THE LOGIC OF SIGNIFICANCE AND CONTEXT

THE LOGIC OF SIGNIFICANCE AND CONTEXT

LEONARD GODDARD
Professor of Logic and Metaphysics
University of St. Andrews

and

RICHARD ROUTLEY
Senior Fellow in Philosophy
Australian National University

VOLUME
ONE

SCOTTISH ACADEMIC PRESS
EDINBURGH AND LONDON
1973

Published by
Scottish Academic Press Ltd
25 Perth Street, Edinburgh 3
and distributed by
Chatto & Windus Ltd
40 William IV Street
London W.C.2

First Published 1973

SBN 7011 1859 8

Printed in Great Britain by R. & R. Clark Ltd, Edinburgh

Contents of Volume 1

PART II

THE PRINCIPLES OF SIGNIFICANCE

An index for both volumes is given in Volume 2.

Acknowledgements

WE began this book in 1966 when we were colleagues at the University of New England, Armidale, NSW. We continued working together, largely by correspondence, when one of us moved to Monash University, later to the Australian National University, and the other to St. Andrews. Because it has been written over such a long period, and in different parts of the world, we have had an unusual opportunity to try out embryonic ideas on a large number of people, especially our colleagues and students. From each of them we have learnt a great deal and to all of them we are extremely grateful. Looking back, we find it difficult to say to what extent we were influenced by any one person since some of our ideas only became clear after discussions involving a number of people, but we wish to record a special debt of gratitude to our former students, Valerie Routley, Ross Brady, Malcolm Rennie and Rowan Rockingham Gill, each of whom has made an independent contribution, in papers or theses, to the development of context or significance logics.

For grants towards the cost of publication, we are much indebted to the Australian Academy of Humanities, the British Academy, the Carnegie Trust, Monash University and the University of St. Andrews.

It remains only to say that each of us is responsible for everything in the book; the order of our names on the title page is simply alphabetical.

R. R.
Canberra

August 1973

L. G.
St. Andrews

CORRIGENDUM

p. 457 rule 4: where R_i is not reflexive, amend the second line to:

$$v(\Phi_i A, \mathrm{H}) = n \quad \text{iff } v(A, \mathrm{H}) = n \ \& \ v(\Phi_i A, \mathrm{H}) \neq t.$$

Correspondingly, J1 (p. 455) is weakened to $\vdash S\ \Phi_i p \rightharpoonup .\ Sp \vee T\ \Phi_i p$. Soundness and completeness proofs then require only trivial modifications.

Introduction

§1 *Historical Background*

Much of the philosophy of this century has depended on appeals to significance. Philosophical theses have been rejected on the grounds that they are nonsense, nonsignificant, absurd, meaningless, unintelligible, incoherent or logically odd. This is not to say that it is only in this century that such claims have been made. There are indications even in Plato that appeals of this kind are legitimate, and in the *Categories* Aristotle made some moves towards a theory of significance. But it is in this century that they have become prominent, especially in the writings of the logical positivists, Russell, Wittgenstein and Ryle.

For Russell, the problem first appears as a technical difficulty in the foundations of mathematics. He was concerned to show that mathematics can be reduced to logic; or equivalently, that the axioms and theorems of mathematics can be derived from the fundamental laws of logic. An essential step in this reduction was to identify numbers as classes of a certain kind. But it was here the difficulty arose. For consider the class of men. Its members are individual men; but the class itself is not an individual man. Therefore the class itself is not one of its own members. On the other hand, consider the class whose members are all those classes which themselves contain, say, more than three members. This class will, presumably, itself contain more than three members. So, being a class of the kind which it contains as members, it will be a member of itself. Similarly, the class of all classes, being a class, will be a member of itself. It is thus possible to distinguish classes according as they are or are not members of themselves. Consider now the class which contains as members all those classes which are not members of themselves. Is this class a member of itself or not? If it is, then it must have the characteristic feature which all its members have, namely that it is one of those classes which are not members of themselves; so it is not a member of itself. If it is not, then it must be a member of the class which contains all those classes which are not members of themselves; and so it is a member of itself. The assumption that it is a member of itself therefore leads to a contradiction, and the assumption that it is not a member of itself leads to a contradiction. This is the standard paradox argument which shows that there is some inconsistency in the naïve conception of a class from which the argument begins. But if there is an inconsistency in the theory of classes then, in terms of

Russell's programme, there is an inconsistency in the foundations of mathematics.

Russell's solution was to claim that the premisses of the argument are meaningless. In this way the paradox is stopped and the inconsistency removed. For the argument can no longer be stated and the contradictory conclusion no longer follows. The argument evaporates.

This claim is supported by an elaborate technical theory, the theory of types, in terms of which individuals and classes are sorted into a hierarchy. Individuals are said to be of type 0. Classes which contain only individuals as members are of type 1. Classes which contain only type 1 classes as members are of type 2; and so on. A criterion of significance then emerges as the condition that a class of type n can only significantly contain elements of type $n-1$ as members. From this it follows that any sentence expressing self-membership, whether positively or negatively, is meaningless. To say that an individual man is a member of the class of men is meaningful and true; to say that an individual man is a member of the class of cats is meaningful and false. Both are meaningful because in each case an individual (of type 0) is being assigned to membership of a class which significantly contains individuals as members (a type 1 class); hence the type rules are satisfied. But to say that the class of men *is* (or *is not*) a member of the class of men is neither true nor false but meaningless, for each assertion puts the class of men on the same typal level as its members; and this is not significant in terms of the type rules. Similarly, the sentences 'The class of all classes which are not members of themselves is (or is not) a member of itself' are nonsignificant. But these are the premisses of the paradox argument.

There are more complicated aspects of the theory of types which need not concern us here. What matters from the present point of view is that certain grammatically impeccable sentences, which on the face of it appear to be significant and therefore either true or false, are held to be meaningless.

There are implications for other branches of logic besides class theory. In particular, because for Russell there is always a predicative sentence equivalent to a given class-membership sentence, so the theory of types transfers to predicate logic. Thus 'Jack is a member of the class of men' is equivalent to 'Jack is a man', or even to 'Jack is human'; hence seemingly innocuous predicative sentences like 'The class of men is not a man' (which is equivalent to 'The class of men is not a member of the class of men') are also meaningless. Properties, as well as classes, have therefore to be arranged in a hierarchy of types. Type 1 properties are those which are significant over individuals, type 2 properties are those which are significant over type 1 properties, and so on. In fact, the class of all individuals, which is a type 1 class, is the *range of significance* of all type 1 properties; and similarly for other types. This is a special version of a general principle which we shall examine in detail in Part III, namely that every predicate has a significance range associated with it, though not necessarily a range consisting of all elements of a given Russellian type.

The formal treatment of the theory of types in *Principia Mathematica* (abbreviated *PM* in what follows), and elsewhere in Russell's writings, is ambiguous. It is not made clear whether type differences are features of individuals and properties, or of the words (names and predicates) which are used to mention them. There are, too, other difficulties which we discuss in detail at various places in Parts II and III. In spite of these difficulties, however, and the later discovery that the contradiction can be removed in other, simpler, ways, the initial success of type-theory made possible the programme of basing a consistent mathematics on logic. Not only this, but the theory seemed to ensure that nonsignificant sentences could be excluded systematically from the formal language of mathematics. And this suggested a much more general programme.

For suppose it is the case that the whole of mathematics can be derived consistently from the primitive assumptions of logic and that this has been achieved in *PM*. It seemed reasonable to suppose that this formal language could be extended to embrace the sciences and indeed all significant discourse. So Russell conceived of a logically perfect language, which was in effect to be the language of formal logic and mathematics with an augmented vocabulary, in which anything significant could be said and anything which could not be said was not significant. This vision of a highly structured, formalized, ideal language is the basis of Russell's philosophy of logical atomism and is characteristic of Wittgenstein's *Tractatus Logico-Philosophicus*.

There is, therefore, a move from the appeal to nonsignificance in terms of a limited technical theory designed to eliminate a formal contradiction in mathematics, to a more general philosophical appeal designed to eliminate spurious theses outside mathematics. But of course in the enlarged programme the limited theory of types could not of itself stand as the only set of significance criteria. A more general theory of meaning and meaningfulness had to be devised. This was, that a sentence is meaningful if its formal logical structure mirrors, or pictures, the structure of a fact. The theory of types can be regarded as a special case of this general theory since the structure of sentences which fail to satisfy the type rules fail to mirror the structure of facts. There is no fact corresponding to the sentence 'The class of men is not a member of the class of men' in terms of which its truth or falsity can be decided. But type theory, restricted as it is to a limited range of sentences, cannot exhaust the general theory; and whereas it is systematically applicable to those sentences to which it does apply, the general theory is not. The philosophy of logical atomism thus implied a large-scale programme of detailed analysis. For given that all those formal sentential structures which mirror the structure of facts are expressible in the ideal language, it follows that any given sentence of natural language must be analysed to see if it can be expressed in terms of them. If it can, it finds its place in the ideal language and is significant; if not, it is nonsignificant.

The shortcomings of the ideal language approach are discussed in detail in Part II. Our concern here is simply to trace the outlines of the part played by significance claims in both logic and philosophy generally, and to emphasise that those sentences which have been dismissed as nonsignificant are neither gibberish nor ungrammatical. They can only be described as nonsignificant in terms of a highly developed and sophisticated theory of language, and it is just this that makes them philosophically important. For the claim is that within a natural language they stand as nonsense masquerading as sense and hence often lead to apparently plausible but in fact absurd conclusions.

The extent of this claim is most obvious in the philosophy of logical positivism. The positivist criterion of meaning is similar to but simpler than Russell's, and the consequences are correspondingly similar but more extensive. For the positivists, a non-mathematical sentence is meaningful, if and only if, it is empirically verifiable. And if we take it, as some positivists did, that a sentence is empirically verifiable if there is some fact in terms of which its truth or falsity may be determined, then the similarity to Russell's criterion is clear. But there is no longer any detailed requirement that the structure of the sentence must mirror the structure of the fact, and hence no prior analysis of the sentence, and a comparison of its structure with those of the basic forms in the ideal language, is necessary. Indeed the ideal language is no longer necessary; it has been supplanted by the language of science. Or more correctly, the ideal language becomes a suitably refined language of science, and expressibility in this becomes the criterion of significance. For it is in science that empirical verification procedures are already most highly developed. So it became possible for the positivists to dismiss as nonsense all the apparently sensible utterances of theologians and metaphysicians. Perhaps, too, this was a consequence of the philosophy of logical atomism since the ideal language was intended to be the language of science and mathematics, but it was never stated explicitly because the ideal language was never constructed, and the detailed analysis never concluded.

The reaction to the over-simple criterion of verification and to the artificiality of the ideal language structures led inevitably to the current interest in natural language and the problem of meaning and meaningfulness as it arises there. This ranges from Wittgenstein's insistence in the *Philosophical Investigations* that the meaning of a word is its use in the language, through Ryle's theory of categories—a much liberalised type theory which draws its strength from ordinary language rather than mathematics—to the transformational grammarians who, like the Russell of the ideal language, are interested in reducing complex linguistic expressions to basic structures but who, unlike Russell, begin with the basic structures of natural languages instead of those of an artificially created ideal. For the philosophers in the story, however, though the criteria of meaning are different, the aim has remained the same. It is to show that some nonsense masquerades as sense and is therefore deceiving; and it is expressed by Wittgenstein in the remark "My aim is to teach you to pass from a piece of disguised nonsense to something that is patent nonsense" (Wittgenstein [2], I, §464).*

To achieve this aim, there must therefore be examples of patent nonsense which are implied by the less obvious but more interesting cases such as 'The class of men is not a man', 'God is in His heaven' and 'The mind is in the body', where, it is said, we have insidious cases of nonsense which purport to be sense. The patent cases of nonsense cannot, however, be gibberish formed from non-words like 'Rubberdock dode', nor flagrantly ungrammatical expressions like 'The old not well because' formed from acceptable words. Though obvious nonsense, they must be grammatical strings of acceptable words otherwise there could be no relation of implication or reduction between them and the less obvious cases. Such are the paradigm cases of category mistakes which Ryle instances. In sentences like 'The number 7 is virtuous' and 'Friday is in bed' we have, it is held, clear cases of absurdity arising from the fact that the subject belongs to a logical kind, type or category over which the predicate is nonsignificant. Numbers are not the kinds of things which can be or fail to be virtuous. This is a type theory of a kind, though different from Russell's. For whereas Russell's individuals all belong to the same type, Ryle's do not. There is, too, a difference in technique. Russell has general rules in terms of which nonsignificance can be decided, Ryle relies on the reduction of doubtful cases to paradigmatic cases. If it can be shown that an apparently meaningful thesis of the kind that the mind is in the body can be reduced to, or shown to imply, paradigmatic cases of category mistakes, then it is not a significant thesis. Exactly this kind of literal *reductio ad absurdum* programme underlies the whole structure of the *Concept of Mind*.

§2 *The Need for a Logic of Significance*

In spite of the central role which the concept of significance has played in recent philosophy, very little has been done to produce a general theory in terms of which the various claims can be evaluated. Both Russell and Ryle go some way towards the development of particular significance theories which they then apply to the examination of philosophical theses, and Russell's theory of types is expressed partly in terms of formal criteria. But the two theories, though similar, are inconsistent with each other, and each is seriously incomplete. Both depend crucially on the acceptance of intuitive and unstated principles.

It is our purpose to develop a general formal theory of significance in terms of which significance claims, and arguments by means of which they

* A name followed by a numeral in square brackets refers to the bibliography.

are made, can be assessed. Thus we aim, eventually, to provide a logic, not previously developed, for much of modern philosophy.

This aim might be challenged in one of three ways. First, even though it is accepted that some sentences are nonsignificant, it might be thought that there is no need of a special logic to take account of them since the ordinary principles of classical two-valued logic can be applied to arguments in which they occur. Secondly, it might be said, given that there is a need for a special logic, the peculiar features of nonsignificant sentences make its development impossible. Thirdly, it has sometimes been suggested, nonsignificance can be wholly explained in terms of other well-understood concepts: in particular, it can either be identified with the ungrammatical, in which case nonsignificant sentences should be excluded from all arguments and the only logic which is necessary is the classical two-valued logic over significant sentences; or it can be identified with the necessarily false, and in this case significance claims and arguments can be handled within classical modal logic.

Consider the first objection. The point of and need for a general formal theory of significance can be justified on general grounds by appealing to the usual reasons for underpinning philosophical investigations by logical studies: namely the clarity and systematisation which result, and the fact that the philosophical theory can be tested for consistency and is generally more readily assessed and open to falsification. But it can also be justified on special grounds since, without it, the assessment of particular significance theories, if not impossible, is at best intuitive. Unless general principles are formulated and justified, it remains obscure which critical arguments are valid within and against a particular theory. Thus, without some systematisation of significance principles, many philosophical arguments, such as those of Ryle and Strawson on the philosophy of mind, are impossible to assess.

As one example of the need to make explicit the principles which are assumed, and the need to evaluate them, consider Russell's assumption, which is nowhere justified, that the paradox argument evaporates and the conclusion no longer follows once it has been shown that the premisses are meaningless. This amounts to saying, in general, that a nonsignificant sentence has no implications. But if this is so, then Ryle's *reductio ad absurdum* technique, and Wittgenstein's declared aim, seem to be illogical. For here, the intention is to show that a given sentence is nonsignificant because it has implications which are nonsignificant. But if it is indeed nonsignificant, then it was so before the actual implications were drawn, and in terms of Russell's principle it cannot have these implications because it has none at all.

In order to resolve problems such as this, it is necessary to construct associated significance logics for particular theories of significance and to ask whether there is an internal inconsistency in either theory, whether there are general principles incompatible with either or both, and whether there is a general theory which can consistently include both. It may be, for example, that the apparent incompatibility of Russell's principle and Ryle's is not real. This would be so if they were adopting different implication connectives; in this case, both principles could consistently appear in the same general theory.

None of these questions can be settled using only a classical two-valued logic, however, since what is in issue is just which relations hold between truth-valued sentences, on the one hand, and nonsignificant sentences, on the other. Similarly, it is inadmissible to evaluate positivist criteria of significance using principles of two-valued logic alone since, given that some sentences are nonsignificant, such a logic cannot hold generally. But the mere recognition that at least a three-valued logic is necessary is of itself inadequate. For unless the *relevant* three-valued logic is specified, the criteria are still not fully assessable.

The complications which arise with respect to the assessment of particular significance theories before the elaboration of their associated logics resemble those difficulties which arose with respect to the assessment of the intuitionistic theory of mathematics before the development, by Heyting and others, of intuitionistic logics: there it was unclear, even at the quantificational level, what was intuitionistically meaningful.

Consider now the second objection. There is a widespread scepticism concerning the very possibility of significance logics. Various reasons have been offered for this. First it is claimed that matters of significance, like matters of context, lie beyond the reach of formal logic, that they belong not to formal but to informal logic. We refute this view by constructing formal theories of both significance and context. Secondly, it has sometimes been thought that significance principles cannot be explicitly stated in a logic but can at best be shown in a metalanguage. Russell, for example, thought that significance range restrictions could not be significantly stated in a formal logic. Connected with this is the belief that a significance logic would be worthless because it would contain no theorems. For such a logic must admit nonsignificant sentences, and thereby nonsignificance as a value of sentential variables. But no compound sentence containing a nonsignificant component can be significant, and no formula containing a variable which takes nonsignificance as a value can be valid. These sceptical arguments, too, will be shown to be unfounded by the construction of significance logics which do just these things which the sceptical arguments claim cannot be done.

Put in its simplest terms, we propose to develop a logic in which both nonsignificant and significant sentences are admitted as values of the variables; to determine the laws which hold between them so that principles such as the literal *reductio* can be assessed; and to state explicitly, and to explore the consequences of, some of the criteria of significance which have been adopted. None of this is possible in classical two-valued logic in which only significant sentences are admitted as values of the variables. We thus intend to construct a general formal theory of significance in terms of which particular theories can be evaluated. We do not intend to advance yet another particular theory. That is, we are not ourselves concerned to make claims of the kind that certain philosophical theses are or are not significant. Our interest is in general laws rather than specific cases and particular applications. We are trying, however, both to find a theory which reflects and systematises significance assignments made in natural language and to clarify the formal relations between various such assignments.

It might yet be said, however, that any such intention is both misguided and pointless. For there are those who hold that there are no nonsignificant sentences. This brings us to the third objection; and it is a criticism on a different level. For whereas the first two objections accept the main thesis that some sentences are nonsignificant, and can be met by showing that two-valued logic is inadequate to deal with arguments in which they occur (the first objection), and by demonstrating that the supposed difficulties are not insurmountable (the second objection), here the problem is different. For the suggestion is that the whole problem of significance is not a problem of significance at all. Quine, for example, seems to take this view:

> the forms concerned would remain still quite under control if admitted rather, like self-contradictions, as false (and false by meaning, if one likes) (Quine [2], p. 229).

Clearly such criticism has to be met at the outset, and indeed Quine's suggestion is not the only possible reductionist thesis which has to be examined. But criticisms of this kind cannot be evaluated fully independently of some of the formal results which are reached as the theory is developed. We shall therefore only note the main points in the next section and hope that the final vindication of the position we take will emerge as the detailed story unfolds.

§3 *Nonsignificance, Falsity and the Ungrammatical*

The claim made by Russell, the earlier and later Wittgenstein, the positivists and Ryle, is that some sentences are both grammatically acceptable yet nonsignificant. And given the further assumption, which is justified by a limitation to certain kinds of indicative use, that nonsignificant sentences are neither true nor false, this amounts to saying that some grammatical sentences are not truth-valued. There are, therefore, two possible lines of attack:

(i) that every so-called nonsignificant sentence is ungrammatical (*cf.* Shorter [1]);

(ii) that every so-called nonsignificant sentence is grammatical but truth-valued (*cf.* Prior [1]).

Quine's suggestion stands as a special case under (ii).

(i) is a much more sympathetic position than (ii); sympathetic, that is, to the claim that there are nonsignificant sentences. For it recognises the oddity of expressions like 'Friday is in bed' but seeks to reclassify them. (ii), on the other hand, represents the direct negative. There are, however, difficulties in both positions.

Consider (i): That the sentences are ungrammatical is obviously not a simple assertion about either text-book grammar or idiomatic grammar. For they are clearly not ungrammatical in terms of criteria drawn from either of these sources. It is only a possible criticism given a much more comprehensive sense of 'grammar'. Nor can the sense of 'grammar' be restricted to what might be called syntactical grammar. The claim that 'Friday is in bed' is nonsignificant is not made solely, or even at all, in terms of its syntactical structure but in terms of the clash of sense between 'Friday' and 'is in bed'. This is generally what is expressed when it is said that Friday is not the kind of thing which could be or fail to be in bed. It is at least, therefore, a semantic claim. This is not of course to deny that structural features are relevant to meaning, but simply to affirm that they do not exhaust it.

In view of this it might perhaps be thought that the criticism is not a real one. In fact, however, it is exemplified by post-Russellian treatments of type-theory which reflect semantic criteria in syntactic structure.

Russell identified types as semantic significance ranges and expressed the type rules in terms of them. A property of a given type can only be significantly affirmed or denied of an element belonging to the next lower type, and this lower type is the significance range of the properties in the higher type. If, however, we introduced different lexicographical sorts of variables and constants, where each sort is restricted in range to a specified type, then the type rules may be expressed as syntactic restrictions on the formation of formulae. Those formulae which satisfy the restrictions are well-formed; those which do not are ill-formed. From the point of view of the underlying semantic range theory it is still possible to say that sentences which contravene the type rules are nonsignificant. But from the point of view of the formal theory it is only necessary to say that they are ill-formed. To the extent that formation rules are grammatical rules, therefore, we may say that sentences which contravene the type rules are ungrammatical. At this point it is possible to throw away the ladder and drop the use of 'nonsignificant'.

Russell's type rules are not of themselves adequate to exclude as ungrammatical many of the sentences which others have wished to exclude. But given an adequate transformational grammar with many more basic forms than Russell envisaged, and a finer list of grammatical categories, much more detailed formation rules could be devised and sentences like 'Friday is in bed' eliminated. Of course the motivation and rationale is still semantic, but the formal rules of construction could be expressed entirely in syntactic terms. So one can imagine the realisation of Russell's dream of an ideal language within which every significant sentence can be expressed and from which every nonsignificant sentence is excluded. In a quite general way, therefore, 'nonsignificant', 'non-truth-valued' and 'non-grammatical', might be claimed to be coextensive over indicative sentences.

Such a position holds no real philosophical threat to theories such as Ryle's which deal heavily in nonsignificance. For it is just as telling to dismiss a philosophical thesis as ungrammatical, in this new extended sense of 'grammatical', as it is to dismiss it as absurd. From this point of view the criticism is unimportant. It might be thought, too, that it is unimportant from the logical point of view and there is no reason why those who wish to adopt it should not replace 'nonsignificant' by 'nongrammatical' in everything that follows, for the distinction between truth-valued and non-truth-valued sentences still remains, and the formal theses which we present are still necessary. There would be a difference only in interpretation, and that of itself is slight.

In fact, however, there are two serious objections to it, one in the form in which it stands, and one in a slightly extended, but not uncommon, form.

The first objection is that to approach the problem of significance in this way is to be committed to the view that nonsignificance is a permanent feature of a sentence, either a permanent semantic feature or a permanent syntactic feature. And this is to neglect entirely the use of sentences in a context. We shall discuss this in much greater detail later, but we simply note at this point that if the sentence 'Friday is in bed', and others like it, were wholly excluded in terms of fixed grammatical categories, either semantic or syntactic, then Robinson Crusoe would not have been able to make the significant remark he no doubt could have made on many occasions.

The second objection arises if it is held that once the comprehensive grammatical rules have been devised, the ungrammatical sentences may then be excluded from further consideration since they are no longer part of the "official" language and cannot enter into logical relations with each other, or with acceptable grammatical sentences. This is the way in which type-transgressing sentences are treated in modern formulations of type theory. The grammatical sentences which remain, being truth-valued, satisfy a classical two-valued logic and this, it is held, is the only logic which is necessary. There is, however, a methodological inconsistency in this position. For the logic of impeccable sentences will not then apply to that part of the language in which the distinction between significance and nonsignificance, or grammatical and ungrammatical, is recognised and in which the separation is carried out. Either all arguments at this level are illogical or else a different logic is being used—one in which the variables range over both significant and nonsignificant sentences, and hence, one in which nonsignificant sentences are well-formed. But it is just such arguments which are philosophically interesting. A restricted logic, in terms of which the formal evaluation of

Russell's paradox argument and Ryle's *reductio* technique is impossible, is inadequate.

Consider, now, (ii): This is a much more telling criticism if it can be sustained, for by identifying nonsignificant expressions as truth-valued sentences of one kind or another, it overcomes the objection that there is no logic of the nonsignificant. On the contrary, it merges with the logic of the significant.

There are various possible versions of this criticism:

(a) that a "nonsignificant" sentence and its negation are both true;

(b) that of a "nonsignificant" sentence and its negation one is true and the other false;

(c) that a "nonsignificant" sentence and its negation are both false.

It is unlikely that (a) could be held seriously. If a sentence and its negation are both true, either there is an inconsistency or some special sense of negation is being proposed. If the latter, then these sentences are being distinguished from others in terms of a negation which cannot consistently operate uniformly over all sentences. In this case nonsignificance can be defined in terms of this negation and no real alternative is being proposed.

The second version (b) is a more likely proposal, but there are a number of difficulties associated with it. First, consider a nonsignificant sentence and its negation: 'Friday is in bed' and 'Friday is not in bed', where it is understood that the context is such that the reference of 'Friday' is to a day of the week and not to a man or to some other physical object. How are we to determine which of these is true and which false? If either one is true, it is clear that it cannot be contingently true because none of the ordinary verification procedures are relevant. It would be pointless to go upstairs to look. Secondly, none of the ordinary contextual implications hold. Told that Friday is not in bed, we cannot ask, "Then where is it? In the garden?"; told that Friday is in bed, we cannot ask, "What time did it go there?" Such sentences, if truth-valued, are as Quine suggests, true or false by meaning. Which is to say that one is analytic, one self-contradictory. This still leaves us with the difficulty, however, of determining which value to give to which sentence.

This aspect of the problem is, surprisingly, often overlooked. There is a tendency simply to say that nonsignificance can be identified with necessary falsehood, and this point is usually made in terms of a few examples. Thus, in the passage quoted from Quine, two examples are given: 'This stone is thinking about Vienna' (Carnap) and 'Quadruplicity drinks procrastination' (Russell): and the suggestion is that these should be taken as necessarily false. But if they are necessarily false, their negations are necessarily true; and in terms of what criteria is the assignment of values made in this way rather than *vice versa*? Those who have wanted to invest in nonsignificance have always been careful to point out that if a sentence is nonsignificant, so is its negation; for it is usually held that the one is just as puzzling as the other. Yet we are now being asked to distinguish between the two without any clear criteria being given of how this is to be done in particular cases. In fact, the need to identify some sentences as nonsignificant arose for Russell simply because the ordinary criteria fail to apply. Thus, one way of showing that a sentence is necessarily false is to show that it implies the paradigm case of a contradiction: *p* and not-*p*. But in the case of the Russell sentence, it can be shown that both it and its negation imply a contradiction, and therefore, on this test, both are necessarily false. It is just this that makes the sentence paradoxical.

One feature of those sentences which have been described as nonsignificant which may account for the temptation to classify them as necessarily false is that they may seem to be less absurd when stated negatively. Thus, 'This stone is thinking about Vienna' is queer by any standards, but 'This stone is not thinking about Vienna' is less so. Similarly, 'The number 7 is not virtuous', and 'Friday is not in bed' might seem to make sense. On the basis of this, it might then be thought that those which contain the 'not' are necessarily true, while those which do not contain it are necessarily false. And it is interesting that when examples are given in support of the claim that nonsignificance can be identified with necessary falsity, as in Quine, it is positively stated sentences which are given as examples. But if this is to be the test, then just this should make the sentences suspect. For it is not a feature of ordinary sentences which are classified as necessarily false. We cannot say that '2+2 is equal to 4' is necessarily false because it fails to contain a 'not'; on the contrary it is the negatively stated negation which is false. And in other cases, of course, the opposite holds. Neither in mathematics nor elsewhere is there a general correlation between necessary truth (or indeed contingent truth) and negative sentences.

It is worth noting too, that if all putative nonsignificant sentences are to be reclassified as false, or necessarily false, if they are positive, and true, or necessarily true, if they are negative, then ordinary equivalences derived from synonymy relations could no longer hold. For example, given that 'in motion' is synonymous with 'not at rest' then in general, 'X is in motion' and 'X is not at rest' are equivalent. Now if we take 'the square root of –1 is in motion' and 'the square root of –1 is not at rest' and accept that these are nonsignificant, then the equivalence still holds in terms of an elementary extension of the definition of equivalence to the three-valued case (§2.12). If, however, nonsignificant negative sentences are reclassified as true and nonsignificant positive sentences as false, then the equivalence fails since 'the square root of –1 is not at rest' and 'the square root of –1 is in motion' then have opposite truth values. One might seek to avoid this by claiming that 'in motion' is implicitly negative, or equivalently, that 'is not at rest' is implicitly positive. But then, if implicit negatives are admitted, the explicit occurrence or absence of 'not' is no longer an absolute guide to the positive or negative character of a given sentence, and it is necessary to have an initial list of positive and negative words in terms of which the assessment can be made. Thus it becomes necessary to invest in an ideal language to make the logic work.

The temptation to describe the negative sentences as true is therefore suspicious; and it can be explained. For it rests on the ambiguity of denials. As we have seen, in order to account for the nonsignificance of sentences like 'Friday is in bed', it is sometimes said that Friday is not the kind of thing which could be in bed. And this denial is significant and true, while its negation 'Friday is the kind of thing which could be in bed' is significant and false. For it is a way of stating the range requirement that Friday is not in the significant range of application of the predicate 'is in bed'. And just because this is significant and true, so the sentence 'Friday is not in bed', which *presupposes* that Friday *is* in the significance range of the predicate 'is in bed', is nonsignificant. But the sentence 'Friday is not in bed' might in some contexts be stated and intended as a shortened idiomatic version of 'Friday is not the kind of thing which could be in bed'. And because there is always a need to force a significant interpretation of any sentence, so there is often an implicit significant reinterpretation of the sentence 'Friday is not in bed' when offered as an example of nonsignificance, which is fostered by the ambiguity of the denial.

Just this ambiguity is present, too, in those formal arguments which seek to show that the truth of certain nonsignificant sentences follows from true premisses. Thus, consider:

The theory of relativity is an abstract object
Abstract objects are not blue
∴ The theory of relativity is not blue
(*cf.* Prior [1] and Drange [1], p. 24).

Here, it may be said, the premisses are true and the argument is valid, so the conclusion is true. Hence the positively stated sentence 'The theory of relativity is blue', which is the negation of the conclusion, is false. Clearly, however, either the second premiss is taken in the sense 'Abstract objects are not the kinds of things which can be blue' (which is the natural way to take it), in which case the premisses are true but the conclusion is 'The theory of relativity is not the kind of thing which can be blue', or else the whole argument begs the question by assuming that the second premiss, when taken as stated and not reinterpreted as a significance condition, is itself significant. For, as stated, its significance is just as much in question as 'The theory of relativity is not blue'.

Different, but related, points are raised by arguments of a similar kind. Consider, for example:

Whatever is in bed is not a day of the week
Friday is a day of the week
∴ Friday is not in bed.

Various assumptions are implicit here. Suppose, first, it is held that the premisses are significant and true. Now there is a presupposed instantiation step from the first premiss to 'If Friday is in bed then Friday is not a day of the week'; and if this is taken to be significant then either the component part 'Friday is in bed' is taken to be significant, in which case the question is begged, or else it has to be accepted that we can have a significant compound with a nonsignificant component. But this raises a question about the significance features of sentential connectives which cannot be answered independently of a logic of significance. On the other hand, if it is accepted that the instantiation case may be nonsignificant, then what are we to say of the first premiss? Can we in fact have true sentences implying nonsignificant sentences, or is it the case that quantified sentences which have nonsignificant instantiation cases are themselves nonsignificant? This, too, raises a question, about quantification, which cannot be answered independently of a logic of significance. Finally, if it is accepted that both the instantiation case and the first premiss may be nonsignificant, it may still be the case that the argument is valid, though this, too, is a matter which can only be decided in terms of an appropriate logic, but even if it is, it clearly cannot be assumed that a valid argument with a nonsignificant premiss guarantees either the truth or the significance of its conclusion.

There is one final objection to the proposal that nonsignificance can be identified with the necessarily false. It is the difficulty which arises in the case of contraries. For suppose it is agreed that 'The number seven likes tomato soup' is necessarily false and its negation, 'The number seven does not like tomato soup' is necessarily true. Presumably it would have to be said also that 'The number seven dislikes tomato soup' is necessarily false and its negation, 'The number seven does not dislike tomato soup' is necessarily true. So far, there is nothing inconsistent in this position. 'The number seven likes tomato soup' and 'The number seven dislikes tomato soup' are contraries, and can be false together; similarly 'The number seven does not like tomato soup' and 'The number seven does not dislike tomato soup' are subcontraries and can be true together. But given that the two negative sentences are true, then so is their conjunction: 'The number seven does not like tomato soup and does not dislike it'. In most ordinary contexts, however, to say that X does not like soup and does not dislike it, is to say that X is indifferent to soup, *i.e.* he can take it or leave it. So we seem committed to the position that 'The number seven is indifferent to tomato soup' is true. But this is a positive ascription of the same kind as the original sentence 'The number seven likes tomato soup' and ought therefore to be classified as necessarily false. Certainly it is of the kind which would be classified as nonsignificant by those who are prepared to accept nonsignificance. And if, in answer to this, it is said that this is not an ordinary context, and hence that the implication from 'does not like and does not dislike' to 'is indifferent to' fails to hold, then it is being conceded that the number seven is of a different logical kind from those items of which it can ordinarily (significantly?) be said that they like, dislike or are indifferent to soup; but in this case the point is made.

This kind of point can be generalised. For it is always possible to form from two contraries, 'X is F' and 'X is non-F', a conjunction of the form 'X is not F and X is not non-F' which is true if both contraries are false; and in many cases, as in the above example, the conjunction can be contracted to a single sentence which attributes to X a determinant of the same kind as those in the original contraries. In this case, however, it should be false, not true.

The proposal (c), like (a), can be dealt with shortly. For if it is held that *all* those sentences which have been claimed as nonsignificant are to be reclassified as false, then any given nonsignificant sentence and its negation are false. In this case, either (i) there is an immediate inconsistency; or (ii), the sentence and its negation are taken to be a sentence and its contrary, in which case there is in at least some instances an implied inconsistency (as above); or (iii) a new sense of negation is being introduced which holds for this special class of sentences alone. But this is enough to distinguish them. That is, we may now simply define nonsignificant sentences as those which satisfy the condition that both they and their "negations" are false.

We do not regard these arguments as conclusive at this stage, and they will reappear in a sharper form at various places in the body of the book. We hope, however, to have made the case sufficiently for it to be accepted that there is a need for a formal theory designed to explore the problems which have been raised.

§4 *The Proposed Logic*

There are a number of semantic problems which have to be resolved before the formal theory can be developed. In particular it is necessary to consider the different ways in which syntactic, semantic and contextual features bear on the evaluation of significance. Some of the problems, mentioned briefly in §3 above, can be set by the questions: In what respects is significance independent of context? and What is significance a property of? Three possible candidates for the last are sentences, statements and propositions, and different authors have made different choices. Answers to these questions are important to us since they control both the semantic interpretation and the formal structure of the proposed theory. We therefore begin in Part I with the semantic basis and from this develop the formal principles of what we propose to call *context logic*. Our conclusions here are largely independent of the special features of significance theory and we hope relevant to any formal system, such as classical sentential and quantification logic, which has a claim to being regarded as a (partial) logic of discourse.

We next turn to the special features of nonsignificant sentences and develop first the logic of principles, in Part II, and then the logic of criteria, in Part III (volume 2). This distinction is not made in terms of different kinds of significance theory, but relates, rather, to the dual features of any given particular theory.

Every particular significance theory must contain criteria in terms of which a given sentence may be determined as either significant or nonsignificant and, as we have already seen, several different kinds of criteria have been proposed by different authors. In general, however, the different particular theories fall under one or other of two main types: they may be classified either as unit theories or factor theories. Unit theories explain nonsignificance in terms of criteria which apply to sentences considered as wholes or units, no matter what their formal structure. Such theories are exemplified by the positivist criterion of verification (though it is unfortunately often called the principle of verification). Factor theories, on the other hand, explain nonsignificance in terms of the incompatibility of sentence parts, and this incompatibility may or may not itself be explained in terms of semantic significance ranges. If it is, the criteria of significance are expressed as range conditions of one kind or another; if not, the criteria are expressed as syntactic conditions. In either case significance is controlled by placing restrictions on the couplings of sentence parts. Both Russell and Ryle favour such theories, as do many modern grammarians.

Unit and factor theorists therefore differ in the kind of answer they give to the question: what are the criteria in terms of which a given sentence may be said to be significant? And different factor theorists, though they agree on the kind of answer, may yet disagree on the actual answer. But given that there are nonsignificant sentences, whatever the criteria in terms of which this decision is made, we may ask a second question: what principles of inference operate in contexts in which they occur? Can we, for example, affirm that any sentence which implies a nonsignificant sentence is itself nonsignificant (the literal *reductio ad absurdum*)? Is it the case that a sentence and its negation are co-significant (either both significant or both nonsignificant)? Is it the case that if two descriptions are about the same item, then any predicate which, when coupled with one of the descriptions yields a nonsignificant sentence, yields a nonsignificant sentence when coupled with the other (the identity law)? Principles which presuppose answers to questions such as these have been adopted either implicitly or explicitly by both unit and factor theorists and there is, therefore, a good deal of common ground between them. Apart from the work of Halldén [1] and other Swedish logicians, however, there has been little serious attempt to state them explicitly and to present them systematically. It is this problem which concerns us in Part II.

In Part III our main concern is to develop a formal logic of factor theories in terms of the significance ranges of sentence parts. We are therefore there concerned with criteria of significance, and in particular, with criteria of a certain kind. And it might perhaps be thought that this is where we should begin, before taking up the question of principles. However, quite apart from the fact that the problem of principles represents an area of ground common to all theories, whether unit or factor, and in that sense forms a convenient starting point, it comes first logically since the use of the principles derived in Part II is essential to the development of range theory. Of course it is not possible entirely to develop the logic of principles independently of some informal consideration of criteria, but we try in Part II to exclude the problem of criteria from the formal development.

PART I

CONTEXT LOGIC

CHAPTER 1

The Semantic Theory

1.1 Initial Problems

Some of the initial problems which confront us are not peculiar to significance theory. They are common to all systems of logic which are intended to have an application to arguments in ordinary discourse rather than in specialised areas of it, such as mathematics, but they become rather more important in relation to an intended logic of significance. They concern the semantic interpretation of the system.

As an example of one such difficulty, consider the lack of agreement among logicians as to whether the ordinary variables of the classical two-valued system, '*p*', '*q*', '*r*', *etc*., stand for sentences, statements or propositions. Even the system in which they occur is sometimes called the sentential calculus, sometimes the calculus of statements and sometimes the propositional calculus. Partly these differences in nomenclature reflect genuine differences of opinion, partly a lazy attitude towards semantics. Classical systems have been developed in spite of this semantic haziness because the decisions taken at the semantic level are thought to be independent of the formal construction of the systems. The syntactic features of a formal system will remain the same, it is supposed, independently of its interpretation. We reject this view in general, and especially in the case of modal and intensional systems, but we do not propose to take up the general issue here. For whether or not such a view can be justified in the case of classical systems, it cannot be supported in the case of significance theory. Here, the semantic decisions have a marked effect on the syntactic features of the formal theory.

Suppose, for example, we wish to affirm or deny of a given expression that it is significant. So to say is to make a remark about that expression, and hence to mention it. But there is a great deal of difference between mentioning a sentence and mentioning a statement, and these differences have to be indicated syntactically. To mention a sentence it is normally necessary to introduce some supplementary device, such as quote marks or italicisation, to indicate that the sentence is being mentioned and not used. To mention a statement, on the other hand, is to use a sentence and not to mention it, and the supplementary devices become unnecessary. Compare, for example, the two different kinds of mentioning in the report:

(1) Jack used the French sentence 'Le monde est rond' to make the statement that the world is round.

It would not matter in which language we wrote the last four words of this report since we are here mentioning the statement he made and not the mode of expression which he used in making it. But the quoted expression must remain in French if he used a French sentence. Thus, if (1) is true, (2) is true but (3) is false:

(2) Jacques a employé la phrase français 'Le monde est rond' pour dire que le monde est rond;

(3) Jack used the French sentence 'The world is round' to make the statement that the world is round.

This is a minor illustration of the sorts of reasons which make it necessary to employ some device such as quote marks together with rules governing their use—that expressions inside them resist translation in some contexts, for example—if sentences are being mentioned. It thus becomes important for us to know whether, when we mention nonsignificant expressions, we are mentioning sentences or statements. If sentences, we have to devise a systematic theory of quotation functions as a preliminary.

Nor is this the only consequence. In a sense, the whole structure of the proposed theory depends on decisions taken at the semantic level. For there are many commonly made assumptions which appear to be inconsistent with each other and which have to be resolved if progress is to be made. It is often said, for example, that if we limit our attention to indicative expressions, neglecting imperatives, interrogatives and the like, then 'nonsignificant' means 'neither true nor false' and 'significant' means 'true or false'. Russell makes this assumption explicitly. A consequence of it, or an equivalent way of stating it, is that every indicative expression is either true, false or nonsignificant. But consider: if we take 'expression' here to mean 'sentence' then we are committed to the view that it is sentences which are true or false; and this might be thought objectionable. Sentences, it has been said, are not candidates for truth or falsity; what is true or false is the statement which the sentence is used to make. On the other hand, if we accept this and take 'expression' to mean 'statement' then we are committed to the view that statements may be nonsignificant. And this might be thought to be self-contradictory. For there are those who would wish to say that statements, by definition, have meaning, make sense, and are true or false.

A good deal hangs on this. If we dismiss the objections and accept the general thesis that every indicative expression is either true, false or nonsignificant, whatever sense is given to 'expression', then we are saying that the basic logic of the proposed significance theory is three-valued; truth, falsity and nonsignificance are, in terms of this, on the same semantic level. They are three comparable values of any given expression. If, on the other hand, we take the view that truth and falsity are characteristic of statements, while non-significance is a feature of sentences, or at any rate of non-statements, then it would seem that the proper development is first to divide sentences into two classes: those which can be used to make statements and are therefore significant, and those which cannot and are therefore nonsignificant; and secondly to subdivide statements into those which are true and those which are false. So if we wish to construct a logic of significance we should first develop a two-valued logic of sentences, whose values are significance and nonsignificance, and then a second two-valued logic of statements (in fact the classical two-valued system) whose values are truth and falsity. In this way we recognise an essential semantic hierarchy; and though we may still feel inclined to define 'nonsignificant' to mean 'not true and not false', the 'not' so used is not the 'not' of negation, instead the description 'not true and not false' means something like 'the words "true" and "false" do not apply'.

To make a decision either way on arguments of this sort, however, would be folly. For the assumptions on which they depend are made intuitively without any clear and explicit indication of what sentences and statements are and in what relations they stand to each other. When it is suggested, for example, that sentences might be divided into two classes, those which are significant and those which are not, there is an implicit assumption that significance is a permanent feature of sentences which is independent of their use and hence independent of contextual considerations. But this, as we have already suggested in the Introduction, is at least doubtful.

To resolve difficulties such as this it is necessary to clarify the semantic presuppositions as a preliminary to the formal development. We do not consider that it is a problem peculiar to significance theory but a general problem which only becomes rather more obvious in relation to significance theory.

1.2 Sentences and their Use

People use written and spoken sentences to communicate with each other: more specifically, to ask and answer questions; to make requests, promises and reports; to give information, advice and orders; to express wishes, fears, hopes and desires; and in general to perform what Austin [1] called an illocutionary act. This is not the only kind of use which can be made of sentences. Written sentences, for example, may be used to decorate wall-paper and fabrics, and in other ways.

To ask a question, to give an order, to give advice, is not to say something which is true or false. The concepts of truth and falsity are only relevant to those illocutionary acts which can be regarded as acts of statement-making. To give information, to make a report, to identify something, to describe something, is to perform a statement-making act. To make a statement is not, for us, a particular and separate illocutionary act itself appearing in the list. Instead, we use the phrase 'statement-making' as a general description of specific acts such as the giving of information. We do not pretend that the list of statement-making acts which we have mentioned is complete; we mean only to mark the general distinction between acts which are statement-making, and to which the concepts of truth and falsity are appropriate, and those which are not.

To use a sentence to ask a question is to use it interrogatively; to give an order, imperatively; to express a wish, optatively; to make a statement, statementally, or as we shall sometimes say, indicatively. These descriptions are adverbial since they qualify the use of the sentence and not the sentence itself. For some purposes, as we shall see, it is useful to transfer descriptions of the use to the vehicles used. So we may come to speak of interrogative, imperative, optative and indicative sentences. But such transferred descriptions can be misleading unless the conditions under which the transfer is made are well understood. The sentence used in an indicative act, for example, need not itself be a grammatical indicative.

The performance of any illocutionary act may be either successful or unsuccessful. The conditions for success are many and varied. They can be classified roughly into those which operate on the physical environment at the time of utterance (*e.g.* that there should not be too much background noise); those which operate on the speaker or writer (*e.g.* that he should not mumble or write with black ink on black paper); those which operate on the sentence itself (*e.g.* that the words in it should belong to the vocabulary of a language common to the participants); and those which operate on the use of the sentence (*e.g.* that its intended use should not be obscure: *cf.* I did not realise you were *asking* me to rearrange the papers when you *said* they were untidy). Our interest is in those conditions which operate on the sentence itself and on its use.

One of the conditions, though as we have wished to emphasise, not the only one, for the successful performance of any illocutionary act is that the sentence used shall be significant. The act need not be statement-making. We can give absurd orders and ask absurd questions. But it is only in the context of intended statement-making that the contrast with truth and falsity can be made. If we seek, therefore, to develop a logic of significance in terms of truth, falsity and nonsignificance, however the relations between these are explained, we must limit attention to intended statement-making, or indicative, acts. This we now do. We emphasise, however, that this is not a limitation to the use of grammatically indicative sentences.

1.3 True and False Sentences

We begin with distinctions made by Strawson ([1], pp. 66 ff.) between:

a sentence;
a use of a sentence;
an utterance of a sentence.

These distinctions are intended to mark, and to provide a vocabulary for talking about, what is thought to be an obvious fact of language: namely that one and the same sentence can be uttered on different occasions and that its use may be different on those occasions. If X and Y both say "Mary is happy", they have both uttered the same sentence. And if one of them is talking about his wife, while the other is talking about his favourite cow, then they have made different uses of the same sentence. Each has used the sentence to make a statement, but they have made different statements. On the other hand, if each is talking about the same individual, they have made two different utterances of the same sentence to make the same statement; that is, they have each made the same use of the same sentence.

It follows immediately from this that we cannot speak of the truth or falsity of *sentences*. For one and the same sentence may be used by X to make a false statement about his wife, and by Y to make a true statement about his cow. Unless, therefore, we allow that the same sentence is both true and false, which presumably we cannot do, we must restrict the application of 'true' and 'false' to the statements made. Alternatively, we may apply them adverbially to the use of the sentence. We may say that X used the sentence 'Mary is happy' truly, on one occasion, while Y used it falsely, on another. But in saying this we are not saying anything new. We are only saying again that X used the sentence to make a true statement while Y used it to make a false one.

If it is also accepted, as it is by Strawson, that it is sentences which are either significant or nonsignificant, then 'significant' cannot mean, as Russell took it to mean, 'true or false'. For sentences are never true or false, but they may be significant. Thus,

> if I ask: 'Is the sentence true or false?' I am asking an absurd question which becomes no less absurd if I add 'It must be one or the other since it's significant'. The question is absurd because the *sentence* is neither true nor false ... (Strawson [1], p. 70).

In terms of this, it is impossible to develop a logic of significance by taking truth, falsity and nonsignificance as the three possible values of any given *sentence*. For truth and falsity are not sentence values at all. Equally, it is impossible to develop significance logic by taking truth, falsity and nonsignificance as the three possible values of any given *statement*. For nonsignificance is not a value of statements. It is therefore impossible to develop a logic of significance as a three-valued logic. The variables will not range uniformly over sentences, on the one hand, nor over statements, on the other. And the matter is not improved by appealing to propositions instead of statements. For if nonsignificance is not a property of statements, it is not a property of propositions. Significant sentences express propositions; non-significant sentences fail to express propositions, they do not express non-significant propositions.

These conclusions seem utterly destructive of any attempt to develop a theory of significance in terms of a three-valued logic. Yet they can be avoided without denying any of the distinctions which Strawson wishes to make. It is only necessary to make the same distinctions in a slightly different way.

The conclusions depend entirely on the assumption that one and the same sentence can be uttered on different occasions. But this is true only if we take 'sentence' to mean 'sentence-type'. It is not true if we take 'sentence' to mean 'sentence-token'.

The standard distinction between types and tokens is meant to mark the fact that although two utterances or inscriptions such as,

Mary is happy

and

Mary is happy

are of the same kind, they are nevertheless different individual items separately locatable. This is often described by saying that they are two different tokens of the same type. Now if the distinction is described in this way, then tokens cannot be uttered or used more than once or by more than one person —except trivially, by cutting the actual token from one page and pasting it on another, or by playing back a tape-recording. So described, however, types can be used more than once and by more than one person. And since the conclusion that sentences are neither true nor false depends on the premiss that *one and the same sentence* may be used by one person to make a true statement in one context and by someone else to make a false statement in another, the most we can in fact conclude is that *type-sentences* are neither true nor false. This leaves it open, however, whether or not it may be appropriate to describe token-sentences as true or false.

To leave the matter there, however, would be misleading, for the distinction between types and tokens is not yet made as clearly as it might be. The belief that types can be uttered or used at all is, to say the least, puzzling. For in an important sense, there are no type sentences. Types are abstractions which do not exist as tokens do. Any given sentence which is used in an utterance has certain features some of which may be used to characterise it as a unique individual (its individuating features) and others of which may be used to classify it as being of the same kind or type as other sentence tokens (its type features). We may, for example, classify it as a unique individual on the basis of its time and place of utterance, or as an individual of a kind in terms of its physical structure. Every individual sentence may therefore be regarded as a token of a type, and to speak of a type-sentence is to speak of a property, or of a class of tokens of a similar structure. Hence any remark about a type-sentence can only be a remark about features of all tokens of that type or, as is more often the case, about an arbitrarily chosen, or typical, token of that type. Individual tokens thus exemplify types, but no token, and indeed no actual item, is itself a type-sentence. To utter or to use a type is just to utter or to use a token which exemplifies a type.

It is now clear that if it is said of a type sentence that it is true, this remark can itself only be true if all tokens of that type are used to make the same true statement. But except in very special cases where we restrict the use of the tokens to a fixed context (as in mathematics), exclude the use of proper names and personal pronouns, and proscribe the possibility of codes, it will never be the case that all tokens of a given type are used to make the same statement, or even that all are used to make true (though perhaps different) statements. We thus reach Strawson's conclusion in the only form in which it is justifiable: that truth and falsity are features of the *use* of token sentences and that, in most cases, it is inappropriate to describe the type, that is, the class of tokens, as true or false.

It is however appropriate to describe some token sentences, namely those which are used to make statements in a specified context, as true or false. For if a token sentence, which is a unique individual, is used to make a true statement on a particular occasion, we may say that it was used truly. We may then transfer this description of the use to the vehicle used. For since the token is unique and the context of its utterance is specified, to say that it is true is simply an abbreviated way of saying that it was used truly, *i.e.* to make a true statement, on that occasion. Similarly, we may say that a given token sentence is false. This use of the words 'true' and 'false' is secondary, being parasitic on their primary use to describe features of statements, and it is a disguised adverbial use. But it is not misleading or inappropriate in the way in which it is both misleading and inappropriate to speak of the truth or falsity of type sentences.

It might perhaps be thought, however, that in certain cases difficulties arise in the case of tokens which are similar to those which arise in the case of types. Suppose, for example, that someone says "Mary is tired" intending this to be a true remark about his wife. Conceivably a hearer might take it to be a false remark about his cow (which he can see jumping around the field). Here, since only one token is uttered, one might draw the conclusion that one and the same token is both true and false, and hence, that it is just as inappropriate to attribute truth-values to sentence tokens as it is to attribute them to sentence types.

What is being neglected in reaching this conclusion, however, is the role of the context. In any real conversation, ambiguity of this sort would become apparent fairly quickly. The speaker might go on to say "Both of her legs are aching" and the hearer would then know that they were at cross-purposes. He might reply "Both? Do you mean all?" At this point the reference of the word 'Mary' is brought into question and the misunderstanding is quickly resolved. Now to establish the reference of subject terms is to take context into account. Indeed we wish to say that, so far as meaning is concerned, the crucial role of the context is simply to establish the actual sense of the sentences used, and most important, what it is that they are being used to mention, *i.e.* their reference, or, more generally, their aboutness.

We adopt the common philosophical use of the word 'reference' which limits its application to existing items (entities) and use the more general word 'aboutness' for non-existent or abstract items, as well as entities. In terms of this, many sentences fail to have a reference. For example 'Pegasus is a winged horse' has no reference; but it is nevertheless about, in a perfectly good sense of 'about', a mythical creature. Hence we may say its aboutness is a mythical creature. (Formal definitions of 'is about' and 'refers to' for words, phrases and sentences, are given in §3.6). When we say, therefore, that the crucial role of the context is that it determines the actual sense and aboutness of the sentences used, it will be clear that we are not restricting the concept of a context to the physical circumstances in which the utterance takes place, nor even to physical circumstances at all. To determine the aboutness of a sentence is not necessarily to seek some actual physical item in the world which the sentence is about; for there may be no such thing. Nevertheless, we can find out what the sentence is about by considering the topic of conversation, the identifications made in the sentences which come both before and after the particular utterance in question (such as "By 'The King of France' I mean 'Louis XIV'"; or "Pegasus is a winged horse in Greek mythology"), the elucidatory side questions which are asked (such as "Both? Do you mean all?"), and so on. All such factors which are relevant to this determination we count as context-determining features. The time and place of utterance, and the physical circumstances in which the utterance takes place, which are sometimes taken to be characteristic of the context, are very often only incidental to this purpose and are only relevant to the extent that they contribute to it.

Now we shall see later (§3.6) that there is a condition of unambiguous aboutness on contexts; that is, subject terms of the same type must have the same aboutness in the same context, or equivalently, subject terms of the same type which have different aboutnesses are being used with respect to different contexts. In terms of this, when X says "Mary is tired" and intends it to be a remark about his wife, he is taking the token with respect to one context, while Y, who hears it as a remark about his cow, is taking it with respect to another. Only one token sentence is involved, and there is only one set of *physical* circumstances of utterance, but there are two contexts; and we have not said that tokens, unqualified, are true or false, but only that we may take tokens in *a specified context* to be true or false. There is, however, no inconsistency in the fact that a given token yields a true statement with respect to one context and a false statement with respect to another. Alternatively, one might wish to say that until the ambiguity is resolved, the context is inadequate to determine what statement has been made and hence that the sentence fails to yield a statement since it is incomplete with respect to aboutness (*cf.* §1.5 below). In either case, an adequate resolution can be given which does not impair the proposal to transfer the descriptions 'true' and 'false' to sentence tokens.

If, therefore, we restrict attention to token sentences uttered in a context, we may take truth and falsity as sentence values. This is the course we propose to adopt. To simplify the discussion we shall use the word 'sentence' to mean 'token sentence', and we shall continue to speak of sentences of a type, or of sentences which exemplify a type, rather than of type sentences.

We do not attempt to state precisely the criteria of identity in terms of which two sentences may be judged to be of the same or different types. We note, however, that they are ordinarily based on sonic or physical similarities and have to be flexible enough to allow two utterances to be counted as being of the same type even though made by speakers with different accents, and two inscriptions to be counted as being of the same type even though written by different hands.

1.4 Significant and Nonsignificant Sentences

We accept with Strawson that it is sentences, rather than statements or propositions, which are significant or nonsignificant. But we do not accept, as Strawson does, that significance is a feature of sentence-types. The difficulties which arise from this supposition are exactly those which arise from the supposition that truth and falsity are features of sentence-types.

If it is held that one and the same sentence can be used by different people then it has to be accepted that one and the same sentence is both significant and nonsignificant, hence that significance and nonsignificance are not features of sentences. For if X uses the sentence 'Rainbows eat flies' to talk about visual phenomena in the sky, then he fails to make a statement; but if Y uses it to talk about trout, then he succeeds in making a statement. In the first case, we may say, the sentence is nonsignificant; in the second case, significant. So if we take it that it is the same sentence which is used, we are committed to saying that one and the same sentence is both significant and nonsignificant, and we must conclude that it is inappropriate to describe sentences as significant or nonsignificant.

Now what is true in this account is that significance and nonsignificance are, like truth and falsity, features of the use of sentences. What is false, is the assumption that we can transfer features of the different uses to features of the one sentence-type. That is why the contradiction arises. But this is not to say that no transfer can be made. The transfer from features of use to the vehicle used can be made unambiguously if the vehicle is unique to the occasion of its use. We may say that X used a token of the type exemplified by 'Rainbows eat flies' nonsignificantly; and that Y used a different token of the same type significantly. And we may then go on to say that the token which X used is nonsignificant while the token which Y used is significant. Or, in terms of our proposed use of 'sentence' (to mean 'sentence token'), we may say that X used a nonsignificant sentence while Y used a significant one. X's sentence and Y's are different, though they are both of the same type.

It might perhaps be thought that we have taken the analogy too far and that significance really is a feature of type-sentences, which is to say a feature of all (token) sentences of a given type, and not at all a feature of their use, as truth and falsity are. For we have simply exploited an ambiguity in the word 'rainbow', and most linguists and lexicographers would wish to explain ambiguity by saying that there is not one ambiguous word but two different unambiguous words which happen to be typographically similar. In terms of this, we might introduce two distinct words 'rainbow_1' and 'rainbow_2' to take the place of the two distinct senses of the old word. We then have to consider two sentences of different types 'Rainbows_1 eat flies' and 'Rainbows_2 eat flies'; and we may say that all tokens of the type exemplified by the first are nonsignificant, while all tokens of the type exemplified by the second are significant. This allows us to say that the two sentence-types are, respectively, nonsignificant and significant.

Such a move, however, introduces a fixed, standard, sense for each of the words employed, and thereby proscribes their use, or the use of sentences in which they occur, in non-standard ways. To take a standard meaning for a word is to limit attention to a standard context, and while this, as we shall see, is necessary for some purposes, it introduces an undesirable limitation on any general theory of language. It means that non-standard uses, which may be quite deliberate and calculated, as in codes, and in the experimental and creative use of language, have to be regarded as incorrect uses. If it were necessary to make such a move in order to formulate a general theory, then the limitation would have to be accepted. But it is quite unnecessary. By taking sentence tokens as the basic elements, any use, whether standard or not, can be given a consistent explanation. In any case, if it were reasonable to limit attention to standard contexts here, it would be equally reasonable to limit it in such a way that truth and falsity could be regarded as features of sentence-types. This indeed is what formal logicians commonly do; and because they do, so they fail to capture important features of natural language, especially the role of context. But it is exactly our purpose to show that this is not a necessary feature of formal logic, but simply a consequence of an unnecessary assumption made by some formal logicians. General laws which allow for the part played by context can be given, whether the context is standard or not.

Even if it were accepted that a limitation to the standard meaning of words is not unreasonable, it would be a quite inadequate restriction. For the significance of many simple sentences cannot be determined independently of their context of use, even if we presuppose a standard unambiguous sense for each of the words employed.

Suppose somebody says "Mary is happy". Just because tokens of the type exemplified by 'Mary' may be used to name people, cows, ships and cyclones, so we cannot, even given a standard sense for 'happy', determine the significance of the sentence independently of its use. Of course, if the context is taken up by identity relations in the sentence itself, there is less difficulty. If somebody says "Mary, my wife, is happy", the use of the proper name, hence the context, becomes irrelevant to the significance of the sentence. A second predicate 'wife' with a presumed standard sense is introduced; and if one knows the standard language, one knows that 'wife' and 'happy' are a significant pair. To make such moves, however, is both to recognise the importance of context and to move a long way from ordinary language in an attempt to preserve the thesis that significance is a permanent feature of sentences. To preserve the thesis adequately we have to presuppose a standard meaning for all predicates and we have to proscribe the use of proper names which are not qualified by some description, or else suppose that every item in the universe has a unique name. We have, that is to say, to presuppose an ideal language.

In general, when significance features depend on naming or mentioning factors, no satisfactory account can be given in terms of standard meaning and sentence types. For the aboutness of a word is not a permanent feature of it; it can only be determined from its use to mention in a context. Thus it would be pointless to explain the fact that Robinson Crusoe could have used a sentence of the type exemplified by 'Friday is in bed' to make a significant remark about Man Friday by claiming that he would then have been using a sentence of a different type, say 'Friday_1 is in bed', from that which is used by someone to make a nonsignificant utterance about a day of the week, say 'Friday_2 is in bed'. For this is not a simple case of ambiguity in the word 'Friday' which can be explained in terms of two distinct senses; it is instead a case of two different mentioning uses, and the number of possible such uses is unlimited. To deal with the problem in this way would be to invoke an indefinitely long list of 'Friday'-types each restricted in use to a predetermined class of aboutness items: people, days of the week, pet dogs, cats, canaries, and whatever. But in ordinary language there are no such lists and no such restrictions. Any proper name or demonstrative pronoun can be used to mention any item, though of course it is the case that some proper names have a standard use to mention items of a certain kind. Standard uses, however, are not the only uses.

We take it then that truth, falsity and nonsignificance may be regarded as three comparable semantic values of sentence-tokens used in a context provided it is recognised that they are not primarily features of the sentences themselves but of their use.

1.5 Incomplete Sentences

If, therefore, we neglect the use of sentences to ask questions, to give orders, and the like, it seems that the way is now open to develop significance theory as a three-valued logic of sentences, where truth, falsity and nonsignificance constitute the set of possible values for any given sentence. For we have shown that the values are comparable, and if we adopt a suitable restriction on the use of sentences (by limiting them to indicative uses), it would seem that they are exhaustive. If a sentence is used in a given context to make a statement, it is either true or false; if it is used in such a way that it fails to make a statement, it is nonsignificant. This would be to adopt Russell's assumption that every (indicative) sentence is either true, false or nonsignificant. It seems, too, that we might adopt Russell's definition: that 'nonsignificant' means 'neither true nor false'. For if a sentence is used in such a way that it fails to make a statement, then it fails to make a true statement or a false one; that is, it is neither true nor false; conversely, if a sentence is used in such a way that it makes neither a true statement nor a false one, *i.e.* is neither true nor false, then it fails to make a statement at all and so is nonsignificant.

Such an assumption, however, would make it impossible for us to take account of sentences which might be held to be significant, which are used indicatively, and yet, because they do not yield statements, are neither true nor false. If, for example, someone utters a sentence of the type exemplified by 'The King of France is wise', when there is no King of France, we could not say, what Strawson says, that a significant sentence has been uttered yet no statement made. We should either have to say that such sentences are nonsignificant or else adopt the Russellian position that a false statement has been made.

Taking 'sentence' to mean 'sentence-type', Strawson points to the fact that anyone uttering 'The King of France is wise' in the reign of Louis XIV would be making a true statement; in the reign of Louis XV, a false statement; and in the present day he would be failing to make a statement at all. Because the sentence can be, or has been, used to make statements, however, Strawson concludes that it is significant: "that if anyone were now to utter it, he would be uttering a significant sentence". This, too, is Russell's view; and because Russell takes it that 'significant' means 'true-or-false', so he is committed to the position that a statement has been made even though there is no King of France. Strawson avoids this conclusion by refusing, as we have seen, to take 'significant' to mean 'true-or-false', since significance is a property of sentences while truth and falsity are not. He is therefore able to say that a significant sentence has been uttered but no statement has been made.

Now we do not wish to say, as Strawson implies, that every sentence token of the type exemplified by 'The King of France is wise' is significant. We wish to provide for the fact which Strawson notes that some tokens of this type have been, or can be, used successfully to make statements and are therefore significant. But we wish also to allow that other tokens of the same type might be nonsignificant (*e.g.* if someone used such a sentence to talk about a ship named *The King of France*), and that yet others might be used in a context where, though 'The King of France' fails to have a reference, the intended reference is of a kind such that a statement would have been made if the item mentioned did in fact exist. In this last case, the sentence (token) is significant even though it is commonly said that it fails to yield a statement.

Sentences which are significant but nevertheless fail to yield statements are, we shall say, incomplete. The incompleteness is semantic, not syntactic; and characteristically, but not essentially, the incompleteness is with respect to reference. As an example of a sentence which is incomplete but not with respect to reference, however, consider 'Everything doubled in size overnight'. Since the criterion for doubling involves some standard which remains constant, there is some breakdown in meaning here which one might wish to recognise yet feel reluctant to classify as a significance failure. Similar problems are raised by examples such as 'This is an East wind' uttered at the North Pole, and 'It is 5 o'clock on the sun'. Again, some, but not all, paradoxical sentences might be regarded as incomplete. Consider, for example, sentences of the type exemplified by 'This sentence is false'. If such a sentence is used to talk about some other sentence which is false, then a true statement has been made; if to talk about another sentence which is true, then a false statement has been made. But if a token of this type is being used to mention itself, then one might wish to say that it fails to yield a statement and is incomplete (see §2.14).

We do not ourselves at this stage wish to say of any particular sentence that it is or that it is not incomplete. The decision in any given case will depend heavily on the context of use. Not only this, but there may be, as we have suggested, different kinds of incompleteness which ought to be distinguished. We do, however, wish to provide for the category of incomplete sentences, and from this point of view we must regard the initial basic logic as four-valued rather than three-valued. That is, any given sentence-token, such as 'Mary is happy', may be used in a context in such a way that it is true (yields a true statement), is false (yields a false statement), is nonsignificant, or is significant but incomplete. Both nonsignificant and incomplete sentences fail to yield statements, and are therefore neither true nor false; but given that there are incomplete sentences, so 'nonsignificant' cannot be taken to *mean* 'neither true nor false'. This definition can be adopted if incomplete sentences are excluded from consideration, and we make this simplifying assumption later (§2.16). Alternatively they can be assimilated as three-valued sentences (§3.9). Initially, however, in the formal development of context logic, we provide for incomplete sentences.

There are of course other alternatives. One might simply adopt the stronger but defective view that incomplete sentences are nonsignificant; or one might hold with Russell that they always yield false statements; or one might claim that they yield statements which are neither true nor false, but this would be to modify considerably the concept of a statement. All but the last of these courses can be expressed within the logic we develop by adopting slightly simpler definitions at certain points.

Finally, it should be noted, completeness and incompleteness, like significance and nonsignificance, are features which apply to all sentences and not simply to those which are used with the intention of making statements. It is only in relation to a statement-making use, however, that they can be compared with truth and falsity.

1.6 Comparable Semantic Values

We conclude, therefore, that if we take token sentences in a context as our basic elements, that is as the items in the range of the sentential variables, then truth, falsity, nonsignificance and incompleteness may be taken as four comparable semantic values. By this we mean that any given sentence token will take one or other of these four values, depending on the context. Compare, for example, the use of sentences of a type exemplified by 'Mary is happy' with respect to the four different contexts: (c_1), in which 'Mary' is used to mention a woman who is in fact happy; (c_2), in which 'Mary' is used to mention a woman who is unhappy; (c_3), in which 'Mary' is used to mention a cyclone; (c_4), in which 'Mary' is used to mention the present Queen of France.

In reaching this conclusion we have taken a good deal for granted. We have assumed some intuitive understanding of the concepts of sentence, statement, context and of the relations between them, and we have expressly introduced a limitation on the kinds of sentential use which are appropriate or admissible (though not a restriction on the kinds of sentences which are admissible). For if a sentence is used to ask a question, it may be both significant and complete yet neither true nor false. To suppose, therefore, that any given sentence takes on one and only one of the four values truth, falsity, nonsignificance and incompleteness is in effect to restrict the application of the proposed logic to indicative acts. This limitation applies to all the significance systems we consider, but we shall show briefly in the next chapter how context logic may be developed to take account of the non-indicative uses of sentences.

Perhaps the most important and fundamental assumption we have made, however, is that it is appropriate to take sentence tokens as the basic elements. And this might be thought to be absurd. For sentence tokens are noises or marks; and in what sense can it be said the noises and marks are true or false? We have of course anticipated this partly by acknowledging that the use of 'true' and 'false' here is derivative; the primary use is in conjunction with statements. To say that a sentence token is true is just to say that it was used on such-and-such an occasion to make a true statement. Even so, it might be said, that which is used to make a statement must have meaning; and in what sense can noises and marks be said to have meaning?

We now try to answer these and related questions.

1.7 The Meaning of Sentences

The meaning of a sentence, or its lack of meaning, when it is used with the intention of performing an illocutionary act, depends both on general features of the language to which it belongs and on features of the context in which it is uttered.

§1. *The Background Language*

Not all utterances are linguistic. We may make noises and marks without intending to perform, or in fact performing, an illocutionary act. To qualify as linguistic the sounds or marks made in an utterance must satisfy certain general conditions which determine features which they have as sounds or marks *of a language*. We shall therefore assume, for the logic we develop, that the values of the variables, whether these are words, phrases or sentences, are tokens of some definite natural language, which we call the background language. This need not be the same for every application of the logic, though we frequently take it to be English, but for any application it must be some specified or specifiable natural language characterised, as a minimum, by a vocabulary, a semantic dictionary and a grammar.

The vocabulary consists of a set of tokens deemed to be acceptable elements of the language and such that any other token of the types which they exemplify are also acceptable elements of the language. The tokens selected in this way from the total set of tokens which it is possible to utter are, we shall say, *words* of the background language. Commonly the vocabulary will be open-ended; that is, it may expand or contract as a result of usage and convention. An account of the way a vocabulary begins, grows, develops and atrophies, is not our concern here, though we take it that there will be historical and causal explanations of various kinds.

The semantic dictionary consists of a set of conventions which associate with some elements of the vocabulary what we propose to call a standard sense, and with all elements a standard, or typical, use.

The standard sense may be given by associating one element of the vocabulary with other elements of it, that is, by an ordinary dictionary entry; or it may be given by a definition which is not usually to be found in an ordinary dictionary (such as the definition of 'well-formed-formula' for a particular formal system); or it may be given by conventions which cannot be written down, but which the user of the language will know, conventions of the kind which are often explained by an appeal to ostensive definition.

The standard use of a word may be given simply by indicating the semantic category—noun, verb, preposition, *etc.*—to which tokens of the type which it exemplifies belong. To say, for example, that a word is a common noun is to say that its standard or typical use is to classify; to say that it is an adjective is to say that its standard use is to describe. Exactly which items or features the word is typically used to classify or to describe is given by specifying its standard sense.

Not all words of a language have a standard sense (*e.g.* some proper names, referring pronouns, and connectives of various kinds, may not have), but all have a standard use. The typical or standard use of a proper name, for example, is simply its use to name; and we may often be more specific and say, *e.g.*, that the typical use of tokens of the type exemplified by 'Mary' is not only to name, but to name female humans.

Some words have more than one standard sense associated with them and more than one standard use. Tokens of the type exemplified by 'rainbow', for example, have more than one standard sense (they may be used to classify trout or visual phenomena in the sky); and tokens of the type exemplified by 'black' have more than one standard use (they may be used as nouns, verbs and adjectives). In the first case we say that an arbitrarily chosen token of the type is ambiguous, and although in most cases such ambiguity can be resolved by distinguishing two types to take the place of the original one, this is not commonly done in natural language and we shall later give a formal account of ambiguity which does not depend on this assumption. In the second case we say that the typical use is indefinite.

So conceived, the semantic dictionary is an ordinary lexicon together with what might be called the working definitions of the language and, most important, those semantic conventions or rules which relate elements of the vocabulary to abstract or actual items which are not themselves elements of the vocabulary. The semantic dictionary is not, therefore, simply a syntactic lexicon which gives rise solely to formal synonymy or equivalence transformations on elements of the vocabulary. It consists also of the whole network of unwritten conventions, learnt in the course of learning a language, which relate words to non-words, whatever these conventions may be and however difficult it may be to explain them or to describe them. That they are unwritten, or indeed cannot be written, is unimportant. What matters is that they can be taught and learnt.

We do not at this point attempt to explain further these conventional relations between words and non-words. At this stage, therefore, the concept of standard sense is a primitive. It can of course be given an explanation in terms of other primitives, and it is just this which philosophical theories of meaning seek to do. The primitives in terms of which the explanation is given may be, for example, universals, properties, abstract ideas, concepts, classes, or, more recently, rules of use. But we wish at this point to avoid these complexities since the logical systems we propose to develop are independent of any such particular theory. Later we shall give a more detailed account of standard sense, though in a way which still leaves it independent of particular philosophical theories.

We shall say that to give the standard sense(s) of a word together with its standard use(s) is to give its standard meaning(s). Hence the semantic dictionary gives the standard meaning(s) of words—as an ordinary lexicon is said to give the meanings of words by listing their semantic classifications and senses. The function of the semantic dictionary, therefore, is the same as that of an ordinary lexicon, for it is simply an augmented lexicon. Our use of the word 'standard' is not meant to mark any difference between the function of an ordinary lexicon and the function of the semantic dictionary: an ordinary lexicon gives standard senses, uses, and consequently, meanings. The word 'standard' is meant to mark a distinction, which we shall consider presently, between features of a word which are characteristic of it simply by virtue of its being an element of a language and features which it has by virtue of its actual use in a context.

Words for us, therefore, are marks and noises. They are not abstract entities, whether types or something else, but neither are they simply marks and noises. They are marks and noises which may be said to "have" a standard meaning because of the conventional associations between tokens of the type which they exemplify, on the one hand, and standard sense (if any) and typical use, on the other. We may, if we wish, describe this by saying that the word-types are related to, or have, a standard meaning, provided it is remembered that this is simply an abbreviated way of talking about a class of tokens, or about the type features of an arbitrarily chosen token of the class.

The grammar of a language consists of a set of conventions which, as a minimum, determine acceptable strings of words, namely phrases and sentences. By 'grammar' we mean idiomatic grammar, not text-book grammar.

Just as words have a typical use, so too, do word strings. Compare for example, the typical use of phrases of the form 'the so-and-so' to refer to unique items; or the typical use of a sentence such as 'The cat is on the mat' to give information and 'Is the cat on the mat?' to ask a question. Word order, verb mood, the use of different articles, and special supplementary devices like question marks, all contribute to the typical use and are part of what we regard as the grammar of the language.

We now extend the notion of sense to word strings and say that the standard sense of phrases and sentences is a function of the standard meaning of the words which comprise them and the grammatical construction. We do not at this point attempt to explain the precise relationship which holds between the standard meanings of the words and the standard sense of the string, but we give a detailed formal account later.

Now although in the case of words we have said that to give the standard sense and use is to give the standard meaning, we cannot say the same thing in the case of word strings. For the string may have a sense and its use may be quite definite, yet it fails to have a meaning. It may, that is to say, be nonsignificant. For example, given the standard meanings of 'purple', 'number' 'one' and 'two' and the fact that phrases of the sort 'The so-and-so' have a typical use to refer, the string 'The purple numbers one or two' has a standard sense and a typical use, but it fails to have a standard meaning. There is in fact a clash of senses; and this is so even though it is a grammatically acceptable string.

It might perhaps be thought that we should say that such strings have no standard sense rather than, as we have said, that they have a standard sense but no standard meaning. But this seems to us to be a less informative way of describing them. For the fact is that each of the words in the string has a standard meaning and the way in which they are put together contributes to the sense of the string as a whole. There is, for example, a difference in sense between 'The purple numbers one or two' and 'The one or two purple numbers', even though both are nonsignificant. It is in fact just because we "see" the standard sense of the phrases that we can say they are nonsignificant. Similarly with sentences. It is because we understand, in one sense of 'understand', 'The colour blue procrastinates furiously', that we can say it fails to have a standard meaning.

To give the standard sense and typical use of a word string is, therefore, to give important information about it even though it is not to give its standard meaning, for it may fail to have one. To mark this, we shall say that it is to give the standard illocutionary force of the string. All phrases and sentences, therefore, that is all grammatically acceptable word strings—as distinct from gibberish (strings which contain some non-words) and garbled strings (all of whose elements are words, but in an ungrammatical sequence: *e.g.* 'if when the ill')—have an illocutionary force but may or may not have a standard meaning.

In saying that some strings fail to have a *standard* meaning we are not saying that *all* tokens of the type which they exemplify are meaningless in all contexts. A given token of such a type may be used with a non-standard sense or in a non-standard way. Phrases of the type exemplified by 'the purple number seven' have no standard meaning. This is true of all tokens of this type. But particular tokens of this type may be used meaningfully with a non-standard sense—*e.g.* to mention the house at 7, Arcadia Avenue, which happens to be painted purple. This is one of the reasons why we use the word 'standard'.

Sentences and phrases, then, like words, are marks and noises; but they are not simply this. They are marks and noises which "have" an illocutionary force derived from the conventions of the language to which they belong. The illocutionary force may be said to give the standard meaning of the string, provided it has a standard meaning at all; *i.e.* provided it is not nonsignificant with respect to its standard sense.

§2. *The Context*

As should now be clear, the typical use of a word, phrase or sentence may be different from its actual use, and its standard sense may be different from its actual sense on the occasion of its use.

We have illustrated this difference by pointing to the fact that all tokens of the type exemplified by 'The purple number seven' fail to have a standard meaning, *i.e.* are nonsignificant, and yet particular tokens of this type may be significant on the occasion of their use. But it is not only with respect to significance that such differences may arise. There are in fact many ways in which standard features may differ from actual features.

First, considered purely as an element of a language and independently of its actual use, a token may be ambiguous or have an indefinite typical use. When it is actually used, however, the use may be, and commonly will be, quite definite, and the ambiguity resolved.

Secondly, the actual use may be quite different from the typical use. By means of changes of intonation or the use of italics and similar devices, it is possible to ask a question or to give a command by uttering a sentence which has a syntactical structure normally employed in sentences used to give information, and to ring the changes in other ways.

Thirdly, the actual sense may be different from the standard sense. Compare, for example, the standard sense of tokens of the type exemplified by 'The cat is on the mat' with the actual sense which a given token of this type would have if someone used it to make a remark about his wife, or again, if a different token of the same type were used to make a remark about the typist who is being reprimanded by the boss. Here the standard senses of the words differ from their actual senses, and consequently the standard sense of the sentences differs from their actual senses. The limiting case of such differences is represented by codes of the kind 'The sky is blue' to mean 'The attack takes place at dawn', which in effect determine new non-standard senses or uses for words and sentences on particular occasions of their use.

Fourthly, even in cases where there is no conflict between the actual sense and use of a sentence and its typical sense and use, the standard meaning cannot always be identified with the actual meaning, even when it is significant. Consider, for example, the difference between the illocutionary force of

'Mary is happy' (*i.e.* its standard meaning since it is presumed significant) and the actual meaning which it would have if it were used by someone to make a remark about his wife. We may know from the general features of the language that 'Mary' is typically used to refer to human females, and the actual use does not conflict with this; but it does enrich it. In such a case the standard meaning of the sentence can only be said to coincide with an attenuated meaning, not with its actual meaning. And this is often the case. It is almost always so when demonstrative pronouns and proper names are used.

Finally, just as a sentence may fail to have a standard meaning and yet be used meaningfully, so a sentence may have a standard meaning and yet be used nonsignificantly. 'Mary is happy' has a standard meaning since 'Mary' is typically used to refer to female humans and the standard sense of 'happy' is appropriate to human beings. But if a token of this type is used to make a remark about a matchbox, then it has been used nonsignificantly. Here there is a clash between the typical use of 'Mary' and its actual use.

These differences should not be over-emphasised, but neither should they be ignored. In pointing to them we do not wish to suggest that every utterance, or even most, is non-standard. Clearly if this were so, there would be no language and no communication. In many cases the standard meaning does coincide with the actual meaning. If a teacher of current affairs utters 'The official residence of the Prime Minister of England is 10 Downing Street' in the course of a lecture, the actual meaning of the sentence will very likely be its standard meaning. But that there can be differences is enough to show that the meaning of linguistic elements (or their lack of meaning) on the occasion of their use, is not determined solely by general linguistic features; other factors have to be taken into account.

Because linguistic elements are linguistic, so their mere utterance creates a presumption with respect to their meaning since they have standard features which anyone who knows the language will recognise. And this presumption is very often correct. That it may on occasion be mistaken, however, means that it may sometimes have to be modified and that it must always be assessed. Those factors in terms of which this presumption is assessed, that is, either confirmed or modified, we shall call features of the context of utterance.

The notion of context is inevitably vague, but we hope to have avoided some of the difficulties by introducing it in terms of its role rather than its features. The kinds of features which are relevant to its role will vary from one context to another. Commonly they will include the time and place of utterance, the names of those involved in the conversation, and a description of the physical circumstances in which the utterance takes place. But on occasion none of these may be necessary. What is important in the specification of a context is that all the information is given which is relevant to the determination of the actual sense, use and aboutness of the sentences which are uttered, insofar as these deviate from, or are not given by, the semantic dictionary of the background language. And this will normally include the topic of conversation, the identifications made (*e.g.* that 'Charlie' is being used to mention the man next door or, simply, that Charlie is the man next door) and the special assumptions and presuppositions in terms of which non-standard senses can be determined, ambiguities removed, and indefinite uses made definite.

An alternative but equivalent way of marking the difference between standard and actual meaning is to say that the standard meaning of a linguistic element is given by that sense and use which it has with respect to a standard context. In terms of this, the vocabulary, semantic dictionary and grammar of a language determine a standard context. Then, an actual utterance may be with respect to this standard context, in which case the actual meaning, or lack of it, coincides with its standard meaning, or lack of it; or it may be that the utterance is made with respect to a non-standard context, in which case there is a difference between the actual linguistic features and standard features. On the occasion of any utterance, however, the presumption is that the context is standard. That a context is standard, however, by no means determines all features of it. The time of utterance, details of speakers, and ambiguities of words, remain to be specified; most important, if pronouns and proper names occur, it has to be determined what items they are being used to mention.

All relevant features of the context, whether standard or not, may be described by using sentences, so that, from a logical point of view, a context may be represented by a set of sentences, namely those which specify the context.

These conclusions shape the development of the logic in the following way:

We introduce variables which take as values words and phrases, and other variables which take as values sentences. Though sameness of type determines sameness of variable, the values of the variables are token inscriptions, elements of the background language. It is always significant to say of them, therefore, that they have, or fail to have, a standard meaning, even though they are tokens. We are, however, concerned with the actual features of tokens on the occasion of their use and it is necessary to take account of those contextual factors which may modify the standard features. We therefore introduce context variables to qualify the initial variables. The complex symbols so constructed then stand for actual utterances in a context. By distinguishing one context from another it is then possible to express facts of the kind that two sentences of the same type may be used to make different statements, or that one may be used to make a statement while the other is nonsignificant, or that two sentences of different types may be used to make the same statement.

1.8 Criteria of Nonsignificance

Given that the values of the variables are limited to words, phrases and sentences of a background language, then we automatically exclude from consideration both gibberish and garbled word-strings. Moreover, words, considered simply as linguistic elements or with respect to their use in a context, are never nonsignificant. As part of the language they always have a standard meaning or meanings, and though they may be ambiguous, and the ambiguity may even remain when they are used (if the context is in some way inadequate), this is not to say that they lack meaning but only that the meaning is indefinite.

These requirements are, as it happens, inessential. At the expense of complicating the logic it would be possible to introduce variables which take as values gibberish, garbled strings, and non-words. But these additional cases are of little philosophical interest and the further complexity is not warranted. Given these conditions, then, the problem of nonsignificance is limited to grammatically acceptable word strings: that is, to phrases and sentences.

Consider, first, sentences. If we limit attention to indicative illocutionary acts, and exclude incomplete sentences, then the over-riding criterion of nonsignificance is that a sentence is nonsignificant if it fails to yield a statement in a given context. And in terms of the presumed restrictions, this can be expressed by adopting the definition that 'nonsignificant' means 'not true and not false'. That a sentence fails to yield a statement, however, can be given different explanations. We may say, for example, that the sentence is unverifiable (a unit account) or that there is a clash of sense in the words employed or that the intended reference is not of the logical kind or type to which the predicate is appropriate (factor accounts). And whatever account is given, it is possible to distinguish the significance or nonsignificance of a sentence with respect to its standard features from its significance or nonsignificance with respect to its actual features.

Now although our main interest is in the actual features of sentences when used, standard features are by no means irrelevant and indeed contribute to the evaluation of significance. Suppose, for example, someone uses the sentence 'Tom is happy' in context c_0 to make a remark about a matchbox. Let us suppose, too, that 'happy' is being used with standard meaning. Then we may say that 'Tom is happy' (c_0) is nonsignificant, and we may give as an explanation that 'Tom' is being used to mention a matchbox and matchboxes are not of the logical kind to which 'happy', with respect to its standard sense, is appropriate. This context-dependent explanation can, however, be made context independent if, on the basis of the context, we affirm that Tom is a matchbox, and hence, that the significance value of 'Tom is happy' with respect to c_0 is the same as the significance value of 'A matchbox is happy' with respect to the standard context. Thus, given the identification of the reference (a matter which can only be determined from the context), significance features can often be evaluated in terms of standard features.

This illustrates a technique which we shall use a good deal, in examples, and for the purposes of illustration, but by itself it is in no way an explanation of nonsignificance. We may say, perhaps, that there is a clash of sense between 'matchbox' and 'happy', but this depends on an intuitive understanding of what constitutes a clash of sense. A full explanation of sense clashes cannot be given, however, until an explanation of sense is given. To the extent that the concept of sense is a primitive, the concept of a sense clash is inadequate. We shall indicate briefly in Chapter 3 a preliminary analysis of sense clashes in terms of significance ranges, but we leave the full development, and the detailed account of sense, to Part III.

The problem of significance for phrases is more difficult and we leave the main discussion of it to Part III. A brief indication of the problem is given at various places in Chapter 7.

CHAPTER 2

Elements of the Sentential Theory

So far we have tried to give an informal account of the features which sentences have considered purely as linguistic elements, and the actual features which they have when used. A sentence, we have said, is a token, in fact a string of marks or noises which satisfies certain syntactic and semantic conditions associated with the type to which it belongs. In terms of these conditions, a sentence may be said to have syntactic and semantic features independently of its use. When the sentence is used, however, these semantic features which are determined by the background language will often be modified by contextual conditions, and its actual semantic features on the occasion of its use will be a function of general linguistic conditions and conditions of the context. Sometimes there will be no difference between the context-free semantic features of a sentence and its actual semantic features. Sometimes the difference will be slight; and in other cases, as we have seen, there will be important differences.

We now try to give formal expression to these ideas. In this chapter we begin the development of context logic by marking the distinction between sentences and statements and concentrating on the use of a sentence in a context to make a statement. In Chapter 3, we extend the analysis to the parts of sentences, in particular to their contextual subjects and predicates, and so take the first steps towards a formal account of referring and saying.

2.1 Sentential Variables and Constants

We first introduce,

sentential variables: $p, q, p', q', p'', \ldots$

The values of these variables are actual sentences, that is, utterances, or tokens, which satisfy the conditions for a sentence imposed by the background language and which have an illocutionary force. No other conditions are imposed on these sentences. They may be either complete or incomplete when used in a context, they may be either significant or nonsignificant, and they may either yield or fail to yield statements. Such possibilities must be admitted if we are concerned not only with successful indicative acts, in which statements are made, but also with unsuccessful acts. We recall, too, that a sentence which is used in an indicative act need not itself be a grammatical indicative. Such limitation as our interest in statement-making imposes is a limitation on the act rather than on the form of the sentence which is used in performing it. In fact, however, even this limitation is unnecessary in the early part and we shall show briefly how other illocutionary acts may be considered formally.

The actual sentences which are values of the sentential variables we call constants in the range of the variables, or briefly, sentential constants or constants. Sometimes, however, it is useful to use an abbreviation for an actual sentence and for this purpose we introduce,

formal sentential constants: $p_0, q_0, p'_0, q'_0, p''_0, \ldots$

Any one of these symbols may be used to abbreviate an actual sentence. We call these symbols, too, sentential constants, or simply constants.

The sentential variables are type variables. By this we mean two things. First, that although the list of variables which has been given is a list of letter tokens, since each exemplifies a type, we adopt the common convention that all letter tokens of these types are likewise variables. Secondly, if two or more letters of the same type occur in a formula, they stand for constants of the same type. Hence, if a given token variable is replaced by a particular sentential constant in a formula, any other occurrence of a token variable of the same type in that formula must be replaced by a sentential constant of the same type. This is the principle of uniform substitution. It does not preclude the replacement of two different variables by sentences of the same type, though the important use of variables of different types is that they can be replaced by sentences of different types. In what follows, therefore, when we speak of the same or different variables we mean variables of the same or different type.

It is sometimes said that although the actual substitution values of the variables are sentences, the *range* of the variables is statements or propositions. This is sometimes supported by claims of the kind that sentences yield statements or propositions, and sometimes, even, that they name them. Such an account is not appropriate here, however. For we are concerned to admit as values of the variables sentences which fail to yield statements as well as those which do. We do not therefore specify a range for the variables distinct from their substitution range. Instead, we give an explicit formal account of what it is for a sentence to yield or to fail to yield a statement in a context. In terms of this account it is then possible to derive a restricted sentential theory of statement-yielding sentences (in fact the classical two-valued logic of sentences) for which it might be said that the substitution range is sentences while the derived "designation" range is statements (see §3.8).

2.2 Connectives and Semi-well-formed Formulae

Sentences may be combined by connectives to obtain compound sentences and we reflect this by introducing,

sentential connectives: $\sim$, &, $\vee$, $\supset$, $\equiv$

interpreted respectively as,

not; and; or (inclusive sense); materially, if ... then ...; materially, ... if, and only if, ...

In classical sentential logic these are not all introduced independently; normally the first two would be taken as primitive and the remainder defined in terms of them. We cannot however assume the adequacy of the standard definitions at this stage since it is not obvious that they hold uniformly over all sentences even though it is accepted that they hold uniformly over all statement-yielding sentences.

We now define the basic semi-well-formed formulae (semi-wffs) by means of the following rules. The symbols Γ and Δ are metavariables which range over the semi-wffs, and the sentential connectives and parentheses are used autonymously. We show later, however, how metavariables may be eliminated in favour of a quotation-function notation (§2.4).

R1 (i) sentential variables and constants are semi-well-formed formulae.
(ii) If Γ is a semi-wff then $(\sim\Gamma)$ is a semi-wff.
(iii) If Γ and Δ are semi-wffs then $(\Gamma \& \Delta)$, $(\Gamma \vee \Delta)$, $(\Gamma \supset \Delta)$ and $(\Gamma \equiv \Delta)$ are semi-wffs.

We take for granted a standard convention concerning the relative strength of the connectives which allow brackets to be omitted (see Kleene [2], pp. 74-75). Further clauses are added to this rule as we proceed.

The formulae defined here are described as semi-wffs rather than wffs because they do not, initially, stand as admissible formulae of context logic but only as well-defined parts of such formulae.

An immediate problem arises concerning compound sentences. For since the sentential variables range over both significant and nonsignificant sentences, by permitting compounds to be formed without restriction, we thereby raise the question of whether a compound with a nonsignificant component is itself significant or nonsignificant. This difficulty is taken up in Part II where various sentential logics are developed in detail. It is unnecessary to take it up here since in context logic we do not *use* formulae which take nonsignificant sentences as values but only *mention* them. This is explained fully below.

2.3 Context Variables and Constants

Since in general the actual meaning of sentence, or its lack of meaning, can only be determined given its context of use, it is necessary to provide for this dependence by introducing context symbols,

context variables: $c, d, c', d', c'', \ldots$
context constants: $c_0, d_0, c'_0, d'_0, c''_0, \ldots$

which qualify sentential formulae, and the further formation rule:

R1 (iv) If θ is a context symbol and Γ is a semi-wff, then $\Gamma(\theta)$ is a semi-wff, provided that none of the sentential symbols in Γ are already qualified by context symbols.

Hence the following are semi-wffs:

$p(c)$, $q_0(c)$, $p(c) \vee q(d)$, $q \supset p(c_0)$, $p(c) \vee q(c)$, $(p \vee q)(c)$.

We now adopt what we propose to call the worlds assumption (by analogy with the logic of situations—see §7.2—according to which all situations are states of affairs or worlds), namely that if a compound semi-wff has the *same* context symbol on all components, the symbol may be removed from each component and written after the whole formula. This assumption can subsequently be expressed by means of equivalence relationships, *e.g.*: $\Gamma(\theta) \vee \Delta(\theta) \equiv (\Gamma \vee \Delta)(\theta)$, *etc.* for each sentential connective. For the present, however, we operate with the assumption as a rule which enables us to replace $\Gamma(\theta) \vee \Delta(\theta)$ by $(\Gamma \vee \Delta)(\theta)$, and conversely; and similarly for each connective. This assumption, that contexts are distributive about the standard connectives, holds for all ordinary contexts, but it fails for some fairly natural extensions of the notion of context: *e.g.* in the context of someone's beliefs; in the context of Restoration manners. The assumption therefore expresses a limitation, though not one which we find restrictive (until §7.2). In terms of it, formulae such as '$(p \vee q)(c)$' and '$p(c) \vee q(c)$' may be regarded as only notationally different; without it, such formulae have to be distinguished and, since they may satisfy different conditions, the logic becomes more complicated.

It is in fact the worlds assumption which underlies the proviso on R1 (iv). The proviso excludes formulae such as '$p(c)(d)$', '$(p(c) \vee q(c))(d)$' *etc.*, and where no compound sentence is involved, as in '$p(c)(d)$', the exclusion can be justified by claiming that if a sentence is evaluated with respect to a given context, the evaluated sentence cannot then itself be evaluated with respect to another context. But without the worlds assumption this explanation is insufficient in the case of formulae such as '$(p(c) \vee q(c))(d)$'. It is only if it is accepted that '$p(c) \vee q(c)$' can be replaced by '$(p \vee q)(c)$' to obtain '$(p \vee q)(c)(d)$', or alternatively that 'd' may be distributed to obtain '$p(c)(d) \vee q(c)(d)$', that it can be said that '$(p(c) \vee q(c))(d)$' is similar to '$p(c)(d)$' because it is implicitly a case of the same sentence(s) being qualified by different contexts at the same time, and hence justifiably excluded. If the context symbols were not distributive, it could be held that although 'p' and 'q' are qualified in '$p(c) \vee q(c)$', '$\vee$' is not, and therefore '$(p(c) \vee q(c))(d)$' is acceptable since 'd' is required to fix the context of '$\vee$'.

We secure the assumption formally by adding a further clause to R1, though the required rule is not strictly a formation rule:

R1 (v) The following pairs of semi-wffs are interchangeable:

$$\begin{array}{lll} \sim(\Gamma)(\theta) & \text{and} & (\sim\Gamma)(\theta) \\ \Gamma(\theta) \vee \Delta(\theta) & \text{and} & (\Gamma \vee \Delta)(\theta) \\ \Gamma(\theta) \,\&\, \Delta(\theta) & \text{and} & (\Gamma \,\&\, \Delta)(\theta) \\ \Gamma(\theta) \supset \Delta(\theta) & \text{and} & (\Gamma \supset \Delta)(\theta) \\ \Gamma(\theta) \equiv \Delta(\theta) & \text{and} & (\Gamma \equiv \Delta)(\theta) \end{array}$$

The analogy with situations, states of affairs, or model sets, can now be made explicit, for, where H is a situation, we could transform these conditions and write them as:

$$\Gamma \notin H \equiv \sim \Gamma \in H$$
$$\Gamma \in H \vee \Delta \in H \equiv (\Gamma \vee \Delta) \in H$$
$$\Gamma \in H \,\&\, \Delta \in H \equiv (\Gamma \,\&\, \Delta) \in H \quad \textit{etc.}$$

In such a case, however, H is a world.

We have chosen to represent the dependence of a sentence on its context by means of the functional notation $\Gamma(\theta)$ in order to emphasise that the semantic features of a sentence, its significance, completeness, meaning and so on, are a function of its use in a context. Thus, by using context symbols we are able to represent the use of sentences of the same type in different contexts by complexes of the sort,

$$p(c); \quad p(d);$$

which are respectively interpreted as,

p with respect to context *c*; *p* with respect to context *d*.

Hence, although the sentential variables stand for any token sentence of a given type, *i.e.* are type variables, when they are qualified with context variables they represent actual tokens on the occasion of their use and we are able to express the different uses of tokens of the same type. Similarly, by using different sentential variables with the same context variable we can represent the use of sentences of different types in the same context by complexes such as,

$$p(c); \quad q(c);$$

which are respectively interpreted as,

p with respect to context *c*; *q* with respect to context *c*.

These distinctions enable us later to consider the use of sentences of the same type to make different statements in different contexts, and the use of sentences of different types to make the same statement in the same or different contexts. If it were necessary to distinguish the utterance of two or more sentences of the same type in the same context, we could use numerical subscripts on the sentential formulae. So, for example,

$$p_1(c) \quad \text{and} \quad p_2(c)$$

represent two different tokens of the same type with respect to the same context. Such subscripts are rarely used, however, and we do not introduce them in the formal theory. They are unnecessary if two different tokens of the same type are uttered in different contexts since our conventions concerning the use of sentential and context variables are sufficient indication of this.

A context is defined by a set of descriptions which give the time and place of utterance, the topic of conversation, the identifications made, and similar detailed information. Hence the context variables range over such sets. That is to say, over sets of sentences which are themselves significant and complete. For example, the specification of a context for a particular utterance may take the following form: The time is 4 p.m. on Wednesday 17 January 1968. The place is the public bar of the Cross Keys Hotel, St Andrews. The speaker is Joe Bloggs. The name 'Bill' in the sentence he just uttered was used to refer to Bill Todd; *etc.* The amount of detail which may be necessary to describe a given context uniquely, however, is unimportant. What matters is whether sufficient indication is or can be given to determine the actual meaning of the sentence on the occasion of its use, and this will depend partly on the identification of the items mentioned by the subject terms (though these need not of course be physically present at the time and place of utterance, and indeed cannot be if they are abstract) and partly on the intended sense of both the subject and predicate terms (which need not be their standard sense). For our purposes, therefore, we may select from a given set of sentences which specify a context those which are relevant to the determination of the actual sense and aboutness of the utterance in the context. From this point of view, a context variable or constant is a vector with two components a and s, the aboutness and sense factors of the context. We could therefore write $\Gamma(a_\theta, s_\theta)$ in place of the simple $\Gamma(\theta)$ to indicate that we are limiting the specification of the context to the specification of its aboutness- and sense-determining factors, but this is an unnecessary symbolic complication and we shall not introduce it formally.

2.4 The Sentential Quotation Function

Since our primary concern is to mention sentences in order to refer to their use, as in,

Jack used a sentence of a type exemplified by 'Le monde est rond' to make the statement that the world is round,

and since we wish to generalise such assertions in order to develop a formal theory, it is necessary to use variables in a quotation context and yet retain their use as variables. In terms of the standard conventions concerning the use of quote marks, however, this is impossible. For if an expression is enclosed in quote marks, we thereby form the name of the expression quoted. Hence, a complex such as,

'*p*'

is the name of the sixteenth letter of the alphabet. If, therefore, we seek to generalise examples of the kind given above by writing,

Jack used a sentence of a type exemplified by '*p*' to make the statement that *q*,

we express the fact that Jack used the sixteenth letter of the alphabet to make a statement. Here the quoted variable no longer functions as a variable and is not open to substitution. There is besides a second difficulty arising from the standard conventions even where we do not wish to introduce variables. For if what we form by the use of quotation marks is the name of the expression inside them, then it is illegitimate to read inside the quotes in order to take into account the sense of the quoted expression since the whole complex including the quote marks is a unity, *i.e.* one name. Such a convention is therefore quite inappropriate if we wish to mention the semantic features of a sentence and is only useful where the mentioning is limited to syntactic features. Most of the mentioning we wish to do, however, is semantic and we therefore have to drop the standard account.

We therefore begin by introducing quote marks which permit semantic mentioning and then solve the problem of generalisation by means of a quotation function which takes quoted expressions as values.

We say that a letter, word, phrase or sentence surrounded by single quotes is a quotation expression and that the use of such expressions is to be in conjunction with arbitrarily chosen tokens of a given type in order to mention them. A quotation expression is not the name of the word, phrase or sentence inside the quotes, instead the quotes function rather as parentheses which permit the expressions inside them to be read so that the sense can be taken into account. Quotation expressions may therefore be used for both semantic and syntactic mentioning.

In order to mention actual tokens, rather than arbitrarily chosen tokens of a given type, we combine the use of quote marks with parenthetical contextual identifications. Thus, if c_0 identifies a definite context, then

The sentence 'Le monde est rond' (c_0) was used by Jack

makes reference to a specific token of a given type even though the actual token-sentence enclosed in quotes is not the very one which Jack used but only of the same type as one which he used.

We now generalise these conventions for use in conjunction with variables. This is still necessary, for even though we now have a looser convention for ordinary quotes than the standard one, in order to admit semantic mentioning, it remains true that if we seek to use them in order to mention an arbitrarily chosen sentence, by placing them round a variable, we only succeed in mentioning an arbitrarily chosen token of the letter type and we lose its use as a variable.

We therefore introduce,

the basic sentential quotation function: $qu(\quad)$

interpreted as,

quotes ( ),

and used in accordance with the following convention:

If Γ is a semi-wff which does not contain a context symbol and if values are substituted for one or more of the variables in Γ, then the value of $qu(\Gamma)$ is the quotation expression of the value given to Γ.

This, together with our conventions concerning quote marks, guarantees that $qu(\Gamma)$ has a unique value for any given value of the argument Γ. The relation is, therefore, appropriately many-one and hence is a function.

So, for example, if

$p =$ The world is round

then,

$qu(p) =$ 'The world is round'.

Similarly, if

$p =$ the world is round; $q =$ the sun is shining

then

$qu(p \& q) =$ 'The world is round & the sun is shining'.

When values are substituted for the variables, therefore, the value of the whole formula in the scope of the basic quotation function is a quoted expression. It is sometimes useful, however, to limit quotation to part of the scope, and for this purpose we may introduce subscripts on the function to indicate that the quotation is only intended to operate with respect to a specified part of a formula.

Thus, given a formula such as,

Jack used p to make the statement that q

then,

qu_p (Jack used p to make the statement that q)

indicates that when values are substituted for the variables, the value of 'p' is a quoted expression. For example, putting 'It is raining' for 'p' and 'rain is falling' for 'q', the value is:

Jack used 'It is raining' to make the statement that rain is falling.

Similarly, the value of

$qu_{p,q}$ (Jack used p and Bill used q to make the same statement)

for the same values of the variables, is

Jack used 'It is raining' and Bill used 'Rain is falling' to make the same statement.

By contrast, the value of,

qu (Jack used p and Bill used q to make the same statement)

is the nonsense,

'Jack used it is raining and Bill used rain is falling to make the same statement'.

Clearly, if f is some sentential function, then

$$qu_{p,q}f(p,q)$$

is equivalent to

$$f(qu(p),qu(q)).$$

For example,

$qu_{p,q}$ (Jack used p and Bill used q to make the same statement),

is equivalent to,

Jack used $qu(p)$ and Bill used $qu(q)$ to make the same statement.

It is always possible, therefore, to achieve these limited cases of quotation in terms of the basic function, and for this reason we shall not introduce the subscripted function in the formal system. It is sometimes useful, however, in the informal metatheory.

The use of the quotation function enables us to generalise assertions which mention arbitrarily chosen tokens of a given type, and their use in the formal theory corresponds to the use of single quotes in conjunction with actual sentences. We can however particularise to the formalisation of assertions which mention specific tokens by using context variables, just as, in the case of single quotes, we particularised to the mentioning of specific tokens by adding definite contextual identifications. Thus, we may now represent the generalised form of assertions such as,

The sentence 'Le monde est rond' (c_0) was used by Jack,

by

The sentence $qu(p)(c)$ was used by Jack.

Here both variables are open to substitution, but the variable 'c' is not quoted. Alternatively we may write,

qu_p (The sentence $p(c)$ was used by Jack).

It is possible to use quotation-function expressions in place of the metavariables, Γ, Δ, *etc.* which occur in the formation rules. For we may say that sentential variables are semi-wffs, that $qu(p \& q)$, $qu(p \vee q)$, *etc.* are semi-wffs, and that if a semi-wff is substituted for a variable throughout a semi-wff, the result is a semi-wff. To keep the symbolism as simple as possible, however, we shall continue to use metavariables in formation rules.

We now provide for the use of the basic quotation function in the formal theory by defining partial well-formed formulae (partial wffs) as follows:

R2 (i) If Γ is a semi-wff not containing a context symbol and $\Gamma(\theta)$ is a semi-wff, $qu(\Gamma)$ and $qu(\Gamma)(\theta)$ are partial wffs.

Later, in Chapter 3, we extend the use of the basic quotation function so that we can generalise on the mentioning of words and phrases: *e.g.* we shall have:

$$f = \text{is red}$$

then

$$qu(f) = \text{'is red'}.$$

One point which must be emphasised about the use of the quotation function is that its sole purpose is to provide for the use of variables in a quotation context. Hence its use must be restricted to those situations in which it is necessary to mention indirectly the values of a variable—which we may describe loosely as the mentioning of a variable while retaining its use as a variable. But this does not mean that all mentioning of variables, or of expressions containing variables, calls for the use of the quotation function. On the contrary, in some cases it would be quite improper to use it. Suppose, for example, we simply wish to say that the letters

$$p, q, p', \ldots$$

are variables. So to say is not indirectly to mention their values and it is unnecessary, and indeed wrong, to retain their *use* as variables in this mentioning. The proper form of such a remark is,

the letters 'p', 'q', 'p'', ..., are variables,

in which the single quotes are used. Similarly, if we simply wish to say that the formula,

$$p \& q$$

is a semi-wff, then the ordinary quote marks are appropriate, as in: the formula '$p\,\&q$' is a semi-wff. By contrast, if we wish to introduce a formation rule which expresses the fact that *any* formula of the form '... &...' is a semi-wff, when the dots are replaced by semi-wffs, then the appropriate formulation is: $qu(p\,\&q)$ is a semi-wff. For given the substitution convention for the quotation function, we may substitute any semi-wffs for the variables and it will follow as a consequence from the rule so expressed that,

'$(p \vee q)\,\&q$' is a semi-wff
'$\sim p\,\&(p' \vee p'')$' is a semi-wff

and so on.

It is not always easy to decide intuitively when the mentioning in the informal metalanguage of an expression containing variables calls for the use of quote marks and when it calls for the use of the quotation function. In the formal theory, however, which is metalinguistic, the mentioning always requires the use of the quotation function since we always wish to retain the use of the variables as variables and provide for substitution on them under quotation. The distinction can best be made, perhaps, by asking in particular cases of mentioning whether what is being formulated is a general law applicable to all sentences even though quoted, or whether a particular formula is being quoted even though it happens to contain variables. If the latter, quote marks are the appropriate mentioning device since we do not, when we use the quotation function, mention the variables at all. We *use* them within the function in order indirectly to mention their values. It is perhaps worth noting that in much of the literature quote marks are used ambiguously both for quote marks of various sorts and for quotation functions. In what follows, to keep the symbolism as simple as possible in cases where no confusion can arise, we sometimes drop the use of quotes entirely in the informal metalanguage and use symbols, especially connectives, autonymously.

There are a number of other problems associated with quotation which need not be pursued here. We have been concerned only to provide a limited account sufficient to carry through the formal theory. A more detailed discussion will be found in Goddard and Routley [1].

2.5 The Relation between Sentences and Statements

It was suggested in §1.6 that if we restrict attention to indicative acts, a sentence is used successfully to make a statement if, and only if, it is both significant and complete in its context of use. Such sentences we call statement-yielding and we introduce special symbols to distinguish them as a particular sub-class of all sentences so that we may investigate their special properties. Strictly speaking the introduction of special symbols is not necessary. For we shall later introduce sentential predicates '*sig*' (is significant) and '*comp*' (is complete) and it is then possible to derive general theorems of the sort:

if *sig qu*(*p*)(*c*) and *comp qu*(*p*)(*c*) then ...;

the consequents of such theorems may then be taken as theses over statement-yielding sentences. However, the technique of using special symbols is initially simpler and we therefore introduce:

restricted sentential variables: $r, s, r', s', \ldots$

which take as values only statement-yielding sentences; and

restricted sentential constants: $r_0, s_0, r'_0, s'_0, \ldots$

which abbreviate actual statement-yielding sentences.

Since these variables and constants are simply a special sub-class of those already introduced, they satisfy the formation rules R1 and R2 and we may therefore construct semi-wffs and partial wffs which contain some of these symbols as elements. In case all the sentential variables and constants in a semi-wff or partial-wff are restricted, we shall say that the semi-wff or partial-wff is itself restricted. To distinguish the different symbols and formulae metatheoretically, we shall adopt the following conventions.

We shall continue to use $\Gamma, \Delta, \Gamma', \Delta', \ldots$ as metavariables which stand for any semi-wff but we shall also introduce $\Phi, \Psi, \Phi', \Psi', \ldots$ as metavariables which stand specifically for restricted semi-wffs. In case we wish to refer to an arbitrary sentential variable, we shall use $P, P', \ldots$. To refer to restricted sentential variables, we shall use $R, R', \ldots$. The subscript zero will be used as it is used in the formal theory. That is, Γ_0, Δ_0, stand for semi-wffs which contain only sentential constants; Φ_0, Ψ_0, stand for semi-wffs which contain only restricted sentential constants; P_0, P'_0 stand for arbitrary sentential constants; R_0, R'_0, stand for arbitrary restricted sentential constants.

Now although a sentence which is complete and significant in its context of use is statement-yielding, it cannot be identified with the statement it is used to make. For sentences of different types may be used to make the same statement, in the same or different contexts, and different sentences of the same type may be used to make different statements in different contexts. Such differences must be recognised and any temptation to identify statements with the sentences used to make them has to be resisted. There is however a relation, and in fact a many-one relation, between statement-yielding sentences in a context and statements. For many different sentences may yield the same statement; and it is also the case that to each statement-yielding sentence *in a context* there corresponds one, and only one, statement. As we saw in the previous chapter (§1.3), certain apparent counter examples to this last condition can be explained in terms of differences in context. It might be thought, for example, that sentences containing proper names or personal pronouns can yield two or more statements in a given context—*e.g.* that 'You are happy' can be used to make a true statement about one person and a false (therefore different) statement about another in the same context. In such cases, however, if the aboutness of the pronoun is vague, ambiguous, or undetermined, then the sentence is incomplete and fails to yield a statement at all. If, on the other hand, the aboutness of the pronoun is established, then a unique statement has been made. It may of course be that X will intend or take the aboutness to be one item and Y will take it to be another; but in this case, since the context is determined, partly at least, by the aboutness factors, X and Y are in fact taking the sentence with respect to different contexts. But that the same sentence may yield one statement in one context and a different one in another, in no way conflicts with the condition that it yields one and only one statement in one context.

We may say, therefore, that since the relation between statement-yielding sentences in context to the statements they are used to make is many-one, there is a function from such sentences to statements. If we represent the value of this function by,

$$\mathbf{st}(\Phi, \theta)$$

then, since the values of the variables in Φ are restricted in range to significant and complete sentences with respect to θ, the values of the function are statements. Implicit reference to this function is being made when it is said, as it often is, that a statement is "given by" a sentence in a context.

Clearly the function is not recursive, in the sense that a finite general procedure could be given for its evaluation in all cases. Nevertheless, the evaluation must be possible in each particular case, otherwise we could not say for a given complete and significant sentence what statement it yields in its context of use. This evaluation, however, will depend heavily on the context, especially on the sense and aboutness factors, and for this reason will not give rise to a recursive procedure for all contexts. Hence in general, although the statement function is not effective, it will always have a definite value for given definite values of the sentential and context variables. It may, of course, have the same value for different values of the arguments. Note, however, that if we allow an unrestricted semi-wff Γ as an argument in place of Φ, then, for certain values of the variables in Γ, the function will have no value. This will be so if the variables in Γ take nonsignificant or incomplete sentences as values, since these fail to yield statements. Such extension would therefore give us a function with singularity points for some of the permissible values of its argument variables.

We now introduce:

statement constants: $a_0, b_0, a'_0, b'_0, \ldots$

as formal abbreviations for elements of the set of values of the statement function; and

statement variables: $a, b, a', b', \ldots$

as variables which take statement constants as values. Thus, the statement variables range over the values of the statement function. We adopt the convention that statement symbols of the same type stand for the same statement, while statement symbols of different types stand for different statements in general, though they may stand for the same statements.

Given a suitable identity connective, we could now express the relation between sentences and statements by means of equalities of the sort,

$$a_0 = \mathbf{st}(r_0, c_0)$$

to show that a significant and complete sentence 'r_0' yields a statement a_0 in context c_0. That different sentences may yield the same statement in the same or different contexts is accounted for by the fact that the statement function may take the same value for different values of the sentential and context variables. Thus, if 'r_0' and 's_0' both yield a_0 in c_0, and 's_0' also yields a_0 in d_0, we shall have,

$$a_0 = \mathbf{st}(r_0, c_0)$$
$$a_0 = \mathbf{st}(s_0, c_0)$$
$$a_0 = \mathbf{st}(s_0, d_0).$$

Again, that sentences of the same type may yield different statements in different contexts will find expression in equalities such as,

$$a_0 = \mathbf{st}(r_0, c_0)$$
$$b_0 = \mathbf{st}(r_0, d_0).$$

Statements so defined, as the values of a function, are abstract items. It should be noted, however, that although this set of abstract items constitutes the *range* of statement variables, the actual *substitution* values which are made for them when the logic is applied are word strings, in fact that-clauses, which are used to mention statements—just as, in arithmetic, the variables range over numbers but the actual substitution values for the variables are words, in fact numerals, which are used to mention numbers. We therefore read the statement symbols, 'a', 'a_0' *etc.*, as 'that-a', 'that-a_0' *etc.*

Given the statement function as primitive, and the statement symbols introduced in this way, it is then possible to introduce formally a two-place relation **Y** (read 'yields the statement') by means of,

$$qu(\Phi)(\theta)\ \mathbf{Y}\ \alpha =_{\mathrm{Df}} (\alpha = \mathbf{st}(\Phi, \theta))$$

where α is a statement symbol and '$=$' is an appropriate identity connective. Thus **Y** takes quoted sentences in a context as substitution values in its first place, and that-clauses as values in its second place. Then, formulae of the sort,

$$qu(r)(c)\ \mathbf{Y}\ a$$

are interpreted as,

qu(*r*) in context *c* yields the statement that-*a*.

From the ordinary connections between functions and relations it will then follow that **Y** satisfies the many-one condition:

$$\text{if } qu(\Phi)(\theta)\ \mathbf{Y}\ \alpha \text{ and } qu(\Phi)(\theta)\ \mathbf{Y}\ \beta \text{ then } \alpha=\beta.$$

We shall, however, develop the formal theory in a slightly different way. We shall initially regard the statement function as an elucidatory device which stands outside the formalism as an explanation of the role of the statement symbols; within the formal theory, however, these are taken as primitive. That is, we shall first operate entirely in terms of the values of the statement function and variables which range over these values. In this way we develop what we call statement logic (**SL**). Later, however, in context logic (**CL**) we introduce **Y** as a further primitive and it is then possible to introduce the statement function by definition. The advantage of developing the formal theory in this way is related to the fact, which we have already noticed, that the statement function fails to have a value for some values of its variables if we extend it to allow unrestricted semi-wffs in its first argument place. Such extension is, however, necessary, if we are to be able to express all that we wish to say.

Suppose, for example, we wish to affirm as a formal thesis, what we have already affirmed informally, that a sentence yields a statement if, and only if, it is both significant and complete. Either one of the formulations:

$qu(\Phi)(\theta)\ \mathbf{Y}\ \alpha$ for some α, if, and only if, $qu(\Phi)$ is significant and complete in θ,

$(\alpha=\mathbf{st}(\Phi,\theta))$ for some α, if, and only if, $qu(\Phi)$ is significant and complete in θ,

is quite inadequate as an expression of the general thesis. For we want to say that *any* sentence satisfies this condition, but we have only managed to say it for the case of statement-yielding sentences since the values of Φ are restricted to them.

Such limitation is clearly unsatisfactory but there is no simple way of avoiding it if we begin with the statement function and use it to introduce the statement symbols formally since its first argument place must, for this purpose, be restricted to statement-yielding sentences. Hence the domain place of **Y**, defined in terms of the function will be similarly restricted. If, on the other hand, we begin with **Y** and the statement symbols as primitive, then we may from the beginning allow any semi-wff within the quotation function in the domain place. That is, we may begin with formulae of the form,

$$qu(\Delta)(\theta)\ \mathbf{Y}\ \alpha$$

where Δ is any semi-wff. For in case Δ takes as values sentences which are either significant or incomplete, the relational formula becomes false for those values. Hence we may affirm quite generally,

$qu(\Delta)(\theta)\ \mathbf{Y}\ \alpha$ for some α, if, and only if, $qu(\Delta)$ is significant and complete in θ, since if Δ takes a nonsignificant or incomplete sentence as a value, both sides of the (material) equivalence are false and the equivalence itself is true. Now if we define the extended statement function by,

$$(\alpha = \mathbf{st}(\Delta, \theta)) =_{\mathrm{Df}} qu(\Delta)(\theta)\ \mathbf{Y}\ \alpha$$

we shall also have,

$\alpha = \mathbf{st}(\Delta, \theta)$ for some α, if, and only if, $qu(\Delta)$ is significant and complete in θ.

That is,

$\alpha \neq \mathbf{st}(\Delta, \theta)$ *for every* α when $qu(\Delta)$ is not statement-yielding.

In other words, we are able to show that **st**(,) fails to have a value for some values of its arguments, and to show what these arguments are.

A related difficulty arises in respect of the values of the statement variables. So far we have simply said that the values are that-clauses. Considered from a purely grammatical point of view, however, it is possible to construct that-clauses from any sentence, whether complete or not, whether significant or not. However, although that-clauses constructed from any grammatical sentence are themselves grammatical, and can form part of significant sentences even though constructed from nonsignificant sentences, only those which are constructed from statement-yielding sentences can stand as admissible substitution values of the statement variables. For these variables are introduced as ranging over statements, and being thus restricted in range their substitution values must be similarly restricted.

This restriction imposes no limitation on the formal development, and in fact is simply a convenience which removes the need for a special postulate. For if we allow the construction of a that-clause from any grammatical sentence, as we may, but allow further that any such clause can be substituted for a statement variable, then it would be necessary to affirm the postulate,

if $qu(\Gamma)$ is either nonsignificant or incomplete, then that-$\Gamma \neq \mathbf{st}(\Phi, \theta)$ for all values of Φ and θ.

By restricting the values of the statement variables to clauses constructed from statement-yielding sentences, we remove the need to state the postulate since it is implicit in the restriction. Thus, formulae such as 'that-Ψ' $= \mathbf{st}(\Phi, \theta)$' will be true for some values of Φ and θ but never false for all. For 'that-Ψ'' will always mention a statement yielded by some sentence in some context, and it may, but need not, mention the actual statement yielded by $qu(\Phi)$ in θ.

The question arises whether that-clauses should be qualified by context symbols. This raises a number of problems. From one point of view such qualification would be both pointless and misleading. If a sentence is uttered, and the sense and aboutness factors of the context taken into account, and if as a result of this evaluation we can say not only that a statement has been made but what the statement actually is, by mentioning it, then we have freed the utterance from its context. This, indeed, is the whole point of the concept of a statement, or for that matter, of a proposition. But of course, in reporting this evaluation we ourselves utter a sentence which is itself context-dependent. And from this point of view it seems that the that-clause, which is part of this utterance, should itself be qualified by a context symbol.

Suppose, for example, we know that the sentence 'The sky is blue' was used as a code to mean the attack takes place at dawn. This may be reported by saying:

X used the sentence 'The sky is blue' in the coding context c_0 to make the statement that the attack takes place at dawn.

Now this whole sentence is uttered with respect to some context, and any of the words in it might be being used in a non-standard way; indeed it may itself be in code. Let us suppose, however, that the structure, or sentence frame,

X used the sentence '—' (c_0) to make the statement ...

itself has a fixed, standard sense with respect to the background language. Then the question of context for such utterances is the question of what context should be supplied for the that-clause which replaces the dots. For although the sentence within quote marks may have a non-standard sense, it presents no problem since it is being mentioned and its context of use is indicated; the that-clause which replaces the dots, however, is being *used* with respect to some context as yet unspecified. Now it may be that this that-clause is in particular cases itself in code (and in a different code from that determined by c_0), but if this were always so we should have a private-language situation involving an endless regress. If evaluations expressed in such sentences are ever to have point, then at some stage we must require the that-clause to be expressed with respect to the *standard context*, which is to say, with standard sense relatively to the background language. It is just this public requirement which gives point to the concept of a statement; or conversely, the concept of a statement demands a public language of standard senses. There is in an important sense no context-free utterance. But utterances which are made with respect to the standard context need no evaluation beyond that which the ordinary user of the background language can automatically supply from his knowledge of the semantic dictionary and grammar. Given that an utterance is made with respect to the standard context, which we shall indicate by using the special constant 'c_S', it is then guaranteed that the words used have standard senses and typical uses, and this will often extend even to some proper names such as 'Socrates' and 'Churchill'. Of course, though we usually take it that the background language is some natural language, it need not be. All that is required is that the parties to a conversation have an agreed public language, that is, an agreed semantic dictionary and grammar, which may nevertheless be private to them.

Unless otherwise indicated, therefore, we shall take it that the that-clauses which are substituted for the statement variables are made with respect to c_S. We shall, however, continue to write formulae of the sort '$a = \mathbf{st}(r, c)$' or '$qu(r)(c)\ \mathbf{Y}\ a$' instead of '$a(c_S) = \mathbf{st}(r, c)$' or '$qu(r)(c)\ \mathbf{Y}\ a(c_S)$', taking it that c_S is an implicit component of a.

We must now ask, however, whether we can, or should, *restrict* the values of the statement variables to that-clauses expressed with respect to the standard context.

Consider again the sentence 'The sky is blue' used as a code. Taking the code into account, we may say it yields the statement that the attack takes place at dawn with respect to c_S. But we may also say it yields the statement that the sky is blue not, certainly, with respect to c_S, but with respect to the coding context, say c_0. For we may well give correct and useful information to those in the know if we affirm that X made the statement that the sky is blue. Of course we only manage to give correct information because those in the know are able to make the proper translation. To those not in the know we should be giving misleading information. It would not, however, be misleading, though it might be pointless, if we qualify our remark by indicating the context; that is, by saying "X made the statement that the sky is blue (c_0)".

It is possible, therefore, to allow the substitution of context-qualified that-clauses for the statement variables in certain circumstances; and it is also useful. To see this, let 'r_0' and 's_0' be respectively the statement-yielding sentences 'The sky is blue' and 'The attack takes place at dawn'. As before, let c_0 be the coding context. We form the clauses 'that-r_0' and 'that-s_0'. Now we may affirm in general, for any sentence s,

$$\mathbf{st}(s, c_S) = \text{that-}s; \quad \textit{i.e.}\ qu(s)(c_S)\ \mathbf{Y}\ \text{that-}s;$$

since if a sentence is uttered with respect to the standard context, and is statement-yielding, the actual statement which it yields can be mentioned by the that-clause formed from it without contextual qualification—in terms of the convention that the absence of a context symbol on a that-clause indicates that its evaluation is with respect to c_S. In particular, therefore,

$$\mathbf{st}(s_0, c_S) = \text{that-}s_0.$$

Again, for any sentence r and any context c, we may affirm

$$\mathbf{st}(r, c) = \text{that-}r(c); \quad \textit{i.e.}\ qu(r)(c)\ \mathbf{Y}\ \text{that-}r(c);$$

for though this may be uninformative, the retention of the context variable on the that-clause prevents it from being misleading. Hence we have the special case,

$$\mathbf{st}(r_0, c_0) = \text{that-}r_0(c_0).$$

Now for the particular sentences under consideration, we know that

$$\text{that-}r_0(c_0) = \text{that-}s_0;$$

i.e. the statement that the sky is blue, made with respect to the coding context, is the same as the statement that the attack takes place at dawn, made with respect to the standard context. Equivalently,

$$\mathbf{st}(r_0, c_0) = \mathbf{st}(s_0, c_s).$$

That is, the two sentences 'r_0' (when uttered with respect to c_0), and 's_0' (when uttered with respect to c_s) yield the same statement. This we express by saying that they are statement-equal.

Hence, by admitting that-clauses qualified by context symbols as values of the statement variables, we are able to express formally the relations which hold between the two sentences and to show that, in general, the problem of determining what statement a sentence makes is simply the problem of translating it into the public language. This is what freeing it from its context amounts to.

This explanation presupposes, of course, some intuitive understanding of statement identity. We shall show in the formal theory, however, how this can be introduced.

2.6 The Logic of Statements, SL

A question which must now be asked is whether statements have a logic at all. For what is often meant by a logic is the set of formal relations which hold between elements of a given kind by virtue of their structure or form. In this sense, sentences certainly have a logic since they have a syntactic structure and we may investigate the formal relations which hold between such structures: *e.g.* we may investigate the relations which hold between sentences of the form $qu(p \vee q)$ and those of the form $qu(p \& q)$. But it is doubtful if statements have a logic in this sense since it is not at all clear that they have a structure. What seems to be clear is that they do not have a syntactic structure if only because they are not syntactic objects. We cannot say, for example, that a given statement is of the form

$$a$$

and another of the form

$$\text{not-}a.$$

For two different sentences, one containing the word 'not' and the other not containing it, may be used to make the same statement. Compare, for example,

He is alive.
He is not dead.

In a suitable context, there are two sentences here but only one statement. Again, we may use a conjunctive sentence to make the same statement as is made by the use of a disjunctive sentence. Compare:

I might go and I might not.
I shall either go or not.

Or consider the use of conjunctive and hypothetical sentences to make the same statement; as in,

I may go; but it would be after seven.
If I go, it will be after seven.

It seems, however, that statements have properties and that there are relations between them. For we may say that statements are true or false, that one implies another, or that one is incompatible with another. And in this sense, it may be said, they have a logic.

What seems to be the case is that we can affirm certain logical relations between statements by *mentioning* them appropriately. For in affirming that one statement implies another, for example, we make a statement about statements. We might express this explicitly in a particular case in some such way as,

The statement expressed by 'Jack is a man' implies the statement expressed by 'Jack is human'.

Here both sentences and statements are being mentioned; and it is assumed that the sentences are significant and complete in the context. Commonly, however, the same thing can be achieved by suppressing the sentential mentioning by means of an indirect speech form containing the word 'that'. We may say,

(the statement) that Jack is a man implies (the statement) that Jack is human.

Here, we use that-clauses constructed from sentences of a type which could be used to make the statements which are being mentioned ('Jack is a man' could be used to make the statement that Jack is a man), without actually making the statement at all. Similarly we may say,

It is true that the cat is on the mat

or,

(the statement) that the cat is on the mat is true.

In each case the whole sentence is used to make a statement about another statement, and the statement which is the subject of the remark is mentioned by using a that-clause constructed from a sentence of a type which could be used to make it, *viz.* 'the cat is on the mat', but which is not in fact being made because it is governed by 'that'.

It seems, therefore, that to the extent that there is a logic of statements, it is a logic of the properties of and relations between statements which can only be expressed as a logic of mentioning. This is how we shall proceed. The important differences between the logic of sentences and the logic of statements is then reflected by the fact that where r and s are sentences, so are $\sim r$, $r \vee s$, $r \& s$, $r \supset s$, and $r \equiv s$, they are in fact compound sentences with a syntactic structure; but where a and b are statements, we do not allow the formation of compound statements such as $\sim a$, $a \vee b$, $a \& b$, *etc.* That such compounds should not be admitted is reinforced by purely grammatical considerations if it is remembered that the substitution values of the statement symbols are that-clauses. For given the clauses 'that it is raining' and 'that it is snowing', the compound 'that it is raining or that it is snowing' is not a clause. By contrast, given the sentences 'It is raining' and 'It is snowing', the compound 'It is raining or it is snowing' is a sentence. Again, clauses are subject terms which stand in need of predicates, whereas sentences are grammatically complete in themselves.

We wish to emphasise, however, that in refusing to allow compounds to be constructed from that-clauses, we are not refusing to admit compound that-clauses as values of the statement variables. Thus, for example, from the compound sentence 'It is raining or it is snowing' we may construct the compound clause 'that (it is raining or it is snowing)'. But this is quite different from the inadmissible compound of clauses 'that it is raining or that it is snowing'. In general, if Φ is a restricted semi-wff, that-Φ is an admissible that-clause, whatever the structure of Φ. Hence we allow clauses such as 'that($\sim r$)', 'that($r \vee s$)', 'that($r \& s$)', *etc.* as values of the statement variables. But if that-Φ and that-Ψ are admissible that-clauses, $\sim$that-Φ, that-$\Phi \vee$ that-Ψ, that-Φ & that-Ψ, *etc.*, are not. Hence, we may allow as special cases of locutions such as,

Jack made the statement a,

sentences such as,

Jack made the statement that if it rains he will attend the lecture,

but not,

Jack made the statement if that it rains then that he will attend the lecture.

Our restriction on the construction of compounds from that-clauses is therefore supported by grammar but it is not simply grammatical. For we take it that the grammar here reflects a fundamental difference between the logic of sentences and the logic of statements, related though these are. What we wish to say is that although someone may use a hypothetical sentence to make a statement, it is improper to conclude that the statement he made itself has a hypothetical structure. For he might have made the *same* statement, and not merely an equivalent one, by using a sentence with a disjunctive structure. So we cannot say of a statement that it has a disjunctive or implicational structure, but we can, by mentioning the statement made, affirm that it stands in the relation of disjunction to another, that it implies another, that it is true, that it is false, and so on.

We now develop the logic of statements as a logic of mentioning and show how to derive the logic of statement-yielding sentences from it.

We introduce two primitives:

the incompatibility connective: $/$

interpreted as,

(the statement) ... is incompatible with (the statement) ...,

which is sentence-forming on clauses. For if the blanks are filled by that-clauses, we have a sentence which can be used to make a statement about statements, namely those mentioned by the that-clauses.

the that-operator: §

interpreted as 'that', which we adopt following a suggestion by W. and M. Kneale ([1], p. 539), and which is clause-forming on sentences.

The statement symbols are taken as primitive and we construct sentential formulae from them, which we call well-formed formulae of statement logic (wffsl), by means of $/$. In view of the fixed interpretation which has been put on $/$, we take it that the sentences which result when values are substituted for the variables are themselves statement-yielding. This allows us to construct from these formulae further clauses using §. Thus,

R3 (i) statement symbols are well-formed clauses (wfc)
(ii) if α and β are wfc, (α/β) is a wffsl
(iii) if A is a wffsl, $\S A$ is a wfc.

Hence clauses, that is statement symbols standing alone and expressions such as $\S A$, are not well-formed formulae since they are not themselves sentences but only well-formed parts of sentences. In terms of this rule, such formulae as the following are wffsl:

$$a/b, \quad a/\S(b/b), \quad \S(a/a)/b, \quad \S(a/a)/\S(b/b).$$

We take the rules for parentheses as understood.

It will be noticed that the sentential formulae such as 'a/b' are not qualified by a context symbol. The reason for this is that given the fixed interpretation which has been put on $/$, all our assertions about the relations between statements are made with respect to the standard context c_s. In effect, therefore, every thesis of **SL** is qualified by c_S. Instead of writing it on each thesis, however, we regard it as a general condition on the whole logic. Similarly, the absence of a context symbol on the statement symbols indicates that, unless otherwise stated, the that-clauses which are taken as values are evaluated with respect to c_S. Later we allow as values that-clauses which are not restricted in this way, but it still remains the case that our theses are all affirmed with respect to c_S. Thus a sentence such as,

$$\S(\text{It is raining})(c_0)/\S(\text{the sky is blue})(c_1)$$

is itself affirmed with respect to c_S, even though its clauses are not, since the interpretation of **/** remains the same.

Next we introduce the following definitions:

D1: where α and β are wfc,

(i) $\mathbf{F}\alpha =_{\mathrm{Df}} \alpha/\alpha$
(ii) $\mathbf{T}\alpha =_{\mathrm{Df}} \mathbf{F}\S\mathbf{F}\alpha$
(iii) $\alpha\ \mathbf{imp}\ \beta =_{\mathrm{Df}} \alpha/\S\mathbf{F}\beta$
(iv) $\alpha\ \mathbf{eq}\ \beta =_{\mathrm{Df}} \mathbf{F}\S(\S(\alpha/\S\mathbf{F}\beta)/\S(\beta/\S\mathbf{F}\alpha))$
(v) $\alpha\ \mathbf{disj}\ \beta =_{\mathrm{Df}} \S\mathbf{F}\alpha/\S\mathbf{F}\beta$
(vi) $\alpha\ \mathbf{conj}\ \beta =_{\mathrm{Df}} \mathbf{F}\S(\alpha/\beta)$.

These are interpreted as:

Fα: (the statement) α is false
Tα: (the statement) α is true
α **imp** β: (the statement) α materially implies (the statement) β
α **eq** β: (the statement) α is materially equivalent to (the statement) β
α **disj** β: (the statement) α stands in the relation of disjunction to (the statement) β
α **conj** β: (the statement) α stands in the relation of conjunction to (the statement) β.

Hence the defined connectives are all quite explicitly mentioning connectives which stand between that-clauses. For example, in particular cases we shall have sentences such as: (the statement) that it is raining is false; (the statement) that it is raining materially implies (the statement) that the clouds are grey; and so on.

We now secure the formal theory by means of the following axioms and rules of a Nicod-type system. The symbol '$\vdash_{SL}$' introduces theses of statement logic; α, β, γ and δ stand for wfc and *A*, *B*, *C* stand for wffsl. A thesis is an axiom or a theorem.

Axioms:

$\vdash_{SL}(1)$ $\alpha/\S(\alpha/\alpha)$
$\vdash_{SL}(2)$ $\S\{\alpha/\S(\beta/\beta)\}/\S\{\S[\S(\gamma/\beta)/\S(\S(\alpha/\gamma)/\S(\alpha/\gamma))]/\S[\S(\gamma/\beta)/\S(\S(\alpha/\gamma)/\S(\alpha/\gamma))]\}$
$\vdash_{SL}(3)$ $\S\mathbf{T}\alpha\ \mathbf{eq}\ \alpha$

Rules:

SL(1) *Definitional replacement:* if a wffsl is replaced by its defined equivalent at any of its occurrences in a thesis, the wffsl so obtained is a thesis.

SL(2) *Equivalence replacement:*
(i) if $\vdash_{SL}\alpha\ \mathbf{eq}\ \beta$ and $\vdash_{SL}A$, where A contains α as a wfc, then $\vdash_{SL}B$, where B is obtained from A by replacing α by β at any of its occurrences in A; similarly β may be replaced by α.
(ii) if $\vdash_{SL}\S A\ \mathbf{eq}\ \S B$ then A may be replaced by B at any point of a thesis to yield a thesis; similarly B may be replaced by A.

SL(3) *Detachment:* if $\vdash_{SL}A$ and $\vdash_{SL}\S A/\S(\S B/\S C)$, then $\vdash_{SL}C$.

These postulates are not independent, but since the minimal representation of a Nicod system is already well known, we here prefer a version which leads to simpler proofs.

Derived Rule SL(4)*:* if $\vdash_{SL}A$ then $\vdash_{SL}\mathbf{T}\S A$ (*The* **T**-*rule*)

Proof: As a case of $\vdash_{SL}(3)$ we have $\vdash_{SL}\S\mathbf{T}\S A\ \mathbf{eq}\ \S A$ for any wffsl A. Suppose $\vdash_{SL}A$. Then by rule 2(ii), $\vdash_{SL}\mathbf{T}\S A$.

Evidently the converse is also derivable, but we state this as a separate rule, SL(6) below.

Theorems:

$\vdash_{SL}(4)$ $\alpha/\S\mathbf{F}\alpha$: D1(i) on $\vdash_{SL}(1)$; rule (1)
(the statement α is incompatible with the statement that α is false)

$\vdash_{SL}(5)$ $\alpha\ \mathbf{imp}\ \alpha$: D1(iii) on $\vdash_{SL}(4)$; rule (1)
(the statement α materially implies the statement α)

$\vdash_{SL}(6)$ $\mathbf{T}\S(\alpha/\S\mathbf{F}\alpha)$: **T**-rule on $\vdash_{SL}(4)$

$\vdash_{SL}(7)$ $\mathbf{F}\S\mathbf{F}\S(\alpha/\S\mathbf{F}\alpha)$: D1(ii) on $\vdash_{SL}(6)$; rule (1)

$\vdash_{SL}(8)$ $\mathbf{F}\S(\S(\alpha/\S\mathbf{F}\alpha)/\S(\alpha/\S\mathbf{F}\alpha))$: D1(i) on $\vdash_{SL}(7)$; rule (1)

$\vdash_{SL}(9)$ $\alpha\ \mathbf{eq}\ \alpha$: D1(iv) on $\vdash_{SL}(8)$; rule (1).

Derived Rule SL(5)*:* if $A =_{Df} B$ then $\vdash_{SL}\S A\ \mathbf{eq}\ \S B$

Proof: For any wffsl A, $\vdash_{SL}\S A\ \mathbf{eq}\ \S A$ by $\vdash_{SL}(9)$ since $\S A$ is a wfc. Let $A =_{Df} B$. Then by rule (1), $\vdash_{SL}\S A\ \mathbf{eq}\ \S B$.

$\vdash_{SL}(10)$ $\S\mathbf{T}\alpha\ \mathbf{eq}\ \S\mathbf{F}\S\mathbf{F}\alpha$: D1(ii) and derived rule (5)

$\vdash_{SL}(11)$ $\S\mathbf{T}\alpha\ \mathbf{eq}\ \S(\S\mathbf{F}\alpha/\S\mathbf{F}\alpha)$: D1(i) on $\vdash_{SL}(10)$; rule (1)

$\vdash_{SL}(12)$ $\S\mathbf{T}\alpha\ \mathbf{eq}\ \S(\S(\alpha/\alpha)/\S(\alpha/\alpha))$: D1(i) on $\vdash_{SL}(11)$; rule (1)

$\vdash_{SL}(13)$ $\S(\alpha/\alpha)/\S(\S(\alpha/\alpha)/\S(\alpha/\alpha))$: from $\vdash_{SL}(2)$ where β and γ are both α, and $\vdash_{SL}(1)$, by detachment

$\vdash_{SL}(14)$ $\S\mathbf{F}\alpha/\S\mathbf{T}\alpha$: from $\vdash_{SL}(13)$ using D1(i) and rule (1), and $\vdash_{SL}(12)$ and rule (2)(i)

(the statement that α is false is incompatible with the statement that α is true)

$\vdash_{SL}(15)$ $\S T\alpha/\S F\S T\alpha$: $\vdash_{SL}(4)$ where α is $\S T\alpha$

$\vdash_{SL}(16)$ $\S F\S F\alpha/\S F\S T\alpha$: D1(ii) on $\vdash_{SL}(15)$; rule (1)

$\vdash_{SL}(17)$ $\S F\alpha\ \mathbf{disj}\ \S T\alpha$: D1(v) on $\vdash_{SL}(16)$; rule (1)

(the statement that α is false stands in the relation of disjunction to the statement that α is true)

$\vdash_{SL}(18)$ $\S F\alpha/\S F\S F\alpha$: $\vdash_{SL}(4)$ where α is $\S F\alpha$

$\vdash_{SL}(19)$ $\alpha\ \mathbf{disj}\ \S F\alpha$: D1(v) on $\vdash_{SL}(18)$

(the statement α stands in the relation of disjunction to the statement that α is false)

$\vdash_{SL}(20)$ $\S T\S(\alpha\ \mathbf{disj}\ \beta)\ \mathbf{eq}\ \S(\alpha\ \mathbf{disj}\ \beta)$: $\vdash_{SL}(3)$ where α is $\S(\alpha\ \mathbf{disj}\ \beta)$

$\vdash_{SL}(21)$ $\S T\S(\alpha\ \mathbf{disj}\ \beta)\ \mathbf{eq}\ \S(\S T\alpha\ \mathbf{disj}\ \S T\beta)$: using $\vdash_{SL}(3)$ twice to make equivalence replacements in $\vdash_{SL}(20)$; rule 2(i)

We shall not give any more proofs. Other useful theses are:

$\vdash_{SL}(22)$ $\S T\S F\alpha\ \mathbf{eq}\ \S F\S T\alpha$

$\vdash_{SL}(23)$ $\S T\S F\alpha\ \mathbf{eq}\ \S F\alpha$

$\vdash_{SL}(24)$ $\S T\S T\alpha\ \mathbf{eq}\ \S T\alpha$

$\vdash_{SL}(25)$ $\S T\S T\alpha\ \mathbf{eq}\ \S F\S F\alpha$

$\vdash_{SL}(26)$ $\S T\S(\alpha\ \mathbf{conj}\ \beta)\ \mathbf{eq}\ \S(\S T\alpha\ \mathbf{conj}\ \S T\beta)$

$\vdash_{SL}(27)$ $\alpha\ \mathbf{imp}\ \S(\alpha\ \mathbf{disj}\ \beta)$

$\vdash_{SL}(28)$ $\S(\alpha\ \mathbf{eq}\ \beta)\ \mathbf{eq}\ \S(\S F\alpha\ \mathbf{eq}\ \S F\beta)$

$\vdash_{SL}(29)$ $F\S(\S T\alpha\ \mathbf{conj}\ \S F\alpha)$

This last is simply an alternative way of writing $\vdash_{SL}(14)$ since we have in general,

$\vdash_{SL}(30)$ $\S(\alpha/\beta)\ \mathbf{eq}\ \S F\S(\alpha\ \mathbf{conj}\ \beta)$

which leads from $\vdash_{SL}(14)$ to $\vdash_{SL}(29)$ by rule 2(ii).

2.7 The Logic of Statement-Yielding Sentences, RSL

We now show how to derive the logic of statement-yielding sentences, which we call restricted sentential logic (**RSL**), from the logic of statements. This is in fact the classical two-valued sentential calculus which is often, in our view mistakenly, taken to be the logic of statements.

The connexion between the two logics depends on recognising explicitly an underlying assumption of classical logic, namely that, if a relation holds between statements, it is reflected by the use of a sentence with an appropriate syntactical structure. In particular, in the two-valued case, the connectives $\sim$, $\vee$, &, $\supset$ and $\equiv$, have a fixed determined sense such that the use of a sentence of a given compound form implies that certain relations hold between the statements yielded by its components. This is sometimes expressed by saying that a compound sentence "shows", but does not state, that the statements yielded by its components stand in certain relations, or that the sentence "reflects" these relations. For example, a sentence of the form $qu(r \vee s)$ shows that the statements yielded by $qu(r)$ and $qu(s)$ stand in the relation of disjunction; a sentence of the form $qu(r \& s)$ shows that the statements yielded by its components stand in the relation of conjunction; and so on. This means in effect that the classical logical constants have a standard sense which is the same for all contexts. In this respect, therefore, they are not the ordinary connectives 'or', 'and', *etc.* of a natural language. It is by virtue of these relations that truth conditions for the various compounds can be given. For a sentence of the form $qu(r \vee s)$ is true (yields a true statement) if, and only if, the statements yielded by $qu(r)$ and $qu(s)$ in fact stand in the relation of disjunction to each other.

These connexions between the sentential connectives and statement relations can be made quite explicit, as we shall see later, by means of the yields relation **Y**. For present purposes, however, it is unnecessary to go so far. All that we need do, to derive sentential logic, is to state explicitly the various truth conditions for statements yielded by compound sentences. Such conditions determine a definite sense for the connectives for all contexts. Connectives used with such a sense, that is in accordance with such conditions, are, we shall say, logically standard. Accordingly, the conditions determine contexts with logically standard features.

To say that a context is logically standard is not to say that it is the standard context of the background language. The standard context will be logically standard; but a context which is logically standard need not be standard in all respects. In a code, for example, the sentential connectives may well have their standard sense even though all other words are used in non-standard ways. Again, if the components of the sentence, 'The number seven is purple and the number three is green' respectively yield the statements that the house next door is painted purple, that the house across the road is painted green, then although the phrases 'number seven' and 'number three' are being used in non-standard ways, the word 'and' is standard if it is accepted that the statements yielded by the components of the compound stand in the relation of conjunction. The utterance is then made with respect to a non-standard context which is nevertheless logically standard.

Before giving the definition which characterises the features of logically standard contexts, it is necessary to extend the class of wfc by admitting clauses formed from restricted sentential semi-wffs. Thus,

R3 (iv) If $\Phi(\theta)$ is a restricted semi-wff, where θ is a context symbol, then $\S\Phi(\theta)$ is a wfc.

Hence we have special wfc which are qualified by context symbols, but this will not affect the fact that all theses of **SL** are asserted with respect to c_s.

The definition is then as follows:

D2 θ is a logically standard context if, and only if,

(i) $\S\mathbf{T}\S{\sim}\Phi(\theta)\ \mathbf{eq}\ \S\mathbf{F}\S\Phi(\theta)$

(ii) $\S\mathbf{T}\S(\Phi\vee\Psi)(\theta)\ \mathbf{eq}\ \S(\S\Phi(\theta)\ \mathbf{disj}\ \S\Psi(\theta))$

(iii) $\S\mathbf{T}\S(\Phi\,\&\,\Psi)(\theta)\ \mathbf{eq}\ \S(\S\Phi(\theta)\ \mathbf{conj}\ \S\Psi(\theta))$

(iv) $\S\mathbf{T}\S(\Phi\supset\Psi)(\theta)\ \mathbf{eq}\ \S(\S\Phi(\theta)\ \mathbf{imp}\ \S\Psi(\theta))$

(v) $\S\mathbf{T}\S(\Phi\equiv\Psi)(\theta)\ \mathbf{eq}\ \S(\S\Phi(\theta)\ \mathbf{eq}\ \S\Psi(\theta))$.

This definition determines a logically standard context by fixing the truth conditions for statements yielded by compound sentences containing the classical two-valued connectives. For although it is expressed entirely in terms of statement equivalence, the connexion between sentences and statements can be made explicit if we follow the suggestion made earlier (§2.5) and adopt as a general postulate,

$$qu(\Phi)(\theta)\ \mathbf{Y}\ \S\Phi(\theta),$$

for any context θ. In terms of this, $qu(\Phi\vee\Psi)(\theta)$ will yield the statement $\S(\Phi\vee\Psi)(\theta)$; and by D2(ii), if θ is logically standard, this statement will be true if and only if, $\S\Phi(\theta)$ stands in the relation of disjunction to $\S\Psi(\theta)$. In other words, the statement yielded by a sentence of the form $qu(\Phi\vee\Psi)$ will be true if, and only if, the statements yielded by its components stand in the relation of disjunction. We shall develop this extended analysis later.

We now introduce special context symbols as follows:

if θ is a context symbol then $\mathrm{L}\theta$ is a context symbol.

We read '$\mathrm{L}\theta$' as 'logically standard θ'. $\mathrm{L}\theta$ is a variable or constant according as θ is a variable or a constant. The rules for the formation of semi-wffs and partial wffs need not be extended since they are expressed in a form which is valid for all context symbols, hence they automatically apply to symbols of the form $\mathrm{L}\theta$.

Now by taking θ in D2 to be $\mathrm{L}\theta$, the condition '$\mathrm{L}\theta$ is a logically standard context' can be affirmed and the set of equivalences asserted unconditionally as theses of SL. Thus,

$\vdash_{\mathrm{SL}}\ \S\mathbf{T}\S{\sim}\Phi(\mathrm{L}\theta)\ \mathbf{eq}\ \S\mathbf{F}\S\Phi(\mathrm{L}\theta)$

$\vdash_{\mathrm{SL}}\ \S\mathbf{T}\S(\Phi\vee\Psi)(\mathrm{L}\theta)\ \mathbf{eq}\ \S(\S\Phi(\mathrm{L}\theta)\ \mathbf{disj}\ \S\Psi(\mathrm{L}\theta))$

$\vdash_{\mathrm{SL}}\ \S\mathbf{T}\S(\Phi\,\&\,\Psi)(\mathrm{L}\theta)\ \mathbf{eq}\ \S(\S\Phi(\mathrm{L}\theta)\ \mathbf{conj}\ \S\Psi(\mathrm{L}\theta))$

$\vdash_{\mathrm{SL}}\ \S\mathbf{T}\S(\Phi\supset\Psi)(\mathrm{L}\theta)\ \mathbf{eq}\ \S(\S\Phi(\mathrm{L}\theta)\ \mathbf{imp}\ \S\Psi(\mathrm{L}\theta))$

$\vdash_{\mathrm{SL}}\ \S\mathbf{T}\S(\Phi\equiv\Psi)(\mathrm{L}\theta)\ \mathbf{eq}\ \S(\S\Phi(\mathrm{L}\theta)\ \mathbf{eq}\ \S\Psi(\mathrm{L}\theta))$.

A condition similar to those in D2, and which in a sense completes the list, is already derivable for any context θ from $\vdash_{\mathrm{SL}}(3)$: *i.e.*

$\vdash_{\mathrm{SL}}\ \S\mathbf{T}\S\Phi(\theta)\ \mathbf{eq}\ \S\Phi(\theta)$.

Hence, in particular,

$\vdash_{\mathrm{SL}}(31)\ \ \S\mathbf{T}\S\Phi(\mathrm{L}\theta)\ \mathbf{eq}\ \S\Phi(\mathrm{L}\theta)$.

Now taking $\vdash_{SL}(3)$ in the form,

$\vdash_{SL}(32)$ §**T**§A **eq** §A,

where A is any wffsl, then, given $\vdash_{SL}$**T**§A, we could use $\vdash_{SL}(32)$ to pass by rule 2(ii) to the thesis $\vdash_{SL}A$. That is, we have in general, for any wffsl, the derived rule:

SL(6) If $\vdash_{SL}$**T**§A then $\vdash_{SL}A$.

By a similar argument, therefore, one might expect to use $\vdash_{SL}(31)$ to obtain the derived rule: (6′) if $\vdash_{SL}$**T**§Φ(Lθ) then $\vdash_{SL}$Φ(Lθ); or indeed simply to assert this as a special case of (6). In terms of it, certain theses would then be in the familiar **RSL** form. However, although clauses formed from restricted semi-wffs have been defined as wfc, and hence formulae such as **T**§Φ(Lθ) are wffsl, and therefore possible theses, restricted semi-wffs standing alone have not been defined as wffsl and (6′) cannot arise as a special case of (6) since Φ(Lθ) cannot be taken as a case of A. Nor is the rule directly derivable from $\vdash_{SL}(31)$ since rule 2(ii), which is required in the derivation, is expressed in terms of wffsl. It would in fact be possible to introduce restricted semi-wffs as wffsl, but we choose not to at this point because their introduction would change the character of the logic. As it now is, when values are substituted for the variables, every well-formed formula of **SL** becomes a sentence which yields a statement about statements. This is so since every wffsl contains a wfc as a part. If, however, we admit restricted semi-wffs such as '$r \vee s$' as wffsl, then when values are substituted for the variables, the sentences which arise, though they will be statement-yielding, need not be such as yield statements about statements. This extension is of course desirable from some points of view, and we shall make it presently, but when it is made, we no longer have just statement logic, but a substantial part of context logic.

For the present, therefore, we keep statement logic, **SL**, distinct from the logic of statement-yielding sentences, **RSL**, and only combine them later when we develop context logic, **CL**. But we secure the link between them by adopting a modified version of (6′), *viz.*

SL(7) If $\vdash_{SL}$**T**§Φ(Lθ) then $\vdash_{RSL}$Φ(Lθ).

Since, in terms of the proposed postulate,

$$qu(\Phi)(\theta)\ \mathbf{Y}\ \S\Phi(\theta),$$

we shall have,

$$qu(\Phi)(\mathrm{L}\theta)\ \mathbf{Y}\ \S\Phi(\mathrm{L}\theta),$$

we may say that the statement §Φ with respect to Lθ is yielded by the sentence *qu*(Φ) in Lθ. Hence, the rule amounts to saying that if it is a thesis of statement logic that the statement yielded by a sentence *qu*(Φ) in Lθ is true, then the sentence *qu*(Φ) is itself a thesis of restricted sentential logic with respect to a logically standard context.

As examples of the derivation of **RSL** theses we show how to obtain $\vdash_{RSL}(\Phi \vee \sim\Phi)(L\theta)$ and $\vdash_{RSL}(\Phi \supset \Phi \vee \Psi)(L\theta)$.

We have as a special case of $\vdash_{SL}(19)$,

$\vdash_{SL}(33)$ $\S\Phi(L\theta)$ **disj** $\S\mathbf{F}\S\Phi(L\theta)$.

Hence,

$\vdash_{SL}(34)$ $\S\Phi(L\theta)$ **disj** $\S\sim\Phi(L\theta)$: from $\vdash_{SL}(33)$ by D2(i), $\vdash_{SL}(31)$ and rule 2(i).

But we also have:

$\vdash_{SL}(35)$ $\S\mathbf{T}\S(\Phi \vee \sim\Phi)(L\theta)$ **eq** $\S(\S\Phi(L\theta)$ **disj** $\S\sim\Phi(L\theta))$: a case of D2(ii).

From $\vdash_{SL}(34)$ and $\vdash_{SL}(35)$, therefore, by rule 2(ii),

$\vdash_{SL}(36)$ $\mathbf{T}\S(\Phi \vee \sim\Phi)(L\theta)$.

Hence, by the link rule (7),

$\vdash_{RSL}(\Phi \vee \sim\Phi)(L\theta)$.

Again, as a special case of $\vdash_{SL}(27)$ we have,

$\vdash_{SL}(37)$ $\S\Phi(L\theta)$ **imp** $\S(\S\Phi(L\theta)$ **disj** $\S\Psi(L\theta))$.

So, by D2(ii) and rule 2(i),

$\vdash_{SL}(38)$ $\S\Phi(L\theta)$ **imp** $\S\mathbf{T}\S(\Phi \vee \Psi)(L\theta)$.

That is,

$\vdash_{SL}(39)$ $\S\Phi(L\theta)$ **imp** $\S(\Phi \vee \Psi)(L\theta)$: by $\vdash_{SL}(3)$ and rule 2(i).

Hence,

$\vdash_{SL}(40)$ $\mathbf{T}\S(\Phi \supset \Phi \vee \Psi)(L\theta)$: from $\vdash_{SL}(39)$ using a case of D2(iv) and rule 2(ii).

Using the link rule, therefore,

$\vdash_{RSL}(\Phi \supset \Phi \vee \Psi)(L\theta)$.

The other **RSL** theses can be derived in a similar way. To show that the whole of **RSL** can be obtained, it is only necessary to derive a standard set of axioms and the rule of detachment in the more familiar form: if $\vdash_{RSL}\Phi(L\theta)$ and $\vdash_{RSL}(\Phi \supset \Psi)(L\theta)$ then $\vdash_{RSL}\Psi(L\theta)$. We shall not go into this in detail here, however, since the proofs simply follow the pattern of those already given, but instead show, in **CL**, how to derive the familiar two-valued matrices (truth tables) for the classical two-valued connectives from the basis which we have here. This provides a more interesting proof that the whole of **RSL** is derivable.

It will be noted that every **RSL** thesis so obtained is qualified by the context symbol 'Lθ'. We may therefore remove it from each thesis and write at the beginning of the logic 'For a given logically standard context ...' as a uniform condition on all theses. That the whole of **RSL** is qualified in this way means that, before the logic is applied, a context must first be specified. Then, all the theses hold for that specified context throughout the application. The context variable thus becomes a parameter rather than a variable. That a context must be fixed before the application, implies, contrary to what has sometimes been supposed, that classical two-valued sentential logic is not context-free. For the context need not be the same for all applications. The only limitation is that the specified context must be kept fixed throughout a given application since none of the theses provide for cross-contextual links. That the chosen context must be logically standard, however, means that the sentential connectives have a uniform standard sense whatever value is specified for θ, though as we have noted, the sentences which are substituted for the variables need not themselves be limited to their standard senses.

2.8 Extended Statement Logic, ESL

Classical sentential logic is sometimes extended to include quantifiers. This, however, raises a number of interpretational difficulties. For formulae such as '(*r*)*r*', read literally as 'for every (sentence) *r*, *r*', are not intelligible or even grammatical. Sense is achieved only by introducing predicates such as 'is true'. Thus, '(*r*)*r*' is paraphrased as 'for every sentence *r*, *r* is true' or as 'every sentence is true'. So to interpret them, however, is to treat the sentential variables as subject variables which stand in need of a predicate in order to make a grammatically complete sentence. Such difficulties must inevitably arise because quantification can only be over subjects, and subjects demand a predicate. To make a general statement is to say something about items of a certain kind—that they are all so-and-so, where 'so-and-so' is a predicate.

These difficulties do not arise if statement logic is extended to include quantifiers. For we here have clear subject terms, the wfc; and every well-formed formula, when values are substituted for the variables, becomes a predicative (or relational) sentence which yields a statement about the items (statements) mentioned by the subject terms. Hence we can make general remarks about these items without encountering interpretational problems since the predicates are already explicit in the wffsl. We therefore extend statement logic to **ESL** as follows:

We add to the primitives,

the universal quantifier: U

interpreted as 'for every'; and supplement the formation rules with the clause:

R3 (v) if A is a wffsl, so is $(U\gamma)A$, where γ is a statement variable. (Hence $\S(U\gamma)A$ is a wfc.)

The particular quantifier P is then defined by:

D1 (vii) $(P\gamma)A =_{\mathrm{Df}} \mathbf{F}\S(U\gamma)\mathbf{F}\S A$

and interpreted as 'for some'. We use the word 'some' here, rather than the more commonly used 'exists' (and hence 'P' rather than '$\exists$') because we do not wish the use of the particular quantifiers to be taken to imply existence. We do not consider that statements exist as individual objects do. Statements, like numbers, are abstractions. Nevertheless, it is possible to make general and particular statements about both statements and numbers, and it is therefore appropriate to use universal and particular quantifiers. In the next chapter we develop the elementary predicate theory of semi-wffs and again use existentially neutral quantifiers, so that it becomes possible to make true affirmations of the sort "Some fairies are ..." even though there are no fairies.

We take for granted here the distinction between free and bound variables and rules for the scope of quantifiers. These are explained in detail in the next chapter where we discuss quantification generally. The formal logic of quantification over statements has practically the same structure as standard two-valued quantification theory. The only difference is that the subject variables have a different range and there are predicate constants (**/**, **T**, **conj**, *etc.*) but no predicate parameters.

The following are examples of formulae which can now be constructed:

(Ua) **T**a: for every statement a, a is true; or, every statement is true.

$(Ua)(a$ **imp** $a)$: for every statement a, a materially implies a; or, every statement materially implies itself.

$(Pa)(Pb)(a$ **disj** $b)$: some statement stands in the relation of disjunction to some statement.

$(Ua)(Pb)(a$ **imp** $b)$: every statement materially implies some statement.

The additional postulates which determine the theses of **ESL** are as follows:

Axioms:

$\vdash_{\mathrm{SL}}$(E1) $\S(U\gamma)(\S A$ **imp** $\S B)$ **imp** $\S(\S A$ **imp** $\S(U\gamma)B)$, where γ does not occur free in A.

$\vdash_{\mathrm{SL}}$(E2) $\S(U\gamma)A$ **imp** $\S\check{S}^{\gamma}_{\alpha} A\,|$, where γ is a statement variable, α is a wfc and $\check{S}^{\gamma}_{\alpha} A\,|$ is the wffsl which results from the substitution of α for all free occurrences of γ in A, provided that no free occurrence of γ is in a well-formed part of A which contains a variable of α bound; and otherwise is A. (This notation is adapted from that of Church [4].)

Rule:

SL(8) If $\vdash_{SL} A$ then $\vdash_{SL} (U\gamma)A$ (Generalisation).

Examples of theses are:

$\vdash_{SL}$(E3) $\S(P\gamma)A$ **eq** $\S\mathbf{F}\S(U\gamma)\mathbf{F}\S A$: D1(vii) by derived rule (5).

The other relations between the quantifiers can be derived from this in the usual way. Thus,

$\vdash_{SL}$(E4) $\S(P\gamma)\mathbf{F}\S A$ **eq** $\S\mathbf{F}\S(U\gamma)\mathbf{F}\S\mathbf{F}\S A$: a case of $\vdash_{SL}$(E3) taking A to be the wffsl $\mathbf{F}\S A$.

$\vdash_{SL}$(E5) $\S(P\gamma)\mathbf{F}\S A$ **eq** $\S\mathbf{F}\S(U\gamma)\mathbf{T}\S\mathbf{T}\S A$: from $\vdash_{SL}$(E4) using $\vdash_{SL}$(25) and rule 2(ii).

$\vdash_{SL}$(E6) $\S\mathbf{T}\S\mathbf{T}\S A$ **eq** $\S\mathbf{T}\S A$: a case of $\vdash_{SL}$(3).

$\vdash_{SL}$(E7) $\S(P\gamma)\mathbf{F}\S A$ **eq** $\S\mathbf{F}\S(U\gamma)\mathbf{T}\S A$: from $\vdash_{SL}$(E5) by $\vdash_{SL}$(E6) using rule 2(ii).

$\vdash_{SL}$(E8) $\S(P\gamma)\mathbf{F}\S A$ **eq** $\S\mathbf{F}\S(U\gamma)A$: from $\vdash_{SL}$(E7) by $\vdash_{SL}$(32) and rule 2(ii).

By similar moves, and by making use of $\vdash_{SL}$(28) where necessary, we also have:

$\vdash_{SL}$(E9) $\S\mathbf{F}\S(P\gamma)A$ **eq** $\S(U\gamma)\mathbf{F}\S A$.

$\vdash_{SL}$(E10) $\S\mathbf{F}\S(P\gamma)\mathbf{F}\S A$ **eq** $\S(U\gamma)A$.

Generalisations of **SL** theses arise in the following manner:

$\vdash_{SL}$(E11) γ **imp** γ, where γ is a statement variable: a case of $\vdash_{SL}$(5) where the wfc α is taken as a statement variable γ (as it may be since statement variables are wfc).

$\vdash_{SL}$(E12) $(U\gamma)(\gamma$ **imp** $\gamma)$: rule (8) on $\vdash_{SL}$(E11).

Special cases of **SL** theses involving quantifiers are obtained by moves such as the following:

$\vdash_{SL}$(E13) $\S(U\gamma)A$ **imp** $\S(U\gamma)A$: a case of $\vdash_{SL}$(5) where the wfc α is taken as the wfc $\S(U\gamma)A$.

We shall not prove any more theses since, as we have noted, the structure of the quantification theory is standard. That is, we have standard theses such as the distribution of the universal quantifier about conjunction, the distribution of the particular quantifier about disjunction, and so on.

It would be possible to derive from **ESL** an extended sentential logic **ERSL** by adding appropriate clauses to D2. We resist this, however, because of the interpretational problems involved. It seems to us that extended sentential logic is simply an inadequate attempt to express a quantified statement logic with proper subject variables and predicates. For this reason we have only defined quantification in conjunction with wffsl; and since we have not defined restricted semi-wffs as wffsl, quantification does not extend to them.

2.9 The Basic Well-formed Formulae of Context Logic, CL

We are now close to being able to formalise assertions such as,

Jack used the sentence 'Le monde est rond' (c_0) to make the statement that the world is round.

To get an exact formalisation, all that is required are

person variables: X, Y, X', Y', ...

which take the names of persons as values:

the use relation: **U**

to abbreviate 'used the sentence'; and

the use modifier: **M**

to abbreviate 'to make the statement'.

With this apparatus, examples such as that above would be represented by formulae of the sort,

$$X\mathbf{U}\ qu(p)(c)\ \mathbf{M}\ a,$$

and it is worth indicating here how the logic could be extended to take into account non-indicative illocutionary acts. The restriction to indicative illocutionary acts is in fact quite arbitrary in terms of the formal requirements and simply reflects our interests since our primary concern is the logic of statement-making. But the extension to other acts could be made by introducing further functions such as

the question function: **quest** (Φ, θ)
the order function: **ord** (Φ, θ)
the request function: **req** (Φ, θ)

and variables which range over their values. For just as many different sentences can be used to make the same statement, but to each statement-yielding sentence in a context there corresponds one and only one statement, so many different sentences can be used to ask the same question but to each question-yielding sentence in a context there corresponds one and only one question. And similarly for sentences which are used to give orders, to make requests, and the like. Hence there are many-one relations between the sentences and the items definable in terms of the illocutionary act which they are used to perform and they can be appropriately represented by functions. The values of these functions, like the values of the statement function, are abstractions; but the substitution values of the variables which range over questions, orders and requests are, respectively, clauses such as 'whether (why) (what) so-and-so', 'to so-and-so', and 'that so-and-so'. Hence if we introduce further modifiers like **M**, *viz.*

Q: which abbreviates 'to ask the question'
O: which abbreviates 'to give the order'
R: which abbreviates 'to request',

we could represent sentences such as,

Jack used the sentence 'Is it raining?' (c_0) to ask the question whether rain is falling;
Jack used the sentence 'Fetch my slippers' (c_0) to give the order to fetch his slippers;
Jack used the sentence 'Please fetch my slippers' (c_0) to request that his slippers be brought;

by means of formulae such as,

$$X\,\mathbf{U}\,qu(p_0)(c_0)\,\mathbf{Q}\,e_0;$$
$$X\,\mathbf{U}\,qu(q_0)(c_0)\,\mathbf{O}\,f_0;$$
$$X\,\mathbf{U}\,qu(p'_0)(c_0)\,\mathbf{R}\,g_0;$$

where e_0, f_0 and g_0 are respectively clause symbols defined in terms of the appropriate functions.

It would then be necessary to develop question logic, imperative logic and request logic along the lines of statement logic as a preliminary to the inclusion of such formulae as '$X\,\mathbf{U}\,qu(p)(c)\,\mathbf{Q}\,e$' into context logic. It is perhaps worth emphasising that the illocutionary act modifiers, **M**, **Q**, **O**, **R**, *etc.*, represent the act performed and not the grammatical form of the sentence used. So, for example, if a sentence of a type which is typically used to ask a question is in fact used to make a request ("Would you like to help with the dishes?"), then the appropriate modifier would be **R** and not **Q**. Sometimes, of course, though a sentence which is used to make a statement only yields one statement on the occasion of its use, and a sentence which is used to make a request only yields one request on the occasion of its use—a feature which is demanded by the introduction of the abstractions as the values of functions —one and the same sentence can be used in a given context to perform two or more different illocutionary acts simultaneously. Thus we might use the sentence 'The door is open' both to give information (*i.e.* to make a statement) and to make a request (Please shut it). Such uses find expression in formulae such as:

$$X\,\mathbf{U}\,qu(p)(c)\,\mathbf{M}\,a \quad \& \quad X\,\mathbf{U}\,qu(p)(c)\,\mathbf{R}\,g$$

and do not destroy the functional basis.

We shall not pursue these developments here, and we shall simplify the statement-making part of context logic by neglecting the use of person variables and by introducing as a formal primitive,

the yields relation: **Y**

which absorbs both **U** and **M**. We then regard formulae such as, '$qu(p)(c)$ **Y** a' ('$qu(p)$ in c yields the statement that-a'), as providing a satisfactory formal analysis, for many purposes, of sentences like "Jack used the sentence 'Le monde est rond' (c_0) to make the statement that the world is round", since it is usually unnecessary to make explicit mention of the user and, in cases where it may be necessary, reference to the user can be regarded as part of the context. The symbolism we adopt is therefore slightly less explicit than it might be, but it has the advantage of simplicity, and for our present purposes this lack of explicitness imposes no limitations and so is acceptable.

The only other primitive which is necessary, in addition to those we already have, is,

the significance predicate: sig

interpreted as 'is significant' and used in conjunction with semi-wffs to make statements about sentences. Alternatively we could choose to take '*comp*' ('is complete') as a primitive, since either one is definable in terms of the other.

We now define the basic well-formed formulae of context logic (wffcl) as follows:

R4 where $qu(\Gamma)$ is a partial wff and α is a wfc
(i) $qu(\Gamma)(\theta)$ **Y** α
(ii) $sig\ qu(\Gamma)(\theta)$
are wffcl.

It should be noted that we do not limit Γ to restricted semi-wffs in (i) (see §2.5).

To these we add wffsl and provide for restricted semi-wffs.

R4 (iii) wffsl are wffcl
(iv) restricted sentential variables are wffcl.

When values are substituted for the variables in wffcl we obtain *sentences* which have a prescribed standard use with respect to the standard context of the background language, namely, in the case of formulae arising from R4(i), to make statements about the relations between sentences and statements; in the case of those arising from R4(ii), to make statements about sentences; and in the case of those arising from R4(iii), to make statements about the relations between statements. The only exception to this is in the case of formulae arising from R4(iv). For we may substitute for the variables in restricted semi-wffs sentences which, though statement-yielding, need not yield statements about either sentences or statements. It is not in fact necessary to introduce restricted semi-wffs as wffcl, though it is necessary to admit wfc formed from them, but their omission leads to minor complications and we accept the slight dilution of purity in the interests of simplicity.

We must now consider how such formulae can be combined and ask what logic they satisfy.

2.10 The Nature of the Basic Formulae

Although formulae defined in terms of R4(i) and (ii) may contain unrestricted sentential variables which take as values nonsignificant or incomplete sentences, we wish to claim that wffcl always give rise to statement-yielding sentences whatever values are substituted for the variables.

Consider, first, formulae defined in terms of R4(i).

If we say of a significant and complete sentence that it yields a particular statement, we thereby make a true-or-false remark about it. The remark we make is true if the sentence happens to yield the statement which we mention, and false if not. So, for example, suppose that the sentence

'It is raining' (c_0)

yields the statement that rain is falling. Then the sentence,

'It is raining' (c_0) yields the statement that rain is falling,

is itself used to make a true statement, while the sentence

'It is raining' (c_0) yields the statement that grass is green,

is itself used to make a false statement.

On the other hand, if we say of a sentence which is either nonsignificant or incomplete that it yields a particular statement, then again we make a true-or-false remark about it. If, for example, a given sentence is used in a context to make a nonsignificant utterance, it is always significant, though false, to say of it that it yields a given statement; and it is always significant, and true, to say of it that it fails to yield a given statement. Thus, the sentence,

'The number seven likes soup' (c_0) yields the statement that the world is round,

itself yields a false statement if it is established that 'The number seven likes soup' is nonsignificant in c_0. For since the sentence which is being mentioned is itself nonsignificant, then it fails to be used in such a way as to make a statement at all, and hence fails to be used to make the statement that the world is round.

Similarly, if we say of any sentence that it is significant, the remark we make is true-or-false. It is true if the sentence is significant and false otherwise. Thus,

'The number seven likes soup' is significant in c_0

is itself statement-yielding even if the sentence which we are mentioning is nonsignificant in c_0. And

'Everything doubled in size overnight' is significant in c_0

is itself significant and complete even though the mentioned sentence is incomplete in c_0.

Hence, although nonsignificant and incomplete sentences cannot be used to make statements in the contexts in which they have these features, sentences in which they are mentioned may themselves be both significant and complete and so can be used to make statements about them. It should be noted, however, that this is a limited conclusion. In particular, we do not assume that every sentence in which other sentences are quoted is statement-yielding. On the contrary, it is possible to make nonsignificant utterances about other sentences whether the sentences mentioned are significant or not, complete or not. If it were said, for example, that Jack used a sentence which likes soup, then a nonsignificant utterance has been made about Jack's sentence even if the sentence he used is itself statement-yielding.

So our conclusion is, quite simply, that sentences constructed by means of the relation-expression 'yields the statement' with a quoted sentence on the left and a that-clause on the right, and those constructed from a quoted sentence and the predicate 'is significant', are always both significant and complete. Hence, all sentences which arise as substitution cases of R4(i) and (ii) formulae are statement-yielding provided that '**Y**' and '*sig*' are used with the fixed sense which has been given to them. That is, provided they occur in assertions made with respect to the standard context c_S.

In the case of formulae arising from R4(iii), since the that-clauses which are used in their construction are formed from statement-yielding sentences, they mention statements. Hence if we relate two such clauses by the relation '... is incompatible with ...', what we say is never nonsignificant or incomplete. Again, if we say that every statement is incompatible with a particular statement, with some statements, or with every statement, the remark we make is true or false. Hence wffsl are statement-yielding sentences when values are substituted for the variables. This, indeed, we have already presupposed by permitting the formation of well-formed clauses from wffsl, since the wfc can only arise from statement-yielding sentences.

Of course, the primitive constants: *qu*, **/**, **Y** and *sig*; and those defined in terms of them, could be used in different contexts with a sense quite different from their intended sense here, and so could be used in the construction of nonsignificant or incomplete sentences. If, however, we specify that their sense is fixed with respect to c_S, and if we affirm all theses of **CL** with respect to c_S, then, since there is a well-defined range for each set of variables, we are assured that formulae constructed in accordance with the rules R4(i)–(iii) never give rise to nonsignificant or incomplete sentences when values are substituted for the variables.

Finally, by definition, sentences substituted for the variables in restricted semi-wffs are always statement-yielding.

2.11 The Formal Basis of **CL**

The basic wffcl thus satisfy the logic of statement-yielding sentences. If, therefore, we take it that any compound formed from complete and significant sentences is itself a complete and significant sentence, we may combine the basic wffcl using the standard two-valued connectives and the formulae so constructed will satisfy the theses of classical logic. This assumption about compound sentences is discussed in Part II. It is of course a standard assumption of classical logic that a compound of two-valued sentences is itself two-valued. So conceived, context logic is a two-valued theory even though some of the sentential formulae which occur in wffcl are not themselves two-valued. This is possible simply because such formulae always occur within a quotation function and their values are therefore never used but only mentioned. For this reason unrestricted semi-wffs are not themselves defined as wffcl but only as well-formed parts of wffcl. To this point, therefore, context logic is essentially a two-valued metatheory of a many-valued object-language. Formally speaking it is an applied classical predicate calculus of first order with special postulates.

We now set up the formal basis as follows:

Consider first basic sentential formulae such as,

$$qu(\Gamma)(\theta)\,\mathbf{Y}\,\alpha;\quad sig\ qu(\Gamma)(\theta);\quad \mathbf{T}\,\alpha;\quad \mathbf{F}\,\alpha;\quad R;\quad S.$$

Combining them by means of the usual connectives, we shall have compounds like,

$$\sim(qu(\Gamma)(\theta)\,\mathbf{Y}\,\alpha);\quad qu(\Gamma)(\theta)\,\mathbf{Y}\,\alpha\,\&\,\mathbf{T}\alpha;\quad (R \vee \sim R)\supset \mathbf{T}\alpha;$$

Now each of these is a sentential formula and can be uttered with respect to a context. And although our interest is in asserting some of them with respect to the standard context, we provide for their utterance in any context by allowing them to be qualified by any context symbol. So we allow, for example,

$$\sim(qu(\Gamma)(\theta)\,\mathbf{Y}\alpha)(\eta);\quad (qu(\Gamma)(\theta)\,\mathbf{Y}\alpha\,\&\,\mathbf{T}\alpha)(\eta).$$

That the formulae already contain context symbols as part of partial wffs such as $qu(\Gamma)(\theta)$, or indeed as part of wfc—*e.g.* $\S\Phi(\theta)$, does not make for ambiguity. For the context symbols already in the formulae indicate contexts which are being mentioned and these contexts need not themselves be the context of utterance of the whole formula in which the mentioning takes place. For example, a sentence such as,

The sentence 'It is raining' (c_0) yields the statement that rain is falling (c_S)

may itself be uttered with respect to any context—c_0, c_S, or some other. Again, being sentential formulae their values (sentences) can be mentioned and it is therefore appropriate to construct partial wffs from them by means of quotation functions. In particular, we can say that they are significant and that they yield statements. So we allow:

$$qu[qu(\Gamma)(\theta)\,\mathbf{Y}\alpha\ \&\ \mathbf{T}\alpha](\eta)\,\mathbf{Y}\,\beta;\quad sig\ qu(\mathbf{T}\alpha)(\eta);$$

and so on. Cases of these are, for example,

The sentence "The sentence 'It is raining' (c_0) yields the statement that rain is falling, and it is true that rain is falling" (c_S) yields the statement that the sentence 'It is raining' (c_0) yields the statement that rain is falling.

The sentence 'it is true that it is raining' (c_S) is significant.

And again these sentences can be uttered with respect to any context. To avoid proliferation of context symbols, however, we adopt the convention that the absence of a context symbol on a sentence indicates that its context of utterance is c_S. We could in fact say all that we want to say by limiting the contextual qualification on wffcl to c_S and then dropping it in terms of this convention so that no context symbol occurs on the whole formula. But in fact it makes for some formal simplicity if we allow general context qualification. The reason for this is that we require restricted semi-wffs to be wffcl but we do not wish to limit their use to the standard context. Hence it would be necessary to assert each postulate and definition twice—once for wffcl which are not restricted semi-wffs and are expressed with respect to the standard context only, and once for restricted semi-wffs. (*Cf.* D2* and $\vdash_{CL}$(P1).)

Finally we allow that statement variables in wffcl may be quantified and so construct formulae such as the following:

$$(Ua)(\mathbf{T}a \vee \mathbf{F}a);\quad (Ua)\mathbf{T}a \supset (Pa)(\mathbf{T}a);\quad (Pa)(qu(\Gamma)(\theta)\,\mathbf{Y}\,a).$$

The formation rules are as follows:

R4 (v) Where A is a wffcl and α a wfc, $A(\theta)$, $sig\ qu(A)(\theta)$, and $qu(A)(\theta)\,\mathbf{Y}\,\alpha$ are wffcl, provided that no part of A, other than the partial wffs or wfc which may occur in it, already contains a context symbol.

(vi) where A and B are wffcl and γ is a statement variable, $(\sim A)$, $(A \vee B)$, $(A \& B)$, $(A \supset B)$, $(A \equiv B)$ and $(U\gamma)A$ are wffcl.

(We adopt the worlds assumption for wffcl, as well as for the semi-wffs: see §2.3.).

It follows from (v) and (vi) that, *e.g.*, $A(\theta) \vee B(\eta)$ is a wffcl but $(A(\theta) \vee B(\eta))(\theta)$ is not. Similarly, $sig\ qu(A \vee B)(\theta)$ is a wffcl but $sig\ qu(A(\theta) \vee B(\eta))(\theta)$ is not. The connectives $\sim$, & *etc.*, and the quantifier, are characterised as two-valued below.

R4 (vii) where A is a wffcl, §A is a wfc.

Since wffsl are wffcl, and since wffsl can be constructed using wfc, the wfc formed here can be used to construct further wffsl which are also wffcl. They can of course also be used to construct other wffcl by R4(v).

The convention that the absence of a context symbol on a wffcl or a wfc indicates that its context of utterance is the standard context of the background language is secured by the following definition, where A is as in R4 (v):

D3 (i) $A =_{\mathrm{Df}} A(c_{\mathrm{S}})$.
It then follows by definitional replacement: (ii) $\S A =_{\mathrm{Df}} \S A(c_{\mathrm{S}})$.

Here A has no context symbol qualifying it (though it may contain context symbols in any partial wffs or wfc which occur in it). For if it had, $A(c_{\mathrm{S}})$ would not be well formed and the definition would fail. Notice that D3 does not provide for the elimination of 'c_{S}' in formulae such as '*sig qu*$(A)(c_{\mathrm{S}})$'. The elimination here would leave us with '*sig qu*(A)' which is not well-formed since R4(v) provides for the well-formation of '*sig qu*$(A)(\theta)$', when A is well-formed, but not '*sig qu*(A)'. Similarly, the context symbol cannot be dropped from formulae such as '*qu*$(A)(c_{\mathrm{S}})$ **Y** α'.

In allowing the formation of wfc from wffcl, we are in effect limiting the application of the logic to contexts in which the wffcl yield statements when values are substituted for the variables. For it is only statement-yielding sentences which give rise to that-clauses which can be used successfully to mention statements, and only such clauses can be taken as values of the statement variables and hence as wfc. Now we tried to show in the previous section that this assumption is justified in the case of the standard context where **Y**, **/**, *etc.* have the fixed interpretation which has been given to them, and it is of course in this interpretation that our interest lies. Nevertheless, there may be other non-standard interpretations of the primitives such that wffcl give rise to statement-yielding sentences when values are substituted for the variables—indeed this is likely to be so unless the logic is categorical—and if so, our logic will apply to the sentences in these non-standard contexts. It will only fail to apply to contexts in which wffcl are not statement-yielding. Hence the limitation is that θ, in $A(\theta)$, must always be such that A yields a statement in θ when values are substituted for the variables. This limitation on the context symbols does not of course apply to those which qualify unrestricted semi-wffs, for the values of the unrestricted semi-wffs are being mentioned, not used, and we can successfully mention their non-statement-yielding uses.

All *theses* of **CL** are asserted with respect to c_{S} and, as in statement logic, instead of writing the contextual qualification on each thesis, we regard it as a condition on the whole logic. This is now explicit in terms of D3 since we have: if $\vdash_{\mathrm{CL}} A$, where A is not qualified by a context symbol, then $\vdash_{\mathrm{CL}} A(c_{\mathrm{S}})$.

The rules allow us to construct formulae such as the following:

qu$(p)(c)$ **Y** a: *qu*(p) in c yields the statement that-a;
qu$(p)(c)$ **Y** a & **T** a: *qu*(p) in c yields the statement that-a, and that-a is true;

$(Pa)(qu(p)(c)\,\mathbf{Y}\,a)$: $qu(p)$ in c yields some statement (*i.e.* $qu(p)$ is statement-yielding in c).

(Strictly speaking the particular quantifier has not yet been defined except for use in conjunction with wffcl which are also wffsl. The extension to all wffcl is given below.)

$(Pa)(qu(p)(c)\,\mathbf{Y}\,a) \supset sig\ qu(p)(c)$: if $qu(p)$ is statement-yielding in c, then it is significant in c;

$(Ua)(\mathbf{T}a \vee \mathbf{F}a)$: every statement is true or false;

$\sim(Pa)(\mathbf{T}a \,\&\, \mathbf{F}a)$: no statement is both true and false;

$qu(p)(c)\,\mathbf{Y}\,\S(Ua)(\mathbf{T}a \vee \mathbf{F}a)$: $qu(p)$ in c yields the statement that every statement is true or false;

$sig\ qu[(Pa)\mathbf{T}a]\,(c_{\mathrm{S}})$: the sentence 'Some statement is true' is significant in c_{S}.

That the formulae of **CL** satisfy classical two-valued logic can be achieved in various ways. We could simply affirm that substitution cases of **RSL**-theses are **CL**-theses, *i.e.*

$$\text{if } \vdash_{\mathrm{RSL}} \Phi(\mathrm{L}\theta) \text{ then } \vdash_{\mathrm{CL}} \mathrm{S}^{\mathrm{R}_1 \ldots \mathrm{R}_n}_{\mathrm{A}_1 \ldots \mathrm{A}_n} \Phi \mid (c_{\mathrm{S}})$$

where $A_1, \ldots, A_n$ are wffcl and $R_1, \ldots, R_n$ are all the restricted sentential variables occurring in Φ. This would amount to saying that since the theses of **RSL** are valid over any logically standard context, and since c_{S} is logically standard, we may take an application of **RSL** over the formulae of **CL**. This would give us the sentential two-valued basis of **CL** in the form of theorem schemata, *e.g.* $\vdash_{\mathrm{CL}} A \supset A \vee B$; $\vdash_{\mathrm{CL}} A \vee A \supset A$ *etc.* However, by taking a variant of D2 we can obtain directly both the sentential and quantificational bases.

Where A and B are wffcl:

D2* θ is a logically standard context if, and only if,

(i) $\S\mathbf{T}\S{\sim}A(\theta)\ \mathbf{eq}\ \S\mathbf{F}\S A(\theta)$

(ii) $\S\mathbf{T}\S(A \vee B)(\theta)\ \mathbf{eq}\ \S(\S A(\theta)\ \mathbf{disj}\ \S B(\theta))$

(iii) $\S\mathbf{T}\S(A \& B)(\theta)\ \mathbf{eq}\ \S(\S A(\theta)\ \mathbf{conj}\ \S B(\theta))$

(iv) $\S\mathbf{T}\S(A \supset B)(\theta)\ \mathbf{eq}\ \S(\S A(\theta)\ \mathbf{imp}\ \S B(\theta))$

(v) $\S\mathbf{T}\S(A \equiv B)(\theta)\ \mathbf{eq}\ \S(\S A(\theta)\ \mathbf{eq}\ \S B(\theta))$.

D2 could not be stated in this general form in **SL** since sentential compounds of wffsl were not defined as wffsl. Hence clauses constructed from them, *e.g.* $\S(A \vee B)$, were not wfc. **SL** could be extended in this way, but it is then very close to being **CL**. Notice that this definition demands immediately that θ be restricted to contexts in which the wffcl are statement-yielding, otherwise clauses such as $\S A$ and $\S(A \vee B)$ would not be statement-mentioning and could not be taken in conjunction with the connectives **T**, **disj**, *etc.*

If we now extend the rule SL(2), equivalence replacement, so that it applies to all wffcl, *i.e.* as a rule of context logic, the derivations given in §2.7 go through as before, though in a more general form. Thus, in place of $\vdash_{SL}(36)$ and $\vdash_{SL}(40)$ we shall have:

$$\vdash_{CL}\mathbf{T}\S(A \vee \sim A)(\mathrm{L}\theta)$$
$$\vdash_{CL}\mathbf{T}\S(A \supset A \vee B)(\mathrm{L}\theta).$$

Taking A and B to be restricted semi-wffs, as we may since these are wffcl, we obtain $\vdash_{SL}(36)$ and $\vdash_{SL}(40)$ in their original form as theses of **CL**: $\vdash_{CL}\mathbf{T}\S(\Phi \vee \sim\Phi)(\mathrm{L}\theta)$, $\vdash_{CL}\mathbf{T}\S(\Phi \supset \Phi \vee \Psi)(\mathrm{L}\theta)$. But in the form in which we now have them, they are much more general than this and give us an application of **RSL** over all wffcl. Thus, since c_s is logically standard, we have as special cases:

$$\vdash_{CL}\mathbf{T}\ \S\ (A \vee \sim A)\ (c_s)$$
$$\vdash_{CL}\mathbf{T}\ \S\ (A \supset A \vee B)\ (c_s),$$

and by D3 we may drop the contextual qualification. Finally, therefore, by rule SL(6)—**T**-elimination—extended to all wffcl, we have:

$$\vdash_{CL}A \vee \sim A$$
$$\vdash_{CL}A \supset A \vee B.$$

The link rule SL(7) is not required here. Rule SL(6) (extended) takes its place since we are not now deriving **RSL** as a separate logic, but **RSL** theorem-schemata for the wffs of **CL**.

The standard two-valued definitions of the connectives come through in the form of equivalence theses: *e.g.* $\vdash_{CL}(A \supset B) \equiv \sim(A \mathbin{\&} \sim B)$, though they could be derived in a definitional form from D1 if we add a rule to the effect that an equivalence replacement in a definition yields a definition. This, however, is unnecessary. But it should be noted that these definitions are only valid for wffcl (which include restricted semi-wffs). In the case of unrestricted semi-wffs, the connectives still have to be regarded as independent until the appropriate many-valued sentential systems are developed in Part II.

We shall take the theses of the two-valued sentential basis for granted without proof, and moves made in accordance with them will be indicated using the phrase 'by **RSL**'.

By cashing θ in D2* as Lθ, and using the worlds assumption, we can obtain versions of the conditions (i)–(v) as equivalence theses of **CL**.

$\vdash_{CL}(1)$ $\S\mathbf{T}\S\sim A(\mathrm{L}\theta)\ \mathbf{eq}\ \S\mathbf{F}\S A(\mathrm{L}\theta)$
$\vdash_{CL}(2)$ $\S\mathbf{T}\S(A \vee B)(\mathrm{L}\theta)\ \mathbf{eq}\ \S(\S A\ \mathbf{disj}\ \S B)(\mathrm{L}\theta)$
$\vdash_{CL}(3)$ $\S\mathbf{T}\S(A \mathbin{\&} B)(\mathrm{L}\theta)\ \mathbf{eq}\ \S(A\ \mathbf{conj}\ \S B)(\mathrm{L}\theta)$
$\vdash_{CL}(4)$ $\S\mathbf{T}\S(A \supset B)(\mathrm{L}\theta)\ \mathbf{eq}\ \S(\S A\ \mathbf{imp}\ \S B)(\mathrm{L}\theta)$
$\vdash_{CL}(5)$ $\S\mathbf{T}\S(A \equiv B)(\mathrm{L}\theta)\ \mathbf{eq}\ \S(\S A\ \mathbf{eq}\ \S B)(\mathrm{L}\theta).$

These in turn give rise to sentential equivalence theses as we shall see presently.

The quantification theory can now be derived as follows:

We may assert $\vdash_{SL}$(E1) and $\vdash_{SL}$(E2) where A, B and $(U\gamma)B$ are taken to be wffcl since they occur in clause form in (E1) and (E2) and clauses formed from wffcl are wfc. So we have, as a case of (E1), where γ does not occur free in A,

$\vdash_{CL}\S(U\gamma)(\S A \textbf{ imp } \S B) \textbf{ imp } \S(\S A \textbf{ imp } \S(U\gamma)B)$.

Now by $\vdash_{CL}(4)$, where $L\theta$ is c_S, and using D3(i) and rule SL(2)(ii) on the antecedent, and SL(2)(i) on the consequent, we may pass to,

$\vdash_{CL}\S(U\gamma)\,\mathbf{T}\,\S(A \supset B) \textbf{ imp } \S\mathbf{T}\S(A \supset (U\gamma)B)$.

By $\vdash_{SL}(3)$, therefore, again using SL(2)(ii) on the antecedent and SL(2)(i) on the consequent, we have,

$\vdash_{CL}\S(U\gamma)(A \supset B) \textbf{ imp } \S(A \supset (U\gamma)B)$.

Hence, by $\vdash_{CL}(4)$, using rule SL(2)(ii),

$\vdash_{CL}\mathbf{T}\S((U\gamma)(A \supset B) \supset (A \supset (U\gamma)B))$.

Finally, therefore, by rule SL(6),

$\vdash_{CL}(U\gamma)(A \supset B) \supset (A \supset (U\gamma)B)$, where γ does not occur free in A.

Similarly, we obtain,

$\vdash_{CL}(U\gamma)A \supset \check{S}^{\gamma}_{\alpha} A\,|$.

The standard definition of the particular quantifier is expressed by the equivalence thesis:

$\vdash_{CL}(P\gamma)A \equiv \sim(U\gamma)\sim A$.

Notice that although $(U\gamma)A$ is a wffcl when A is a restricted semi-wff, this does not give us extended sentential logic, but instead formulae of the sort '$(Ua)r$' in which the quantification is vacuous.

Thus the quantificational theory is simply an applied form of standard two-valued first-order quantification theory, which we call **RQL** (restricted quantificational logic). Theses of it will be taken for granted without proof and we indicate moves made in accordance with them by the phrase 'by **RQL**'.

For example, we have as a case of $\vdash A \vee \sim A$,

$$\vdash_{CL}\mathbf{T}a \vee \sim\mathbf{T}a.$$

Hence, by generalisation,

$$\vdash_{CL}(Ua)(\mathbf{T}a \vee \sim\mathbf{T}a);$$

and by **RQL**,

$$\vdash_{CL}(Pa)(\mathbf{T}a \vee \sim\mathbf{T}a);$$

hence, again by **RQL**,

$$\vdash_{CL}(Pa)\,\mathbf{T}a \vee (Pa)\sim\mathbf{T}a.$$

and so on.

The derivations of the sentential and quantificational bases presupposes that we take over theses of **SL** as theses of **CL** and that the rules of **SL** are extended appropriately and taken to be rules of **CL**. This we now do, though with two variations in the rules. The first variation is that we write the rule of detachment in its more familiar form since we now have the normal connectives. The second variation is more fundamental. It concerns the application of rule SL(2)(i), equivalence replacement.

If we state the rule in the form:

if $\vdash_{\mathrm{CL}}\alpha\ \mathbf{eq}\ \beta$ and $\vdash_{\mathrm{CL}}A$, where A contains α as a wfc, then $\vdash_{\mathrm{CL}}B$, where B is obtained from A by replacing α by β at any of its occurrences in A,

where A and B are *any wffcl*, then it will follow that if we have a thesis of the form,

$$\vdash_{\mathrm{CL}} qu(A)(\theta)\ \mathbf{Y}\alpha$$

and we also have,

$$\vdash_{\mathrm{CL}}\alpha\ \mathbf{eq}\ \beta$$

we shall be able to pass to,

$$\vdash_{\mathrm{CL}} qu(A)(\theta)\ \mathbf{Y}\beta.$$

Hence,

$$\vdash_{\mathrm{CL}} qu(A)(\theta)\ \mathbf{Y}\alpha\ \ \&\ \ qu(A)(\theta)\ \mathbf{Y}\beta$$

by the ordinary compounding rule of **RSL**. Now if we assert the many-one property of **Y** by affirming,

$$\vdash_{\mathrm{CL}}(qu(\Pi)(\theta)\ \mathbf{Y}\alpha\ \&\ qu(\Pi)(\theta)\ \mathbf{Y}\beta) \supset (\alpha = \beta),$$

where Π is either a semi-wff or a wffcl, we shall be able to pass to

$$\vdash_{\mathrm{CL}}\alpha = \beta.$$

Hence, whenever we can prove that two statements are equivalent and that a sentence yields one of them, we shall be able to prove that the two statements are identical. That is, we shall have a derived rule of the form,

$$\text{if } \vdash qu(A)(\theta)\ \mathbf{Y}\alpha \text{ and } \vdash\alpha\ \mathbf{eq}\ \beta \quad \text{then } \vdash\alpha = \beta;$$

and even if we do not affirm the many-one character of **Y**, we shall still have:

$$\text{if } \vdash qu(A)(\theta)\ \mathbf{Y}\alpha \text{ and } \vdash\alpha\ \mathbf{eq}\ \beta \text{ then } \vdash qu(A)(\theta)\ \mathbf{Y}\beta;$$

that is, a sentence yields every statement which can be proved to be materially equivalent to any one which it yields. Taking A to be any unrestricted semi-wff Φ, these results hold for any statement-yielding sentence.

Now we shall always be able to prove that clauses formed from two theses are equivalent. That is, if $\vdash A$ and $\vdash B$ then $\vdash \S A\ \mathbf{eq}\ \S B$ (derived rule CL(10) below); and we wish also to affirm that statement-yielding sentences uttered with respect to the standard context yield statements which can be mentioned by prefixing them with a that-clause: *i.e.* $\vdash qu(A)(c_S)\ \mathbf{Y}\ \S A$. Hence if A and B are theses we shall have both

$$\vdash qu(A)(c_S)\ \mathbf{Y}\ \S A \text{ and } \vdash \S A\ \mathbf{eq}\ \S B.$$

Consequently $\vdash \S A = \S B$. That is,

$$\text{if } \vdash A \text{ and } \vdash B \text{ then } \vdash \S A = \S B.$$

So, since we have $\vdash r \vee \sim r$ and $\vdash r \supset r \vee s$, we should have $\vdash \S(r \vee \sim r) = \S(r \supset r \vee s)$. For example, that it is raining or it is not, is the same statement as, that if it is raining then it is raining or the sun is shining. In general, that is to say, every tautology makes the same statement with respect to the standard context.

These conclusions are not unconnected with Wittgenstein's remark in the *Tractatus* "a proposition asserts every proposition which follows from it" (5.124)—in our terminology the slightly weaker form that a sentence yields every statement which is logically equivalent to any one which it yields. They are even closer to the positivist doctrine that the meaning of a proposition is the set of all its implications. For we should be able to say that every logical equivalent of a statement has the same meaning as it (since they are all the same statement) and hence that the meaning of a statement is the set of its logical equivalents. Finally, if we take a thesis version of either one of the "rules",

$$\vdash qu(A)(\theta)\,\mathbf{Y}\alpha \text{ and } \vdash \alpha\,\mathbf{eq}\,\beta \text{ then } \vdash \alpha = \beta$$
$$\vdash qu(A)(\theta)\,\mathbf{Y}\alpha \text{ and } \vdash \alpha\,\mathbf{eq}\,\beta \text{ then } \vdash qu(A)(\theta)\,\mathbf{Y}\beta,$$

i.e.

$$\vdash [qu(A)(\theta)\,\mathbf{Y}\alpha \;\&\; (\alpha\,\mathbf{eq}\,\beta)] \supset [\alpha = \beta]$$
$$\text{or } \vdash [qu(A)(\theta)\,\mathbf{Y}\alpha \;\&\; (\alpha\,\mathbf{eq}\,\beta)] \supset [qu(A)(\theta)\,\mathbf{Y}\beta],$$

then we have Frege's doctrine that there are only two statements, the True and the False.

For some, perhaps, these consequences are, or have been, acceptable. For others, though unacceptable, it might be thought that the most damaging of them could be stopped by refusing to affirm the many-one property of **Y**. We point out, however, that if this course is adopted, then it has to be accepted that every statement-yielding sentence in a given context yields an infinite number of statements in the same context. For any statement has an infinite number of logical equivalents.

We find all of these conclusions unacceptable. It seems to us that there is a clear difference between two different statements' being equivalent and two, apparently different, statements' being the same. To say that two apparently different statements are the same is at least to imply that they have the same meaning. To say that two statements are equivalent, even logically equivalent, is not to imply this at all. Equally, there is a clear difference between a sentence's yielding one and only one statement which may be logically equivalent to an infinite number of others, and a sentence's yielding an infinite number of statements. For when we use a sentence to say something, what we say may be equivalent to many other things which we could also say, but we do not in fact say them.

In general, an extensional connection such as material equivalence will only licence a replacement in extensional positions, not in non-truth-functional positions such as those governed by **Y** or by intensional predicates (see Carnap [2] and §7.2). We therefore accept an extended form of SL(2) for **CL** but in effect restrict its application to truth-functional positions by refusing to accept the replacement of equivalents in the second position of '**Y**', *i.e.* by refusing to accept that a sentence which yields a statement also yields every logically equivalent statement. Later (§2.13), we extend this restriction to identity formulae.

The initial rules we adopt are as follows:

CL(1) *Definitional Replacement* for wffcl (extension of SL(1))

CL(2) *Equivalence Replacement* (modified SL(2)).

(i) if $\vdash_{CL}\alpha$ **eq** β and $\vdash_{SL}A$, where A contains α as a wfc, then $\vdash_{SL}B$, where B is obtained from A by replacing α by β at any of its occurrences in A which do not appear in well-formed parts of the form $qu(\Pi)(\theta)$ **Y**α, where Π is either a semi-wff or a wffcl.

(ii) if $\vdash_{CL}\S A$ **eq** $\S B$ then A may be replaced by B at any point of a thesis to yield a thesis, subject to the same restrictions as in (i).

This restriction on rule (2) does not affect any of the derivations necessary to obtain the sentential and quantificational bases since the application of the rule is there carried out entirely on wffcl which are also wffsl, *i.e.* no replacements are made in formulae of the sort $qu(\Pi)(\theta)$ **Y**α.

CL(3) *Detachment:* if $\vdash_{CL}A$ and $\vdash_{CL}A \supset B$, then $\vdash_{CL}B$ (modified SL(3))

CL(4) *The* **T**-*rule* for wffcl (extension of SL(4))

CL(5) *Equivalence representation of definitions* (extension of SL(5))

CL(6) **T**-*elimination*—the converse of CL(4) (extension of SL(6))

CL(7) *The Link Rule* (extension of SL(7))

CL(8) *Generalisation* (extension of SL(8)).

We also have the following additional derived rules:

CL(9) $\vdash_{CL}\S A$ **eq** $\S B$ if, and only if, $\vdash_{CL}A \equiv B$

Proof: Suppose $\vdash_{CL}\S A$ **eq** $\S B$. Then by $\vdash_{CL}(5)$, using rule CL(2)(ii), $\vdash_{CL}$**T**$\S(A \equiv B)$. Hence, by rule (6), $\vdash_{CL}A \equiv B$. Conversely, if $\vdash_{CL}A \equiv B$, then $\vdash_{CL}$**T**$\S(A \equiv B)$ by rule (4). Hence, again by $\vdash_{CL}(5)$ and rule (2)(ii), $\vdash_{CL}\S A$ **eq** $\S B$.

It follows immediately that,

CL(10) if $\vdash_{CL}A$ and $\vdash_{CL}B$ then $\vdash_{CL}\S A$ **eq** $\S B$.

Proof: For by a derived rule of **RSL**, if $\vdash A$ and $\vdash B$ then $\vdash A \equiv B$. Result by rule (9).

We can now affirm the statemental-equivalence theses derivable from D2* in the form of sentential equivalences. For we have, from $\vdash_{\mathrm{CL}}(1)$ using a case of $\vdash_{\mathrm{SL}}(3)$,

$$\vdash_{\mathrm{CL}} \S\mathbf{F}\S A(\mathrm{L}\theta)\ \mathbf{eq}\ \S{\sim} A(\mathrm{L}\theta),$$

which, by rule CL(9) leads to,

$$\vdash_{\mathrm{CL}}(6)\ \mathbf{F}\ \S A(\mathrm{L}\theta) \equiv {\sim} A(\mathrm{L}\theta).$$

Similarly,

$\vdash_{\mathrm{CL}}(2)$ leads to, $\vdash_{\mathrm{CL}}(7)\ \mathbf{T}\S(A \vee B)(\mathrm{L}\theta) \equiv (\S A\ \mathbf{disj}\ \S B)(\mathrm{L}\theta)$
$\vdash_{\mathrm{CL}}(3)$ leads to, $\vdash_{\mathrm{CL}}(8)\ \mathbf{T}\S(A \,\&\, B)(\mathrm{L}\theta) \equiv (\S A\ \mathbf{conj}\ \S B)(\mathrm{L}\theta)$
$\vdash_{\mathrm{CL}}(4)$ leads to, $\vdash_{\mathrm{CL}}(9)\ \mathbf{T}\S(A \supset B)(\mathrm{L}\theta) \equiv (\S A\ \mathbf{imp}\ \S B)(\mathrm{L}\theta)$
$\vdash_{\mathrm{CL}}(5)$ leads to, $\vdash_{\mathrm{CL}}(10)\ \mathbf{T}\S(A \equiv B)(\mathrm{L}\theta) \equiv (\S A\ \mathbf{eq}\ \S B)(\mathrm{L}\theta)$.

And finally, we have from $\vdash_{\mathrm{SL}}(3)$,

$\vdash_{\mathrm{CL}}\S\mathbf{T}\ \S A(\mathrm{L}\theta)\ \mathbf{eq}\ \S A(\mathrm{L}\theta)$, which leads to $\vdash_{\mathrm{CL}}(11)\ \mathbf{T}\ \S A(\mathrm{L}\theta) \equiv A(\mathrm{L}\theta)$.

So we are able to represent the truth conditions for statements yielded by compound sentences.

Instead of adding the postulates of **SL** and the original rule of detachment, we simply affirm that every thesis of **SL** is a thesis of **CL**:

CL(11) If $\vdash_{\mathrm{SL}} A$ then $\vdash_{\mathrm{CL}} A$.

This rule has been taken for granted in the above derivations.

2.12 Properties of Sentences

Several sentential properties can be defined for all sentences, whether statement-yielding or not. In particular, where A is a wffcl and Γ is an unrestricted semi-wff, it is possible to give two identical definitions of certain features, one involving A and the other involving Γ. To avoid this duplication we use new metavariables Π and Θ which stand for both unrestricted semi-wffs and wffcl. As before, γ stands for any statement variable.

D4 *Completeness:* a sentence is complete in its context if, and only if, when it is significant it yields a statement:
$$comp\ qu(\Pi)(\theta) =_{\mathrm{Df}} [sig\ qu(\Pi)(\theta) \equiv (P\gamma)(qu(\Pi)\ (\theta)\ \mathbf{Y}\gamma)].$$

D5 *Truth:* a sentence is true in its context if, and only if, it yields a true statement:
$$tru\ qu(\Pi)(\theta) =_{\mathrm{Df}} (P\gamma)(qu(\Pi)(\theta)\ \mathbf{Y}\gamma\ \&\ \mathbf{T}\gamma).$$

D6 *Falsity:* a sentence is false in its context if, and only if, it yields a false statement:
$$fal\ qu(\Pi)(\theta) =_{\mathrm{Df}} (P\gamma)(qu(\Pi)(\theta)\ \mathbf{Y}\gamma\ \&\ \mathbf{F}\gamma).$$

D7 *Statement Equality:* two sentences are statement equal, in the same or different contexts, if, and only if, there is some statement which they both yield:

$$qu(\Pi)(\theta)\ steq\ qu(\Theta)(\eta) =_{\mathrm{Df}} (P\gamma)(qu(\Pi)(\theta)\ \mathbf{Y}\gamma\ \&\ qu(\Theta)(\eta)\ \mathbf{Y}\gamma).$$

D8 *Strong Extensional Equivalence:* two sentences are extensionally equivalent in the strong sense if, and only if, they are both true or both false or both nonsignificant or both incomplete:

$$\begin{aligned} qu(\Pi)(\theta)\ stee\ qu(\Theta)(\eta) =_{\mathrm{Df}} & [(tru\ qu(\Pi)(\theta)\ \&\ tru\ qu(\Theta)(\eta)) \\ & \vee (fal\ qu(\Pi)(\theta)\ \&\ fal\ \mathrm{qu}(\Theta)(\eta)) \\ & \vee (\sim sig\ qu(\Pi)(\theta)\ \&\ \sim sig\ qu(\Theta)(\eta)) \\ & \vee (\sim comp\ qu(\Pi)(\theta)\ \&\ \sim comp\ qu(\Theta)(\eta))]. \end{aligned}$$

To simplify the symbolism we introduce a number of absorption principles which enable us to drop the writing of quotation functions in many cases. We write:

S	for	*sig qu*
C	for	*comp qu*
T	for	*tru qu*
F	for	*fal qu*
$\doteqdot$	for	*qu*( ) *steq qu*( )
$\simeq$	for	*qu*( ) *stee qu*( )

(In general, we use bold face type to indicate properties of and relations between statements, or relations between sentences and statements or words and items, and ordinary italics to indicate properties of and relations between sentences or words).

So, for example, taking Π to be the semi-wff 'p', Θ to be 'q', and θ and η both as 'c', we can write an equivalence thesis arising from D8 in the following form:

$$\vdash_{\mathrm{CL}}\{[p \simeq q] \equiv [(Tp\ \&\ Tq) \vee (Fp\ \&\ Fq) \vee (\sim Sp\ \&\sim Sq) \vee (\sim Cp\ \&\sim Cq)]\}\ (c).$$

Since the same context symbol occurs on each variable, with this particular application of D8, it can be removed from each and written after the whole formula.

D4 provides us with a minimal test of completeness. It tells us only that a sentence is complete if, when it is significant, it yields a statement. But this is in no sense an analysis of the concept of completeness determining criteria in terms of which the completeness or incompleteness of a sentence can be assessed. For that matter, of course, we do not at this stage have any formal criteria in terms of which significance can be assessed. The concept of significance is at present a formal primitive and is only given a more detailed analysis in terms of range theory in Part III. However, in the next chapter we take the first steps towards providing criteria of both completeness and significance.

D5 and D6 are the foreshadowed definitions of the truth and falsity of sentences. They can only be applied to a sentence in its context and are parasitic on the corresponding properties of statements.

Given that **Y** is many-one, it will follow from D7 that if two sentences are statement equal, they yield one and the same statement. For since each sentence can yield one and only one statement in a specified context, the statement which they both yield must be the same.

D8 gives a definition of equivalence which is a simple extension of the ordinary two-valued material equivalence. The definition reduces to that of '$\equiv$' for statement-yielding sentences. Notice that although it can hold between any sentences, it is a two-valued relation since $\Pi \simeq \Theta$ is a wffcl even when Π and Θ are both unrestricted semi-wffs. This does not, however, impinge on the general feature of **CL** that unrestricted semi-wffs are not used but only mentioned, for '$\simeq$' absorbs a quotation function. On the other hand, it can be given a use, rather than a mentioning, interpretation, *viz* 'if, and only if' (see §5.11).

We now explore some of the consequences of these definitions. First we characterise the yields relation more precisely by means of special postulates. This enables us to introduce statement identity and so leads to the introduction of the statement function.

2.13 THE STATEMENT-YIELDING RELATION, STATEMENT IDENTITY AND THE STATEMENT FUNCTION

We first affirm a generalised version of the postulate suggested earlier (§2.5 and §2.7) that a statement-yielding sentence in a context yields the statement mentioned by the that-clause formed from it and qualified by the same context. Thus,

$\vdash_{CL}$(P1) $qu(A)(\theta)\ \mathbf{Y}\ \S A(\theta)$.

As special cases we then have:

$\vdash_{CL}$(12) $qu(\Phi)(\theta)\ \mathbf{Y}\ \S\Phi(\theta)$

$\vdash_{CL}$(13) $qu(A)(c_s)\ \mathbf{Y}\ \S A$ (using D3(ii)).

This gives us the limiting, trivial, sense of 'statement'. In terms of it we may say that if the sentence 'the number seven is purple' is statement-yielding in some context c_0, then the statement it yields can be mentioned by the that-clause 'that the number seven is purple' if this is also uttered with respect to c_0. The interesting concept of a statement, however, is that which takes us from a sentence in a non-standard context to a that-clause in the standard context, so that we may say, for example, 'The number seven is purple' (c_0) yields the statement that the house next door is painted purple. However, as we shall see, the limiting case is useful in that it enables us to relate non-standard contexts to c_s. Even if $\vdash_{CL}$(P1) were rejected, it would still be necessary to affirm $\vdash_{CL}$(13) independently, since the special case of it,

$\vdash_{CL}(14) \quad qu(\Phi)(c_s)\,\mathbf{Y}\,\S\Phi,$

which expresses the fact that the statement yielded by any statement-yielding sentence in the standard context can be mentioned simply by forming the that-clause formed from it and dropping the contextual qualification, is the only general way we have of showing under what conditions a sentence can be freed from its context and the statement it makes identified.

It follows immediately from (P1) by **RQL** that every wffcl is statement-yielding: *i.e.*

$\vdash_{CL}(15) \quad (P\gamma)(qu(A)(\theta)\,\mathbf{Y}\gamma);$

and the special case,

$\vdash_{CL}(16) \quad (P\gamma)(qu(\Phi)(\theta)\mathbf{Y}\gamma).$

The first of these expresses the main methodological assumption we have made that the formulae of **CL** are only being considered with respect to statement-yielding contexts. The second expresses the fundamental use of restricted semi-wffs. Since it holds, it would be inconsistent to substitute for the variables in them sentences which do not yield statements.

It is a feature of the yields relation that its first place is not limited to quoted wffcl, for $qu(\Gamma)(\theta)\,\mathbf{Y}\,\alpha$ is well-formed where Γ is an unrestricted semi-wff. And we now give postulates which express in a more explicit and general form the truth conditions for any compound sentences in a logically standard context. As in the definitions D4–D8, Π and Θ stand for either unrestricted semi-wffs or wffcl.

$\vdash_{CL}(P2) \quad (qu(\Pi)(L\theta)\,\mathbf{Y}\alpha\ \&\ \mathbf{F}\alpha) \supset T{\sim}\Pi(L\theta)$
$\vdash_{CL}(P3) \quad (qu(\Pi)(L\theta)\,\mathbf{Y}\alpha\ \&\ qu(\Theta)(L\theta)\,\mathbf{Y}\beta\ \&\ \alpha\ \mathbf{disj}\ \beta) \supset T(\Pi \vee \Theta)(L\theta)$
$\vdash_{CL}(P4) \quad (qu(\Pi)(L\theta)\,\mathbf{Y}\alpha\ \&\ qu(\Theta)(L\theta)\,\mathbf{Y}\beta\ \&\ \alpha\ \mathbf{conj}\ \beta) \supset T(\Pi\,\&\,\Theta)(L\theta)$
$\vdash_{CL}(P5) \quad (qu(\Pi)(L\theta)\,\mathbf{Y}\alpha\ \&\ qu(\Theta)(L\theta)\,\mathbf{Y}\beta\ \&\ \alpha\ \mathbf{imp}\ \beta) \supset T(\Pi \supset \Theta)(L\theta)$
$\vdash_{CL}(P6) \quad (qu(\Pi)(L\theta)\,\mathbf{Y}\alpha\ \&\ qu(\Theta)(L\theta)\,\mathbf{Y}\beta\ \&\ \alpha\ \mathbf{eq}\ \beta) \supset T(\Pi \equiv \Theta)(L\theta).$

None of these could be stated generally for both unrestricted semi-wffs and wffcl if we did not have the concept of truth over sentences. If, for example, we write $\vdash_{CL}$(P3) in the form,

$$(qu(\Pi)(L\theta)\,\mathbf{Y}\alpha\ \&\ qu(\Theta)(L\theta)\,\mathbf{Y}\beta\ \&\ \alpha\ \mathbf{disj}\ \beta) \supset \mathbf{T}\S(\Pi \vee \Theta)(L\theta),$$

with a wfc in the consequent, it would not hold for unrestricted semi-wffs. For if Π or Θ or both, fail to yield a statement, $\S(\Pi \vee \Theta)$ fails to mention a statement and is not a wfc. Hence $\mathbf{T}\S(\Pi \vee \Theta)$ is not well-formed. However, $T(\Pi \vee \Theta)(L\theta)$ is well-formed since we have

$$T(\Pi \vee \Theta)(L\theta) =_{Df} (P\gamma)(qu(\Pi \vee \Theta)(L\theta)\,\mathbf{Y}\gamma\ \&\ \mathbf{T}\gamma)$$

even if $\Pi \vee \Theta$ is an unrestricted semi-wff which fails to yield a statement.

The difference between the conditions under which '$T\Pi$' and '$\mathbf{T}\S\Pi$' are well-formed is not accidental but a consequence of an important difference between the concept of truth as applied to sentences and as applied to statements, even though the one is defined in terms of the other. For we may significantly say of a sentence that it is true when it fails to yield a statement. So to say is simply to make a false remark. But we cannot say of a statement that it is true if no statement has been made. Hence if Π fails to yield a statement, '$T\Pi$' is statement-yielding, and in fact '$\sim T\Pi$' is true; for 'not true' embraces 'not significant' and 'not complete' as well as 'false'. So '$\sim T\Pi$' may be true just because Π is not statement-yielding. By contrast, if Π fails to yield a statement, '$\mathbf{T}\S\Pi$' is not statement-yielding and '$\sim\mathbf{T}\S\Pi$' is not true. For in the case of statements, '$\sim\mathbf{T}$' embraces 'false' only, in fact is equivalent to 'false', as we shall see. But when Π fails to yield a statement, it is not the case that $\S\Pi$ is false any more than it is the case that it is true. For $\S\Pi$ does not mention a statement at all, and therefore mentions neither a true one nor a false one. The concepts of truth and falsity over sentences, therefore, though parasitic on the corresponding properties of statements, are by no means identical with them except when applied to statement-yielding sentences. There is much that cannot be said if we limit ourselves to the concept of truth over statements.

A thesis corresponding to (P2), and which completes the list, is immediately derivable by **RQL**. For taking A in $\vdash_{\mathrm{CL}}\check{\mathrm{S}}^{\gamma}_{\alpha}A\,|\supset(P\gamma)A$ to be '$qu(\Pi)(\theta)\,\mathbf{Y}\gamma\ \&\ \mathbf{T}\gamma$', the proviso, that no free occurrence of γ is in a well-formed part of A which contains a variable of α bound, can be satisfied for every α. Hence,

$$\vdash_{\mathrm{CL}}\quad (qu(\Pi)(\theta)\,\mathbf{Y}\alpha\ \&\ \mathbf{T}\alpha)\supset(P\gamma)(qu(\Pi)(\theta)\,\mathbf{Y}\gamma\ \&\ \mathbf{T}\gamma).$$

That is, by D5,

$$\vdash_{\mathrm{CL}}(17)\quad (qu(\Pi)(\theta)\,\mathbf{Y}\alpha\ \&\ \mathbf{T}\alpha)\supset T\Pi(\theta),$$

and consequently the special case for $\mathrm{L}\theta$.

We now introduce statement identity by taking '$=$' as a further primitive and by adding to the formation rules:

R4 (viii) where ρ and σ are either wfc or values of the statement function (see D9 below), $(\rho=\sigma)$ is a wffcl.

We then take the usual postulates for identity restricted to wfc (this restriction is explained below):

$\vdash_{\mathrm{CL}}$(P7) (i) $(\alpha=\alpha)$

(ii) $(\alpha=\beta)\supset(A(\alpha)\supset A(\beta))$

where α and β are wfc, $A(\alpha)$ is a wffcl which contains the wfc α with variables in α not in the scope of quantifiers binding variables in α or β, and $A(\beta)$ is

A with β in place of α. This determines **CL** as an extended predicate logic in which the wfc are the terms. Hence, statement identity has all the usual features such as symmetry and transitivity. We shall take these for granted.

We now postulate that **Y** is a many-one relation:

$\vdash_{CL}$(P8) $(qu(\Pi)(\theta)\,\mathbf{Y}\alpha \;\&\; qu(\Pi)(\theta)\,\mathbf{Y}\beta) \supset (\alpha = \beta)$;

that is, any sentence (token) which yields a statement in a given context yields a unique statement in that context. This enables us to define the statement function:

D9 $(\alpha = \mathbf{st}(\Pi, \theta)) =_{\mathrm{Df}} qu(\Pi)(\theta)\,\mathbf{Y}\,\alpha$.

Notice that the function is defined even though Π is not statement-yielding, for we may take an unrestricted semi-wff Γ as a case of Π. This has certain advantages as we have mentioned (§2.5). In particular, we can show that in such cases the function has no value. Thus, from D9, by **RQL**, we have immediately,

$\vdash_{CL}$(18) $(P\gamma)(qu(\Pi)(\theta)\,\mathbf{Y}\gamma) \equiv (\mathrm{P}\gamma)(\gamma = \mathbf{st}(\Pi, \theta))$.

Hence, by **RQL**, and writing $\rho \neq \sigma$ as an abbreviation for $\sim(\rho = \sigma)$, we have:

$\vdash_{CL}$(19) $\sim(P\gamma)(qu(\Pi)(\theta)\,\mathbf{Y}\gamma) \equiv (U\gamma)(\gamma \neq \mathbf{st}(\Pi, \theta))$.

Thus, if 'p_0' is a definite sentence which fails to yield a statement in c_0, we shall be able to affirm $\sim(Pa)(qu(p_0)(c_0)\,\mathbf{Y}a)$, hence $(Ua)(a \neq \mathbf{st}(p_0, c_0))$. That is, $\mathbf{st}(p_0, c_0)$ fails to have a value.

This result does not of course conflict with the many-one postulate which, using D9, can be written as,

$\vdash_{CL}$(20) $((\alpha = \mathbf{st}(\Pi, \theta)) \;\&\; (\beta = \mathbf{st}(\Pi, \theta))) \supset (\alpha = \beta)$,

since this only tells us that the function has at most one value; it does not imply that it always has at least one value. On the other hand, if Π is taken to be a statement-yielding sentence, or more generally a wffcl, then we can show that the function always has at least one value. Thus from $\vdash_{CL}$(18) we have:

$\vdash_{CL}$(21) $(P\gamma)(qu(A)(\theta)\,\mathbf{Y}\gamma) \equiv (P\gamma)(\gamma = \mathbf{st}(A, \theta))$.

But the LHS of this is a thesis: $\vdash_{CL}$(15). Hence,

$\vdash_{CL}$(22) $(P\gamma)(\gamma = \mathbf{st}(A, \theta))$.

If, therefore, the first argument place in the statement function is restricted to wffcl (which include all statement-yielding sentences Φ) then it has at least one value and at most one value. In such cases it always "designates" a unique statement and so may be taken as a wfc. We affirm this formally in the rule:

R4 (ix) $\mathbf{st}(A, \theta)$ is a wfc.

We call this the restricted statement function.

We are now in a position to explain the restriction on the identity postulates. If we allowed the postulates to apply to all well-formed identity formulae, including such as $\mathbf{st}(\Pi, \theta) = \gamma$, so that α and β in the postulates stood for all statement function expressions as well as wfc, then we should have undesirable results. In particular, since we can obtain by **RQL** from (P7)(i) the thesis,

$$\vdash_{\mathrm{CL}}(23) \quad (P\gamma)(\alpha = \gamma),$$

we should have immediately: $\vdash(P\gamma)(\mathbf{st}(\Pi, \theta) = \gamma)$. But this would be inconsistent with the interpretational requirements since it amounts to saying that every sentence is statement yielding: *i.e.* by D9, $(P\gamma)(qu(\Pi)(\theta)\ \mathbf{Y}\gamma)$. On the other hand, taking the restricted statement function as a case of α in $\vdash_{\mathrm{CL}}(23)$ simply gives us $\vdash_{\mathrm{CL}}(22)$ again, and leads to no difficulty. Hence the limitation on the identity postulates. They are restricted in application to wfc, and not all statement-function formulae are wfc.

Even this restriction might seem to be inadequate, however, for the use of identity theses involving the restricted statement function enables us to derive the two rules which were rejected in §2.11, *viz.*:

if $qu(A)(\theta)\ \mathbf{Y}\alpha$ and $\vdash\alpha\ \mathbf{eq}\ \beta$ then $\vdash\alpha = \beta$
and, if $qu(A)(\theta)\ \mathbf{Y}\alpha$ and $\vdash\alpha\ \mathbf{eq}\ \beta$ then $\vdash qu(A)(\theta)\ \mathbf{Y}\beta$.

For suppose,

$$\vdash qu(A)(\theta)\ \mathbf{Y}\alpha \text{ and } \vdash\alpha\ \mathbf{eq}\ \beta$$

then by D9,

$$\vdash\alpha = \mathbf{st}(A, \theta) \text{ and } \vdash\alpha\ \mathbf{eq}\ \beta$$

By CL(2)(i), therefore,

$$\vdash\beta = \mathbf{st}(A, \theta).$$

Hence, by symmetry and transitivity,

$$\vdash\alpha = \beta.$$

Consequently, $\vdash qu(A)(\theta)\ \mathbf{Y}\beta$.

Now this argument could be stopped by refusing to apply the transitivity step; and this could be done by denying that the statement function is a wfc even when it is restricted. In this case the identity postulates do not apply to the statement function at all. In our view, however, this would be a mistake. For it is clear that the unacceptable step occurs before transitivity is applied. The use of CL(2)(i) to replace α by β in $\vdash\alpha = \mathbf{st}(A, \theta)$, by virtue of $\vdash\alpha\ \mathbf{eq}\ \beta$, though it is not prevented by the existing restriction on CL(2)(i), in effect bypasses the restriction. For the existing proviso would prevent the use of CL(2)(i) at the previous step and disallow the replacement of α by β in $qu(A)(\theta)\ \mathbf{Y}\alpha$. Yet at the point where the replacement is made, the formula which is operated on, $\alpha = \mathbf{st}(A, \theta)$, is the *defined equivalent* of $qu(A)(\theta)\ \mathbf{Y}\alpha$. But if a restriction is to be effective on a given formula, it must apply not only to that formula but to its defined equivalent and to all consequences of the definition.

That the problem is independent of the transitivity step, and hence of taking the restricted statement function as a wfc, can be shown more generally as follows:

For suppose we have $\vdash \alpha$ **eq** β. Then since we also have $\vdash \alpha = \alpha$, we could use CL(2)(i) to pass to $\vdash \alpha = \beta$. That is, in general: if $\vdash \alpha$ **eq** β then $\vdash \alpha = \beta$. This clearly, is absurd; and it has nothing to do with the identity postulates as such, nor with the special fact that we have taken $\mathbf{st}(A, \theta)$ to be a wfc.

The difficulty is, quite simply, that wfc, unlike the terms of ordinary extended predicate logic, stand in both truth-functional and identity relations to one another, and the two relations must be kept distinct since statement identity is not an extensional relation. It is only because they have sometimes been confused that odd philosophical conclusions such as Frege's, that there are only two statements, the True and the False, seem to be plausible.

We therefore propose to extend the proviso on CL(2) by refusing to allow the replacement of equivalent wfc in identity formulae. On the other hand, of course, the converse rule, that identical wfc may be replaced in equivalence theses, is perfectly satisfactory, and is indeed a consequence of (P7)(ii). The revised rule is as follows:

CL(2) modified: Add to the end of CL(2)(i):

, or $\alpha = \rho$, where ρ is either a wfc or a statement function formula.

We now show the relation between the statement-forming connective §, and the statement function.

From D9 we have, for wffcl,

$\vdash_{CL}(24)$ $qu(A)(\theta)\, \mathbf{Y}\alpha \equiv (\alpha = \mathbf{st}(A, \theta))$.

Hence the particular case,

$\vdash_{CL}(25)$ $qu(A)(\theta)\, \mathbf{Y}\S A(\theta) \equiv (\S A(\theta) = \mathbf{st}\,(A, \theta))$.

By (P1), therefore,

$\vdash_{CL}(26)$ $\S A(\theta) = \mathbf{st}\,(A, \theta)$.

Hence, § may be regarded as the restricted statement function. In particular, for any statement-yielding sentence $qu(r)$,

$\vdash_{CL}(27)$ $\S r(c) = \mathbf{st}(r, c)$;

i.e. the statement which $qu(r)$ yields in c is $\S r(c)$.

Using $\vdash_{CL}(26)$ to make a replacement in $\vdash_{CL}(24)$, we have,

$\vdash_{CL}(28)$ $qu(A)(\theta)\, \mathbf{Y}\alpha \equiv (\alpha = \S A(\theta))$.

This is a fundamental theorem of context logic, a reducibility thesis of a kind, which characterises **Y** as a special kind of relation which we call a contraction relation. Another example is the referring relation **R**. The features of these relations are discussed in the next chapter (§3.6).

As a special case of $\vdash_{CL}(26)$, taking θ to be c_S, we have:

$\vdash_{CL}(29)$ $\S A = \mathbf{st}(A, c_S)$;

and similarly, as a special case of $\vdash_{CL}(28)$ we have:

$\vdash_{CL}(30)$ $qu(A)(c_S)\,\mathbf{Y}\alpha \equiv (\alpha = \S A)$.

A number of interesting and useful theses follow from the many-one postulate. Thus, taking β in (P8) to be a statement variable γ, we have:

$$\vdash_{CL} \quad (qu(\Pi)(\theta)\,\mathbf{Y}\alpha \,\&\, qu(\Pi)(\theta)\,\mathbf{Y}\gamma) \supset (\alpha = \gamma);$$

and by the symmetry of identity this leads to

$$\vdash_{CL} \quad (qu(\Pi)(\theta)\,\mathbf{Y}\alpha \,\&\, qu(\Pi)(\theta)\,\mathbf{Y}\gamma) \supset (\gamma = \alpha).$$

Now from (P7)(ii) we also have

$$\vdash_{CL} \quad (\gamma = \alpha) \supset (\mathbf{T}\gamma \supset \mathbf{T}\alpha).$$

Hence, by **RSL**,

$$\vdash_{CL} \quad (qu(\Pi)(\theta)\,\mathbf{Y}\alpha \,\&\, qu(\Pi)(\theta)\,\mathbf{Y}\gamma) \supset (\mathbf{T}\gamma \supset \mathbf{T}\alpha).$$

That is, by **RSL**,

$$\vdash_{CL} \quad (qu(\Pi)(\theta)\,\mathbf{Y}\alpha \,\&\, qu(\Pi)(\theta)\,\mathbf{Y}\gamma \,\&\, \mathbf{T}\gamma) \supset \mathbf{T}\alpha.$$

By **RQL**, therefore,

$$\vdash_{CL} \quad (qu(\Pi)(\theta)\,\mathbf{Y}\alpha \,\&\, (P\gamma)(qu(\Pi)(\theta)\,\mathbf{Y}\gamma \,\&\, \mathbf{T}\gamma)) \supset \mathbf{T}\alpha;$$

provided γ does not occur free in α; but this condition can always be satisfied by a change of variable.

By D5, therefore, we have:

$$\vdash_{CL} \quad (qu(\Pi)(\theta)\,\mathbf{Y}\alpha \,\&\, T\Pi(\theta)) \supset \mathbf{T}\alpha;$$

i.e.

$\vdash_{CL}(31)$ $qu(\Pi)(\theta)\,\mathbf{Y}\alpha \supset (T\Pi(\theta) \supset \mathbf{T}\alpha)$.

That is: if a sentence yields a statement then, if it is true, the statement it yields is true.

Now from $\vdash_{CL}(17)$ by **RSL** we have,

$$\vdash_{CL} \quad qu(\Pi)(\theta)\,\mathbf{Y}\alpha \supset (\mathbf{T}\alpha \supset T\Pi(\theta)).$$

And combining this with $\vdash_{CL}(31)$ gives:

$\vdash_{CL}(32)$ $qu(\Pi)(\theta)\,\mathbf{Y}\alpha \supset (T\Pi(\theta) \equiv \mathbf{T}\alpha)$.

Taking Π to be a wffcl A, α to be $\S A$ and θ to be c_s, this leads to,

$$\vdash_{CL} \quad qu(A)(c_s)\,\mathbf{Y}\S A \supset (TA \equiv \mathbf{T}\S A),$$

using D3. And since the antecedent is a thesis, $\vdash_{CL}(13)$, we have:

$$\vdash_{CL}(33) \quad TA \equiv \mathbf{T}\S A$$

for any wffcl A. But we also have $\vdash_{CL}(11)$ $\mathbf{T}\S A \equiv A$. Hence by **RSL**

$$\vdash_{CL}(34) \quad TA \equiv A.$$

In spite of the generality of $\vdash_{CL}(32)$, holding as it does for unrestricted semi-wffs as well as wffcl, the steps after that only hold in terms of wffcl. Indeed if we take A to be an unrestricted semi-wff, say 'p', in $\vdash_{CL}(34)$, then we have '$Tp \equiv p$' which is not well formed since 'p' is not well formed. It should in any case be obvious that '$Tp \equiv p$' is unlikely to hold in general. For if we substitute a nonsignificant sentence 'p_0' for the variable, then 'Tp_0' will be false (since we are saying of a nonsignificant sentence that it is true, and thus making a false remark) and we shall have an equivalence between a false sentence and a nonsignificant sentence. Whether such an equivalence can be accepted, therefore, will depend on the properties of '$\equiv$' in the many-valued system in which it occurs. It is precisely because of difficulties such as this that unrestricted semi-wffs are not at this stage accepted as wffcl. In Part II, however, we admit unrestricted semi-wffs as well formed and investigate the properties of the connectives in various many-valued systems. In some of them, '$Tp \equiv p$' is a thesis, in others not. By contrast, '$Tr \equiv r$' is a thesis in all systems we consider (being a case of $\vdash_{CL}(34)$). It expresses the T-elimination principle for statement-yielding sentences. This emphasises the point made earlier that $\mathbf{T}$ and T only have the same properties when T is applied to statement-yielding sentences. Indeed $\vdash_{CL}(33)$ says precisely this.

As one might expect, there are close connexions between statement identity and the relation of statement equality between sentences. To show this, we first prove some preliminary theorems involving '$\doteq$'.

We have, as a case of (P5),

$$\vdash_{CL} \quad (qu(A)(c_s)\,\mathbf{Y}\alpha \;\&\; qu(B)(c_s)\,\mathbf{Y}\alpha \;\&\; \alpha \;\mathbf{imp}\; \alpha) \supset T(A \supset B).$$

Hence by **RSL**, using $\vdash_{SL}(5)$,

$$\vdash_{CL} \quad (qu(A)(c_s)\,\mathbf{Y}\alpha \;\&\; qu(B)(c_s)\,\mathbf{Y}\alpha) \supset T(A \supset B).$$

Taking α to be a statement variable γ which does not occur free in A or B, we then have by **RQL**,

$$\vdash_{CL} \quad (P\gamma)(qu(A)(c_s)\,\mathbf{Y}\gamma \;\&\; qu(B)(c_s)\,\mathbf{Y}\gamma) \supset T(A \supset B).$$

By D7, therefore,

$$\vdash_{CL} \quad (A \doteq B) \supset T(A \supset B)$$

and by $\vdash_{CL}(34)$,

$\vdash_{CL}(35)$ $(A \doteq B) \supset (A \supset B)$.

Again, from D7, taking Θ to be Π and η to be θ, we have:

$\vdash_{CL}$ $(\Pi \doteq \Pi)(\theta) \equiv (P\gamma)(qu(\Pi)(\theta)\ \mathbf{Y}\gamma\ \&\ qu(\Pi)(\theta)\ \mathbf{Y}\gamma)$.

That is, by **RQL**,

$\vdash_{CL}(36)$ $(\Pi \doteq \Pi)(\theta) \equiv (P\gamma)(qu(\Pi)(\theta)\ \mathbf{Y}\gamma)$.

Hence, a sentence is statement-equal to itself if, and only if, it yields a statement. This is one of several relations we define (in Chapter 3 especially) which are only reflexive under a certain condition (*cf.* also *PM**14.28, $((\imath x)\phi x = (\imath x)\phi x) . \equiv . E!(\imath x)\phi x)$.
For example, we shall have:

$\vdash_{CL}$ $(p \doteq p)(c) \equiv (Pa)(qu(p)(c)\ \mathbf{Y}a)$

for any sentence, but

$\vdash_{CL}$ $(r \doteq r)(c)$

for the special case of statement-yielding sentences since we can affirm $\vdash_{CL}(Pa)(qu(r)(c)\ \mathbf{Y}a)$ as a case of $\vdash_{CL}(16)$.

Taking Π in $\vdash_{CL}(36)$ to be a wffcl A, we can use $\vdash_{CL}(15)$ to pass to,

$\vdash_{CL}$ $(A \doteq A)(\theta)$.

In particular, therefore, for the standard context,

$\vdash_{CL}(37)$ $A \doteq A$.

The connexion between '$=$' and '$\doteq$' now follows from the identity postulate (P7)(ii). For we may take $A(\alpha)$ to be $A(\alpha) \doteq A(\alpha)$, and $A(\beta)$ to be $A(\alpha) \doteq A(\beta)$. This gives:

$\vdash_{CL}$ $(\alpha = \beta) \supset (A(\alpha) \doteq A(\alpha) . \supset . A(\alpha) \doteq A(\beta))$.

Now let A in $\vdash_{CL}(37)$ contain the wfc α. Then we have

$\vdash_{CL}(38)$ $A(\alpha) \doteq A(\alpha)$.

Hence, by **RSL**,

$\vdash_{CL}(39)$ $(\alpha = \beta) \supset (A(\alpha) \doteq A(\beta))$.

In other words, a sentence which contains a that-clause which can be used to mention a statement is statement-equal to the sentence which is obtained by replacing the that-clause, at any of its occurrences, by another which can be used to mention the same statement.

The thesis $\vdash_{CL}(39)$ can in fact be taken as a postulate in place of (P7)(ii) since the latter is deducible from it. Thus, taking A in $\vdash_{CL}(35)$ to be $A(\alpha)$ and B to be $A(\beta)$ we have:

$$\vdash_{CL} \quad (A(\alpha) \overline{\overline{\cdot}} A(\beta)) \supset (A(\alpha) \supset A(\beta)).$$

(P7)(ii) now follows immediately from this and $\vdash_{CL}(39)$ regarded as a postulate.

There are in fact even closer connexions between statement equality and statement identity. For suppose that the statement-yielding sentences 'r_0' and 's_0' are statement-equal: *i.e.* $r_0 \overline{\overline{\cdot}} s_0$. Then they yield the same statement. Now we know that 'r_0' yields $\S r_0$ and 's_0' yields $\S s_0$. Hence we must have $\S r_0 = \S s_0$. Conversely, suppose that $\S r_0 = \S s_0$. Since $\S r_0$ is yielded by 'r_0' and $\S s_0$ is yielded by 's_0', 'r_0' and 's_0' yield the same statement. Hence they are statement equal. We conclude that $\S r_0 = \S s_0$ if, and only if, $r_0 \overline{\overline{\cdot}} s_0$.

This informal argument neglects the role of context, but we prove the corresponding formal theses below. They are of particular importance in elucidating the role of statement equality in ordinary discourse. They show that the relation of statement equality is a translation relation since it preserves meaning. For if two sentences yield the same statement, they have the same meaning. The translation may be from a non-standard context to the standard context of the background language, or from the standard context of one natural language to the standard context of another. For if we choose a background language, say English, then c_S is the standard context of English, and from the point of view of this context, the standard context of another natural language, say French, can be regarded as a non-standard context, say c_1. For the words of French, though they translate into a standard English sense, do not *have* a standard English sense. Some indeed, have a non-standard sense from the point of view of English: *e.g.* 'mince'. An alternative reading of '$\overline{\overline{\cdot}}$' is, therefore, 'translates into'.

This enables us to interpret some of the theorems in such a way as to show that the problem of freeing a sentence from its context and determining what statement it yields, or what it means, is simply the problem of translating it into a sentence of the background language with respect to the standard context. This is so whether we are interested in deciphering codes, an "odd" use of an apparently standard sentence, or a foreign language. A thesis which expresses this is:

$$\vdash_{CL} \quad (r(c) \overline{\overline{\cdot}} s) \equiv qu(r)(c) \mathbf{Y} \S s.$$

For example,

> 'Le monde est rond' uttered with respect to the standard context of French, say c_1, translates into 'The world is round' when uttered with respect to the standard context of English, if, and only if, 'Le monde est rond' yields, in c_1, the statement that the world is round.

Similarly,

> 'The sky is blue' uttered with respect to the coding context, say c_0, translates into 'The attack takes place at dawn' in the standard context, if, and only if, 'The sky is blue' in code means that (yields the statement that) the attack takes place at dawn.

To the extent that we understand translation from one language into another, therefore, to that extent we understand the concept of a statement.

The thesis is obtained as follows:

Taking α to be a statement variable γ in $\vdash_{CL}(28)$,

$\vdash_{CL}(40)$ $qu(A)(\theta)\ \mathbf{Y}\gamma \equiv (\gamma = \S A(\theta))$. Similarly,
$\vdash_{CL}(41)$ $qu(B)(\eta)\ \mathbf{Y}\gamma \equiv (\gamma = \S B(\eta))$.

Hence by **RSL**,

$\vdash_{CL}$ $(qu(A)(\theta)\ \mathbf{Y}\gamma\ \&\ qu(B)(\eta)\ \mathbf{Y}\gamma) \supset ((\gamma = \S A(\theta))\ \&\ (\gamma = \S B(\eta)))$.

So, by symmetry and transitivity,

$\vdash_{CL}$ $(qu(A)(\theta)\ \mathbf{Y}\gamma\ \&\ qu(B)(\eta)\ \mathbf{Y}\gamma) \supset (\S A(\theta) = \S B(\eta))$.

By **RQL** we may bind the variable in the antecedent with a particular quantifier and then by D7 we have:

$\vdash_{CL}(42)$ $(A(\theta) \doteq B(\eta)) \supset (\S A(\theta) = \S B(\eta))$.

This thesis is one half of the principle assumed in the informal argument above. The other half is proved below.

By **RQL** we have:

$\vdash_{CL}(43)$ $(qu(A)(\theta)\ \mathbf{Y}\S B(\eta)\ \&\ qu(B)(\eta)\ \mathbf{Y}\S B(\eta)) \supset (P\gamma)(qu(A)(\theta)\ \mathbf{Y}\gamma\ \&\ qu(B)(\eta)\ \mathbf{Y}\gamma)$.

Now as a case of $\vdash_{CL}(28)$ we have:

$\vdash_{CL}(44)$ $qu(A)(\theta)\ \mathbf{Y}\S B(\eta) \equiv (\S B(\eta) = \S A(\theta))$.

And by **RSL**, since $qu(B)(\eta)\ \mathbf{Y}\S B(\eta)$ is a thesis, (P1), we have from this,

$\vdash_{CL}$ $(qu(A)(\theta)\ \mathbf{Y}\S B(\eta)\ \&\ qu(B)(\eta)\ \mathbf{Y}\S B(\eta)) \equiv (\S B(\eta) = \S A(\theta))$.

So by $\vdash_{CL}(43)$,

$\vdash_{CL}$ $(\S B(\eta) = \S A(\theta)) \supset (P\gamma)(qu(A)(\theta)\ \mathbf{Y}\gamma\ \&\ qu(B)(\eta)\ \mathbf{Y}\gamma)$.

Hence, by D7,

$\vdash_{CL}(45)$ $(\S B(\eta) = \S A(\theta)) \supset (A(\theta) \doteq B(\eta))$.

The two theses $\vdash_{CL}(42)$ and $\vdash_{CL}(45)$ together give:

$\vdash_{CL}(46)$ $(\S A(\theta) = \S B(\eta)) \equiv (A(\theta) \doteq B(\eta))$.

We can now obtain the desired thesis immediately. For we have from $\vdash_{CL}(44)$ and $\vdash_{CL}(46)$,

$\vdash_{CL}(47)$ $(A(\theta) \mathrel{\underline{\sqsubset}} B(\eta)) \equiv qu(A)(\theta)\ \mathbf{Y}\S B(\eta)$.

Hence the special case,

$\vdash_{CL}$ $(r(c) \mathrel{\underline{\sqsubset}} s) \equiv qu(r)(c)\ \mathbf{Y}\S s$.

It should be noticed the $\vdash_{CL}(47)$ does not provide a criterion for translating under quote marks. For example, we cannot take, say, '$qu(r)$' as a special case of A since it is not a wffcl. So it would not follow from either $\vdash_{CL}(46)$ or $\vdash_{CL}(47)$ that because, say, 'Le monde est rond' translates into 'The world is round' so "'Le monde est rond'" translates into "'The world is round'". And of course we should not want this to be so. On the other hand, there is nothing to prevent us from replacing A by a sentence which contains a quoted sentence. Thus, '$sig\ qu(r)(c)$' is a wffcl which contains a quoted expression and can be taken as a case of A. So we might take as a special case of $\vdash_{CL}(47)$,

$\vdash_{CL}$ $((sig\ qu(r)(c))(c_S) \mathrel{\underline{\sqsubset}} s) \equiv qu(sig\ qu(r)(c))(c_S)\ \mathbf{Y}\S s$,

where 's' need not itself contain a quotation. For example, "The sentence 'it is raining' is significant in c_1" uttered with respect to c_S, translates into s_0 if, and only if, "The sentence 'it is raining' is significant in c_1" yields, in the standard context, the statement that-s_0. Now we might take 's_0' to be 'it makes sense to say, in c_1, that it is raining', in which no quotation occurs. Or we might take it to be "The sentence 'it is raining' is truth-valued in c_1", which does contain a quotation. In neither case, however, is the quoted sentence itself being translated. What is being translated is the whole sentence in which the quotation occurs. Whether we can, in some circumstances, translate under quotation signs without change of meaning, is therefore not determined by $\vdash_{CL}(47)$. Clearly there are some cases in which we cannot—*e.g.* in locutions such as "Jack used the French sentence 'Le monde est rond' to make the statement that ...", though by $\vdash_{CL}(47)$ the possibility of paraphrasing it into a sentence which does not contain a quotation is not excluded. In other cases, perhaps, translation under quote marks is possible, but that it is possible is not a consequence of $\vdash_{CL}(47)$.

We show finally that **Y** has the detachment property of an implication relation by proving the derived rule:

CL(12) if $\vdash_{CL}$ $qu(A)(c_S)\ \mathbf{Y}\S B$ and $\vdash_{CL} A$ then $\vdash_{CL} B$.

Thus if we introduced, say, '$\rightarrow$', as a yields connective which absorbs a quotation function on the left and a statement function (*i.e.* a that-operator) on the right, the rule would take the more familiar form : if $\vdash A \rightarrow B$ and $\vdash A$ then $\vdash B$.

Proof: Suppose $\vdash_{CL}$ *qu*$(A)(c_S)$ **Y**§B. Then by $\vdash_{CL}(47)$, we have $\vdash_{CL} A \doteq B$. So by $\vdash_{CL}(35)$, $\vdash_{CL} A \supset B$. Hence, by detachment, if $\vdash_{CL} A$ then $\vdash_{CL} B$.

As one would expect, this rule can be strengthened to show that **Y** is an equivalence relation. That is, we have

CL(13) if $\vdash_{CL}$ *qu*$(A)(c_S)$ **Y**§B and $\vdash_{CL} B$ then $\vdash_{CL} A$.
i.e. if $\vdash A \to B$ and $\vdash B$ then $\vdash A$.

The proof of this is similar but depends on a thesis which expresses the symmetry of '$\doteq$'.

$\vdash_{CL}(48)$ $(A \doteq B) \equiv (B \doteq A)$: from D7 by the commutative law for &.

Now as a case of $\vdash_{CL}(35)$ we have,

$\vdash_{CL}$ $(B \doteq A) \supset (B \supset A)$.

So, by $\vdash_{CL}(48)$,

$\vdash_{CL}(49)$ $(A \doteq B) \supset (B \supset A)$.

Proof of CL(13)*:* Suppose $\vdash_{CL}$ *qu*$(A)(c_S)$ **Y**§B. Then by $\vdash_{CL}(47)$, $\vdash_{CL} A \doteq B$. So by $\vdash_{CL}(49)$, $\vdash_{CL} B \supset A$. Hence, if $\vdash_{CL} B$ then $\vdash_{CL} A$.

Of course, $\to$ is an implication relation with very special properties. In particular, from the many-one theorem we have:

$$((A \to B) \;\&\; (A \to C)) \supset (\S B = \S C).$$

2.14 Truth and Falsity

We now derive some of the main theorems which characterise the properties of truth and falsity over both statements and sentences.

The following derived rule is immediate from $\vdash_{CL}(34)$ by **RSL**:

CL(14): $\vdash_{CL} TA$ if, and only if $\vdash_{CL} A$.

Now we saw in §2.11 that $\vdash_{CL}(Ua)(\mathbf{T}a \vee \sim\mathbf{T}a)$ is derivable just by **RSL** and **RQL**. To strengthen this to the more familiar form $\vdash_{CL}(Ua)(\mathbf{T}a \vee \mathbf{F}a)$, every statement is either true or false, we use (P3). For we have, as a case of (P3),

$\vdash_{CL}$ $($*qu*$(\mathbf{T}\alpha)(c_S)$ **Y**§$\mathbf{T}\alpha$ & *qu*$(\mathbf{F}\alpha)(c_S)$**Y**§$\mathbf{F}\alpha$ & $\mathbf{T}\alpha$ **disj** $\mathbf{F}\alpha) \supset T(\mathbf{T}\alpha \vee \mathbf{F}\alpha)$.

But each component of the antecedent is a thesis. The first two are special cases of $\vdash_{CL}(13)$, and the third is $\vdash_{SL}(17)$. So by detachment $\vdash_{CL} T(\mathbf{T}\alpha \vee \mathbf{F}\alpha)$; and by derived rule CL(14),

$\vdash_{CL}(50)$ $\mathbf{T}\alpha \vee \mathbf{F}\alpha$.

Taking α to be a statement variable, we then have by generalisation:

$\vdash_{\text{CL}}(51)$ $(Ua)(\mathbf{T}a \vee \mathbf{F}a)$.

Again as a case of (P2) we have:

$\vdash_{\text{CL}}$ $(qu(\mathbf{T}\alpha \mathbin{\&} \mathbf{F}\alpha)(c_{\text{S}})\ \mathbf{Y}\S(\mathbf{T}\alpha \mathbin{\&} \mathbf{F}\alpha) \mathbin{\&} \mathbf{F}\S(\mathbf{T}\alpha \mathbin{\&} \mathbf{F}\alpha)) \supset T{\sim}(\mathbf{T}\alpha \mathbin{\&} \mathbf{F}\alpha)$.

Hence by **RSL** and $\vdash_{\text{CL}}(13)$,

$\vdash_{\text{CL}}$ $\mathbf{F}\S(\mathbf{T}\alpha \mathbin{\&} \mathbf{F}\alpha) \supset T{\sim}(\mathbf{T}\alpha \mathbin{\&} \mathbf{F}\alpha)$.

Now by $\vdash_{\text{CL}}(3)$ we have, $\vdash_{\text{CL}}$ $\S\mathbf{T}\S(A \mathbin{\&} B)\ \mathbf{eq}\ \S(\S A\ \mathbf{conj}\ \S B)$; and by $\vdash_{\text{SL}}(3)$ we may make an equivalence replacement to get $\vdash_{\text{CL}}$ $\S(A \mathbin{\&} B)\ \mathbf{eq}\ \S(\S A\ \mathbf{conj}\ \S B)$. Hence we may replace $\S(\mathbf{T}\alpha \mathbin{\&} \mathbf{F}\alpha)$ in the above thesis by $\S(\S\mathbf{T}\alpha\ \mathbf{conj}\ \S\mathbf{F}\alpha)$ and so obtain,

$\vdash_{\text{CL}}$ $\mathbf{F}\S(\S\mathbf{T}\alpha\ \mathbf{conj}\ \S\mathbf{F}\alpha) \supset T{\sim}(\mathbf{T}\alpha \mathbin{\&} \mathbf{F}\alpha)$.

But the antecedent of this is the thesis $\vdash_{\text{SL}}(29)$. So by detachment and rule CL(14), we have:

$\vdash_{\text{CL}}(52)$ ${\sim}(\mathbf{T}\alpha \mathbin{\&} \mathbf{F}\alpha)$; that is, by generalisation and **RQL**,
$\vdash_{\text{CL}}(53)$ ${\sim}(Pa)(\mathbf{T}a \mathbin{\&} \mathbf{F}a)$: no statement is both true and false.

From $\vdash_{\text{SL}}(10)$ by rule CL(9) we have:

$\vdash_{\text{CL}}$ $\mathbf{T}\alpha \equiv \mathbf{F}\S\mathbf{F}\alpha$; and by $\vdash_{\text{CL}}(6)$ this leads to:
$\vdash_{\text{CL}}(54)$ $\mathbf{T}\alpha \equiv {\sim}\mathbf{F}\alpha$.

Hence, 'not false' over statements is equivalent to 'true'. And of course, 'not true' is equivalent to 'false': for we have immediately from $\vdash_{\text{CL}}(54)$ by **RSL**.

$\vdash_{\text{CL}}(55)$ ${\sim}\mathbf{T}\alpha \equiv \mathbf{F}\alpha$.

Other similar theses which arise in this way are:

$\vdash_{\text{CL}}(56)$ $\mathbf{FT}\alpha \equiv \mathbf{F}\alpha$; $\vdash_{\text{CL}}(57)$ $\mathbf{TF}\alpha \equiv \mathbf{F}\alpha$;
$\vdash_{\text{CL}}(58)$ $\mathbf{FF}\alpha \equiv \mathbf{T}\alpha$; $\vdash_{\text{CL}}(59)$ $\mathbf{TT}\alpha \equiv \mathbf{T}\alpha$; *etc.*

To show that 'true not' is equivalent to 'false', and 'false not' to 'true', we have to derive theses in a slightly different form. We cannot, for example, affirm $\mathbf{T}{\sim}\alpha \equiv \mathbf{F}\alpha$ since '${\sim}\alpha$' is not well-formed (*e.g.* if α is a statement variable). However, $\S{\sim}A$ is well-formed where A is a wffcl, and we can show that $\mathbf{T}\S{\sim}A \equiv \mathbf{F}\S A$. As follows:

$\vdash_{\text{CL}}$ $\mathbf{T}\S{\sim}A \equiv {\sim}A$: a case of $\vdash_{\text{CL}}(11)$. So by $\vdash_{\text{CL}}(6)$,
$\vdash_{\text{CL}}(60)$ $\mathbf{T}\S{\sim}A \equiv \mathbf{F}\S A$.

Similarly,

$\vdash_{\text{CL}}(61)$ $\mathbf{F}\S{\sim}A \equiv \mathbf{T}\S A$.

The properties of truth and falsity over statement-yielding sentences are identical, and follow immediately from $\vdash_{CL}(34)$ $TA \equiv A$ and $\vdash_{CL}$ $FA \equiv \sim A$ which is proved below.

From (P2) we have by **RQL**,

$\vdash_{CL}$ $(P\gamma)(qu(\Pi)(L\theta)\,\mathbf{Y}\gamma\;\&\;\mathbf{F}\gamma) \supset T{\sim}\Pi(L\theta)$. So, by D6,
$\vdash_{CL}(62)$ $F\Pi(L\theta) \supset T{\sim}\Pi(L\theta)$.

Hence, for wffcl,

$\vdash_{CL}(63)$ $FA(L\theta) \supset T{\sim}A(L\theta)$.

Now $\vdash_{CL}(34)$ can be derived with a context qualification (the proof parallels that of $\vdash_{CL}(34)$) and we may affirm:

$\vdash_{CL}(64)$ $TA(L\theta) \equiv A(L\theta)$. Hence,
$\vdash_{CL}(65)$ $T{\sim}A(L\theta) \equiv {\sim}A(L\theta)$.

From $\vdash_{CL}(63)$ and $\vdash_{CL}(65)$, therefore,

$\vdash_{CL}(66)$ $FA(L\theta) \supset {\sim}A(L\theta)$.

We obtain the converse as follows:

As a case of the **RQL** thesis, $\vdash \check{S}^{\gamma}_{\alpha}A \mid \supset (P\gamma)A$, we have,

$\vdash_{CL}$ $(qu(\Pi)(\theta)\,\mathbf{Y}\alpha\;\&\;\mathbf{F}\alpha) \supset (P\gamma)(qu(\Pi)(\theta)\,\mathbf{Y}\gamma\;\&\;\mathbf{F}\gamma)$.

That is, by D6,

$\vdash_{CL}$ $(qu(\Pi)(\theta)\,\mathbf{Y}\alpha\;\&\;\mathbf{F}\alpha) \supset F\Pi(\theta)$. Hence,
$\vdash_{CL}$ $(qu(A)(\theta)\,\mathbf{Y}\S A(\theta)) \supset (\mathbf{F}\S A(\theta) \supset FA(\theta))$.

By (P1) and detachment, then,

$\vdash_{CL}(67)$ $\mathbf{F}\S A(\theta) \supset FA(\theta)$.

Now we have from $\vdash_{CL}(11)$, $\vdash_{CL}$ $\mathbf{T}\S{\sim}A(L\theta) \equiv {\sim}A(L\theta)$. So by $\vdash_{CL}(60)$,

$\vdash_{CL}(68)$ ${\sim}A(L\theta) \equiv \mathbf{F}\S A(L\theta)$.

Hence, by $\vdash_{CL}(67)$,

$\vdash_{CL}(69)$ ${\sim}A(L\theta) \supset FA(L\theta)$.

Together, $\vdash_{CL}(66)$ and $\vdash_{CL}(69)$ give:

$\vdash_{CL}(70)$ $FA(L\theta) \equiv {\sim}A(L\theta)$.

As special cases of $\vdash_{CL}(64)$ and $\vdash_{CL}(70)$ we then have,

$\vdash_{CL}(71)$ $T\Phi(L\theta) \equiv \Phi(L\theta)$.
$\vdash_{CL}(72)$ $F\Phi(L\theta) \equiv {\sim}\Phi(L\theta)$.

Using these equivalences to make replacements in **RSL** theorems we shall have, for any logically standard context, theses such as:

$\vdash_{CL} Tr \equiv r$; $\vdash_{CL} Fr \equiv \sim r$; $\vdash_{CL} Tr \vee Fr$; $\vdash_{CL} \sim(Tr \,\&\, Fr)$; $\vdash_{CL} T(r \,\&\, s) \equiv Tr \,\&\, Ts$; $\vdash_{CL} T(r \vee s) \equiv Tr \vee Ts$; $\vdash_{CL} TTr \equiv Tr$; $\vdash_{CL} TFr \equiv Fr$; *etc.*

Now although, as we have suggested, the unrestricted analogue of '$Tr \equiv r$', *viz* '$Tp \equiv p$', cannot be a thesis of **CL** since it is ill-formed, and in any case is unlikely to hold in general in many-valued systems in which it can occur, it might perhaps be thought that in cases where all the variables in the above theses are covered by 'T' or 'F', there will be analogues in terms of unrestricted variables. Thus one might perhaps expect '$T(p \,\&\, q) \equiv Tp \,\&\, Tq$' and '$T(p \vee q) \equiv Tp \vee Tq$' to be theses since these express distributive properties of 'T' which are independent of the kind of sentence substituted. Such formulae are well-formed, but we cannot at this stage derive any of them as theorems, and in fact, as we shall see not all of them are acceptable. For all the many-valued systems which we consider in Part II, '$T(p \,\&\, q) \equiv Tp \,\&\, Tq$' is a thesis, but '$T(p \vee q) \equiv Tp \vee Tq$' is not. One half of this latter holds, namely '$T(p \vee q) \supset Tp \vee Tq$'; but the converse does not.

We can, however, represent theses such as '$T(r \,\&\, s) \equiv Tr \,\&\, Ts$' in terms of unrestricted variables if we write them with an antecedent condition which expresses the fact that the variables in them are limited to statement-yielding sentences. For example, the formula,

$$[(Pa)(qu(p)(Lc)\, \mathbf{Y}a) \,\&\, (Pb)(qu(q)(Lc)\, \mathbf{Y}b)] \supset [T(p \,\&\, q) \equiv Tp \,\&\, Tq](Lc).$$

expresses in terms of unrestricted variables exactly what '$[T(r \,\&\, s) \equiv Tr \,\&\, Ts](Lc)$' expresses in terms of restricted variables. To take these more general formulations as theses would not add anything new, and in particular would not lead to the conclusion that unrestricted variants of the restricted formulae are theses, but it is sometimes an advantage to represent them in this way and we introduce a rule for this purpose. The rule has to be expressed in such a way that we do not get unrestricted variables which are not covered by 'T' or 'F' in the consequent. For example, the rule must not allow a move from $\vdash (Tr \equiv r)(Lc)$ to $\vdash (Pa)(qu(p)(Lc)\, \mathbf{Y}a) \supset (Tp \equiv p)(Lc)$ since the consequent is not well formed. This limitation can be achieved simply by requiring that the restricted variables which are to be replaced occur within a quotation function. For since 'T' and 'F' both absorb quotation functions, the replacement will be permissible for variables covered by 'T' and 'F' but not for those occurring uncovered. The rule is as follows:

> CL(15): if $\vdash_{CL} A(\theta)$ then $\vdash_{CL} (P\gamma)(qu(P)(\theta)\, \mathbf{Y}\gamma) \supset \mathsf{S}^{R}_{P}A \,|\, (\theta)$ provided every occurrence of R in A is within the scope of the quotation function, and $\mathsf{S}^{R}_{P}A \,|$ is the formula which results from A by replacing every occurrence of the restricted variable R by an unrestricted variable P.

The rule is not derivable but is nevertheless a consequence of our conventions concerning the use of restricted and unrestricted variables. For a thesis expressed in terms of restricted variables indicates that it is limited in application to statement-yielding sentences. Hence if we affirm this as an antecedent condition, we can express the same thesis in terms of unrestricted variables. So, for example, from $\vdash_{CL} Tr \vee Fr$ we shall be able to pass to,

$\vdash_{CL} \quad (Pa)(qu(p)(c_S)\,\mathbf{Y}a) \supset (Tp \vee Fp).$

As it happens, we can obtain this independently of CL(15) in a much more general form ($\vdash_{CL}$(74) below), but this is not always the case. The main use of the rule is that it facilitates the derivation of a number of theses involving '*T*' and '*F*' with unrestricted variables. For example, if we can show that a formula holds for sentences which do not yield statements, and if we know that it also holds for sentences which do, then it will hold for all sentences. Thus, suppose we can establish:

$$\sim(Pa)(qu(p)(Lc)\,\mathbf{Y}a) \supset \sim(Tp \,\&\, Fp)$$

i.e. non-statement yielding sentences are not both true and false. Then since we also have,

$$\sim(Tr \,\&\, Fr),$$

which expresses the fact that statement-yielding sentences are not both true and false, we can use the rule to pass to,

$$(Pa)(qu(p)(Lc)\,\mathbf{Y}a) \supset \sim(Tp \,\&\, Fp);$$

and by **RSL** this will lead to $\vdash\sim(Tp \,\&\, Fp)$.

We now show how to derive a more general version of $\vdash(Pa)(qu(p)(c_S)\,\mathbf{Y}a) \supset (Tp \vee Fp)$ independently of CL(15). The theorem is the most fundamental of those which express the properties of '*T*' and '*F*' in conjunction with unrestricted variables. The proof is as follows:

Where Γ is an unrestricted semi-wff, we have by D5 and D6:

$\vdash_{CL} \quad (T\Gamma \vee F\Gamma)(\theta) \equiv (P\gamma)(qu(\Gamma)(\theta)\,\mathbf{Y}\gamma \,\&\, \mathbf{T}\gamma) \vee (P\gamma)(qu(\Gamma)(\theta)\,\mathbf{Y}\gamma \,\&\, \mathbf{F}\gamma);$

which leads by **RQL** to,

$\vdash_{CL}(73) \quad (T\Gamma \vee F\Gamma)(\theta) \equiv (P\gamma)(qu(\Gamma)(\theta)\,\mathbf{Y}\gamma \,\&\, (\mathbf{T}\gamma \vee \mathbf{F}\gamma)).$

Now as a special case of the **RSL** thesis $A \supset (B \equiv B \,\&\, A)$, we have:

$\vdash_{CL} \quad (\mathbf{T}\gamma \vee \mathbf{F}\gamma) \supset (qu(\Gamma)(\theta)\,\mathbf{Y}\gamma \equiv qu(\Gamma)(\theta)\,\mathbf{Y}\gamma \,\&\, (\mathbf{T}\gamma \vee \mathbf{F}\gamma));$

So by detachment, using $\vdash_{CL}$(50), and generalisation:

$\vdash_{CL} \quad (U\gamma)(qu(\Gamma)(\theta)\,\mathbf{Y}\gamma \equiv qu(\Gamma)(\theta)\,\mathbf{Y}\gamma \,\&\, (\mathbf{T}\gamma \vee \mathbf{F}\gamma)).$

Hence, by **RQL**,

$\vdash_{CL} \quad (P\gamma)(qu(\Gamma)(\theta)\,\mathbf{Y}\gamma) \equiv (P\gamma)(qu(\Gamma)(\theta)\,\mathbf{Y}\gamma \,\&\, (\mathbf{T}\gamma \vee \mathbf{F}\gamma)).$

Finally, therefore, using this to make a replacement in $\vdash_{CL}$(73),

$\vdash_{CL}(74) \quad (T\Gamma \vee F\Gamma)(\theta) \equiv (P\gamma)(qu(\Gamma)(\theta)\,\mathbf{Y}\gamma).$

In particular,

$\vdash_{CL}(75)\quad (Tp \vee Fp)(c) \equiv (Pa)(qu(p)(c)\ \mathbf{Y}a).$

This expresses the fact that a truth-valued sentence in a context is statement yielding, and conversely. We explore the consequences of this in the next section.

We show finally how the liar paradox argument may be formalised.

It was suggested in §1.5 that one way of resolving the paradox might be to claim that the sentence 'This sentence is false' used in a context c_0 in such a way as to refer to itself, fails to yield a statement. For if we write it in the form, 'This sentence yields a false statement', then the reference is to a pair consisting of a sentence and a statement, and while one element of this pair, the sentence, is known, the other element, the statement, is not. In terms of this analysis, the sentence is incomplete in that context in which reference is made to itself. By contrast, if the sentence is used in a different context to make a remark about some other sentence, then it is complete. For if the second sentence being referred to yields a statement, then a remark has been made which is either true or false—true if the second sentence yields a false statement, false if it yields a true one; if on the other hand, the sentence which is being mentioned fails to yield a statement, then it fails to yield a false statement, and the remark is then false. Hence tokens which exemplify the type 'This sentence yields a false statement' are incomplete if self-referring and statement-yielding if the reference of 'This sentence' is to some other sentence.

If this analysis is justifiable we should therefore expect to be able to show that any token which exemplifies the type 'This sentence yields a false statement' cannot yield the statement that it itself is false. We should not expect to be able to show that all tokens of this type fail to yield statements, for in other contexts they may.

We have as a case of $\vdash_{CL}(47)$,

$\vdash_{CL}\quad (r(c) \mathrel{\underline{\underline{\sqsubset}}} Fr(c)) \equiv qu(r)(c)\ \mathbf{Y}\S Fr(c).$

But by $\vdash_{CL}(35)$ and $\vdash_{CL}(49)$, using **RSL**,

$\vdash_{CL}\quad (r(c) \mathrel{\underline{\underline{\sqsubset}}} Fr(c)) \supset (r(c) \equiv Fr(c)).$ Hence,
$\vdash_{CL}(76)\quad qu(r)(c)\ \mathbf{Y}\S Fr(c) \supset (r(c) \equiv Fr(c)).$

Now for any logically standard context Lc, we have, by $\vdash_{CL}(70)$,

$\vdash_{CL}\quad Fr(Lc) \equiv {\sim} r(Lc).$

So by **RSL** and $\vdash_{CL}(76)$,

$\vdash_{CL}\quad qu(r)(Lc)\ \mathbf{Y}\S Fr(Lc) \supset [(r(Lc) \equiv Fr(Lc))\ \&\ (Fr(Lc) \equiv {\sim} r(Lc)].$

Hence,

$\vdash_{CL}\quad qu(r)(Lc)\ \mathbf{Y}\S Fr(Lc) \supset (r(Lc) \equiv {\sim} r(Lc)).$

But we have the **RSL** thesis,

$\vdash_{CL}$ $\sim(r(Lc)\equiv\sim r(Lc))$. Hence,
$\vdash_{CL}$ $\sim(qu(r)(Lc)\ \mathbf{Y}\S Fr(Lc))$.

That is, no statement-yielding sentence can yield, in a logically standard context, the statement that it is itself false. Now by rule CL(15), we then have:

$\vdash_{CL}(77)$ $(Pa)(qu(p)(Lc)\ \mathbf{Y}a)\supset\sim(qu(p)(Lc)\ \mathbf{Y}\S Fp(Lc))$.

But we also have, by **RQL**,

$\vdash_{CL}$ $qu(p)(Lc)\ \mathbf{Y}\S Fp(Lc)\supset(Pa)(qu(p)(Lc)\ \mathbf{Y}a)$.

Hence by $\vdash_{CL}(77)$ and **RSL**,

$\vdash_{CL}$ $qu(p)(Lc)\ \mathbf{Y}\S Fp(Lc)\supset\sim(qu(p)(Lc)\ \mathbf{Y}\S Fp(Lc))$.

Consequently,

$\vdash_{CL}$ $\sim(qu(p)(Lc)\ \mathbf{Y}\S Fp(Lc))$.

i.e. whatever sentence we take, it does not yield, in a logically standard context, the statement that it is false. Or, by D9,

$\vdash_{CL}$ $\S Fp(Lc)\neq \mathbf{st}\,(p,Lc)$;

the statement yielded by $qu(p)$ in Lc is not the statement that $qu(p)$ is false.

Take, then, a sentence 'p_0' and make the supposition that it does yield the statement that it is false. Writing $\vdash_{CL}(77)$ in an equivalent form, we shall have, for 'p_0',

$$qu(p_0)(Lc)\ \mathbf{Y}\S Fp_0(Lc)\supset\sim(Pa)(qu(p_0)(Lc)\ \mathbf{Y}a).$$

Hence the supposition will lead us to the conclusion that 'p_0' is not statement-yielding in a logically standard context. This of course leaves it open whether it is nonsignificant or incomplete. In favour of the view that it is incomplete, however, we may note that since it purports to say that the statement which it itself yields is false, and since we have shown in general that no sentence can yield the statement that it is itself false, then if it yields a statement at all, it must be some other statement. In that case, however, we do not know what the statement is, and therefore we cannot identify what the sentence is about.

Notice that if 'This sentence' in 'This sentence is false' is used in a context Lc to mention not the sentence in which it occurs, but some other sentence, say 'q_0', we then have, $qu(p_0)(Lc)\ \mathbf{Y}\S Fq_0(Lc)$ in place of $qu(p_0)(Lc)\ \mathbf{Y}\S Fp_0(Lc)$. In this case none of the conclusions follow.

2.15 The Valuation Function and the Derived Sentential Range

The initial range of the sentential variables is a set of token sentences, but we now show how this may be replaced by a derived range consisting of the four values t (true), f (false), n (nonsignificant) and i (incomplete), which reduces, in the case of statement-yielding sentences to the two values t and f. We call t, f, n and i value constants and we introduce a valuation function $v(\ )$ which takes any wffcl or unrestricted semi-wff in its argument place and one of the value constants as a value. We show below that for all values of the variables in Π, $v(\Pi)$ is identical with one, and only one, of the value constants. Hence the function is appropriately many-one. Value constants and $v(\Pi)$ are value terms.

R4(x) if μ and ω are value terms, $\mu=\omega$ is a wffcl.

We thus use the same symbol '=' for value identity as for statement identity but there should be no confusion since it joins different symbols. In particular, μ and ω are not wfc.

We take the ordinary identity postulates to characterise value identity:

$\vdash_{\mathrm{CL}}$ (P9) (i) $\mu=\mu$
(ii) $\mu=\omega\supset(A(\mu)\supset A(\omega))$.

The function is defined successively for its various values as follows:

D10 (i) $(v(\Pi)=t)(\theta)=_{\mathrm{Df}}T\Pi(\theta)$
(ii) $(v(\Pi)=f)(\theta)=_{\mathrm{Df}}F\Pi(\theta)$
(iii) $(v(\Pi)=n)(\theta)=_{\mathrm{Df}}\sim S\Pi(\theta)$
(iv) $(v(\Pi)=i)(\theta)=_{\mathrm{Df}}\sim C\Pi(\theta)$.

Now from D4 we have,

$\vdash_{\mathrm{CL}}$ $C\Pi(\theta)\equiv[S\Pi(\theta)\equiv(P\gamma)(qu(\Pi)(\theta)\,\mathbf{Y}\gamma)]$; and by **RSL** this gives,
$\vdash_{\mathrm{CL}}(78)$ $(C\Pi(\theta)\ \&\ S\Pi(\theta))\supset(P\gamma)(qu(\Pi)(\theta)\,\mathbf{Y}\gamma)$,

i.e. if a sentence is both significant and complete, then it is statement yielding. This theorem is in fact uninteresting where Π is a restricted semi-wff or a wffcl for by $\vdash_{\mathrm{CL}}(15)$ the consequent is a thesis and it can therefore be established by the **RSL** law, $B\supset(A\supset B)$. In what follows, therefore, we shall only consider its consequences where Π is taken to be an unrestricted semi-wff Γ. Taking it as such, and using $\vdash_{\mathrm{CL}}(74)$, we have:

$\vdash_{\mathrm{CL}}$ $C\Gamma(\theta)\ \&\ S\Gamma(\theta)\supset T\Gamma(\theta)\vee F\Gamma(\theta)$; *i.e.*
$\vdash_{\mathrm{CL}}(79)$ $(\sim C\Gamma\vee\sim S\Gamma\vee T\Gamma\vee F\Gamma)(\theta)$. In particular,
$\vdash_{\mathrm{CL}}(80)$ $(\sim Cp\vee\sim Sp\vee Tp\vee Fp)(c)$.

This is the analogue, for any sentence, of $\vdash_{\mathrm{CL}}Tr\vee Fr$ over statement-yielding sentences. It expresses the fact that every sentence is either true, false, nonsignificant or incomplete in its context of use.

We now fix finally the sense which we give to 'statement-yielding' by adopting the converse of $\vdash_{\mathrm{CL}}(78)$ as a postulate, *viz*:

$\vdash_{\mathrm{CL}}(P10)$ $(\mathrm{P}\gamma)(qu(\Pi)(\theta)\,\mathbf{Y}\gamma)\supset(C\Pi(\theta)\ \&\ S\Pi(\theta))$.

That is, a statement-yielding sentence is both significant and complete. This makes explicit the basic assumptions we have made throughout about the relations between the concepts of statement-yielding, completeness and significance; or what amounts to the same thing, as we shall see presently, about the relations between the four values, *t*, *f*, *n* and *i*. For taking the postulate in conjunction with $\vdash_{CL}(78)$ we have:

$$\vdash_{CL} \quad (P\gamma)(qu(\Gamma)(\theta)\,\mathbf{Y}\gamma) \equiv (C\Gamma \,\&\, S\Gamma)(\theta),$$

for the special case of unrestricted semi-wffs, *i.e.* for all sentences, whether truth-valued or not. But we also know, from $\vdash_{CL}(74)$, that statement-yielding sentences are truth-valued. Hence,

$$\vdash_{CL}(81) \quad (T\Gamma \vee F\Gamma)(\theta) \equiv (C\Gamma \,\&\, S\Gamma)(\theta);$$

i.e. truth-valued sentences are both significant and complete, and conversely.

Now from $\vdash_{CL}(79)$ by D10 we have,

$$\vdash_{CL}(82) \quad [(v(\Gamma)=t) \vee (v(\Gamma)=f) \vee (v(\Gamma)=n) \vee (v(\Gamma)=i)](\theta).$$

That is, every sentence has the value *t*, *f*, *n* or *i*.

We call the set of values $\{t, f, n, i\}$ the derived sentential range. It is of fundamental importance in Part II for we are there concerned to investigate the relations which hold between sentences when they occur in a non-quotation context, *i.e.* when they are used rather than mentioned. We wish to know, for example, whether standard laws such as '$(r \supset r \vee s)(Lc)$' are valid when the variables range over all sentences and not simply over those which yield statements. Is it the case, for example, that '$(p \supset p \vee q)(Lc)$' is a thesis valid over all sentences? Now this amounts to asking whether it is a theorem of four-valued logic when the values are *t*, *f*, *n* and *i*, and one way of approaching the problem is to examine the possible four-valued matrices which may be constructed for the connectives &, $\vee$, *etc*. Our purpose here, however, is not to examine such matrices (though we show below how to derive them for the two-valued case) but simply to show that it is a legitimate approach; and this is what $\vdash_{CL}(82)$ establishes. It has to be recognised, however, that this derived range can only be introduced with respect to a context: in a given context a particular token sentence is either true, false, nonsignificant or incomplete; this is the content of $\vdash_{CL}(82)$. If, therefore, we wish to develop the logic of sentences by means of matrices, the same fixed context must be taken for all variables before any application of the logic. This is a feature which, as we have already seen, characterises **RSL**. And it can be met, in the way we have already adopted for **RSL**, by limiting the logic to formulae qualified by the same context variable and then removing it from each and writing "For given context *c*" as a general condition which qualifies all theorems. This then gives us a sentential calculus the theorems of which are valid for an arbitrarily chosen context but which is such that the context must be kept fixed for any particular application. The theorems of such a logic are not valid for an application in which sentences are taken from different contexts. The chosen context will, however, be logically standard if the matrix definitions of the ordinary connectives &, $\vee$, etc., are such as include their standard two-valued definitions. This is so for all the many-valued systems we consider. Hence they are all controlled by the prefix "For a logically standard context...." No thesis of any many-valued system in Part II is context free. If they are taken to be unqualified by a context symbol (unless with the special assumption that this indicates an application in c_S) they are incomplete symbols in exactly the sense defined in *PM* (Chapter III, p. 66).

That the four value-constants are exhaustive, and thus constitute the total range of the value function, is given by $\vdash_{CL}(82)$. We now show that they are also exclusive. This implies that the value function cannot take more than one of the constants as a value, for any given argument Γ, and is, therefore, many-one.

Since, by **RSL**, we have:

$\vdash_{CL} T\Gamma(\theta) \supset (T\Gamma \vee F\Gamma)(\theta)$ and $\vdash_{CL}(C\Gamma \;\&\; S\Gamma)(\theta) \supset C\Gamma(\theta)$,

we have immediately from $\vdash_{CL}(81)$,

$\vdash_{CL} T\Gamma(\theta) \supset C\Gamma(\theta)$; *i.e.* by **RSL**,
$\vdash_{CL}(83) \sim(T\Gamma \;\&\; \sim C\Gamma)(\theta)$.

This is the first of the exclusive conditions we require. Writing it in terms of the valuation function, using (D10), it is:

$\vdash_{CL} \sim [(v(\Gamma) = t) \;\&\; (v(\Gamma) = i)](\theta)$.

That is, the value of Γ cannot be both t and i in the same context.

Three of the other conditions, *viz* $\vdash_{CL} \sim(T\Gamma \;\&\; \sim S\Gamma)(\theta)$, $\vdash_{CL} \sim(F\Gamma \;\&\; \sim C\Gamma)(\theta)$, and $\vdash_{CL} \sim(F\Gamma \;\&\; \sim S\Gamma)(\theta)$, are proved in a similar manner. Hence t and n, f and i, and f and n, are mutually exclusive pairs.

The proof of $\vdash_{CL} \sim(\sim C\Gamma \;\&\; \sim S\Gamma)(\theta)$ which determines i and n as an exclusive pair, is slightly different, but again depends on $\vdash_{CL}(81)$; as follows. From $\vdash_{CL}(81)$ using $\vdash_{CL}(C\Gamma \;\&\; S\Gamma)(\theta) \supset S\Gamma(\theta)$, a case of an **RSL** thesis, we have:

$\vdash_{CL}(T\Gamma \vee F\Gamma)(\theta) \supset S\Gamma(\theta)$.

Hence, by the **RSL** thesis $B \supset (A \supset B)$ and detachment,

$\vdash_{CL} \sim S\Gamma(\theta) \supset ((T\Gamma \vee F\Gamma)(\theta) \supset S\Gamma(\theta))$.

Again, as a case of the **RSL** thesis $\sim A \supset (A \supset B)$ we have,

$\vdash_{CL} \sim S\Gamma(\theta) \supset (S\Gamma(\theta) \supset (T\Gamma \vee F\Gamma)(\theta))$.

Putting these together, therefore, gives:

$\vdash_{CL} \sim S\Gamma(\theta) \supset (S\Gamma(\theta) \equiv (T\Gamma \vee F\Gamma)(\theta))$; *i.e.* by D4 and $\vdash_{CL}(74)$,
$\vdash_{CL} \sim S\Gamma(\theta) \supset C\Gamma(\theta)$; and by **RSL**,
$\vdash_{CL}(84) \sim(\sim S\Gamma \ \& \sim C\Gamma)(\theta)$.

There is of course a sense of 'incomplete' in terms of which one might wish to say that incomplete sentences may be nonsignificant, and in this case $\vdash_{CL}(84)$ fails. But the sense we have taken, which rules out the possibility of a "double" failure in a sentence, has the strong advantage that it implies the distinctness of the values *i* and *n* and so justifies a matrix approach in four-valued sentential logic. It is not necessary, however, to reject $\vdash_{CL}(84)$ if another sense of 'incomplete' is preferred. For context logic can easily accommodate both senses. Thus, we could define an additional constant C' such that $\sim(\sim S\Gamma \ \& \sim C'\Gamma)(\theta)$ is not a thesis; but then, of course, the corresponding value i' would not be independent of *n* and could not be taken as a matrix value since some sentences would take both i' and *n* as a value in the same context. We return to this point in §3.9.

The final condition on the exclusiveness of the value constants, *viz.* $\vdash_{CL} \sim(T\Gamma \ \& \ F\Gamma)(\theta)$ is derivable as follows:

From $\vdash_{CL}(31)$, where Π is Γ, we have by **RSL**,

$\vdash_{CL} \quad qu(\Gamma)(\theta)\, \mathbf{Y}\alpha \supset (\sim \mathbf{T}\alpha \supset \sim T\Gamma(\theta))$.

Hence, by $\vdash_{CL}(55)$,

$\vdash_{CL} \quad qu(\Gamma)(\theta)\, \mathbf{Y}\alpha \supset (\mathbf{F}\alpha \supset \sim T\Gamma(\theta))$.

That is, by **RSL**,

$\vdash_{CL} \quad (qu(\Gamma)(\theta)\, \mathbf{Y}\alpha \ \& \ \mathbf{F}\alpha) \supset \sim T\Gamma(\theta)$.

Where α is a statement variable γ, then by **RQL** this leads to:

$\vdash_{CL} \quad (P\gamma)(qu(\Gamma)(\theta)\, \mathbf{Y}\gamma \ \& \ \mathbf{F}\gamma) \supset \sim T\Gamma(\theta)$.

By D6, therefore,

$\vdash_{CL} \quad F\Gamma(\theta) \supset \sim T\Gamma(\theta)$.

Hence, by **RSL**,

$\vdash_{CL}(85) \quad \sim(T\Gamma \ \& \ F\Gamma)(\theta)$.

Our assumptions therefore imply that every many-valued system we consider satisfies the law of non-contradiction: *i.e.* *t* and *f* are mutually exclusive.

We now show how to derive the familiar matrices (truth tables) for **RSL**. From,

$\vdash_{CL} \quad (T\Phi \vee F\Phi)(\mathrm{L}\theta)$, which arises from $\vdash_{CL}(71)$ and $\vdash_{CL}(72)$, we have,

$\vdash_{CL}(86)\quad (v(\Phi)=t)\vee(v(\Phi)=f))(L\theta);$

and using a special case of $\vdash_{CL}(85)$, we also have,

$\vdash_{CL}(87)\quad \sim[(v(\Phi)=t)\ \&\ (v(\Phi)=f)](L\theta).$

Hence, every restricted semi-wff has either the value t or the value f, and none has both. Again, from $\vdash_{CL}(T{\sim}\Phi\equiv F\Phi)(L\theta)$, another consequence of $\vdash_{CL}(71)$ and $\vdash_{CL}(72)$, we have,

$\vdash_{CL}(88)\quad (v({\sim}\Phi)=t)\equiv(v(\Phi)=f))(L\theta);$

and from $\vdash_{CL}(T\Phi\equiv F{\sim}\Phi)(L\theta)$,

$\vdash_{CL}(89)\quad ((v(\Phi)=t)\equiv(v({\sim}\Phi)=f))(L\theta).$

Hence, if a restricted semi-wff Φ has the value t, ${\sim}\Phi$ has the value f, and if Φ has the value f, ${\sim}\Phi$ has the value t, and conversely in each case.

These results we may therefore sum up by writing a matrix:

Φ	${\sim}\Phi$
t	f
f	t

Here, each restricted semi-wff Φ and ${\sim}\Phi$ has the value t or the value f and neither has both. Hence $\vdash_{CL}(86)$ and $\vdash_{CL}(87)$ are satisfied. Again, Φ has the value t when, and only when, ${\sim}\Phi$ has the value f; and conversely. So $\vdash_{CL}(88)$ and $\vdash_{CL}(89)$ are satisfied.

Now from $\vdash_{CL}(71)$ and $\vdash_{CL}(72)$ by **RSL** we have for $L\theta$,

$\vdash_{CL}T\Phi\ \&\ T\Psi\supset T(\Phi\supset\Psi);\ \vdash_{CL}F\Phi\ \&\ T\Psi\supset T(\Phi\supset\Psi);$
$\vdash_{CL}T\Phi\ \&\ F\Psi\supset F(\Phi\supset\Psi);\ \vdash_{CL}F\Phi\ \&\ F\Psi\supset T(\Phi\supset\Psi).$

So, by D10,

$\vdash_{CL}(90)\quad [(v(\Phi)=t)\ \&\ (v(\Psi)=t)]\supset(v(\Phi\supset\Psi)=t);$
$\vdash_{CL}(91)\quad [(v(\Phi)=f)\ \&\ (v(\Psi)=t)]\supset(v(\Phi\supset\Psi)=t);$
$\vdash_{CL}(92)\quad [(v(\Phi)=t)\ \&\ (v(\Psi)=f)]\supset(v(\Phi\supset\Psi)=f);$
$\vdash_{CL}(93)\quad [(v(\Phi)=f)\ \&\ (v(\Psi)=f)]\supset(v(\Phi\supset\Psi)=t)$; for $L\theta$.

The following matrix summarises these results under the conditions imposed by $\vdash_{CL}(86)$ and $\vdash_{CL}(87)$:

Φ	Ψ	$\Phi\supset\Psi$
t	t	t
f	t	t
t	f	f
f	f	t

i.e.

$\supset$	t	f
t	t	f
f	t	t

The remaining matrices can be derived in a similar way, or immediately from the equivalence theses corresponding to the standard definitions, and we shall have:

&	*t*	*f*
t	*t*	*f*
f	*f*	*f*

∨	*t*	*f*
t	*t*	*t*
f	*t*	*f*

≡	*t*	*f*
t	*t*	*f*
f	*f*	*t*

The technique can be applied to formulae containing '*T*' and '*F*' in conjunction with restricted variables, *i.e.* to formulae of **RSL** extended by '*T*' and '*F*'. For we have, by $\vdash_{CL}(64)$, $\vdash_{CL}(65)$ and $\vdash_{CL}(70)$,

$$\vdash_{CL} T\Phi \equiv TT\Phi;\ \vdash_{CL} F\Phi \equiv FT\Phi;\ \vdash_{CL} T\Phi \equiv FF\Phi;\ \vdash_{CL} F\Phi \equiv TF\Phi.$$

Hence, by D10, for Lθ,

$\vdash_{CL}(94)$ $(v(\Phi)=t) \equiv (v(T\Phi)=t)$
$\vdash_{CL}(95)$ $(v(\Phi)=f) \equiv (v(T\Phi)=f)$
$\vdash_{CL}(96)$ $(v(\Phi)=t) \equiv (v(F\Phi)=f)$
$\vdash_{CL}(97)$ $(v(\Phi)=f) \equiv (v(F\Phi)=t)$.

From the first two of these we obtain the matrix for '*T*', and from the second two the matrix for '*F*'. Thus:

Φ	$T\Phi$
t	*t*
f	*f*

Φ	$F\Phi$
t	*f*
f	*t*

We cannot at this point derive matrices over unrestricted semi-wffs because we have no theses which determine the values of compound unrestricted semi-wffs given values for their components. It is possible, however, to derive a set of matrices for statement logic, **SL**, if we introduce statement value constants **t** and **f** and a statement value function $\mathbf{v}(\alpha)$ such that $(\mathbf{v}(\alpha)=\mathbf{t}) =_{Df} \mathbf{T}\alpha$ and $(\mathbf{v}(\alpha)=\mathbf{f}) =_{Df} \mathbf{F}\alpha$. The technique is exactly the same as that for **RSL** and leads to the following matrices:

	t	**f**
t	**f**	**t**
f	**t**	**t**

imp	**t**	**f**
t	**t**	**f**
f	**t**	**t**

conj	**t**	**f**
t	**t**	**f**
f	**f**	**f**

disj	**t**	**f**
t	**t**	**t**
f	**t**	**f**

eq	**t**	**f**
t	**t**	**f**
f	**f**	**t**

α	$\mathbf{T}\alpha$
t	**t**
f	**f**

α	$\mathbf{F}\alpha$
t	**f**
f	**t**

Hence, as one would expect, there is a matrix isomorphism between the restricted sentential connectives $\supset$, &, $\vee$, $\equiv$, T, F and, respectively, the statemental relations and properties **imp**, **conj**, **disj**, **eq**, **T** and **F**.

2.16 The Reduced Sentential Range

Since our primary interest is in the problem of nonsignificance, we shall for many purposes neglect incomplete sentences and deal only with those which are either true, false or nonsignificant in a given context. In making this simplification, however, we do not wish it to be thought that the theory of incomplete sentences is unimportant. On the contrary, it seems to us that it is relevant to many problems and, as we have suggested, some paradoxes might be resolved in terms of this concept rather than nonsignificance.

The simplification can be achieved formally by reclassifying incomplete sentences as either truth-valued or nonsignificant (see §3.9) or by affirming an additional postulate which expresses the fact that all sentences are complete. Since it follows in any case from (P1) that all wffcl (hence all restricted semi-wffs) are complete when values are substituted for the variables, since they are statement-yielding, the new postulate need only be affirmed for unrestricted semi-wffs Γ:

$\vdash_{\mathrm{CL}}$(P11) $C(\Gamma)(\theta)$.

To distinguish theorems which depend on this postulate from the more general theses, we add an extra 'c' to the gate sign and write '$\vdash_{\mathrm{CCL}}$'. The set of such theorems may then be regarded as a sub-theory of **CL** which applies to complete sentences only. Similarly, sub-theories applying to significant sentences only, and to statement-yielding sentences only, could be developed.

From (P11) and an equivalence thesis arising from D4 we have immediately:

$\vdash_{\mathrm{CCL}}$(98) $S\Gamma(\theta) \equiv (P\gamma)(qu(\Gamma)(\theta)\ \mathbf{Y}\gamma)$,

i.e. a sentence is significant in a context if, and only if, it yields a statement. This is an important criterion of significance in terms of which most other criteria can be evaluated. From this and $\vdash_{\mathrm{CL}}$(74) we have:

$\vdash_{\mathrm{CCL}}$(99) $S\Gamma(\theta) \equiv (T\Gamma \vee F\Gamma)(\theta)$.

This corresponds to Russell's definition of significance and illustrates the implicit assumption made by Russell, and others, that incomplete sentences are excluded or else absorbed in the class of nonsignificant or false sentences. It illustrates, too, that non-indicative acts are excluded from consideration, otherwise a definition in terms of truth and falsity would be inappropriate. Of course, we too have accepted this exclusion, though we do not accept that it implies a limitation to sentences which are grammatical indicatives.

From $\vdash_{CL}(79)$ and (P11) we have,

$\vdash_{CCL}(100)$ $(T\Gamma \vee F\Gamma \vee \sim S\Gamma)(\theta)$;

and by D10 this determines the derived range as the set of values $\{t, f, n\}$. It is this set which we shall use for much of the development of the many-valued systems in Part II. That is, for the most part, we consider three-valued rather than four-valued sentential systems. We have already shown that these values are exclusive in the set $\{t, f, n, i\}$ and the elimination of i does not affect this conclusion. Since the above theorem tells us that the three values are exhaustive when i is deleted, we therefore have the necessary conditions satisfied for the matrix development of the three-valued systems. We call t, f and n the three significance values.

2.17 Strong Extensional Equivalence

An equivalence relation which is used extensively in Part II is strong extensional equivalence defined in D8. It has the feature that it holds between two sentences if, and only if, they have the same value: *i.e.* if, and only if, both have the value t, both f, both n, or both i. By D10 it has the following matrix:

$\simeq$	t	f	n	i
t	t	f	f	f
f	f	t	f	f
n	f	f	t	f
i	f	f	f	t

It is therefore a two-valued connective, as it must be by D8 since $\Gamma \simeq \Gamma$ is a wffcl, even though it relates many-valued sentences. If Γ is two-valued, '$\simeq$' reduces to '$\equiv$'.

Writing D8 as an equivalence thesis we have,

$$\vdash_{CL}(\Gamma \simeq \Delta)(\theta) \equiv [(T\Gamma \,\&\, T\Delta) \vee (F\Gamma \,\&\, F\Delta) \vee (\sim S\Gamma \,\&\, \sim S\Delta) \vee (\sim C\Gamma \,\&\, \sim C\Delta)](\theta)$$

and taking Δ to be Γ, the special case,

$\vdash_{CL}(\Gamma \simeq \Gamma)(\theta) \equiv [T\Gamma \vee F\Gamma \vee \sim S\Gamma \vee \sim C\Gamma](\theta)$, by **RSL**.

Hence, by $\vdash_{CL}(79)$,

$\vdash_{CL}(\Gamma \simeq \Gamma)(\theta)$.

The same conclusion follows for the reduced sentential range since $(\sim C\Gamma \,\&\, \sim C\Delta)$ is then a contradictory conjunction and can be dropped. Hence '$\simeq$' is reflexive; and it is easily shown that it is also symmetric and transitive.

2.18 Further Developments

We have been concerned in this chapter to examine the main semantic features of statement-yielding, but we now extend the analysis to other semantic concepts. In all of the theses which have been derived, the sentential variables, both restricted and unrestricted, stand for sentences of any form. In particular, therefore, they stand for sentences of the form 'fx', '$g(x, y)$', '$(Ux)fx$', *etc.*, whether many-valued or two-valued, and by considering the semantic function of the various parts of sentences, we are able to give a preliminary analysis in **CL** of such notions as sense, aboutness and reference. This is the purpose of the next chapter. We then turn, in Part II, to the development of various many-valued sentential and quantificational systems. Until this is done, context logic remains seriously incomplete. For we have as yet no logic of the unrestricted semi-wffs. Until we know what general laws hold between sentences, we cannot manipulate any of the unrestricted formulae which may occur in a quotation function. For example, unless we know that, say, '$(\sim p \vee q) \equiv (p \supset q)$' is a law of an acceptable three- or four-valued sentential logic, or that '$\sim(Px)\sim fx \equiv (Ux)fx$' is a law of an acceptable many-valued quantification theory, we cannot make even such elementary moves as passing from a thesis such as '$T(\sim p \vee q)(c) \supset (Pa)(qu(\sim p \vee q)(c)$ **Y**$a)$' to '$T(\sim p \vee q)(c) \supset (Pa)(qu(p \supset q)(c)$ **Y**$a)$'; or from '$T(Ux)fx(c) \equiv T(Ux)fx(c)$' to '$T\sim(Px)\sim fx(c) \equiv T(Ux)fx(c)$'. From this point of view, although we investigate various many-valued logics as independent systems of significance logic in Part II, they may be regarded as extensions of **CL**. Given a set of postulates for a suitable many-valued system, they can be taken as additional postulates of **CL**. If the system is four-valued, they may be taken as postulates of the full **CL**; if three-valued, as postulates of the sub-system **CCL**.

CHAPTER 3

Elements of the Predicate and Quantification Theory

So far we have considered sentences as unanalysed wholes and concentrated on their use to make statements and the conditions they must satisfy in order to yield statements. It is time to examine the analysis of sentences into parts. This is necessary for a variety of reasons. It is primarily in terms of the coupling of sentence parts that the significance, completeness, and statement-capability of many sentences (in their contexts) is assessed. Again, as in classical logic, an analysis of sentences into parts is essential for the development of quantification theory.

In this chapter we shall provide the semantic basis for the quantified significance logics of Chapter 7, but this is not our only concern. Part of our business is to relate significance theory to grammar, on the one hand, and to the world, on the other. Hence the preliminary theories of subject and predicate, of aboutness and items, and of reference and entities, which we develop.

3.1 Syntactical and Semantical Preliminaries

For most of this volume our analysis of the *syntactical* structure of sentences follows classical lines. Thus, we introduce two sorts of variables corresponding to the classical individual and predicate variables, and similarly two sets of constants. We then consider the class of sentences which can be syntactically represented in terms of subject-predicate kernel sentences using connectives, quantifiers and description-forming operators. We also adopt the usual conventions for symbolising multiple-subject or relational sentences. For example, a sentence such as 'Tom loves Mary' will normally be represented as 'Loves (Tom, Mary)' where '(Tom, Mary)' is an ordered subject-pair and 'loves' is the (relational) predicate, rather than as, what is also correct but less revealing, 'Tom (loves Mary)' where 'Tom' is the subject and 'loves Mary' is the predicate. Nothing we do, however, prevents the enrichment of the syntax by the addition of grammatical transformations; such transformations can be handled in a similar way to definitional additions.

We introduce, then, corresponding to the individual variables of classical logic:

subject arguments: $x, y, z, x', y', z', x'', \ldots,$

which take as values any word or phrase which can stand as a *singular* grammatical subject. Subject arguments of the same type take as values words and phrases of the same type throughout a formula. Subject arguments are quantifiable variables. The admitted ranges of these arguments, or variables, will be progressively clarified in the next sections.

Corresponding to the individual constants of classical logic we introduce singular subject constants 'x_0' 'y_0' *etc.* by means of the general rule:

If X is a subject argument then X_0 is a *singular subject constant.*

Here 'X' is a metavariable which stands for any subject argument. Hence the rule expresses the convention that any letter of a type which is classified as a variable and to which is appended a subscript zero, is a formal constant of the same grammatical category as that variable and is a substitution value for all variables of that category. In applications of the logic these formal constants often stand as abbreviations of actual constants (*i.e.* actual words and phrases) of the background language, just as the formal sentential constants 'p_0', 'q_0' *etc.* often stand as abbreviations of actual sentences of the background language.

Much as in classical quantification logic we introduce:

predicate parameters: $f, g, h, f', g', h', f'', \ldots$

which take as values any linguistic expression which can serve as a grammatical predicate. Predicate parameters of the same type take as values expressions of the same type throughout a formula. As usual, such expressions may be simple or complex, and they need not be one-place. They may be self-contradictory or even nonsignificant in a given context. So, for example, the following are all admissible values of the predicate parameters: 'is red', 'are believed to be men', 'is a red hat', 'loves', 'runs', 'is probably running', 'owes', 'is a round square', 'is a purple number', *etc.* These values are English predicate constants. The logic will have corresponding formal predicate constants:

If F is a predicate parameter then F_0 is a *predicate constant.*

Here, 'F' is a metavariable which stands for any predicate parameter. Hence 'f_0', 'g_0', *etc.* are formal predicate constants, and we may in applications take, say, 'g_0' to stand for an actual constant, *e.g.* 'is green', or for an arbitrarily chosen constant.

In case it is necessary to indicate the number of places which a predicate has, we use indices. Thus, 'f^m' is a predicate parameter of m-places and '$f_0{}^m$' is a predicate constant of m-places. In general,

If F is a predicate term (*i.e.* parameter or constant), then F^m is a predicate term of m-places.

The new metasymbolism which we use is illustrated above. Thus, we take X, Y, Z, *etc.*, to stand for any subject argument and F, G, H, *etc.* to stand for any predicate parameter. Zero subscripts and rank indices are used as on the arguments and parameters themselves. As before, Φ, Ψ stand for any restricted semi-wff; Γ, Δ, for any unrestricted semi-wff; *A*, *B*, for wffcl; and Π, Θ for both wffcl and semi-wffs.

Despite the surface syntactical similarity there are important semantic differences between the classical approach and ours, particularly in respect of the specified value ranges of the arguments and parameters, and accordingly in the interpretation of the quantifiers. Indeed a serious obstacle in the way of the development of specifically significance and intensional logics is the classical account of constants, variables, variable ranges and quantifiers. For the usual semantics has been tailored to fit classical designation-style logics, and thereby, as we shall explain, to exclude significance and intensional logics. Accordingly we have to reconsider such basic notions as those of constant, variable and quantifier. Our aim, however, is not so much to define these concepts rigorously as to draw attention to relevant features of them and their roles, and to point out the special interpretations or reinterpretations of them which prove most advantageous for the development of significance logics. Some of the account we give is foreshadowed by Carnap ([2], pp. 44-46).

To elucidate semantical notions the theory of quotation has to be extended to words and phrases. We have already provided in §2.4 for the use of single quotes in conjunction with words and phrases (as well as sentences) and we now extend the use of the quotation function to variables which range over them by defining quasi-wffs analogously to partial wffs.

R5 (i) Where X is a subject argument (or subject vector or set, in the sense defined below (§3.5)), F is a predicate parameter and θ a context variable, *qu*(X), *qu*(X)(θ), *qu*(F) and *qu*(*F*)(θ) are quasi-wffs.

In an expression like *qu*(F)(θ), the scope of context variable θ is *qu*(F).

Quotation functions of subjects and predicates are explained quite analogously to quotation functions of sentences. As before, we specify that whatever value is given to the variable in a quasi-wff, the value of the function is the quotation expression of the value given to the variable. If a constant occurs in a quotation function, the value of the function is the quotation expression of the constant. Hence, if

x = This cat

then, *qu*(*x*) = 'This cat'

Similarly, if

f = is happy

then, *qu*(*f*) = 'is happy'

and *qu*(is happy) = 'is happy'.

3.2 Constants, Subjects and Predicates

A constant is a definite sentence, or part of such a sentence, which has a fixed role and meaning determined with respect to a given context. By a subject constant, we mean a subject element of a sentence; and by a predicate constant, a predicate element of a sentence. More generally, a constant of a given grammatical sort is a linguistic expression of that sort which has, with respect to a definite context, a constant or invariable role.

A constant is, thus, a word, a well-formed word string or a whole sentence. It is a verbal or symbolic expression of a given grammatical kind. Sentential constants, that is constant sentences, have already been introduced. Subject constants, which we now need to introduce, are not restricted to proper names or even to names; any constant subject expression is a subject constant, including definite and indefinite descriptions, abstract nouns or property names (*e.g.* 'Happiness' as in 'Happiness is a virtue') and so on.

We cannot, however simply take traditional grammatical classifications for granted, or we shall be in trouble; in particular we should encounter well-known logical difficulties over quantified subjects such as 'everyone' and 'nobody'. We distinguish, then, *singular subjects* from such *quantified subjects* as 'all men', 'every man', 'men', 'some men' and 'no men', and we exclude quantified subjects as singular subject constants, and accordingly as values of subject arguments, preferring instead to guarantee cross-reference and proper scope by a quantificational analysis, as in classical logic. Serious ambiguities arise and logical fallacies result if such phrases are not re-analysed (*cf.* Geach [1] and [2]). These fallacies arise, of course, because the subjects are already quantified and so, if admitted within the range of quantifiers, could be given inconsistent assignments. Thus for example, that every man is tall or not would entail, by instantiation, that no man is tall or not. And ambiguities occur because the scope of quantifiers in multiply quantified sentences is not uniquely determined and can fluctuate.

On the other hand we do not find sufficient reasons for the classical strictures on the class of logical subjects, and we find good reasons, which will appear, for pursuing a much more liberal course. As subject constants, and hence as acceptable values of subject arguments, we allow each of the following *singular subjects*:

Demonstrative and personal pronouns: *e.g.* this; that; it; I; he; she; we;
Proper names: *e.g.* Tom; Mary; Harold Wilson;
Indefinite descriptions: *e.g.* a man; a certain cat;
Definite singular descriptions: *e.g.* the King of France; the man next door;
Definite general descriptions: *e.g.* the lion (as in 'The lion is tawny'); the aeroplane (as in 'The aeroplane has a long history');
Plural descriptions: *e.g.* the inhabitants of the moon;
Class descriptions: *e.g.* the class of men; the class of cats;

Singular common nouns: *e.g.* man (as in 'Man is a rational animal');
Abstract nouns: *e.g.* whiteness; triangularity;
Adjectival nouns: *e.g.* red (as in 'Red is a colour');

Elements of the list are not exclusive; for example definite singular descriptions may be analysed in terms of indefinite descriptions and identity, and plural descriptions in terms of classes (*cf PM*, *37). Nor is the list exhaustive, but merely illustrative. Classically the most striking feature of the list is that we do not exclude as subject constants, or subject argument-values, expressions which have no existing reference in a given context, *e.g.* 'Pegasus' and 'The present King of France'; nor expressions which are self-contradictory in a standard context, *e.g.* 'the round square'; nor phrases such as 'Wittgenstein's happy stove' and 'the purple number 7', which are nonsignificant in a given context. All these are admissible, and indeed not easy to separate off. Their admissibility leads to the introduction of very wide quantifiers and entails the rejection of common tests such as reference, naming and designation tests as criteria for determining the subjects of sentences. Even if 'the round square' may be said to designate, in a wide and misleading sense of 'designate', an impossible item, no such tenuous claim can be made in the case of nonsignificant subject constants such as 'the purple number' or 'Wittgenstein's happy stove'. Such nonsignificant constants cannot, however, be excluded from any general theory, nor need they be since we have ways of handling them.

Subject constants are closed singular subjects (*i.e.* singular subjects which contain no free variables as parts) and predicate constants are likewise closed predicates, evaluated with respect to a fixed, suitably circumscribed context of application. The distinguishing feature of constants, as distinct from variables which are likewise verbal expressions, lies naturally in their constancy of role; they are closed and accordingly invariant in a given context. In case they are significant and have a sense, they have a definite sense with respect to their context. Of course subject constants need not have a sense (as logically proper names which presumably have a completely fixed reference are supposed not to have) and they need not have any reference, but whatever sense they do have must be fixed before they can be regarded as constants. There are certain subjects, usually regarded as constants, which have a fixed sense but a variable reference: examples are words like 'I', 'now', 'here', 'it', *etc.* This variability occurs when different contexts are taken, and even in single contexts which are large enough to permit changes of (person) reference. It can be eliminated by taking a small enough context. Again, predicate constants such as 'cardinal' which have an ambiguous sense with respect to the semantic dictionary of the background language only become true constants by reference to a fixed context which determines a constant sense. However, we do not exclude expressions which remain indefinite or ambiguous until contextual considerations are taken into account.

We show later, moreover (Part III), how a fixed context formal theory can be extended to admit ambiguous predicates.

The logical rules associated with constants arise directly from their constancy features. So, for example, a constant cannot be generalised upon. In $\vdash Tfx_0 \vee \sim Tfx_0$, it is not asserted that an arbitrary item satisfies a certain logical principle, but only that a particular one, x_0, does. And what holds for a given item need not hold for all. By contrast, $\vdash Tfx \vee \sim Tfx$, where 'x' is an argument, is a sufficient basis for generalising to: $\vdash$ for all (items) x, $Tfx \vee \sim Tfx$. For according to the premiss, the principle holds not for a fixed item, but for an arbitrarily chosen one. Of course, even here the generalisation is restricted by the presuppositions of the logic: *e.g.* in classical logic (but not for us) 'all items' means 'all existing items'.

The class of subjects admitted as subject constants, and accordingly as values of subject arguments, is specified so far (and will be determined in context logic) just by an enumeration of sorts of singular subjects. We aim however to provide a more general characterisation of subjects, both of singular subjects and of quantified subjects. Although the quantification packed into quantified subjects calls for logical analysis, both quantified and singular subjects count as subjects or subject expressions. Both fulfil typical subject roles. Both, for example, provide answers to such requests for subjects of sentences as "What (is, are)?", "Which one(s)?"; that is, both quantified and singular subjects are picked out by these question forms of, what is more general, quantificational tests for subject position. Both classes of subjects too are selected by aboutness tests—related special quantificational tests for subject position. Indeed (following a suggestion of V. Routley), we may characterise a subject in terms of its accessibility to quantification. This characterisation is, of course, only plausible given wide quantifiers.

A connected expression (which may occur more than once) is then *a subject* of a sentence if it is open to quantification, *e.g.* by particularisation, as singular subjects are, or it includes already (implicit) quantification, as quantified subjects do, and hence may be replaced by a singular subject, the resulting sentence being connected either by instantiation or by particularisation with the original. In other words, a subject is either a singular subject that is open to quantification, or it is the result of such quantification. To illustrate: 'Pegasus' is a subject in 'Pegasus is a winged horse', since the position is open to quantification and the sentence can be derived from 'Everything is a winged horse' and yields 'some item is a winged horse'; and by the same connections 'Everything' and 'some item' are subjects—quantified subjects. Similarly 'Nobody' is a subject of 'nobody is in sight' because it entails, by instantiation of the negative quantifier 'Jack is not in sight'; and 'the happy stove' is a subject in 'The happy stove sits smiling in the corner' since (see **QS**$_6$, §7.7) this implies that some item sits smiling in the corner. The criterion admits, of course, multiple subject sentences, but it (decently) leaves it as a question for decision whether the impersonal 'it' occurs as a subject in such sentences as 'It (the season) is winter' and 'It (the distance) is ten miles to Bungaree'. The decision turns on whether one wants, as we do, to quantify over expressions about such abstractions as seasons, distances and so forth, or not—where one *quantifies over* an expression if it is a value of a subject variable.

In admitting multiple subject sentences we are departing from the traditional grammar, enforced by traditional logic, according to which each sentence has just one subject. There are good reasons for this departure. First, the subject required by traditional grammar was never satisfactorily marked out; in fact only conflicting criteria, which failed to survive simple transformations such as the active-passive conversion or translation to other languages, were offered (see Lyons [1]). We can, however, place an order on the subjects of a sentence in terms of their order of (first) occurrence (in left to right order in English) in the sentence; and frequently the first subject in our sense coincides with *the* subject of traditional grammar. Secondly, multiple subject or relational sentences are essential for logic, and in order to avoid certain metaphysical impasses, as Russell has explained in detail (in Russell [1] and [5]).

The common distinction between logical and grammatical subject accordingly disappears; and likewise the distinction between logical and grammatical predicate. We use logic to define the grammatical subject.

Strawson ([3], pp. 156-7), has objected to Quine's characterisation of singular subjects in terms of quantification that it adds nothing to Strawson's earlier characterisation and indeed seems to rest upon it. This objection does not apply to our characterisation of subjects because subjects are not restricted in the narrow way that most of Strawson's functions (*e.g.* in his A_1 list) curtail his class of subjects. Strawson's preliminary characterisation is inadequate, since even if liberally construed (*i.e.* we count a subject in if it passes just one of his tests) it excludes too many subjects. Moreover the objection is a poor one, since the quantificational characterisation both adds to Strawson's characterisation in precision and in settling doubtful cases, and also has the usual advantages of a logical characterisation which starts from a very limited subclass, certain quantified subjects, of the expressions to be characterised. Finally the quantificational characterisation does not rest on Strawson's characterisation: his way of explaining quantificational discrepancies between subject and predicate positions (Strawson [3], pp. 236-7), hardly adequate for non-contingent predicates, or as a reason for giving up higher-order predicate logic, breaks down completely once subjects which do not refer and nonsignificant subjects are admitted.

Given the notions of sentence and subject, a *predicate* of a sentence may be defined as the remainder of the sentence when one or more subjects are deleted. It follows that multiple-subject sentences can be analysed in various ways to give different predicates; *e.g.* 'Captain Haddock was looking for a bottle of whisky' gives the predicates '... was looking for a bottle of whisky', 'Captain Haddock was looking for ...' and '... was looking for ...'. The analysis made in any given case will depend typically on the context in which the sentence occurs; for the context frequently elicits a particular subject set. Hence, full grammatical analysis depends typically both on logic (quantificational selection, subject deletion) and on context.

The criteria we have offered for distinguishing subject and predicate sharpen and improve on the familiar explanation that a subject of a sentence is what the sentence is *about*, and an associated predicate is what the sentence *says about* its subject (or subjects). The familiar test has to be handled with caution, and not merely because it confuses subjects and predicate, which in the sense concerned should be linguistic items, with what they are about. For one thing, what the sentence is about cannot be identified with what the supposed subject refers to, for the sentence may be about Pegasus and hence refer to nothing. For another, nonsignificant sentences certainly have subject-predicate analyses, but only in a tenuous sense of 'say something about' does the predicate say something about the subject: does 'sleep furiously' say something about colourless green ideas?

Predicates should be distinguished from functions. A predicate is sentence-forming on subjects, but a function combined with a subject yields a further subject. So, for example, over the field of real numbers, '... is prime' is a predicate constant which yields a true or false statement when the blank is replaced by a substantival constant—in this case a numeral, but '... + ...' and '$(\quad)^2$' are function constants which yield further subject constants, not statements, when the blanks are replaced by subject constants. A function (expression) may be defined, analogously to a predicate, as the remainder of a complex subject when one or more subjects within it are deleted. Thus there are close connexions between predicates and functions. In particular, given an equality relation of some sort (=) and a one-place function constant, it is possible to define a two-place predicate constant, or *vice versa*. Thus, given the one-place function constant '$\mathbf{B}_0$' which abbreviates 'the only brother of ...', we may define the two-place predicate constant 'B_0', '... is the only brother of ...', by means of: $yB_0x =_{\mathrm{Df}} y = \mathbf{B}_0x$. Function constants, like predicate constants, therefore usually have a sense determined by the fixed context of application.

Quotation expressions, and that-clauses constructed from sentences, are singular subject constants in the required sense. For both quote marks and the that-operator are devices which enable us to mention particular items—in the one case words, phrases and sentences, in the other, statements. Even in the case of a quantified sentence, surrounding quote marks turn it into a singular subject. Thus, the sentence "'Everyone loves someone' is true" does not itself have a quantified subject, rather it expresses a fact about a particular item which happens to be a quantified sentence. Similarly, 'That everyone loves someone is true' has a singular subject, namely the that-clause 'that everyone loves someone'.

Given our conventions concerning the use of a quotation function, which permit the quantification of variables in its scope, quotation expressions are open to quantification and so satisfy our criteria for subjects. Thus, "'Everyone loves someone' is true" implies 'for some sentence $qu(p)$, $qu(p)$ is true', and can be derived from 'for all sentences $qu(p)$, $qu(p)$ is true'. Or again, "'Short' is short" implies 'for some subject $qu(x)$, $qu(x)$ is short', and is yielded by 'for all subjects $qu(x)$, $qu(x)$ is short'. So, too, as we have seen in **SL**, that-clauses pass the tests for subjects since variables which range over them may be quantified.

That quotation expressions and that-clauses are subject terms is guaranteed formally in **CL** by the fact that neither are well-formed formulae but instead, respectively, partial (or quasi) wffs and wfc which stand in need of a predicate. Predicates which are significantly sentence-forming on such subjects are sometimes called sentence functors, but this is too limited a description and we use the more general terms 'sentence predicates' and 'sentence connectives', taking the former to include word, sentence and statement predicates. Some of these predicates are limited in their significant application to certain kinds of subjects. For example 'contains ten letters' is restricted to linguistic items—quoted words, phrases, sentences *etc.* Others are applicable to both statements and sentences (*e.g.* 'is true', 'is false' *etc.*), while others include some of these subjects in their significant range but are not limited to them. Thus 'is short' is appropriate to linguistic items and to grass; the predicate 'Tom believes' may be completed by 'the ten commandments' and 'that there is snow at the North Pole'; 'Bill sees' by 'the tree' and 'that the tree is a stringy-bark'. Again, connectives like 'and' are sentence-forming on sentences and predicate-forming on both sentence predicates as well as ordinary predicates. In general, a predicate or connective is a sentence predicate or connective when the subjects with which it significantly combines *includes* quotation expressions or that-clauses.

For the purposes of statement logic and the formal development of **CL** in Chapter 2, we limited the construction of that-clauses to those obtained from statement-yielding sentences. These are the wfc of **CL**. But we now have to consider the formation of many-valued sentences of the form FX, *i.e.* semi-wffs rather than wffcl, where the subject X may be a that-clause and F an appropriate sentence predicate; and from this point of view it is no longer necessary to restrict the construction of that-clauses. For given a predicate such as 'Jack said' or 'Jack believes', the subject may well be a that-clause obtained from a sentence which is not statement-yielding. Thus, if 'Mary kissed my thoughts' is nonsignificant in c_0, we may form the sentence 'Jack said that Mary kissed my thoughts' which is itself significant in c_0 even though its subject, the that-clause, is not. To restrict the formation of that-clauses to those constructed from significant sentences, therefore, would be to restrict the application of the logic in a serious way. We take it, then, that if Π is either a semi-wff or a wffcl, §Π is a subject term and F§Π is a semi-wff, whether or not Π is significant. Such subject terms include the previously defined wfc but some of them are not wfc.

We therefore admit quotation expressions and that-clauses as values of the subject arguments, and sentence predicates as values of the predicate parameters. The sentences which can be constructed by means of these expressions satisfy the same general principles of quantification logic as those constructed from other subjects and predicates; in particular, such principles as the distribution of quantifiers, instantiation and generalisation. The admission of these subjects and predicates does not introduce any new problems. The subjects differ from those already admitted only in the way that other sortal subjects differ from each other (for example in the way that numerals as special subjects differ from the names of people), and since we already have such sortal differences there is nothing new here. So far as quantification logic is concerned, all such differences can be isolated in the instantiation scheme. Again, the predicates may differ from others in their significant range of application, though as we have seen there may often be overlap, but this is true of many different sorts of predicates which are commonly accepted as values of the predicate parameters. Of course it will be possible to select a subject constant 'x_0' and a predicate constant 'f_0' such that 'f_0x_0' is non-significant (*e.g.* 'That Tom is tall is happy') but again, the construction of nonsignificant sentences is already possible and it is not in any case our concern to exclude them but to handle them explicitly within the logic. In general, sortal differences are significance matters and should be treated as such in significance logics. Since sortal subjects can be treated unproblematically it is logically advantageous to adopt an analogous treatment in the case of quotation subjects and that-clauses even though it may be a borderline issue whether sentences like 'That Tom is tall is happy' are grammatically ill-formed or just nonsignificant in standard contexts.

There are, then, no disadvantages in extending the range of the subject arguments and predicate parameters in this way; but there are clear advantages. For the quantification theory will now apply to sentences such as 'That Tom is tall is interesting' as well as to those like 'The book is interesting'. In this way the application of the logic is extended well beyond the limits of application of classical quantification theory.

The decision to take 'that' as an explicit part of the subject implies a departure from the usual treatment of what are sometimes called intensional predicates. The usual analysis of a sentence like 'Tom knows that there is no King of France' is to take 'knows that' as the predicate and 'Tom' and the *sentence*, 'There is no King of France' as subjects; or alternatively, 'Tom knows that' as the predicate and 'There is no King of France' as subject. This leads to problems in the formalisation of identity postulates, as we shall see presently, since predicates like 'knows that' satisfy different conditions from ordinary (extensional) predicates like 'is red'. It also entails that 'knows' and 'knows that' are different predicates. For 'knows' may be applied significantly to non-sentential predicates such as 'Bill Smith'—as in 'Tom knows Bill Smith', while 'knows that' can only be applied significantly to sentences. Our approach, as we hope to show, has many advantages for the theory of identity, and it also enables us to say that the same predicate, 'knows', occurs in 'Tom knows Bill Smith' and 'Tom knows that there is no King of France'. The difference, for us, is in the subjects. In the one case 'knows' applies to "ordinary" subjects, in the other, to that-clauses; and in each case it applies significantly and satisfies the same conditions. There is, however, nothing in our approach which prevents the adoption of absorption principles in given cases. Thus, for example, just as '*T*' absorbs '*qu*' by abbreviating '*tru qu*', so too we could take '*K*' to absorb '§' by abbreviating '*kn*§' where '*kn*' is 'knows'. Then, for example, although the correct representation of 'Tom knows that there is no King of France' is 'Tom *kn* §p_0', we could *write* it as 'Tom Kp_0', and there is sometimes an advantage in doing so. But this is not to take '*K*' as a primitive intensional predicate and 'Kp_0' as a semi-wff with a sentence as subject. In general, formulae such as '*fp*' are not semi-wffs unless '*f*' can be expanded, by an absorption principle, and the formulae written as '*g*§*p*' or '*g qu*(*p*)'.

The use of absorption principles enables us to make explicit the role of 'that' in natural language. For one of its main features is its apparent dispensability. By transforming English sentences in which it occurs, it can almost always be eliminated. Thus, 'It is true snow is white' and 'Tom believes the number seven is happy' are perfectly acceptable idiomatic variants of 'It is true that snow is white' and 'Tom believes that the number seven is happy'. In such cases, '§' has been absorbed by the predicate. Hence we may formalise them respectively as '$\mathbf{T}'r_0$' and 'Tom Bp_0', where '**T**'' is '**T**§' and '*B*' is 'believes §'. The absorption principles then stand as (grammatical) rules on which the transformations depend.

In general, then, we take it that sentences, as opposed to quoted sentences and that-clauses constructed from them, are not values of the subject arguments. That is, sentences are not subjects; and variables which range over them are not open to quantification. For sentences are grammatically complete in themselves and do not stand in need of a predicate. They can however be *converted* to subjects, *i.e.* subjects can be constructed from them, by writing them within the scope of '*qu*' or '§'. The complex expressions so formed then stand in need of a predicate and are quantifiable. It follows that we completely reverse the classical position by refusing to accept quantification over unquoted sentences and by permitting quantification under quotation. Hence, as foreshadowed in §2.8, we do not accept an extended sentential logic which allows formulae such as '$(Up)(p \vee \sim p)$', since these contain no predicates. We do, however, allow '$(Uqu(p))\,(Tp \vee Fp)$'. For here there is both a subject, '*qu*(*p*)', and predicates, '*tru*' and '*fal*', since it may be written as '$(Uqu(p))(tru\, qu(p) \vee fal\, qu(p))$' and regarded as a special (restricted variable) case of '$(Ux)(trux \vee falx)$', or more generally, of '$(Ux)(fx \vee gx)$'.

Several restrictions on the way we formulate and develop logic result at once from the characterisation of subjects and predicates. In particular, contrary to classical higher-order logics:

(i) *No expression can be at once a subject and a predicate.*

Otherwise the same expression would have to be both accessible to quantification and not. A subject may be a proper part of a predicate, but since subject and predicate have different and exclusive sentence roles, an expression cannot be both at once. A predicate cannot be a subject, or we should end with such gibberish as 'chimes (chimes)' and 'is hard is hard'. By a similar substitution argument, a subject cannot be a predicate, or we should have as grammatical sentences such unlinked strings of names and subjects as 'Julius Caesar Rome Tuesday', 'the lion the unicorn' and 'Julius Caesar resilience fortitude', and such quantificational garbled strings as (Frege's) 'for some *x*, *x* Julius Caesar'. It follows that expressions of the form X(Y) and F(G), and in particular of the form X(X) and F(F), are ungrammatical and accordingly not significant. This may appear to put us in agreement with the requirements of type theory, but the appearance is shortlived. For on the one hand expressions like $F_{n+1}(F_n)$, where the typal subscripts indicate that the one-place predicates are properly typed, are likewise grammatical nonsense. On the other hand there is nothing to stop us forming, by sentence transformations, subjects from predicates or predicates from subjects. By doing so we shall arrive at sentences the type theorist equally rules out, not just 'Transitivity is transitive' and 'Redness is red' but the true 'Abstractness is abstract' and the debatable 'Superficiality is superficial'.

(ii) *Predicates cannot be quantified over.*

For quantification is just what selects and applies to subjects, and predicates are the remainder after subject selection. Of course predicates may contain proper parts, subjects, which are accessible to quantifiers; and by quoting a predicate '*f*' a new subject '*qu*(*f*)' may be formed. The fact that we hereby apparently rule out all classical predicate logic beyond first-order quantification theory, though it may please the nominalist, will be turned as an objection against us. For cannot sentences like 'Every item has some property' and 'Napoleon had all the features of a great general' occur in arguments which can be logically appraised? Certainly, but we do not need or want quantification of the predicate place to handle them. The effects that logicians think are obtained by quantification of the predicate are in fact almost always obtained by replacing the predicate by a property-expression joined by the expression 'has' or 'instantiates' with the former subject of the predicate, and by quantifying over this newly obtained *subject*. Thus, in the cases in question '$f(x)$' is transformed (converted) to 'x has f-ness', which we symbolise '$x\ i\ \lambda f$'. The resulting sentence has *two subject* places both of which are open to quantification; and quantifying both appropriately we obtain '$(Ux)(P\lambda f)(x\ i\ \lambda f)$' *i.e.* 'for every item x and for some property λf, x has λf'. Now compare what is done in understanding such misleading second-order notation as '$(x)(\exists f)f(x)$': '$f(x)$' is always paraphrased so that the predicate is replaced by a property-expression which occupies a newly-found subject position; it is always read 'x has (or instantiates) f-ness (or the property of being f)' , and it seems impossible to represent the quantified expression in English without making the conversion. Related points hold good against standard formulations of extended sentential logic, which contain wff of the form '$(p)p$', read literally 'for every (sentence)p, p', but rendered grammatical and intelligible only by insertion of the predicate 'is true' or by paraphrasing as 'every proposition is true'. In order to bring 'p' into a quantificational position it has to be treated as a subject; but without a predicate to complete it no well-formed sentence on which to operate has been provided.

Restrictions (i) and (ii) help decide the way quantified significance logic should be developed. The development will be analogous to developments of first-order axiomatic set theory, as distinct from that of higher-order type theory, characteristic features of which (i) and (ii) prohibit. Corresponding to quantification logic, *i.e.* first-order predicate logic, which is admissible according to requirements (i) and (ii), we shall develop various significance quantification logics. But we cannot intelligibly extend these logics to higher-order significance logics; these formalisms at best represent a thoroughly confused attempt to introduce further subject terms into quantification logic. Instead we extend quantified significance logic, first by admitting sentence predicates which satisfy additional postulates, and second, and independently, by admitting further subject expressions and by developing the logic of subject conversion, *i.e.* the principles which govern the formation of subjects by property abstraction, by quotation, and so on. In particular, we may add to quantified significance logic the two-place instantiation predicate 'i', read 'has', and the attribute abstraction operator 'λ' and develop the significance theory of attributes and instantiation, much as set theory elaborates the theory of classes and membership. This method has the advantage, for philosophical application, not only of staying much closer to discourse, as we have already observed, but of facilitating the expression of sentences that are either excluded or incorrectly rendered nonsignificant under type logic. Since among the latter are many sentences of important philosophical theories, such as Plato's theory of ideas where self-predication is fundamental, this is no minor matter. (On this see also Fitch [1]).

Within extended quantified significance logics we obtain, nevertheless, a syntactical analogue of second-order logic. For second-order logic can be viewed syntactically as adopting a semantically confused abbreviated notation: roughly, λ-abstraction notation is deleted from λ-restricted quantifiers where they occur, and the combination $(X_1, ..., X_n)\ i\ \lambda F^n$ is transformed wherever it occurs to $F^n(X_1, ..., X_n)$. Thus second-order logic abbreviates '$(Ux)(P\lambda f)(x\ i\ \lambda f)$' as '$(Ux)(Pf)f(x)$'. The two logics do not however keep step: *e.g.* the restricted subject variable 'λf' can be replaced by a general subject variable, 'y', yielding the unproblematic '$(Ux)(Py)(x\ i\ y)$', whereas replacement of 'f' by 'y' gives only the gibberish '$(Ux)(Py)\ y\ (x)$'.

Since we can express unambiguously within subject-enriched quantified significance logic everything that can be grammatically expressed in higher-order significance logics, essentially by reversing the rough mapping above, and since we can express as well assertions that cannot be expressed in higher-order typed logics, we shall drop the confused higher-order notation altogether.

The remainder definition of *predicate* carries no commitment to a property abstraction thesis of the usual sort: in symbols,

$$(P) \qquad (Pf)(Ux_1, ..., x_n)[f(x_1, ..., x_n) \simeq \Gamma(x_1, ..., x_n)],$$

where 'f' is not free in $\Gamma(x_1, ..., x_n)$. Indeed this thesis contravenes (ii) by assuming quantification over predicates. We are committed to saying that every sentence with subject places deleted provides a predicate *qu*(f), a claim often conflated with and erroneously used to justify a property abstraction thesis, But our thesis is, in standard logical parlance, a metalogical thesis concerning syntax, and it does not reduce to object language thesis (P), since *qu*(*qu*(f)) is never the same as *qu*(f). Nor are we in any way committed to what (P) tries to assert, namely a *property* abstraction or conversion principle,

$$(Q) \qquad (Pz)(Ux_1, ..., x_n)[(x_1, ..., x_n)\ i\ z \simeq \Gamma(x_1, ..., x_n)],$$

or to (Q) with the bound subject variable 'z' replaced by the restricted subject variable 'λf'. Given any sentence the remainder upon deletion of subject constants is a predicate constant; but in this there is no commitment to the wholesale generation of property constants, by conversion principle (Q), from predicate constants. Neither principle (Q) nor the further shuffles made in the impredicativity paradox argument are supported by our account of predicates.

Our subject-predicate theory resolves several traditional philosophical puzzles, the so-called problems of predication, which the subject-predicate relation has generated. In particular, the problem of the link that combines subjects and predicates and the "glue" that holds sentences together is resolved at once as a consequence of the definition of 'predicate'. Because the predicate is the remainder of a whole sentence the predicate itself links the sentence subjects together. As the predicate is unsaturated, the remainder, it holds the sentence together: no further glue, no further relation is needed—in contrast to the traditional case of two subject terms, *e.g.* 'men' and 'mortal', where a predicate is needed to link the subjects. Nor is there any mystery about Frege's idea that predicates are unsaturated: a predicate is unsaturated just because it is in need of supplementation (Frege [1]), and it is in need of supplementation just because it is the remainder of a sentence.

3.3 Place-holders, Arguments, Parameters and Variables: Their Values and Ranges

Some recent accounts of variables and variable values (*e.g.* by Quine in [2] esp. p. 192 ff.) are simply tailored to fit extensional two-valued logics. For reference assumptions concerning both existence and identity are built into the accounts. Not surprisingly the accounts do not extend satisfactorily beyond the Establishment systems to other systems; but rather surprisingly, this has sometimes been used as a reason against the development of non-extensional or many-valued systems. Those who attempt to set up such logics using standard accounts of variables arrive at bizarre conclusions about the notions they seek to elucidate (as, for example, with the modal and significance essentialisms explained below); and those who attempt to set up non-classical systems using different accounts of the basic concepts are either said to be unintelligible or at best to be misusing and not understanding such concepts as those of a variable and a variable value. This self-supporting case for limiting attention to classical logic is quite insidious. By offering our own explanations of key semantical notions we aim to anticipate and avoid this sort of objection.

Our account takes the following directions: We first explain place-holders; place-holders hold or link places for expressions of a given grammatical sort. Place-holders, like pronouns, also serve as place-linkers. Among place-holders we distinguish those that hold subject places and those that do not, but hold predicate places. Those that do not hold subject places we call *parameters*, those that hold subject places *arguments* or *proper variables*. If an argument is open to quantificational operations it is *free*; otherwise, if it is already part of a quantified or descriptive subject at one of its linked places, it is *bound*. Bound variables are the only bound place-holders; we retain such inapposite terms as 'bound variable' because they are so well entrenched.

Arguments and parameters are introduced, then, by way of place-holders. Any sentence expression or component, not excluding the whole sentence, is called a sentence factor (the terminology is Ryle's [2], pp. 68-69). The remainder of the sentence, if any, when the sentence factor is deleted and replaced by a blank symbol or a symbol is called a sentence frame, an open sentence, or a sentential form or function; what fills the gap, including blank signs such as '—' and '...', are called *gap signs*. A similar procedure can also be adopted when expressions which are not sentences are taken as a starting point. In this case we obtain sentence factors and expression frames or forms. This extends the analogous account of forms given by Church ([4], p. 10).

For example, given the expression 'the man who broke the bank at Monte Carlo', we obtain under one particular division the sentence factor 'broke the bank at Monte Carlo' and the form 'the man who *f*', choosing '*f*' as the gap sign, or place-holder for factors. The same terminology is used when several sentence factors are extracted from a sentence or expression. Thus, in this example, the factor 'man' could also be extracted leaving the form 'the *x* who *f*'. When several factors are extracted, the extraction may be carried out in one stage instead of single successive stages, provided certain factors do not occur as components of others and the role of gap signs in cross-indexing is preserved. Furthermore, the same factor may be extracted from more than one place in a given sentence. In the resulting frame, sameness of factor is shown by sameness of gap sign. In this way gap signs serve to link places. A gap sign is not so far a place-holder.

A gap sign which holds a place for a range of, or certain of a range of sentence factors, is a *place-holder*, and it *links* the places it holds. Consider for instance, the sentence form

(1) *x* was founded by Plato

obtained from the sentence,

(1a) Plato's academy was founded by Plato,

by extracting the factor 'Plato's academy'. What sentence-factors does the gap sign '*x*' hold places for other than the factor extracted from (1a)? If '*x*' holds a place only for 'Plato's academy' then it is a constant-symbol which abbreviates the factor which it replaces. But if '*x*' also holds a place for other factors as well, factors such as 'Hernando's hideaway' and 'slavery', then the letter is called a place-holder, and in this case a subject argument, and the various factors constitute its constants-range. Though a place-holder or variable, the symbol itself is fixed and does not vary. What it does, in case it is not bound, is to stand in for, or hold a place for, each and every one of the factors in its constants-range.

A place-holder that holds places for singular subjects (in places none of which are proper parts of quantified or descriptive subjects) is a free subject argument or variable: it is characterised by its constants-range. A place-holder that holds places for predicates is a predicate parameter. An argument, or a parameter, of a given grammatical sort is, like a constant, a linguistic expression of that grammatical sort. The sort may be made specific in the notation, as in the case of place-holding expressions like 'the first item' and 'the other', or may be separately specified as when we stipulate that '*x*' and '—', but not '*f*', are subject variables.

A parameter differs from an argument in logical role. A free argument or variable is open to quantificational operations such as generalisation and particularisation, those operations in fact which distinguish subjects; but a parameter is not, because it does not hold subject places. The *OED* gives the point of difference when it characterises a 'parameter' as a 'quantity constant in the case considered, but varying in different cases'. A parameter holds a place for any one of its factors in a given context, but not for every one. For a predicate parameter is not open to subject-forming operations such as quantification, without conversion from its predicate role: 'every *f*' like 'every runs' is gibberish *i.e.* ungrammatical in a logically serious way. Thus for parameters the notion of freedom applicable to subject variables—essentially that of being open to quantificational transformations—is not significant. A sentence form containing free subject variables which is asserted as true is normally understood as an implicit generalisation, as asserting the sentence for any factor for which the variable holds a place. Hence when the form is asserted as true any arbitrarily chosen factor from the range of the variable can be substituted for the variable. It is this too which justifies a principle of generalisation. For since a free variable stands in for every constant in its range, any assertion of the form containing the variable amounts to an assertion of every sentence which can be obtained by replacing the variable by a constant. Satisfying this condition in fact serves to characterise the notion of freedom of a variable.

Once bound up as part of a quantified subject however, it is no longer permissible to substitute an element of the constants-range of a variable for the variable, that is to replace a place-holder by one of the factors for which it holds a place. Though we can *instantiate* 'Every *x* such that *x* is a natural number is less than some *y* such that *y* is a natural number', we cannot *substitute* constants, such as 'five' for any occurrence of the variables (including those which occur in the quantifiers) without destroying the sentence, any more than we can substitute 'Socrates' for 'man' and its linked pronoun 'his' in such sentences as 'Every man is an island and his home is his castle'. Subject variables can perfectly well serve as place-holders for subject parts of quantified subjects and subject parts of complex subjects, and for subject stand-ins such as pronouns which are linked with these subject parts. In other words, variables may operate as cross-indexing or place-linking devices which link place-holders, which are specifically bound by operators, with other occurrences of the same place-holder in a sentence. In this respect variables replace and extend the function of pronouns and relatives of ordinary language. For example, in 'the item *x* such that *x* is green' and in 'every car *x* is such that *x* is green', the variable '*x*', though it has appropriate constants-ranges, is not open to replacement by factors in the ranges—

consider the gibberish that would result in each case upon simultaneous substitution of a constant for the two occurrences of '*x*'—but the first occurrence of the variable can be deleted and the second replaced by the linking particle 'it'. Such variables, that are controlled by quantifiers and descriptors or that replace linked pronouns, are said to be *bound*, and to bind the places they hold; they are bound *by* the controlling operators.

The differences between arguments and parameters reflect the differences between subjects and predicates. It is not usual to insist on the differences between arguments and parameters, and often it is not important to do so, *e.g.* at the purely sentential level. It is in fact usual to call all place-holders and place-linkers *variables*. We find advantages in sticking to customary usage though we wish to avoid customary associations, such as the usual ideas of values and ranges. We shall, however, only call place-holders and place-linkers either filling subject-roles, as arguments do, or which can be given a subject-role without conversion, *proper variables*. As to the last, a logic may use proper variables without assigning them a full argument role, as happens in zero-order logics, *i.e.* predicate logics without quantifiers.

The account given has some minor and inessential special features. First, the system variables and parameters we shall use are place-holders for grammatical or well-formed factors; they are not just string variables. Second, all the factors for which they hold places derive (ultimately when other variable blanks, if any, are replaced by factors) from grammatical or well-formed expressions. This second feature is intended to have the consequence that all factors for which a place-holder stands are of the same grammatical class, according to the background language, and that the place-holder may therefore be assigned to the same grammatical class. So a place-holder which holds places for subjects is a subject argument, and, where unrestricted, it has a constants-range consisting of all subject expressions. These two limitations are not essential to the concept of a place-holder. Since place-holders are distinguished initially according to the strings or factors for which they hold places, it is possible to introduce general place-holders which stand for any string, whether well-formed or not, and whether or not the factors for which they hold places are derived from well- or ill-formed strings. Extra-systematic place-holders (*e.g.* metavariables of various kinds) are like this. Indeed we could, it seems, begin with general string place-holders, and then distinguish, using logical features of the sort we have noticed (*e.g.* significance of quantification of a component), such notions as that of a subject variable. This approach could form part of a programme for handing grammar back to logic.

Unlike a subject constant, a subject argument never has a sense and is not about a particular item; and it certainly does not have a fixed meaning. The distinguishing feature of a variable and of a parameter, as distinct from a constant, is that it has various ranges of values, its main or primary range being its constants-range which is given by the class of constants of the same grammatical sort as it, and is such that any member of the appropriate class of constants may be substituted for it. By a somewhat artificial extension, one may perhaps speak of the meaning of a variable being constituted by the meaning of the constants which may be substituted for it; but this is unnecessary, and, in case nonsignificant constants occur, perverse.

An element of the constants-range of an argument, variable or parameter is called a *value* or argument-value of the argument, variable or parameter. This generally coincides with what Carnap ([2], p. 42) calls a value expression; it differs from what Quine [2] insists should be a value, namely an "object" designated by the expression. Thus, given that 'is green', 'is five feet tall', *etc.*, are constants in the primary range of a predicate parameter '*f*', then these *linguistic expressions* are values of the parameter '*f*'. The constants-range of a variable or parameter is then the class of its values. A variable or parameter is only grammatically restricted if its class of values consists of all expressions of its grammatical sort of the background language. In other words the range of such place-holders is controlled only by grammatical requirements. There is of course nothing to prevent the introduction of grammatically unrestricted place-holders, as we mentioned, only they will satisfy few postulates of immediate interest. We consider first subject arguments restricted only by the requirement (of §3.1) that their values are singular subjects; such arguments we call *unrestricted* subject arguments, even though they are not completely unrestricted even grammatically. The constants-range of these unrestricted subject arguments is called their *primary range*. Subject arguments whose values form a proper subclass of the primary range are called *restricted* subject arguments. We shall see that the subject arguments of classical logic are heavily, and unnecessarily, restricted.

What we have so far spoken of as the values of a subject argument are substitution values, that is, words or phrases of a certain sort which can be substituted for the arguments; and we have called the set of values the constants-range of the arguments. It is customary in classical logic, however, to introduce instead the designation range, which consists of the set of entities designated by the constants. We do introduce such ranges, as *derived* ranges, below. But we find that there is no simple way of introducing them immediately, or without a good deal of preliminary adjustment and scene-setting. First, because the reference of a subject may be different in different contexts, so we cannot regard any word or phrase as having a fixed designation pinned on to it. Secondly, a designation range, to be comprehensive enough, would have to "contain" non-existing items like Pegasus, impossible items like the round square, and nonsignificant items like the purple number seven. We meet this problem by defining explicitly what it means to say that a given subject term designates or has a reference in a context. The specification of a reference range can then be made in terms of the use of a constant to refer; that is, in terms of the class of constants and the logic of reference. The specification of an independent reference range is only necessary as an artificial device in logics which make no provision for the use of a constant to refer in a context.

There are, then, fundamental differences between the classical account of values of *subject* arguments and our account. For substitutions for the free individual variables of classical logic are usually restricted to proper names which name existing items, and the values of individual variables are not constants at all but the entities denoted by constants in a standardised context. Unlike the individual variables of classical predicate systems, subject arguments are not restricted to the (proper) names of individual entities. They stand in for any singular subject factor including, as we have noted, descriptions, abstract nouns, noun phrases *etc.*; and such elements, though they must be grammatically well-formed, need not be significant. Even if we restricted them to significant expressions, however, we still would have a quite different set of constants from those of classical logic. There are several reasons why we regard our divergence from the classical account as both necessary and justifiable.

Because of the standard limitations it has to be assumed classically, in order to treat some of that vast collection of sentences which otherwise would lie beyond the scope of logic, that any sentence with a "complex" subject can be analysed in such a way that the complex subject disappears and the new sentence so obtained can be represented entirely by means of proper names, individual variables restricted in range to proper names, predicate variables and constants, standard sentential connectives, the quantifiers, and (sometimes) identity. This, in effect, is the assumption of an ideal language, which is discussed in Chapter 4. The paradigmatic classical analysis is provided, for the particular case of singular definite descriptions, by Russell's Theory of Descriptions. According to this, the sentence 'The present King of France is bald' becomes on analysis, 'There exists exactly one individual who is the present King of France and is bald'. The simple sentence with a complex subject has been replaced by a complicated compound sentence and the initial subject has become a predicate. Since the compound sentence implies that the present King of France exists, it is false; and indeed every primary statement about the present King of France, except the statement that he does not exist, is false. The analysis therefore fails if this implication is not accepted; yet it cannot be in the case of all singular definite descriptions. Compare, for example, 'The winged horse is a mythical creature', or even 'Pegasus is a mythical creature', since 'Pegasus' would have been taken by Russell to be a disguised definite description rather than a proper name. Are we to say that these sentences are false because on analysis they imply the existence of Pegasus and it is known that it does not exist? (For a detailed criticism of Russell's theory of descriptions and related theories, see Routley

[7] and Linsky [1]). An alternative proposed by Strawson [1] is that 'The present King of France is bald' does not imply but presupposes the existence of the present King of France. It cannot therefore yield a true statement if there is no such person, but neither does it yield a false one. Rather, it fails to yield a statement at all, true or false, though it is significant; it is, that is to say, incomplete. This too, however, fails to do justice to 'The winged horse is a mythical creature'. For it does yield a true statement just because Pegasus is a mythical creature. But this does not mean that it simply affirms that Pegasus does not exist, though it does imply it. For that Pegasus is a mythical creature is not implied by 'Pegasus does not exist'; not all non-existent items are mythical creatures.

These facts are well known, and certain of the difficulties can be overcome by introducing a particular quantifier which does not have existential commitment. We cite them here simply to emphasise the danger present in the assumption that there is always a uniform analysis, in terms of the standard apparatus of quantification theory, for every sentence containing a complex subject of a given sort. The danger is, that since the truth conditions for the standard connectives can be precisely given, so a uniform analysis may imply that every sentence containing the same complex subject has the same truth conditions. But of course, complex subjects can be used, as proper names can, in sentences used to make true statements, false statements and non-significant utterances, as well as in sentences which, though significant, fail to yield statements. Thus,

'The winged horse is a mythical creature' is true;
'The winged horse is a mythical elephant' is false;
'The winged horse is factorisable by twenty-seven' is nonsignificant;
'The winged horse is heavier than every unicorn' is incomplete.

Any uniform analysis which brings each of these out as having the same truth conditions (*i.e.* such that all are false or all are incomplete) is inadequate. This does not of course totally preclude the possibility of some, as yet unrevealed, non-uniform analysis; but it makes a satisfactory analysis seem very unlikely.

The lack of such an analysis need not, however, impede logical investigation. For we can treat singular definite descriptions equally with proper names as values of subject arguments in their own right (without any disasters: see Church [3]). The only difference is that in case the description has no existing reference, questions about the completeness, significance and truth and falsity of sentences in which they occur are settled entirely in terms of the signification of the sentence and are reference-independent. But this is so for some proper names: *e.g.* 'Pegasus'.

Yet if this is true of singular descriptions, which have received so much logical attention, it is equally true of other complex subjects which have received so little. There is for example a large class of sentences which take as subjects general definite descriptions, abstract nouns, singular common nouns and adjectival nouns, which resist uniform analysis in terms of the standard connectives of quantification theory. Thus consider:

Red is a colour
Whiteness is a property
The lion is tawny
Man is a rational animal.

It has sometimes been suggested that such sentences are merely idiomatic variants of ordinary universal sentences (Ryle [1]) or logically necessary universal sentences (Sellars [1]). In certain cases, this may be so. But it is not true in general. Does 'Red is a colour' mean the same as 'All red things are coloured'? The singular statement implies the universal, but it is not clear that the converse holds. For given the plausible assumption that all material objects are coloured, it will be true that all square things are coloured, but this will not imply that square is a colour. To get the inference from 'All red things are coloured' to 'Red is a colour' we need to know that 'red' is a colour word; that is, we need to know that red is a colour. And clearly, 'Whiteness is a property' neither implies nor is implied by 'All white things are properties'. Again, 'Man is a rational animal' does not mean the same as 'All men are rational animals', though it implies it, for it tells us something about biological classifications which the universal sentence does not. In the case of sentences which have general descriptions as subject terms, the discrepancy is as great. 'The lion is tawny' does not mean the same as 'All lions are tawny' since it affirms a characteristic feature of lions which need not be satisfied by all lions. Compare 'The lion is dangerous' and 'All lions are dangerous': the latter would be falsified by the existence of a tame pet lion, the former not. And it should be clear that 'The aeroplane has a long history' is not equivalent to 'All aeroplanes have a long history'.

In fact the main reasons for insisting on some or other of these inadequate analyses are removed given our account of subjects, arguments and quantifiers. For then descriptions can be taken unproblematically as values of arguments.

Quite apart from the fact that it seems to be impossible satisfactorily to analyse complex subjects away in other than a quite piecemeal way, and at the same time preserve truth let alone meaning, there are further reasons why it is important for us to admit them directly as values of the subject arguments. For the significance of a sentence which contains a proper name or a pronoun as subject cannot be fully assessed except by way of identifications made in terms of descriptions. Thus, consider 'Tom is happy'. This is significant in a given context if 'Tom' is being used to identify a human being, or perhaps an animal, otherwise not. To assess the significance, therefore, it is necessary both to be given the sentence and the aboutness of the subject term; and this last can either be given by an explicit statement of what the aboutness is (*e.g.* by means of a statement of the sort: 'Tom' is being used to refer to the man next door), or, equivalently, by means of an identity statement (*e.g.* Tom is the man next door). And this is so with all cases of context-dependent significance. The procedure is not, however, restricted to cases in which the proper name has an actual reference. For the identification in terms of a description can be made even though, in the context, neither the name nor the description has an actual reference. If, for example, 'Tom' is being used to mention the one and only man who is twenty feet tall, we may conclude that the sentence is significant (though it is no doubt incomplete). By contrast, if 'Tom' is being used to name the match box on the table or the real square root of -1, the sentence is not significant independently of the fact that the matchbox actually exists and the real square root of -1 does not. For the determination of significance, what matters is what the subject term is about; it does not matter whether what it is about exists. What is required is an identification in terms of some description; and indeed it is not even necessary to have a definite description, an indefinite description is usually enough. The fact that a description of some sort is required, however, means that the procedure for assessing the significance of a sentence containing a proper name or pronoun as subject really amounts to an assessment of the significance of a sentence which contains a description as subject. Given 'Tom is happy' we use the identity 'Tom is the man next door' to obtain the further sentence 'The man next door is happy', and the question of whether the original sentence is significant is now the question of whether there is a significance clash between the complex subject 'the man next door' and the predicate 'is happy'. Sentences with complex subjects must therefore be admitted in their own right if context-dependent significance is to be evaluated; and from this point of view, they are much more informative, and indeed more basic, than sentences which have pronouns or proper names as subjects. Of course this procedure would not of itself preclude some analysis of the sentence with complex subject in favour of a sentence without it, but from the point of view of significance evaluation such analysis is unnecessary and merely complicates what is initially a simple technique.

The fact that the existence of an item corresponding to the subject term is irrelevant to the problem of significance explains in part why we are willing to admit as values of the subject variables, names and descriptions which have no (existing) reference. In any case, given our starting point of the use of sentence tokens in a context, it is impossible to specify in advance which words and phrases have a reference and which do not. To assume that 'Pegasus' has no reference is to assume some sort of standard context in which 'Pegasus' is always and only used to name a winged horse. But there is no reason whatever why, in ordinary contexts, 'Pegasus' should not be used to name, say, a living horse or some other creature. Similarly, words like 'Tom', though ordinarily used to refer to actual items, might equally well be used of items which do not exist. And this is true also of descriptions. Even a phrase like 'The man next door' fails to have an actual reference in a given context if there is no one man living next door, but this is not a reason for excluding it as a subject or for limiting the application of the logic to contexts in which it has a reference. And as much is true of "impossible" descriptions and nonsignificant descriptions. Both must be admissible values since no description is uniformly self-contradictory or nonsignificant in every context. 'The purple number seven' might well be used to refer to the house across the road, and 'the round square' might be the name of a city square which happens to be circular. But even given standard senses, it is still necessary to admit such phrases as values of the subject arguments since they contribute to the assessment of significance.

We have concentrated so far on unrestricted arguments which admit all values in their primary range. Also important, however, are *restricted arguments* or *variables* (and likewise restricted parameters) which have as constants-range only a proper subclass of the range an unrestricted argument occupying the same place would have; *i.e.* restricted arguments are positively restricted by some further requirement or condition which restricts their constants-range. Thus a subject argument might be restricted to that subclass of its primary range for which it yields only significant sentences with respect to a certain predicate or predicates (its s-range), or to that subclass of its primary range which are, in the context, about entities; and so on.

The general distinction drawn here between restricted and unrestricted variables should not be confused with that drawn in *PM*. The unrestricted variables there introduced are a very special kind of restricted variables, namely those which are restricted to their significant values (s-ranges). Subsequently (in Chapter 7) such significance-restricted variables will be introduced in the systems we consider, and significance-restricted sentential variables will be introduced in weak sentential S-logics, extending their use in Chapter 2. These *S*-restricted variables '*r*', '*s*' *etc.* have a constants range restricted to significant sentences. In stronger S-logics a way of eliminating these restricted variables will be demonstrated.

A restricted place-holder may be either *implicitly restricted* by requirements of a context or *explicitly restricted* on each occasion of its use. For example, our sentential variable '*r*' is implicitly restricted in the context of this book by the requirement that it holds places just for significant sentences. A variable that is normally unrestricted may also be restricted in a particular context, for example for a given system or model or in the application of a system to some example. An explicitly restricted variable, in contrast, carries its restriction with it. For instance in place of '*r*' we may choose sometimes to use the symbolism '$p \mid Sp$', read 'p restricted by the requirement that Sp', or '$p \ni Sp$' read 'p such that (for which) Sp'. In these uses the constants-ranges are sometimes said to be *conditionally restricted*—in the example given, by a significance requirement.

Of course the explicitness of explicit restrictions is a relative matter; in all cases of use of variables there is scene-setting. Implicit restrictions can often be eliminated by way of explicit restrictions, though it is often not convenient to make the elimination; and explicit restrictions using special connectives like Peano's *such that*, symbolised '$\ni$', can sometimes be eliminated in favour of more familiar connectives. For example, it is commonly contended that restricted variables can be generally eliminated within the framework of quantification logic as follows: take the universal closure of any formula containing explicitly /-restricted variables; then replace each occurrence of '$(Ux \mid B(x))\, A(x \mid B(x))$' by '$(Uy)(B(y) \supset A(y))$' and each occurrence of '$(Px \mid B(x))\, A(x \mid B(x))$' by '$(Py)(B(y)\ \&\ A(y))$'. Thus it is claimed that '$(Uy \mid \text{greek}(y))$ mortal $(y \mid \text{greek}(y))$' representing 'Every Greek is mortal', can be correctly analysed as '$(Ux)(\text{greek}(x) \supset \text{mortal}(x))$'. But it is far from clear that it can. First, 'Every Greek is mortal' is apparently about Greeks whereas in the classical analysis the sentence is not about Greeks (see §3.6); but if they are about something different they surely cannot mean the same? (For much more on this topic see Passmore [1].) Secondly, they appear to support different entailments; if so they have a different logical content, and once again do not mean the same. For instance that every Greek is mortal certainly seemed, before the classical analysis was enforced, to entail that some Greek is mortal; but $(Ux)(\text{greek}(x) \supset \text{mortal}(x))$ of course does not entail $(Px)(\text{greek}(x)\ \&\ \text{mortal}(x))$. The constant temptation of novices to express sentences like 'Some Greeks are mortal' as '$(Px)(\text{greek}(x) \supset \text{mortal}(x))$' also points to the deficiency in standard recipes: such a formulation would be satisfactory given a proper conditionally-restricted operation in place of material implication.

All this suggests that conditionally restricted variables and operations should be given a separate logical treatment, rather than the classical elimination treatment. Such treatments can be given (see Behmann [1], Belnap [1]) and are important for significance theory. Accordingly we shall take the case of restricted variables seriously and not regard explicitly restricted variables as automatically eliminable in favour of unrestricted variables. Implicitly restricted variables may likewise be uneliminable either in favour of explicitly restricted variables or in favour of unrestricted variables, because the logic in which they feature is insufficiently rich; for just this reason the S-restricted variables of weak sentential significance logics (*e.g.* S_1 of Chapter 5) are uneliminable.

Since restricted variables are sometimes formed from other place-holders by logical operations they may exhibit notational complexity. An example is provided by attributive arguments such as 'λf' and '$\lambda y\ g(y)$' which are obtained by abstraction from predicate parameters. For exactness we specify:

> Where F is a predicate parameter, λF and *qu*(F) are (restricted) subject variables; and where P is a sentence parameter, §P and *qu*(P) are (restricted) subject variables, namely assertoric and quotational variables respectively.

If an expression containing a free unrestricted variable is asserted (as true, as a thesis) this amounts, as we have noticed, to the assertion of every sentence derivable from it by substitution of *any* value from the constants-range of the variable. Of course, in sentential systems, a variable like '*qu*(p)' does not read 'any sentence'; but assertion of, say *qu*($Tp \vee \sim Tp$), is tantamount to the assertion of 'for any sentence *qu*(p), *qu*($Tp \vee \sim Tp$) is true'. The two assertions are "equivalent" in the sense that we are entitled to infer either one from the other. (Such significance unrestricted sentence variables have then a generality interpretation in Kleene's sense (Kleene [3], p. 149)). Variables restricted by the requirement that they satisfy certain conditions are common not just in significance logics but also in algebra. Consider, for example, the variable 'x' so restricted that the equation ($x^2+2=3x$) holds. (Such restricted variables have a conditional interpretation in Kleene's sense.) An expression containing such free variables is asserted for any value from the constants-range which satisfies the condition.

It is important to distinguish an implicitly restricted variable which is restricted to a certain range α from an unrestricted variable which always leads to the nonsignificance of expressions in which it occurs upon substitution of values from its range outside α. To illustrate, consider the sentence frame,

(1) x is prime

where 'x' is a restricted variable whose range is limited to natural number expressions. Then 'Socrates is prime' is ill-formed since 'Socrates' is not a permissible substitution value. By contrast, given that the variable in

(2) x is prime

is unrestricted, then 'Socrates' is a permissible substitution value and the resultant sentence 'Socrates is prime' is nonsignificant. Thus (1) is significant for all substitution values of its variable while (2) is not. Again, given that the variable in

(3) x is green

is restricted to natural number expressions, then all substitution instances obtained by replacing the variable by its values are nonsignificant, but 'Plato's Academy is green' does not arise as the result of a significant substitution; on the contrary, it is ill-formed since 'Plato's Academy' is not a permissible value for 'x'. Elimination of an implicitly restricted variable 'x' depends not on extending the class of well-formed expressions containing '*x*' but on replacing '*x*' by a new variable '*y*', not so implicitly restricted, whose restriction is made explicit.

We give no ontological, as distinct from logical, preference to any particular kind of place-holder, subject or predicate, restricted or not. Subject variables, for instance, have no priority or advantage over such restricted variables as sentential variables. Our separation of variables from parameters, though reminiscent of Quine's distinction of variables and schematic letters, is differently motivated, differently drawn and differently defended. In particular, we are not concerned to draw ontological conclusions of any kind. It should be clear from the explanation of constants-ranges that adoption of a particular kind of variable does not entail ontological commitment to certain kinds of things. For example, the adoption of sentential variables does not commit us to the existence of all sentences, or even to the existence of some sentences, and it does not commit us to the existence or subsistence of "items" (propositions, statements, *e.g.*, or the Fregean "The True" and "The False") which might be said by some to stand as the designata of sentences. We know independently of our use of sentential variables that some token sentences occur (*e.g.* this one), that not all possible sentences have actual tokens, and we have ways of introducing statements and truth values by reference to the use of actual sentences in a context (Ch. 2). Type sentences, which in a sense might be said to stand as the values of the sentential variables, are abstractions. In a good sense of 'exist' they do not exist. What is substituted for a sentential variable, indeed for any variable, is a token which exemplifies a type; and insofar as the thesis in which the substitution is made is independent of the token features of the sentence actually substituted, so we may say that the thesis holds for types and that the values substituted are types. But however we may describe what happens, it is undeniable that what in fact happens is the substitution of a token.

3.4 Operators: Quantifiers and Descriptions

Our account of the syntax of operators and quantifiers follows that of Church ([4], pp. 39-46) for the most part. But there are major semantical divergences from Church, first because of the different account given of variables, and second because the explanation of quantifiers in terms of the two truth values truth and falsity has to be abandoned.

By an operator we mean a combination of symbols used together with one or more arguments or variables, called the operator arguments or variables, and one or more constants or forms, called the operands, to produce a new constant or form. For example, in '$(Ux)(x$ is happy)', the universal quantifier '*U*' is the operator, '*x*' is the operator variable, and '*x* is happy' is the form.

The operator together with its operator variables is said to be applied to the forms or constants or both on which it operates. For example, '(Ux)' is applied to the form 'x is happy', and 'x is happy' is the form on which it operates.

The scope of an operator, together with its operator variables, consists of its operands. Often the scope of an operator is shown by bracket pairs which open immediately following the operator, together with its variables, and close at the end of the scope. But though such an indication of scope can be guaranteed by appropriate formation rules, it is not always adopted or advantageous. For example, the formal implication operator, symbolised together with its operator variable 'x' by '$\supset_x$', is normally sandwiched between components of its operands as in '$(— \supset_x ...)$'.

An occurrence of a variable in a form is bound (by an operator) if either it occurs as an operator variable or is typographically the same as an operator variable and occurs in the scope of an operator. Otherwise the occurrence of the variable is free.

In applications of an operator, the operator variables will typically occur as free variables in some of the operands and only become bound when the operator is applied, not being already within the scope of and bound by another operator. Where they do not, or where they do not occur at all in the operands, the operator has a vacuous occurrence (or occurs vacuously).

Bound variables have, as already explained, distinct logical roles from free variables. Substitution of values from its constants-range is a relevant logical procedure in the case of a free variable, but not in the case of a bound variable. Bound variables are linked place-holders, or place-linkers, which may not be replaced by constants in their range.

An expression in which all the variables are bound, called a closed expression, has what meaning it has independently of the particular variables in it. Accordingly it is not affected by alphabetic changes of the bound variables in it, provided the alphabetic replacement is made at every occurrence of a given variable in both the operator and the scope. Thus the rule for the change of bound variables is simply an outcome of the features of bound variables much as the rule of substitution is an outcome of the logical features of free variables.

An expression containing bound variables only is thus a constant expression. It does not depend for its content on values of its variables, and in particular is not a function of its variables. For this reason Russell called bound variables, apparent variables; and free variables, real variables.

An operator is called *m*-ary-*n*-ary if, and only if, it is used with *m* distinct operator variables and *n* operands. A 1-ary-1-ary operator is called a simple operator. Most of the operators we introduce are simple.

We distinguish two important classes of operators: descriptors and quantifiers. A quantifier is an operator for which both the operands and the resulting form are sentence forms. A descriptor, on the other hand, is an operator for which the operands are sentence forms (including sentences) and the resulting form, produced by application of the operator, called a description, is a subject form.

Most of the logics we consider will contain only simple quantifiers as operators. The most important are:

the universal quantifier: U, read 'for every'
the particular quantifiers:
the classical: P, read 'for some out of all'
the non-classical: q, read 'for some'
the existential quantifier: $\exists$, read 'there exists a'.

The quantifiers are not all independent. Given either U or P as primitive, the other may be defined in terms of it in the usual way (below); and in sufficiently rich logics, q may be defined in terms of U (see §7.7). It is necessary to distinguish a particular quantifier from an existential quantifier if we are to be able to represent adequately some ordinary speech forms. Thus, in 'Some things do not exist', the word 'some' cannot have existential implications. Again, we may wish to speak of some abstract items, some numbers, some fairies, some propositions, and so on, without in any way wishing to imply that there exist items satisfying these descriptions. The existential quantifier may either be taken as primitive or defined in terms of P and a new primitive predicate constant 'E' (read 'exists'). We prefer the latter course, in spite of the long tradition that existence is not a predicate. For whatever the philosophical reasons which may be given in support of this, it can scarcely be denied that it is a grammatical predicate; and our interest is in acceptable grammatical expressions. There are in any case very many real advantages in adopting 'E' as a predicate constant and no obvious philosophical difficulties.

We need to use quantifiers in conjunction with both semi-wffs and wffcl. For we wish to consider the features of many-valued sentences such as '$(Ux)(x$ is happy)' in formulae like '$qu((Ux)(x$ is happy$))$ (c_0) $\mathbf{Y}a_0$'. Hence we require $(Ux)\Gamma$ to be a semi-wff where Γ is a semi-wff. In this use the quantifiers are many-valued. At the same time, in order to develop the full theory of context logic, we need to quantify over subject arguments which occur as parts of wffcl and which may either be in the scope of the quotation function or not. For example, we shall introduce the referring relation $\mathbf{R}$ (read 'refers to') and construct formulae of the sort '$qu(x)$ $\mathbf{R}y$'. In such formulae, both subject arguments are open to quantification. Thus, we allow, say, '(Px) $(qu(x)$ $\mathbf{R}y)$' (read, 'some (subject) expression refers to y') and '$(Px)(Py)$ $(qu(x)$ $\mathbf{R}y)$' (read, 'some (subject) expression refers to some item'). Hence, in general, we also require (Ux) A to be a wffcl where A is a wffcl. In such uses the quantifiers are two-valued. It would be possible to use a different symbolism to mark this difference, but this is both unnecessary, since the positions of quantifiers in formulae serves to distinguish them (those in the scope of a quotation function are many-valued, those outside are two-valued), and undesirable, since a change in notation would destroy important logical connexions shown by identity of symbolism types. Moreover the quantifiers, whether two-valued or many-valued, satisfy similar definitions and have the same readings.

We give a full specification of semi-wffs and wffcl in the next section, but at this stage, given the universal quantifier U and the predicate constant 'E' (exists) as new primitives, we can define P and $\mathcal{I}$ for both semi-wffs and wffcl as follows:

(D11) where Π is either a semi-wff or a wffcl and X is a subject variable:
(i) $(PX)\Pi =_{Df} \sim(UX)\sim\Pi$
(ii) $(\mathcal{I}X)\Pi =_{Df} (PX)(\Pi \ \& \ EX)$.

We shall not at this stage define $\mathcal{Q}$. It will be taken as a primitive for the construction of semi-wffs and is not needed for use in conjunction with wffcl since, in the two-valued case, it reduces to P. Other, restricted, quantifiers can be defined using the primitive / or ∋ ('such that', §3.3) and the relations between universal, particular and existential restricted quantifiers are like those defined in (D11). Sentential quantifiers, restricted in application to sentential variables within the scope of the quotation function or the that-operator, can either be introduced as restricted subject-quantifiers or taken as primitive.

For the explanation of quantifiers in context and significance logics the account of arguments and argument-values we have offered becomes important. Quantifiers of classical predicate logics are usually explained in terms of designation ranges and the two truth-values, truth and falsity. Thus '$(x)(x$ is happy)' is read 'for every entity x, x is happy' (where 'entity' usually means 'existing entity') or 'everything is an object x such that x is happy'; and its truth-values are explained in terms of the truth or falsity of sentences of the form *qu*(x is happy) for designation values of the variable 'x'. The reading is a consequence of the designation-value account since the standard designation of a referring expression is the entity it designates. Thus, 'for all designation values of *qu*(x)' amounts to saying 'for all entities x'.

From our point of view, however, this account is deficient in two ways. First, because expressions in the range of our subject variables may not designate anything. In the standard context, 'The round square' designates no entity, except in a very extended and tenuous sense of 'entity', and 'the purple number' does not designate at all, given standard senses for 'purple' and 'number', even if skeletal senses of 'designate' and 'entity' are introduced to admit 'the round square' as a designating expression. Consequently, we cannot adopt a designation-value account of the role of variables in quantified sentences.

The simple reading we adopt for '$(Ux)(x$ is happy)' is first, 'for every x, x is happy', or sometimes, 'for every item x, x is happy', where the word 'item' has no existential implications, but such an interpretation is simply a contraction for 'for every constant value of *qu*(*x*), *x* is happy'. Similarly, the reading 'for some *x*, *x* is happy', which we adopt for '(ᑫ*x*)(*x* is happy)' may be regarded as a contraction for 'for some constant value of *qu*(*x*), *x* is happy'. The constant values are, of course, expressions; they are not quoted expressions, even though the variable here appears quoted. In order to mention them, we have to quote them, but as values they are not quoted. Thus, suppose a suitable constant value of '*x*' for which '(ᑫ*x*)(*x* is happy)' is true is 'Bill'. Then it is the truth of 'Bill is happy', not the truth (!) of "'Bill' is happy" which guarantees the truth of '(ᑫ*x*)(*x* is happy)'. Similarly, given a sentence form containing a free variable, what is substituted for the variable is neither an entity nor a quoted expression but simply an expression; for it is expressions which occur as constants in the range of the variable. In case the quantification is over a variable within the scope of a quotation function, the value of the variable is still an expression, not a quoted expression, but the value of the function is a quoted expression. For example, the truth of '(ᑫ*x*)(*qu*(*x*) contains four letters)' is guaranteed by '*qu*(Bill) contains four letters', not by '*qu*('Bill') contains four letters'. But the value of *qu*(Bill) is 'Bill' so that '*qu*(Bill) contains four letters' becomes "'Bill' contains four letters". The range of the variables is thus exactly the same, whether they appear in quotation functions or not; hence the explanation and interpretation of the quantifiers is the same. But since, in such cases, quoted subjects occur in the formula, it is sometimes convenient to make an informal distinction by reading, say, '(ᑫ*x*)(*qu*(*x*) contains four letters)' as 'some subject contains four letters' and '(ᑫ*x*)(*x* is happy)' as 'some item is happy'. Hence the reading suggested earlier of 'some subject refers to some item' for '(*Px*)(*Py*)(*qu*(*x*) **R***y*)'; the correct interpretation, however, is: for some constant values of *qu*(*x*) and *qu*(*y*), *qu*(*x*) refers to *y*. To obtain an instantiation case from this, the whole of the sentence up to the comma must be deleted and unquoted expressions substituted for '*x*' and '*y*' in the remainder.

The second way in which the standard account is deficient, from our point of view, arises because quantified sentences may be neither true nor false but nonsignificant or incomplete for some values of the variables. Hence we cannot adopt a semantic analysis of quantifiers in terms of truth and falsity alone. The semantic analysis of many-valued quantifiers is discussed in detail in §7.3; at this stage, as with sentential connectives, we can only consider and only need to consider their two-valued features for use in conjunction with wffcl. However, the general account, which we give later, spans both the two-valued and many-valued cases and the main features relevant here are summed up in the equivalences:

(*U*X) ΠX is true if ΠX is true for every value of *qu*(X);
(ᑫX) ΠX is true if ΠX is true for some value of *qu*(X).

For the two-valued case, ᑫ may be taken to be *P*.

We turn now to the other main type of operator, necessary if the logic of complex subjects is to be adequately investigated, descriptors. These are subject-forming on sentential forms, whereas of course, quantifiers are sentence-forming on sentential forms. Simple descriptors of logical relevance include the following:

the indefinite descriptor κ: read, 'a' or 'an'
the definite descriptor ι: read, 'the'
the choice operator ξ: read 'an' or 'any'
the abstraction operator λ: (see below for interpretation).

As with quantifiers, we need to use descriptors in conjunction with both semi-wffs and wffcl. In either case, simple descriptions are formed according to the following formation rule: where τ is a simple descriptor and Π is either a semi-wff or a wffcl, $(\tau X)\Pi$ is a description (and thus a subject factor) and X is a bound argument. In subjects like $(\tau X)FX$, in which the only variable is bound by a descriptor, the argument X is a dummy and we can simplify the notation without loss by omitting the argument and writing τF. In case τ is λ, we read this as F-ness. For example, if 'f_0' is 'is white', then '$(\lambda x)f_0x$' *i.e.* 'λf_0', is the subject term 'whiteness'. More generally, writing '$X^{(n-1)}$' as an abbreviation for '$X^{(n-1)\text{primes}}$' we condense $(\tau X)(\tau X')$ ... $(\tau X^{(n-1)})\ F^n(X, X', ..., X^{(n-1)})$, where all the descriptors are the same, to τF^n: for example, '$(\iota x)(\iota y)g_0(x, y)$' is written '$\iota g_0{}^2$'. Again, if τ is λ we read τF^n as F^n-ness. This reading is not always natural, of course, since not all abstract nouns have the suffix 'ness', but this is unimportant. What is important is that, grammatically speaking, λ transforms a sentence form to an abstract noun.

Descriptors are of value because they enable us to emphasise and to study the special features of certain kinds of subjects, subjects which exhibit important logical features in their own right, and which have quite distinctive logical roles. Part of their distinctiveness is due to the fact that they contain predicates as syntactical parts, a feature displayed in notation such as '$(\iota x)\ g_0x$' and '$(\lambda x)(\lambda y)\ h_0(x,y)$'. Since descriptions (*i.e.* subjects formed by means of a descriptor) are admitted to all subject positions, we can express sentences such as 'Whiteness is brighter than blackness' directly in the notation '$f_0{}^2((\lambda x)\ g_0x, (\lambda y)\ h_0y)$', or '$f_0{}^2(\lambda g_0, \lambda h_0)$', where '$g_0$' is 'is white', and '$h_0$' is 'is black', and '$f_0{}^2$' is 'is brighter than'. Similarly ,we can represent directly sentences such as 'The lion is tawny' and 'Whiteness and abstractness are both abstract'.

Formulae such as '$f((\lambda x)\ fx)$', in which the same predicate term occurs both as the predicate of the sentence and as part of the subject, are not excluded. Such formulae commonly are excluded in type-theory extended to predicate logic in order to avoid the paradox of impredicability—the predicative version of the class paradox. For if we call a property predicable if it applies to itself, and non-predicable if not (the property of being abstract is itself abstract; the property of being coloured is not itself coloured), we are faced with a contradiction when we ask if the property of being non-predicable is itself predicable or non-predicable. This question cannot be asked, however, and the paradox cannot be stated, if we exclude formulae of the form '$f((\lambda x) fx)$', just as the class paradox cannot arise if we exclude from the list of well-formed formulae those expressions which stand for self-membership sentences. From our point of view, such a solution is not in point. If we are to evaluate the paradoxes, and solutions such as those embodied in type-theory, we must admit the formulae in terms of which the paradox is expressed. For only then can a solution of the type-theoretical kind—that *e.g.*, sentences of the form '$f((\lambda x) fx)$' are nonsignificant [*i.e.* $\sim Sf((\lambda x) fx)$]—be stated and examined explicitly.

3.5 The Syntax of Context Logic, and Identity

We turn now to the formal development of **CL** by extending it to include the conventions introduced in the preceding sections.

We take as new primitives in the construction of both semi-wffs and wffcl:

the universal (*subject*) *quantifier*: U
the existence predicate constant: E
the descriptors: κ, ι, ξ, λ (and such others as may be added from time to time).

As new primitives used in the construction of semi-wffs only, we also require, initially:

the non-classical particular quantifier: $\mathcal{q}$
the subject-identity predicate constant: $=$

It is not necessary to take $\mathcal{q}$ as a primitive of **CL** since, as we have noted, it reduces to P for the two-valued case and hence is definable by (D11). In Chapter 7 we show how to eliminate it as a primitive in the construction of the many-valued semi-wffs. Nor is it necessary to introduce '$=$' as a primitive in the construction of wffcl since, as we show below, a two-valued identity connective can be defined. This enables us to give identity postulates for **CL** but we do not at this stage develop a many-valued theory of identity; this, too, is given in Chapter 7.

For many purposes it is necessary to distinguish simple and complex subjects. We therefore define subject terms and subject factors, as follows:

R6 (i) Predicate parameters and predicate constants are predicate terms, and sentence parameters and sentence constants are sentence terms.
(ii) Subject arguments and subject constants which do not contain predicate terms or sentence terms, are subject terms.

(iii) Where Π is either a semi-wff or a wffcl, X is a subject variable, and τ is a descriptor, $(\tau X)\,\Pi$ is a description and § Π is a that-clause (also a description).
(*Note*: This does not extend the class of wfc. We retain the earlier definition in terms of which § Π is a wfc if, and only if, Π is a wffcl. Hence not all that-clauses, as now defined, are wfc.)

(iv) Subject terms, descriptions, that-clauses, quasi-wffs and partial wffs, are subject factors.
(*Note*: subject factors are within the range of subject arguments.)

In order to represent sentences with two or more subject factors, it is advantageous to introduce a vector notation. A sentence such as 'John loves Margaret' is written as 'Loves (John, Margaret)' where the expression '(John, Margaret)' is an ordered pair of subject constants such that 'John' fills the first place and 'Margaret' the second place of the two-place relational predicate 'loves'. Similarly 'Jack owes Bill £5' is represented as 'Owes (Jack, Bill, £5)' where '(Jack, Bill, £5)' is an ordered triple of subject constants. Such ordered sets of subject constants are called subject vectors; and in general, any ordered set of subjects, whether the elements are variables, constants, descriptions, *etc.*, or a mixture of these, is a subject vector.

R6 (v) An ordered set of subject factors is a subject vector.
(vi) An unordered set of subject factors is a subject set.
(vii) A term vector (set) is a subject vector (set) such that all the subject factors in it are subject terms.

We adopt the standard device of using round brackets or angle brackets to indicate ordered sets and curly brackets to indicate unordered sets. It is convenient to abbreviate the vector notation by using the ordinary algebraic convention of writing one variable with a tilde below it, *e.g.* $\underset{\sim}{X}$, in place of an ordered set; and it is sometimes useful to add an index to indicate the number of places in an ordered set. Thus, $\underset{\sim}{X}^m$ is a vector of m-places, or as we shall say, of rank m. Likewise we abbreviate the set notation by writing a single variable with a tilde above it: *e.g.* $\tilde{X}$ in place of an unordered set or, if the number of places is important, $\tilde{X}^m$ for a set of m elements.

Vector and set variables: if X is a subject variable then $\underset{\sim}{X}$ ($\tilde{X}$) is a vector (set) variable standing for any subject vector (subject set) with an arbitrary number of places, and $\underset{\sim}{X}^m$ ($\tilde{X}^m$) is a vector (set) variable standing for any subject vector (subject set) with m places. $\underset{\sim}{X}^m$ ($\tilde{X}^m$) is a vector (set) of rank m. Subject vector (set) variables are values of the subject arguments.

Hence, we may write '$\underset{\sim}{x}$' instead of '(x,y)', '(y,z,x)', *etc.*, if the number of places is unimportant; or, if we wish to indicate the number of places: '$\underset{\sim}{x}^2$', say for '(x,y)', '(y,z)', '(x',y')' *etc.*; '$\underset{\sim}{z}^3$', say, for '(x,y,z)', '(x',y,x'')', *etc.*; and so on. It is not necessary that there should be a typographical similarity between the vector variable and the elements of the vector, but unless otherwise indicated in the text, we shall adopt throughout the useful convention that '$\underset{\sim}{x}^m$' stands for the ordered set $(x, x', x'', ..., x^{(m-1)\text{primes}})$'; '$\underset{\sim}{y}^m$' stands for '$(y, y', y'', ..., y^{(m-1)\text{primes}})$'; and '$\underset{\sim}{z}^m$' stands for '$(z, z', z'', ..., z^{(m-1)\text{primes}})$'. A single bracketed subject variable is a subject vector or subject set of rank 1.

Finally, if $\underset{\sim}{X}$ is a subject vector, then $\tilde{X}$ is the subject set obtained by dropping the ordering of $\underset{\sim}{X}$. Thus $\underset{\sim}{X}$ and $\tilde{X}$ have the same elements.

We are now in a position to define further semi-wffs and wffcl. The distinction between atomic and elementary semi-wffs is needed for the theory of aboutness in the next section.

R1 (vi) If $\underset{\sim}{X}^m$ is a subject term-vector of rank m and F^m is a predicate term of m-places, $F^m\underset{\sim}{X}^m$ is an atomic semi-wff.

(vii) If $\underset{\sim}{X}^m$ is a subject vector with subject factors other than terms in any of its places, then $F^m\underset{\sim}{X}^m$ is an elementary semi-wff.

(viii) Atomic semi-wffs and elementary semi-wffs are semi-wffs.

(ix) If Γ is a semi-wff and X a subject variable then $(U\text{X})\,\Gamma$ and $(\text{ꝗX})\,\Gamma$ are semi-wffs.

It follows from these rules:

(x) If X is a subject factor, then EX is a semi-wff;

(xi) If X and Y are subject terms, X = Y is an atomic semi-wff, where $\text{X} = \text{Y} =_{\text{Df}} = (\text{X}, \text{Y})$;

(xii) If either X or Y (or both) is a subject factor other than a term, then X = Y is an elementary semi-wff.

(xiii) If Γ is a semi-wff and $\S P$ or $qu(P)$ is a restricted subject variable, then $(U\S P)\Gamma$, $(ꝗ\S P)\Gamma$, $(Uqu(P))\,\Gamma$ and $(ꝗqu(P))\,\Gamma$ are semi-wffs.

(*Note*: This allows '$(U\S p)f\S p$' and '$(ꝗqu(q))g(qu(q), x)$' but not, *e.g.* '$(Up)p$'. On the other hand, '$(U\S p)p$' and '$(Uqu(p))p$' are admissible since 'p' is a semi-wff and the quantifiers are appropriate (restricted) subject quantifiers; in such cases, however, the quantification is vacuous since the formulae are special cases of $(U\text{X})\,\Gamma$ where Γ does not contain X.)

When m takes its smallest value, 1, F^m is one-place and $\underset{\sim}{X}^m$ is a one-element vector. In such cases we frequently omit the brackets; e.g. we write 'fx' for '$f(x)$'. [Admitting the case where m takes the value 0, so that sentences could be accommodated as a special case with vectors of rank 0 (in effect semi-wffs without vectors) would confuse sentences with both subjects and unsaturated predicates.] Except to avoid ambiguity, or to emphasise well- or ill-formedness, it is unnecessary to write the rank index. The subject vector may of course be written out in full. Hence the following are all semi-wffs

atomic: $f(x)$ [or fx]; $f(x,y)$; $g^2(x,y)$; $g^3\underset{\sim}{x}^3$; $h\underset{\sim}{x}$; $f_0(x_0)$; $x = y$.

Where an expression such as '$h\underset{\sim}{x}$' is cited as a semi-wff, the definition ensures that whatever the rank of '$\underset{\sim}{x}$', 'h' must be a predicate of the same rank, *i.e.* with a number of places equal to the rank of the vector.

elementary: $g((\lambda x)(fx))$; $f(x,(\imath x)\,hx)$; $g((\lambda x)(fx \,\&\, h(x,z)),z)$; $g((\imath y)((Ux)\,f(x,y)))$; $f(\S(p \,\&\, q))$; $f(qu(fx))$; $h(qu(x),qu(f))$; $x = (\imath x)\,fx$.

Special cases of each of these elementary semi-wffs are, respectively, as follows:

whiteness is abstract; John is taller than the King of France; the property of being red and heavier than the moon, is more interesting than the moon; the item such that every item is larger than it, is green; that John stayed and accepted his punishment is commendable; 'Grass is short' is short; 'The man in the moon' contains more letters than 'is red'; Mao is the Emperor of China.

We have chosen here to take significant sentences as examples, but of course nonsignificant cases can arise. Thus a nonsignificant sentence arising from '$h(qu(x), qu(f))$' is " 'The man in the moon' loves 'is red' ".

The quantifiers satisfy the definitions already given, (D11)(i) and (ii), and restricted quantifiers will be introduced as they are needed. Strictly speaking $(U\S P)$ is a restricted subject quantifier, though it has not been defined explicitly as such. To define it, and other restricted quantifiers, it is necessary to add the further primitive: $\ni$ (or /). When $\ni$ is introduced as a new primitive, we adopt the rules:

If F is a one-place predicate constant, X $\ni$ FX is a (restricted) subject variable. Then $(U\mathrm{X} \ni \mathrm{FX})$ and $(\mathfrak{q}\mathrm{X} \ni \mathrm{FX})$ are (restricted) quantifiers, and $(\tau\mathrm{X} \ni \mathrm{FX})\,\Gamma$ is a restricted description.

It likewise follows, by admitting restricted subject variables as subjects, that

$$\Gamma(\mathrm{X} \ni \mathrm{FX}) \text{ is a semi-wff where } \Gamma(\mathrm{X}) \text{ is a semi-wff.}$$

To obtain the restricted variables (of §3.3) that we have used, specific predicate constants, such as the constant h_0 (read, approximately, '$qu(...)$ is a that-qualified sentence term'), have also to be introduced. Then we can define:

$$(U\S P)\,\Gamma(\S P) =_{\mathrm{Df}} (U\mathrm{X} \ni h_0\mathrm{X})\,\Gamma(\mathrm{X} \ni h_0\mathrm{X}); \text{ and so on.}$$

The method is explained in detail in §7.9 and §7.12.)

In case one or more quantifiers of the same kind occur juxtaposed in a formula, the vector notation may be used to shorten the string. Thus, we may write '$(U\underset{\sim}{x}^3)$' to abbreviate '$(Ux)(Ux')(Ux'')$'; or again, if the length of the string is arbitrary, we may simply write '$(U\underset{\sim}{x})$'. Similarly for strings of particular and existential quantifiers.

The previous compounding rules R1(i)-(iii), given in §2.2, apply, as do the rules for introducing context symbols, R1(iv) and (v) of §2.3. Hence the following are all semi-wffs:

$f(x,y)(c)$; $f\underset{\sim}{x}^m(c)$; $(Ux)f(x,y)$; $(U\underset{\sim}{x}^m)\,g\underset{\sim}{x}^m$;
$(Px)f(x,(\lambda y)\,gy)(c)$, $(Px)(\exists y)(g(x,y,z)\,\&\,f(qu(x)))$; $(U\S p)f(\S p)$;
$[(U\S p)f(\S p)\,\&\,(\exists x)\,g(x,qu(y))](c)$.

Notice, too, that that-clauses and descriptions constructed from wffcl can occur as parts of semi-wffs (by R6(iii) and (iv), and R1(vii)). Thus we have as semi-wffs:

$$f(\S\mathbf{T}a);\quad g((\iota x)(qu(fx)(c)\,\mathbf{Y}a_0),\S r);\quad etc.$$

Not only this, but certain wffcl can occur directly as semi-wffs. For example, suppose we have a formula of the form FX. We may take X to be the partial wff '*qu*(*p*)', since partial wffs are subject factors, and we may take F to be 'is true', *i.e.* '*tru*', since any grammatical predicate constant is within the range of the predicate parameters. Hence as a special case of the semi-wff we have, '*tru qu*(*p*)'; that is, '*Tp*'. Of course, wffcl can always arise as values of the restricted sentential variables in semi-wffs.

The class of partial wffs is now automatically extended since R2 (§2.4) is a rule for constructing partial wffs from any semi-wff. Hence it applies to the new semi-wffs. Again, the class of wffcl is automatically extended since the clauses of R4 (§2.9 ff.) provide for the construction of wffcl from any semi-wff.

When constants are substituted for the variables in semi-wffs, we have sentences which have illocutionary and sense features determined componentwise by the semantic dictionary of the background language. In general, they are three- or four-valued sentences where the values are t, f, n and i (in case i is taken as the fourth value). Hence all those features of semi-wffs determined in Chapter 2 automatically accrue to those now defined. For example, as a special case of

$\vdash_{CL}(74)\quad (T\Gamma \vee F\Gamma)(\theta) \equiv (P\gamma)(qu(\Gamma)(\theta)\,\mathbf{Y}\gamma)$,

we have immediately,

$\vdash_{CL}(74.1)\quad (TG\underset{\sim}{X} \vee FG\underset{\sim}{X})(\theta) \equiv (P\gamma)(qu(G\underset{\sim}{X})(\theta)\,\mathbf{Y}\gamma)$,

taking the semi-wff $G\underset{\sim}{X}$ as a case of Γ.

So, for example, if $G\underset{\sim}{X}$ is '$f_0(x_0,y_0)$', where 'f_0' is 'to the left of', 'x_0' is 'the bed' and 'y_0' is 'the chair', we shall have:

$\vdash_{CL}(74.2)\quad (Tf_0(x_0,y_0) \vee Ff_0(x_0,y_0))(c_0) \equiv (Pa)(qu(f_0(x_0,y_0))(c_0)\,\mathbf{Y}a)$;

i.e. 'The bed is to the left of the table' is true or false in c_0 if, and only if, it yields a statement in c_0. But we might equally well take $G\underset{\sim}{X}$ to be '$h_0qu(p_0)$', where 'h_0' is '*tru*', and we shall then have:

$\vdash_{CL}(74.3)\quad (TTp_0 \vee FTp_0)(c_0) \equiv (Pa)(qu(Tp_0)(c_0)\,\mathbf{Y}a)$;

i.e. 'Tp_0' is true or false in c_0 if, and only if, it yields a statement. There is however no *requirement* that 'h_0' be an appropriate sentence predicate even though we take a quoted sentence as a subject. We might, for example, take 'h_0' to be 'is happy' and 'p_0' to be 'The moon is round'. Then the thesis,

$$\vdash_{\text{CL}}(74.4) \quad (Th_0qu(p_0) \vee Fh_0qu(p_0))(c_0) \equiv (Pa)(qu(h_0qu(p_0))(c_0)\ \mathbf{Y}a),$$

has the interpretation: " 'The moon is round' is happy" is true or false in c_0 if, and only if, it yields a statement. That the sentence being mentioned, *viz.* " 'The moon is round' is happy" is in fact nonsignificant, is irrelevant. The thesis that it is true or false provided it yields a statement is itself significant and true. Again, taking $G\underset{\sim}{X}$ in (74.1) to be '$f_0\S p_0$', where 'p_0' is the nonsignificant sentence 'Wittgenstein's stove is happy', it is possible to choose 'f_0' in such a way that '$f_0\S p_0$' is significant (*e.g.* take 'f_0' to be 'Jack said') or nonsignificant (*e.g.* take 'f_0' to be 'is elastic-sided'); in either case, the thesis that it is true or false if, and only if, it yields a statement, remains significant and true.

Clearly we do not in this way obtain any new results since the application of theorems already derived does not depend in any way on the special features of the predicate semi-wffs. We hope to have illustrated by the above examples, however, the wide range of application to sentences of all kinds which arise from the unrestricted constants range of the subject arguments and predicate parameters.

Just as in the previous chapter we did not attempt to say which particular many-valued logic sentential semi-wffs satisfy, so here we do not attempt to say which particular many-valued quantification theory the predicate semi-wffs satisfy. These problems are taken up, in the case of standard contexts in Chapters 5 and 6 (for many-valued sentential logics) and Chapter 7 (for many-valued predicate logics). At this stage, the construction of semi-wffs continues to be a means to an end. Our main purpose, still, is to make two-valued statements about them in **CL** by incorporating them in quotation functions. As we have seen, this has already been achieved to some extent simply because predicate semi-wffs are admissible values for the unrestricted sentential variables, but this is a rather trivial extension. Given the special features of predicate expressions, especially the fact that they exhibit the parts of sentences, it is possible to affirm two-valued theses which cannot be stated when only sentential variables are employed. So, for example, given the relation constant **R** ('refers to'), we can express the fact that a given subject or a given sentence is used to refer to a given item in a context by means of formulae such as '$(qu(x)\ \mathbf{R}y)(c)$' and '$(qu(fx)\ \mathbf{R}y)(c)$'; or again, that a given expression refers to some item by, '$(Py)(qu(x)\ \mathbf{R}y)(c)$', and so on. We shall introduce these in the next section as two-valued wffcl even though they may contain, as in the sentential theory, many-valued expressions (*e.g.* 'fx') inside the quotation function, or a symbol (*e.g.* 'x') which is not itself even a semi-wff. In the former respect they do not differ from formulae constructed from Y, *e.g.* '*qu*(*fx*)(*c*) Y*a*'.

Such wffcl are introduced progressively as new relation constants are admitted. As a preliminary, however, we need to provide for quantification over the subject variables which may occur in wffcl and, since most of the relation constants we require are defined in terms of identity, to develop a two-valued theory of subject identity. We first provide for quantification:

R4 (xi) If X is a subject factor, *E*X is a wffcl.
(xii) If *A* is a wffcl and X is a subject variable, then (*U*X)*A* is a wffcl. (Thus, as in the case of semi-wffs, we have as special cases formulae such as '(*Uqu*(*p*))*Tp*' but, again, not '(*Up*)*p*').

The previous compounding rules of R4 apply to the new wffcl, as do the rules for introducing context variables. The quantifiers satisfy D11.

The two-valued use of the quantifiers is secured by the standard postulates affirmed for logically standard contexts :

(P12) (i) $\vdash_{\mathrm{CL}}[(U\mathrm{X})(A \supset B) \supset (A \supset (U\mathrm{X})B)](\mathrm{L}\theta)$, where X is a subject argument which does not occur free in *A*.
(ii) $\vdash_{\mathrm{CL}}((U\mathrm{X})A \supset \check{\mathrm{S}}^{\mathrm{X}}_{\mathrm{Y}}A|)(\mathrm{L}\theta)$, where X is a subject argument, Y is a subject factor, and $\check{\mathrm{S}}^{\mathrm{X}}_{\mathrm{Y}}A|$ is the wffcl which results from the substitution of Y for all free occurrences of X in *A* provided that no free occurrence of X is in a well-formed part of *A* which contains a variable of Y bound; and otherwise is *A*.

It follows from this and D11 that the two-valued use of the quantifiers *U* and *P* satisfies, in standard logical contexts, the principles of classical quantification logic, with *U* behaving like the universal quantifier and *P* like the classical particular quantifier. The two-valued logical properties of our existential quantifier *Ǝ* are determined by its definition in terms of *P* and the predicate constant *E*. In fact *Ǝ* behaves like the existential quantifier of free logic. We emphasise that the two-valued quantifiers introduced here for use in quantifying the subject variables of wffcl are different from the many-valued quantifiers which may occur in the quotation function as parts of semi-wffs. Thus, in '(*Px*)(*qu*((*Py*)*f*(*x*,*y*))(*c*) Y*a*)', '(*Px*)' is two-valued while '(*Py*)' is many-valued.

Two-valued identity formulae are now introduced as wffcl by means of the following definitions:

(D12) (i) $\mathrm{X} =^{+} \mathrm{Y} =_{\mathrm{Df}} T(\mathrm{X} = \mathrm{Y})$
(ii) $\underset{\sim}{\mathrm{X}}^{m} =^{+} \underset{\sim}{\mathrm{Y}}^{m} =_{\mathrm{Df}} (\mathrm{X} =^{+} \mathrm{Y})\ \&\ (\mathrm{X}' =^{+} \mathrm{Y}')\ \&\ \ldots\ \&\ (\mathrm{X}^{(m-1)\mathrm{primes}} =^{+} \mathrm{Y}^{(m-1)\mathrm{primes}})$

where $\underset{\sim}{\mathrm{X}}^{m}$ is (X, X′, ..., $\mathrm{X}^{(m-1)\mathrm{primes}}$) and $\underset{\sim}{\mathrm{Y}}^{m}$ is (Y, Y′, ..., $\mathrm{Y}^{(m-1)\mathrm{primes}}$). Thus, vector identity is only defined for vectors of the same rank. The identity of unordered sets, *i.e.* $\tilde{X}^m = \tilde{Y}^m$, cannot be defined unless we introduce $\in$ (membership) or $\subset$ (containment) as a further primitive. To keep the formalisation as simple as possible, however, we shall regard **CL** as being extended by such elementary two-valued set theory as we need without actually adding further primitives and postulates to achieve this. Elementary set-theoretic theses are therefore taken for granted without proof, in much the way that theses of **RSL** and **RQL** are taken for granted. We have given the necessary postulates to ensure the use of **RSL** and **RQL** since these two logics are so much an integral part of **CL**, and in any case arise out of **SL**, but to provide a full axiomatic set-theory within **CL** is an unnecessary complication because we need so little at this stage. Later, we shall consider many-valued set theory in detail. In general, then, we continue to regard **CL** as an extendible logic, open to extension by the introduction of new primitives and postulates, and where these are standard we take them over without providing a full formalisation.

It is clear that nonsignificant or incomplete identity formulae of the form $X = Y$ can arise in a variety of ways. We may, for example, choose subject factors of different kinds (*e.g.* one a that-clause and one an abstract noun) to obtain sentences such as, 'Whiteness is identical with (the statement) that it is raining' and 'Tom is redness'. Or again, we may choose factors of the same kind which are themselves nonsignificant or fail to have a reference. Thus, suppose 'The purple number 7' is nonsignificant in c_0 and 'the King of France', though significant, fails to have a reference in c_0; nevertheless, we can construct the identity sentence 'The purple number 7 is the King of France' (c_0). The effect of the first clause of D12, however, is to introduce two-valued identity formulae by means of a transformation on corresponding four-valued formulae. For we know that $T\Gamma$ is two-valued (being a wffcl) even though Γ is many-valued. The nature of the transformation which is made arises from the truth conditions for $T\Gamma$. These have not yet been given, but we shall see in Part II that if Γ is true, so is $T\Gamma$; if Γ is false, so is $T\Gamma$; and if Γ is either nonsignificant or incomplete, $T\Gamma$ is false. Hence the value of true and false sentences remains unaffected by the transformation; the important effect which it has is that all sentences which are either nonsignificant or incomplete are converted to false sentences.

It might perhaps be thought that a two-valued identity for use in the construction of wffcl could be defined by appropriate restrictions on formation rules, *e.g.* of the sort: $X =^{+} Y$ is a wffcl provided X and Y are subject factors of the same kind. This would eliminate sentences like 'Whiteness is the King of France', but by itself it would be too weak since there is nothing, so far, which would prevent us from substituting nonsignificant or non-referring phrases for the subject arguments in formulae in which only subject factors of the same kind are related. Hence substitution restrictions would be necessary too. Even then it is doubtful if an effective sieve is created. We resist this approach, not only because it may be ineffective, but mainly because our concern throughout is to admit nonsignificant and incomplete sentences and deal with them explicitly. The general inadequacy of the approach by way of formation and other syntactic restrictions is discussed in Chapter 4. We may note here, however, that at the very least it limits investigation to a fixed, definite, context (usually the standard context of the background language); for a sentence may be significant in one context, yet nonsignificant in another. Hence, if such a sentence passes through the syntactic sieve, and is thereby deemed to be significant, investigation of its use in those other contexts is automatically excluded.

The technique of covering a sentence by 'T', which is exploited in D12, is a useful and simple way of ensuring two-valuedness without making any false assumptions about the use of sentences in non-standard contexts. Indeed it can often be shown that those who wish to insist that every sentence in every context is true or false are implicitly using this transformation. The rationale behind the definition is that there is *an* identity relation—by no means the only one, but one which can always be defined given *some* identity relation—with which one can, in a given context, always make a statement that one item is or is not the same as another. One might enlarge on the reasonableness of introducing such an identity in the following, not implausible, way: In an identity claim, all that is being said, in effect, is that, in the context, two different words or phrases are being used to mention the same item; and whether the phrases are significant, or whether any such item exists, is irrelevant. That the phrases are or are not being used to mention the same item in the context is something that can always be determined (given adequate contextual conditions) and the claim is therefore either true or false. Of course, were this argument accepted without qualification, then we could introduce $X = Y$ directly as a two-valued wffcl. This would be to deny in general that nonsignificant or incomplete identity sentences can arise. In this case, the transformation has no effect, though it still holds, in fact is derivable, as a special case of $\vdash_{CL}(22)\ TA \equiv A$.

With context introduced it is important to distinguish intra-context identity, which we represent $(X = Y)(\theta)$ from inter- or trans-context identity, which we represent $X(\theta) = Y(\eta)$. It is perhaps helpful to compare intra- and trans-context identity respectively with intra- and inter-linguistic synonymy; the problems of *defining* inter-linguistic synonymy and trans-context identity have a good deal in common. We consider at this stage only intra-context identity.

We maintain that there is not a single identity relation, but a myriad of identity relations: these relations are, however, all determinates under the same determinable relation which simply requires coincidence of *at least* all extensional properties of identified items. (The position is explained fully in Routley [9]; and the theory of sameness as a determinable is elaborated in

§4.6.) If two items are identical in all extensional respects, then they are one and the same in fact; but it by no means follows that they share all intensional properties—one may be thought to be explosive, the other not. Hence stronger identity relations require that identified items share further intensional features. Later, we shall use the symbol '*I*' to stand in for determinants of the logical identity relation; and we use the familiar symbol '=' both to represent the logical determinable 'is identical with' and to symbolise the basic identity relation, that lowest common denominator, extensional identity. We therefore begin in **CL** with extensional intra-context subject identity as the basic notion.

Subject identity overlaps statement identity. For X $=^+$ Y is well-formed if X and Y are both that-clauses; hence, if we choose that-clauses which are also wfc, we have a statement-identity expression. Of course not all identity formulae involving that-clauses are statement-identity expressions, for we may choose an unrestricted semi-wff Γ which is nonsignificant and form $\S\Gamma =^+ \S\Gamma$. Conversely, not all statement-identity formulae are subject-identity formulae; for we have not defined statement variables as subject factors, hence '$a = a$', is not a subject-identity formula. It is in fact possible to define statement variables as a restricted class of subject arguments, but it is not necessary at this stage and for many purposes it is useful to keep them distinct.

The admission of that-clauses and quotation expressions as subjects simplifies the usual treatment of identity by enabling us to characterise precisely the problem of intension and to show that the usual distinction between intensional and extensional predicates can be made explicit in terms of § and *qu*.

To see this, consider first the ordinary two-valued postulates without context qualification as they occur in extended predicate logic:

(i) $x = x$
(ii) $(x = y) \supset (\Phi(x) \equiv \Phi(y))$

where Φ is any wff of extended **RQL** and $\Phi(y)$ is like $\Phi(x)$ except for containing 'y' for 'x'. Here and elsewhere, we drop the two-valued marker '+' when it is clear from the discussion, or the formulae used, that we are concerned only with two-valued expressions.

The second of these postulates is said to hold for extensional predicates such as 'is red' but not for intensional predicates like 'Bill knows that ... is red'. Indeed the distinction between the two kinds of predicates is sometimes (mistakenly) made by defining extensional predicates to be those which satisfy (ii) (see, *e.g.* Carnap [2]). For example, if we take the subject constant 'The thing in the corner' as a value of 'x' and 'Tom's ball' as a value of 'y', we shall have as a case of (ii),

if the thing in the corner is Tom's ball, then the thing in the corner is red if, and only if, Tom's ball is red;

and so for all extensional predicates. But taking the intensional predicate 'Bill knows that ... is red', we shall have,

> if the thing in the corner is Tom's ball, then Bill knows that the thing in the corner is red if, and only if, Bill knows that Tom's ball is red;

and the consequent of this may be false when the antecedent is true. For Bill may know that Tom's ball is red yet not know that the thing in the corner is red, even though the thing in the corner is in fact Tom's ball. For he may not know that the thing in the corner is Tom's ball.

Similar problems arise with any predicate which involves an intensional operator such as 'believes that', 'thinks that', *etc.*, and they arise also with predicates which contain quotation. To take an example from Quine ([3], Ch. VIII),

$$\text{Cicero} = \text{Tully},$$

but not

$$\text{`Cicero' contains six letters} \equiv \text{`Tully' contains six letters.}$$

Hence, the predicate,

$$qu(\quad) \text{ contains six letters},$$

is intensional. So, too, are the modal predicates 'it is necessary (that)', 'it is possible (that)', and so on.

It seems, therefore, that Φ in (ii) must be restricted in application to sentences which contain no intensional predicates; but simply to state this restriction is not enough. It fails to give us an effective theory of identity unless we have a pre-determined list of intensional (or extensional) predicates, or a general criterion in terms of which it can be decided whether or not a given predicate is intensional and therefore an improper case of Φ. But it cannot be (ii) itself which provides us with the test; for *qua* test, it is not effective since we have to draw on outside knowledge to determine whether substitution cases of (ii) are true or false.

Since we admit, and find no good reason to exclude, predicates like 'knows' and 'believes' and sentences such as 'Bill knows that Tom's ball is red' the problem confronts us at once. Thus let 'x_0' and 'y_0' be, respectively, 'the thing in the corner' and 'Tom's ball', and let 'R_0' be 'is red' and 'K' be 'knows that'. Then we can symbolise the two cases above as follows:

$$(x_0 = y_0) \supset (R_0 x_0 \equiv R_0 y_0),$$

the acceptable extensional case; and

(iia) $(x_0 = y_0) \supset (\text{Bill } KR_0 x_0 \equiv \text{Bill } KR_0 y_0)$,

the unacceptable intensional case. But since, for us, 'Bill KR_0x_0' is not well formed unless 'K' absorbs a that-operator, *i.e.* is '*kn* §', where '*kn*' is 'knows', the second of these expands to, (iib) $(x_0 = y_0) \supset (\text{Bill } kn\S R_0 x_0 \equiv \text{Bill } kn\S R_0 y_0)$.

Now it might be thought that our account of subject and predicate provides us with an immediate and simple solution. For we have said that the subject of a sentence such as 'Bill $kn\S p_0$' is the that-clause '$\S p_0$' while the predicate is 'Bill *kn*' (assuming that 'Bill' is not extracted as a further subject, but this does not affect the general point which is being made). This we have insisted on in opposition to an analysis in which '§' is taken to be a part of the predicate. For if the predicate is taken to be 'Bill *kn*§', then the *sentence* 'p_0' has to be taken as the subject; but it is our contention that while that-clauses stand in quantifiable (*i.e.* subject) positions, sentences do not. It might therefore be argued that in the sentence 'Bill $kn\S R_0 x_0$', since it is a special case of 'Bill $kn\S p_0$', the subject is '$\S R_0 x_0$' and the predicate is 'Bill *kn*'. If this is so, however, then (iib) is not a case of (ii). For in order to get the left-hand side of (iib), we have to take 'x' and 'y' of (ii) to be, respectively 'x_0' and 'y_0'; and if the substitution is uniform throughout, this entails that 'x_0' and 'y_0' are the subjects on the right also, and hence that the predicate is 'Bill $kn\S R_0$' not, as it should be, 'Bill *kn*'. Alternatively, if we take the correct subjects '$\S R_0 x_0$' and '$\S R_0 y_0$' for 'x' and 'y' on the right of (ii), and the correct predicate 'Bill *kn*' as a case of Φ, then the substitution is not uniform. Hence (iib) cannot arise as a case of (ii): either we have an incorrect predicate or non-uniform substitution. So the problem does not arise. Moreover, if we do take the correct subjects and predicate, then what we get as a case of (ii) is,

(iii) $(\S R_0 x_0 = \S R_0 y_0) \supset (\text{Bill } kn\S R_0 x_0 \equiv \text{Bill } kn\S R_0 y_0)$;

and this is a perfectly acceptable consequence which expresses the fact that if $\S R_0 x_0$ *is the same statement as* $\S R_0 y_0$, then Bill knows $\S R_0 x_0$ if, and only if, Bill knows $\S R_0 y_0$. More generally we have:

(iiia) $(\S R_0 x_0 = \S R_0 y_0) \supset (G(\S R_0 x_0) \equiv G(\S R_0 y_0))$,

for any predicate G. Each of these follows correctly and innocuously from (ii). Nor are they new. For they arise directly from the postulate we have already adopted for statement identity:

$\vdash_{CL}$(P7) (ii) $(\alpha = \beta) \supset (A(\alpha) \supset A(\beta))$, in §2.13,

which implies

$\vdash_{CL}$ $(\alpha = \beta) \supset (A(\alpha) \equiv A(\beta))$,

hence (iiia), where α and β are the wfc '$\S R_0 x_0$' and '$\S R_0 y_0$' and A is G, in case '$R_0 x_0$' and '$R_0 y_0$' are statement-yielding sentences, as we are assuming them to be. Correct instantiation of (ii) therefore confirms the postulate for statement identity; and *vice versa*, the statement identity postulate indicates how (ii) should be instantiated correctly.

Unfortunately, however, the matter is not so simple as this. It is true that *one* correct analysis of 'Bill $kn\,\S R_0x_0$' is to regard it as being of the form 'Bill $kn\,\S p_0$'; in which case we may extract '$\S R_0x_0$' as the subject and take 'Bill *kn*' to be the predicate. Then (iib) does not follow as a case of (ii). At the same time, our account of subject and predicate allows for the possibility of more than one analysis of a given sentence, and in particular, that analysis in terms of which 'x_0' is taken to be the subject and 'Bill $kn\,\S R_0$' the predicate, is not ruled out. For although it would be incorrect, for the reasons already given, to analyse 'Bill $kn\,\S p_0$' in such a way that the subject is the sentence 'p_0' and the predicate is 'Bill $kn\,\S$', the matter is different when 'p_0' is itself given a structure; for then other possible subjects appear. Thus there is nothing inconsistent in affirming that 'Bill $kn\,\S$' is an incorrect predicate while accepting that 'Bill $kn\,\S$ ... F' is correct. In fact we can extract subject 'x_0' from sentences like 'Bill $kn\,\S R_0x_0$'. For unlike 'p_0' in 'Bill $kn\,\S p_0$', 'x_0' does stand in a quantifiable position, since our account of quantification certainly does not exclude quantification into intensional sentence frames. Of course one cannot quantify into intensional sentence frames, and thereby extract subjects, if one insists on making quantifiers transparent (and existentially loaded). Quine [2] imposes just such requirements on the quantifiers of his canonical language. But our quantifiers, like those of a natural language, are not so regimented. There is nothing to prevent our quantifying into intensional sentence frames; and indeed there are many precedents in ordinary pre-analytic arguments for so doing, and great advantages in so doing. But given such quantification, we may take 'x_0' and 'Bill $kn\,\S R_0$' to be the subject and predicate of 'Bill $kn\,\S R_0x_0$'; in which case (iib) does follow as a case of (ii).

Similarly, if we consider a modal sentence such as '$\Box\S p_0$' then the subject is '$\S p_0$' and the predicate is '$\Box$' ('is necessary'). But in the case of a sentence like '$\Box\,\S f_0x_0$' we may extract either '$\S f_0x_0$' as subject or 'x_0' since both expressions stand in quantifiable positions. On the first analysis the predicate is again '$\Box$'; on the second, '$\Box\,\S f_0$'.

The situation with sentences containing quotation expressions is the same. In a sentence like "'Cicero' contains six letters", the only term in a quantifiable position is the *quoted expression* "'Cicero'". Hence we cannot accept,

(iic) (Cicero = Tully) ⊃ ('Cicero' contains six letters ≡ 'Tully' contains six letters),

as a case of (ii). For in order to get the left-hand side, the unquoted expressions 'Cicero' and 'Tully' have to be taken as cases of 'x' and 'y' respectively, but then (iic) can only be regarded as a case of (ii) if '*qu*( ) contains six letters' is taken to be the predicate, and for us this is not possible. The subjects on the right are "'Cicero'" and "'Tully'" while the predicate is 'contains six letters'; and taking these as cases of 'x', 'y' and Φ, respectively, what does arise correctly from (ii) is:

(iv) ('Cicero' = 'Tully') ⊃ ('Cicero' contains six letters ≡ 'Tully' contains six letters);

and this presents no problem. The antecedent, as well as the consequent, is false since 'Cicero' ≠ 'Tully'. A generalised form of (iv) which also follows correctly from (ii) is:

(iva) $(qu(x) = qu(y)) \supset (G(qu(x)) \equiv G(qu(y)))$,

for any predicate G (assuming the significance of G, $(qu(x))$, *etc.*, as we are at this stage). This provides a criterion of identity for quotation expressions.

However, if we take a sentence like,

'Cicero was a Roman' contains fifteen letters,

we have a case exactly analagous to 'Bill $kn\S R_0 x_0$'. For we may abstract either the quoted expression "'Cicero was a Roman'" as subject, in which case the unacceptable,

(iid) (Cicero = Tully) ⊃ ('Cicero was a Roman' contains fifteen letters ≡ 'Tully was a Roman' contains fifteen letters)

presents no problems since it does not then arise as a case of (ii), while

('Cicero was a Roman' = 'Tully was a Roman') ⊃ ('Cicero was a Roman' contains fifteen letters ≡ 'Tully was a Roman' contains fifteen letters)

which does follow correctly from (ii) is, like (iv) innocuous. Or we may extract the unquoted 'Cicero' from "'Cicero was a Roman'" since it stands in a quantifiable position. Then (iid) does follow. In other words, although '*qu*( ) contains six letters', like 'Bill $kn\S$' is not an acceptable predicate, '*qu*(F ...) contains six letters', like 'Bill $kn\S R_0$', is; and these last raise the problem of intensionality immediately by direct instantiation of (ii).

In general, our wide quantificational criterion for determining subjects and predicates at present allows us to say that if 'x_0' is a subject of 'p_0', and '$\S p_0$' (or '$qu(p_0)$') is a subject of 'q_0', then 'x_0' is a subject of 'q_0'. It might however be questioned whether the subject terms within subject factors should themselves be counted as subjects of the sentences of which the factors are subjects, and hence whether some restriction should be put on the quantificational test (though not on the quantification itself since it is essential to permit quantification into intensional frames). For the present, however, we leave this question on one side and accept the consequences of the wide test. It cannot be answered satisfactorily here since the answer to it turns partly on a theory of aboutness for sentences with complex subjects and this, in turn, raises problems in range theory. In the next section we give a preliminary account of aboutness but deliberately exclude from consideration sentences with subject factors other than terms. Thus we do not raise the question of whether a sentence like 'Jack said that Tom is happy' is about both the statement, that Tom is happy, and about Tom. Of course the answer to this might be negative even though we accept that both the clause, 'that Tom is happy', and the name, 'Tom', are subjects of 'Jack said that Tom is happy'; it will depend on the criteria of aboutness. We take up the whole question of complex subjects in detail in volume 2.

As it happens, even if we were prepared to restrict the quantificational test in such a way that 'Bill $kn\,\S R_0$' is not an acceptable predicate, we still do not avoid the problem of intensionality. For (iib) still arises as a consequence of (ii); under such a restriction it only fails to arise as a direct instantiation of (ii). Hence from this point of view there is nothing to be gained by restricting the test. This suggests, indeed, that attempts to solve the problem by restricting the class of predicates are mistaken; and in fact the alternative derivation of (iib) (and (iic)), which we give below, indicates an immediate general solution which does not require expressions like 'Bill $kn\,\S R_0$' to be excluded as predicates.

Thus, the objectionable (iib), or generalised forms of it, *e.g.*

$$(x=y)\supset(\mathrm{G}(\S\mathrm{F}x)\equiv\mathrm{G}(\S\mathrm{F}y))$$
$$(x=y)\supset(\Phi(\S\mathrm{F}x)\equiv\Phi(\S\mathrm{F}y))$$

will follow from the acceptable (iii), or its generalised form (iiia), if the following holds:

$$(x=y)\supset(\S\mathrm{F}x=\S\mathrm{F}y).$$

For using this, and some version of (iii), we obtain the unwanted conclusion by modus ponens. Similarly, the objectionable (iic) would follow from (iv) if

$$(x=y)\supset(qu(x)=qu(y))$$

holds. And in fact, each of these is a consequence of (ii) unless suitable restrictions are adopted.

Thus, from (ii) we have, correctly:

$$(x=y)\supset(\Phi(x,x)\equiv\Phi(x,y));$$

and taking $\Phi(x,x)$ to be $\S\mathrm{F}x=\S\mathrm{F}x$, and $\Phi(x,y)$ to be $\S\mathrm{F}x=\S\mathrm{F}y$, as we may, since each of these is a wff of the correct form, we have,

$$(x=y)\supset(\S\mathrm{F}x=\S\mathrm{F}x\equiv\S\mathrm{F}x=\S\mathrm{F}y).$$

But when Fx is statement-yielding,

$$\S\mathrm{F}x=\S\mathrm{F}x$$

is a thesis, a case of (P7)(i). Hence by **RSL**,

$$(x=y)\supset(\S\mathrm{F}x=\S\mathrm{F}y).$$

Similarly, we may derive,

$$(x=y)\supset(qu(x)=qu(y)).$$

That we must find some way of preventing these derivations is clear not only from the fact that the derived formulae lead immediately to others which we wish to reject, but also because the formulae are themselves plainly false. Indeed they crystallise the problem of intensionality. For '$(x=y)\supset(\S Fx=\S Fy)$' says in effect that if two items are the same, then any remark about one yields the same statement as a similar remark about the other. This fails, however, for almost every case of F which we may care to take; and given a suitable choice for F, it fails for every case of x and y. Thus, taking F to be 'is identical with x', then we have,

$$(x=y)\supset(\S(x=x)=\S(y=x));$$

but that x is identical with x is evidently not *the same statement as* that y is identical with x, even given that x is identical with y. Given the antecedent that x is identical with y, then the two *sentences* 'x is identical with x' and 'y is identical with x' are truth-functionally equivalent, but this certainly does not imply that they yield the same statement. And this means that (ii) is unacceptable, and that it ought to be separated from, and certainly should not imply, the non-thesis about statement identity. It is just because this implication is implicit in classical logic that the problem of intensionality arises. For by failing to distinguish formally between sentences and statements, statement identity is willy-nilly conflated with sentential equivalence.

Similarly, the "thesis"

$$(x=y)\supset(qu(x)=qu(y))$$

must be rejected as a plain falsehood. For that two items are the same cannot in any circumstances *imply* that the expressions used to mention them are the same.

We therefore restrict (ii) by adding the condition:

> provided that no replacement for 'x' (or 'y') is made for any occurrence within the scope of § or *qu*.

This prevents the derivation of $(x=y)\supset(\S Fx=\S Fx)$ from (ii) since if we take $\S Fx=\S Fx$ as a case of $\Phi(x,x)$ we cannot now take $\S Fx=\S Fy$ as a case of $\Phi(x,y)$. Similarly, the derivation of $(x=y)\supset(qu(x)=qu(y))$ is prevented. The restriction does not, however, prevent the derivation of the acceptable (iii) and (iv); for here, we are not replacing an occurrence of 'x'('y') within the scope of § or *qu*, but instead taking, respectively, §-clauses and *qu*-expressions as subject factors uniformly throughout (ii).

The adoption of this proviso means that we do not require any list of intensional predicates before (ii) is applied. Indeed we do not have intensional predicates at all, rather we have two intensional, or opaque, operators § and *qu* which resist the substitution of identities within them. This is not to say, however, that § and *qu* are the only intensional operators, but rather that these are the only two which have so far been introduced in **CL**. Others occur in different kinds of locutions. Thus, for example, the question-forming operator 'whether', like the statement-forming operator 'that', is opaque. Given that $x = y$, it does not follow that the question whether y is x, is simply the question whether x is x. That is, in general, $(x = y) \supset (?Fx = ?Fy)$ fails. In the case of indicative logic, there may be others besides § and *qu*, but we regard these as the most important since by grammatical transformations most opaque forms can be represented in terms of them. For example 'x is thought to be explosive' can with little loss be expressed as 'that x is explosive is thought to be true'; and similarly for 'believed to be', 'known to be', *etc*. Again, the classical modal operators 'it is necessary that' and 'it is possible that' find expression in terms of the predicates 'is necessary' and 'is possible' in conjunction with that-clauses taken as subjects. Hence the opacity of the modal operators is simply a special case of the opacity of §, just as the opacity of 'knows that', 'believes that', 'said that', and so on, are.

Of course, the assumption that (ii), with the proposed restriction, is sufficient to deal with all cases of intension, is really two. First, that all intensional predicates can be resolved into non-intensional (sentence/statement) predicates in conjunction with an intensional operator, often as part of the subject; and second, that the only intensional operators necessary for indicative logic are § and *qu*. The second assumption is less important than the first, for given that there are other independent operators, they could simply be added to the restriction. The first assumption, however, is more far-reaching and is discussed in more detail below and in Chapter 7. Given for the present that it is justified, however, then although we adopt (ii) (modified) and have a logic rich enough to allow the formulation of sentences containing intensional predicates, we avoid the standard associated problems. At the same time, we can express truths which cannot be stated classically. Some of these are instanced above. Others arise because predicates like 'knows' and 'believes' are often, in ordinary language, applied to subjects other than that-clauses. We can say "Tom knows Bill Smith" and "Bill knows the ten commandments"; and this use finds expression as a consequence of (ii). For example, take as a value for 'x' the subject constant 'the ten commandments' and for 'y' 'the laws of Moses'. Then taking 'Bill knows' or 'Bill believes' for Φ, we shall have as a case of (ii),

> if the ten commandments are the laws of Moses, then Bill knows (believes) the ten commandments if, and only if, Bill knows (believes) the laws of Moses.

And this is unobjectionable, for the classical difficulty associated with 'knows (believes) that' does not arise. If the antecedent is true, so is the consequent. For even if Bill does not know (believe) that the ten commandments are the laws of Moses, it is nevertheless true that if they are in fact the laws of Moses, then if he knows (believes) the commandments, he knows (believes) the laws of Moses. In order to get this as a case of (ii), it is not necessary for us to distinguish two different predicates, 'knows' and 'knows that' and hold that (ii) is acceptable for the first but not for the second; for 'knows that' is not in any case a predicate (though 'knows that F' is). We have one predicate 'knows' which is sometimes applied to that-clauses and sometimes to subjects which are not that-clauses.

To simplify the exposition we have made these points in terms of the standard two-valued extended **RQL**, but it is worth pointing out that the solution which follows as a consequence of taking that-clauses as subject factors is not possible in a two-valued predicate theory. Given the predicate 'Bill knows that ... is red', it can be claimed, though we think implausibly, that its significance range is the same as that of ordinary (extensional) predicates like '... is red'. Thus, 'the ball' will significantly fill the blanks in either. But given the predicates, 'Bill knows ...' and '... is red', then they do not have the same significance range, since the first will take that-clauses as well as ordinary names and descriptions as subjects, while the second will not significantly take that-clauses as subjects (*cf.* 'that it is raining is red'). Hence, by admitting that-clauses as subjects, nonsignificant sentences are bound to arise as cases of formulae of the form FX. So, for some substitution cases of (ii), the right-hand side will contain nonsignificant components and it is therefore necessary to have a many-valued predicate theory. In our view, however, it is in any case implausible to suppose that nonsignificance is avoided by always taking 'that' as part of the predicate. For if 'Bill knows that ...' and 'Bill said that ...' are allowed as predicates, as they are classically, we certainly cannot significantly fill the blanks with ordinary subjects like 'the ball' or 'Tom'.

It remains only to express the solution in terms of many-valued sentences and to introduce intra-contextual qualifications. We propose to adopt versions of (i) and (ii) as postulates of **CL** but expressed in terms of any wffcl or semi-wff Π instead of a two-valued Φ. It follows that the **CL** version of (ii) must be such that the right-hand side is a wffcl even though it embraces semi-wffs. We cannot, therefore, adopt,

$$x =^{+} y \supset (\Pi(x) \equiv \Pi(y)),$$

for that would give as a special case,

$$x =^{+} y \supset (\Gamma(x) \equiv \Gamma(y));$$

and the right-hand side is not a wffcl. This last (without the '+') would in fact be the postulate in a many-valued extended predicate logic corresponding to the two-valued (ii). This difficulty is easily overcome, however, by taking the equivalence to be $\simeq$ rather than $\equiv$. Secondly, as with the other logical connectives, we only characterise the use of '$\simeq$' in a logically standard context. Hence the postulates are:

(P13) (i) $\vdash_{CL}(X =^{+} X)(L\theta)$

(ii) $\vdash_{CL}(X =^{+} Y)(L\theta) \supset (\Pi(X) \simeq \Pi(Y))(L\theta)$,

provided that (a) no replacement for X (or Y) is made for any occurrence within the scope of § or *qu* (or quote marks), and (b), the usual proviso on identity, X is not in the scope of operators binding X or Y.

Here, taking Π to be Γ, $\Gamma(X)$ and $\Gamma(Y)$ may be nonsignificant or incomplete for given values of X, Y and Γ, but $\Gamma(X) \simeq \Gamma(Y)$ is nevertheless two-valued since it expresses the fact that $\Gamma(X)$ and $\Gamma(Y)$ have the same value. Taking Π to be a wffcl A, $A(X) \simeq A(Y)$ reduces to $A(X) \equiv A(Y)$ where $\equiv$ is two-valued.

We call the identity determined by (P13), extensional identity. It follows in the usual way that it is symmetric and transitive, and therefore, since it is also reflexive, (P13)(i), it is an equivalence relation.

The assumption we have made throughout this discussion is that, for all the sentences which we wish to consider, all quotation can be explicitly isolated and represented by quotes or *qu*, and all non-quotational intensionality can be isolated, transformed by grammatical transformations into that-locutions, and be explicitly represented by §. Otherwise the proviso on (ii) is ineffective. If this assumption is denied, then the alternative is to suppose that it is possible to distinguish between primitive (extensional) predicates and others, and then adopt in place of (ii),

$$(X =^{+} Y) \supset (F(\ldots X \ldots) \simeq F(\ldots Y \ldots)),$$

where X and Y may be replaced by subject factors which contain only primitive predicates and F is any primitive predicate. This restriction is necessary, for given our account of predicates, if we adopt

$$(X =^{+} Y) \supset (F(X) \simeq F(Y))$$

for *every* predicate F, we have immediately

$$(X =^{+} Y) \supset ((\ldots X \ldots) \simeq (\ldots Y \ldots)),$$

for *every* sentence frame. The ordinary problems of intensionality then follow at once. Now although the restriction on X, Y and F prevents this derivation, it does raise immediately the problem of how we can be sure that a given predicate can be taken as primitive. But this, after all, may simply be the same problem as the one which arises from the solution we have adopted, *viz.* how can we be sure that all intensionality, other than quotation, can be isolated and represented explicitly by means of §? We discuss these difficulties, and examine in more detail the alternative approach suggested here, in Chapter 7.

Extensional identity provides a satisfactory basis on which to develop a theory of aboutness and reference (§3.6), but to develop a theory of sense, a stronger identity base is needed. In particular, we require an identity relation which permits substitutivity in all non-quotational sentence frames. That is, if it is accepted that all non-quotational intensionality can be represented in terms of §, an identity which permits substitutivity under §. Here full Leibnizian identity, which we represent by '$\approx$', suggests itself.

For a theory of sense, then, we add the new primitive: $\approx$ (read 'is Leibnizian identical with'). New semi-wffs are then defined by the derived formation rule:

R1 (xiv) If X and Y are subject terms, $X \approx Y$ is an atomic semi-wff; if either X or Y (or both) are subject factors other than terms, $X \approx Y$ is an elementary semi-wff (where $X \approx Y =_{\mathrm{Df}} \approx(X, Y)$).

Then, as before, we secure a two-valued form of $\approx$ by means of the definition:

(D12) (iii) $X \approx^{+} Y =_{\mathrm{Df}} T(X \approx Y)$

(iv) $\underset{\sim}{X}^{m} \approx^{+} \underset{\sim}{Y}^{m} =_{\mathrm{Df}} (X \approx^{+} Y) \;\&\; (X' \approx^{+} Y') \;\&\; \ldots \;\&(X^{(m-1)\mathrm{primes}} \approx^{+} Y^{(m-1)\mathrm{primes}})$

where $\underset{\sim}{X}^{m}$ and $\underset{\sim}{Y}^{m}$ are as in D12 (ii).

The identity postulates are then as follows:

(P14) (i) $\vdash_{\mathrm{CL}}(X \approx^{+} X)(L\theta)$

(ii) $\vdash_{\mathrm{CL}}(X \approx^{+} Y)(L\theta) \supset (\Pi(X) \simeq \Pi(Y))(L\theta)$

provided that no replacement for X (or Y) is made for any occurrence within the scope of *qu* (or quote marks), and X is not in the scope of operators binding X or Y.

Now by the previous argument, it will follow immediately from this, that,

$$\vdash_{\mathrm{CL}}(x \approx^{+} y)(L\theta) \supset (\S Fx =^{+} \S Fy)(L\theta);$$

and as special cases,

$$\vdash_{\mathrm{CL}}(x \approx^{+} y)(L\theta) \supset (\S(x =^{+} x) =^{+} \S(y =^{+} x))(L\theta)$$

and,

$$\vdash_{\mathrm{CL}}(x \approx^{+} y)(L\theta) \supset (\S(x \approx^{+} x) =^{+} \S(y \approx^{+} x))(L\theta).$$

These indicate the strength of $\approx$. We rejected them earlier in the case of extensional identity because in a good sense of 'is the same as', we can accept that x is the same as y yet deny that $\S Fx$ is the same statement as $\S Fy$, and in particular, that $\S(x =^{+} x)$ is the same statement as $\S(y =^{+} x)$. What we are now claiming, however, is that there is *an* identity, Leibnizian identity, such that if x is Leibnizian the same as y, then $\S Fx$ is the same statement as $\S Fy$, or in particular, $\S(x =^{+} x)$ is the same statement as $\S(y =^{+} x)$. Evidently, if two words stand in such a relation, they have the same sense; for substitution of one for another in a sentence will not change the statement made. An identity stronger than Leibniz identity which permits replacement under '*qu*' would define type identity of expressions.

The theory of identity is not only important for the analysis of sense and reference. Context can often be taken up, and contextual differences eliminated, by using identity connexions. Consider, for example, the sentence 'Tom is happy' uttered in a context c_0 for which it is known that Tom is the butcher. Formally speaking, c_0 is a set of sentences, and the sentence 'Tom is the butcher' or 'Tom = the butcher' can be represented as an explicit part of the context by means of,

Tom is happy ($c_0 \cup \{$Tom = the butcher$\}$).

Similarly we may write,

Bill is unhappy ($c_0 \cup \{$Bill = the baker$\}$)

to make explicit the fact that 'Bill = the baker' is one of the context-determining sentences with respect to the utterance 'Bill is unhappy'. Now if the set c_0 does not contain 'Tom = the butcher' or 'Bill = the baker', then $c_0 \cup \{$Tom = the butcher$\}$ and $c_0 \cup \{$Bill = the baker$\}$ are different sets, *i.e.* different contexts. By taking up the identity into the sentence, however, a common context can be obtained, thus,

Tom is happy where Tom is the butcher (c_0)
Bill is unhappy where Bill is the baker (c_0).

Then, since the context is common, the sentences may be combined to give:

Tom is happy where Tom is the butcher, but Bill is unhappy where Bill is the baker (c_0);

and under special conditions, *e.g.* everything is nicely referential, the identities may be substituted to give:

The butcher is happy but the baker is unhappy (c_0).

We have, then, a way of representing formally shuffles that are commonly made in bringing discourse into a manipulable logical form. It remains, however, to determine the precise conditions under which each of these steps, whereby context is taken up and made explicit, can be carried through. In the process of obtaining a common context in which different segments of discourse can be jointly assessed, translation and substitutions may also be needed, *e.g.* a change or elaboration of proper names may be necessary to avoid clashes and to permit quantificational operations.

3.6 Aboutness and Items and Other Contractions

How sentences connect with reality is typically explained through connexions of their subject and predicate parts with things and the relations things stand in. Thus the subject of a sentence is about an item—very often it refers to an entity, *i.e.* an existing individual—and the predicate says something about or comments on the item. Where contextual analysis brings a sentence out as multi-subject, the subjects will be about, or have as topic, a set of items, and the comment will concern all these items.

The aboutness relation which connects subjects of a sentence with items may be extended to apply to sentences. What a sentence is about is what its subjects are about. In fact the subjects of a sentence are often picked out in just this way, by asking what the sentence is about and then determining which phrases of the sentence are about these items.

Both what the subject is about and the item it is about can be defined using identity. For what a subject is about is that item which it identifies; and the item is what is so identified. An identity relation, being an equivalence relation, partitions its field into exhaustive and exclusive sets. Each of these sets, so distilled, identifies an item, and the relation between a subject and the item, that is the equivalence class to which it belongs, is the aboutness relation relating subject to item. Algebraically then, aboutness composes quotation with a membership relation.

We explain the theories of aboutness and reference in an intuitive way first, before going on to details and qualifications. Subsequently we generalise on the aboutness relation; for aboutness is but a special *contraction* relation between expressions and items. The generalisation has the advantage of revealing how little metaphysical sting there is in our aboutness relation. Nonetheless the aboutness relation is a fundamental one for classical-looking semantics of significance predicate logics. In such semantics, given in Chapter 7, our *defined* aboutness relation supplants the primitive denotation or reference relation of the usual semantics. With aboutness, and more generally with contractions, we lay the basis of the contractional semantics we shall provide for quantificational significance logics. The generalisation of aboutness also reveals a simple way of beginning on theories of sense and content.

To say, in a given context, that the subject 'Tom' of the sentence 'Tom is only 4 feet tall' is about the man next door is simply to say that Tom is the man next door. This sense of 'about', in terms of identity, rests on the fact that what a subject is about depends on some identification having been made in the context. Thus, Tom has been identified as the man next door, or equivalently, the man next door has been identified as the item which 'Tom' is being used to mention. Since our subject-variables take definite descriptions and other complex expressions as values, as well as proper names and pronouns, we may capture this sense by affirming:

$$(qu(x) \textit{ is s-about } y)(c) \equiv (x = y)(c).$$

For we may substitute a proper name for 'x' and a description for 'y', or *vice versa*, or a description for both. Thus, we may say, for example:

'Pegasus' is s-about a winged horse in context c_0, if, and only if, Pegasus is a winged horse.

or

> 'The purple number 7' is s-about the house next door in context c_0 if, and only if, the purple number 7 is (has been identified in c_0 as) the house next door.

By a simple extension it is then possible to define the s-aboutness of a singular subject-predicate sentence by affirming as expected, that the sentence is s-about whatever its subject is s-about, *i.e.*

$$(qu(f(x))\ \textit{is s-about}\ y)(c) \equiv (qu(x)\ \textit{is s-about}\ y)(c).$$

To achieve a similar result for multiple subject sentences it is only necessary to introduce subject vectors and their subject sets. Then

$$(qu(\underset{\sim}{x})\ \textit{is s-about}\ \tilde{y})(c) \equiv (\tilde{x} = \tilde{y})(c)$$

and $$(qu(f\underset{\sim}{x})\ \textit{is s-about}\ \tilde{y})(c) \equiv (qu(\underset{\sim}{x})\ \textit{is s-about}\ \tilde{y})(c).$$

Here, and in what follows, since we are only concerned with theses of **CL**, all identity formulae are two-valued. We therefore write the simple '$=$' in place of '$=^+$'.

The rank of the vectors and subject sets in theses such as those above is unimportant, but the rank of $\underset{\sim}{x}$ must initially be the same as the rank of $\tilde{y}$ in order to secure a one-one link between subjects and items and so provide an aboutness for each subject which is extracted in the context. This does not mean, however, that the subject sets cannot be reduced by ordinary set-theoretic procedures. For example, where 'Tom' and 'himself' are both s-about Mr Bloggs in a suitable context, the sentence 'Tom hates himself' is s-about the set {Mr Bloggs, Mr Bloggs}; but this then reduces to {Mr Bloggs}, *i.e.* on the theory we shall use, just Mr Bloggs. We shall take over *that feature* of Quine's system **ML** (Quine [1]) according to which individuals, and non-sets generally, coincide with their unit sets. Again, that the rank of $\underset{\sim}{x}$ must initially be the same as the rank of $\tilde{y}$, imposes no restrictions on the number of subject factors which are extracted in a given context. Thus we may extract just 'Mary' from the sentence 'Mary loves Tom', or the ordered pair '(Mary, Tom)', or just 'Tom', depending on the analysis required in the context. In the first case we represent the sentence as being of the form 'fx', in the second case as being of the form '$g(x,y)$', and in the third case as being of the form 'hy'. The conditions on the rank then require that the subject sets are, respectively, a unit set, an unordered pair, and a unit set. In general, because sentences of the same type need not have the same subject-predicate analysis uniformly over all contexts, or even within the same context, so they do not necessarily have the same kind of aboutness sets. The kind of aboutness sets which they have will depend on the analysis into subject-predicate constituents, and this in turn will depend on the reason for the analysis. Not only this, but even if two sentences of the same type have the same kind of aboutness sets in different contexts, the sets themselves may have different members. For what a subject is about depends crucially on the context; 'Tom' may be about a person in one context and a dog in another. Hence, though two different tokens of the type exemplified by 'Mary loves Tom' each has a rank 2 aboutness set in different contexts, the sets may be quite different.

The motivation for using unordered sets is as follows:—'Mary is loved by Tom' is but a transformation of 'Tom loves Mary'; hence given a context in which they both occur, and which is such that 'Mary' and 'Tom' have the same respective s-aboutness in the two sentences, then if one is s-about a given pair, so is the other; *i.e.* the order of the subjects does not affect the s-aboutness. Again, 'Tom and Mary hate the butcher' is presumably s-about whatever 'Tom hates the butcher and Mary hates the butcher' is s-about (given that 'Tom', 'Mary' and 'the butcher' have the same respective s-aboutness in the two sentences). But these can only easily be s-about the same things, and s-about what 'The butcher is hated by Tom and Mary' is s-about, if they are all s-about the same set, say {Tom, Mary, the butcher} in case all the subjects are extracted. (In contrast, 'Jones and Brown are two', though s-about {Jones, Brown} = {Jones}∪{Brown}, does not transform to 'Jones is two and Brown is two' since the predicate is not distributive.)

Finally, the s-item $qu(x)$ is s-about is $\{y : y = x\}$. Where x is an individual, this s-item (subject item) is just $\{x\}$; *i.e.* on the theory we shall use, just x itself.

We now give a formal definition which secures all the theses we want for the case of term vectors and atomic semi-wffs. Later we shall extend the analysis to compound and quantified semi-wffs whose component parts are atomic. We shall not, however, develop a theory of s-aboutness for elementary semi-wffs at this stage. The necessary extension is not difficult, but it does introduce a complexity into what is, initially, a very simple theory, and for most interpretational applications it is unnecessary. Some of the complications arise from the fact already noted, that different subjects may be extracted in different contexts.

To see this, consider a sentence which contains a fairly complex definite description, *e.g.* 'The man who owes £5 to Bill is greedy'. There are a variety of ways in which subject terms could be extracted from this. The simplest is to take 'f_0' as 'is greedy' and 'x_0' as the single subject 'the-ower-of-£5-to-Bill'. Then the form of the sentence is 'f_0x_0'. The most complex is to extract three subjects and represent the sentence as being of the form '$f_0(\iota x)g_0(x, y_0, z_0)$', where '$y_0$' is '£5', '$z_0$' is 'Bill', and '$g_0$' is 'owes'. Depending on the analysis, therefore, the s-aboutness sets will differ—just as in the cases we have already considered. But there is an important difference between this case and those mentioned above. For, given a sentence like 'Tom loves Mary', it can be represented as being of the form 'fx', '$g(x,y)$' or 'hy' but each of these is absorbed in the theory of aboutness for sentences of the form $\mathrm{F}^m\mathfrak{X}^m$, since each is a predicate-vector sentence. Hence by giving a general aboutness condition for $\mathrm{F}^m\underset{\sim}{\mathrm{X}}^m$, we thereby give the different conditions corresponding to all possible subject-predicate analyses of 'Tom loves Mary', in whatever context. Now if we regard 'The man who owes £5 to Bill is greedy' as being of the form 'f_0x_0', the theory for predicate-vector sentences comprehends it; but if 'x_0' is itself regarded as analysable into, say, '$(\iota x)h_0(x, z_0)$', where 'h_0' is 'owes £5', or '$(\iota x)g_0(x, y_0, z_0)$', where '$g_0$' is 'owes', then the subject set is not clearly defined: it may be either a singleton, a pair or a triple. And such analyses are not included as special cases of $\mathrm{F}^m\underset{\sim}{\mathrm{X}}^m$ since, *e.g.*, '$f_0(\iota x)g_0(x, y_0, z_0)$' is not of the form '$f(x,y,z)$'. Similar difficulties arise, as we noted in the previous section, in the case of sentences which contain '§' or '*qu*'. To extend the theory to such cases, it is therefore necessary to add special conditions for subject factors other than terms. This we do in volume 2.

The limitation of the theory to subject terms is less restrictive than it might appear to be. For as we have noted, it is always possible to represent a description, say, as a single subject. To this extent the theory applies to contexts in which the extraction of different subjects within the same factor is unnecessary or unimportant. Conversely, at the level of interpretation we can always substitute a subject factor constant for a subject argument (since the variables range over all subject constants), regarding the factor constant as a single subject. Hence there is no real restriction in application. What is missing, only, is an extended analysis, which could be added to the one we give, in terms of which more detailed information about subject sets is made explicit in certain cases.

The initial definitions we adopt, then, for atomic semi-wffs, are as follows:

(D13) Where $\underset{\sim}{\mathrm{X}}^m$ and $\underset{\sim}{\mathrm{Y}}^m$ are subject term vectors of rank m, $\tilde{\mathrm{X}}^m$ and $\tilde{\mathrm{Y}}^m$ are the corresponding subject sets, F^m is an m-place predicate term and θ is a context symbol:

(i) $(qu(\underset{\sim}{\mathrm{X}}^m)\ is\ s\text{-}about\ \tilde{\mathrm{Y}}^m)(\theta) =_{\mathrm{Df}} (\tilde{\mathrm{X}}^m = \tilde{\mathrm{Y}}^m)(\theta)$

(ii) $(qu(\mathrm{F}^m\underset{\sim}{\mathrm{X}}^m)\ is\ s\text{-}about\ \tilde{\mathrm{Y}}^m)(\theta) =_{\mathrm{Df}} (qu(\underset{\sim}{\mathrm{X}}^m)\ is\ s\text{-}about\ \tilde{\mathrm{Y}}^m)(\theta)$

(iii) $(the\ s\text{-}item\ set\ qu(\underset{\sim}{\mathrm{X}}^m)\ is\ s\text{-}about)(\theta) =_{\mathrm{Df}} \{\tilde{\mathrm{Y}}^m : (\tilde{\mathrm{Y}}^m = \tilde{\mathrm{X}}^m)(\theta)\}$.

Thus, by taking $m = 1$ we have, for individuals, the theses mentioned above:

$\vdash_{\mathrm{CL}}(101)$ $(qu(x)\ is\ s\text{-}about\ y)(c) \equiv (x = y)(c)$;

$\vdash_{\mathrm{CL}}(102)$ $(qu(fx)\ is\ s\text{-}about\ y)(c) \equiv (qu(x)\ is\ s\text{-}about\ y)(c)$

We may say, for example,

'Tom' is s-about the man next door, in c_0, if, and only if, Tom is the man next door; and

'Tom is happy' is s-about the butcher, in c_0, if, and only if, Tom is the butcher.

Here we have taken descriptions as values of the subject arguments, as we may; and in general any constant subject factor, provided it is regarded as a single subject, can be taken as a value for the arguments. Notice, however, that since s-aboutness is defined in terms of extensional identity, it is undefined for intensional sentence frames. We cannot, for example, conclude from the fact that the morning star is the evening star that the sentence 'Bill knows that the morning star is Venus' is s-about the evening star. Even allowing the substitution of any factor constant for the variables, this cannot arise as a case of $\vdash_{CL}(102)$. Thus, if 'B_0' is 'Bill knows', 'x_0' is 'the morning star', 'y_0' is 'the evening star' and 'z_0' is 'Venus', we do not have,

$$(\text{`}B_0\S(x_0 = z_0)\text{'}\ \textit{is s-about}\ y_0)(c_0) \equiv (\text{`}x_0\text{'}\ \textit{is s-about}\ y_0)(c_0),$$

as a case of $\vdash_{CL}(102)$, since the substitution is not uniform; hence not,

$$(\text{`}B_0\S(x_0 = z_0)\text{'}\ \textit{is s-about}\ y_0)(c_0) \equiv (x_0 = y_0)(c_0).$$

There is of course a sense in which 'Bill knows that the morning star is Venus' is extensionally s-about the evening star. And this we could show if, for example, we specified that the subject factor '$\S(x_0 = z_0)$' is s-about $\{x_0, z_0\}$. For then, given that $x_0 = y_0$ and $z_0 = y_0$, the set $\{x_0, z_0\}$ reduces to $\{y_0\}$, *i.e.* to y_0 itself. But this is to make a number of assumptions which are not always justified.

By taking $m = 2$ in (D13) we have:

$\vdash_{CL}$ $[qu(x,x')\ \textit{is s-about}\ \{y,y'\}](c) \equiv [\{x,x'\} = \{y,y'\}](c)$; and
$\vdash_{CL}$ $[qu\,f(x,x')\ \textit{is s-about}\ \{y,y'\}](c) \equiv [\{x,x'\} = \{y,y'\}](c)$;

and similarly for $m>2$. As examples we may take:

'(Mary, Tom)' is s-about {the waitress, the baker} in c_0 if, and only if, {Mary, Tom} = {the waitress, the baker}; and
'Mary loves Tom' is s-about {the waitress, the baker} in c_0 if, and only if, {Mary, Tom} = {the waitress, the baker}.

Again, under the same conditions of s-aboutness for '(Mary, Tom)', the sentences, 'Tom loves Mary', 'Tom is loved by Mary', 'Mary is loved by Tom', 'Tom hates Mary', *etc.*, are s-about the same s-item set {the waitress, the baker} that 'Tom loves Mary' is s-about, in contexts in which both 'Tom' and 'Mary' are extracted as subjects.

Now by standard set theory, if two ordered sets are identical, the unordered sets corresponding to them have the same members: $\vdash(\underset{\sim}{X}^m = \underset{\sim}{Y}^m) \supset (\tilde{X}^m = \tilde{Y}^m)$. Consequently, by (D12)(ii) and (D13) (i) we shall have:

$$\vdash_{CL}[(X = Y)\ \&\ (X' = Y')\ \&\ \ldots\ \&\ (X^{(m-1)\text{primes}} = Y^{(m-1)\text{primes}})]\ (\theta)$$
$$\supset [qu(\underset{\sim}{X}^m)\ \textit{is s-about}\ \tilde{Y}^m]\ (\theta).$$

In particular, therefore, taking $m = 2$,

$$\vdash_{CL}[(x = y)\ \&\ (x' = y')]\ (c) \supset [qu(x,x')\ \textit{is s-about}\ \{y,y'\}]\ (c);$$

and similarly for sentences:

$\vdash_{CL}[(x = y) \; \& \; (x' = y')] \, (c) \supset [qu \, f(x,x') \textit{ is s-about } \{y,y'\}] \, (c).$

Thus,

if Tom is the baker and Mary is the waitress, in c_0, then '(Tom, Mary)' and 'Tom loves Mary' are s-about {the baker, the waitress}.

Similarly,

if Tom is the baker and Mary is the waitress, in c_0, then 'Mary is loved by Tom', 'Tom hates Mary', 'Tom is the father of Mary' *etc.* are all s-about {the baker, the waitress}.

It is evident, then, that the s-aboutness of a sentence, the topic, is independent of what is said, the comment.

Now again by standard set theory we shall have: $\vdash \tilde{\mathrm{X}} = \tilde{\mathrm{X}}$. Alternatively we may regard this as a special case of (P13)(i) since set variables are within the range of subject arguments. But from (D13) (i) we have:

$\vdash_{CL}(qu(\underset{\sim}{\mathrm{X}}) \textit{ is s-about } \tilde{\mathrm{X}})(\theta) \equiv (\tilde{\mathrm{X}} = \tilde{\mathrm{X}})(\theta).$

Hence,

$\vdash_{CL}(103)$ $(qu(\underset{\sim}{\mathrm{X}}) \textit{ is s-about } \tilde{\mathrm{X}})(\theta)$; and similarly,

$\vdash_{CL}(104)$ $(qu(\mathrm{F}\underset{\sim}{\mathrm{X}}) \textit{ is s-about } \tilde{\mathrm{X}})(\theta).$

This gives us the trivial sense of 'about' in terms of which we may say, in every context, 'Tom' is about Tom, '(Tom, Joe)' is about the pair {Tom, Joe}, and 'Tom is taller than the butcher' is about {Tom, the butcher}. Because it holds, however, it follows by quantification logic that every subject and every elementary sentence is s-about something: *i.e.*

$\vdash_{CL}(105)$ $(P\underset{\sim}{\mathrm{Y}})(qu(\underset{\sim}{\mathrm{X}}) \textit{ is s-about } \tilde{\mathrm{Y}})(\theta)$

$\vdash_{CL}(106)$ $(P\underset{\sim}{\mathrm{Y}})(qu(\mathrm{F}\underset{\sim}{\mathrm{X}}) \textit{ is s-about } \tilde{\mathrm{Y}})(\theta).$

We read these as '$qu(\underset{\sim}{\mathrm{X}})$ [$qu(\mathrm{F}\underset{\sim}{\mathrm{X}})$] has an s-aboutness'. But it is important to remember here that the particular quantifier is not existential. In this sense of 'about' we may say that 'Pegasus', 'the purple number 7' and 'Wittgenstein's happy stove' are all about something—either in the trivial sense (*e.g.* 'the purple number 7' is about the purple number 7) or more informatively (*e.g.* 'Pegasus' is about a winged horse)—without being committed to the existence of the item mentioned or the significance of the description used.

We say that two subjects or two elementary sentences are s-about the same thing, or have the same s-aboutness, which we represent by '$=_{sa}$', if, and only if, they are both s-about the same item:

(D14) (i) $(qu(\underset{\sim}{\mathrm{X}}) =_{sa} qu(\underset{\sim}{\mathrm{Y}}))(\theta) =_{Df} (P\underset{\sim}{\mathrm{Z}})((qu(\underset{\sim}{\mathrm{X}}) \textit{ is s-about } \tilde{\mathrm{Z}})(\theta)$
$\& \; (qu(\underset{\sim}{\mathrm{Y}}) \textit{ is s-about } \tilde{\mathrm{Z}})(\theta))$

(ii) $(qu(\mathrm{F}\underset{\sim}{\mathrm{X}}) =_{sa} qu(\mathrm{G}\underset{\sim}{\mathrm{Y}}))(\theta) =_{Df} (qu(\underset{\sim}{\mathrm{X}}) =_{sa} qu(\underset{\sim}{\mathrm{Y}}))(\theta).$

Thus, 'Bill is taller than Fred' and 'Mr Bloggs is heavier than Mr Roll' have the same s-aboutness in c_0 if, and only if, '(Bill, Fred)' and '(Mr Bloggs, Mr Roll)' have the same s-aboutness; and this is so if, and only if, $(Pz)(Pz')$ ('(Bill, Fred)' is s-about $\{z,z'\}$ in c_0 and '(Mr Bloggs, Mr Roll)' is s-about $\{z,z'\}$ in c_0). Evidently this will be satisfied if, say, Bill Bloggs is the butcher and Fred Roll is the baker. Here, too, we emphasise that the particular quantifier is not existential so that we may say two subjects have the same s-aboutness even though they are not about existing items.

By (D13)(i) we have:
$\vdash_{\mathrm{CL}}$ $(qu(\mathrm{X})\ \textit{is s-about}\ \mathrm{Z})(\theta) \equiv (\mathrm{X} = \mathrm{Z})(\theta)$; and
$\vdash_{\mathrm{CL}}$ $(qu(\mathrm{Y})\ \textit{is s-about}\ \mathrm{Z})(\theta) \equiv (\mathrm{Y} = \mathrm{Z})(\theta)$.

Hence, by **RSL**,

$$\vdash_{\mathrm{CL}} ((qu(\mathrm{X})\ \textit{is s-about}\ \mathrm{Z})(\theta)\ \&\ (qu(\mathrm{Y})\ \textit{is s-about}\ \mathrm{Z})(\theta)) \equiv (\mathrm{X} = \mathrm{Z}\ \&\ \mathrm{Y} = \mathrm{Z})(\theta).$$

By the transitivity of identity, therefore,

$$\vdash_{\mathrm{CL}} ((qu(\mathrm{X})\ \text{is s-about}\ \mathrm{Z})(\theta)\ \&\ (qu(\mathrm{Y})\ \text{is s-about}\ \mathrm{Z})(\theta)) \supset (\mathrm{X} = \mathrm{Y})(\theta).$$

More generally, by **RQL,**

$$\vdash_{\mathrm{CL}} (P\mathrm{Z})((qu(\mathrm{X})\ \textit{is s-about}\ \mathrm{Z})(\theta)\ \&\ (qu(\mathrm{Y})\ \textit{is s-about}\ \mathrm{Z})(\theta)) \supset (\mathrm{X} = \mathrm{Y})(\theta).$$

For example: if there is some item which the phrases 'the morning star' and 'the evening star' are both s-about, then the morning star is the evening star.

The converse follows immediately from the properties of identity. For since we have $\vdash_{\mathrm{CL}}(\mathrm{X} = \mathrm{X})(\theta)$, then by **RSL**:

$$\vdash_{\mathrm{CL}} (\mathrm{X} = \mathrm{Y})(\theta) \supset ((\mathrm{X} = \mathrm{X})(\theta)\ \&\ (\mathrm{Y} = \mathrm{X})(\theta)).$$

Hence by **RQL,**

$$\vdash_{\mathrm{CL}} (\mathrm{X} = \mathrm{Y})(\theta) \supset (P\mathrm{Z})((\mathrm{X} = \mathrm{Z})(\theta)\ \&\ (\mathrm{Y} = \mathrm{Z})(\theta)).$$

That is, by (D13)(i),

$$\vdash_{\mathrm{CL}} (\mathrm{X} = \mathrm{Y})(\theta) \supset (P\mathrm{Z})((qu(\mathrm{X})\ \textit{is s-about}\ \mathrm{Z})(\theta)\ \&\ (qu(\mathrm{Y})\ \textit{is s-about}\ \mathrm{Z})(\theta)).$$

For example: if the morning star is the evening star then there is some item which 'the morning star' and 'the evening star' are both s-about.

Putting the two results together gives the equivalence thesis:

$$\vdash_{\mathrm{CL}} (\mathrm{X} = \mathrm{Y})(\theta) \equiv (P\mathrm{Z})((qu(\mathrm{X})\ \textit{is s-about}\ \mathrm{Z})(\theta)\ \&\ (qu(\mathrm{Y})\ \textit{is s-about}\ \mathrm{Z})(\theta)).$$

Consequently, by (D14)(i),

$$\vdash_{\mathrm{CL}} (\mathrm{X} = \mathrm{Y})(\theta) \equiv (qu(\mathrm{X}) =_{\mathrm{sa}} qu(\mathrm{Y}))(\theta).$$

Similarly for subject vectors and atomic sentences.

This result seems to capture leading features of the concept of extensional identity, namely that to say that two items are the same is simply to say that the phrases which are used to mention them are both s-about the same item: *e.g.* the morning star is the evening star if, and only if, there is some item which 'the morning star' and 'the evening star' are both s-about in the same context. Hence identity theses can always be represented as theses about subject terms or factors rather than theses about items. Put intuitively, when we say that two items are the same, we do not imply that there are two distinct items which are nevertheless one, but rather that there is one item which can be mentioned successfully by means of two distinct phrases. An alternative approach to identity, therefore, would be to take s-aboutness as primitive and to define '$X = Y$' in terms of '$qu(X) =_{sa} qu(Y)$'.

Taking Y to be X in the last result we have immediately:

$\vdash_{CL}$(107) $(qu(X) =_{sa} qu(X))(\theta)$.

In a given context, that is, every subject has the same s-aboutness as itself. In other words, a subject term, or set of subjects must be used univocally throughout a context. This is the condition of unambiguous s-aboutness on contexts.

It may be claimed that this condition is too strong, on the ground that proper names, especially, may be used to mention different items in the same context. On reflection, however, the condition is unobjectionable. For if we use, say, the name 'Tom' to mention different items, without first making it clear that we are doing so, then ambiguity does arise and contexts overlap. Commonly, ambiguity of this sort is avoided by distinguishing, say, Tom Jones from Tom Sawyer; that is, by introducing two distinct names. In the case of demonstrative pronouns the ambiguity is frequently resolved by—what can be represented as part of the context—pointing. If someone points to x_0 and says "This is red", and then to y_0 and says "This is red", the two uses of 'this' are effectively the uses of two distinct words since here both the noise and the pointing together make up a complex mentioning, and the two pointings are different because two different items are pointed out.

Our condition on unambiguous s-aboutness does not exclude disagreement and contradiction but it does exclude and so isolate certain cases of being at cross purposes. For example, if a person says, in a context, 'x_0 is red' and then 'x_0 is not red', then he contradicts himself by making inconsistent comments on one and the same s-item x_0; similarly if he says 'x_0 has the property of redness' and 'x_0 does not have the property of redness' then the subjects 'x_0' and 'the property of redness' are the same but the predicate, or comment, is different. On the other hand if someone says 'x_0 is red' in a context where he refers with 'x_0' to Petra; and someone also objects with 'x_0 is not red' referring to Leptis Magna, then they are at cross purposes, and the contexts are different and have to be separated. The gain that use of context offers over classical logic, where the unambiguity of subjects has *always* to be assumed, is clear.

The condition is necessary in order to ensure intracontext quantification; but because the condition may not be satisfied from one context to the next inter, or trans, context quantification is only sometimes permissible, namely when the relevantly same subject occurs in the contexts quantified across. *When* intercontext quantification is significant sameness of s-aboutness can be defined for different contexts, as follows:

$$\left.\begin{array}{l} qu(\underset{\sim}{\mathrm{X}})(\theta) =_{\mathrm{sa}} qu(\underset{\sim}{\mathrm{Y}})(\eta) \\ qu(\mathrm{F}\underset{\sim}{\mathrm{X}})(\theta) =_{\mathrm{sa}} qu(\mathrm{G}\underset{\sim}{\mathrm{Y}})(\eta) \end{array}\right\} =_{\mathrm{Df}} (P\underset{\sim}{\mathrm{Z}})[(qu(\underset{\sim}{\mathrm{X}}) \textit{ is s-about } \tilde{\mathrm{Z}})(\theta) \mathbin{\&} (qu(\hat{\mathrm{Y}}) \textit{ is s-about } \tilde{\mathrm{Z}})(\eta)].$$

But one cannot conclude using $\vdash_{\mathrm{CL}}(105)$ that $qu(\underset{\sim}{\mathrm{X}})(\theta) =_{\mathrm{sa}} qu(\underset{\sim}{\mathrm{X}})(\eta)$, and therefore that context does not matter; for the conditions for well-defined *i.e.* significant, quantification are frequently violated. In fact the definition is not much help, since from our absolute (or logically standard) position as regards contexts it is apparent that the requirement that quantification be well-defined is at least that $\tilde{\mathrm{Z}}_\theta = \tilde{\mathrm{Z}}_\eta$ (where $\tilde{\mathrm{Z}}_\theta$ takes up sufficient of the θ-context, and $\tilde{\mathrm{Z}}_\eta$ sufficient of the η-context, for us to make an identification in a new context).

Because nonsignificant descriptions may be used, or non-existent items mentioned, so it is necessary to introduce narrower senses of 'about' in order to distinguish the different cases and obtain such familiar notions as those of reference and entity, and to derive the usual ranges of variables.

First, however, we extend this wide sense of 'about' to compound and quantified sentences constructed from atomic semi-wffs. Our basic definition defines aboutness for atomic sentences. We take this definition as the inductive basis and specify a set of rules, rather after the pattern of formation rules, in terms of which the s-aboutness of more complicated sentences can be determined.

Consider first negated sentences. We shall say, what seems customary for sentences which do not contain quantifiers, that the negation of a sentence is s-about whatever the original sentence is s-about. Thus, if Γ is a semi-wff, then in a given context,

R7 (i) if $qu(\Gamma)$ *is s-about* $\tilde{\mathrm{X}}$, then $qu(\sim\Gamma)$ *is s-about* $\tilde{\mathrm{X}}$.

So, for example, if 'Tom is happy' is s-about the butcher in some context, then, in that same context, 'Tom is not happy' is s-about the butcher also. Again, if 'John loves Margaret' is s-about the pair {the man across the road, the girl next door}, then so, too, is 'John does not love Margaret'. Different things are being said, evidently, but what is being said is being said about the same items; the comment is different, but the topic the same.

Consider now the case of conjunction. If 'Tom is happy' is s-about the butcher and 'Bill is sad' is s-about the baker, then 'Tom is happy and Bill is sad' is s-about both the butcher and the baker. The conjunction, therefore, is about the union of the sets which the two sentences separately are s-about.

Since the set union imposes no ordering, Γ & Δ is s-about what Δ & Γ is s-about, as we want. For the ordinary sense of 'and', as distinct from the tensed connective 'and then', guarantees a commutative law which in most contexts allows us to say that Γ & Δ and Δ & Γ are equivalent. Since we wish to retain this general principle, we are committed to the commutativity of & in any many-valued sentential system which we adopt for semi-wffs. Since the conjunction of sentences is s-about the union of what its conjuncts are about, it is about the united set of these elements; but it is not, in general, about the elements of this union.

The components of a conjunction may themselves contain *m*-place vectors. Consider, for example, 'Tom loves Mary and Tom owes £5 to Bill'. If 'Tom loves Mary' is s-about {the butcher, the waitress} and 'Tom owes £5 to Bill' is s-about {the butcher, £5, the baker} then 'Tom loves Mary and Tom owes £5 to Bill' is, in this context, s-about {the butcher, the waitress, £5, the baker}. Thus s-aboutness has a certain stability under sentence transformation and does not depend critically on the choice of atomic sentences. We should for instance have the same s-aboutness in the given context for 'F(Tom, Mary, £5, Bill)'. Generally then,

R7 (ii) if $(qu(\Gamma)$ *is s-about* $\tilde{X})(\theta)$ and $(qu(\Delta)$ *is s-about* $\tilde{Y})(\theta)$ then $(qu(\Gamma \mathbin{\&} \Delta)$ *is s-about* $\tilde{X} \cup \tilde{Y})(\theta)$.

The condition for a conjunction with any number of components follows immediately; and conditions for sentences of the form $\Gamma \vee \Delta$, $\Gamma \supset \Delta$, $\Gamma \equiv \Delta$ either follow analogously, or from definitions of their connectives in terms of '∼' and '&'. It follows that the ordering of components in all compounds is irrelevant to their s-aboutness. Further, any two compound sentences with the same components (whether or not these appear negated) have the same s-aboutness independently of the connective employed. But of course this in no way implies that they mean the same. As in the case of negation, different things are being said, but about the same items.

The recipe for quantified sentences is not quite so simple. At this stage we limit attention to unrestricted quantifiers and take up the problem of restricted quantifiers in volume 2.

If we conceive a simple universally quantified sentence with a single one-place predicate on the analogy of an infinite conjunction, then, where the constants of the logic are 'x_0', 'y_0', 'z_0', 'x_0'', ..., we can represent

$$(Ux)f(x) : f(x_0) \mathbin{\&} f(y_0) \mathbin{\&} f(z_0) \mathbin{\&} f(x_0') \mathbin{\&} \ldots .$$

Now if 'x_0', 'y_0', ..., are respectively s-about x^*, y_0^*, ..., then $qu(fx_0)$, $qu(fy_0)$, ..., are respectively s-about x_0^*, y_0^* ..., and by the condition for conjunction, $qu((Ux)fx)$ will be s-about the set $\{x_0^*, y_0^*, \ldots\}$. That is, writing this universal set of all items (as fixed by the given context) as $\mathbf{U}$, $qu((Ux)f(x))$ is s-about $\mathbf{U}$. In fact $\mathbf{U}$ is implicitly contextually qualified since what we define is s-aboutness in a given context. Since 'x_0' is s-about x_0, 'y_0' s-about y_0, *etc.*, we may set $\mathbf{U} = \{x_0, y_0, ...\}$. Similarly conceiving of $(Px)f(x)$ on the analogy of an infinite disjunction (or as the limit of increasingly long disjunctions) it follows that $qu((Px)f(x))(c)$ is s-about $\mathbf{U}$.

In terms of this, therefore, the two sentences 'Everything is red' and 'Something is red' would have the same s-aboutness. Again, of course, this is not to say that they have the same meaning or even the same truth conditions.

Generalising this account, a sentence containing a non-vacuous quantifier is, in standard contexts, about $\mathbf{U}$, since the union of $\mathbf{U}$ with any set is $\mathbf{U}$. Thus, where Q is a non-vacuous quantifier binding some variables in a semi-wff Γ,

R7 (iii) $(qu(Q\Gamma)$ *is s-about* $\mathbf{U})(\theta)$.

It follows then that sentences, however compounded from '&', '$\vee$' and '$\sim$' using the same components, are s-about the same s-items; and that every sentence containing one non-vacuous unrestricted quantifier is about $\mathbf{U}$, the universe of discourse. This provides a complete recipe for the closed wff of the usual quantification logic; for vacuous quantifiers may be deleted. It is apparent from the account that the s-aboutness of *sentences* is quite worthless from the point of view of distinguishing sentences as to their meaning or content. For different compounding and different quantification both make differences in meaning and content. Since, as we shall explain, reference is but a special case of s-aboutness, reference likewise is worthless as an explanation of meaning or content of sentences. It is of course possible to give different recursion rules from those we have given here for s-aboutness, rules which make further discriminations. But these rules tend, our investigations seem to show, to have overwhelming drawbacks: first, what a sentence is s-about comes to depend critically on its logical form and is sensitive to variations that appear not to affect content; and second, the rules are not just complex but rather arbitrary, suggesting that there is no stable logical notion to explicate. Why s-aboutness and reference of sentences *taken on their own* are inadequate to provide an explication of content is, of course, that though s-aboutness and reference specify the topic they do not discriminate sentences which make different comments on the same topic. In order to discriminate such sentences more than reference is needed. In fact even ardent reference theorists do not try to work entirely in terms of reference. To reach an account of *content* it is common to supplement an account of s-aboutness (or reference) by sets of truth conditions. However, truth conditions are inadequate unless they are supplemented by, or taken to include, significance and completeness conditions; and when so supplemented s-aboutness proves otiose, since whatever can be done using the expanded truth conditions and s-aboutness can be done with the expanded truth conditions alone (see §7.3, where domains are eliminated). Finally, even expanded truth conditions are inadequate in the case of sentences containing intensional connectives unless the extended truth conditions are relativised to classes of contexts or set-ups (see §7.2).

For related reasons the thesis (of Ryle) that the meaning of an expression can be explained in terms of its significance couplings is mistaken; for different expressions of the same (minimal) categories will have the same significance couplings. Just as truth conditions have to be supplemented by significance conditions, so significance conditions have to be supplemented by truth conditions.

The theory of s-aboutness also reveals once more (thus adding to §3.3) the inadequacy of the standard paraphrase of 'all' and 'some' sentences in logical notation. For sentences like 'Everyone loves someone' and 'All the books on Moore's shelf are blue' are not both about the total universe of discourse, but the first is about all people and the second about books on Moore's shelf. This problem can be solved by a proper use of restricted variables, which are not in turn eliminable by many-valued extensional connectives. For example, 'All the books on Moore's shelf are blue' will be rendered 'For every x such that x is a book on Moore's shelf, x is blue', and this will be s-about every s-item identical with a book on Moore's shelf, and not s-about **U**. (The full analysis of traditional categorical sentences will be undertaken when range theory is developed in volume 2; but see §7.9, §6.)

Aboutness relations vary as the identity relation used in the basic definition. As we have remarked, where we have an intensional sentence frame such as 'It is logically necessary (that ...)' or 'Bill believes (that ...)', it is no longer satisfactory in general to use extensional identity in the determination of s-aboutness, *given* that one wants to guarantee that when a sentence yields a true statement, its predicate is true of the s-item which the sentence is s-about. Thus 'the morning star is necessarily a morning phenomenon' is not s-about Venus, though Venus is (extensionally) the morning star, since it is not true of Venus that it is necessarily a morning phenomenon. For this reason we did not extend the rules for s-aboutness to sentences containing intensional sentence predicates, though it is not difficult to see more or less how they would go, namely each intensional connective would require the use of a suitably strengthened identity relation (one at least strong enough and preferably just strong enough to ensure intersubstitutivity) in the initial clause defining s-aboutness of subjects.

We turn now to the problem of distinguishing more restricted aboutness relations. 'Jack' may indeed be s-about the purple number seven in a given context: because, however, the phrase 'the purple number seven' is nonsense, there is, in a narrower and more familiar sense of 'about', no item which 'Jack' is really about. Again, in literal contexts, although the phrase 'the present king of France' is significant and is accordingly about some item, there is an even narrower sense of 'about' and a correlated sense of 'thing' in which the phrase is about nothing since there is no such *thing* as the present king of France. In particular, we wish to distinguish s-aboutness from reference. We shall say that subjects which are s-about s-items which exist have *a reference*, and have as *referent*, or *entity* they *refer to*, an existent s-item. In terms of this, both the significant phrase 'the present king of France' and the nonsignificant phrase 'the purple number seven' have no reference, but 'the present Prime Minister of England' has a reference. Thus a subject has a reference in a given context iff it is s-about an item which exists; and similarly for sentences. Thus, in terms of the primitive predicate 'E' read 'exists', we define the relation **R**, read 'refers to', as follows:

(D15) (i) $(qu(\mathrm{X})\ \mathbf{R}\mathrm{Y})(\theta) =_{\mathrm{Df}} (qu(\mathrm{X})$ *is s-about* $\mathrm{Y}\ \&\ E(\mathrm{Y}))(\theta)$.
(ii) $(qu(\mathrm{FX})\ \mathbf{R}\mathrm{Y})(\theta) =_{\mathrm{Df}} (qu(\mathrm{X})\ \mathbf{R}\mathrm{Y})(\theta)$.

For each context θ, **R** is a function, since $(qu(\mathrm{FX})\ \mathbf{R}\mathrm{Y})(\theta)$ and $(qu(\mathrm{FX})\ \mathbf{R}\mathrm{Z})(\theta)$ imply $(\mathrm{Y} = \mathrm{Z})(\theta)$, by the properties of extensional identity. Hence the converse relation $\mathbf{R}^c$ may be read 'is *the* reference of'. It follows from the definition that

$\vdash_{\mathrm{CL}}(108)$ $(qu(\mathrm{FX})\ \mathbf{R}\mathrm{X})(\theta) \equiv E(\mathrm{X})(\theta)$ [*i.e.* $\equiv (qu(\mathrm{X})\ \mathbf{R}\mathrm{X})(\theta)$].

Thus reference may alternatively be taken as a primitive and existence defined, but postulationally this alternative course is more costly since the connexion of reference with identity, for example, has then to be assumed. Once the logic is enriched by predicate connectives, existence can be defined.

In a standard context the significant phrase 'the present king of Australia' has no reference, since where c_8 is a standard context the present king of Australia does not exist. If, however, in a non-standard, but literal, context someone uses 'the present king of Australia' to mention Mr Gorton, then it has a reference since 'Mr Gorton' has a reference. This sort of example, like that of metaphor, emphasises the important part played by context, and shows, to the detriment of simple reference theories, why it is impossible to regard any word or phrase as having a permanent reference associated with it.

Commonly remarked features of reference follow from the account given. For example it follows that reference is an identifying relation and that it identifies an existent item. Using the theory of descriptive functions (outlined in §7.9 *ff*) we can study the properties of the reference of $qu(\mathrm{X})$; for

(D16) $\mathbf{R}`qu(\mathrm{X})(\theta) =_{\mathrm{Df}} (\iota \mathrm{Y})(qu(\mathrm{X})\ \mathbf{R}\mathrm{Y})(\theta)$.

The definition of reference can be extended in turn to, (1) subject term vectors—a subject term vector $qu(\underset{\sim}{\mathrm{X}})$ refers to $\tilde{\mathrm{Y}}$ in a given context iff $qu(\underset{\sim}{\mathrm{X}})$ is s-about $\tilde{\mathrm{Y}}$ there and also *each* element of $\tilde{\mathrm{Y}}$ exists; otherwise only some or none of the members of $qu(\underset{\sim}{\mathrm{X}})$ will refer to the s-items they are s-about; (2) multi-subject atomic sentences; (3) compound sentences—for example, if $(qu(\Gamma)\ \mathbf{R}\tilde{\mathrm{X}})(\theta)$ and $(qu(\Delta)\ \mathbf{R}\tilde{\mathrm{Y}})(\theta)$ then $(qu(\Gamma\ \&\ \Delta)\ \mathbf{R}\ \tilde{\mathrm{X}} \cup \tilde{\mathrm{Y}})(\theta)$; and (4), quantified sentences—whatever $qu(\Gamma)(\theta)$ refers to, $(qu(Q\Gamma)\ \mathbf{R}\ E\mathbf{U})(\theta)$, where $E\mathbf{U}$ is the universal class of entities with respect to given context θ. An *entity* is an s-item that exists.

Two subjects or two sentences have the same reference in a given context iff they refer to the same entity, *i.e.*

(D17) (i) $(qu(\mathrm{X}) =_{\mathrm{r}} qu(\mathrm{Y}))(\theta) =_{\mathrm{Df}} (P\mathrm{Z})(qu(\mathrm{X})\ \mathbf{R}\mathrm{Z}\ \&\ qu(\mathrm{Y})\ \mathbf{R}\mathrm{Z})(\theta)$.
(ii) $(qu(\mathrm{FX}) =_{\mathrm{r}} qu(\mathrm{GY}))(\theta) =_{\mathrm{Df}} (qu(\mathrm{X}) =_{\mathrm{r}} qu(\mathrm{Y}))(\theta)$.

In particular then,

$$\vdash_{\mathrm{CL}} (qu(x) =_{\mathrm{r}} qu(x))(c) \equiv [(Pz)(x = z\ \&\ E(z))](c) \equiv E(x)(c).$$

Thus for a subject to have the self-same reference the item must exist; and $=_{\mathrm{r}}$ is an "identity" relation which is reflexive only when an existence condition is met. 'The morning star' and 'the evening star' have the same reference; but 'Pegasus' and 'the famous winged horse', though s-about the one item, do not have the same reference, since they neither have a reference. In this respect what we say contrasts with the usual, but invalid (given our account of quantifiable subjects), Alice-in-Wonderland argument according to which 'Pegasus' and 'the famous winged horse' have the same reference (*e.g.* the non-entity) since they both have no reference. Defining a relation which absorbs the quotation functions,

(D18) $\mathrm{X} =_{\mathrm{d}} \mathrm{Y} =_{\mathrm{Df}} qu(\mathrm{X}) =_{\mathrm{r}} qu(\mathrm{Y})$

we obtain a relation, $=_{\mathrm{d}}$, which behaves, for each context, like Russell's identity relation with respect to described items and classes; for example,

(the x which $f =_{\mathrm{d}}$ the x which f) $\equiv$. the x which f exists.

The apparatus we have developed for defining referents and reference relations for subjects and for sentences can be readily adapted to define other kinds of s-items. In fact it is only necessary to couple with the contraction defining s-items a relevant special predicate sorting out the kinds of s-items in question. Suppose, for example, we wished to distinguish possibly existing or, as they are sometimes called, *subsistent* items. Given the modal predicate '$\Diamond$' (of §7.2), read 'is possible', we define subsistence:

$$\Diamond(\mathrm{X})(\theta) =_{\mathrm{Df}} (\Diamond\S E(\mathrm{X}))(\theta).$$

By replacing in the definitions of '**R**', '$=_{\mathrm{r}}$', *etc.*, 'E' by '$\Diamond$', we obtain definitions of '**D**', '$=_{\mathrm{des}}$', *etc.*, which are read 'designates', 'has the same designation as' *etc.* Then in a standard context c_{s}, 'the round square' and 'the seat of the bed' (Smart's example, which is discussed in §7.9) have no designation, and the items, the round square, and the seat of the bed, fail to subsist; but 'Pegasus', though it has no reference, does designate, and designates the subsistent item, Pegasus. Since existence implies subsistence, whatever has a reference in a given context also, thereby, designates. Though the subsistence and designation terminology can be given a technically impeccable introduction, it needs to be handled with care since the mere use of th terminology in the sense we have given—only one of the varying senses in which the expressions are commonly used—leads too many philosophers to conclude, quite invalidly, that some kind of existence and determinacy, as distinct from the mere logical possibility of existence, is thereby ascribed to subsistent items.

Given the further notion of a significant subject (discussed in §7.9 §2) we can define a full aboutness relation and associated items. For example,

$$(qu(\underset{\sim}{\mathrm{X}})\ \textit{is about}\ \tilde{\mathrm{Y}})(\theta) =_{\mathrm{Df}} (\tilde{\mathrm{X}} = \tilde{\mathrm{Y}}\ \&\ S\tilde{\mathrm{Y}})(\theta),$$

where a subject set is significant iff each subject in the set is. Thus, in standard contexts the eye of a needle is an item but the liver of 4 o'clock is not. Then the field of s-items comprises the following subsets:

s-items, items, subsistents, entities,

with each set including the next.

One of our aims has been to clarify the concepts of subject-aboutness and reference; and we can then distinguish different kinds of mentionings. Given the trip-over sense of 'exists', to mention abstract items is not the same as to mention cats and dogs. We may refer to the latter but not the former. The philosopher Socrates is now only designated when he is mentioned, but Quine is referred to (in 1968). This is not to say, however, that all sentences which have no reference thereby fail to have a truth value. On the contrary, we can make true statements about both numbers and Socrates. The truth procedures, however, are different (though a uniform truth definition can be provided). It is not in point to look for a designation in order to observe its properties but it is always in point, in principle, to look for a reference.

The contraction relations defined are special cases of a more general relation, *K-I-contraction*, where K is a one-place predicate and I an identity predicate:

(D19) $(\mathrm{Y}\ \textit{is the K-I-contraction of qu}(\mathrm{X})(\theta)) =_{\mathrm{Df}} (\mathrm{X}I\mathrm{Y}\ \&\ K(\mathrm{Y}))(\theta).$

I-contraction is that case where the further property holds of all subject items, *e.g.* where the property is self-identity. Not only reference and designation relations are special contraction relations: the *yields* relations of Chapter 2 is another special contraction relation, where the subjects are quoted sentences, the identity is that of statement-equality among sentences, and the special property is that of being truth-valued. The subject item defined by the contraction is, of course, the statement. A contraction is, then, a function of three elements: the class of subjects on which the contraction is made, the identity relation used in forming equivalence sets, and the special property.

Contraction relations are not limited to reference and designation relations in the usual philosophers' senses. Contraction can perfectly well be carried out with stronger identity relations than those used in determining reference, and without supposing that the item identified exists. On the other hand it is clear why many philosophers have wanted to assimilate all contraction relations to reference, designation or naming relations. For it is a consequence of any reference theory that there is just one sort of identity, identity of reference, and that whatever is talked about successfully is referred to and exists.

By contracting on stronger subject identity relations we can define such notions as *sense* and *concept*. The central relation needed, in terms of which other notions can be defined, is that of synonymy of expressions. We define *complete* synonymy in terms of Leibnitz identity, thus:

(D20) $(qu(X)\ csyn\ qu(Y))(\theta) =_{Df} (X \approx Y)(\theta)$.

Thus *csyn* is an equivalence relation on subjects, weaker than type-identity of expressions, which warrants intersubstitutivity in all *non-quotational* sentence frames. The emphasis on 'non-quotational' is important; for many objections to a *salva veritate* (plus *salva significatione*) synonymy principle appeal to "counterexamples" using quotational sentence frames. The account also seems to capture what is right in the meaning-is-use thesis. This thesis is not the thesis that the meaning of a subject consists of its actual uses, because, as Wittgenstein pointed out, meaning is not distributed over time as uses are, because many uses are irrelevant to meaning, and because actual uses are often not enough. It is rather that the meaning of a subject is given by the class of its relevant correct uses. Now the relevant declarative uses of a subject can be given through the class of sentence frames in which the subject occurs grammatically, but the correct uses are not so easily characterised. What seems clear, however, is that *qu*(X) has the same declarative use as *qu*(Y) in context θ iff for every non-quotational sentence frame *qu*(F), $(F(X) \simeq F(Y))(\theta)$. But then, under generally accepted assumptions, it follows that *qu*(X) is completely synonymous with *qu*(Y) in a context iff *qu*(X) has the same declarative use as *qu*(Y) in that context.

As a preliminary to defining *sense* it helps to distinguish two ways in which one may *give* the sense of a subject such as 'Bert's brother'; one way is to give a synonymous expression, such as 'Bert's male sibling'; another, the Frege-Church way, is to specify a non-linguistic item, *e.g.* the man who is sibling of Bert. We shall use 'sense' in the second way which puts sense on a par with reference, so that when a subject *expresses* a sense (Church's terminology [4], p. 6) a linguistic expression is related to an item. But we can cater for the first way, as follows:

$(a\ synonym\ for\ qu(\mathrm{X}))(\theta) =_{\mathrm{Df}} \xi qu(\mathrm{Y})(qu(\mathrm{Y})\ csyn\ qu(\mathrm{X}))(\theta)$,

(where ξ is the arbitrary selection operator featured in §7.9 *ff.*). As to the second, we define

(D21) (i) $(qu(\mathrm{X})\ expresses\ sense\ \mathrm{Y})(\theta) =_{\mathrm{Df}} (\mathrm{X} \approx \mathrm{Y})(\theta)$, and
(ii) $\mathrm{S}`qu(\mathrm{X})(\theta) =_{\mathrm{Df}} (\iota \mathrm{Y})(\mathrm{X} \approx \mathrm{Y})(\theta)$.

The concept of X, which is the same as the sense of *qu*(X), in one sense of 'concept', is similarly defined.

It is common, especially in translating and providing analyses, to work with a less exacting notion of sense than that characterised in terms of complete synonymy. What this amounts to is preparedness to work with an identity relation which guarantees intersubstitutivity not in all but only in some intensional sentence frames. Once the relevant class of sentence frames is marked out, by no means a simple task usually, the corresponding identity *I* can be defined or axiomatised and the definitions we have just given can be reapplied, with '*I*' supplanting '$\approx$', to define the new notions. In particular the definitions we present are mostly weaker than complete synonymy relations. The logicist definition of number we offer (in §7.13) for example, of the number *n* as a certain property of classes, though providing at least a strict identity, does not, and is not intended to preserve truth-values under substitutions in such frames as 'x_0 learned (that ... $=n$)'. We should emphasise too that significance logics and significance theory are not tied to one particular account of *statement* and *sense*; they are compatible with various different accounts, accounts which may vary as regards the identity relations used in characterising the notions contractionally.

A further dimension has to be allowed for in a full theory of contraction. For the results of contraction are typically context-independent, but what is contracted upon, subjects, are not. Such s-items as statements and entities are stable items that ought to escape the vagaries of context; they ought to be invariants which do not change when contexts change. In fact part of the point of introducing statements at all, and of founding the semantics of logics on items and restricting the ranges of variables to s-items, just is to make the notions context- and language-invariant. But the idea of a total escape from context in doing logic is illusory. For even extensional logic deals with expressions and their features, in particular, sense features; and no expression has these features *per se* in a context-independent way. Moreover, so far as books are concerned, expressions in context are the only general way we have of getting at items by contraction relations. Earlier, in the use of statements, we may have appeared to escape the problem of context by use of context-invariant statement variables; but in fact we fixed the contexts so that variables '*a*' '*b*' *etc.* were restricted in *those* contexts to ranging contraction-wise over statements. Thus statement variables were introduced through a function whose arguments were sentence and context variables.

The fact remains, however, that described s-items are independent of context in a way that descriptions are not, and that this difference should be reflected in the characterisation of s-items. The relevant features can be brought out using a trans-context identity. Described s-items are independent of context inasmuch as different descriptions in different contexts may describe the same s-items. Thus an s-item itself may be characterised as the contraction of its descriptions in different contexts. More generally,

Y(θ) *is the K-I-contraction of qu*(X)(η) iff X(η) *I* Y(θ) & (*K*(X))(η).

For example, where d_0 is the context determined by the Post Office coding and c_0 is a suitable context determined by our conversation,

the purple number seven (d_0) is the *E*-contraction (reference) of 'Mrs. Jones's house' (c_0), since, in fact, (Mrs. Jones's house exists) (c_0) and Mrs. Jones's house is, according to the Post Office coding, the purple number seven.

Once trans-context connexions—which can be built up from trans-context covalue conditions, *e.g.* from such facts as that $\Gamma(\theta)$ has the same truth-value as $\Delta(\eta)$—are admitted, various of the notions we have introduced split in two. Take the case of sentences. Within a single context 'I am happy' is synonymous with 'I am happy', and they each yield the same statement. But consider these type-identical sentences in different contexts, say in c_1 where the speaker is person x_1, and in c_2 where the speaker is x_2. It is tempting to say that the sentences remain synonymous, provided that the contexts are standard English contexts, but that the statements yielded are different: at least something is different and what better way to mark the distinction than to deploy the synonymy/statemental identity terminology to do it? However, to make the differences between trans-context and purely intra-context notions formally sharp and to formalise trans-context synonymy we should have to inflate greatly the resources of context logic. This is an important task but one we prefer to postpone.

Contraction relations and their refinements provide the relations, among other things, between language and the world. They do not provide the only such relations, since they do not include, without much further ado, relations like extension, but they provide a logically very important class. To guard against misunderstandings over our views on the relations between language and the world that the theory of contractions may generate, we want to stress that we are not saying that language determines the world (what is the case; the totality of truths, not of things); we are not saying that what exists depends on how we talk. For whether the basic identity relations, in terms of which reference and so forth are determined, hold or not depends on what is the case, on what features items do actually have.

The contrast between subjects and their contractions clarifies an important reason why disciplines like significance and intensional logics are obliged to take variables as ranging over linguistic expressions and not over their contractions (a point further explained in §3.8). For contractions by identifying differently described items may blur or eliminate features of descriptions which are important for the assessment of significance, necessity and other intensional properties. Indeed switching to contractions as values of variables results in modal and significance paradoxes (see §3.8). For a number of different descriptions with different modal and significance properties have the same contractions. In a logic where primary values of variables are subject-items any property that does not transfer over the identity used in the contraction concerned generates difficulties, and can at best be approximated in the logic. Such a logic can typically only cope with, what is defective, an essentialist account of significance, according to which significance features are ascribed to s-items no matter how they are described. Working as we do, however, with a much richer logical apparatus, in terms of which finer discriminations can be made, problems like those of significance essentialism (§3.8) on the one side; and conventionalism (§4.5) on the other, are avoided; yet, through contraction, items can be introduced. Using identity we can take up as much of the context as reference theorists do. Naturally the more elaborate theory introduces problems, like those tied up with introducing context-independent items in terms of context-dependent descriptions, which do not arise for classical logic.

3.7 Predicates, Properties and Universals

Predicates, unlike subjects, are not about anything; so we shall not try to define an aboutness relation for predicates. That predicates are not about anything follows from the fact (argued in §3.2) that predicate places are not open to quantification; for only if they were so open could we instantiate to obtain items the predicates are about. (Such operations as negation also point up the differences between predicates and subjects: *cf.* 'is not red' with 'not redness'.) Conversely, to suppose that predicates were significantly about something would be to suppose that predicates are subjects of some sort, and hence that predicate positions are open to quantificational moves. But as a predicate is an incomplete, unsaturated expression, how could it hold a subject place, how could it be about something?

What are we to make then of the common claim that predicates are about properties, relations, activities and so on; that the predicate 'is white' is about whiteness, both 'runs' and 'is running' about the activity of running, and 'is believed to be between ... and ... gardening' about some complex activity? These claims combine by composition relevantly distinct relations.

First there are important elementary transformations which convert predicates to subjects. Subjects may be formed from predicates either by quotation of the predicate or by conversion of the predicate (as remarked by

Frege [1]). By the second method an attributive, *i.e.* an attribute *expression*, is formed from a predicate—a technique similar to the English transformation whereby attributives are formed from predicates by means of such suffixes as '-ness', '-ity' and '-hood'. The method generalises on the grammatical transformations, however, because given a many-place predicate with distinct subject places held by distinct variables, conversion may be applied to some but not all variables; given for example the predicate 'x is between y and z' with its three places marked by subject variables, we may form in turn '$(\lambda x)(x$ is between y and $z)$', read 'the property of x such that x is between y and z', '$(\lambda y)(\lambda x)(x$ is between y and $z)$' and '$(\lambda z)(\lambda y)(\lambda x)(x$ is between y and $z)$'. Only the last represents the attributive 'betweenness'. (The logic of this λ-conversion, a variation of that developed by Church [1], is discussed in Chapter 7.) An attributive is then obtained from a predicate by λ-conversion of all its places, *i.e.* by *full conversion* of the predicate. In brief, where F^m is an m-place predicate, its full conversion is $(\lambda \underset{\sim}{X}^m)\, F^m \underset{\sim}{X}^m$, which we write as λF, and the relation,

$$(\alpha){:}\ \ qu(\lambda F)\ f\text{-}conv\ qu(F)$$

holds; '*f-conv*' is read 'is the full conversion of'. Full conversion is a function.

The condition (α) follows immediately from the following definition,

(D22) $(qu(Y)\ f\text{-}conv\ qu(F))\ (\theta) =_{Df} (Y \approx \lambda F)(\theta)$,

by taking Y to be λF and using (P14)(i). We drop the two-valued marker '+' on '≈' since we are only concerned with two-valued formulae.

Because attributives are subjects, they are s-about certain s-items, attributes typically. Composing the s-aboutness function with *f-conv* yields a relation, the predicate-property relation, between predicates and s-items. The predicate-property relation is analogous to a contraction relation. Although the relation is of course not an aboutness function, the connexions suffice to explain the common claim that predicates are "about" properties and relations:

(D23) [X *is the property determined by (associated with)* qu(F)](θ) $=_{Df}$ $(PY)[qu(Y)\ f\text{-}conv\ qu(F)$ & X *is the contraction of* qu(Y)](θ).

Since the relation defined is many-one, the reading '*the* property ...' is warranted. The definiendum may also be read '(*Predicate* qu(F) *specifies* X)(θ)' or, at a risk, '(qu(F) *connotes* X)(θ)'. Only strong identity contractions are usually tolerated in defining properties; otherwise part of the point of introducing properties—to solve puzzles concerning subjects which are s-about the same item but nonetheless have different roles—is defeated. For simplicity we suppose that the contraction is made using Leibniz identity; then (D23) becomes:

(D23*) $(qu(F)\ connotes\ X)(\theta) =_{Df} (PY)(qu(Y)\ f\text{-}conv\ qu(F)\ \&\ (X \approx Y))(\theta)$.

This, as the consequences suggest, seems to capture one, but only one, of the common notions of a property. For we have, by (D22) and **RSL**,

$\vdash_{CL}$ $((qu(\mathrm{Y})\ f\text{-}conv\ qu(\mathrm{F}))\ \&\ (\mathrm{X} \approx \mathrm{Y}))(\theta) \equiv ((\mathrm{X} \approx \mathrm{Y})\ \&\ (\mathrm{Y} \approx \lambda\mathrm{F}))(\theta)$.

Hence, by the transitivity of $\approx$ and **RQL**,

$\vdash_{CL}$ $(P\mathrm{Y})(qu(\mathrm{Y})\ f\text{-}conv\ qu(\mathrm{F})\ \&\ (\mathrm{X} \approx \mathrm{Y}))(\theta) \supset (\mathrm{X} \approx \lambda\mathrm{F})(\theta)$.

Conversely, since $\vdash_{CL}(\mathrm{X} \approx \mathrm{X})(\theta)$, we have by **RSL**,

$\vdash_{CL}$ $(\mathrm{X} \approx \lambda\mathrm{F})(\theta) \equiv (\mathrm{X} \approx \mathrm{X}\ \&\ \mathrm{X} \approx \lambda\mathrm{F})(\theta)$.

Hence by **RQL**,

$\vdash_{CL}$ $(\mathrm{X} \approx \lambda\mathrm{F})(\theta) \supset (P\mathrm{Y})(\mathrm{X} \approx \mathrm{Y}\ \&\ \mathrm{Y} \approx \lambda\mathrm{F})(\theta)$.

By (D22), therefore,

$\vdash_{CL}$ $(\mathrm{X} \approx \lambda\mathrm{F})(\theta) \supset (P\mathrm{Y})(\mathrm{X} \approx \mathrm{Y}\ \&\ qu(\mathrm{Y})\ f\text{-}conv\ qu(\mathrm{F}))(\theta)$.

Putting these two results together, we have the equivalence thesis:

$\vdash_{CL}$ $(P\mathrm{Y})(qu(\mathrm{Y})\ f\text{-}conv\ qu(\mathrm{F})\ \&\ (\mathrm{X} \approx \mathrm{Y}))(\theta) \equiv (\mathrm{X} \approx \lambda\mathrm{F})(\theta)$.

Hence by (D23*),

$\vdash_{CL}$(109) $(qu(\mathrm{F})\ connotes\ \mathrm{X})(\theta) \equiv (\mathrm{X} \approx \lambda\mathrm{F})(\theta)$.

Thus, suppose we have a code such that 'is prime' means 'is dangerous'. Then, 'is prime' connotes dangerousness *in that context*; which is to say that, for that context, primeness is (is taken to be) dangerousness. Since further,

$\vdash_{CL}$(110) $(qu(\mathrm{F})\ connotes\ \lambda\mathrm{F})(\theta)$,

follows immediately from $\vdash_{CL}$(109) by taking X to be λF and using (P14)(i), the connotation relation has some resemblance to Mill's notion of connotation. It is that concept in terms of which we can say, in every context, that 'is red' connotes redness, and 'is white' (not Mill's 'white') connotes whiteness.

Again, by **RQL** from $\vdash_{CL}$(110) we have:

$\vdash_{CL}$(111) $(P\mathrm{Y})(qu(\mathrm{F})\ connotes\ \mathrm{Y})(\theta)$.

Hence, if we define

(D24) $(qu(\mathrm{F})\ has\ a\ connotation)(\theta) =_{\mathrm{Df}} (P\mathrm{Y})(qu(\mathrm{F})\ connotes\ \mathrm{Y})(\theta)$,

it follows that every predicate, even 'is polywobble' and 'is a purple number', has a connotation, *viz*. polywobbleness and being a purple number. This conclusion, for every predicate, follows since '*qu*(*f*)', unlike '*f*', is a subject and therefore open to quantification.

To say that a predicate has a connotation is not, of course, to say or to imply that there is, let alone exists, any item which has the property which it connotes. Hence $\vdash_{CL}$(111) is innocuous so far as existence implications are concerned. For certain purposes, however, there are distinct advantages in using a loaded contraction so as to exclude predicates like 'is a purple number'. This may bc achieved by requiring, for example, that the connoted attribute be consistent or that the connoting predicate be significant. In this way specialised senses of 'connotation' can be defined, in much the way that specialised senses of 's-aboutness', such as 'refers to' and 'designates', are defined.

Much as we defined the aboutness of sentences in terms of the aboutness of subject sets, so we may define the connotation of sentences in terms of the connotations of their predicate sets. For example, 'The man next door loves the girl across the road', when represented in the elementary form '$h_0((\iota x)f_0 x, (\iota y)g_0 y)$', has connotation, $\{\lambda f_0, \lambda g_0, \lambda h_0\}$. Connotations of complex sentences may then be defined by recursion; but, as in the case of aboutness, the merit of defining connotation for sentences is dubious. Puzzles such as the morning star paradox can be resolved, by a theory like Mill's, without the extension to sentences. 'The morning star' has the same reference as 'the evening star', but the predicates 'is a morning star' and 'is an evening star' have different connotations. Hence the informativeness of the extensional identity: the morning star is the evening star. (Solution of identity paradoxes is also discussed in §7.9 ff.)

Full λ-conversion is not, of course, the only sort of subject conversion we have considered. All closed descriptors convert, and §-conversion converts a sentence into a subject. By composing §-conversion with the aboutness relation defined for subjects we obtain a derivative relation between sentences and items, in usual parlance between sentences and propositions. Once again the derivative relation is not an aboutness relation. It is important not to restrict subjects so as to exclude those formed by conversion from other parts of speech; otherwise, since many familiar natural language predicates can be significantly coupled both with subjects formed by conversion as well as with subjects not so introduced, many important connexions are lost. Thus, we may state in the one context: Pegasus is interesting; the word 'Pegasus' is interesting; equinity is interesting; that the book about Pegasus is long, is interesting; and our notation enables us to reflect both the similarities and the distinctive logical features of these sentences. We shall later have to consider, as well as conversion relations, their trickier inverse relations, which take us back from subjects to other sentence parts.

The instantiation relation just is the relation between subject vector $\underset{\sim}{X}$ and subject λF which holds when, and only when F$\underset{\sim}{X}$. In short, we may define *i*, read 'instantiates' or 'has', by:

(D25)(i) $(\underset{\sim}{X}\ i\ \lambda F)(\theta) =_{Df} F\underset{\sim}{X}(\theta)$.

In fact this concretion scheme could serve to define jointly both *i* and λ (see §7.9). It is useful to have also a two-valued instantiation relation, i^+ defined by:

(D25)(ii) $(\underset{\sim}{X}\ i^{+}\ \lambda F)(\theta) =_{Df} T(\underset{\sim}{X}\ i\ \lambda F)(\theta)$.

Then, of course,

$\vdash_{CL}(112)\quad (\underset{\sim}{X}\ i^{+}\ \lambda F \equiv TF\underset{\sim}{X})(\theta)$

—a fundamental connexion despite its trivial derivation.

Since $\vdash_{CL}(110)(qu(F)$ *connotes* $\lambda F)(\theta)$ and $\vdash_{CL}(103)$ $((qu(X^{j})$ *is s-about* $X^{j})(\theta))$, for each *j*, where *j* is the number of primes, it then follows, by **RQL**, writing $X^{(m-1)}$ for $X^{(m-1)\text{ primes}}$,

$$\vdash_{CL}(113)\quad TF(X, X', \ldots, X^{(m-1)})(\theta) \equiv (PZ)(PZ') \ \ldots\ (PZ^{(m-1)})(PY)(qu(X) \text{ is s-about } Z \,\&\, \ldots \,\&\, qu(X^{(m-1)}) \text{ is s-about } Z^{(m-1)} \,\&\, qu(F) \text{ connotes } Y \,\&\, (Z, Z', \ldots, Z^{(m-1)})\ i^{+}Y)(\theta)$$

or, more succinctly,

$$\vdash_{CL}\quad TF\underset{\sim}{X}(\theta) \equiv (P\underset{\sim}{Z})(PY)(\underset{\sim}{X} = \underset{\sim}{Z} \,\&\, qu(F) \text{ connotes } Y \,\&\, \underset{\sim}{Z}\ i^{+}\ Y)(\theta).$$

This theorem captures leading features of the correspondence theory of truth, according to which an atomic sentence $F(X, X', \ldots, X^{(m-1)})$ is true if, and only if, what each $qu(X^{i})$ in turn is about stands in the relation, or has the attribute, that *qu*(F) connotes. For example, 'Tom is happy' is true iff what 'Tom' is about has the property 'is happy' connotes; 'Tom loves Mary' is true iff the items which 'Tom' and 'Mary' are about stand in the relation connoted by 'loves'.

We rely on analytical connexions like $\vdash_{CL}(113)$ to underpin the inductive definition of truth we use in developing the semantic theory of quantified significance logics (in §7.7 ff.). A model for quantified significance logic assigns contractions or s-items to subjects and connotations or attributes to predicates, and stipulates for atomic formulae which sets of contractions truly (or significantly) instantiate which connotations. We thus stretch classical model theory in several directions, in which it needs extension, at once; namely in assigning s-items and not merely references to subjects, in assigning connotations and not simply more references to predicates, and in providing a many-valued instantiation relation. Our models then, unlike usual classical models, build in neither existence assumptions nor referential identity assumptions, nor two-valued assumptions; and they are much better equipped to serve as bases for models for intensional logics. Given truth and significance assignments for *atomic* formulae, assignments for compound formulae can then be specified, inductively, for each connective and quantifier using theses of the system expressed in terms of the valuation function. Thus, from theses such as,

$$(T(\Gamma \,\&\, \Delta) \equiv T\Gamma \,\&\, T\Delta)(\theta),$$

we have the value specification for conjunctive formulae:

$$v(\Gamma \,\&\, \Delta) = t \text{ iff } v(\Gamma) = t \text{ and } v(\Delta) = t;$$

while theses such as,

$$(T(UX)\Gamma \equiv (UX)T\Gamma)(\theta);\ (S(Ux)\Gamma \equiv (UX)S\Gamma)(\theta);$$
$$(T(\text{ɿ}X)\Gamma \equiv (\text{ɿ}X)T\Gamma)(\theta);\ (S(\text{ɿ}X)\Gamma \equiv (\text{ɿ}X)S\Gamma)(\theta);$$

give rise immediately to corresponding value specifications of the sort:

$$v((UX)\Gamma) = t \text{ iff } v(\Gamma) = t \text{ for every X.}$$

The use of descriptions in the construction of elementary formulae adds a further complication which is treated in Chapter 7.

This method is by no means the only way of basing an inductive (not recursive) characterisation of truth. One alternative is to specify the truth and significance conditions for atomic formulae by introducing truth and significance ranges. Thus, where 'TD‘ F' and 'SD‘ F' abbreviate 'the truth range (truth domain) of F' and 'the significance range of F', we may define:

$$\text{(i) } \underset{\sim}{X} \in TD\text{‘ F} =_{Df} TF\underset{\sim}{X},$$
$$\text{(ii) } \underset{\sim}{X} \in SD\text{‘ F} =_{Df} SF\underset{\sim}{X};$$

i.e. $\underset{\sim}{X}$ belongs to the truth range (significance range) of F iff $TF\underset{\sim}{X}$ ($SF\underset{\sim}{X}$). Then, instead of specifying instantiation relations in models, we require that a model assigns to each primitive predicate a truth and significance range. As before the conditions for compound and quantified formulae are then specified inductively. The definition (i), or its thesis-equivalent, $\underset{\sim}{X} \in TD$‘ F $\equiv TF\underset{\sim}{X}$, called the *range theorem*, is also important in the significance range theory given in Part III. (See Goddard [2] for a preliminary discussion.)

Other standard methods of specifying truth and significance conditions over a model, such as that of first defining satisfaction inductively and then defining truth in terms of satisfaction, also adapt. Moreover each of these methods can be consistently combined with, and serves to enhance, the neutral theory of truth of Chapter 2 expressed in such laws as $\vdash_{CL}$**T**a **eq** a and $\vdash_{CL} T\Gamma(\theta) \equiv (Pa)(qu(\Gamma)(\theta)\ \mathbf{Y}a\ \&\ \mathbf{T}a)$. Though the neutral theory is informative, it hardly asserts much that could be disputed, and it does not suffice, without supplement, for the semantic theory of quantified significance logic.

The thesis $\vdash_{CL}(113)$ not only provides us with an explicit analysis of one version of the correspondence theory of truth, and hence the basis for the formal model theory of quantified logics, but it stands also as an explanation of the meaning of general words (predicates). Traditionally such an explanation was given by a theory of universals, and $\vdash_{CL}(113)$ can be interpreted as the condition which any such theory must satisfy. More generally, it can be interpreted as a neutral condition which any theory of meaning must satisfy. To see this, consider the simpler version,

$\vdash_{CL}(114)$ $\quad TF\underset{\sim}{X}(\theta) \equiv (PY)(qu(F)$ *connotes* $Y\ \&\ \underset{\sim}{X}\ i^{+}\ Y)$,

which arises from $\vdash_{CL}(113)$ by taking the trivial s-aboutness of subject terms $qu(X^{j})$ for each j. It was characteristic of traditional theories of universals that an account of this form stood not only as an analysis of truth (How can things be as they are?), but also as an analysis of the meaning, or naming function, of predicate terms (How can we name things as we do?).

For suppose we ask how it is that two different (predicate) tokens of the same type can have the same meaning on different occasions of their use if they are used to describe different items. Within the framework of the denotation (naming) theory of meaning, the answer had to be that there is some one thing to which the tokens and the items are related. This unifying element was called the universal, and by analogy with the paradigmatic naming situation which *seems* to prevail in the case of proper names the tokens were said to mean because they named a universal. (For a fuller discussion see Goddard [5].)

The main point of a theory of universals is thus to give some (denotational) account of the relation between a predicate term and those items which it may be used truly to describe or to classify. And the characteristic form of the solution is to introduce some third item which links the words to the items and so stands in relations to both. The tokens mean the same because they are related in the same way to the constant universal; while the items, though different individuals, may be said to be the same in a given respect if they are related in the same way to the unchanging universal. Thus, if N and M are (for the moment) unspecified relations, then the universals thesis may be expressed quite generally by affirming:

$$\text{(i)}\ (P\mathrm{Y})(qu(\mathrm{F})N\mathrm{Y}\ \&\ \mathrm{Y}M\underset{\sim}{\mathrm{X}}),$$

where $\underset{\sim}{\mathrm{X}}$ takes as values just those subject terms for which $\mathrm{F}\underset{\sim}{\mathrm{X}}$ is true. Evidently, this condition may be stated explicitly by affirming:

$$\text{(ii)}\ T\mathrm{F}\underset{\sim}{\mathrm{X}} \equiv (P\mathrm{Y})(qu(\mathrm{F})N\mathrm{Y}\ \&\ \mathrm{Y}M\underset{\sim}{\mathrm{X}}).$$

Traditional theories of universals typically take the relation N to be the relation of naming. Under such theories the clause '$qu(\mathrm{F})N\mathrm{Y}$' may be taken to be an analysis of the standard meaning of the predicate term, given, what the theories typically assumed, that the (standard) meaning of $qu(\mathrm{F})$ is given by the fact that it names Y. If, therefore, we can say what Y is, we can say what the standard meaning of $qu(\mathrm{F})$ is.

So far, the choice of what to take as Y is open, and there are various possible solutions, but the most elementary choice is to take Y to be a property. Then, we may say, $qu(\mathrm{F})$ names $\lambda\mathrm{G}$ and $\lambda\mathrm{G}$ stands in some relation M to X. Evidently, the converse of M, *viz.* M^{c}, is the relation of having, or being an instance of; or more directly, M is the relation of application.

Were we to equate 'names' with 'connotes', then, $\vdash_{\mathrm{CL}}(114)$ would express a standard, and old-fashioned, theory of universals. Now the objectionable features of theories of universals arise from the choice made for Y along with the existential interpretation insisted upon for quantifiers, and from the interpretations of the relations N and M. They do not arise from the form of the analysis itself. For one thing, $\vdash_{CL}(114)$, of which (ii) is a generalised form, is an analytic connection. For another, however the meaning of predicates is to be explained, it must, on a denotation theory, be in terms of some relative constant which stands in relation to different tokens of the same type and in some other relation to items which, though different individuals, are nevertheless the same in some respect. To say that a predicate gets its meaning by naming some property, or indeed that the property is the meaning, may well be philosophically objectionable. But to say that the predicate stands in some constant relation N to a constant, is not. On the contrary, if the sameness of meaning and items is to be explained when the tokens are different from each other *qua* individuals, and the items are different from each other *qua* individuals, there must be recourse to something which remains the same throughout the differences. Every theory of the meaning of predicates, that is to say, must satisfy the condition (i). That this condition is satisfied by modern theories of meaning we shall see presently.

If, for the moment, however, we retain the interpretation of 'N' as 'names', then many different choices may be made for Y and this in turn determines the interpretation of 'M'. Thus, we may take Y to be a Platonic form and M^c to be the relation of participation; or we may take Y to be a concept and M to be the relation of comprehension; or Y to be a Russellian class, and M^c to be the relation of class-membership. Or the account may be more complicated, as in Locke, where *qu*(F) names a nominal essence, this in turn is related to a real essence, and this to the items which satisfy the predicate. Here, Y is analysed into two universals and M into two relations.

Each such account attempts to explicate properties, and what it is for an item to have a property; and is supported by details of the underlying metaphysics. Of course, whether or not one accepts that Platonic forms, or any of the other notions, are less mysterious than properties, will depend on the persuasive power of the metaphysical argument. But given that one can be persuaded, so properties can be eliminated in favour of other items. For example, should one be persuaded that the notion of class is more comprehensible than that of property (it isn't, and the identity conditions that fix properties make them much more satisfactory for intensional analysis), then one may reduce '*qu*(F) N λG' to '*qu*(F) $N\{Z:G(Z)\}$'; in which case, to say that X has the property λG is just to say that X is a member of $\{Z:G(Z)\}$. Our account of connotation in terms of properties, therefore, is not inconsistent with, indeed is quite compatible with, an analysis of λG in terms of some other primitive. Formally speaking, all that is required is a further reduction assumption (of the sort $X\, i^+ \lambda G \equiv .\, T(X \in \{Z:G(Z)\})$) which embodies the reduction thesis of the favoured analysis.

What distinguishes one traditional theory of universals from another is the different determination given to Y while the interpretation of 'N' remains essentially the same (though the interpretation of '*M*' will depend on the choice of Y). What distinguishes all traditional theories from modern theories is that the latter demand an interpretation of '*N*' different from 'names'. So, for example, a rule theory of meaning would require that Y be taken as a set of rules and the relation *N* as some (usually unspecified) relation between them and the predicate. This relation is sometimes expressed by saying that the rules "give" the meaning of the predicate. But whatever the relation is, it is clear that it cannot be the relation of naming. For the predicate does not name the rules. The relation M^c, on the other hand, is a relation of satisfaction: the items which the predicate is truly used to describe satisfy the rules.

Such an account appears to differ markedly from traditional accounts. The different traditional accounts simply replace properties by items of a similar logical kind (forms, concepts, classes, *etc.*). But the substitution of rules or criteria for properties seems to introduce items of a different logical kind. But the appearance is deceptive; for the rules turn out to be *sets* of rules, while criteria as *tests* or *formulae* are not unlike Plato's forms under one interpretation of the theory of forms. Though the analyses of properties given by modern theories are more like those of traditional theories than is often supposed, modern theories nonetheless make improvements: by connecting properties with tests, criteria and sets of rules they help scotch the idea that predicates name or are about properties, that properties are further individuals, and that properties need exist or have independent existence.

As an example of a more modern theory (which goes back to Wittgenstein [2]), consider how it might be explained what the property of being well-formed is, with respect to some formal system. A set of rules is given in terms of which it may be discovered whether or not any formula is well formed. Those formulae which pass the test, *i.e.* satisfy the rules, are well formed; those which fail to satisfy them are not. Whether or not anyone can, after sufficient practice, perceive the property of being well-formed in a given formula is irrelevant. For he has only come to this ability by way of the rules, and, it may be said, to say that he perceives the property is simply to say that he is applying the rules automatically. To know the property just is to know the rules; and to know the meaning of the predicate 'is well-formed' is, again, just to know the rules. The rules thus stand both as an explanation of meaning and as an explanation of 'having the same property'. That is, they stand as a universal; for they remain the same throughout the different uses of tokens of the type 'is well-formed', and independently of the fact that different formulae are being described. The predicate and the items are both related to the same one item, a set of rules. The rules cannot, however, be identified with the meaning nor with the property. Instead, it may be said, they "give" both: in that to know what the meaning is, and to know what the property is, is simply to know the rules.

This example is of course favoured since formation rules are commonly recursive and give rise to a fixed finite procedure. But this is not a necessary feature. If we extend our notion of rules to include criteria, and include tests among criteria, a similar account can be given even though the tests are not recursive. Suppose, for example, we teach someone how to heft two objects to compare them in respect of weight. Then, though the test is crude and will not always yield the same result for different people, or even for the same person on different occasions, the test itself stands as a criterion which both "gives" the meaning (however blurred) of 'has the same weight as' and at the same time "gives" the relation of having the same weight. For someone who has learnt only this test, to know the test is to know both the meaning and the relation; and there is no difference in principle if the crude test is replaced by a set of overlapping scientific procedures.

To make such a theory formally explicit we could introduce a criterion predicate, '*Cr*', in terms of which we define criteria variables 'μ', 'v' *etc.* as restricted variables, and a determining predicate '*D*' for which we assume: $(U\mathrm{Y})(P\mu)\mu D\mathrm{Y}$ and $\mu DY \supset (P!Y)\mu DY$ *i.e.* every property is determined (or given) by some criterion. Then, defining a giving relation: $qu(\mathrm{F})N\mu =_{\mathrm{Df}} (P\mathrm{Y})(qu(\mathrm{F})$ *connotes* $\mathrm{Y}\ \&\ \mu D\mathrm{Y})$, and an application relation: $\mu M\underset{\sim}{\mathrm{X}} =_{\mathrm{Df}} (P\mathrm{Y})(\mu D\mathrm{Y}\ \&\ \underset{\sim}{\mathrm{X}}\ i^{+}\ \mathrm{Y})$, it follows: $T\mathrm{F}\underset{\sim}{\mathrm{X}} \equiv (P\mu)(qu(\mathrm{F})N\mu\ \&\ \mu M\underset{\sim}{\mathrm{X}})$—in traditional normal form. Such a theory has of course to be underpinned by an account of criteria, or a definition of the predicate '*Cr*', and by an account of determining relations. As to the last, if $\mu D\mathrm{Y}$ then relevant features of μ transfer to Y; for example, to know μ is to know Y; by no means all features transfer however, and *D* is not an identity relation.

We make no attempt to develop the details of such a theory here. Our concern is to show that the account we have given of connotation in terms of properties, *i.e.* in terms of attributives and their aboutness, is not incompatible either with other more favoured traditional theories of meaning or with more recent treatments. Since we introduce attributives as primitives, satisfying a λ-conversion scheme, our account is neutral as between theories of universals. Attributives in turn may be regarded as dummies which, like intervening variables, can be interpreted or explained in different ways. We have given the barest sketch of a preferred way of explaining—but not reducing—properties.

3.8 Derived Ranges and Significance Essentialism

Given a standard context for which univocal senses have been specified for all expressions, secondary ranges of variables and parameters, including the usual reference-ranges of logicians, are readily explained. They are obtained by contraction from subclasses of the *primary* or constants-range of arguments and parameters.

Under the limitations already discussed, the variables and parameters of our systems each have as their primary ranges a class of sentence factors, of substituenda which are such that well-formation or grammaticalness is preserved upon replacement of a variable or parameter by any factor of the range. In terms of this limitation, the primary ranges of our variables and parameters are also their *grammatical ranges*. Other properties of the expression in which the replacement is made, such as significance, sense *etc.* (if relevant) are not in general preserved. With a general extra-systematic variable as opposed to a system variable, not even grammaticalness need be guaranteed, though it can be. Compare, for example, the difference between 'Let Δ be any formula ...' and 'Let Δ be any wff ...'.

We consider first certain subclasses of the primary ranges:

§1 *Significance Ranges*

Consider again the expression,

(1) Plato's Academy *f*.

For some values of '*f*', for example, 'was founded by Plato' and 'is covered with ivy', (1) is significant. For other values, for example, 'eats Gouda cheese for breakfast', (1) is, given a standard context, not significant. The class of values of parameter '*f*' for which (1) is significant is called the *significance range of* 'Plato's Academy'. More generally, the class of values for which $qu(fx_0)$ is significant, where 'x_0' is a subject constant, is called the significance range of (the constant) 'x_0'. More generally still, where 'x' is a subject parameter, not a variable, the class of values of '*f*' for which $qu(f(x))$ is significant is the *significance-range* of *parameter* 'x'. Significance ranges (*s-ranges*) are defined with respect to given constants, or generally parametrically, *not* with respect to variables. For even if both 'x_0 is white' and 'x_1 is white' are both significant, the s-range of 'x_0' may differ from the s-range of 'x_1'. This will be so if, for example, 'x_0' is 'snow' and 'x_1' is 'Socrates' or 'what I am thinking of now'. It follows that the expression 's-range of the variable 'x'' is not well defined without explicit indication of the sort of 'x'.

Consider next the expression,

(2) x is green.

The s-range of 'is green' is defined as the class of (substitution-) values of $qu(x)$ for which (2) is significant. Thus, 'the roof' belongs to the s-range of 'is green', but 'the number two' does not. More generally, the *s-range of a predicate constant* 'f_0' is defined as: the class of values of variable $qu(x)$, for which $qu(f_0x)$ is significant. Similarly for predicate parameter '*f*'. The non-s-range of a predicate constant 'f_0' is the class of values of $qu(x)$ for which $qu(f_0x)$ is nonsignificant. Thus, the non-s-range of 'f_0' is the complement of the s-range of 'f_0' with respect to the class of its primary values. The non-s-range of a subject constant 'x_0' may be defined similarly.

A one-place predicate constant 'f_0' belongs to the s-range of a subject constant 'x_0' if, and only if, 'x_0' belongs to the s-range of 'f_0'. Thus it is not really necessary to define s-ranges both for subjects and for predicates. Reasons for concentrating on the s-ranges of unsaturated expressions, like predicates, rather than on s-ranges of subjects emerge as soon as many-place predicates are considered.

The s-range of a given constant is a subclass of its grammatical range (in English usually a proper subclass of its grammatical range). Thus in general, significance ranges are subclasses of primary ranges of a given grammatical kind. Since, therefore, the primary ranges of our variables are grammatical ranges, significance is not in general preserved upon substitution of a value from the primary (grammatical) range. This is in marked contrast to classical type-logics in which, generally speaking, primary ranges are defined or assumed to be coincident with significance ranges. Given a formula of the form $qu(fx)$ a legitimate substitution value for us would be 'The number two is happy' even though this is nonsignificant since 'The number two' does not belong to the s-range of 'is happy', but this could not count as a legitimate substitution in classical type-logic since there are no formal principles defining the relations between nonsignificant and other expressions. The classical restriction has many disadvantages, as we shall see (Ch. 4), and it is difficult to justify from a semantic point of view. The classical restrictions may, however, be written into the formation rules, as in modern treatments of type theory, so turning significance criteria into grammatical criteria, or they may be ensured by the introduction of sortal variables and restricted quantification. Either way is unsatisfactory, since both presuppose a large-scale informal analysis of significance outside the formal system. Sortal ranges $K_0, K_1, ..., K_m$ for the individual variables have to be associated, extra-systematically, with each predicate constant 'f_0', 'f_1', ..., 'f_m' such that $qu(f_i x)$ is significant if, and only if $qu(x) \in K_i$. In the case of type theory, of course, we have the unrealistic requirement that if 'f_0', 'f_1', ..., 'f_m' are all type-1 predicates, there is only one class K (the type of individuals) from which the value of $qu(x)$ may be chosen in order to ensure the significance of $qu(f_i x)$. (Similarly for all types in the type-hierarchy). The class K is thus both the grammatical range (total range) and the significance range of $qu(x)$ with respect to any type-1 predicate constant.

The problem of how best to deal logically with nonsignificance arose with modern logic: it confronted Frege (in [1]). If we begin with the distinction between complete expressions such as 'Tom', '7+5', '7 is greater than 5' which designate "objects", and incomplete or unsaturated expressions such as '... +5' and '... is greater than 5' which designate "functions", the question arises whether any restrictions are to be imposed on the substitution of complete expressions for the gaps in unsaturated expressions. Frege's answer to this is "No", so that nonsignificant expressions such as 'Tom+5' and

'Tom is greater than 7' may arise. Since some of these are (grammatical) sentences, and since the proposed logic is two-valued (every sentence designates 'The True" or "The False") it is necessary to ask which truth-value should be assigned to these special cases. Frege's answer seems to be that an arbitrary stipulation is satisfactory in any given case; alternatively one might stipulate that every such expression designates "The False".

The artificiality and unsatisfactoriness of such solutions (see Introduction §3, and §4.6), force one to consider the two possible alternatives: (a) of restricting substitution in such a way that nonsignificant expressions cannot arise, and writing these restrictions into the syntactic and semantic theory; (b) of increasing the number of values from two to three (in Frege's terminology, allowing that some sentences and some non-sentential expressions (phrases, clauses, *etc.*) may designate "The Non-Significant").

We have chosen course (b) believing it to be closer to the ordinary facts of language as well as being both simpler and more comprehensive. In spite of the fact that we have three values as against two, the logics we investigate are simpler because we only require in general two types of parameters: subject and predicate, restricted only by their grammatical ranges. We therefore gain the advantage of Frege's unrestricted individual variables which range over all "objects", and we do not suffer the disadvantage of having many sorts of predicate variable as in type or sortal systems. At the same time, the three-valued logics are much more comprehensive, both because we are able to affirm principles which cannot be affirmed in two-valued classical systems, and because we can derive many-sorted systems from them as special cases (§7.5 and §7.6). Quite apart from this, we avoid the inadequacies of type-theory in terms of which all individuals belong to the same sort, and all type-m predicates, for a given m, belong to the same sort. Such loose restrictions are quite incapable of preventing the occurrence of nonsignificant sentences such as 'The tree is envious' and 'The lake is sailing' and the failure of type theory in this respect is no accident. It is a consequence of the ill-conceived attempt to force significance criteria into grammatical (syntactic) criteria.

These facts become even more important in relation to two- or more place predicates. For the s-range of a subject variable with respect to one place of a many-placed predicate constant may be different from its s-range with respect to another place. Thus, consider

(3) x is larger than y.

Each subject place has a significance range. In the case of this particular predicate constant, the s-ranges of the first and second places are the same, but in the case of a two-place predicate constant like 'admires', the s-ranges of the first and second places differ.

A complication arises from some predicates like 'is larger than' which exhibit some sort of ambiguity or heterogeneity. For since '9 is larger than 7' is significant, '9' belongs to the s-range of the first place. Also, since 'Plato is larger than Socrates' is significant, 'Plato' belongs to the s-range of the first place. But the sentence 'Plato is larger than 7' is not significant, even though replacements for the variables of (3) are chosen from the respective s-ranges. Thus, for certain polyadic predicates, substitutions for the variables by values chosen from the appropriate s-ranges do not always result in significant sentences. This problem can be overcome either by limiting the context so that, for example, 'is larger than' is restricted to a numerical sense and the s-ranges of the places restricted to numerals, or by distinguishing homogeneous predicates (those which yield significant sentences for an arbitrary choice of values from the s-ranges of their places) from heterogeneous or ambiguous predicates (which fail to satisfy this condition). Then, either the logic can be restricted to homogeneous predicates, or the special relations between predicates of the two kinds can be investigated. The problem of the zeugma can be dealt with in this way. (This general difficulty is mentioned below in another context (§5.12) and a fuller discussion will be found in Goddard [2] and [3]. The question will be raised in detail in Part III).

The definition of the s-range of the first place of a two-place predicate constant 'f_0' (such as 'is larger than') is: the class of expressions from the primary constants range of $qu(x)$ such that, for *some* value of $qu(y)$, $qu(f_0(x, y))$ is significant. The s-range of the second place is similarly defined.

In general, the definition for the ith place of a m-place predicate constant (or parameter) 'f_0', where $1 \leqslant i \leqslant m$, is a direct extension of the two-place case. An m-place predicate constant has m places each of which can be occupied by a variable. Each of these place-holding variables will have a constants-range. The places can be ordered, and in general must be so if ungrammaticalness is to be avoided. Suppose, then, that the places are ordered from 1 to m. Each place of such a predicate has an s-range. A subject constant 'x_0' belongs to the s-range of the ith place of an m-place predicate constant 'f_0' if, and only if, for some values 'x_1', ..., 'x_{i-1}', 'x_{i+1}', ..., 'x_m' of the variables of the remaining places, '$f_0(x_1 \ldots x_{i-1}, x_0, x_{i+1} \ldots x_m)$'is significant. The s-range of the ith place of 'f_0' is the class of such expressions, *i.e.* the class of expressions such that, for *some* values of the other places, the resulting sentence is significant. An ordered set $\langle$'a' ... 'a^i' ... 'a^m'$\rangle$ of values, 'a^1' from the first s-range, ..., 'a^i' from the ith s-range, ..., 'a^m' from the mth s-range of 'f_0' for which '$f_0(a^1 \ldots a^i \ldots a^m)$' is significant is called a *cosignificant set* for 'f_0'. Thus $\langle$'Socrates', 'Plato'$\rangle$ is a cosignificant set for 'is larger than'; so, too is $\langle$'7', '9'$\rangle$; but $\langle$'Socrates', '9'$\rangle$ is not.

§2 *Truth and Falsity ranges*

Just as the significance range of a given constant (with respect to a given context) is a set of expressions which constitute a subclass of its grammatical range, so the truth and falsity ranges may be defined as subclasses of the significance range. Thus, the truth range of 'is red' is the class of expressions for which sentences of the form *qu*(x is red) are true when the value for the variable 'x' is selected from its grammatical range. Similarly the falsity range is the set of expressions for which sentences of the form *qu*(x is red) are false. Again, these definitions can be extended to polyadic predicates in the manner outlined for significance ranges. A *cotruth set* for a polyadic predicate 'f_0' is then an ordered set of expressions which satisfy 'f_0'.

Significance, truth and falsity ranges are not introduced explicitly, except occasionally for illustration, until §7.8 (and then only in virtual form). Their explicit introduction requires an abstraction thesis of some sort (cf. §3.7 and §7.8); but equivalent formulations of assertions such as "'x_0' belongs to the s-range (t-range, f-range) of 'f_0'" do of course occur since whenever 'x_0' belongs to the s-range of 'f_0' we may affirm 'Sf_0x_0', and whenever 'x_0' belongs to the t-range or f-range of 'f_0' we may affirm, respectively, 'Tf_0x_0' and 'Ff_0x_0'. Range assertions are therefore *represented* in the predicate systems.

§3 *Secondary Item Ranges*

So far we have concentrated on ranges which are classes of expressions. But corresponding to each such range there are various ways in which we might define associated item and reference ranges. Indeed each contraction on primary ranges yields secondary ranges.

Suppose, to illustrate, we have with respect to a fixed standard context c_s a class of subject constants which are values of the subject variables. Now since the reference of these constants, where applicable, can be established with respect to c_s, we may take the *reference range* to be the class of items which stand as the references of these constants. This range will be restricted to those entities referred to using the constants in the given context (together, perhaps, with an artificially inserted null item or set). Similarly any contraction relation can, in a fixed context, be used to define a secondary item range; it defines the items which are contractions under that relation of the primary range.

The richest item-range arises when we contract on subjects using just Leibniz identity: for any "contraction" on a stronger relation would have to be partly quotational and would therefore not yield items. There are said to be serious difficulties about such a wide item-range: in what sense, for example, can an item-range "contain" mythical and nonsignificant items as well as actual individuals? In the ordinary sense of set-membership: for there is nothing in the elementary theory of sets which restricts its application to entities. There are many other moves however which try to create difficulties for non-existent and nonsignificant items: they are designed in fact to force us back to reference ranges for variables. They typically take us to what may be called the *standard* reference range. For there are many words and phrases of English which, if there are no special contextual limitations, are taken to have a standard reference. Thus, in ordinary discourse, 'Socrates' would be taken to refer to the Athenian philosopher who drank hemlock, *etc.* If, therefore, we take a fixed semantic grammar and a supposed *standard* context, rather than just a fixed context, we may take, say, the class of singular unambiguous proper names and define the common standard reference for each of these as the actual item they did denote at some time, if any, and the null class otherwise. Under this prescription, 'Moses' has as its standard reference the individual Moses who led the Jews out of Egypt and did numerous other things related in the Bible, while 'Pegasus' has as a standard reference the null class. Given standard references for all expressions of a constants range of a subject variable, a standard reference range can be defined, for that variable, as the class of standard references of its constants-range. Again, it has to be stipulated that nonsignificant descriptions have no reference.

Such highly artificial reference ranges not only have little value, but, in significance, context and intensional logics, are a very serious handicap. Given an "ideal extensional language" in which tokens of the same type are always used to refer to the same individual, and such that no individual has more than one referring phrase associated with it, and words like 'this' and 'that' are excluded, then a standard reference range may have point. But in an actual language different individuals can be referred to by tokens of the same type (and however restricted the context, it must be possible to refer to three different individuals named 'Tom'), some individuals have more than one name, and demonstrative elements like 'this' and 'that' are present. Besides this, there may be contexts in which 'Pegasus' is used to refer to a pet cat, 'Socrates' to a pet canary, and 'the purple number 7' to a house. In such cases these constants have actual references. Using only reference ranges is also a serious handicap because truth and significance are assessed in terms of values taken by variables; thus exclusive use of reference ranges leads to modal and significance paradoxes (see §7.3 ff.) and to the problems of modal and significance essentialism. (How refusal to tolerate values and ranges other than reference ones can destroy intensional logic is well illustrated by Quine [2]).

These problems also sabotage the classical enterprise of working out reference ranges associated with predicate constants. For very often a good deal more than reference classes must be ascertained in order to assess truth and significance. Nonetheless, given the sense of a predicate such as 'is a horse' and a fixed context in terms of which there is no ambiguity, we may distinguish various reference classes associated with the predicate—an actual reference class of current horses; a total reference, which is time-invariant and consists of all horses which have been, are or will be actual (at some time); and a possible "reference" class (sometimes called the comprehension) which includes the total reference but also contains elements like Pegasus.

Similarly, given a two-place predicate such as 'heavier than', we may distinguish an actual reference class of ordered pairs of current items, a total reference class of ordered pairs of actual items some of which may not now or may not yet exist, and a possible reference class of ordered pairs of actual and possible items. Now if for each predicate constant three such reference classes can be distinguished, then we may define three reference ranges which may be associated with the primary range of the predicate parameters, namely the class of all actual reference classes, the class of all total reference classes, and the class of all possible-reference classes.

What is the point of defining these various ranges? Not much; the usual reasons given, that (some of) these ranges are needed in giving an account of the meaning of predicates and of the truth and significance of subject-predicate sentences, do not stand up. For accounts can be given without these ranges (see §3.7); and accounts given using just these ranges are quite inadequate for much of discourse. In fact what is important is not the construction of reference ranges but the evaluation of a sentence in a context; and for this purpose we do not want, or need, to be bound by imposed artificialities such as those of associated reference ranges, and deciding *which* of the various ranges is appropriate. In any case, as is well known, some parameters simply do not have reference ranges associated with their primary range. Consider, for example, the one-place sentence functor κ which has as a constants-range one-place sentence connectives such as '*T*', '*F*', '*D*', '$\sim$', '$\neg$' ('*D*' and '$\neg$' are defined in §5.13 and §5.6 respectively), and one-place predicates such as 'Bill knows (that)' and 'it is thought (that)'. Each element of the constants-range of κ when applied to a grammatical sentence yields another grammatical sentence. But the sentence to which the functor applies need not be significant, nor need the sentence which results from the application be significant (*cf.* the application of connective '*D*' of chapter 5 to a true sentence and of the Slupecki-$\dot{T}$ to any sentence). The sense of 'refers' would have to be stretched beyond reasonable limits if it were required of, say, '*D*', that it should designate. Despite the lack of a clear-cut reference range, however, such parameters as κ are perfectly satisfactory, and their role, explained in terms of place-holding and constants alone, is quite intelligible.

When a secondary item range is associated with a primary range, subclasses of it may be associated with subclasses of the primary range. For example, the *v-range* of a monadic predicate 'f_0' is defined as: the class of standard references of elements of the s-range of 'f_0'. The v-range thus consists of a set of items which may significantly be said to have the property λf_0. Similarly, subclasses of the v-range may be associated with the truth and falsity ranges. The definition of a v-range given here can be extended immediately to derive Halldén's [1] definition of the v-range of a property λf_0 since this last is the v-range of the predicate 'f_0', where property λf_0 is the connotation of 'f_0'.

An item from the standard reference or designation range of a variable is called a reference or designation value of that variable. Under suitable choice of standard designations, a designation value of a variable is one of Carnap's value extensions of that variable. Similarly, we may define an associated intension range, where intensions are defined in Carnap's way. This range will be the class of Carnap's value intensions of the relevant variable.

§4 *Significance Essentialism*

It is a common presumption of classical logic that the specification of reference ranges is essential to the semantic theory. Thus Frege held that individual variables range over objects, while Russell took the view that they ranged over individuals. Such specification, however is far too vague. Immediately one wants to ask "What counts as an individual?" or "What is an object?" And why impose such restrictions? They are neither necessary nor desirable. Partly to meet this difficulty, modern treatments depend on the specification of a well-defined set *d*, which determines the domain of a model, and the laws of the formal system are restricted to an application over *d*. It is no good, however, saying that *d* may be any set, if the semantics is supposed to furnish illumination as to meaning; and in any case applications of the logic to given subject matter force specific choices of *d*. A different specification of *d* will result in a different interpretation. So, for example, we may define *d* to be the class of numbers, for one application, or the class of living animals for another, and so on. Such a restriction is impossible, however, if we wish to represent assertions which cross the boundaries of two or more specifications of *d*. If we restrict *d* to numbers or to living beings, then we have no way of dealing with sentences of the form 'John multiplied the number seven by the number three and got the answer twenty four'. And in general, to accommodate predicates of the kind 'John is thinking of ...', the class *d* has to be extended at least to all entities or objects and we are back at the Frege-Russell position. (Actually the theory applies as well to items which are not clear and distinct entities and which appear ineligible for membership of well-defined Cantorean sets, to items such as clouds, rainbows, shades of colour, chimeras and round squares; but current logical dogma would confine quantification logic in its application to entities).

No such difficulty—as to the characterisation of the relevant semantical objects or entities—arises if we work in terms of primary ranges with respect to the determined grammar of a given language. For then, the "individual" variables, *i.e.* subject arguments, may simply be said to range over the subjects of that language, whatever these may be said to designate. And it is interesting that Frege begins in this way, by distinguishing grammatically complete and incomplete (unsaturated) *expressions*, and then introduces derivatively objects as the designata of complete expressions and functions as the designata of incomplete expressions. It is interesting, too, that Russell, though in his atomistic period he defines types as classes of "things" later repudiates this view and defines them as classes of expressions in order to avoid the difficulty, for his *informal* account of types, that we cannot say significantly that a given "thing" does not belong to a given type (see §4.3).

Nevertheless, the limitation to primary ranges might be thought to have an overwhelming defect, namely that we have no semantic theory at all unless we supply reference ranges. As a special case of this objection, it might be said that in terms of Russell's later version of the theory of types, though we can say that the word 'Socrates' belongs to a type different from those which can significantly fill the blank in the sentence factor '... is divisible by 2', and so explain the nonsignificance of 'Socrates is divisible by 2', we can give no explanation of *why* the word 'Socrates' belongs to a different type from those words which can significantly fill the blank. In particular, we cannot say that it is because people and numbers belong to different logical kinds. So the whole theory of significance is thrown into language, and this is a ridiculous position to take because the language is as it is because "things" are as they are.

Such criticisms are misplaced. For in terms of primary ranges we can define reference and other secondary ranges, and thereby link the theory with things. It is not that such secondary ranges cannot be introduced; it is the idea that one has to work exclusively with reference ranges and entities that is pernicious. The *logic* of quantification theory, for example, imposes no such restrictions; and *semantics* as a study is by no means confined to a theory of reference or of referentially characterisable notions. Semantics is not rendered syntactical by our method, since the constants in our primary ranges are not simply syntactic strings. They are descriptions and phrases (except in special cases) some of which are used to refer in the context of application. Hence even though we concentrate formal attention on primary ranges we do not thereby divorce language from real situations and consider the relations between linguistic expressions in an idealised logical vacuum. On the contrary, we wish to claim that far from lacking a semantic theory, we have through the theory of contractions, a more complete and much more realistic semantics than is classically secured by the specification of artificial reference ranges.

Indeed concentrating on constants-ranges and s-ranges of variables rather than reference ranges, sense ranges and v-ranges, has big advantages. For example, the problems of selecting standard references and of deciding whether 'Pegasus' and 'the Red Queen' have the same reference or not, and the problems of ambiguity, can be evaded. Moreover, once intensional functors and the various special functors of significance logic are introduced, it becomes practically impossible to develop an adequate theory if *only* reference ranges for variables are permitted. For expressions purportedly referring to items which do not exist or even cannot exist may quite well belong to constants-ranges and s-ranges—since sentences such as 'Pegasus likes grass' and 'The round square is round' are significant—but the admission of "items" corresponding to them in standard reference ranges is usually prohibited. Even worse, expressions which are not significant may belong to the constants-ranges of variables, but since they do not significantly designate, they cannot even be represented in the way 'Pegasus' can in a wide "reference" range. Indeed, any attempt to develop significance logic in terms of reference ranges appears to presuppose some sort of significance essentialism thesis, that items, things, properties, and so forth have significance features irrespective of how they are described, characterised, *etc.* (We owe this point to V. Routley; see R. and V. Routley [1]).

Arguments similar to those assembled by Quine and others against modal essentialism can be marshalled against significance essentialism. In particular, some damaging paradoxes reminiscent of modal and intensional paradoxes can be constructed to discredit significance essentialism.

If essentialism were correct, then it should not make any difference how we describe, or what expressions we use to refer to, a given item. But suppose that what a person Tom is thinking about at seven o'clock is the number seven. Then the number seven can be alternatively described by 'the number seven' and 'what Tom is thinking about at seven o'clock'. Now consider,

The number seven likes soup

and, What Tom is thinking about at seven o'clock likes soup.

The first is nonsignificant, given a standard context, but it is arguable whether the second is. There are various ways in which the second sentence might be analysed. If the reference of 'What Tom is thinking about ...' is supplied, the sentence is complete in the given context. But is it nonsignificant? One might say that its significance cannot be assessed until the reference is supplied; alternatively, that it is significant before the reference is given, but that, after the reference is known, it has a kind of absurdity. But in either case, contrary to essentialism, the way in which the number seven is described seems to make a difference to the significance properties of the sentences used. The reason for this is that the expression 'the number seven' has a standard sense which singles out a particular item as what it is about and there is therefore an obvious and immediate significance clash in 'the number seven likes soup'. The expression 'what Tom is thinking about at seven o'clock', however, though it has a fixed sense does not of itself single out a particular item as what it is about, and one cannot say, in terms of the standard sense alone, that there is or is not a significance clash. Certainly there is no sense-sense clash in 'What Tom is thinking about at seven o'clock likes soup' as there might be said to be in the case of 'The number seven likes soup'. Of course in non-standard contexts even 'The number seven likes soup' need not be nonsignificant.

Since the way in which an item is described makes a difference to its significance properties, essentialism is not correct. And other more striking examples which, like that given, depend on identity claims of some sort, can be constructed. Consider, for example, cross-category identifications such as,

(a) material objects are structured collections of microphysical particles, and
(b) the odour of violets is the molecule irone.

In the case of (a), there are predicates which can be significantly coupled with the left-hand side of the identity sentence but not the right-hand side. Examples are 'have solid parts', 'are of uniform material throughout' and 'are sat on sometimes'. In the case of (b) there are predicates which can be significantly coupled with the left-hand side but not the right—*e.g.* 'is sweet', 'is sharp' and 'is pungent'—as well as predicates which can be coupled with the right but not the left—*e.g.* 'has shape' and 'weighs ...'. If, as seems certain, some heterogeneous identities such as (a) and (b) are true, then significance cannot be a property of items irrespective of how they are characterised, and significance essentialism is false.

If significance essentialism is to be evaded, either ranges other than reference and designation ranges must be introduced or else familiar reference ranges must be replaced by "reference" ranges which correspond in effect to these other ranges. In either case, the entity or pure reference account of the value of a variable must be abandoned, and the whole classical treatment of the semantics of a formal theory in terms of a model must be reshaped. The obvious and least metaphysical way out of the difficulty is to introduce at the outset constants-ranges as primary.

In fact the idea of a model of a formal system, and the implied significance essentialism, owes much to the philosophy of logical atomism and the belief in an ideal language (positions which we shall go on to reject, for related reasons, in §4.2–§4.4). For a model is constituted by a set or sets of entities (the reference ranges of the system variables), and relations between these items, which together stand as a set of "facts" in terms of which the truth or acceptability of the interpreted theses of the formal system are assessed. Further, the structure of the formal system "mirrors" or "maps" the structure of the model. Such a philosophy of formal systems is simply an application of that philosophy of "ordinary" language, which is based on belief in an ideal or canonical language; and the resulting formal semantics is as artificial with respect to formal languages as the non-formal semantics is with respect to natural language. Our relocating of reference ranges, model theory and standard semantics is, therefore, simply an extension of our insistence throughout that any formal system which has claims to being regarded as the logic of discourse must be tied to the real semantics of grammar and contexts.

3.9 Separating Incompleteness from the Varieties of Nonsense

The nonsense ragbag has been stretched wide, in philosophy and without, to catch not merely absurd and incomplete sentences but also many *statements*, for example patently false statements, inconsistent and contradictory statements, and contextually self-refuting statements—such as that we are not now writing in English about nonsense. The nonsense bag has also been adjusted to catch some non-empirical and non-extensional statements, whether true or false.

It will be clear that we do not subscribe to these attempts to "write off" inconvenient truth-valued assertions as nonsense. Instead, as we hope to have shown, we make explicit provision for sentences which do not fit conveniently within the framework of classical logic and there is, therefore, no need to reduce them to the nonsignificant. Conversely, as we indicated in the Introduction and as we shall try to show in more detail later (§4.6), nonsignificance is not reducible to other categories which have sometimes been favoured. In particular, we take it that truth, falsity and nonsignificance are three independent sentence-values.

Our position with regard to incompleteness is, however, different. We have suggested (§2.14) that the liar sentence is both significant and incomplete, and we shall argue in detail (in volume 2) that the key sentences in logical and semantical paradoxes, the paradox-generating sentences, have this property. This suggests that nonsignificance and incompleteness are independent values. But at this stage we wish to leave open the question of how best to treat incomplete sentences. Especially we wish to leave open whether incompleteness is best treated (1) as a further *independent* value alongside truth, falsity and nonsignificance, or (2) as in some way reducible or assimilable under these values.

Now it might be thought that our treatment of incompleteness so far, in terms of which i is taken as a fourth value, commits us to (1). In fact, however, it does not. To take it initially as anything other than a fourth value would be to beg the question in favour of (2); to take it initially as a fourth value leaves it open whether it is independent, or, when the necessary apparatus is developed, whether it is later reducible to or definable in terms of, the other three values. Partly for this reason, and partly for reasons of simplicity, we restrict attention in Part II to many-valued systems interpreted over just the three independent values t, f and n.

There are various ways in which incompleteness might be reduced. In terms of a Russellian analysis, sentences like 'The present King of France is bald', 'The distance from Pluto to Vulcan is very great', 'There is a west wind blowing at the North Pole', and negations of these, where the description has a primary occurrence, are all taken to yield false statements. In some cases, this assignment is reasonable and one which many English speakers are inclined to make; in other cases it is less reasonable. In particular, if the paradox-generating sentences are admitted as incomplete, then a Russellian analysis, in terms of which both a sentence and its negation are assigned the value *f*, fails unless some modification is made. This can be done by extending the theory of primary and secondary occurrences, or by using predicate connectives (Routley [7]). In such cases, incompleteness really amounts to a cross-classification of truth-valued sentences (see §5.4).

An alternative, very simple, reduction would be to take all incomplete sentences as nonsignificant. More favoured by some philosophers are treatments which reduce some incomplete assertions, *e.g.* those about possible items such as the present King of France, or those about reasonable (non-defective) items, to false assertions, and the remainder, including usually paradox-generating sentences, to nonsignificant sentences.

Neither of these last seems to us satisfactory, however, and though we leave open the best treatment of incomplete sentences until volume 2, we need here to mark out the class of incomplete sentences rather more precisely and to separate them from the sorts of nonsignificance which we shall later distinguish in more detail (§7.14).

Given a sentence Γ_0, which may contain a number of subject vectors whose places are filled by factors of various kinds, we may distinguish one variety of nonsignificance as that which depends on what the subject factors are about, in a given context c_0: we call this, *absurdity* (with respect to c_0). A second kind of nonsignificance does not depend on what the subject factors are about but only on the senses which the expressions in Γ_0 have in c_0: this we call aboutness-independent nonsignificance, or *nonsignificance proper*. With a knowledge of proper nonsignificance we can determine, using the identities made in the context, all absurdities in that context, simply by making appropriate substitutions. For example, given that 'The table was only slightly amused' is nonsignificant in c_0, because of the clash of senses involved, it follows that if 'Tom' is s-about the table in c_0, then 'Tom was only slightly amused' is an absurdity in c_0. This approach, which takes over significance proper, with respect to a fixed context, as primary, we shall use later in Chapter 7. But for most of the logical developments we make, '$\sim S$' may be interpreted either in terms of nonsignificance proper or in terms of absurdity.

Absurdity and nonsignificance proper do not exhaust the nonsense notions we think are of philosophical importance. Another such notion is that according to which an assertion is *nonsense* if it entails some assertion that is absurd.

Now what distinguishes incompleteness from all such nonsense notions is the way the former is bound up with existence. It makes no difference to the nonsignificance or absurdity of a sentence whether its subject factors refer or not, whether what the sentence is about exists or not. But it makes all the difference to incomplete sentences. If the present King of France exists, then 'The present King of France is bald but still handsome' is not incomplete. Should Vulcan exist, the distance from Pluto to Vulcan would be a determinate, measurable, one and statements about the distance would not be incomplete. Should, *per impossible*, the least rapidly convergent sequence exist, incomplete assertions about it would be resolved. Notice, however, that the condition for completeness is not necessarily the existence of some item which the subject term is about, though it usually is in the simple cases. But given the sentence 'It is five o'clock on the sun', the condition for completeness is that time discriminations exist (or are determinate) on the sun; again 'Everything doubled in size overnight' is complete if absolute space exists. In general, the features of incompleteness, unverifiability, irresolvability, arbitrariness and so forth, are all removed given existence of the appropriate items. This is not the case, however, with nonsignificance features.

In terms of this we can give a rough criterion for separating incomplete and nonsignificant sentences, though not one which applies directly to sentences where the condition for completeness is the existence of items which are presupposed rather than mentioned by the subject factors (*e.g.* 'Everything doubled in size overnight'). To give a criterion for such sentences, full determinacy, rather than its special case, existence, appears to be the proper test, but we do not yet have the apparatus adequate to express this. The limited test we therefore propose is as follows:

Let Γ_0 be such that it fails to yield a (determinate) statement in c_0. Now suppose that all the items Γ_0 is about in c_0 exist. Does $\Gamma_0(c_0)$ then yield a (determinate) statement? It it does, Γ_0 is incomplete in c_0, otherwise nonsignificant. The criterion can be dressed up formally. Let $X_1, ..., X_n$ be all the subject factors of Γ. Let θ be a context such that some of $X_1, ..., X_n$ do not exist, and let θ' be a context like θ except that $E(X_1)$ & ... & $E(X_n)$. Then,

> $\Gamma(\theta)$ is incomplete if, and only if, $\Gamma(\theta')$ yields a (determinate) statement but $\Gamma(\theta)$ does not.

In case incompleteness is taken as an independent value, the bracketed 'determinate' can be deleted. If, however, incomplete sentences are taken to yield statements (as on a Russellian, or similar, analysis), then the statements they yield have to be distinguished, *e.g.* as not determinate, or as secondary.

Notice that the criterion does not imply that every sentence about non-existing items is incomplete. We are not saying, for example, that because the number two does not exist, in the trip-over sense of 'exist', so 'The number two is prime' is incomplete in the standard arithmetical context. For in that context, the sentence does yield a statement and the final conjunct of the criterion fails to be satisfied. It is only when it is known that a sentence fails to yield a statement in a given context that the criterion can be applied. It is essentially, therefore, a criterion for distinguishing different kinds of non-statement yielding sentences, *viz*. those which are incomplete from those which are nonsignificant.

The criterion does assume, however, against Hume, that we can suppose that impossible items exist or that we can evaluate strictly impossible contexts. But this is not unreasonable. Admittedly, if we suppose, in order to decide whether 'Meinong's round square is yellow' is incomplete, that Meinong's round square exists in Quine's backyard, we have to take the consequences of this assumption: we are committed to supposing that the item has both a radius and a square shape, as well as a definite colour. Geometers, however, regularly make such assumptions in the course of constructing *reductio* proofs.

One useful consequence of the criterion is that it helps to explain important differences in the compounding rules for incomplete sentences and nonsignificant sentences. Given classical connectives (§§5.1–5.3), nonsignificance dominates over truth and falsity. For example, if 'Jack is a sailor' is truth-valued and 'The rainbow is miserable' is nonsignificant in c_0, then the conjunction 'Jack is a sailor and the rainbow is miserable' is nonsignificant. But the result of conjoining 'The present King of France is living in exile in Nova Scotia' and 'The present King of France is bald' is not incomplete. Again, intensional attitudes held towards assertions which on their own are incomplete are not so exotic as similar attitudes held of nonsignificant assertions: one may believe that Vulcan is near to Pluto simply because one believes that Vulcan exists, but mere belief in the existence of the *planet* Vulcan would not of itself explain the belief that Vulcan is singularly unhappy.

Given that we can so separate incompleteness and nonsignificance, we can characterise a sentence as nonsignificant in its context if it is complete but fails to yield a statement.

PART II

THE PRINCIPLES OF SIGNIFICANCE

CHAPTER 4

Towards a Logic of Significance

4.1 THE METHOD TO BE ADOPTED

THE main theorems of the context logic so far developed treat of four-valued sentences as mentioned, *i.e.* as covered by quotation functions. The theorems therefore express relations between the properties of sentences and not between the sentences themselves. Because of this limitation to the mentioning of sentences, it was possible to develop context logic as a two-valued system and to adopt the laws of **RSL** and **RQL** to carry the deduction. Context logic, that is to say, as it has so far been developed, is simply an applied classical two-valued system with special postulates. But if we ask the question: what general relations hold between sentences when used?, the problem is quite different.

This question is often taken as the starting point of formal logical systems and the answer to it is given by laws which express such relations. So, for example, under the hypothesis that all sentences are truth-valued, the various laws of **RSL** and **RQL** such as,

$$r \,\&\, s \supset r \vee s$$
$$\text{and} \quad (Ux)fx \supset (Px)fx$$

can be shown to hold as consequences of postulate-rule sets or, in appropriate cases, of matrix (*i.e.* truth-table) definitions of the logical constants.

It is analogues of such laws we now wish to examine under the hypothesis that any given sentence is three-valued, in the sense that, depending on the context, it may yield a true statement, a false statement, or be nonsignificant. We thus move from the full derived range to the reduced range. And since we are to consider such formulae in non-quotational contexts, it is necessary now to regard our earlier semi-wff as wff. It is necessary, too, to examine the matrix and postulational bases of several different three-valued systems which might be adopted so that we can decide which is most appropriate to our purposes.

If we wished to develop a full logic of sentences over the unrestricted range which includes incomplete sentences, with incompleteness taken as a further value, then it would be necessary to examine the matrix and postulational bases of various four-valued systems. For the time being, however, we leave this development on one side, partly to simplify the problem and partly to focus attention on the special problem of significance.

We shall say that a sentential or predicate system is a significance logic if it is three-valued where the values are t (truth), f (falsity) and n (nonsignificance). As one would expect, however, since the choice of three-valued systems is so wide, there are many which could qualify as significance logics in this sense and the real problem is to examine the assumptions implicit in the various matrix and postulational bases and to assess their value. We therefore develop several systems in some detail before choosing a particular one.

We shall call the laws which follow from a matrix or postulational basis of a significance logic, *principles* of significance. So, for example,

$$p \,\&\, q \supset p \vee q$$

$$\text{and} \quad (Ux)fx \supset (Px)fx$$

will qualify as principles of significance if they can be shown to hold as consequences of a postulate or matrix set of a significance logic. Principles are therefore always system-relative, but a given set of principles may be more important than another because one system may be more important than another—in terms of its application, for example.

Significance logics, though they can be set up as closed independent systems with respect to their own bases, are more correctly regarded as subsystems of context logic either under the assumption that incomplete sentences can be assimilated (§3.9) or under the further hypothesis $\vdash Cp(c)$. By this last we mean that, if we take $\vdash_{\mathrm{CL}}$(P11) $C(\Gamma)(\theta)$ as a postulate of context logic (*see* §2.16) and add the postulates and rules of a significance logic, then all the principles of significance with respect to that particular basis are $\vdash_{\mathrm{CCL}}$-theorems of a three-valued context logic. If, on the other hand, we developed a four-valued sentential and predicate theory with values t, f, n and i, then the postulates and rules of such a system, when added to context logic, yield theorems which are independent of the hypothesis $\vdash_{\mathrm{CL}}C(\Gamma)(\theta)$, *i.e.* they are full theorems of four-valued context logic.

Such development of context logic is essential if it is not to be seriously incomplete (*cf.* the concluding remarks of §2.16) and we therefore regard the discussion in Part II as a continuation of Part I rather than as an independent investigation of significance. This is true also of Part III. We shall, however, simplify the development by restricting the principles of significance to application in a fixed context. Thus, we shall not consider formulae such as,

$$p(c) \supset p(c) \vee q(d),$$

but only those which have the same context variable qualifying the whole formula *e.g.*

$$(p \supset q \vee q)(c).$$

We then make the assumption that the same context variable applies to all formulae, so that it becomes unnecessary to write it. That is, we represent the above formula, parametrically, as

$$p \supset p \vee q;$$

and similarly for all others. This amounts to saying that for any given application of a significance logic, or for that part of context logic which consists of a set of significance principles, it must be restricted to a fixed context. This is not to say, of course, that the whole logic cannot then be reapplied in a different context. Rather, the restriction is that the logic cannot be applied in such a way as to relate different contexts. We therefore take it that the condition,

For a fixed context *c*,

is an extra-systematic assumption of any given significance logic (this corresponds to the technique adopted for **RSL** in Ch. 2). The fixed context chosen for any given application need not be the standard context c_S, though it will in general be logically standard. From the viewpoint of context logic then, the formulae of significance logic are like operator equations of the calculus. Thus we can abstract from context without invoking the so-called "ideal" of contextless sentences. In fact, contrary to familiar criticisms (*e.g.* those of §4.5), formal logic has never really required for its application absolutely contextless sentences. Rather, at the beginning of an application or a problem the context is fixed; the domain of discourse (or of individuals in the model) is specified, the (reference) values of individual constants are laid down and so on. Often this is done informally, but there is provision in the standard quantificational semantics for much of it to be done in a more formal way, and this formalisation can be increased.

A significance logic modelled on the classical two-valued system, such that all the principles express relations between sentences when used, is the simplest of its kind. But it is necessary to consider also more complicated systems containing principles which express relations between sentences which are both used and mentioned. This is so because many arguments appear to yield a conclusion of the sort, "'p_0' is nonsignificant", where 'p_0' is used but not mentioned in the premisses.

One might argue, for example, that if a sentence implies a nonsignificant sentence, it is itself nonsignificant. Or that if a sentence implies its negation, and its negation implies it, it is nonsignificant. Again, one might conclude from the fact that two subjects have the same aboutness and that a nonsignificant sentence can be formed by taking one of them with a given predicate element, that a nonsignificant sentence is formed by taking the other with the same predicate element.

Arguments such as these can only be assessed satisfactorily by representing the putative principle formally, as in the following formulae where the negations and implications remain to be determined:

$((p \supset q) \mathbin{\&} \sim Sq) \supset \sim Sp$ (the literal *reductio ad absurdum*);
$((p \supset \sim p) \mathbin{\&} (\sim p \supset p)) \supset \sim Sp$ (the paradox law);
$(x = y) \supset (\sim Sfx \supset \sim Sfy)$ (the identity law);

and considering significance logics in which they are or fail to be derivable principles. For if there are strong extra-systematic reasons why a given significance logic should be adopted and if such formulae, though well-formed, fail to be derivable, this is evidence of their unsuitability. On the other hand, if they are derivable in a well-supported system, this is evidence of their suitability. Of course, one might argue that they ought to be principles of any satisfactory system; and this decision would then fashion the development of it.

So it is necessary, in order to assess arguments in which sentences are both used and mentioned, to admit formulae in which both three-valued and two-valued expressions are well-formed, to allow compounds of these, and to consider relations between them. For otherwise it would be impossible to represent formulae such as those above in which, *e.g.*, '$(p \supset q)$' occurs as a three-valued element and 'Sp' occurs as two-valued, and hence impossible to evaluate arguments which depend on them as principles. And it is necessary, too, to consider formulae which are wholly two-valued. For there are many principles which have been adopted implicitly or explicitly by various authors and which it is necessary to come to a decision about. Such principles, for example, as that a sentence and its negation are co-significant, *i.e.*

$$Sp \equiv S{\sim}p;$$

or that a compound sentence is significant if, and only if, each of its components is significant, *i.e.*

$$S(p \mathbin{\&} q) \equiv Sp \mathbin{\&} Sq,$$
$$S(p \vee q) \equiv Sp \mathbin{\&} Sq, \quad \textit{etc.}$$

These principles, of course, could arise in that part of context logic developed in Part I, given suitable postulates, since they are wholly two-valued.

If, therefore, we regard the basic significance logics as being like the classical systems, in that all variables are uncovered, though of course differing in having three values instead of two, then there are extended significance logics, rather like the classical modal systems, which contain operators of various kinds. This analogy must not be taken too far, however, for modal formulae such as,

$$\Box p, \ \Diamond p, \quad \textit{etc.}$$

(which are usually interpreted, respectively, as 'it is necessary that *p*' and 'it is possible that *p*') are, like the initial variables, two-valued, while "significance" formulae such as,

$$Sp, \ Tp, \quad etc.$$

are two-valued even though the initial variables are three-valued. There is nothing in classical modal logics corresponding to the mixture of values which occurs in significance logics.

Our method then, in Chapter 5 and subsequent chapters, will be to examine the matrix and postulational bases of various sentential significance logics some of which contain both two-valued and three-valued formulae, as well as formulae which are compounded from these two kinds. We shall then consider corresponding predicate systems.

4.2 The Standard Method

The standard approach to significance has been different. Instead of constructing special logics containing variables which can take nonsignificant sentences as values, the technique has been to exclude nonsignificant sentences by means of appropriate formation rules so that only truth-valued sentences remain as possible values for the variables of the logic. In terms of such an approach, **RSL** and **RQL** are the only (extensional) logical systems necessary. The standard method may be seen at work not only in mathematical logic, but also in most modern theories of grammar, and in recent attempts to salvage the empiricist theory of significance by design of an empiricist language into which all significant discourse is supposed to be translatable.

The approach is well displayed in treatments of type theory, and it is important to look at some of the details here since our reasons for offering a different method derive largely from our belief that the standard approach is inadequate.

The theory of types is a set of criteria designed to separate nonsignificant sentences from others; and the motive for developing the theory is the well-known contradiction which arose in Frege's arithmetical system (Russell [1], Appendices A and B). Indeed a whole series of paradoxes, the most notable of them Russell's paradox concerning the class of all classes which are not members of themselves (outlined in the Introduction §1) arose to threaten the foundations of mathematics and, in particular, class theory.

Russell's solution of this version of the problem, as we have noted, consists in dividing items into different logical kinds or types. In particular, individuals are said to be items of type 0, classes of individuals are items of type 1, classes of classes of individuals are items of type 2, and so on. It is then specified that the membership relation can only hold significantly between x and y if the type of x is exactly one less than the type of y. Thus, if x is of type 2, then $qu(x \in y)$ is significant if, and only if, y is of type 3. It is a consequence of this specification that all self-membership expressions, such as $qu(x \in x)$, are nonsignificant. And since the premisses of the paradox argument are of this form, the argument can no longer be stated and the contradiction, so it is claimed, is resolved.

Because the paradox can be formulated entirely in terms of properties and independently of the concept of a class, however, the solution has to be extended and a typal structure imposed on the predicate calculus. Thus, if we say that a property is predicable when it has its own property (the property of being abstract is itself abstract), otherwise impredicable (the property of being concrete is not concrete), then we obtain a contradiction when we ask whether the property of impredicability is itself impredicable. Hence properties have to be sorted into types, and predicative sentences are held to be significant if, and only if, the type of the subject element is exactly one less than the type of the predicate element. In fact, given the class abstraction thesis,

$$x \in \hat{z}(\phi z) \equiv \phi x,$$

where $\hat{z}(\phi z)$ is the class generated by the predicate $qu(\phi)$ [or the property $\lambda\phi$], the typal hierarchy of classes implies the typal hierarchy of properties; or conversely, given the property hierarchy, the class hierarchy follows since classes and properties are formally related.

There are complications which need not concern us here. First, some technique has to be developed which enables the type of a relation to be determined when it relates elements of the same or different types. Secondly, types themselves have to be structured internally by a hierarchy of orders to prevent the occurrence of certain vicious circle fallacies which are not removed by the typal hierarchy. There are, too, several well-known difficulties which we shall take up later; though we look briefly at one of them, that the theory cannot be consistently stated, in the next section. In general, however, these complications and difficulties are related to the question of whether or not type theory constitutes a set of adequate or reasonable criteria in terms of which the significance of a given sentence can be decided, and we postpone the discussion of that to Part III. But what does concern us here are the formal techniques to be adopted for dealing with nonsignificance when it has been recognised; when, that is, the theory of types has been applied as a set of (presumed adequate) criteria and the nonsignificant sentences have been isolated.

Russell's practice is ambiguous in this respect because he did not make distinctions which were later made between a formal system and its metatheory. There are places where it seems clear that the type rules are regarded as extra-systematic and used at the pre-formal level to exclude nonsignificant sentences from the logic since the only admissible formulae are "predicative" (in Russell's special sense) and only predicative formulae are significant (see, *e.g.*, Russell [2], V and VI). But types are nevertheless defined within the formal system, and theses about them are derived as theorems (*e.g. PM* *63). More important, from the present point of view, several theorems are clearly principles of significance, and others have explicit significance conditions imposed on them, though the conditions are not expressed in the symbolism of the system. Thus, for example,

*9.14 If "ϕx" is significant, then if x is of the same type as a, "ϕa" is significant, and vice versa (*PM*, p. 133).

*10.39 ... This proposition is only true when the conclusion is significant; the significance of the hypothesis does not insure that of the conclusion (*PM*, p. 147).

The question to be asked here is how such conditions should be taken up in a complete formalisation of the logic of *PM*. They could be represented explicitly as formal parts of the formal theses, and this is the method we propose to adopt. It requires, however, that the sentential and predicate formulae be three-valued. For it must be possible to say, in certain cases, that a given sentence is nonsignificant; and if such a remark is ever to be true, the sentence itself cannot be truth-valued and any formula of which it is a case must therefore be three-valued.

The alternative, which we propose to reject, is to distinguish the metatheory from the formal theory, to regard the type rules as metatheoretic, and to express their consequences as formation rules designed to exclude nonsignificant expressions as well-formed formulae. This is the method adopted in more recent treatments of type theory, and there are various ways of doing it. We may, for example, introduce numerical typal subscripts on the variables and specify that a predicate or class formula is well-formed if, and only if, the typal subscript of its subject element is exactly one less than the typal subscript of the predicate element (with appropriate complications for relational formulae).

In terms of such restrictions, formulae which represent nonsignificant sentences cannot arise within the formal system and it is therefore two-valued, *i.e.* it is like the classical **RQL**. It is this technique of eliminating nonsignificance from the formal system which is opposed to our approach and which we call the standard method. Theses such as *9.14 and *10.39 are now no longer required. The principle expressed by *9.14, for example, is secured by the typal suffixes and the formation rules; and the condition on *10.39 becomes unnecessary since the significance of the conclusion is assured if it is well-formed, as it must be to occur at all.

4.3 Deficiencies of the Standard Method

We develop three main objections to the standard method. The first is that it is only appropriate to highly artificial and technical theories of significance, which is what Russell's theory of types is, and that it fails completely for any theory, such as Sommers' theory of types ([1] and [2]) which comes closer to satisfying the conditions of a natural language. The second is that any set of significance criteria, including the theory of types, both presupposes and implies a number of principles of significance which can never be stated formally if nonsignificant sentences, or formulae which represent them, are excluded. And the third is that it becomes impossible to evaluate arguments in which nonsignificance is an issue, including those arguments which are advanced in favour of a particular set of significance criteria, such as the rules of type theory. We shall discuss each of these in turn.

It is a feature of the theory of types that significance is wholly determinable as a syntactic feature of sentences. This is not obvious in terms of its initial formulation since types were assigned to individuals, classes, properties and relations, rather than to the words which denote or refer to them. But it became necessary, in order to avoid certain difficulties in the theory, to re-interpret types as characteristic of the words themselves rather than as characteristic of the items denoted by the words. In particular, if we take a realist view of types, we can never say significantly that a given item does not belong to a given type. For suppose we wish to say of an individual of type 0 that it does not belong to a class of type 2. This assertion has to be expressed by a formula of the sort,

$$\sim(x \in \alpha);$$

but since, *ex hypothesi*, x is of type 0, and α is of type 2, the sentence,

$$\sim(x \in \alpha),$$

is not significant. And yet, of course, if the theory is true, the assertion has not only to be significant but true, since individuals do not, indeed cannot, belong to type 2 classes. Russell's proposed solution, which appears in several places, is that the theory of types is really a theory of symbols, not of things. For we can say significantly of the symbols that there is a type difference between them, but we cannot say this of the items which the symbols denote (see, for example, *PM* vol. ii, p. 34; Russell [3] VII p. 267; Schilpp [1], pp. 601–2).

When the theory is interpreted in this way, therefore, we may say that a single-subject sentence is significant (to take the simplest case) if, and only if, its subject element is exactly one type less than its predicate element. It is thus admirably suited to syntactical representation in terms of sorted variables and formation restrictions on syntactic structure. And indeed it is only when it is represented as a syntactic theory that it can be encapsulated in formation rules. Conversely, and more generally, because formation rules are wholly syntactic rules, they can only succeed as a sieve designed to strain out nonsignificant sentences if the criteria in terms of which significance is assessed are themselves wholly, or wholly reducible to, syntactic criteria. And this means that if the standard approach, which is peculiarly suited to type theory, is adopted as a general method, then we are committed to the view that the significance or nonsignificance of a sentence is a permanent feature of it, because it is a feature of its structure, and this is permanent; and secondly that it is a feature which is independent of both semantic and contextual considerations. Once semantic and contextual features are admitted, however, any structural theory of significance is hopelessly inadequate.

The failure of the theory of types at the semantic level is apparent. For if we take it, as Russell does, that all individuals are of the same type, or in later versions, that all proper names are of the same type no matter what items they refer to, we are immediately faced at the semantic level with the problem of who or what to admit as individuals, or of what symbols are admissible as proper names. If we admit 'Margaret' and 'Saturday' to the list, and take 'is in bed' as a type 1 predicate significant over individuals, then we have to accept that both 'Margaret is in bed' and 'Saturday is in bed' are significant. But of course, if 'Saturday' is taken in its standard sense, then 'Saturday is in bed' is not significant. Again, since all individuals are of the same type, any one individual variable can be substituted for another without affecting significance. Hence given a thesis in which a relational formula of the sort '*xRy*' occurs, it is usually possible to derive a further thesis containing the converse of such a formula, namely '*yRx*'. And this means that if significant theses always lead to a significant theses, the second thesis is significant whenever the first is. At the level of interpretation, however, there are many relations which fail to satisfy the condition that if *qu*(*xRy*) is significant, so is *qu*(*yRx*). If we take *qu*(*R*) to be 'loves', then we may significantly say that Margaret loves Saturdays, but the converse, 'Saturdays love Margaret', is not significant. It is a further consequence of type theory that because the type of predicate is irrevocably a permanent feature of it, so no predicate which is significant over an individual can also be significant over a class of individuals. Hence, since a club is a class of a different type from its members, we cannot significantly say that both a club and Jones, who is a member of it, donated £5 to a worthy cause.

These criticisms are not simply criticisms of classical type theory, which of course has a number of quite special defects. They are criticisms also of linguistically more sophisticated type theories such as cumulative type theories; and they can be extended to Chomsky's programmatic revitalisation of type theory. For they are criticisms which arise directly from the fact that type theory is solely a syntactic theory. And though it may be the case that not exactly the same difficulties would arise within other syntactical theories, other similar difficulties would. For it is always possible to construct nonsignificant sentences at the semantic level which violate criteria of significance which are wholly syntactic. In this respect Ryle's account of categories stands in marked contrast to the theory of types and other syntactic theories, for Ryle insists that "absurdities result from the improper coupling not of expressions but of what the expressions signify, though the coupling and mis-coupling of them is effected by operating upon their expressions" (Ryle [2], p. 77).

But even the admission of semantic criteria is not enough. For unless we also admit contextual features, we shall still remain committed to the position that significance is a permanent feature of the sentence, even though it is now regarded as a semantic sentence. This is again obvious in terms of the theory of types which we might now seek to modify by introducing semantic conditions. Thus, suppose we decide to limit the range of the individual variables in order to avoid the occurrence of nonsignificant sentences such as 'Saturday is in bed' and 'Saturdays love Margaret'. This we might do by excluding 'Saturday' from the range. In this case, however, we thereby restrict the application of the logic, since it is no longer applicable to *any* sentence which contains the word 'Saturday', and we eliminate nonsignificance at the semantic level only at the price of eliminating much that is significant. We shall no longer be able to say, for example, that Margaret loves Saturdays. Hence the logic can no longer claim to stand as the logic of discourse.

Such tinkering is valueless simply because it is the case that, just as the rules of a syntactic grammar (which is what formation rules are) are inevitably inadequate as criteria of nonsignificance, so the rules of a semantic grammar (which is what range specifications are) are inadequate. If semantic restrictions are to work, it has to be supposed that words are always used in a standard sense, or with a standard aboutness. But we may use 'Saturday' to mention a boy, in which case 'Saturday is in bed' is significant; or we may use 'Margaret' to mention a cyclone, in which case 'Margaret is in bed' is nonsignificant. Are we then to admit 'Saturday' to the range and exclude 'Margaret'?

This is simply to re-emphasise that significance is neither a permanent feature of a syntactic sentence nor yet of a semantic sentence. For it is not a permanent feature of a sentence at all. It is a feature of its use; which is to say that, in most cases, significance cannot be evaluated independently of context. Any given sentence of a given structure and illocutionary force may be such that it is truth-valued in one context and nonsignificant in another. And this means that if we attempt to exclude nonsignificant sentences by means of grammatical rules, whether syntactic, semantic or both, we must inevitably fail. We shall be committed to excluding too much or too little. This becomes obvious in terms of a later doctrine of Russell's that the only genuine proper names, which can be substituted for an individual variable, are demonstrative pronouns such as 'this' and 'that'. But this entails that every admissible formula of the form $qu(\phi x)$ yields a significant sentence if $qu(\phi)$ is replaced by a type 1 predicate and $qu(x)$ is replaced by a demonstrative pronoun. Hence every sentence of the form qu(This is ϕ), where $qu(\phi)$ is a type-1 predicate, is significant in every context, a consequence which is a *reductio ad absurdum* of the theory and of any analogous inclusive type theory, since *every* description *qu*(the item which is ϕ is ψ) is significant where *qu*(ϕ) and *qu*(ψ) are type-1 predicates.

Our second criticism, which we shall again illustrate by way of the theory of types, but which we believe has general force, turns on the fact that any set of significance criteria both presupposes and implies a number of principles of significance which must be accepted if the theory is itself acceptable. If it is the case, however, that the formal system, developed by reference to the significance criteria, is designed to exclude nonsignificant sentences, then the implied principles cannot be stated formally.

As an example of a principle which is presupposed by the theory of types, consider again the assumption which provides the motive for the theory, namely that the contradiction in the paradox argument evaporates if it can be shown that the premisses are nonsignificant. The point of the theory is to establish a set of criteria designed to show that the premisses are nonsignificant, but this assumption about the argument is independent of the theory of types, and indeed is independent of any other set of criteria which might be put up in justification of the claim that the premisses are nonsignificant.

It is not clear exactly how to formulate the principle implicit in this assumption. We might, for example, represent it as,

$$\sim Sp \supset \sim [((p \supset \sim p) \ \& \ (\sim p \supset p)) \supset (p \ \& \ \sim p)],$$

or

$$\sim [((p \supset \sim p) \ \& \ (\sim p \supset p) \ \& \sim Sp) \supset (p \ \& \ \sim p)],$$

but the widespread belief that the self-references paradoxes in general can be resolved by dismissing the paradoxical sentences as nonsignificant, often seems to depend on the stronger assumption that any paradox argument virtually implies the nonsignificance of its own premisses. That is,

$$((p \supset \sim p) \ \& \ (\sim p \supset p)) \supset \sim Sp.$$

But however it is represented, the point we wish to make here is that it cannot be formulated at all in any system which excludes nonsignificance as a value. For if *qu*($\sim Sp$) is true in any particular case, then *qu*(p) must take the value n throughout. Hence the principle cannot be expressed in a two-valued system.

As an example of a principle which is implied by the theory of types, we need only note that if *qu*($x \in y$) is significant and *qu*($y \in z$) is significant, then if *qu*(x) is of type n, *qu*(y) must be of type $n+1$; and if *qu*(y) is of type $n+1$, then *qu*(z) must be of type $n+2$. But from this it follows that *qu*($x \in z$) is nonsignificant. Hence, in general,

$$S(x \in y) \ \& \ S(y \in z) \supset \sim S(x \in z).$$

This principle is itself two-valued since *qu*($S(x \in z)$), *qu*($S(y \in z)$) and *qu*($\sim S(x \in z)$) are all two-valued, but the elements *qu*($x \in y$), *qu*($y \in z$) and

$qu(x \in z)$ must themselves be three-valued. For if the consequent is true in any particular case, $qu(x \in z)$ has to take the value *n*. While if the consequent is false, so is the antecedent; but in this case one or both of $qu(x \in y)$ and $qu(y \in z)$ takes the value *n*. Hence this principle, and it is only one of many, cannot be stated in any formal system which excludes as ill-formed nonsignificant membership formulae.

Our third criticism derives from the second. For if nonsignificance is to be relegated to a pre-formal level, how are we to evaluate arguments in which it is in issue? We have seen that Russell's assumption about the paradox argument depends upon the acceptance of a significance principle. The question is, however, whether this principle is true. For it is only by answering this question that we can evaluate the argument. To determine whether or not it is true, however, depends on showing whether or not it is a principle of an acceptable three-valued system. The evaluation of Russell's argument is impossible, therefore, unless we have a formal theory which admits, rather than excludes, nonsignificant sentences.

Many such principles have been adopted implicitly, or more or less explicitly, by various authors. Ryle, for example, seems to claim the acceptability of the literal *reductio ad absurdum*,

$$((p \supset q) \mathbin{\&} \sim Sq) \supset \sim Sp$$

(see Ryle [2] and [3]) and adopts it implicitly as part of a major theme since it is his purpose to show that the doctrine of dualism is absurd because it has absurd consequences (Ryle [4], Introd. & Ch. 1). If, therefore, we wish to ask whether Ryle's arguments are valid, we are in effect asking whether this principle is true, and this cannot be answered unless we construct appropriate significance logics. So it is in the case of other principles which have frequently been assumed but never demonstrated.

In sum, we conclude, first, that any formal system which has claims to being regarded as the logic of ordinary discourse cannot exclude nonsignificance either by means of formation rules or by means of semantic range restrictions since this is to commit the fallacy of supposing that significance is a permanent feature of sentences. Secondly, that in order to examine the presuppositions and consequences of any particular significance theory, it is necessary to represent explicitly and formally the principles which are adopted. And finally, that arguments in which nonsignificance is an issue, though they may occur in the metatheory of a given formal system and are pre-formal with respect to that system, must not be pre-formal with respect to all systems if they are to be evaluated. If, for example, the principles are significantly statable metatheoretically then they are formalisable, and are accordingly, from the point of view of the meta-metatheory, formalisable in a two-level formal system.

4.4 Ideal Languages

We have made most of our criticisms in terms of the theory of types, but it seems to us that they are general and not limited to the particular example we have chosen. The reason for this is that our objection to Russell's approach to significance is not, at this point anyway, an objection to particular details of type theory but instead to the general presupposition that the proper way to deal with nonsignificance at the formal level is to exclude it, rather than to admit it as an explicit value and then to explore the consequences of so doing. And this general presupposition underlies many other theories of significance. It is evident in levels-of-language solutions of the semantic paradoxes, which not only follow type theory in the hierarchical structure but which also, like type theory, use the hierarchy to impose syntactic restrictions on what can and cannot be said significantly. It is evident, too, in the theory of semantic categories developed by the Polish logicians, and in the writings of logical positivists. The absence of a formal logic of significance in positivist writings is especially surprising, for without explicitly formulated principles many arguments and claims which they make cannot be assessed. In particular, it remains obscure which arguments for and against the criterion of verifiability are sound (*cf.* Presley [1]).

There was perhaps a half-recognition that there is here an obvious gap in formal logic since, as we shall see in the next chapter, both Russell and the positivists here and there make assumptions which commit them to a special set of three-valued sentential connectives, which we call the classical significance connectives. And there is besides, as we have already mentioned, a partially explicit treatment of type theory in *PM* as well as explicitly stated significance conditions on some of the theorems. In spite of this, the full recognition and the necessary detailed development was not made; and one important reason for this is that it was inhibited by the generally accepted belief, which is also Russell's, that the proper task of analysts and linguists is to construct an ideal language rather than to investigate the actual features of real ones.

For an ideal language will only contain sentences whose parts are univocal and whose sense, and even aboutness, is entirely independent of context. Illocutionary force will coincide with actual sense; and meaning will be independent of use. All sentences of an ideal language must express genuine propositions, and from this it follows that nonsignificant sentences are to be excluded. The logic of an ideal language will therefore be the logic of propositions, so that significance is something to be assessed before the real business of logic is begun. Logic is concerned only with what is expressed, and since nonsignificant sentences do not express propositions, they are outside the scope of logic.

This attitude appears in Russell's review of MacColl's logic, where he criticises MacColl's introduction of the "unmeaning" as a separate class of proposition. This introduction, says Russell, "again illustrates the fact that his [MacColl's] system is concerned only with the verbal expression, not with what is expressed. For what is unmeaning is merely a phrase; it is by no means nothing, on the contrary, it is a definite form of words. In logic, we ought to adopt such a language, and such rules for its employment, that unmeaning phrases shall not occur." (Russell [4], p. 257).

The idea of an ideal language later appears in a much more elaborate form in the philosophy of logical atomism (Russell [3], pp. 175–282). It is there described as the logically perfect language, and is in effect to be the system of *PM* with an augmented vocabulary. Anything that can significantly be said can, Russell thought, be said in this language. If we take any complex expression of natural language and want to know what it means, we have first to analyse it to see if it can be translated into the logically perfect language. If it can, then it is significant; otherwise not. The point reappears in the thesis expressed by Russell in his Introduction to Wittgenstein's *Tractatus*: "a logically perfect language has rules of syntax which prevent nonsense"; and is echoed by Wittgenstein's epigram "Whereof one cannot speak, thereof one must be silent" (Wittgenstein [1], §7).

Later attempts by positivists to formulate the criterion of verifiability satisfactorily also take the ideal-language direction, though it is now to be the system of *PM* together with the language of physics. For anything which can be said significantly is expressible either in pure logic (which is meant to include mathematics) or in physics. Carnap [1] and Hempel [1], for example, in effect proposed an ideal language the rules of which would act as a sieve to strain out meaningless (*i.e.* metaphysical, theological, *etc.*) utterances. Meaningless sentences must not be formulable within the language. A logic of significance is thus excluded; and what talk about significance remains, will belong to the unformalised metalanguage of the ideal language. By this means we shall achieve an unadulterated physicalist language of science (Popper [1], pp. 206–7).

Still more recently new ideal language programmes have been insidiously introduced, for example by Quine and Davidson—where what cannot be formulated in an exclusive but fashionable canonical notation does not survive (see Quine [2] and Davidson [1]).

There is now of course a considerable literature devoted to a critical rejection of the ideal-language programme, and we scarcely need to add to it here (see *e.g.*, Wittgenstein [2], esp. Part 1, §§1–46; Popper [1], pp. 183–226; and Black [1], pp. 229–55). Whether such languages are possible depends crucially on exactly what conditions they are supposed to satisfy and it is therefore singularly unfortunate that the character of ideal languages has never been completely specified. More important, it has never been explained what language and what logic are to be used in the initial investigations required to set up the criteria for an ideal language.

If a language is to be constructed in such a way that its rules exclude non-significant sentences, the actual construction must be carried out in a language in which such sentences appear. Rules for their exclusion cannot be formulated unless we can recognise them for what they are and make significant arguments about them. This problem is always left untidily as something to be taken care of in an unformalised metalanguage. But formalised or not, the conclusions reached in the metalanguage, if they are to be convincing, must be arrived at in terms of criteria and principles which are applicable to nonsignificant sentences. That a problem, an argument or a conclusion is pre-formal does not imply that it is pre-logical. It is in fact an unstated presupposition of every ideal language programme that its justification and development is carried out in a non-ideal language which contains an unformalised significance logic.

The curious identification of logic with the logic of an ideal language, which is to say the logic of genuinely significant sentences, is inexcusable. It implies that all arguments which contain or are about nonsignificant sentences are non-logical. Yet such arguments, in the metalanguage, are crucial; and are in fact the most interesting arguments associated with the programme. It is not surprising that advocates of an ideal language never persevered in pushing the construction beyond the early stages; for they were more concerned with avoiding assessment of the promised language and with justifying the programme in the philosophically interesting, but on their own premisses, improper, metalanguage. If we were restricted to working within an ideal language, then the theses of type theory, the rules for the semantic categories of the Polish logicians and Carnap's provisions for the physicalist language, could not be stated, much less justified. The implication of the programme, that the existence of an ideal language is a necessary condition for philosophical investigation (see Copi [1], pp. 65–70, for an elaboration of this view), is mistaken. For no language can be known to be ideal independently of a prior philosophical investigation in a non-ideal language.

What is especially puzzling in all of this is that Russell, Tarski and others, when faced with the fact that classical two-valued logic fails in certain contexts, drew the surprising conclusion that natural language is wrong, or at best misleading, instead of taking the blow that classical logic is inadequate. It is only in terms of this false conclusion that the ideal-language programme, as a substitute for an investigation of natural language, has any credence whatever. Such an ideal language programme is typically a programme for covering up the deficiencies of classical logic, for saving classical logic as a philosophically indispensible theory. If, however, we insist on looking at language—where philosophical problems are formulated—as it is, and if we recognise that meaning, truth, falsity and significance are not permanent features of sentences but instead features of their use, then we have to accept the need for a logic of tokens, or types in contexts, the need to recognise explicitly the role played by context, and the need to accept nonsignificance as a possible value along with truth and falsity. In this way, in particular, we are led to the conclusion that classical logic is inadequate.

4.5 The Scope of Logic and the Anti-Logic Campaign

The ideal-language programme implies restrictions on the directions in which formal logic should be developed. In particular, it limits logical investigation to the two-valued analysis of relations between significant sentences when used. But even though we reject the ideal-language programme, it may still be argued that the limitation is acceptable, or compulsory, for reasons which are independent of the programme.

One reason for this is the belief that the matters we are concerned to discuss and formalise, context, significance, tense and reference, for instance, lie beyond the scope of formal logic. Such indeed is a thesis of the Two-Kinds-of Logic doctrine espoused by Ryle ([5] Ch. VIII) and Strawson. Thus Strawson asserts ([2], pp. 213–16):

> . . . many features of the use of ordinary speech which are sufficiently general to deserve consideration under such a title as 'The Logic of Language' are necessarily omitted from consideration under the narrower title 'Formal Logic'.

The same mistaken assumption, that classical logic exhausts formal logic, which was part of the case for building ideal languages around classical logic, underlies the Two-Kinds-of-Logic doctrine. (The assumption also appears to be an invariant throughout Wittgenstein's thinking). It was concluded from the correct observation that classical logic is inadequate to take account of much of the logic of language, using the mistaken assumption that formal logic just is classical logic, that formal logic is inadequate to the task. Formal, or symbolic, logic is not, however, a fixed and complete discipline crystallised in some version of modern quantification theory. In fact there is not a single authorised version or interpretation of quantification logic any more than there is of class or description theory: there are various competing systems. It is true that none of these present systems yet gives a completely adequate map of the logic of language in its statement-making aspect (Strawson [2], p. 230). Nevertheless better and more comprehensive symbolic systems are being designed, some of which already suffice to show that the examples of tense and significance as logical topics beyond the reach of formal logic fail.

Another reason for the limitation is that the development of satisfactory significance logics requires the enlargement and modification of the standard theory of quotation, in directions we have already outlined. It is necessary, for example, to acknowledge a difference between syntactic and semantic mentioning, to reject the standard theory of quotation in terms of which a quotation expression is the proper name of the expression being quoted, and to admit quotation functions to the formal system. Anyone, therefore, who insists on the correctness of the standard theory of quotation will object to significance logics just because they include a non-standard quotation theory, quite apart from the separate fact that nonsignificance is admitted as a third value. Such objections we believe to be mistaken; rather, we should insist that the standard quotation theory is unnecessarily artificial and restrictive. But the general objection is given a slightly different emphasis by Tarski. In this form it bears closer on the special question of the development of significance logics.

Tarski claims that if we admit quotation functions, we are thereby forced to admit "certain linguistic constructions whose agreement with the fundamental laws of syntax is at least doubtful, *e.g.* meaningful expressions which contain meaningless expressions as syntactical parts (every quotation-name of a meaningless expression will serve as an example)" (Tarski [1], p. 162). What is being suggested here is that if a given expression is nonsignificant, then it is at least doubtful that any other expression in which it appears quoted is in agreement with the fundamental laws of syntax. What is not clear, however, is what is meant by "the fundamental laws of syntax" nor yet how it could be determined whether or not a given expression "agrees" with them. Since such expressions as those which Tarski wishes to exclude occur frequently in natural languages, they cannot be the syntactical laws of any natural language; and if they are the laws of some unspecified ideal language, they are irrelevant to our purpose. We can and do significantly say that a given sentence is nonsignificant by quoting it in the course of making a statement; and there is no reason why such assertions cannot be treated formally. But even if it were the case that to mention a nonsignificant sentence is always to utter another nonsignificant sentence, we could still express this formally. In any case, as we have already argued, whether or not a given sentence is in agreement with syntactical rules, even if these are in some sense fundamental, is never of itself sufficient to determine significance. Semantic and contextual considerations are essential.

This brings us to the second of the possible objections to the development of significance logics. For it might be said that just because significance and related notions are partly semantic and metalinguistic, so it is inappropriate to deal with them formally. The strength of such an objection is not clear. It is certainly true that detailed formalisation of semantic metalanguages has been repeatedly postponed despite their importance. It is true also that there is a constant tendency to push into the metalanguage whatever cannot readily be carried out in the object language, and to forget it thereafter. It is as if semantic problems, and metalinguistic problems generally, furnished an old neglected attic. But none of this provides us with what the positivists would have called a "logical" reason for the neglect, leading to a conclusion of the sort that it is wrong, inappropriate, improper or impossible to formalise a semantic metatheory. And it is clear that there can be no such reasons. For since, as we shall show, significance logics can be formalised, to the extent that they are metalinguistic semantic systems so metalinguistic semantic systems can be formalised. In any case, as will become clear, we do not accept the rigid distinction between an object language and its metalanguage.

Finally, it might be said, classical logics have the merit of simplicity while significance logics require a good deal of additional apparatus and are generally more complex in structure. But such an objection could only have weight if classical logics were adequate. Not only this, but the complexity is certainly not apparent in many significance logics. Sentential significance systems often seem very simple, and type-free predicate significance logics are in fact simpler than classical predicate logic. But it seems to us pointless to take this objection seriously. For the merit of object-language simplicity is not clear to us, especially since it is never an infallible guide to correctness.

In general, philosophers and logicians who have taken significance to be a central notion, but who have yet failed to develop the necessary logical apparatus essential to the assessment of their own arguments and theses, have been inhibited more by outlook than good reasons. Their special views about the nature of quotation, syntax, semantics, metalinguistic problems, and in general the nature of logic, derive from a philosophical outlook in terms of which nonsignificant sentences *have* to be excluded. But the scope of logic, as conceived both traditionally and more recently, has always been much wider than this; and certainly wide enough to include the assessment of arguments in which significance claims are made. If we were to adopt Russell's restrictive view about the nature of logic, then it would be incorrect to describe his own arguments about significance, in his discussion of the paradoxes, as either valid or invalid. Indeed it would be improper to describe them as arguments. It is true that, except in the paradox argument itself, nonsignificant sentences do not occur uncovered (*i.e.* in non-quotational contexts) in Russell's discussion of significance; but the fact that they are covered by quotation does not in any way remove them from the scope of logic. For nonsignificant sentences must be admitted to the range of the variables in order to say truly, in any particular case, that a given sentence is nonsignificant.

In fact, in terms of Curry's three senses of 'Logic' (Curry [1], pp. 1–3; see also Church [4], pp. 1–2), significance logics qualify as logic in each case. In the first sense, logic is the analysis and criticism of thought, the assessment of arguments as valid or invalid, and the study of the principles of valid reasoning. In this sense, certainly, significance logics count as logic. In the second sense, logic is mathematical or symbolic logic, the study of logical systems; and as we shall show, significance logics can certainly be developed as symbolic systems. In the third sense, a logic is a system or theory of the sort considered in mathematical or philosophical logic. That significance logics are logics in this sense goes without saying. Finally we may note that the formal systems we consider qualify as symbolic logics in Carnap's sense (Carnap [6]) since, as we shall see, they may be presented as purely formal languages which have a semantics of the standard sort. But even if it were still held that the real business of logic is the investigation of those relations which hold between significant sentences, this of itself is no hindrance to the development of significance logics. For most of them include classical logic (at least at the sentential and predicate levels). Hence the "real business" of logic can carry on just as well if we take a significance logic as a basis. We ask only that the real business should not be identified with the total business.

We are perhaps tilting at windmills here; for it seems unlikely that one could take the various restrictive notions of logic and formal logic we have discussed seriously except as part of an ideal-language or anti-logic programme. But it is perhaps important to show that the restrictive views have no independent support, so that if the programmes are jettisoned, there remains no reason to accept the restrictions.

4.6 The Irreducibility of Nonsignificance

Nonsignificance is a further logical dimension not reducible to, or adequately explainable in terms of, two-valued truth or falsity on the one hand or ill-formedness or ungrammaticalness on the other. It is often claimed however, as we explained in the Introduction (§2 and §3), that we can dispense with nonsignificance by reducing it in one way or another, by reclassifying examples of nonsignificant sentences either as truth-valued sentences of certain sorts or as ungrammatical sentences. But by noticing how nonsignificant sentences differ from truth-valued and ungrammatical sentences one can see at the same time the point of the classification, *nonsignificant*, the need for it, and some of the philosophical and logical purposes for which it is wanted.

Nonsignificant sentences are distinguished from false and logically false sentences in particular by the behaviour of their classical negations. In this respect nonsignificance resembles ill-formedness or non-grammaticalness. The importance of this feature, namely that the classical negation of a nonsignificant sentence is nonsignificant whereas the negation of a false (logically false) sentence is true (necessary), is that we are thereby enabled to put aside from truth-valued and modal-valued consideration a whole class of possibly troublesome sentences, both the sentence and its negation and very many compounds of these. Thus, for example, rejection of 'the mind is inside the head' on the grounds that it is nonsignificant does not commit us to acceptance of 'the mind is outside the head'; significance rejection of the existence of properties or statements does not commit us thereby to their non-existence;

we can deny that the number seven dislikes dancing without commitments to the number seven's indifference to or preference for dancing; we can escape endless issues as to the shape and location of Virtue, the colour of Tuesday, the observing of sensations. and so on. Moreover, compounding of sentences raises serious problems for any reclassification of *prima facie* nonsignificant sentences as truth-valued, *e.g.* as false, because expected nonsignificance assignments and falsidal classifications diverge.

Certain tempting routes to falsidal and L-falsidal positions, *i.e.* those positions which reclassify (positive) nonsignificant sentences as false or logically false, should however be blocked. Otherwise, with a little sophistry which English abets, falsidal positions are easily reached. Consider the slide: if the sentence 'This stone is thinking about Vienna' is not significant, then this stone cannot significantly be thinking about Vienna, *i.e.* it cannot be thinking about Vienna, and therefore the statement that it is thinking about Vienna is false. But the argument is invalid; the qualification 'significantly' on 'cannot' cannot be dropped or replaced by 'possibly'. To avoid such sophistical arguments, as well for other reasons, it is important to distinguish negations, clustered under 'not', and impossibilities, clustered under 'cannot'. It is equally important, in avoiding sophistical arguments, to separate certain implications and conditionals clustered under 'implies' and 'if ... then ...'. The crucial point is that several logically distinct but related connectives are grouped under the head *negation*, another set is grouped under *conditional*, just as several logically distinct quantifiers are grouped as *universal* and as *particular quantifiers* (see also Chapter 3). Unfortunately members of a set are frequently conflated, *e.g.* different negation criteria are identified; and when this happens the consequences for significance theory can be disastrous. In particular, arguments for the two-valued thesis, according to which every grammatical sentence has truth-value true or false, get started. (See Toms [1] for sample arguments). But once connectives are properly distinguished the invalidity of these arguments stands out. The main falsidal arguments, designed to show that paradigmatic nonsignificant sentences are false, or L-false, because their negations are (necessarily) true, are also of this sort. Consider a typical falsidal argument and its more general form:

Abstract items do not dislike dancing:	No f's g
The number seven is an abstract item:	x is an f
∴The number seven does not dislike dancing:	∴x does not g

But the negations, 'not' and 'no', occurring are not the classical negation '$\sim$' but the wide negation '$\urcorner$' (equivalent to '$\sim T$'). (The distinctive features of '$\sim$' and '$\urcorner$' are elaborated in Chapter 5). The argument is not correctly formalised using a classical negation, for then the first premiss, as well as the conclusion, is nonsignificant and detachment is unwarranted. It can be correctly formalised using a wide negation but then the conclusion does nothing to support the two-valued thesis, or the falsidal claims, since the negation of the conclusion is not a paradigmatic nonsignificant sentence at all.

How best to describe the situation with regard to key logical notions, with regard to negation, implication, quantification and identity, itself constitutes a minor problem. To say that implication, for example, is ambiguous is misleading; for it is not that there are several unrelated senses of 'implies'; on the contrary it is possible to give a general characterisation of implication as providing a universal guarantee over a class of situations. To say that implication is systematically ambiguous is somewhat better, only it suggests the technical sense given to systematic ambiguity in the ramified theory of types, and in any case needs explanation. We shall say that such words as 'everything', 'some', 'same', 'possible', 'not' and 'implies' are *logical determinables.* In each case there is a common *covering* sense, and under this a class of distinct senses, the determinate senses. For example, the word 'everything' has as a covering sense that of universality of the class of things (taken distributively), but specification of the nature of the contraction class leads to distinct senses under the cover. In one such sense, that demanded by most classical logicians, "everything exists" is true, but in another it is false; thus *salva veritate* may be used to show that the senses under the cover are distinct. Sometimes the context settles the contraction class, the extent of the quantifier, and sometimes it does not, as in the case of generalisations which *may* support counterfactuals; consider the different ranges of 'every' in 'Every raven is black'. The determinable 'not' has as covering function an exclusion or negativing function (see *OED*; the issue of the general characterisation of negation is further discussed in Chapter 5), but out of context the specific features of the exclusion are not indicated. A generalised *salva veritate* principle shows that the specific senses are distinct; thus the determinates '$\neg$' and '$\sim$' differ in sense since '$\sim$(7 dislikes dancing)' is nonsignificant but '$\neg$(7 dislikes dancing)' is true. For very many everyday linguistic purposes a single logical determinable suffices; the determinate in question in a given context is specified with sufficient precision by the scope of the determinable, or the scope of the determinable in combination with a modifier, *e.g.* 'not' followed by 'significantly' as distinct from 'possibly'. Where scoping brackets or particles are unsuitable for the task, as in spoken English, variant forms with an appropriately built-in scope may take over; hence such negative particles as 'in-', 'un-', 'dis-', 'non-' *etc.* It is not surprising that logicians keen on economy of notation and ideas prefer to reduce all differences between determinates to differences of scope, and scope to bracketing scope. Given appropriate auxiliaries this can be done; a determinate is replaced by its determinable together with distinguishing auxiliaries and scopes. Thus all negations can be reduced by occurrences of '$\sim$', which can then be taken to symbolise the determinable, together with scopes, *e.g.*

sentence scope, predicate scope, and auxiliaries, *e.g.* '*T*', '*S*', '◇'. Such a reformulation does not however eliminate determinate senses, and it is no concession to the mistaken *one word—one determinate sense* assumption. For many logical purposes, moreover, the clarity obtained by using distinct logical symbols for different determinates is more important than the apparent economy got by relying on scoping devices and modifiers.

The doctrine of logical determinables appears to subvert all popular arguments for falsidal and other truth-valued reclassifications of nonsignificant sentences: it also provides a basis for meeting many standard objections impeding the development and application of significance logics. Thus using it assists in meeting Russell's objection that significance ranges of quantifiers cannot be explicitly stated (met in §5.11), and Pap's objection that significance ascriptions cannot be significantly explained and defended (answered later in the section). Both Pap's objection and the falsidal arguments rely, in different ways, on one's failing to notice the determinability of 'not'. Yet the determinability of 'not' between various determinates, already evident from alternative logics when these are combined, shows up strikingly when 'not' is combined with modal expressions, *e.g.* as in 'cannot'. Consider, for example, 'Redness cannot be hairy' and 'A raven cannot be white'. The sort of incompatibility of nonsignificance can be made determinate by introduction of appropriate modifiers, though sometimes a speaker may intend only to assert the determinable and not be prepared to go on to a particular determinate: compare 'the object was coloured', where the observer is not prepared to go on to specify the determinate colour, with 'the class of lions is not a lion' where the speaker is not prepared to go on and determine the negation.

The thesis of traditional grammar (Lyons [1], p. 424), that "the significance of grammatically well-formed sentences is accounted for in terms of certain general principles of compatibility between the "meanings" of their constituent elements", does not offer a satisfactory reduction of significance distinctions. For compatibility and incompatibility are likewise logical determinables; and when resolved the thesis is either mistaken or else not a reduction. The chief rival determinates of 'incompatible' are, first, 'logically false' and second, 'nonsignificant'. Though the first would indeed provide a reduction, the reduction is but that of an L-falsidal theory; and the second provides no reduction. In fact significance incompatibility is tantamount to the notion of significance clash, and gives the same sort of illumination.

Symmetry with respect to classical negation is a feature nonsignificance shares with non-grammaticalness, and similarly we want to appeal to non-grammaticalness in order to set aside from consideration—but from *logical* consideration, not just truth-valued and modal consideration—a whole class of sentences and their compounds. But the appeal is different, and its point is different. The difference is that in the case of non-grammaticalness we want to put aside from *all* contexts of language use *all* (unquoted) compounds of a non-grammatical sentence. That is, non-grammatical sentences are much more thoroughly excluded from a place in the language than are non-significant ones. Several features which distinguish non-grammatical from nonsignificant sentences soon emerge. For a grammar, or at least a traditional grammar, should meet the requirements that it is

(i) context independent,
(ii) a syntactic matter,
(iii) homogeneous, and
(iv) recursive.

In all these respects significance differs. To explain:— According to (i) one and the same English (type) sentence cannot be grammatical in one context and ungrammatical in another; in contrast an English sentence may be significant or absurd in one context but not in another. (The relativisation to a language is essential: without it the contrast can be eroded.) Nonsignificance is not a permanent feature of sentences of a language in the way that ungrammaticalness is (a point already developed in Introduction §3 and in §4.3.) For similar reasons there is no grammatical counterpart of context-dependent nonsignificance where nonsignificance depends on reference, as with 'That is happy' where 'that' refers to an inkbottle. Non-grammaticalness can never be rectified just by change of reference of a subject. This is part of the force of (ii), according to which the grammatical status of a sentence, unlike its significance, is unaffected by semantic changes such as changes in senses and aboutness of component expressions. Of course changes which alter the parts of speech of words affect grammatical status, and these parts of speech may ultimately be semantically determined—through standard uses (see §1.2).

Another important distinguishing feature of grammatical sentences, feature (iii), is that they satisfy the following *homogeneity* requirement: any sentence compound of an unquoted non-grammatical *sentence* is likewise non-grammatical. The requirement is not adequately expressed by saying (as in Tarski [1], p. 162) that a grammatical sentence may not contain a non-grammatical *component* as a syntactical part; for 'Socrates is identical with Aristotle' contains the non-grammatical part 'Socrates is identical'. But given, say, Husserl's 'green is or' we cannot form grammatical sentences such as 'Husserl mistakenly believed that green is or'; the sentence does not even provide a basis for a possibly mistaken opinion. Fulfilment of a well-formedness requirement is a pre-condition for something's being even mistakenly maintainable. Thus where $\triangle$ is non-grammatical $f\triangle$ is non-grammatical, unless f sets $\triangle$ in suitable quote marks. In contrast, significance does not satisfy such a homogeneity requirement (see also §5.1 and §7.2). Significant sentences may contain nonsignificant component sentences which are not quoted, *i.e.* significant sentences may contain nonsignificant component sentences as *genuine* parts. For example, certain philosophers have believed, or thought they believed, that the Absolute was green, that the mind was a sort of box, that the mind is located near the pineal gland, that universals are certain sorts of individuals, though the sentences 'The Absolute is green', 'The mind is a kind of box' and so on are nonsignificant. Again Pythagoras might have mistakenly thought that Virtue is triangular, though 'Virtue is triangular' is not significant. It is unsatisfactory to try to reconstrue these sentences as involving quotation, for example to reconstrue the sentence concerning Pythagoras as of the form "Pythagoras might have mistakenly thought 'Virtue is triangular' ..." for the usual reasons, namely that the sentences need not concern a specific form of words of a particular language. Indeed the standard objections to analyses of intensional sentences as involving quotation apply with equal force here: the analysed sentences typically do not concern a particular form of words and accordingly their truth- and modal-values are invariant under substitution of synonyms within the sentences and translation of the sentences, whereas the analysing sentences, involving quotation marks or quotation functions, are not. (For the standard objections see *e.g.* Church [2].)

These features of significance on their own reveal the unsatisfactoriness of attempts to reduce nonsignificance to ill-formation. For many philosophical purposes, as well as in the interests of accuracy, the distinction is important —otherwise beliefs and opinions philosophers have held could not even represent opinions, and many of their alleged arguments in favour of their positions could not even be counted as such.

Another argument for the significant applicability of sentence predicates to nonsignificant sentences starts from the fact, already noticed, that 'Virtue is triangular' has a different sense from 'The mind is a complicated kind of signal box'. Since they are different there must, by the converse of *salva veritate*, be sentence predicates, not involving quotation, which separate them in value; and hence there must be a predicate '*f*' such that the result of applying '*f*' to one or other of the sentences is significant. In fact the predicate 'Pythagoras may have believed' does just such a job. Hence, that *P* is nonsignificant by no means guarantees that *f*§*P* is nonsignificant.

As against Hume, what is intelligible may not be possible, or even significant. Nonsignificant sentences which are grammatical can represent opinions, they can convey information, as they do in some continental philosophy and in fairy stories for instance, and they can be understood in the way that sentences of a story and contradictions can be understood. Non-grammatical sentences however serve none of these functions; they convey no information. In support of the controversial thesis that nonsignificant sentences may convey *some* information we appeal to examples; and first to the examples of nonsignificant theses that philosophers have held, and the fact that such nonsignificant theses may be assessed and argued about and may very well be the object of intensional attitudes. Nor is it true that the story of the sorcerer's apprentice talking with the broom conveys nothing to the listener. Stories which contain nonsignificant sentences nonetheless convey genuine information, and they do this without changing the senses of expressions and without conventional associations of nonsignificant sentences with other significant sentences. It may well be that some transference theory can be invoked to explain how we understand the conversation with the broom in terms of how we understand conversations between people, but this does not affect the point. For the story of the apprentice is generally taken at face value; it conveys information on its own without conventional associations of the sentences with significant ones; and indeed there is no ready-made way of getting rid of the fact that the apprentice talks with what is unmistakably a broom. In this respect fairy tales differ from the gimmicks of poetry and advertising, of which metaphors are an important subclass, where significant messages are conveyed through recognised associations of significant sentences with the metaphorical ones. An eggboard only invests heavily in the advertisement 'Happiness is eggshaped' because it thinks that this nonsignificant sentence will in a striking way convey, by readily understood associations, a message which will help increase its sale of eggs. Finally a convincing logical theory, which assigns content to nonsignificant sentences, can be devised. Since content is *a* measure of information, nonsignificant sentences with non-null contents will convey some information. At this stage we simply outline relevant features of the theory, explaining certain features of a typical model and, where interpretation is needed, of the canonical model of the theory. (A substantial part of the logical theory is developed in §7.2, and many remaining details of a full theory are given and defended in R. and V. Routley [2].)

A typical model consists of a class K of set-ups, among which is a select designated set-up G, and of various relations between set-ups. A set-up is, like a context, formally a set of sentences. In fact contexts are set-ups with special sorts of elements relating to a given sentence, or group of sentences. Set-ups are, in effect, generalisations of state descriptions. In the canonical model G is the set of *true* sentences; and among the set-ups are those marked out by intensional functors. Thus one set-up H_R will be given by the class of Goddard and Routley's beliefs, another set-up H_P by Pythagoras's opinions, and so on. A sentence *P* holds in a set-up H, does not (truly) hold, or does not significantly hold, according as *TP* belongs to H, *FP* belongs to H, or $\sim SP$ belongs to H. For usual set-ups H, *P* holds in H iff *P* belongs to H. Now for a given model, *P* is true, false or nonsignificant according as *P* holds, does not hold, or does not significantly hold in designated set-up G of that model. Moreover the *content* of *P* in that model is given (compare Popper [2] and Carnap [5]) by the class of set-ups in which *P* fails to hold, that is in which *P* either does not truly hold or does not significantly hold. Now if we consider the canonical model, which reflects actual content, we will see, given plausible assumptions as to what is significant and believed, that the sentence 'Virtue is triangular', though nonsignificant, has a non-null content because it fails to hold in G and H_R in particular; but, equally important, it does not have total content because it does hold in H_P. Fairy tales are accommodated under the same logical account, because lead-in or introductory functors like 'It is told that' or 'Once upon a time' which cover the story are but further (intensional) functors which mark out further set-ups; and sentences which are in fact (in G) nonsignificant may hold in the marked-out set-ups.

The logical theory roughed out also shows how to meet certain commonplace objections to the *idea* of a significance logic, such as: How can logical relations obtain between significant and nonsignificant sentences? How can nonsignificant sentences be logically equivalent or synonymous (we asserted in the Introduction that they could)? For example, a logical equivalence relation can be defined in terms of sameness of content. Since, for instance, 'Either virtue is triangular or virtue is square' has exactly the same set-up behaviour as 'Either virtue is square or virtue is triangular' given the expected symmetric *or*-rule, these sentences have the same content and are logically equivalent. Moreover given sufficiently strict conditions on set-ups and their relations, synonymy can be similarly defined. Just as inclusion of content relations may hold between sentences one or more of which is nonsignificant —*e.g.* given the expected *and*-rule 'virtue is square and vice is cool' includes in its content 'vice is cool'—so an analogous entailment relation can be defined such that, that virtue is square and vice is cool entails that vice is cool. Classing a sentence as nonsignificant does not, then, assign it to oblivion, or ban it from important logical roles.

Not only are nonsignificant sentences admissible as genuine grammatical parts of significant sentences, but also there is no guarantee that significant sentences properly combined grammatically yield a significant sentence. In contrast, well-formed sentences always combine *grammatically* to provide a well-formed sentence. Consider such sentence predicates as 'and', '*Q*' (of §5.13), and 'Rover smells (that)'. Though two cubed is eight, the well-formed sentence 'Rover smells that two cubed is eight' is not significant. Or consider the sentences 'Arthur is heavy and Arthur is immaterial' and 'The thing I am thinking about is a number and it is blue'. Though both sentences of each taken separately are (context-independently) significant the grammatical conjunctions are not, in the ordinary (not classical) sense of 'and'. Such sentence predicates and connectives which do not preserve significance are discussed in detail later (§5.13 and §7.2). Since significance ranges are often needed in order to assess the significance of combinations where connectives and predicates which do not preserve significance occur, and since determination of s-ranges typically requires meaning considerations, the occurrence of such connectives and predicates provides still further evidence for the fact that significance is not simply a syntactical matter and for the fact that well-formation is no substitute for significance. Unlike the grammatical case there are not, in general, recursive compounding rules based on componentisation which preserve significance. Often only use of significance range theory is going to give requisite results; often, whether even a sentential connective combines significantly with a sentence demands assessments at the predicate level.

Finally, according to (iv), the class of grammatical sentences, or of well-formed sentences of a system (*any* logistic system according to Church [4], p. 51 ff.), is recursive, *i.e.*, roughly, the grammatical sentences are effectively decidable from an effective set of rules. In contrast, for certain systems the class of significant sentences is not recursive. For consider, for example, the formal system S for the relation $R(x,z)$, *i.e.* $z = \mu y\, T_1(x,x,y)$, representing Kleene's function $\mu y\, T_1(x,x,y)$. S will be adequate for formalising $R(x,z)$ iff system S has a formal relation $\mathscr{A}(\mathbf{x},\mathbf{z})$ such that:

$\vdash_S T\mathscr{A}(\mathbf{x},\mathbf{z})$ iff $R(x,z)$ is true, and
$\vdash_S S\mathscr{A}(\mathbf{x},\mathbf{z})$ iff $R(x,z)$ is significant.

Taking $R(x,z)$ as *significant* iff $\mu y\, T_1(x,x,y)$ is defined, it follows from Kleene ([2], p. 325) that there is no recursive algorithm for deciding whether $R(x,z)$ is significant or not. Hence the class of demonstrably significant wff of S is not recursive. Consequently a recursive grammar cannot provide an adequate theory of significance; and expressibility in a recursively-specified language of science, say, cannot provide adequate necessary and sufficient conditions for significance.

The non-recursiveness of significant sentences explains in part why we resort to the axiomatic-semantic method in formalising significance theory and do not follow a grammatical pattern.

The context-independence, syntacticalness, homogeneity and recursiveness of grammaticalness, but not of significance, shows that significance is not reducible to grammaticalness; for grammaticalness just does not have the right properties. But may not the sense of 'grammar' be stretched, the scope of grammar widened, as is happening with modern grammars? Yes, in certain ways, as we noticed in the Introduction. But not only does no grammar violate homogeneity and recursiveness requirements; it seems that no grammar should violate these requirements, for the sentences a grammar excludes as ill-formed should only be those thoroughly excluded from linguistic and logical roles. If neglecting these points someone, Wittgenstein for instance, insists on extending 'grammar' and its associates so that nonsignificant sentences are non-grammatical, then this constitutes no threat at all. Our investigation remains a logical one, but becomes called a grammatical one. Instead of describing our enterprise as a study of significance we could have described it as a study of logical grammar; but the description would have been far less apposite because of the hard distinction between traditional grammar and significance marked out by the necessary conditions (i)–(iv) above. By liberalising 'significance', significance theory could in turn swallow grammar; but going to this opposite extreme would likewise blur important distinctions.

The details which reveal that significance is not reducible to grammaticalness equally reveal that significance is not reducible to well-formation in an empiricist language (and why accept the restriction to an *empiricist* language?). Recent empiricist theories of significance which test for significance in terms of expressibility in an empiricist language are but special cases of the misguided attempt to eliminate significance in favour of grammar.

Given that significance distinctions are made amongst grammatical sentences, there will be sentences P and P' such that SP and $\sim SP'$. There is nothing to stop us from considering, and much to encourage us to consider, the logic of these expressions of significance distinctions. Given our previous arguments the logic will have at least these features: since nonsignificance is not reducible to a truth-value, and classical negation is symmetrical with respect to significance but not with respect to truth or falsity, P and P' will have at least values, true, false and nonsignificant, the negation '$\sim$' will invert values true and false but leave nonsignificance intact, and the connective 'S' will assign value false when what it applies to is nonsignificant and value true otherwise. It follows (for main details see §5.9) that significance is not classically definable, *i.e.* 'S' cannot be expressed in classical predicate logic. The argument can be extended to apply to *usual* quantified intensional logics where intensional predicates Φ satisfy the condition: $S\Phi P$ iff SP. (On these logics see §7.2.) But a logic would only provide a reduction of significance if it were a usual intensional 2-valued, logic. Hence *the logic of significance is irreducible.*

Though so irreducible, nonsignificance is of course not undefinable in terms of other notions of the same order; for example 'S' can be defined in terms of significance incompatibility, in terms of connective 'T' as $SP =_{\mathrm{Df}} TP \vee T{\sim}P$, and in terms of coincidence of negations thus: $SP =_{\mathrm{Df}} {\sim}P \simeq \neg P$. What is important is that there is no way of breaking out of the circle: there is a cluster of interrelated notions grouped about significance which cannot be absorbed into prevailing logical theories.

4.7 Degrees of Significance

The observation of similarities between significance matters and grammatical matters coupled with the observation that some nonsignificant sentences are somehow worse, less intelligible or interpretable, than others, has suggested (to Ziff, Chomsky and others) that what is really required is a degrees of significance, or degrees of grammaticalness theory. The deviance theory of nonsignificance—the theory that all nonsignificant sentences deviate (reprehensibly) from some accepted norm, *e.g.* of logical grammaticalness, leads at once to a degrees theory, because there are of course degrees of deviance from the norm. The breakdown of attempts to assimilate nonsignificance to ungrammaticalness also motivates the degrees of grammaticalness idea: even if these similar syntactical matters cannot be identified at least, so it is claimed, they can be synthesised in a common theory by use of degrees. (And Chomsky envisages such a synthesis with his sketchy theory of degrees of grammaticalness, built on the theory of types: see Chomsky [1].)

First, however, such degrees are otiose; for we can account for all the features which lead to the idea without introducing degrees. Significance logic, early on, accounts for the behaviour of classical connectives, *etc.* As for the rest, consider again the logical modelling case sketched in defining a content measure for nonsignificant sentences. The same modelling can be used to define a significance spread for each sentence: in a given model the *significance spread* of a sentence is the class of set-ups in which it significantly holds. If now we assess ill-formed as well as well-formed sentences in terms of the modelling, then an ill-formed sentence will have null significance spread in every model; but, given standard contexts, certain significant sentences will have total significance spread. Moreover that one nonsignificant sentence is less intelligible than another can be explicated in terms of the significance spread of one being included (or perhaps largely included) in that of the other. The modelling does not however provide a *degrees* of significance notion; for the inclusion ordering of significance spreads is not connected, and accordingly not appropriately linear.

More generally, since remote nonsignificant sentences are not comparable as to intelligibility, there is no connected ordering of degrees correlated with intelligibility, and the degrees idea breaks down. Of course it is possible to define, rather arbitrarily, a degree of significance for sentences. Consider, just to illustrate, a rough and artificial grading defined along the following lines: Call any significant sentence P^* obtained from a (nonsignificant) sentence P by replacing a minimum number of words a *significant transform* of P and the number of word changes made in obtaining P^* the *transform number* of P. If P is significant it will be one of its own transforms and its transform number will be zero. Call the number of words in P the *word-number* of P. Then the *degree of nonsignificance* of P may be defined as: the ratio of the transform number of P to the word number of P, and the *degree of significance* as: $1 -$ (the degree of nonsignificance). A significant sentence has degree of significance 1; a completely nonsignificant sentence, composed of non-words, degree 0. The sentence 'The mountain watched the rainbow slowly' a significant transform of which is 'The man watched the rainbow intently' has degree of nonsignificance $\frac{1}{3}$ and degree of significance $\frac{2}{3}$. Naturally much elaboration would be wanted before this primitive account would begin to stand up; *e.g.* decisions would have to be made about hyphenated words, prefixing, word inflexion, and so on. Quite independently of the elaboration however, such an account of degrees has a fatal drawback: it assigns different degrees of significance to synonymous nonsignificant sentences. It has other drawbacks too, *e.g.* it assigns certain long, traditionally ungrammatical, sentences a very high degree of significance and a much higher degree than some philosophically interesting nonsignificant but grammatical sentences. And the measure lacks point. It is as if, having noticed resemblances between identity and similarity, those resulting from their both being defined in terms of sharing of features, we tried to combine them in a degrees of identity doctrine. There would be analogous drawbacks; and just as such a move would leave identity intact as a viable notion, so the degrees of significance idea leaves significance intact as a viable notion.

4.8 Conventionalist and Contingency Objections

Although significance is irreducible, it is often claimed, on a variety of grounds, that significance distinctions are dispensable. Dispensable for what purposes? Although significance distinctions, and many other important distinctions, can be dispensed with in limited language games and formal languages, like that for set theory, significance distinctions are not, we believe, dispensable in sufficiently rich languages. They are certainly indispensable in languages adequate for the formalisation, or expression, of philosophical theories. Philosophers of almost every persuasion appeal to significance considerations; they dismiss such and such sentences as nonsignificant. That is one reason why, as we have explained, significance logic is a philosophically important matter. These philosophical purposes cannot be served by reclassification of nonsignificant sentences as false or as non-grammatical.

What then are the grounds for preferring to adopt, or insisting that philosophers should adopt, a manifestly weaker and philosophically inadequate language where significance distinctions do not appear? Significance distinctions are said to be conventional, artificial, arbitrary, merely contingent and "not worth bothering about"—to list a few complaints. We shall tackle some of these complaints briefly. To do so more simply and more convincingly we first introduce the notion of an assertion.

Enough features of the logic of nonsignificant sentences are now visible to make it plain that the statement contraction defined for truth-valued sentences can be extended to nonsignificant indicatives. For consider the sorts of reasons given for introducing linguistic invariant statements: all those have extensions which apply as well to contents of nonsignificant sentences. In order not to confuse earlier points (especially in Chapters 1 and 2) we do not extend the term 'statement', but instead introduce the term 'assertion' as a technical term. Statements are assertions which are either true or false; but an assertion may be nonsense. Sentences which are absurd in their contexts will yield nonsense assertions; assertions are contractions of declarative sentences unrestricted by the requirement that the contraction be true or false. Different sentences, perhaps in different languages, will yield the same assertion. The same (type) sentence in different contexts will yield different assertions. An assertion can be the object of intensional attitudes; assertions can stand in logical relations such as entailment, incompatibility; and so. Assertions are then a three-valued analogue of statements. The statement notion does not however extend to non-grammatical sentences, even though these sentences may appear to share in some statemental features of grammatical sentences, *e.g.* Husserl's 'grun ist oder' can be given a (word by word) translation into English. For non-grammatical sentences lack content entirely, they do not provide contractions which stand in logical relations or are the objects of intensional attitudes, and so forth.

An assertion is nonsense, *i.e.* neither true nor false, when it exhibits significance incompatibilities, much as an assertion is logically false when it is false because of logical incompatibility. Nonsignificant sentences are not senseless, in the sense that they do not have senses or contents, and different nonsignificant sentences may have different senses and different contents.

Appeal to features of assertions is useful in rebutting some superficial objections to significance theory, for instance the vague charges that significance distinctions are a merely linguistic matter and hence irrelevant to the facts and to science and something one need not bother about in philosophy. Since assertions, which are linguistic invariants, do have distinct significance values significance distinctions are not a merely linguistic matter. Nor are any of the other large claims correct. Linguistics, grammar and syntax are sciences and present a body of facts of considerable philosophical interest. Even physicists employ significance distinctions and make use of a nonsense filter, as an examination of standard textbooks on quantum theory will confirm. The sentence/assertion distinction also pays off in rebutting the conventionalist charge.

The conventionalist charge (levelled by Popper [1], but thoroughly confused there with the essentialist issue) is that significance distinctions are conventional, a result of conventional rules laid down by people, and not natural, or a matter of fact. But Popper's natural *versus* conventional distinction, like the referential *versus* quotational distinction from which it is often derived [either a property is a property of a reference, the entity, and factual and natural, or a property is a property of a word, and conventional] poses a false dichotomy. The ambiguity and general inadequacy of rude contrasts of this sort is well known, from parallel conventionalist charges that used to be levelled against logical necessity. The fact that significance is not a thing-property, invariant under changes of description of the thing, does not entail that it is quotational, arbitrary or man-made, or that it is an epistemic or psychological matter. That it is not quotational follows from the interchange principles that hold for nonsignificant sentences, as we have explained. The fact that some categorical distinctions are universal, *i.e.* language invariant, casts substantial doubt on the remaining, usually unsupported, claims. Since there is very considerable agreement that it is nonsense to say that numbers enjoy horseriding or that virtue is a fireshovel, in whatever language is taken, some significance distinctions are not arbitrary or a merely psychological matter. Admittedly there is a conventional element in significance claims, as there is in necessity claims, because there is a conventional element in the senses assigned to words; but a parallel conventional element appears also in truth claims (*cf.* Austin [2] Ch. 4; Schilpp [2], p. 918) without undermining such claims as conventional. Once the senses of expressions are determined, significance claims are no more conventional than truth claims, as the three-valued equivalence $SP \simeq TP \vee FP$ reveals. The only way to avoid these conventional elements is to avoid language, to get down (by contraction) to the facts themselves. But to describe these facts adequately both significance and intensionality are essential. It is as much a matter of fact, and due to the nature of Socrates, and certainly not features of 'Socrates', that it is nonsignificant that Socrates is prime as it is that it is false that Socrates is an abstraction and false that Socrates had a hook nose. Popper claims to have completely destroyed such a naturalistic theory of meaninglessness. By a naturalistic theory he means (Popper [1], p. 190):

> the doctrine that every linguistic expression purporting to be an assertion is either meaningful or meaningless; not by convention, or as a result of rules which have been laid down by convention, but as a matter of actual fact, or due to its nature, just as a plant is, or is not, green in fact, or by nature, and not by conventional rules.

Not only is a misconceived dichotomy between convention and fact basic in the formulation, worse, there is a shift from linguistic expressions to the statements made, or facts asserted using them, which is employed to advantage on one side of the argument and neglected on the other. For there are conventional elements in the linguistic expressions 'That plant is not green' and 'Pigs fly' in so far as these expressions typically yield one statement and not another, though the facts stated are free of these elements. Likewise the fact that Socrates is not significantly prime is free of these, and other, conventionalistic elements. Since it is not unlikely that Popper's claimed demolition of naturalistic theories has some currency it is worth pointing out that most of Popper's arguments apply only to very special naturalistic theories and by no means to most, or more plausible naturalistic theories; and the same goes for his alleged criticisms of the requirements—ones we have adopted—on a meaningful sentence that (a) all words which occur in it have meaning, and (b) all words which occur in it fit together properly. Nor does Popper's criticism of (b)—that an expression like *qu*(x is not a member of x) is not absolutely or intrinsically meaningless because there are rival set-theoretical languages which assign the expression truth-values (for values of *qu*(x))—carry any weight. Not even truth is an absolute notion in this sense; for what *expresses* a truth in one set-theoretical language may express a falsehood or nonsense in another. It is quite unnecessary to meet Popper's challenge to provide a proof of intrinsic meaninglessness, *i.e.* of meaninglessness of expressions with respect to every language; for, first, such a proof is irrelevant so far as criticising philosophical *assertions* as meaningless is concerned, and second, neither naturalism nor (b) implies that meaninglessness is such an absolute notion. Once again the shift from sentences which in the context of different (set-theoretical) languages express different assertions, to the assertion or fact expressed has been exploited to discredit meaninglessness but not truth and falsehood. But the facts expressed—those of nonsignificance along with those of truth—are absolute, in the sense of being language invariant and independent. Nor are the sentences of the rival set theories incompatible; the sentences have different logical contents and '$\in$' has a different sense in the different contexts, as *salva veritate* reveals: class membership is a logical determinable.

A common argument to show that significance is a conventional matter relies on these premisses: if significance is not to be a conventional matter then it must be essentialist; but it is not, as we have ourselves contended (§3.8). The first premiss is false; "essentialist or conventionalist" simply re-presents the false dichotomy already escaped. An item typically has intensional and category features; that these depend on how the item is described does not show that the features are conventionally ascribed. Not only are intensional and significance features not in fact conventionally ascribed, but the nature of an item is not exhausted by its referentially transparent features, as the argument of §3.7 reveals.

An objection, which often forms the basis of a conventionalist charge, is that significance and nonsignificance assertions are merely contingent assertions. But if they are contingent how are they contingent? And how are nonsignificance claims explained and justified? Since empiricists claim to find no evidence of significance distinctions in things in the world—if *they* did significance would of course be essentialistic—they leap to the conclusion that significance claims are either matters of linguistic habit or else psychological or pragmatic matters. There is more than one way in which these scattered charges can be met; but the elaboration of these different ways leads to diverse significance theories. The dividing issue is as to whether true significance claims are necessarily true or are only contingently true, or whether they are sometimes necessary and sometimes contingent, and analogously for false significance assertions. Given, however, as we shall argue, that true assertions of significance proper (§3.9) are logically necessary and false assertions of significance proper are logically impossible, since only sense features are involved, the whole mere-contingency objection collapses since its first assumption is repudiated. Moreover, the contingency of some significance assertions, whether true or false, can be explained in terms of the theory of absurdity (§3.9), since absurdities depend on aboutness features, and therefore on contingent identities.

But whatever line is taken the separable problem as to how significance and nonsignificance assertions are justified should be met. Consider then how we resolve Pap's puzzle (posed in Pap [1], pp. 43–47) concerning the significant vindication of significance assertions. To illustrate we show that Socrates is not significantly prime by the following argument (A):

(1) $\sim T$(Socrates is a natural number)
(2) S(Socrates is prime) $\equiv T$(Socrates is a natural number)
$\therefore$(3) $\sim S$(Socrates is prime).

The argument generalises to the following form (B):

(1′) $\sim T\hat{f}(x)$
(2′) $Sf(x) \equiv T\hat{f}(x)$
$\therefore$(3′) $\sim Sf(x)$

where $qu(\hat{f})$ is the superpredicate of $qu(f)$. Superpredicates, which mark out significance ranges, are introduced and discussed in Routley [3].

To meet the objection that the difficulties have only been postponed—for now vindicate (1), or what entails it: $\sim S$(Socrates is a natural number)—step (1)–(3) are iterated with new premisses like,

(4) $\sim T$(Socrates is a number)
(5) $\sim T$(Socrates is an abstract item)

supplanting (1) in each iteration. The regress is not endless. For since,

(6) Socrates is a concrete individual

is true, indeed analytic;

(7) Socrates is an abstract item

is false, indeed logically false.

Since also (7) is significant, (6) entails (5). (5), together with an instance of superpredicate theorem (2′), entails (4); (4) with another instance of (2′) entails (1); and finally (3) follows. Further, since (6) and (2′) are both analytic, that (3) is logically necessary follows by distributivity of necessity. The argument can of course be shortened by using a bridging premiss such as that (6) entails (1). What is important about the quite representative full argument is that it shows both how nonsignificance assertions and claims can be significantly vindicated and that, and why, assertions of nonsignificance like (3) are logically necessary. Analogous arguments can be designed to show that it is logically necessary that it is significant that Socrates is happy. These arguments, justifying our thesis that assertions of significance proper are non-contingent, capture what is correct, from a significance point of view, in the L-falsidal arguments for the logical falsity of 'Socrates is prime'. The regress in the justifying arguments from a predicate to its superpredicate always terminates; for ultimate predicates like 'may be thought of' and 'is an item' are significant of every subject, whether it is significant or not. The particular argument we have examined assumes a fixed context where 'Socrates' has its standard sense, namely as a name of a person; the context dependence could of course be reduced by replacing 'Socrates' throughout by 'the philosopher Socrates'. Although such nonsignificance assertions are non-contingent, absurdity claims like 'It is happy' in a context where 'it' refers to a garbage can are contingent since, as we have noted, the reference identities are contingent.

Although we have explained how non-contingent assertions of non-significance and significance are significantly vindicated, and to this extent have given an explanation of significance distributions, we have not told the full story. In particular we have not explained the analytic connexion provided by the superpredicate theorem (2′). The problem of explanation here merges with the problem of explaining analytical statements; it is but a special case of this more general problem. We owe such an explanation; for significance analyticities, no more than analyticities in general, are not just given, either as innate notions or as notions that somehow come down from heaven, and which we just intuit. The explanations are in both cases the same. Analytic statements, and analytic connexions of predicates with their super-predicates, are true in virtue of the senses of expressions in suitably fixed contexts. In volume 2 we shall provide a logical model which, we hope, will satisfactorily verify a version of the widely disputed linguistic theory of analyticity. Indeed we shall defend a linguistic theory of significance, to the effect that a sentence is absurd in a context, and the assertion it yields is nonsense, in virtue of the senses and aboutness of the expressions in the sentence in that context, and, very roughly, that an assertion is sensible or nonsense in virtue of contingent linguistic and contextual matters. Thus the senses of expressions, contingent matters, explain non-contingent significance and nonsignificance assertions. It is important to realise that explanation is not reduction. Although significance- and nonsignificance-assertions cannot be reduced to linguistic contingencies—those of sense—they can be explained in terms of these contingent matters. It is very important to notice that a principle commonly used to attack linguistic theories of analyticity, the distributivity of contingency principle, according to which contingent statements cannot entail or support necessary ones and that necessary statements cannot hold in virtue of contingent matters, is false (see R. and V. Routley [3]). In particular, nonsignificance analyticities, though necessary, are true in virtue of linguistic contingencies, namely those of a semantic dictionary.

Since context-independent significance is a logical matter, it is not just a psychological or pragmatic matter. Indeed the logical account of significance assertions we have sketched refutes at once the diverse charges that significance distinctions are solely psychological matters of no logical concern, are assertions of linguistic conventions or habits, are arbitrary, non-compulsory distinctions, and are without basis.

Though we have opted for a non-contingency account of significance proper in rebutting the mere-contingency objection, practically nothing of what follows in Chapters 5, 6 and 7 depends on this account; the subsequent theory is equally compatible with accounts according to which some aboutness-independent significance claims are contingent. Only in §7.2 do we take up the matter again and there we provide semantical analyses of both contingency and non-contingency accounts.

CHAPTER 5

Sentential Significance Logics I The Matrix Approach

WE turn now to an examination of various sentential significance logics; first by means of matrices, and later (Ch. 6), by means of an axiomatic approach.

By the matrix approach we mean the following:

Values (in this case *t*, *f*, *n*) are assigned to the sentential variables, operations or functions on the values are assigned to the connectives, a subset of the values is selected as designated, and theoremhood is decided by computing whether the value of the complex function corresponding to the formula belongs to the set of designated values for all assignments of values to its variables (see, *e.g.*, Prior [2], Church [4] and Rosser and Turquette [1]). The matrix approach to two-valued sentential logic is represented by the familiar truth-table technique where the values are *t* and *f*, and *t* is selected as the only designated value; a theorem is then defined as a formula whose value is *t* for all assignments of values to its variables.

The initial difficulty associated with a matrix approach to significance logic, that nonsignificance is a different sort of value, not comparable to the two truth-values, we have met in Part I. The important remaining difficulties are first, what connectives should appear in sentential significance logics; secondly, what matrices should be assigned to these connectives; and third, what values should be selected as designated.

We discuss each of these questions in turn and then, in terms of the decisions made, develop a series of logics through several stages. Those of Stage 0 are what might be called the basic systems; they are simple three-valued extensions of the classical two-valued systems which contain no new operators. We then extend these systems by adding, successively, new connectives of various kinds and by introducing other modifications. Our strategy is to begin with a weak and quite unobjectionable three-valued logic and to convince the reader, stage by stage, of the need for more comprehensive significance logics. In Chapter 6 we parallel the matrix treatment by axiomatising the various logics at different stages. The extension to predicate logic is given in Chapter 7 when matrices are replaced by models or sets of matrices.

5.1 Extensional Connectives

We require of the various sentential logics that they contain logical constants interpretable as the extensional sentence-connectives, 'and', 'or', 'not' and others definable in terms of these. For in general we wish the logics to cover the same ground as classical logic and to incorporate what seems to be unassailable there. We wish, too, to be able to formalise Russell's theory of significance, including the elementary part which includes significance principles for the extensional connectives, and to symbolise and assess similar significance principles adopted by various other authors. Hence our basic logics must contain the familiar constants '&' (extensional conjunction), '$\vee$' (extensional disjunction) and '$\sim$' (extensional negation) and be susceptible of their normal interpretation, even though three-valued.

Of course '&', '$\vee$' and '$\sim$' are by no means the only, or even the most important, components which emerge when the determinable connectives 'and', 'or' and 'not' of a natural language are analysed. But these components do often occur in discourse, especially in technical and mathematical discourse, and they are especially easy to handle within a matrix approach. For at least in their classical two-valued part, the value of $qu(p \,\&\, q)$ is determined just by the value of the components and not by any further relations which may hold between $qu(p)$ and $qu(q)$. In fact it is to these, and only to these, sorts of connectives that the matrix approach is really suited. The approach breaks down rather badly for connectives which depend not only on the values of the components but also on other relations between, or properties of, the components. Just this sort of difficulty arises with intensional connectives such as '... implies ...' and 'it is logically necessary (that ...)'. With such connectives, various extensions of the matrix approach may sometimes be adopted. For example, the operation assigned to the connective '$\Box$' (it is logically necessary (that ...)) depends both on the value assigned to the variable and on a further relation which connects the value so assigned to other values. Extended matrix methods may then be determined by providing rules for the deletion of certain value assignments (see *e.g.* Feys [1], pp. 109–111). But such elaborations are extensions of the matrix approach *simpliciter*. Similarly, though the matrix approach can be extended to certain infinitely-valued logics, without the introduction of further relations or deletion techniques, its direct and immediate application is to finitely-valued logics.

In order to keep within the bounds of the matrix approach *simpliciter* for as long as possible, and without undue artifice, we concentrate initially on three-valued extensions of the purely truth-functional connectives '&', '$\vee$',

'$\sim$' of classical logic. We therefore wish these extensions to be, like the two-valued connectives they extend, extensional in the sense of 'matrix-accessible'.

Definition: An n-place connective Φ^n connecting variables $P_1, P_2, ..., P_n$ is *matrix-accessible* over a set of values $v_1, ..., v_k$ if, and only if, for any assignments of values from $v_1, ..., v_k$ to variables $P_1, ..., P_n$, $\Phi^n(P_1, ..., P_n)$ takes an assigned value from $v_1, ..., v_k$, where this assignment depends solely on the initial values given to $P_1, ..., P_n$ and a recursive set-closing function ψ.

Thus, Φ^n, connecting variables $P_1, ..., P_n$, is matrix-accessible over the set $V = \{v_1, ..., v_k\}$ if, and only if, there is a set-closing function $\psi : V^n \rightarrow V$, such that for any assignment of values from V, say $v_{i_1}, ..., v_{i_n}$, to $P_1, P_2, ..., P_n$ respectively, the value of $\Phi^n(P_1, P_2, ..., P_n)$ is $\psi(v_{i_1}, ..., v_{i_n})$, and ψ is an extensional function which depends only on values from V; *i.e.* if $v_{i_j} = v_{i_k}$ then $\psi(v_{i_1}, ..., v_{i_j}, ..., v_{i_n}) = \psi(v_{i_1}, ..., v_{i_k}, ..., v_{i_n})$.

The set of assignments can be specified by a Cartesian *tensor* array for Φ^n, where the co-ordinate values for each of the n co-ordinates are the distinct values $v_1, ..., v_k$, and the values for given values of the co-ordinates are values of ψ for the co-ordinates. Where Φ^1 is a 1-place connective, this reduces to the familiar *vector* for a 1-place connective; where Φ^2 is a 2-place connective, the array is a familiar square *matrix*. A tensor giving the assignments for Φ^n over $v_1, ..., v_k$ may or may not give correct assignments depending on the role Φ^n plays. If Φ^n is a connective given in an axiomatic logic, for example, the correctness of a tensor assignment will not be decidable independently of the other connectives of the logic, or independently of the question of whether the tensors are required to be characteristic—*i.e.* such that all and only theorems of the axiomatic logic are satisfied. If, however, Φ^n is *defined*, with respect to $v_1, ..., v_k$, by a tensor, then the assignment is *a fortiori* correct. What distinguishes the simple matrix approach is just that each connective is defined by a tensor, and hence is logically characterised by a tensor, without reference to the requirements of an axiomatic system.

We now specify:

Condition I The constants '&', '$\vee$', '$\sim$' are connectives over the set of values $Val = \{t, f, n\}$;

Condition II The connectives '&', '$\vee$', '$\sim$' are matrix-accessible over the set of values $Val = \{t, f, n\}$.

The requirement I ensures that the connectives are three-valued and that their values are those of the variables. That this should be so follows simply from our earlier assumptions that all sentences are three-valued and that if P and Q are sentences, so are $\sim P$, $P \,\&\, Q$, $P \vee Q$, *etc.*

The requirement II ensures that the value of a compound sentence depends solely on the values assigned to its components and not on any further features of, or relations between, them. Hence, no matter what sentences P and Q happen to be, given that they have definite values, then $\sim P$, $P \& Q$, $P \vee Q$ have definite and determinable values.

It is of course possible to introduce *ad hoc* connectives which satisfy these two conditions but we wish to claim more than this. In particular, that connectives which satisfy these requirements approximate well to those used in ordinary discourse and to those adopted by various authors, especially Russell ([2] and [3]) and Halldén [1].

It follows from I and II that '&', '∨' and '∼' are characterised by tensors of the following kind:

&	*t*	*f*	*n*
t	1	2	3
f	4	5	6
n	7	8	9

∨	*t*	*f*	*n*
t	10	11	12
f	13	14	15
n	16	17	18

∼	
t	19
f	20
n	21

where the values 1–21 are each one, and only one, of *t*, *f* and *n*. Hence, corresponding to the connective '&', for example, is a two-place function & from the pairs, $Val \times Val$ into Val; i.e. $Val \times Val \rightarrow Val$ in the usual algebraic notation. Thus: $\&(f,n) = 6$ where 6 is in the set $\{t,f,n\}$.

It is on the satisfaction of I and II that a matrix approach to significance logic hinges. Though these requirements are by no means sufficient, they are necessary conditions for a matrix approach. If one of them fails, the logic must be approached differently; though of course the matrix technique may still figure in some matters, *e.g.* in establishing consistency and independence.

5.2 Extended Classical Connectives

In order to ensure that the three-valued connectives '&', '∨' and '∼' are not only extensional, in the sense of 'matrix-accessible', but also that they are extensions of, rather than alternatives to, the classical two-valued connectives, we require the further condition:

Condition III Classical and-or-not sentential logic holds with the usual matrix assignments for all true-or-false sentences, *i.e.* for all significant sentences.

Logics which satisfy this requirement may be said to "contain" classical logic in the sense that, if the value *n* is eliminated, the matrices reduce to the classical matrices.

In terms of this requirement, therefore, many of the matrix values 1–21 are fixed as follows:

&	t	f	n
t	t	f	3
f	f	f	6
n	7	8	9

∨	t	f	n
t	t	t	12
f	t	f	15
n	16	17	18

~	
t	f
f	t
n	21

The numbered places are yet to be determined, however, and there are 3^{11} ways in which they could be filled and yet satisfy I, II and III.

Of course there are some fairly simple ways in which the number of choices can be reduced. Thus, for example, since the intended interpretation of '&' and '∨' is as the extensional connectives 'and' and 'or', and since these are symmetric, it follows that for given sentences *P* and *Q*, *P*&*Q* must have the same value as *Q*&*P*, and similarly, the value of *P*∨*Q* must be the same as the value of *Q*∨*P*. Hence the matrices, too, should be symmetric. Even after symmetry has been taken into account, however, there remain 3^7 possible sets of matrices and further requirements therefore have to be introduced if the connectives are to be uniquely determined.

What these additional requirements should be depends on what features significance is claimed to have; and relatedly, on which significance principles are held to be correct. At this point, therefore, a number of decisions have to be made and these to some extent depend on which examples are taken, as initial data, as correct examples on which to generalise.

Now we do not wish to claim that there is a unique set of matrices to which we are inevitably forced. On the contrary, we hold that there is no such unique set. There are for example, three possible and different negation-type connectives, namely

~	
t	f
f	t
n	n

~	
t	f
f	t
n	f

~	
t	f
f	t
n	t

all of which have interest and merit, and all of which have occurred in philosophical discourse at least.

Consider, for example, the last of these. We should be committed to this matrix if it were held that although some predicates are nonsignificant of certain items, *e.g.* 'is the highest mountain in the world' is not significant over numbers, nevertheless it is always significant and true to deny of an item that it has the property determined by a predicate which is nonsignificant over it. So whereas '3 is the highest mountain in the world' is nonsignificant, '3 is not the highest mountain in the world' is true. Such a view, apparently Wittgenstein's later view, is tenable, and derives its attraction from the apparent significance of negative assertions (a fact which can also be used, as we saw in the Introduction, to support the view that there are no non-significant sentences, all those which appear to be so in fact being false). But of course there are difficulties associated with it. In particular, if we consider sentences containing a double negation, *e.g.* '3 is not not the highest mountain in the world', then we must either reject the general law: $\sim\sim p \simeq p$, or else reject the condition III. For if we hold that double negation always cancels out, then '3 is not not the highest mountain the world' is equivalent to '3 is the highest mountain in the world', and since this is *ex hypothesi* nonsignificant, so must be '3 is not not the highest mountain in the world'. But this last is the negation of a true sentence, *viz.* '3 is not the highest mountain in the world'. Hence, in some cases, at least, the negation of true sentences leads not to false ones but to nonsignificant ones; and this would cause us to reject III. Alternatively, if we insist on retaining III, then we cannot accept in general that $\sim\sim p \simeq p$. Either one of these courses can be consistently adopted, however, so there are no immediate *a priori* objections which can be made on logical grounds alone, though there are philosophical objections since we are then committed to the view that there are positive and negative predicates and important differences, perhaps indeed ontological differences, between them.

Our claim, then, is not that any one set of matrices is correct, but the much weaker claim that a given set may be more reasonable than, or more philosophically interesting or important than, others. In particular, we hold that there is one set, those which determine the classical significance connectives, which we believe to have been adopted at least implicitly by many authors and which have played an important part in the history of philosophy. For this reason alone, systems based on them must figure prominently in any general theory of significance.

5.3 Classical Significance Connectives

The connectives which we consider to have been of central importance are singled out by the condition:

Condition IV (a) Any compound sentence with a nonsignificant component is nonsignificant:

(b) Any compound sentence in which all the components are significant is itself significant.

IV(b) is clearly entailed by III, but we here state it separately in order to emphasise it since there are arguments which can be brought against it in this form.

We shall call IV the principle of component homogeneity. It is expressed equivalently by saying that any sentence compounded from components by means of the connectives '&', '$\vee$' and '$\sim$', is significant if, and only if, each of its components is significant. Hence, the matrices determined by the requirements I–IV are as follows:

&	t	f	n
t	t	f	n
f	f	f	n
n	n	n	n

∨	t	f	n
t	t	t	n
f	t	f	n
n	n	n	n

~	
t	f
f	t
n	n

If we now adopt the classical definitions of '⊃' and '≡' in terms of '~' and '&', or '~' and '∨', then we also have:

⊃	t	f	n
t	t	f	n
f	t	t	n
n	n	n	n

≡	t	f	n
t	t	f	n
f	f	t	n
n	n	n	n

These connectives, component-homogeneous extensions of the classical two-valued connectives, we call the classical significance connectives.

The historical importance of classical significance connectives derives from the fact that they occur repeatedly in unformalised arguments about significance, as any cursory examination of the literature will reveal. Indeed some, at least, of these connectives have been adopted by Plato [1], Aristotle [2] and [3], Hume [1], Kant [1], Mill [1] and Husserl [1]; and more recently by the logical positivists, the Cambridge philosophers in the early part of this century, especially Russell, and the Oxford philosophers of ordinary language. Details of some recent significance arguments and appeals in which the classical connectives are used will be found in Sommers [1] and [2]. That their adoption has been widespread is thus obvious. The most important sources, however, are Russell and the positivists.

Not only does Russell argue informally using the connectives 'and', 'or', and 'not' of natural language as being in accord with the classical significance connectives, but the symbolic constants '.', '∨', '~' which occur in *PM* and elsewhere are often so construed as to satisfy the above matrices. For we know, first, that the two-valued section of these matrices, where the components have values *t* and *f*, were adopted by Russell (*cf. PM* *4, p. 115; and Russell's Introduction to Wittgenstein [1]). It remains to consider, therefore, what values are assumed for compound sentences when one or other of the components takes the value *n*. That *n* is regarded as a possible value in many contexts is clear from several discussions in *PM* (*e.g.* pp. 77, 95, 129, 133, 137, 166). There is no doubt, therefore, that at many places the connectives are treated as three-valued. It is also true, however, that they are taken to be the classical significance connectives. In the case of '~', it is affirmed repeatedly that a sentence ~*P* is nonsignificant if, and only if, *P* is nonsignificant (see

PM, p. 77, pp. 43–47, p. 133; and Russell [2], pp. 72–75 and p. 79). In the case of '$\vee$' and '$\supset$', which may be considered together, given the matrix for '$\sim$', since they are then interdefinable, it is affirmed that if $P \supset Q$ (or $P \vee Q$) is significant then P is significant and Q is significant (*PM*, p. 95; Russell [2], p. 72); while the converse, that if P is significant and Q is significant then $P \vee Q$ is significant is adopted in various places (*PM*, p. 97, p. 133, pp. 136–137; *cf.* also Russell [2], pp. 86–87). Hence there is an explicit commitment to the classical matrices for '$\vee$' and '$\sim$' at least, and the commitment to the remainder follows by definition.

These facts are of particular importance in relation to the question of how a complete formalisation of the logic of *PM* should be carried out. We have already rejected the commonly accepted solution of this problem, namely that significance principles should be regarded as metatheoretic and significance criteria be exemplified by appropriate formation rules. In our view, quite apart from the philosophical objections which can be made to any programme designed to exclude nonsignificance, a good deal of the logical richness of *PM* is lost if this technique is adopted. The significance riders on *10.39, *10.4 and *10.41 disappear, as do the principles expressed by *9.14, *9.15 and parts of *63. To retain such conditions explicitly in a fully formalised system, however, involves the recognition of nonsignificance as a possible value and the adoption of the classical matrices. In terms of such an approach various postulates, which do not seem obviously to be significance principles, are best expressed as such. In particular, *1.7–*1.72, which define elementary propositions, are in effect significance principles since elementary propositions are, in our terminology, significant sentences. When so construed, *1.7–*1.72 are postulate-equivalents of some of the conditions which determine the classical significance matrices. This will be shown in detail in Stage 0 of Chapter 6.

That the logical positivists were in general committed to the classical significance connectives is again evident. For it is clear, first, that classical matrices were adopted for significant sentences (see, *e.g.* Carnap [3], p. 20; and Wittgenstein [1]) and hence that the two-valued sections of the three-valued matrices were assumed to hold. Thus it was supposed in general that truth-functional operations on true and false sentences did not lead out of the class of true or false sentences, and in particular never led to nonsense. It was also supposed that extensional operations, such as conjoining and negating, could not lead from nonsignificant sentences to significant sentences. This is apparent from the fact that significance is often conflated with well-formation. For suppose $P \& Q$, being nonsignificant, is not well-formed; then one of P or Q at least is not well-formed. That they were committed also to the remaining values of the classical significance matrices, however, is clear from the explicit affirmation of various principles. In particular it is claimed that a sentence P is nonsignificant if, and only if, $\sim P$ is nonsignificant (see, *e.g.*,

Schilpp [1], p. 868; and Ayer [1], p. 36); and it is implied by the belief that significance and well-formation are identical notions (*cf.* Popper [1], p. 209, *n.* 57a) *i.e.* whenever P is nonsignificant or Q is nonsignificant, then $P \& Q$ and $P \vee Q$ are nonsignificant, for whenever P is not well-formed or Q is not well-formed, $P \& Q$ and $P \vee Q$ are not well-formed (see, *e.g.*, Carnap [1], pp. 419–471; Schilpp [2], p. 881; Popper [1], pp. 207–209; and Reichenbach [1]).

It seems in any case apparent that the positivists were committed to the classical significance connectives by their commitment to the criterion of verification as a criterion of significance. For if we adopt,

(*CV*) A sentence is significant if, and only if, it is empirically verifiable (assessable as true or false),

then, without going into details as to how the notion of empirical verifiability should be explicated, we may ask what consequences the criterion has in terms of compounding principles.

Consider, first, simple indicative sentences without quantifiers, *e.g.*

(1) The Absolute is green.
(2) Snow is green.

If (1) is not empirically verifiable, then

(3) The Absolute is not green

is not empirically verifiable. For if it were, we could assess (3) by observation, experiment, etc., as true or false. In this case, however, (1) is false or true, as the case may be, so contradicting the assumption. It seems in general, therefore, that we may conclude that P is empirically verifiable if, and only if, $\sim P$ is empirically verifiable, and indeed this is often conceded (see, *e.g.* Reichenbach [1]). Hence it follows from *CV* that, for a large class of sentences, P is significant if, and only if, $\sim P$ is significant.

Again, consider the conjunction (1) and (2). If (1) is not empirically verifiable, neither is (1) and (2), while if (1) and (2) is not, either (1) is not or (2) is not. Hence, in general, we may say: $P \& Q$ is empirically verifiable if, and only if, P is empirically verifiable and Q is empirically verifiable. That is, by *CV*, $P \& Q$ is significant if, and only if, P is significant and Q is significant. And similarly for disjunction.

It has to be recognised, however, that in arguing thus we have mapped empirical verifiability from a statemental property to a (derived) sentential property, and extended it to cover all declarative-type sentences. In the case of empirical verifiability, intended to be significant but false of nonsignificant sentences, this mapping on to sentences is important. Otherwise the notion could not be applied to weed out nonsignificance in many of the cases for which it was devised.

It is true that Popper, on the strength of examples involving multiple quantifiers, has objected to the principle that *P* is empirically verifiable if, and only if, $\sim P$ is empirically verifiable. However, Popper regularly disassociates himself from the positivists and his rejection of this principle is one of the grounds of his disassociation. It is far from clear, however, that Popper is right in rejecting the principle of negation since it can only be maintained given, what is doubtful, that the negation of an empirical statement may not be empirical. If, however, Popper is concerned with significance, as he appears to claim but Carnap denies (see Popper [1] and Schilpp [2], pp. 880–881), it would be interesting to know what matrices he is committed to for his connectives. As it is, it seems that his theory may be open to the criticism which he has made of others, that it is not open to logical assessment or falsifiability.

There is indeed something of a case for arguing that the positivists were really committed to the classical significance matrices in advance of, or independently of, their adoption of the verifiability (or later confirmability) criterion. For in view of their acceptance of the classical two-valued logic for significant sentences and a powerful principle of extensionality, they were committed to the two-valued sections of the matrices. And they were committed to a homogeneous extension of the two-valued connectives since they took it as a central thesis that many sentences, especially those of traditional metaphysics, were nonsense no matter how compounded using classical connectives.

Our conclusion that Russell and the positivists were committed to the classical significance connectives supports Halldén [1] and other Swedish logicians who have developed significance logics based on these connectives in order to explicate the theory of types and the principles of positivism. More recently, however, there has been a departure from classical significance connectives by Åqvist [1] (pp. 138–157) who claims that his "calculus **Å** presents an alternative to Halldén's logic **C** which deserves serious consideration". As we shall see, however (§5.12), it is not clear that Åqvist's system is in fact an alternative to Halldén's, rather it contains it. Halldén's connectives '$\sim$', '&' and '$+$' (our '*S*') can be defined, preserving matrix values, in **Å**. In fact it is possible to define '$\sim$', '$\vee$', '&', '$\supset$', '$\equiv$', '*S*' and '$\simeq$' (strong extensional equivalence) in **Å** in such a way that the following are theorems:

$$S(p\,\&q)\simeq Sp\,\&\,Sq;\; S(p\vee q)\simeq Sp\,\&\,Sq;\; S{\sim}p\simeq Sp;\; S(p\supset q)\simeq Sp\,\&\,Sq;\; S(p\equiv q)\simeq Sp\,\&\,Sq.$$

These theorems effectively express our assumption IV and are the principle-correlates of the classical matrices. It is not clear, therefore, that Åqvist has succeeded in rejecting IV, which he calls "the doctrine of the atheoretical element", as he claims to do. Rather, he has simply chosen a different set of basic connectives for which the assumption does not hold. Hence his way of challenging certain suppositions of **C**, that it is impossible to decide in any convincing way between the conflicting "intuitions" which underly **C** and those which underly **Å**, is misleading. For he has not succeeded in producing a logic which stands in opposition to **C**, and the intuitive bases of the two are not therefore in conflict. What we may conclude, however, is that there is a case for the view that there are different senses of '&', '∼', '∨' *etc.* which are captured by different matrices and are perhaps appropriate for dealing with different kinds of nonsignificance. From this point of view **Å** is a more complete logic than **C** since it contains both those connectives which are given a sense by the classical matrices and those which are given a sense by what we propose to call the strong regular matrices, and which satisfy Åqvist's conditions. There is, however, another important difference between **Å** and **C** which will concern us later, namely that *t* is the only designated value with respect to **Å** while both *t* and *n* are designated with respect to **C**. Even this difference can be minimised in finitely-valued matrix logics, however, by setting up mapping relations between the two systems.

5.4 Strong Regular Significance Connectives

There is a strong case for the introduction of non-classical significance connectives, as we shall argue subsequently. For even with just one interpretation of significance in focus, a range of non-classical connectives is formally advantageous, indeed almost indispensable. More important, however, there is no *prima facie* reason to assume that there is just one sort of nonsignificance. And if there are different kinds of nonsignificance, or at least of non-truth-valuedness, then it might be expected that these differences will be reflected by different sets of connectives.

If we reject the requirement IV, for whatever reason, there are many different possible sets of matrices which extend the classical two-valued matrices and which have advantages of one kind or another, but we here introduce only those which satisfy Åqvist's basic connectives. These are the strong regular matrices, which were adopted by Kleene ([2], pp. 333–336) for a different purpose.

The condition which singles out these matrices, and which is opposed to IV, is:

Condition V (a) A conjunction with one false component is false, no matter what the values of its other components; is true when, and only when, all components are true; and is nonsignificant otherwise:

(b) A disjunction with one true component is true, no matter what the values of its other components; is false only in case all components are false; and is nonsignificant otherwise:

(c) A sentence is true if its negation is false; false if its negation is true; otherwise nonsignificant.

It will be seen that if we neglect the last clause of each of (a), (b) and (c), we have exactly the conditions for the ordinary two-valued matrices.

The strong regular negation as defined by (c) is the same as the classical significance negation; and it follows, too, that the strong regular equivalence is the same as the classical significance equivalence. The distinguishing connectives are, therefore, those which represent 'and', 'or' and 'materially, if ... then ...' and we shall use the symbols'$\cap$', '$\cup$' and '$\Supset$' to mark them off from the classical '&', '$\vee$' and '$\supset$'. The symbols '$\equiv$' and '$\sim$' remain the same.

The distinguishing strong regular matrices are therefore as follows:

$\cap$	t	f	n
t	t	f	n
f	f	f	f
n	n	f	n

$\cup$	t	f	n
t	t	t	t
f	t	f	n
n	t	n	n

$\Supset$	t	f	n
t	t	f	n
f	t	t	t
n	t	n	n

As Åqvist emphasises, these connectives are a natural extension of the familiar two-valued ones since they preserve the classical prescriptions: namely that, a disjunction is true if, and only if, one of its components is true; a conjunction is true if, and only if, each of its components is true; and a material implication is true if, and only if, either its antecedent is false or its consequent is true. But of course, the strong regular extensions are not the only ones which satisfy these prescriptions; matrices with *f* in place of *n*, in some positions, will satisfy them.

Kleene considers these tables in the context of partially recursive functions. For if Qx is a partial recursive predicate defined over such a function, there is an algorithm on its range of definition which enables us to say, for each x, whether Qx is t or f. But there may be no algorithm for deciding, for a given x, whether Qx is defined or not. When Qx is undefined it is said to take the value u. In such a case, however, Qx is still either true or false; to say that it has the value u is simply to say that its actual truth-value is not determined by the algorithm; it is not to say that it has no truth-value. If Qx is u, therefore, the value of any compound containing it, say $Qx \vee Px$, cannot be arrived at by utilising information about Qx, since none is available. Kleene therefore takes it that the value range is the set $\{t, f, u\}$; and he concludes that the appropriate matrices which give the sense of the connectives in such a context are the strong regular matrices above (with 'u' for 'n'); that is, in our symbolism, $P \vee Q$ is $P \cup Q$; $P \& Q$ is $P \cap Q$; and $P \supset Q$ is $P \Supset Q$. We note, however, that although the intended initial interpretation of the third value is 'undefined', rather than 'nonsignificant', there is a suggestion that it might be interpreted as 'meaningless' (*i.e.* 'nonsignificant') (Kleene [2], example 1, p. 335).

We now wish to argue, first, that the strong regular matrices are correct under an extended Kleene-type interpretation which includes, besides algorithmic indeterminacy, other kinds of undecidability or incompleteness. Secondly, that such an interpretation is distinct from the interpretation of '*n*' as 'nonsignificant', and indeed that it excludes this interpretation unless further, and questionable, assumptions are added. Third, and against Åqvist, that if '*n*' is interpreted as 'nonsignificant', the connectives so defined are not those commonly used in philosophical and ordinary discourse where significance is an issue.

That the strong regular matrices are correct for certain other kinds of statement-incapability or sentential incompleteness, besides that which is related to algorithmic undecidability, is not established either by Åqvist, whose arguments are unsatisfactory, or by Kleene, whose arguments are inconclusive. Kleene's arguments can be extended, however, in such a way that '*u*', interpreted as 'undefined' may be replaced by '*i*', interpreted as 'not truth-value determinable with respect to a given procedure'.

Kleene's arguments, reinforced by Theorem XXI and subsequent examples 6–8 (Kleene [2], pp. 339–340), support the general thesis, independently of taking the third value as '*u*', that the matrices chosen for negation, conjunction and disjunction over the set $\{t, f, i\}$ must be regular to the extent that a given column (row) which contains *t* in the *i*-row (column) should consist entirely of *t*'s, and similarly for *f*. This commits us to matrices for '∩' and '∪' selected from the following:

∩	*t*	*f*	*i*
t	*t*	*f*	$\bar{i}$
f	*f*	*f*	–
i	$\bar{i}$	–	$\bar{i}$

∪	*t*	*f*	*i*
t	*t*	*t*	–
f	*t*	*f*	$\bar{i}$
i	–	$\bar{i}$	$\bar{i}$

where $\bar{i}$ represents either value *f* or value *i*, and the blank places may as yet be filled by any one of the three values.

That the tables chosen should be strong regular, however, follows from a consideration of the generalisable features of algorithms. We suppose, as Kleene does, that assignments of values for compound sentences are always made on the basis of assignments for the components. And as Kleene observes, the information that *Q*, say, is *i*, is not utilisable by the algorithm, or indeed by any procedure in terms of which truth and falsity are determined. Hence any assignment of value *t* or *f* to a compound containing *Q*, say $P \cup Q$, cannot depend on information about *Q* since there is none. From this it follows that $\sim P$, $P \cap Q$, and $P \cup Q$ must be assigned value *i* when both *P* and *Q* are assigned value *i*. Thus, '∼' will have the same matrix as the classical significance negation, and the (*i*, *i*) place in the '∩' and '∪' matrices will be occupied by *i*.

There is at this point a gap in Kleene's arguments, but we may complete them and generalise them as follows: In terms of the intended interpretation, any given sentence is either *t* or *f*, but we are to assign the value *i* when we lack information enabling us to decide which of these values it has. Consider, then, $P \cap Q$ when *P* is assigned the value *i* and *Q* has value *f*. Now *P* in fact has value *t* or value *f*, but it is assigned the value *i* since the algorithm, or whatever the procedure might be, does not enable us to make a decision. We may say, however, that since it is known that *Q* is *f*, then $P \cap Q$ is *f*. For we can reach this information independently of information about *P*. For if *P* were *t*, we have (*t*,*f*) which yields *f*; and if *P* were *f*, we have (*f*,*f*) which again yields *f*. So our lack of information concerning the value of *P* does not affect our information concerning the value of $P \cap Q$; it will always be *f* even though *P* has to be assigned the value *i*. When, instead, *Q* has value *t*, $P \cap Q$ has value *t* when *P* has value *t*, and value *f* when *P* has value *f*. Here, therefore, the lack of information concerning the value of *P* transfers to a lack of information concerning the value of $P \cap Q$. So it, too, must be given the value *i*. In this way the matrices for '$\cap$' and '$\cup$' may be completed, and the strong regular matrices result (with '*i*' for '*n*').

There are two points to note about these arguments. First the dependence on an algorithmic procedure for determining the values *t* and *f* is unimportant. The arguments depend only on there being a lack of information, whether in relation to an algorithmic or any other kind of procedure or technique. And this justifies the extended interpretation from *u* to *i*. Secondly, the undecidable sentences are taken to be significant and truth-valued; it is simply that the procedure adopted for determining which truth-value is appropriate, whatever this procedure might be, is inadequate in some cases. And this really *modifies* the interpretation of '*t*' and '*f*'. For we cannot interpret them as 'true' and 'false' but only as 'determinable as true by the procedure' and 'determinable as false by the procedure'. In view of this, the strong regular tables may be "read off" the classical two-valued {*t*,*f*}-matrices using only fairly weak assumptions such as: that if *P* is assigned value *t* by the procedure, then it is true; and that the procedure determines the value of compounds classically when it determines the values of all components; *etc.* It will be seen that the above arguments in fact depend on such assumptions.

Viewed in this way, the third value of the strong regular matrices is not properly interpretable as 'non-truth-valued' when established by means of such arguments, but only as 'not truth-value determinable under the procedure'. *A fortiori*, therefore, the third value cannot be read as 'nonsignificant' in relation to such arguments, since nonsignificant sentences are non-truth-valued, and not simply indeterminate.

When the arguments for the matrices are in the form given, therefore, the matrices themselves provide a cross-classification of two-valued sentences and not a classification of three-valued sentences. This condition is satisfied by Kleene's initial interpretation of '*u*' as 'undefined', and by his proposed extended interpretation of '*t*' as 'known to be true', '*f*' as 'known to be false', and '*u*' as 'truth-value unknown'. But it is not satisfied by the interpretation 'verifiable as true' (for '*t*'), 'verifiable as false' (for '*f*') and 'not verifiable' (for '*i*') if we adopt the criterion of verifiability, *CV*, for then 'not verifiable' is equated to 'meaningless' and the arguments will not go through in the form in which they have been given since they depend on the fact that sentences which have value *i* are nevertheless *t* or *f*, though which value they take is unknown. Clearly, however, under the criterion *CV*, this assumption, that sentences which take the value 'meaningless' are nevertheless true or false, is itself false. And in general, any arguments designed to establish the strong regular matrices for the set $\{t, f, n\}$ must be different from those which have been given.

There are other interpretations of '*i*', related to its basic sense of epistemic indeterminacy, which satisfy the general condition that the matrices provide a cross-classification of true-or-false sentences and not a genuine classification of three-valued sentences. Two obvious such interpretations are:

(a) 'provable' (in a given formal system L), for '*t*',
'disprovable' (or 'refutable') in L, for '*f*',
'undecidable' (or 'irresoluble') in L, for '*i*';

(b) 'necessary', for '*t*',
'impossible', for '*f*',
'contingent', for '*i*';

though in each of these cases slight modifications have to be made to the matrices to satisfy the special interpretation. That the general condition is satisfied, however, follows given that all sentences of L are true or false under a given interpretation of L, whether or not they are provable, disprovable or undecidable in L; or again, that all sentences which are true or false are either necessary, impossible or contingent. This amounts to saying, what is commonly accepted, that provability and modal classifications are alternative classifications of truth-valued sentences and not classifications of three-valued sentences.

To show that (a) is a possible application, let L be Peano arithmetic. Then in view of Gödel's incompleteness theorem, L will contain at least one formula G which is neither provable nor disprovable (irresoluble) but which is true under a given interpretation. Now if P and Q both take the value t (*i.e.* are provable), then P & Q takes the value t, by the rule of adjuction in L; if either P or Q is refutable (f), then P & Q is refutable, by applying the thesis of L, $\sim P \supset \sim(P \,\&\, Q)$. To complete the matrix, it remains to show that if one of P and Q is irresoluble (i) and the other is either irresoluble or provable (i or t), the compound P & Q is irresoluble. Now under such conditions, P & Q cannot be provable; for it if were, by the L-thesis, $P \,\&\, Q \supset P$, both P and Q would be separately provable, so contradicting the assumed irresolubility of at least one of them. But nor can P & Q be refutable when one of them, say P, is provable and the other, Q, is irresoluble. For then we should have $P \mathbin{\&} {\sim}(P \mathbin{\&} Q)$ provable, by adjunction; and hence Q is refutable, by disjunctive syllogism, so contradicting the hypothesis that Q is irresoluble. Hence, when one of P or Q takes the value t, and the other the value i, P & Q takes the value i. It is not, however, true in general that when both P and Q are irresoluble, P & Q is irresoluble. For Q may be ${\sim}P$, in which case P & Q is $P \mathbin{\&} {\sim}P$ and is refutable. But when P and Q are refutationally independent, so that P & Q is refutable only if one of P and Q is refutable, then P & Q is irresoluble if both P and Q are irresoluble.

We therefore conclude that the sense of '&' here, with a crucial variation in the (i,i) place, is that given by the matrix for '$\cap$' (with 'i' for 'n').

A similar pattern appears when the application is as under (b). Using standard results of modal logic we easily arrive at these matrices, which are also satisfied by (a):

$\cap$	t	f	i
t	t	f	i
f	f	f	f
i	i	f	i or f

$\cup$	t	f	i
t	t	t	t
f	t	f	i
i	t	i	i or t

$\sim$	
t	f
f	t
i	i

In each case, if we restrict the application to logically independent sentences, the matrices reduce to the strong regular matrices.

There are, then, good reasons why matrices other than those for the classical significance connectives might be taken as basic. But, as we have shown, in none of the cases so far considered is it possible to interpret the third value as 'nonsignificant'. For the arguments which establish the tables over the set $\{t,f,i\}$ depend on assumptions which are false if i is replaced by n. This distinction is especially important in cases where it might be thought the sense given to the values are interchangeable with other senses—as 'known to be true', 'known to be false', 'unknown' seem to be interchangeable with 'verifiable as true', 'verifiable as false' and 'unverifiable'. This would be the case in the ordinary sense of 'verifiable', but not in that sense which is given to it by *CV*, so that these arguments cannot be used to support the view that the positivists were committed to the strong regular tables. Similarly, such intuitionistic principles as might support the adoption of the strong regular matrices are inconsistent with the view that the intuitionists have discerned a sort of meaninglessness in classical mathematics. On the contrary, they have merely introduced a cross-classification of significant mathematical sentences by means of the interpretation:

(c) 'provable by intuitionistic means' for 't'
'refutable by intuitionistic means' for 'f'
'undecidable by intuitionistic means' for 'i'.

Given an intuitionistic verifiability criterion,

(*ICV*) Intuitionistically undecidable sentences of classical mathematics are meaningless,

then it seems to follow that '*i*' may be interpreted as '*n*' in the strong regular matrices. But this involves the same fallacy as before. For if *ICV* is adopted, and if 'meaningless' implies 'non-truth-valued', then the arguments which establish the strong regular matrices in terms of *i* fail.

We conclude, therefore, that if the strong regular matrices are adopted, and '*n*' is taken to be 'nonsignificant', independent arguments consistent with this interpretation must be offered. They are not automatically established, indeed not established at all, by arguments arising from the features of indeterminate sentences of one kind or another. There is no overlap, and can be no overlap, simply because indeterminate sentences are truth-valued and nonsignificant sentences are not.

There are other, independent, reasons in favour of the strong regular matrices containing *n*, however, though none are really decisive. A technical, but negative, reason, is that which arises from the mistaken belief, which we discuss in detail in the next section, that logics based on the classical matrices can have no theorems. A second, more important reason, is that any adequate significance logic should yield principles which have been commonly adopted in the formulation of significance arguments. One such is the *reductio* variant:

If under the assumption that *Sp*, *qu*(*p*) implies a contradiction, then ~*Sp*,

which is attributable, with some justification, to Russell and Whitehead [1], p. 77; Quine [3]; and Ryle [4]. As we shall see, however, this fails to be a principle if the connectives are either classical or strongly regular, so though it might be held to count against the classical connectives, it cannot be used to support the strong regular connectives. But in our view it ought in any case not to appear as a principle in this form, since in general, the fact that *qu*(*p*) implies a contradiction at most establishes that *qu*(*p*) is not true, not that it is nonsignificant. This is so since if *qu*(*p*) is any contradiction then it implies a contradiction, and except on the very strong view that a contradiction is meaningless, we cannot draw the conclusion ~*Sp*. In any case, if we took this strong view then *qu*(*p*) could not be assumed significant in the antecedent.

This principle can in fact be reformulated, and in such a way that it reflects better Russell's treatment of the paradox argument, as follows:

If under the assumption that *Sp*, the truth of *qu*(*p*) implies a contradiction and the falsity of *qu*(*p*) implies a contradiction, then ~*Sp*.

In this form it can be formalised using classical significance connectives and the predicate '*T*' and it is then a theorem of logic $\mathbf{S}_2$ below. But the principle in this form can also be satisfactorily formalised using strong regular connectives, and it is again a theorem. Hence it provides no way of choosing between them. There is however an important conclusion to be drawn from this fact. For since the principle is valid in terms of strong regular matrices, and since the third value of such matrices can be interpreted as 'indeterminate' rather than 'nonsignificant', *reductio* arguments such as Russell's do not establish without further argument or assumption, the nonsignificance of the premisses. Thus, a principle of the sort:

> If under the assumption that *qu*(*p*) is truth-value determinable, the truth of *qu*(*p*) implies a contradiction and the falsity of *qu*(*p*) implies a contradiction, then *qu*(*p*) is not truth-value determinable,

holds. Hence even though the corresponding significance principle is also valid, we cannot conclude from the paradox argument that it is nonsignificant to say of a class that it does or does not belong to itself. For we might instead invest in the alternative indeterminacy principle and conclude simply that it is indeterminate but truth-valued to say of a class that it does or does not belong to itself. For similar reasons, Quine's alleged demonstration of the nonsignificance of certain quantified modal expressions (Quine [3], p. 148; and [2], p. 147) is inadequate, as are Ryle's arguments purporting to show the absurdity of certain accounts of minds and of mental states and processes (Ryle [4]), unless supplemented by criteria designed to show that, in these special cases, either the classical matrices should be taken or the strong regular matrices are correct with *n* as the third value.

The requirement that an adequate significance logic should bring out as valid certain argument forms which have been commonly adopted, will not withstand much examination. The only requirement that can reasonably be imposed is that the logic should not bring out as invalid intuitively valid arguments. It is not enough to show that a given argument, or something like it, has been accepted: for it is well known that philosophers, and even logicians, sometimes argue invalidly. But of course, even with this weaker requirement, there remains the minor problem of showing that an intuitively valid argument should be formalised in one way rather than another; and the far from minor obstacle of showing that the argument is, indeed, intuitively valid.

The weaker requirement is, however, very important. It is on this basis that we shall argue that classical significance connectives, though central, are not exhaustive. Adoption of classical significance connectives alone for sentential significance logic is not nearly sufficient for the formalisation of all valid sentential significance arguments. We thus wish to make it perfectly clear that we are not advocating the classical connectives as the only important significance connectives; we are only advocating them as important enough to be included as initially basic in any formalisation of significance logics which aims to have historical relevance and to be applicable to everyday and philosophical significance arguments. At the same time, however, we find little to justify the adoption of the strong regular connectives in this regard. Rather, we feel, that the direct arguments in favour of them justify the third value as *i* and not *n*. That is, in general, that their adoption as primary indicates that some notion other than nonsignificance is being explicated. Their adoption in fact suggests that really a cross-classification of truth-values is being made, a cross-classification better and more honestly achieved by the explicit introduction of a new connective (as is done by Prior using the intensional connective 'it is the case (that)' and by Routley [7] with predicate connectives). Our discussion of strong normal connectives also suggests that incompleteness can be treated as a cross-classification of truth-values rather than as a further value, a matter that can be accomplished using a new connective (*cf.* also §3.9).

The main non-classical 3-valued connectives which we do find it necessary to introduce, though they enable us to define strong regular connectives, are not themselves strong regular; but we postpone the justification of their use since it depends on the examination of intuitively acceptable principles which they validate but which are not validated either by the classical connectives or by the strong regular connectives.

5.5 Designated Values: **S** and **C** Logics

We propose now to develop a number of logics in which the classical connectives are taken as initially basic; but even though we adopt certain matrices, we have no logic at all until we decide which values are designated or thesis-determining.

It might perhaps seem obvious that *t* and only *t* can be adopted as a designated value. For clearly *f* cannot be taken as designated without undermining the whole purpose of the logic since this would commit us to the adoption of theses which take the value *f* for some or all assignments of values to the components. And it would seem that there is a similar obnoxious consequence if we adopt *n* as a designated value, since we are then committed to asserting theses which turn out to be nonsignificant for some or all values of the components.

However, to take the obvious course of adopting *t* as the only designated value is often supposed to have an overwhelming defect: namely that, a system containing only classical significance connectives, and which has *t* as the only designated value, has no theorems (see *e.g.* Presley [1]). For let $\Gamma(p_1, ..., p_m)$ be a formula containing variables $p_1, ..., p_m$, each of which ranges over sentences taking one of the three values *t*, *f* and *n*, and just classical significance connectives. Then it follows, by induction over the number of connectives in $\Gamma(p_1, ..., p_m)$, that $\Gamma(p_1, ..., p_m)$ has the value *n* whenever any one of $p_1, ..., p_m$ has the value *n*. That is, because of the way the classical matrices are constructed, every compound has the value *n* whenever one or more of its components has the value *n*; and since all components range over *n* as a value, every compound takes the value *n* for some value(s) of its components. Hence no compound (and certainly no single variable) can be a thesis if *t* is the only designated value.

There are however two assumptions in the argument each of which can be amended to lead out of the difficulty.

(1) $\Gamma(p_1, ..., p_m)$ contains *only* classical connectives. By enlarging the *total* class of connectives, even though the classical ones are taken as basic, theorems containing classical connectives (as well as others) can be obtained.

(2) The variables $p_1, ..., p_m$ *all* range over the set $\{t,f,n\}$. By introducing restricted variables, in addition to the ordinary variables, which range only over the values $\{t,f\}$, theorems containing both kinds of variables and classical connectives can be obtained. This is the solution we shall adopt in weak significance logics.

The alternative, adopted by Halldén [1] in spite of its intuitive unacceptability, is to take both *t* and *n* as designated values, though now the word 'designated' seems inappropriate and should be replaced by 'thesis-determining' or simply 'determining' value. Quite apart from the fact that this course commits us to the affirmation of nonsense in some cases, it has other drawbacks at the axiomatic level. For it amounts to saying that a formula is a thesis provided it does not take the value *f* for any assignment of values to its variables and this means that modus ponens does not always lead from true premisses to true conclusions. Thus, for example, when both *t* and *n* are designated values, and the connectives are classical, the schema $P \supset P \vee Q$ is valid since the main column of its valuation table does not contain *f*. But given that *P* has the value *t*, it does not follow in general that $P \vee Q$ has the value *t*, but only that it has the value $\bar{f}$ (not false). In fact, if we substitute in such a way that *P* has the value *t* and *Q* has the value *n*, *e.g.* by putting 'Saturday is a day of the week' (assumed to be *t*) for *P*, and 'Saturday is in bed' (assumed to be *n*) for *Q*, then we may affirm,

> If Saturday is a day of the week then Saturday is a day of the week or Saturday is in bed,

as "true" (designated), being a special case of a thesis, and affirm the antecedent, given as true, and so conclude by modus ponens:

> Saturday is a day of the week or Saturday is in bed;

which, in terms of the classical connectives, is *n*. (cf. Halldén [1], Presley [1] and Goddard [3]).

Halldén concedes that an unqualified modus ponens rule cannot be retained and accordingly weakens modus ponens in his systems. But is there much of a case for this amendment, or in general for treating n as a designated value? The only real case is that by taking n as designated we obtain theses comparable to classical two-valued theses, while if we do not take n as designated, we have no theses at all unless other amendments are made. But "if a formula for certain values becomes meaningless, then how can it be true?" (Halldén [1], p. 47). Halldén's answer to his question, that a formula P is to be taken as asserting something only under significant replacements, effectively concedes the point at issue. For his answer amounts to this: assertion or theoremhood of P is really only assertion of P for its significant values, and we do not have a genuine three-valued logic at all since we have in effect reduced it to the two-valued case. At best, P is designated in some degenerate way when it has value n. How does this differ from saying that P is not strictly designated at all when it has such a value, but is rather a substitution instance of a wff designated for its significant values?

It should be noticed here that, within an axiomatic approach, the initial difficulty over theorems containing only classical connectives—that there are none—only emerges given that both uniform substitution and modus ponens are retained as inference rules. It needs only a few examples to show that whatever credence is given to the rule of uniform substitution in ordinary logics should be withheld when uniform substitutions of nonsignificant sentences in theses is considered. Consider some examples: substituting in $qu(p \supset p)$, we obtain such nonsignificant sentences as 'If the Absolute is green then the Absolute is green'; 'If quadruplicity drinks procrastination, then it does'; and by substituting in $qu((Ux)fx \supset fy)$, we have 'If all numbers like soup, then the number six likes soup.' Hence, significance is not preserved under uniform substitution. Consequently, in the main weak axiomatic **S**-logics which we consider, the rule of uniform substitution will be restricted so as to allow uniform substitution of significant wff only for certain of the variables. Once this restriction is introduced, modus ponens can be reinstated. Now the semantical equivalent of restricted substitution in the axiomatic system is restricted assignments of values to the sentential variables, and the particular restriction suggested amounts to specifying that certain variables should vary only over the set $\{t,f\}$ and not the full set $\{t,f,n\}$. Such variables we call S-restricted variables (see §5.6). In this way, by abandoning assumption (2) above, we overcome the difficulties which arise initially when t is taken as the only designated value.

We shall concentrate on significance logics where t is taken as the only designated value, called **S**-logics, but we shall also develop parallel **C**-logics in which both t and n are taken as designated. For although we regard the designation of n as counterintuitive, in terms of the intended application, we do not consider this an adequate reason for dismissing **C**-logics from the start. Quite apart from the fact that a **C**-logic was adopted by Halldén, and in that sense they have at least historical interest, it seems to us important to judge them in terms of their theses and to compare these with the theses of corresponding **S**-logics in order to see precisely what their advantages and disadvantages are. In the event this leads us to a compelling reason for rejecting them entirely, for we show that no **C**-logic can be made functionally complete without at the same time being made inconsistent with respect to classical negation.

5.6 The Problem of Consistency for Matrix Logics

In two-valued matrix logics consistency does not raise a serious problem, for reasons which we shall see, and in consequence the idea of consistency is usually only formulated for axiomatic systems. In many-valued logics, however, the matter is different. It is known that many-valued matrix logics which have more than one designated value may be inconsistent with respect to some monadic connective κ (κ-inconsistency), in the sense that both Δ and $\kappa\Delta$ are theses for some wff Δ, where κ has a matrix which justifies its interpretation as a negation (see Rosser and Turquette [1], p. 43). Hence for **C**-logics in particular, since these have more than one designated value, it has to be shown that they are not inconsistent with respect to classical negation. As it turns out, this is only possible for weak **C**-logics.

There are various kinds of inconsistency which may be distinguished:

(a) *Absolute inconsistency*: a logic is said to be absolutely inconsistent if every wff is a thesis, otherwise absolutely consistent.

A logic which has this property is valueless since the notion of a thesis coincides with that of a well-formed formula, but this kind of inconsistency is really only applicable to axiomatic systems. A matrix logic is always absolutely consistent provided at least one value is undesignated and at least one set of its variables ranges over all values. For then, any such variable standing alone will have a matrix which contains an undesignated value and hence will not be a thesis. All the matrix logics which we consider satisfy these two conditions and they are, therefore, absolutely consistent.

(b) *Post-inconsistency*: a logic is said to be Post-inconsistent if there is a thesis which consists of an unrestricted variable standing alone, otherwise Post-consistent.

Here, too, inconsistency of this kind is only properly applicable to axiomatic systems. For given a matrix logic which has at least one undesignated value (and if it did not, there would be no point to it) and variables which range over all the values, then, as we have seen, no such variable can be a thesis. Again, each system we consider is Post-consistent, and indeed we used the property of Post-consistency to establish the absolute consistency in (a) above.

There is clearly a close connexion between Post-consistency and absolute consistency, but they are not the same. Given an axiomatic system which has a rule of uniform substitution, a Post-inconsistent system will always be absolutely inconsistent. For since we may substitute any wff for a variable in a thesis, we may substitute any wff in that variable which is itself a thesis. Hence every wff is a thesis. But an axiomatic system which does not have a rule of uniform substitution may be Post-inconsistent but not absolutely inconsistent. Again, as we have seen, a matrix logic which is Post-consistent is absolutely consistent. But since a matrix logic may not contain a (derived) rule of uniform substitution, the fact that it is Post-inconsistent will not entail that it is absolutely inconsistent, and indeed it may be that a variable standing alone is a thesis when some other wff is not.

(c) *κ-inconsistency*: a logic is said to be inconsistent with respect to a monadic connective κ (briefly κ-inconsistent) if, for some wff Δ, both Δ and $\kappa\Delta$ are theses, otherwise the logic is κ-consistent.

So defined, however, inconsistency may be trivial and uninteresting, and certainly not indicative of a defect in the logic. If, for example, we take κ to be the identity operator, defined over our three values *t*, *f*, *n*, by,

Γ	*t*	*f*	*n*
$\kappa\Gamma$	*t*	*f*	*n*

then, necessarily, given any wff Δ which is a thesis, $\kappa\Delta$ is also a thesis. Again, let κ be the significance operator *S* defined by

Γ	*t*	*f*	*n*
$S\Gamma$	*t*	*t*	*f*

If Δ is a wff which takes only the value *t* for all values of the variables in it, then both Δ and $S\Delta$ are theses.

The kind of κ-inconsistency which does indicate a defect of the system is that for which κ is interpretable as a negation. For then, the fact that both Δ and $\kappa\Delta$ are theses commits us to the assertion of both Δ and not-Δ. We therefore define,

(d) *simple inconsistency*: a logic is said to be simply inconsistent if (i) it is κ-inconsistent and (ii) κ is interpretable as a negation.

The difficulty here, however, is in laying down conditions which express precisely what is meant by saying that κ is interpretable as a negation. This is no great difficulty in two-valued logics for we simply rely on the intuitive criterion that 'not' converts a true sentence to a false one, and *vice versa*. Hence simple consistency is κ-inconsistency where κ is '$\sim$', defined by,

Γ	*t*	*f*
$\kappa\Gamma$	*f*	*t*

In the case of many-valued logics, however, there may be several κ's which are intuitively interpretable as a negation, all with different matrices. And in any case, though we may set up matrix conditions under which a given κ may be regarded as a negation, such criteria will not be directly applicable to axiomatic systems, and not applicable at all to an axiomatic system which has no finite characteristic matrices satisfying the axioms. We must therefore seek both matrix conditions and formal conditions to elucidate (d)(ii) above. We first consider matrix conditions.

Let us suppose that we have an *m*-valued logic with a finite set of values {1, 2, 3, ..., M}, where at least one value is designated and one undesignated. Since the set of values is finite we may without loss of generality order them in such a way that the first value is always designated and the last undesignated; and indeed, if there are *s* designated values we may order them in such a way that the first *s* are designated and the last $m-s$ are undesignated. We take it, therefore, that if the values are 1, 2, ..., M, then 1 is always designated and M is always undesignated. Further, since our main interest is in a *truth-value* interpretation, we shall take it that 1 is *t* (interpretable as 'true') and M is *f* (interpretable as 'false').

There are various criteria which might be demanded of a monadic connective κ in order that it be interpretable as a negation. One very weak criterion would be to require of κ that it changes at least one designated value to an undesignated value. In terms of this, a contradiction operator is a negation: *e.g.* in the two-valued case, taking *t* as designated and *f* as undesignated, κ defined by,

Γ	t	f
$\kappa\Gamma$	f	f

is a negation. Equally weak by itself is the converse criterion that κ changes at least one undesignated value to a designated value. Then, a tautology operator is a negation: *e.g.* in the two-valued case:

Γ	t	f
$\kappa\Gamma$	t	t

There is no over-riding reason why such operators should not be regarded as negations for special purposes, though they have obvious drawbacks. The contradiction operator, for example, is an identity operator over contradictions, while the tautology operator is an identity operator over tautologies. We get closer to the concept of an intuitive negation, however, by requiring that at least both of these conditions be satisfied. That is, we usually require of a negation connective that it converts some designated value to an undesignated value and some undesignated value to a designated value. In the two-valued case, where the values are *t* and *f*, this condition would in fact uniquely specify the classical '$\sim$' defined by,

Γ	t	f
$\sim\Gamma$	f	t

But in the many-valued case, a connective which satisfies just this requirement is still very weak. For example, if we have the set of values {1, 2, 3, 4,} such that 1 and 2 are designated and 3 and 4 are undesignated, then κ defined by,

Γ	1	2	3	4
$\kappa\Gamma$	4	2	2	4

satisfies the condition since it converts 1 (designated) to 4 (undesignated) and 3 (undesignated) to 2 (designated). It is characteristic of the two-valued classical '$\sim$', however, that it not only converts some designated value to some undesignated value, and some undesignated value to a designated value, but that it *interchanges* a designated value with an undesignated value. This, of course, is inevitable, since there are only two values, one designated and one undesignated, so that the application of the weak conversion principle automatically results in an interchange. An interchange of values is, however, a separate feature distinguishable from weak conversion in the many-valued case and is, from many interpretational points of view, an essential feature of a negation. In terms of this, κ as defined above over the set of values {1, 2, 3, 4,} would not be a suitable negation since it does not result in an interchange. Over the same set of values, where again 1 and 2 are designated and 3 and 4 are undesignated, the following κ is an interchange negation:

Γ	1	2	3	4
$\kappa\Gamma$	1	3	2	4

Even so, it is arguable whether it is a suitable negation. For if we are taking 1 to be t and 4 to be f, such a κ, which leaves t and f unchanged is certainly not a standard negation. This feature of interchanging t and f, which is indistinguishable from both weak conversion and simple interchanging in the two-valued case, is again a separately distinguishable feature in the many-valued case and seems to be the minimum characterising feature of a standard negation. We therefore define:

(i) A monadic connective κ is minimally interpretable as a standard negation if, and only if, it interchanges 1 (t) and M (f).

Thus, given that the values are {1, 2, ... ,M} with 1 (t) designated and M (f) undesignated, κ is minimally interpretable as a negation if it has as matrix:

Γ	1 2 ... M
$\kappa\Gamma$	M ... 1

where the complete row opposite Γ is the full ordered set (1, 2, ..., M) while the row opposite $\kappa\Gamma$ may be completed by an arbitrary choice of values from the set {1, 2, ..., M}.

Evidently, in the two-valued case, where the values are 1 and 2 only, of the four possible monadic connectives,

Γ	$\kappa^1\Gamma$	$\kappa^2\Gamma$	$\kappa^3\Gamma$	$\kappa^4\Gamma$
1	1	2	1	2
2	1	2	2	1

only one, κ^4, satisfies the condition for a minimally interpretable negation. Writing '*t*' for '1' and '*f*' for '2', this is in fact '$\sim$'. κ^1 is the tautology operator, κ^2 the contradiction operator and κ^3 the identity operator.

In the three-valued case, where the values are 1, 2, 3, there are 27 possible monadic connectives, three of which satisfy the minimal condition for a negation. For the matrix must be of the form,

Γ	$\kappa\Gamma$
1	3
2	
3	1

where the vacant space may be filled by any one of the three values. Thus taking the three values as t, f, n, ($t = 1, n = 2, f = 3$), we have,

Γ	$\kappa_1\Gamma$	$\kappa_2\Gamma$	$\kappa_3\Gamma$
t	f	f	f
n	t	n	f
f	t	t	t

κ_1 is a connective which we later represent by '$\urcorner$', κ_2 is '$\sim$' and κ_3 is 'F'.

Much stronger conditions could be required of a negation. The strongest would be to require that every designated value is interchanged with an undesignated value. Clearly this can only be satisfied in logics which have an even number of values and is such that the number of designated values is the same as the number of undesignated values. A less strong condition (which as we shall see is nevertheless too strong for many purposes), which can be made independently of whether the number of values is odd or even, is the requirement that every designated value be converted to some undesignated value and *vice versa*. We therefore define:

(ii) A monadic connective κ is maximally interpretable as a negation if, and only if, it converts every designated value to some undesignated value and every undesignated value to some designated value.

In the two-valued case, κ^4 is maximally interpretable as a negation. Hence the classical two-valued '$\sim$' is both maximally and minimally interpretable as a negation. In the three-valued case, however, only κ_3 (*i.e.* '*F*') is maximally interpretable as a negation if both *t* and *n* are designated values, as they are in **C**-logics. If, on the other hand, *t* is the only designated value and *n* is undesignated, as in **S**-logics, then κ_1 (*i.e.* '$\neg$') is the only maximally interpretable negation connective. In no case is κ_2 (*i.e.* '$\sim$') a maximally interpretable negation.

It is clear from this that to insist of a monadic connective κ that to be interpretable as a negation it must be maximally interpretable, is too strong a requirement. In fact, for most applications of three-valued sentential logics, κ_2 would be the natural choice for a negation symbol. If the third value is *n*, then κ_2 has to be chosen in order to satisfy the condition that the negation of a nonsignificant sentence is itself nonsignificant. Similarly, if the third value is *i* (indeterminate, incomplete or unknown), then again κ_2 has to be chosen in order to satisfy the condition that the negation of an indeterminate sentence is indeterminate.

These remarks, however, make appeal to intuitive requirements formulated in terms of an intended application and are not expressible as conditions which must be satisfied by the matrix of a negation connective. We can, however, introduce such a condition by noticing that one general feature which negation is usually required to satisfy is that double negation functions as an identity operator. That is, not-not-*P*, where *P* is a sentence, should be strongly equivalent to *P*, in the sense that they always have the same value. In terms of the matrix approach, this amounts to saying that κ should satisfy the condition that Δ and $\kappa\kappa\Delta$ have the same value for all possible sets of values for the variables in Δ. Such a condition cannot, however, be regarded as completely general. For the main negations of some logics, most notably intuitionistic logic (with main negation *absurdity* so-called), do not satisfy this condition.

Two points should be noted about such a condition. The first is that by itself it is insufficient to characterise κ as a negation since it is satisfied also by the identity operator. For if κ is an identity operator, so is $\kappa\kappa$ and hence Δ and $\kappa\kappa\Delta$ always have the same value. The second is that the equivalence which is being demanded is '$\simeq$' (see §2.12) not the three-valued '$\equiv$'. For in view of the matrix for '$\equiv$', we may have a thesis of the form $\Delta \equiv \Gamma$ when Δ and Γ do not take on the same values for all values of the variables. Thus, if Γ has the value *n* for all values of the variables, then by the matrix for '$\equiv$', so does $\Delta \equiv \Gamma$. Hence in a **C**-logic, where *n* is a designated value, $\Delta \equiv \Gamma$ will be a thesis whatever the values of Δ, provided Γ is always *n*. For given values of the variables, therefore, Δ and Γ may have different values even though $\vdash \Delta \equiv \Gamma$. Using '$\simeq$', however, $\Delta \simeq \Gamma$ can only be a thesis if Δ and Γ have exactly the same values.

We therefore define a normally interpretable negation as follows:

(iii) A monadic connective κ is normally interpretable as a negation if, and only if, (a) it is minimally interpretable as a negation and (b) for any wff Δ, Δ and $\kappa\kappa\Delta$ have the same value for all values of the variables in Δ (equivalently, for logics in which '$\simeq$' is definable, $\Delta \simeq \kappa\kappa\Delta$ is a thesis).

We see that the two-valued '$\sim$' is a normally interpretable negation. So, too, is the three-valued '$\sim$', κ_2; but κ_1 and κ_3 are not normally interpretable negations. This is not to say, however, that they are not interpretable as negations at all. On the contrary, for some applications κ_1 or κ_3 may be preferred to κ_2.

The main purpose of these remarks is to establish that to be an adequate negation, a monadic connective need not be a maximally interpretable negation, and in particular, that the most favoured negation in the three-valued case is never maximally interpretable. Again, it would be wrong to insist that a negation connective must be normally interpretable. In the case of 'F' (κ_3), for example, we should regard a logic as inconsistent under the intended application if both Δ and $F\Delta$ were theses even though we do not have $\Delta \simeq FF\Delta$; and the same is true of κ_1.

We therefore take it that in the definition of simple consistency, (d) above, the second condition, (ii) that κ be interpretable as a negation, is satisfied if κ is minimally interpretable as a negation.

It is now easy to show:

I No two-valued matrix logic which has one designated value and one undesignated value is simply inconsistent.

Proof: For since '$\sim$' is the only monadic connective which is interpretable as a negation, and since the matrix for '$\sim$' interchanges the two values, whenever a wff Δ has the designated value (t), $\sim\Delta$ has the undesignated value (f), and *vice versa*. Hence there is no wff Δ such that both Δ and $\sim\Delta$ take only designated values.

It is for this reason that the simple inconsistency of two-valued matrix logics is never a problem, and indeed consistency of any kind is never a problem. For as we have seen, every two-valued matrix logic which has one value undesignated and variables ranging over both values is always absolutely consistent and Post-consistent. As we shall see, these facts are important in consistency proofs for corresponding two-valued axiomatic systems.

A similar result can be shown to hold for three-valued S-logics in which t is the only designated value:

II No three-valued matrix logic which has one designated value and two undesignated values is simply inconsistent.

Proof: For each of the three possible negation connectives κ_1, κ_2, κ_3, is such that whenever a wff Δ has the designated value (t), $\kappa_i\Delta$ takes an undesignated value (in fact f always); conversely, if $\kappa_i\Delta$ takes the designated value (t), then Δ takes an undesignated value (f in the case of κ_2 and κ_3; f or n in the case of κ_1). Hence there is no wff Δ such that both Δ and $\kappa_i\Delta$ take the designated value.

Three-valued matrix logics, therefore, like two-valued matrix logics, are always absolutely, Post and simply consistent provided they have *only one* designated value, and variables which range over all the values. These conditions are satisfied by **S**-logics.

In the case of three-valued matrix logics which have more than one designated value, however, the result fails to hold. To see this, observe that if we choose κ_2 as a negation and take a wff Δ which takes the designated value n for all values of the variables in it, then by the matrix for κ_2, $\kappa_2\Delta$ always takes the designated value n and hence both Δ and $\kappa_2\Delta$ are theses. Similarly, if we choose κ_1 as a negation and take a wff Γ which always takes the designated value n, then by the matrix for κ_1, $\kappa_1\Gamma$ always takes the designated value t. Hence again, both Γ and $\kappa_1\Gamma$ are theses. Only κ_3 avoids this difficulty since it converts each of the designated values to the undesignated value. Now κ_3 is the only maximally interpretable negation when two values are designated, but as we have seen, it is unreasonable so to define simple consistency that a matrix logic is simply inconsistent when, and only when, it is κ-inconsistent with respect to a κ which is maximally interpretable as a negation. Such a course would exclude a great many applications for which minimally, or at least normally, interpretable negations are the natural choice.

In general,

III A three-valued matrix logic which has two designated values and one undesignated value and which contains,

(i) a negation connective which converts one designated value v_i to the same or another designated value,

(ii) a wff which takes the value v_i for all values of the variables in it,

is simply inconsistent with respect to that negation. And conversely, a simply inconsistent system contains a negation which has these properties.

Proof: Let Δ be the wff which takes the value v_i for all values of the variables in it. Then since v_i is designated, Δ is a thesis. Let κ be the negation connective which converts v_i to v_i or to another designated value. Then $\kappa\Delta$ is a thesis. Conversely, let Δ and $\kappa\Delta$ be theses. Then Δ takes a designated value, say v_i, and $\kappa\Delta$ takes a designated value. Hence the matrix for κ must convert the designated value v_i to the same or another designated value.

The converse shows that matrix logics are never simply inconsistent with respect to maximally interpretable negations, since these never convert designated values to designated values. In particular, therefore, **C**-logics are not simply inconsistent with respect to '*F*'.

But each **C**-logic satisfies clause (i) of III; in particular, each **C**-logic contains the classical three-valued '$\sim$' which converts the designated value *n* to *n*. What has to be shown, therefore, to establish the simple consistency of the **C**-logics, is that no wff takes the value *n* for all values of the variables in it (the same would have to be shown for a **C**-logic which contains '$\neg$'). Now this can be shown by inductive proofs for certain **C**-logics, but beyond a certain point inconsistency with respect to '$\sim$' (and '$\neg$') is inevitable. In fact any functionally complete **C**-logic must be simply inconsistent.

A matrix logic is said to be *functionally complete* if, and only if, given any matrix over the set of values, a connective τ which has this matrix can be defined in terms of the primitive connectives (this concept is discussed more fully in §5.11 ff.). Suppose, then, we have a functionally complete **C**-logic, then it is possible to define a connective τ which has the matrix,

Δ	t	n	f
$\tau\Delta$	n	n	n

i.e., for all values of the variables in Δ, $\tau\Delta$ always takes the value *n*. This particular τ is the Słupecki Ṫ-operator (Słupecki [1]). Now since $\tau\Delta$ always takes the value *n*, it is a thesis. But so, too, is $\sim\tau\Delta$ (and $\neg\tau\Delta$).

Any matrix **C**-logic which contains '$\sim$' (or '$\neg$') will therefore be inconsistent if it also contains the Słupecki 'Ṫ', and any functionally complete system will contain 'Ṫ'. Conversely, matrix **C**-logics which contain '$\sim$' (or '$\neg$') and are inconsistent will also contain the Słupecki 'Ṫ' if they have a rule of uniform substitution. For a **C**-logic will be inconsistent if it contains any connective or set of connectives τ such that, for some wff Δ, $\tau\Delta$ takes the value *n* for all values of the variables in Δ (this is what III establishes). Now let Δ be $\Delta(p, p', p'' \ldots)$. Since this has the value *n* for all values of the variables in it, so also does $\Delta(p)$, obtained by substituting '*p*' for each of the variables '*p*'', '*p*''' *etc.*; and $\Delta(p)$ can then be taken as Ṫ*p*. It should not be supposed from this, however, that a **C**-logic must be functionally complete in order to be inconsistent. All that has been shown (and what is formally established in IV below) is that a functionally complete system must be inconsistent. But this represents the upper limit, the point at which we know inconsistency is inevitable. However weaker logics are also inconsistent. Indeed as we have seen, a **C**-logic containing *only* the connectives '$\sim$' and 'Ṫ' is inconsistent. But this logic is not functionally complete; the classical '&', for example, cannot be defined.

In general, then,

IV A functionally complete three-valued matrix logic which has two designated values and one undesignated value is simply inconsistent with respect to a monadic connective κ which is minimally interpretable as a negation.

Proof: For let the designated values be 1 and 2, and the undesignated value be 3. Then by the property of functional completeness, connectives κ and τ with the following matrices are either primitive or can be defined in terms of the primitive connectives:

Γ	$\kappa\Gamma$	$\tau\Gamma$
1	3	2
2	2	2
3	1	2

Since 2 is a designated value, $\tau\Delta$ is a thesis for all wff Δ. So, too, is $\kappa\tau\Delta$ since this always takes the value 2. But κ is a minimally interpretable negation. Hence the logic is simply inconsistent.

The results III and IV can be further generalised to m-valued logics which contain more than one designated value. Thus,

V A many-valued matrix logic with values 1, 2, ..., M, which has two or more designated values and at least one undesignated value and which contains

(i) a negation connective which converts one designated value v_i to the same or another designated value,

(ii) a wff which takes the value v_i for all values of the variables in it,

is simply inconsistent with respect to that negation. And conversely, a simply inconsistent system contains a negation which has these properties.

Proof: As for III.

VI A functionally complete many-valued matrix logic with values 1, 2, ..., M, which has two or more designated values and at least one undesignated value, is simply inconsistent with respect to a monadic connective κ which is minimally interpretable as a negation.

Proof: The theorem follows as a corollary from V. For since the logic is functionally complete, connectives can be defined within it such that conditions (i) and (ii) of V are satisfied.

It is important to notice that a matrix logic may be simply inconsistent even though it is Post-consistent and absolutely consistent. And in fact functionally complete **C**-logics are both absolutely and Post-consistent. Thus, as in (a) and (b) above,

VII A [many-valued] matrix logic which has at least one undesignated value, and variables which range over the total set of values, is both Post-consistent and absolutely consistent.

Proof: Let P be such a variable; then its matrix contains at least one undesignated value. Hence it is not a thesis. Since no unrestricted variable is a thesis, the logic is Post-consistent. Since every variable is a wff, there are wff which are not theses and the logic is absolutely consistent.

Corollary: A functionally complete many-valued matrix logic which has two or more designated values, at least one undesignated value, and variables which range over the total set of values (*i.e.* a **C**-logic), is both Post-consistent and absolutely consistent.

Proof: For the proof of VII is independent of the number of designated values and the functional completeness of the logic.

This result, which together with VI entails that a many-valued matrix logic may be absolutely consistent even though simply inconsistent, is in marked contrast to axiomatic logics which satisfy certain general conditions. For,

VIII If an axiomatic logic is κ-inconsistent with respect to some monadic connective κ [which is interpretable as a negation], then it is absolutely inconsistent and Post-inconsistent provided it contains,

(i) the rule of adjunction: if Δ and Γ are theses, so is $\Delta \mathrm{x} \Gamma$;

(ii) the modus ponens rule: if Δ and $\Delta \rightarrow \Gamma$ are theses, so is Γ;

(iii) the thesis: $(\Delta \mathrm{x} \kappa \Delta) \rightarrow \Gamma$, for all Δ and Γ; where 'x' is a two-place (conjunction) connective, '$\rightarrow$' is a two-place (implication) connective, and κ is a monadic (negation) connective.

[Depending on how the logic is formulated, whether with axioms or axiom schemata, so it may be necessary to include the rule of uniform substitution as an explicit condition. This, however, is implicit in the rider on (iii) which requires the thesis to hold for all Δ and Γ.]

Proof: If the logic is simply inconsistent with respect to κ, then for some wff Δ, both Δ and $\kappa\Delta$ are theses. Hence by (i) $\Delta \mathrm{x} \kappa \Delta$ is a thesis. By (ii) and (iii), therefore, Γ is a thesis for all Γ.

Corollary: If an axiomatic logic satisfying the conditions in VIII is absolutely consistent, it is simply consistent.

Perhaps because of this well-known connexion between the simple consistency and absolute consistency of axiomatic systems, together with the fact that matrix logics which satisfy the most elementary requirements are absolutely consistent (VII), it has sometimes been taken for granted that matrix logics which satisfy the conditions of VII are simply consistent also. But this, as we have seen, is not in general so. Two-valued matrix logics, and many-valued matrix logics with only one designated value, are always simply consistent. But this is for special reasons arising from the matrix structure of the definable negation connectives, and these conditions no longer hold if there is more than one designated value. Sometimes, too, it has been supposed that a set of matrices provides a "concrete" interpretation, and no concrete interpretation can be simply inconsistent. The important point to notice, however is that because there is no immediate connexion between the simple consistency and absolute consistency of a matrix logic, so simple inconsistency is not so serious a matter in a matrix logic as it is in the usual axiomatic systems. It does not, as it does in the case of an axiomatic system, obliterate the distinction between the concept of a thesis and the concept of a wff.

Now although in the statement of VIII κ is parenthetically specified as a negation connective, this is not an essential requirement. Provided the axiomatic system is κ-inconsistent and satisfies condition (iii), it will be absolutely inconsistent whether or not κ is interpretable as a negation. Of course it might be said that it is just the thesis (iii) which does determine κ as a negation connective. But this thesis is not a feature of every sentential axiomatic logic which contains a negation (certain entailment and paradox-free implicational systems do not contain (iii) for example) and the question of how to characterise a monadic connective formally in such a way that it is interpretable as a negation becomes important. This problem, however, though it can be solved at the matrix level, admits of no easy general solution for axiomatic systems. Suppose, for example, by analogy with the matrix condition, we require of κ that to be a normally interpretable negation it should satisfy a non-intuitionistic thesis of the form $\Delta \leftrightarrow \kappa\kappa\Delta$, where '$\leftrightarrow$' is some equivalence connective which itself satisfies the replacement rule: If $g(\Delta)$ is a thesis and $\Delta \leftrightarrow \Gamma$ then $g(\Gamma)$ is a thesis. Such a rule corresponds to the matrix requirement that the equivalence relation should express the fact that formulae related by it always have the same value, for it is only if they have the same value that one is replaceable by the other without change of value. The rule is satisfied by the two-valued '$\equiv$' and by '$\simeq$', but not by the three-valued '$\equiv$'. In effect, therefore, we should be requiring that κ satisfy $\Delta \simeq \kappa\kappa\Delta$ in axiomatic systems containing '$\simeq$'. But as in the matrix case, this of itself fails to distinguish κ from an identity operator. And if we go further and require of κ that it does not satisfy, say, a thesis of the form $\Delta \simeq \kappa\Delta$, as the identity operator does, then we may beg the question of simple inconsistency. For in many simply-inconsistent logics, κ does satisfy $\Delta \simeq \kappa\Delta$. But we can hardly require of a monadic connective that to be a negation operator at all, the logic should be consistent with respect to it. This is a feature we wish the logic to have, but not a condition which can be laid down from the start.

In the case of axiomatic systems which have a matrix interpretation, however, the problem is much simpler since we need not seek formal conditions but simply take over the matrix conditions. Most of the axiomatic systems which we consider in Chapter 6 do have a matrix interpretation, indeed they are axiomatic formulations of the matrix logics developed in this chapter, and we shall therefore limit our attention to these. The relations between a matrix logic and an axiomatic formulation of it can then be characterised as follows:

Let **LM** be a sentential matrix logic and **LA** a sentential axiomatic logic such that:

(i) **LM** and **LA** have the same variables, the same connectives primitive and defined, and the same formation rules. In short, **LM** and **LA** have the same morphology.

(ii) In addition to its morphology, **LM** is characterised by a finite set of m values, 1, 2, 3, ..., S, S + 1, ..., M, described as the range of the variables, the first s of which are designated, the remaining $m - s$ undesignated. At least one value must be designated and at least one value must be undesignated. Hence there must be at least two values (i.e. $m \geqslant 2$), and since at least one must be designated, $s \geqslant 1$. In particular we take it that 1 is always designated and M is always undesignated.

(iii) The primitive connectives of **LM** are matrix accessible: which is to say that if Δ is a wff and κ a primitive monadic connective, the value of $\kappa\Delta$ is wholly determined by a defining matrix for κ for all possible sets of values for the variables in Δ; similarly, if Δ and Γ are wffs and Φ is a primitive two-place connective, the value of $\Phi(\Delta, \Gamma)$ is wholly determined by a defining matrix for Φ for all possible sets of values for the variables in both Δ and Γ; similarly for m-place connectives. (We limit attention to one- and two-place connectives, but this is not essential.)

(iv) A wff containing a defined connective has, for any given set of values for the variables, the same value as the wff which can be obtained from it by eliminating the defined connective and writing the whole in primitive notation. Hence all defined connectives of **LM** are matrix accessible. (This is not of course to say that for any given matrix a connective can be defined which has that matrix; *i.e.* (iii) and (iv) do not imply the functional completeness of **LM**, though a given **LM** may in fact be functionally complete.)

(v) A wff Δ of **LM** is a thesis or theorem of **LM**, briefly $\vdash_{\mathbf{LM}}\Delta$, if, and only if, it takes only designated values for all possible sets of values for the variables in Δ.

(vi) In addition to its morphology, **LA** is characterised by a set of postulates (axioms, or axiom schemata, and rules of inference).

(vii) A wff Δ of **LA** is a thesis or theorem of **LA**, briefly $\vdash_{\mathbf{LA}}\Delta$, if, and only if, Δ is either an axiom or follows by rule of inference from an axiom or a previously derived formula. (Here the description could be made more rigorous, but for present purposes we take the notion of derivation as intuitively understood.)

(viii) **LA** is complete with respect of **LM** if, and only if, every thesis of **LM** is a thesis of **LA**. (In the familiar two-valued case for classical sentential logic: if, and only if, every matrix tautology is a theorem of the axiomatic system.)

(ix) **LM** is complete with respect to **LA** [alternatively, **LA** is plausible with respect to **LM**], if, and only if, every thesis of **LA** is a thesis of **LM**. (In two-valued sentential logic: if, and only if, every theorem of the axiomatic logic is a matrix tautology).

(x) **LM** and **LA** are deductively equivalent if, and only if, they are complete with respect to each other; *i.e.* if, and only if, every thesis of one is a thesis of the other, and hence every non-thesis of one is a non-thesis of the other.

(xi) If **LM** and **LA** are deductively equivalent, **LA** is a complete axiomatic formulation of **LM** and **LM** is a complete matrix interpretation of **LA**.

(xii) A monadic connective κ which is matrix-wise interpretable as a negation (either minimally, normally or maximally) with respect to **LM** is interpretable as a negation (minimally, normally or maximally) with respect to **LA**.

IX If a matrix logic which contains at least one undesignated value and variables which range over the total set of values is deductively equivalent to an axiomatic system which contains the rules of adjunction, modus ponens, and the thesis $\Delta \text{ x } \kappa\Delta \rightarrow \Gamma$ for all Δ and Γ (as in VIII), then

(i) the axiomatic system is simply consistent

(ii) the matrix logic is simply consistent.

Proof: (i) Since the matrix logic satisfies the conditions of VII, it is absolutely consistent. Since the axiomatic system is deductively equivalent to the matrix logic, every non-thesis of the matrix logic is a non-thesis of the axiomatic system. Hence the axiomatic system is absolutely consistent. But it satisfies the conditions of VIII. So it is simply consistent.

(ii) Since the axiomatic system is simply consistent, for every wff Δ, either Δ or $\kappa\Delta$ is a non-thesis. But by the deductive equivalence, every thesis (non-thesis) of the axiomatic system is a thesis (non-thesis) of the matrix logic. So in the matrix logic, for every wff Δ, either Δ or $\kappa\Delta$ is a non-thesis according as Δ or $\kappa\Delta$ is a non-thesis of the axiomatic system. That is, the matrix logic is simply consistent.

Corollary 1. If a matrix logic is simply inconsistent, there is no simply consistent axiomatic system satisfying the conditions of VIII which is deductively equivalent to it.

Corollary 2. Any simply consistent axiomatic formulation (satisfying the conditions of VIII) of a simply inconsistent matrix logic is incomplete with respect to the matrix basis.

Proof: For there will be a valid formula in terms of the matrices (one of Δ or $\kappa\Delta$ for some Δ) which is not provable in the axiomatic system.

These last results have a special importance for us since all the matrix logics which we consider satisfy the conditions of VII and most of the axiomatic formulations of them in Chapter 6 satisfy the conditions of VIII (in weak **C**-logics, the modus ponens rule is modified but not in such a way that would affect the proof of VIII). Hence in cases where deductively equivalent axiomatic formulations of the matrix logics can be given, the simple consistency of both the axiomatic system and the matrix logic follow immediately by IX. In the case of matrix logics which are simply inconsistent, however, the corollaries of IX entail that no axiomatic system which is simply consistent and satisfies the conditions of VIII can be complete with respect to them. It is however possible to set up complete axiomatic formulations of matrix **C**-logics which are simply inconsistent but absolutely consistent by ensuring that one or other of the conditions in VIII fails to be satisfied. If, for example, modus ponens is modified appropriately, then a simply inconsistent axiomatic system which is not absolutely inconsistent, and is deductively equivalent to a matrix **C**-logic which has these same features, can be set up.

The general position with regard to **C**-logics is as follows:

We have already mentioned that Halldén adopts a modified modus ponens rule (MMP) for his axiomatic **C**-logic. This rule is expressed in terms of the classical significance connective '$\supset$' and is designed to overcome the difficulty that an unqualified modus ponens rule will not always lead from true premisses to true conclusions. For since both *t* and *n* are designated, if the rule were expressed in the form:

if Γ and $\Gamma \supset \Delta$ are both designated, then Δ is designated,

it would be satisfied when Γ is *t* and Δ is *n*, for then $\Gamma \supset \Delta$ is *n*. The modification which is required, however, and a variant of which we adopt for our initial **C**-systems, is not necessary from the point of view of consistency. For the weak matrix systems $\mathbf{C}_0$, $\mathbf{C}_1$ (which is Hallden's system **C**) and $\mathbf{C}_2$(§5.8–§5.10) are in any case simply consistent, and axiomatic versions of them (§6.1–§6.3) are also simply consistent.

The next strongest matrix system $\mathbf{C}_3$ (§5.11) is, like the earlier ones, simply consistent, so that again no special difficulty arises with its axiomatisation (§6.4). It does, however, contain new connectives which makes it necessary to introduce two modus ponens rules in the axiomatic version: the existing

MMP restricted in application to formulae which do not contain the new connectives and a special modus ponens (T – ed MP) for those formulae which do. Each of these detachment rules is formulated in terms of '$\supset$'.

The strongest matrix systems $\mathbf{C}_5$ and $\mathbf{C}_6$ (§5.13 and §5.14) are deductively equivalent and functionally complete; and being functionally complete, they are simply inconsistent with respect to both '$\sim$' and '$\neg$'. It happens, however, that the two-tier modus ponens rule already adopted for the axiomatic version of $\mathbf{C}_3$ is sufficient to stop the simple inconsistency of the axiomatic versions of $\mathbf{C}_5$ and $\mathbf{C}_6$ (§6.6 and §6.7) from spreading into absolute inconsistency. That is to say, the axiomatic $\mathbf{C}_5$ and $\mathbf{C}_6$ are, respectively, deductively equivalent to the simply inconsistent but absolutely and Post-consistent matrix systems $\mathbf{C}_5$ and $\mathbf{C}_6$. This means that the two-tier modus ponens rule does not satisfy the conditions of VIII; in fact it fails to satisfy the condition that $\vdash\Delta$ and $\vdash\sim\Delta$ (or $\vdash\Delta$ and $\vdash\neg\Delta$) yield $\vdash\Gamma$, for all Γ, though $(\Delta \,\&\! \sim\Delta)\supset\Gamma$ and $(\Delta\,\&\neg\Delta)\supset\Gamma$ are both theses.

The axiomatic versions of $\mathbf{C}_5$ and $\mathbf{C}_6$ are, however, cumbersome; and as an alternative we present a system $\mathbf{D}_6$ (§6.7) which is deductively equivalent to both but which has only one modus ponens rule with no restrictions on it since it is formulated in terms of an implication connective '$\rightharpoondown$' different from the classical '$\supset$'. Here, though we have unrestricted modus ponens for '$\rightharpoondown$', we do not have either $(\Delta\,\&\!\sim\Delta)\rightharpoondown\Gamma$ or $(\Delta\,\&\neg\Delta)\rightharpoondown\Gamma$ as theses. Hence again, this system fails to satisfy the conditions of VIII and is both simply inconsistent yet absolutely consistent.

The fact that such systems can be developed is not really surprising. It is only in standard classical systems which feature laws such as $(\Delta\,\&\!\sim\Delta)\supset\Gamma$, together with an unqualified modus ponens rule, that simple inconsistency spreads to absolute and Post-inconsistency. But there is in general no reason why this should be so, and in fact it is possible to construct near-classical systems in which a simple inconsistency does not spread. As an example, consider the following connexive implicational system $\mathbf{I}'$ with axioms,

I′(1) $(p\rightarrow q)\rightarrow((q\rightarrow r)\rightarrow(p\rightarrow r))$
I′(2) $(p\,\&\,q)\rightarrow(q\,\&\,p)$
I′(3) $(p\,\&\,q)\rightarrow p$
I′(4) $(p\rightarrow q)\rightarrow\sim(p\rightarrow\sim q)$

and uniform substitution and unrestricted modus ponens for '$\rightarrow$' as rules. Then $\mathbf{I}'$ is simply inconsistent. For,

I′(5) $(p\,\&\!\sim p)\rightarrow(\sim p\,\&\,p)$: from I′(2) by subst.
I′(6) $(\sim p\,\&\,p)\rightarrow\sim p$: from I′(3) by subst.
I′(7) $(p\,\&\!\sim p)\rightarrow\sim p$: from I′(5), I′(6) and I′(1) by modus ponens
I′(8) $(p\,\&\!\sim p)\rightarrow p$: from I′(3) by subst.
I′(9) $\sim((p\,\&\!\sim p)\rightarrow\sim p)$: from I′(4) and I′(8) by modus ponens.

Since I′(7) and I′(9) are both theorems, **I′** is simply inconsistent with respect to '$\sim$'. Nevertheless **I′** is absolutely consistent. For by mapping '$\sim$' into the identity operator, **I′** is mapped into a subsystem of classical two-valued sentential logic. Since this logic is absolutely consistent, so is **I′**.

An even more trivial system which satisfies these conditions is **I″** with axioms:

I″(1) $(p \rightarrow p)$
I″(2) $\sim(p \rightarrow p)$

and uniform substitution as the only rule. That **I″** is simply inconsistent is obvious; that it is absolutely consistent follows as for **I′**, *i.e.*, in effect by using the matrices,

p	$\sim p$
t	t
f	f

$\rightarrow$	t	f
t	t	f
f	t	t

I′ is rendered absolutely inconsistent by the addition of the spread law

$$p \rightarrow (\sim p \rightarrow q) \quad \text{(or adjunction and: } (p \,\&\! \sim p) \rightarrow q)$$

but **I″** only by the addition of both this and modus ponens.

Neither simple inconsistency nor absolute inconsistency need be wholly undesirable features. Given a standard definition of well-formed formulae (wff), simple inconsistency which does not spread to an absolute inconsistency which renders all wff as theorems, is certainly not a serious formal defect, though it may be a serious matter at the level of interpretation. Again, given a non-standard definition of wff, absolute inconsistency may be a trivial matter. If, for example, we redefine the concept of a wff for the standard two-valued sentential logic in such a way that a formula is only well-formed in case it satisfies the usual formation rules *and* is a tautology, then the axioms are well-formed, the rules of inference preserve this property, but there is no wff which is not a theorem.

The interesting feature of axiomatic systems like $\mathbf{C}_5$, $\mathbf{C}_6$, $\mathbf{D}_6$, **I′** and **I″** which have the usual definition of wff and for which a simple inconsistency does not spread to an absolute inconsistency is that inconsistencies appear as localised and isolated singularity points, rather like singularities in the complex number field. This gives point to Wittgenstein's remark "'Contradiction destroys the calculus'—what gives it this special position? With a little imagination, I believe, it can certainly be demolished" (Wittgenstein [3], p. 170). There is in fact no special reason why a simple inconsistency should in general destroy a calculus; but of course, it may destroy its application.

The general results established here simplify the metatheory of the particular logics we consider. For we have shown,

(1) that every two-valued matrix logic over $\{t,f\}$ is absolutely and Post-consistent, and simply consistent with respect to '$\sim$';
(2) that every three-valued matrix logic over $\{t,f,n\}$ is absolutely and Post-consistent;
(3) that every matrix **S**-logic is simply consistent with respect to '$\sim$', '$\neg$' and '*F*';
(4) that every matrix **C**-logic is simply consistent with respect to '*F*';
(5) that every matrix **C**-logic is simply consistent with respect to both '$\sim$' and '$\neg$' *provided* it contains no wff which takes the value *n* for all values of the variables in it (hence the consistency proofs for these systems consists only in establishing this proviso);
(6) that every axiomatic system satisfying the conditions of VIII which is deductively equivalent to a matrix system satisfying the conditions of VII, is simply consistent with respect to all monadic connectives interpretable as negations, and absolutely and Post-consistent;
(7) that any complete axiomatic formulation of a simply-inconsistent matrix system must itself be simply inconsistent but absolutely and Post-consistent and cannot, therefore, satisfy the conditions of VIII.

5.7 Sorted Variables

For sentential **C**-logics, it is enough to have just one sort of variable, namely *S*-unrestricted variables,

$$p, q, p', q', p'', \ldots$$

a denumerably infinite set recursively defined by the following rules:

V is an *S*-unrestricted variable if, and only if,
(i) *V* is one of *p*, *q*, or
(ii) for some *W*, *V* is *W*′ and *W* is an *S*-unrestricted variable.

As before, these variables initially range over finitely many or infinitely many constant sentence-tokens. Since, however, under present assumptions, the context is fixed for any given application of the logic, tokens of the same type have the same values, *t*, *f* or *n* (neglecting incomplete sentences) and we may therefore regard the primary range of the variables as constituted by type-sentences, and the derived range (§2.15, §2.16) by the value-set $\{t,f,n\}$. If we call *t*, *f* and *n* the significance values of a sentence, then we may define the derived values of *V* as the significance values of the primary values of *V*.

In the case of *weak* **S**-logics, as opposed to **C**-logics, it proves best, as we suggested in the previous section, to introduce two sorts of sentential variables, namely the *S*-unrestricted variables defined above together with the *S*-restricted variables,

$$r, s, r', s', r'', \ldots$$

again a denumerably infinite set recursively specified by:

V is an *S*-restricted variable if, and only if,
(i) *V* is either *r* or *s*, or
(ii) for some *W*, *V* is *W'* and *W* is an *S*-restricted variable.

S-restricted variables, under the assumptions as above, have as their primary range significant sentences, and as their derived range the value-set $\{t,f\}$.

In general, therefore, a variable is *S*-unrestricted if its primary range consists of complete sentences which are either significant or nonsignificant; and a variable is *S*-restricted if its primary range consists of significant sentences only.

In what follows, to avoid an over-prolific use of quotation functions in the metatheory, we shall continue to use metavariables:

$R, R_1, R_2, \ldots$ or $R, R', R'', \ldots$ to range over *S*-restricted variables;
$P, P_1, P_2, \ldots$ or $P, P', P'', \ldots$ and sometimes $Q, Q_1, Q_2, \ldots$ or $Q, Q', Q'', \ldots$ } to range over both sorts of variables.

The two sorts of variables are only essential in weak S-logics which contain insufficient apparatus for imposing restrictions of the required sort. Thus, for example, if we wish to assert a thesis over significant sentences only, say $R \supset R$, we can do so by affirming,

$$r \supset r;$$

but if we had the operator '$\rightharpoonup$' ('provided that ... then ...'), we could instead affirm,

$$Sp \rightharpoonup (p \supset p).$$

Such an operator cannot be satisfactorily formulated using, *e.g.* '$\supset$', or any other classical connective. For if we write, say,

$$Sp \supset (p \supset p),$$

which is interpreted as: If *P* is significant then $P \supset P$; the whole formula becomes nonsignificant when the variable takes nonsignificant values. In stronger S-logics, '$\rightharpoonup$' can be introduced, however, and the need for two sorts of variables disappears (§5.11). Even so, there are some advantages to be gained by keeping the two sorts.

Two-sorted sentential logics can be generalised to *m*-sorted logics in much the way that three-valued matrix systems may be generalised to *m*-valued, or even infinite-valued, matrix logics. There is in fact room for a theory of *m*-sorted sentential logics which provides general conditions under which the number of sorts may be reduced, as the two-sorted system may be reduced to a one-sorted system containing '$\rightharpoonup$'.

5.8 Stage 0: the Logics $\mathbf{S}_0$ and $\mathbf{C}_0$

We have now reached the point where we can present matrix significance logics containing just classical connectives and no others. We call these 0-logics. From these we progress by successive stages through $\mathbf{S}_1$ and $\mathbf{C}_1$ (which contain classical connectives together with '*S*'), $\mathbf{S}_2$ and $\mathbf{C}_2$ (which contain classical connectives together with '*S*' and '*T*'), ..., to the functionally complete systems $\mathbf{S}_6$ and $\mathbf{C}_6$. At each stage it is possible to formulate various versions of the semantic (*i.e.* matrix) system under consideration. This arises because the system may be more or less formalised. For example, in stating a valuation rule we may say,

For any *S*-unrestricted variable *qu*(*p*), either the value of *qu*(*p*) is *t*, or the value of *qu*(*p*) is *f*, or the value of *qu*(*p*) is *n*:

or we may represent it more formally as,

$$(Uqu(p))(qu(p) \in SUV \rightarrow v(p) = t \ \vee \ v(p) = f \ \vee \ v(p) = n),$$

where '*SUV*' (*S*-unrestricted variable) is a metapredicate, '(*Uqu*(*p*))' a restricted subject quantifier, '$\rightarrow$' an implication, '*v*' a valuation function which absorbs '*qu*', and '$=$' identity. Such variants, though the same in content, differ greatly in that the former depends on an intuitive understanding of natural language while the latter requires a much richer formal basis for its development. In most cases we present only one such variant, usually the intuitively based version, but occasionally we introduce more formal versions. We distinguish these by labels of the sort, $\mathbf{S}_2-1$, $\mathbf{S}_2-2$, *etc.*, indicating variants of the system $\mathbf{S}_2$ which have the same set of tautologies but differ in their degree of formalisation. Similarly, in the next chapter, where we present axiomatic formulations of the various systems, it is possible to develop variants, labelled *e.g.* ${}_1\mathbf{S}_2$, ${}_2\mathbf{S}_2$, *etc.*, which are deductively equivalent but may differ in the axioms and/or rules chosen as primitive.

The morphologies of the semantic systems coincide with that of the related axiomatic system. By the morphology of a system we mean that part of the primitive frame or specification which comprises the vocabulary and formation rules. The morphology of $\mathbf{C}_0$ is the same as that of classical two-valued sentential logic; that of $\mathbf{S}_0$ is almost the same, and does not differ in form provided the formation rules are stated using metavariables. We first specify a matrix system for $\mathbf{C}_0$ since it is simpler.

The Logic $\mathbf{C}_0$

(a) *Morphology*

We take just '$\sim$' and '$\vee$' as primitive.

Improper primitive symbols: $\sim$, $\vee$, (,).
Variables: p, q
The Symbol rule: If P is a variable, then P' is a variable.

Formation rules:
(1) A variable standing alone is a wff.
(2) If Γ and Δ are wff then $(\sim\Gamma)$ and $(\Gamma\vee\Delta)$ are wff.
(3) Defined equivalents of wff are wff.
(4) These are the only wff.

We often drop brackets using standard conventions and sometimes use dots instead.

$$\text{Definitions: } \Delta\supset\Gamma =_{\text{Df}} \sim\Delta\vee\Gamma$$
$$\Delta\&\Gamma =_{\text{Df}} \sim(\sim\Delta\vee\sim\Gamma)$$
$$\Delta\equiv\Gamma =_{\text{Df}} (\Delta\supset\Gamma)\ \&\ (\Gamma\supset\Delta).$$

Metavariables may be eliminated in favour of quotation functions. The symbol rule is then:

If $qu(p)$ is a variable, then $qu(p')$ is a variable;

and the second formation rule is re-written as:

(2a) If $qu(p)$ and $qu(p')$ are wff, then $qu((\sim p))$ and $qu((p\vee p'))$ are wff.

Similarly for the definitions.

The *variables* of our sentential systems are not proper variables, *i.e.* not arguments. Though they *can* be construed as holding subject-places, except in the isolated case of a sentential variable standing alone, they are not open to quantification. At the sentential level these variables in fact function as no more than parameters. However the morphology can be reshaped so that the variables do assume a full argument role (see §7.12 and the statement logic of chapter 2), but then their values have to be alternatively interpreted as assertions.

To specify $\mathbf{C}_0$ semantic (matrix) systems we must provide, as well as the morphology, valuation rules. There are various ways in which this can be done, more set-theoretic or less set-theoretic, more formal or less formal, depending on the degree to which the underlying predicate logic is formalised. But any formulation must be implicitly set- or predicate-theoretic since what we have to define are effectively calculable functions over classes of values. The matrices given simply specify these valuation functions. Our presentation of the valuation rules, though in the style of many modern semantic systems, is not highly formalised and relies heavily on extra-systematic terminology.

(b) *Valuation Rules*

(1) t is a value, f is a value, n is a value; these values are distinct and are the only values.
(2) A sentential variable takes as a value either t, f or n.

We symbolise 'The sentential variable $qu(p)$ takes as value the value j' by '$v(p,j)$' or '$v(p)=j$' ('the value of $qu(p)$ is j'), where j is one of t, f or n. Then we may re-formulate (2) as:

(2a) For any sentential variable $qu(p)$,
$v(p)=t$ or $v(p)=f$ or $v(p)=n$, where 'or' is taken in its exclusive sense.

(3) For any wff Δ,
if $v(\Delta)=t$ then $v(\sim\Delta)=f$; if $v(\Delta)=f$, then $v(\sim\Delta)=t$;
and if $v(\Delta)=n$, then $v(\sim\Delta)=n$

where '$v(\Delta)=j$' represents 'the wff Δ takes the value j'. Thus Rule (3) is summarised by the classical significance matrix for '$\sim$', *viz.*:

Δ	t	f	n
$\sim\Delta$	f	t	n

This extension of notation from sentential variables to all wff could be done recursively over rules for wff, and in this way further metavariables could be eliminated in favour of explicit quotation devices.

(4) For any wff Γ and Δ,
if $v(\Gamma)=t$ and $v(\Delta)=t$ or $v(\Delta)=f$, then $v(\Gamma\vee\Delta)=t$;
if $v(\Delta)=t$ and $v(\Gamma)=f$, then $v(\Gamma\vee\Delta)=t$;
if $v(\Delta)=f$ and $v(\Gamma)=f$, then $v(\Gamma\vee\Delta)=f$;
if $v(\Delta)=n$ or $v(\Gamma)=n$ then $v(\Gamma\vee\Delta)=n$;
where 'or' is taken in the inclusive sense.

Rule (4) is thus summarised by the matrix:

	$\vee$	t	f	n	$v(\Delta)$
	t	t	t	n	
$v(\Gamma)$	f	t	f	n	
	n	n	n	n	

where the body of the matrix gives $v(\Gamma\vee\Delta)$.

The valuation rules for '$\supset$', '&' and '$\equiv$' follow by the definitions and determine the matrices as the classical significance matrices.

These rules only provide valuations of wff over the derived range.

Carnap [4] typically breaks up valuation rules into rules of designation and rules of truth. So, for example, rule (1) would appear as a rule listing designation values, and rule (2) would be represented as a rule of truth, connecting truth-values with designation values, as follows:

(2b) A sentential variable takes the truth-value t if, and only if, it designates designation value t, *etc.*

Rules (3) and (4) would be treated as further rules of truth. But the designation-truth terminology has a serious disadvantage from our point of view.

For even if 'designation' is taken in its weakest sense to mean 's-aboutness', to say that a sentence is s-about a truth-value, rather in the way that a proper name 'Scott' is s-about Scott, is to treat sentences, not quoted sentences, as if they were subject terms. But the connexion between true sentences and being true, or the truth-value truth, is not at all like the connexion between a proper name and the individual which it is s-about. Thus, 'Scott', 'The author of *Waverley*' and 'x_0' are s-about the same individual, Scott, if x_0 is extensionally identical to the author of *Waverley*. But the notion of extensional identity in terms of which s-aboutness is explicated, and which is tied to stronger notions like designation by way of s-aboutness (since $qu(x)$ designates y if, and only if, x is extensionally identical with y and it is possible that y exists (§3.6)) is not applicable to the relation between sentences and their values. 'There is snow at the North Pole' and 'London is the capital of England' are not in any sense extensionally identical sentences, though they have the same derived value in the standard context, *viz. t*. This is not of course to deny that a sentence may be said to *have* an s-aboutness in the sense defined in §3.6, namely that it is s-about whatever its subject terms are s-about, or, if it is quantified, it is s-about the total set of items which constitutes the universe of discourse. But this is a quite different concept from that of treating the whole sentence as if it were itself a name which names (*is* s-about) whatever truth-value it happens to have *qua* sentence.

These difficulties extend to significance logics since we should be led to the position of having to say that, in a suitable context, 'The number 7 likes soup' and 'The picture admires its painter' are both s-about, or designate nonsense; and this again would commit us to saying that the sentences are in some sense extensionally identical. Less serious, the Carnapian phrase 'rules of truth' is no longer apposite for prescriptions including nonsignificance since nonsignificance assignments preclude truth assignments.

The valuation rules could easily be more fully formalised. For example, quantifiers and sentential connectives could be used to formalise rule (2). If the semantic system were more fully formalised, however, it would outstrip the axiomatic system whose interpretation it is later intended to provide. In fact, since it would contain predicates and quantifiers, it would no longer be simply a sentential system.

(c) *Tautologies*

Given the valuation rules, the notion of a semantical thesis of $\mathbf{C}_0$, or $\mathbf{C}_0$-tautologousness, may be defined as follows:

A wff Δ of $\mathbf{C}_0$ is a $\mathbf{C}_0$-tautology if, and only if, $v(\Delta) = t$ or $v(\Delta) = n$ for all values of the variables of Δ.

Thus both t and n are taken as designated values, and in consequence the classical notion of a tautology is stretched, as is common in many-valued logics, to embrace all wff which take only thesis-determining values irrespective of whether all thesis-determining values are truth-values. Hence we may say, for example, that $qu(p \vee \sim p)$ is a tautology even though it has the value n when $v(p) = n$.

(d) *Valuation Tables*

A valuation table for a wff Δ of $\mathbf{C}_0$ is a linear array of 3^m lines, where m is the number of distinct variables in the formula, one line for each distinct logically possible combination of values for the variables of Δ, such that, in each line, the resulting values of all sub-wff of Δ, as determined by valuation rules for the variable values of that line, are written.

A valuation table is thus a simple many-valued extension of the familiar classical truth-tables for the two-valued sentential logic.

(e) *Formal Theorems*

We now specify that a wff Δ is a theorem (or thesis) of $\mathbf{C}_0$ if, and only if, it is a $\mathbf{C}_0$-tautology. This specification formally determines t and n as designated (thesis-determining) values.

Among the theorems are the axioms of classical two-valued logic, *e.g.* the set

$$p \supset (q \supset p);\ (p \supset (q \supset p')) \supset ((p \supset q) \supset (p \supset p'));\ (p \supset q) \supset (\sim q \supset \sim p).$$

These may be established by means of valuation tables.

(f) *Metatheorems*

Theorem (1): If Δ is a $\mathbf{C}_0$-tautology, so is $S^P_\Gamma\Delta|$, where $S^P_\Gamma\Delta|$ is the formula which results from Δ when the wff Γ is substituted for each occurrence of a variable P contained in Δ.

The proof is by induction, and we shall not give it in full since it simply follows the standard proof for classical two-valued sentential logic. It depends essentially on noting that Γ takes only a (perhaps improper) subclass of the values of P so that, since Δ takes thesis-determining values always for P, it does the same for Γ in place of P.

Theorem (2): If Δ and $\Delta \supset \Gamma$ are $\mathbf{C}_0$-tautologies, so is Γ.

To prove this result the following lemma is required:

Lemma: A wff Δ takes the value n when, and only when, at least one variable of Δ takes the value n.

This is proved, using the valuation rules, by induction over the number of connectives in Δ. Again we omit the full proof since it is obvious from the nature of the classical significance matrices defined by the valuation rules.

To prove (2), suppose that Δ and $\sim \Delta \vee \Gamma$ take thesis-determining values, but that Γ takes the value f for some assignment of values to the variables. By the valuation rules, this cannot happen if Δ takes the value t. Thus it remains to consider the case where Δ, and thus $\sim\Delta \vee \Gamma$, takes the value n. Then by the lemma, some variables, say just $P_1, P_2, \ldots, P_m$ of Δ take the value n and these variables do not occur in Γ. Since Δ is a $\mathbf{C}_0$-tautology, Δ will take the value t or n when variables $P_1, \ldots, P_m$ are assigned values t or f instead of n. Suppose, then, that these variables are assigned value t but that all other variables in $\sim\Delta \vee \Gamma$ are assigned the same values as under the initial hypothesis. This can happen since $P_1, \ldots, P_m$ do not occur in Γ and no other variables than $P_1, \ldots, P_m$ in Δ then take the value n. Applying the lemma again, Δ must take the value t since Δ is a $\mathbf{C}_0$-tautology, and Γ must take the value f under the hypothesis. Hence $\sim\Delta \vee \Gamma$ takes the value f, by the valuation rules, contradicting the hypothesis that $\sim\Delta \vee \Gamma$ is a $\mathbf{C}_0$-tautology. In other words, if there is a line of the joint valuation tables for Δ and $\sim\Delta \vee \Gamma$ under which both take the value n while Γ takes the value f, there is another line of this joint valuation table which shows that not both Δ and $\sim\Delta \vee \Gamma$ are $\mathbf{C}_0$-tautologies.

It now follows by induction over proofs in classical sentential logic:

Theorem (3): Every theorem of classical two-valued sentential logic (**RSL**) is a $\mathbf{C}_0$-tautology.

For we have shown that the axioms of **RSL** are $\mathbf{C}_0$-tautologies and that the rules of proof of **RSL** preserve this feature.

Since every classical two-valued tautology is a theorem of **RSL**, it follows that every classical two-valued tautology is a $\mathbf{C}_0$-tautology. This establishes one arm of the next theorem.

Theorem (4): Δ is a $\mathbf{C}_0$-tautology if, and only if, Δ is a classical two-valued tautology.

For the valuation rules for $\mathbf{C}_0$-tautologies include those for **RSL**-tautologies, and the classical valuation for any $\mathbf{C}_0$-tautology Δ may be effectively obtained from the valuation table for Δ by deleting all lines on which a variable of Δ takes the value n. Thus if Δ is a $\mathbf{C}_0$-tautology, by the valuation table it is also a classical tautology. This establishes the other arm.

We see, therefore, that though $\mathbf{C}_0$ is three-valued, its set of theses coincides exactly with those of **RSL**. And the main reason for this is that by taking both t and n as designated, we effectively reduce the three values to two, *viz.*, f and $\bar{f}$ (not-f), where $\bar{f}$ is either t or n. By writing $\bar{f}$ everywhere for t or n, and deleting repeated lines in the valuation tables, we reduce $\mathbf{C}_0$ to the classical **RSL** with values $\bar{f}$ and f where $\bar{f}$ is designated. This simple reduction is not, however, possible for richer **C**-logics.

Conversely, by reinterpretation as **C**-logics classical axiomatic logics can provide a (seriously inadequate) formalisation of nonsignificance or non-truth-valuedness (see for example van Fraassen [1]).

C-*logics Consistency Theorem: Part* 0

Matrix $\mathbf{C}_0$-logics are simply consistent with respect to all primitive and definable monadic connectives κ which are minimally interpretable as negations.

Proof: We show that there is no wff Δ of $\mathbf{C}_0$ which takes the value n for all values of the variables in it, by showing that Δ never takes the value n when all its variables are assigned value t.

We establish this by induction over the number m of connectives in Δ. If $m=0$, then Δ is P, a variable standing alone, to which t is assigned.

For $m>0$, we show that if the theorem is true for all wff which contain m or fewer connectives (the induction hypothesis) then it holds for all wff which contain $m+1$ connectives. By induction, the theorem then holds for all m. Since any wff can be written in primitive notation, it is enough to show that the theorem holds for '$\sim$' and '$\vee$'. There are thus two cases to consider when m 0:

(i) Δ is of the form $\sim\Delta_1$, where Δ_1 contains m connectives. By the induction hypothesis, Δ_1 does not take the value n for assignment t to the variables in it. Therefore, by the matrix for '$\sim$', ncither does Δ.

(ii) Δ is of the form $\Delta_1 \vee \Delta_2$, where the total number of connectives in Δ_1 and Δ_2 taken together is m. By the induction hypothesis, neither Δ_1 nor Δ_2 take the value n for the given assignment to the variables in them. Hence by the matrix for '$\vee$', neither does Δ.

The result follows by III of §5.6.

Now it is easily shown that of the three possible negation connectives '$\neg$', '$\sim$' and 'F', only '$\sim$' can occur in $\mathbf{C}_0$-logics since neither '$\neg$' nor 'F' are definable, *i.e.* there is no wff $\Delta(P)$ of $\mathbf{C}_0$-logics such that $\Delta(P)$ has the same matrix as $\neg P$ or FP. For when P has the value n, both $\neg P$ and FP have the value $\bar{n}$ (t or f), but $\Delta(P)$ can never lead out of the value n to the value $\bar{n}$. This is again proved by induction over the number m of connectives in Δ(P).

If $m=0$, then Δ(P) is P which has value n when P has value n.

If m>0, there are two cases:

(a) Δ(P) is of the form $\sim\Delta_1(P)$. By the induction hypothesis $\Delta_1(P)$ never has value $\bar{n}$ when P has value n. Therefore by the matrix for '$\sim$', neither does $\Delta(P)$;

(b) $\Delta(P)$ is of the form $\Delta_1(P) \vee \Delta_2(P)$. By the induction hypothesis, $\Delta_1(P)$ and $\Delta_2(P)$ never have value $\bar{n}$ when P has value n. Therefore by the matrix for '$\vee$', neither does $\Delta(P)$.

It follows that the part 0 consistency theorem is equivalent to the assertion that $\mathbf{C}_0$-logics are simply consistent with respect to '$\sim$'.

The Logic $\mathbf{S_0}$

(a) *Morphology*

Again we take just '$\sim$' and '$\vee$' as primitive.

Improper primitive symbols: $\sim$, $\vee$, (,).

Variables of two sorts:

S-unrestricted variables: p, q

S-restricted variables: r, s.

The symbol rule: If P is a variable then P' is a variable of the same sort (*S*-restricted or *S*-unrestricted).

Formation rules:

(1) A variable standing alone is a wff.

(2) If Γ and Δ are wff, so are $(\sim\Gamma)$ and $(\Gamma \vee \Delta)$.

(3) Defined equivalents of wff are wff.

(4) These are the only wff.

Definitions:

$\Delta \supset \Gamma \quad =_{\mathrm{Df}} \; \sim\Delta \vee \Gamma$

$\Delta \,\&\, \Gamma \quad =_{\mathrm{Df}} \; \sim(\sim\Delta \vee \sim\Gamma)$

$\Delta \equiv \Gamma \quad =_{\mathrm{Df}} \; (\Delta \supset \Gamma) \,\&\, (\Gamma \supset \Delta)$.

Again metavariables may be eliminated in favour of quotation functions. Then the symbol rule is re-stated:

If *qu*(p) is an *S*-unrestricted variable, then *qu*(p') is a variable of the same sort;

If *qu*(r) is an *S*-restricted variable, then *qu*(r') is a variable of the same sort.

The formation rules are re-phrased:

(1a) An *S*-unrestricted variable standing alone is a wff.

(1b) An *S*-restricted variable standing alone is a wff.

(2) If *qu*(p) and *qu*(p') are wff, then *qu*($(\sim p)$) and *qu*($(p \vee p')$) are wff.

Here the variable in '*qu*(p)' may take as values both restricted and unrestricted variables. The symbol rule could be written in this way also, given this condition. The apparent ambiguity caused by using the same letters both as system variables of a special sort and as extra-systematic variables standing for either sort, could be avoided by adopting different symbols in the meta-theory. But in practice there is no need to do this.

(b) *Valuation Rules*

(1) t, f and n are values, and the only values.

(2) An *S*-unrestricted variable takes as its value either t, f or n.

(3) An *S*-restricted variable takes as its value either t or f.

(4) As for valuation rule (3) of $\mathbf{C_0}$.

(5) As for valuation rule (4) of $\mathbf{C_0}$.

It should be noticed that although these last two rules are identical with those of $\mathbf{C}_0$, and so specify conditions determining the same matrices for '$\sim$' and '$\vee$' (the classical significance matrices) some of the rows or columns will be otiose if a wff contains restricted variables. In fact, for a wff Δ compounded entirely from restricted variables, the conditions under which $v(\Delta) = n$ are unfulfilled and the matrices reduce to the standard two-valued matrices.

(c) *Tautologies*

A wff Δ of $\mathbf{S}_0$ is a $\mathbf{S}_0$-tautology if, and only if, $v(\Delta) = t$ for all values of the variables of Δ.

(d) *Valuation Tables*

A valuation table for a wff Δ of $\mathbf{S}_0$ with k restricted variables and m unrestricted variables has $2^k \times 3^m$ lines, one line for each distinct possible combination of values of the variables of Δ. (Δ is a $\mathbf{S}_0$-tautology if, and only if, the main value of each line of its valuation table is t.)

(e) *Formal Theorems*

A wff Δ is a theorem of $\mathbf{S}_0$ if, and only if, it is an $\mathbf{S}_0$-tautology.
This establishes t as the only designated value.

(f) *Metatheorems*

Theorem (5): A wff Δ is an $\mathbf{S}_0$-tautology only if all its variables are *S*-restricted.

For it can be shown by induction over the number of connectives in Δ that if Δ contains an unrestricted variable, there are assignments of values such that Δ has value n.

Theorem (6): A wff Δ is an $\mathbf{S}_0$-tautology if, and only if, it is a tautology of classical two-valued sentential logic (**RSL**).

For the valuation rules for classical tautologies coincide with those for $\mathbf{S}_0$-tautologies for wff of $\mathbf{S}_0$ containing only *S*-restricted variables.

The set of theorems of $\mathbf{S}_0$ thus coincides with the set of theorems of **RSL**. It should be noticed, however, that although the set of $\mathbf{C}_0$ theorems also coincides with the set of **RSL** theorems, there is an important difference between $\mathbf{S}_0$ and $\mathbf{C}_0$, namely that the $\mathbf{C}_0$ theorems are three-valued while the $\mathbf{S}_0$ theorems are, like those of **RSL**, two-valued. The only difference between $\mathbf{S}_0$ and **RSL** is that $\mathbf{S}_0$ contains wff which are three-valued and some which contain both two- and three-valued parts, as well as two-valued formulae. However, since only two-valued formulae can be theorems, and since those which are, are exactly those of **RSL**, $\mathbf{S}_0$ has little independent interest. Its value is simply as a preliminary to the development of more complex S-systems.

Like all the logics we develop, 0-logics are susceptible of alternative interpretations. So, for example an alternative interpretation of S_0 interprets value '*n*' as 'non-grammatical'; another interpretation, taken over all sentences rather than just those used in the performance of indicative acts, interprets '*n*' as 'non-indicative'. The main intended interpretation of '*n*' is, of course, as 'nonsignificant' in the classical sense which coincides, at the sentential level, with the chief sense given to it by type-theorists and positivists. But the logics developed can be taken as the formalisations of many distinct, though related, theories in the history of philosophy, and of many different theories of significance. For the logics of many theories coincide at the sentential level and the formal differences only begin to appear in the predicate logic, or even later. For example, the real difference between the free type theory we develop, and the fixed type theory of Russell, only emerges at the predicate level. In particular, *t*, *f* and *n* may be construed as assertion values, *i.e.* as contractions on sentence values, instead of as sentence values; and then *n* may be alternatively interpreted as non-truth-valuedness, as unverifiableness, *etc.*

It may perhaps seem strange that we are prepared to speak of alternative interpretations, or even the interpretation, of a semantic system such as S_0. For are not such systems already interpreted? Not necessarily, from the way S_0 *e.g.*, is presented, and from the way the expression 'semantic system' is, unfortunately, commonly used. A semantic system, when more fully formalised, is simply a formal system of a certain sort open to various interpretations. For, in the valuation rules, *e.g.* for S_0, no interpretation for '*t*', '*f*', '*n*' need be specified and the rules could be expressed formally in predicate logic. The symbols '*t*', '*f*' and '*n*' may simply be specified as constants of a certain sort. Hence these symbols are open to a variety of interpretations. Only when sufficiently many primitives of the formalism are tied down by meaning rules or postulates (see *e.g.* Carnap [2]) which relate these expressions to entrenched expressions of an already intelligible language, usually a natural language, or are tied down directly in much the way that expressions of a natural language are themselves tied down, can the formalism be said to be fully interpreted. Usually, when interpretations in terms of familiar languages are provided, more than extensional relations are guaranteed. In the case of referring expressions, for example, more than the references of the primitive expressions are provided; and in such cases the semantics given through the meaning rules is an intensional semantics (see Church [3] and [4]).

Strictly, the terminology should be rectified. In the case of the sentential systems with which we are concerned, this could be done by calling them 'matrix systems'. Matrix systems would then be contrasted with genuine semantic systems containing meaning rules or postulates, for a matrix system is a formal system open to different interpretations even though it may be theorem-wise equivalent to an axiomatic system and, under a given interpretation, stand as the semantic theory of that axiomatic system.

5.9 Stage 1: The Logics $\mathbf{S_1}$ and $\mathbf{C_1}$

Despite the formal completeness of 0-significance logics, they are seriously incomplete in terms of the intended application—so much so that it is only by courtesy we call them *significance* logics at all. They do not even provide us with the means of saying that a given sentence is or is not significant. For the interpretation to be extended in this way, it is necessary that a logic contain the significance operator '*S*' explicitly, or else provide means of defining '*S*' in terms of the other connectives.

That 0-logics do not contain '*S*' explicitly is obvious; but neither is it possible to define '*S*' within them. In order to see this, consider what valuation rules are necessary to characterise '*S*'. If it is to be interpretable in the required sense, then it must satisfy the condition:

Condition VI To say of a true sentence that it is significant, is true; to say of a false sentence that it is significant, is also true; while to say of a nonsignificant sentence that it is significant, is false.

Thus '*S*' is a two-valued connective and is matrix-accessible with matrix:

Γ	t	f	n
$S\Gamma$	t	t	f

Theorem (7): '*S*' is not definable in 0-logics.

Given only the one condition on '*S*' that $v(Sp) = f$ when $v(p) = n$, it is easily shown that it is not definable in 0-logics. For consider any such proposed definition of $qu(Sp)$. It must be of the form $\Delta(p)$ where Δ is a wff containing only classical connectives, or is a variable standing alone. But if $\Delta(p)$ is to stand as a definition of $qu(Sp)$, then $qu(Sp)$ must take the same values as $\Delta(p)$ for all values of $qu(p)$. Now when $v(p) = n$, $v(Sp) = f$; but by induction over the number of connectives in Δ, $v(\Delta(p)) = n$. Hence the proposed definition is impossible.

It is therefore necessary to extend the 0-logics by taking '*S*' as a further primitive. This gives us 1-logics. We also change the basic morphology slightly by taking '$\sim$' and '$\supset$' as primitives instead of '$\sim$' and '$\vee$'. This makes for simpler proofs of metatheorems in some cases.

The Logic $\mathbf{C_1}$

(a) *Morphology*

Add to the morphology of $\mathbf{C_0}$, with $\sim$ and $\supset$ taken as primitives instead of $\sim$ and $\vee$:

the improper primitive symbol: *S*

the formation rule: if Δ is a wff, then so is $S\Delta$.

(b) *Valuation Rules*

Add to the valuation rules of $\mathbf{C}_0$,

If $v(\Delta) = t$, then $v(S\Delta) = t$; if $v(\Delta) = f$, then $v(S\Delta) = t$; if $v(\Delta) = n$, then $v(S\Delta) = f$.

(c) *Tautologies*

A wff of $\mathbf{C}_1$ is a $\mathbf{C}_1$-tautology if, and only if, it takes the value t or the value n for all assignments of values to its variables.

It follows at once that all $\mathbf{C}_0$-tautologies are $\mathbf{C}_1$-tautologies, but not conversely; and that all classical theses are $\mathbf{C}_1$-tautologies.

(d) *Valuation tables* are defined precisely as for $\mathbf{C}_0$, except that values for wff of the form $S\Gamma$ must also be determined.

(e) *Formal Theorems*

A wff Δ is a theorem of $\mathbf{C}_1$ if, and only if, it is a $\mathbf{C}_1$-tautology. Among the theorems are all those of $\mathbf{C}_0$ together with:

$$SSp;\ p \supset Sp;\ Sp \equiv S{\sim}p;\ S(p \vee q) \equiv Sp \,\&\, Sq;\ S(p \,\&\, q) \equiv Sp \,\&\, Sq;$$
$$S(p \vee q) \equiv S(p \,\&\, q);\ Sp \supset (Sq \supset S(p \,\&\, q));\ ((p \supset q) \,\&{\sim}\, Sq) \supset {\sim}\, Sp;$$
$$((p \supset q) \,\&{\sim}\, Sq) \supset Sp;\ ((p \supset q \,\&{\sim} q) \,\&\, ({\sim}p \supset q \,\&{\sim} q)) \supset {\sim}\, Sp;$$
$$((p \supset q \,\&{\sim} q) \,\&\, ({\sim}p \supset q \,\&{\sim} q)) \supset Sp.$$

It will be seen that among these are theorem-equivalents of some of the valuation rules, *e.g.* $Sp \equiv S{\sim}p$ and $S(p \vee q) \equiv Sp \,\&\, Sq$.

It is easy to prove that a wff Δ is a $\mathbf{C}_1$-theorem iff the main column of the valuation table for Δ contains no occurrence of value f. Hence each of the theorems may be established by its valuation table and we here give the table for $Sp \,\&\, Sq \equiv S(p \,\&\, q)$ to indicate the method of construction. There are two distinct variables, therefore $3^2 = 9$ distinct lines.

p	q	$p \,\&\, q$	Sp	Sq	$S(p \,\&\, q)$	$Sp \,\&\, Sq$	$Sp \,\&\, Sq \equiv S(p \,\&\, q)$
t	t	t	t	t	t	t	t
f	t	f	t	t	t	t	t
n	t	n	f	t	f	f	t
t	f	f	t	t	t	t	t
f	f	f	t	t	t	t	t
n	f	n	f	t	f	f	t
t	n	n	t	f	f	f	t
f	n	n	t	f	f	f	t
n	n	n	f	f	f	f	t

C-*logics Consistency Theorem: Part 1*

Matrix $\mathbf{C}_1$-logics are simply consistent with respect to all primitive and definable monadic connectives κ which are minimally interpretable as negations.

Proof: It is necessary only to add one further case to the Part 0 theorem (§5.8):

(iii) Δ is of the form $S\Delta_1$, where Δ_1 contains m connectives. By the induction hypothesis, Δ_1 does not take the value n if all the variables in it are assigned value t. By the matrix for 'S', therefore, neither does Δ.

The result follows by III of §5.6.

We show in *Theorem* (8) of §5.10 that 'T', hence 'F', cannot be defined in 1-logics, and a similar argument will show that '$\neg$' cannot be defined since $\neg\Delta$ has the same matrix as $\sim T\Delta$. Hence, as in Part 0, the Part 1 consistency theorem is equivalent to the assertion that $\mathbf{C}_1$-logics are simply consistent with respect to '$\sim$'.

The Logic $\mathbf{S}_1$

(a) *Morphology*

Add to the morphology of $\mathbf{S}_0$, with $\sim$ and $\supset$ taken as primitives instead of $\sim$ and $\vee$:

the improper primitive symbol: S

the formation rule: if Δ is a wff, then $S\Delta$ is a wff.

(b) *Valuation Rules*

Add to the valuation rules of $\mathbf{S}_0$,

If $v(\Delta) = t$, then $v(S\Delta) = t$;
if $v(\Delta) = f$, then $v(S\Delta) = t$;
if $v(\Delta) = n$, then $v(S\Delta) = f$.

In case Δ contains only S-restricted variables, the last clause is not applicable.

It can be shown by induction over the number of connectives in Γ, that if all the variables in Γ are either S-restricted or occur in sub-wff of the form $S\Delta$, then $v(\Gamma) = t$ or $v(\Gamma) = f$.

(c) *Tautologies*

A wff of $\mathbf{S}_1$ is a $\mathbf{S}_1$-tautology if, and only if, it takes the value t for all assignments of values to its variables.

It follows that all $\mathbf{S}_0$-tautologies are $\mathbf{S}_1$-tautologies, but not conversely. Hence all classical two-valued tautologies are $\mathbf{S}_1$-tautologies, but not conversely.

(d) *Valuation Tables*

If a wff Δ contains k S-restricted variables only, its valuation table has 2^k independent lines (a classical truth-table) even though it contains sub-wff of the form $S\Gamma$, provided Γ contains only S-restricted variables. If Δ contains k S-restricted variables and m S-unrestricted variables, its valuation table contains $2^k \times 3^m$ distinct lines (a $\mathbf{S}_0$-table) whether or not some of the S-unrestricted variables occur in sub-wff of the form $S\Gamma$. If Δ contains m S-unrestricted variables, its valuation table contains 3^m distinct lines, whether or not it contains sub-wff of the form $S\Gamma$, provided Γ contains S-unrestricted variables.

(e) *Formal Theorems*

A wff Δ is a theorem of $\mathbf{S}_1$ if, and only if it is a $\mathbf{S}_1$-tautology. Among the theorems, as well as all $\mathbf{S}_0$-theorems, we have:

$$SSp; Sp \equiv S{\sim}p; S(p \vee q) \equiv Sp \,\&\, Sq; S(p \,\&\, q) \equiv Sp \,\&\, Sq; S(p \vee q) \equiv S(p \,\&\, q);$$
$$Sp \supset (Sq \supset S(p \,\&\, q)); \; Sp \supset S(p \,\&\, r); \; r \supset Sr; \; Sr; \; Sr \equiv (r \supset r).$$

$\mathbf{S}_1$ clearly has many advantages over $\mathbf{S}_0$. For its set of theorems no longer simply coincides with classical two-valued logic, and although all its theorems are two-valued, and t the only designated value, many of them, such as '$Sp \equiv S{\sim}p$', contain S-unrestricted variables. Hence the general difficulty, of having no theorems at all which contain S-unrestricted variables, has been overcome. Moreover, in $\mathbf{S}_1$, unlike $\mathbf{S}_0$, there are theorems, such as '$Sp \supset S(p \,\&\, r)$', which contain both S-restricted and S-unrestricted variables.

The theorem $qu(Sr)$ expresses the fact that all values for S-restricted variables are significant sentences; while the theorem $qu(Sr \equiv (r \supset r))$ shows that the significance operator 'S' can be defined in classical two-valued sentential logic where only S-restricted variables are admitted.

It will be seen that $qu(SSp)$, which expresses the fact that it is always significant to say of any sentence that it is significant, is a thesis of both $\mathbf{C}_1$ and $\mathbf{S}_1$. Surprisingly, Halldén has claimed that this theorem is a "far-reaching assertion which should be met with some doubt" (Halldén [1], p. 56), but he does not provide any satisfactory reasons for the doubt. Halldén suggests that 'S' may be systematically ambiguous and that, because significance ranges can only be regarded as vaguely determined, it cannot be automatically guaranteed that they are so determined as to yield $qu(SSp)$ as a theorem.

The second objection carries no weight unless it depends on the first. For otherwise, even if significance ranges are vaguely determined, a predicate is either significant of a given individual or it is not. And if it is significant of the individual, it is significant to say that it is; while if it is not significant of the individual, then it is false to say that it is significant, and therefore significant that it is significant of the individual. In any case, the intuitive notion of significance we are seeking to reflect by 'S' is not so vaguely determined as to undermine $\vdash SSp$. Suppose, however, we accept that 'S' is systematically ambiguous. Let us resolve it into a set of non-ambiguous components of which 'S_j' is typical. We may suppose that this resolution may always be carried out unless 'S' has no systematic application at all, a possibility which may clearly be rejected. Now for the ambiguity to be such as to refute $qu(SSp)$ for some value, it would have to be such that, for some component 'S_j', $qu(S_jp)$ is sometimes false or nonsignificant when $qu(p)$ is significant, or else such that $qu(S_jp)$ is nonsignificant when $qu(p)$ is. In none of these cases, however, is 'S_j' a satisfactory significance operator. For either 'S_j' does not specify a significance operator at all, because $qu(S_jp)$ is not true when $qu(p)$ is significant, or it really only specifies a significance operator over a restricted class of sentences, *e.g.* over a given Russell-type. This last is the case which should occur if 'S' is systematically ambiguous; but in this case, 'S_j' can be extended to a satisfactory significance operator 'S_1', where 'S_1' has the same value as 'S_j' over S_j's type and value f otherwise (effectively, $S_1p \simeq TS_jp$). For **C**-logics, Halldén's position is worse still. For 'S_j' must then be false for a given class which includes sentences of the form $qu(S_kp)$. Once again, however, 'S_j' is not a significance operator at all since it assigns value f when $qu(S_jp)$ has value t or f, *i.e.* when $qu(p)$ is significant. In any case, unless some type restrictions are imposed in advance, we can always define an operator 'S', which is matrix-accessible and has the matrix as assigned above, and which thereby reflects the preferred notion. There is no reason why such an operator should not be introduced, and there are good reasons, already indicated, why it should be introduced. In contrast to 'S', systematic ambiguity is a theory-saving and theory-complicating device which is best omitted.

There are however other features of $\mathbf{C}_1$ and $\mathbf{S}_1$ which are less desirable. In $\mathbf{C}_1$, for example, the modus ponens rule:

If Δ and $\Delta \supset \Gamma$ are $\mathbf{C}_1$-tautologies, then Γ is a $\mathbf{C}_1$-tautology,

does not hold generally. The argument for metatheorem (2) of $\mathbf{C}_0$ fails at the assumption that the variables of Δ which take value n do not occur in Γ (otherwise they would also take value n in Γ). For now, with the introduction of 'S', we may have $qu(Sp)$ in Γ which takes value f when $qu(p)$ takes value n. So, for example, take Δ to be $qu(p \vee \sim p)$ and Γ to be $qu(S(p \vee \sim p))$. Then Δ is a $\mathbf{C}_1$-tautology and so is $\Delta \supset \Gamma$, but Γ is not.

In $\mathbf{S}_1$, though the modus ponens rule is valid, the rule of uniform substitution fails to hold for all substitutions of wff on S-restricted variables: if, for example, we substitute the unrestricted variable $qu(p)$ for the restricted variable $qu(r)$ in the theorem $qu(Sr)$, we obtain the non-theorem $qu(Sp)$.

The failure of modus ponens is a definite weakness of $\mathbf{C}_1$, and of logics which include it. It means that, under the intended interpretation, tautologousness can no longer be compared to truth, since truth is preserved under modus ponens. This is in any case obvious from the role which n plays with respect to $\mathbf{C}_1$-tautologousness, together with the intended interpretation for *n*. Because of this separation of tautologousness and truth, **C**-logics in general fail to provide only logical truths which involve only classical connectives in combination with significance; and in an important sense, therefore, they fail to do the job required of them. For it is just the logical truths which we require significance logics to capture.

For similar reasons, several theorems of **C**-logics which include $\mathbf{C}_1$ do not admit of plausible interpretations. In particular, substitution instances of,

$$p \supset Sp,$$

when the value substituted for *qu*(*p*) is a nonsignificant sentence, are especially implausible. Thus, let Γ_0 be a particular nonsignificant sentence and consider the substitution instance,

$$\Gamma_0 \supset S\Gamma_0$$

which asserts that Γ_0 implies its significance. How can this be? How can material implications relating nonsignificant sentences with other sentences, so resulting in a nonsignificant compound, be affirmed in any plausible way? Should we not simply insist that a system is incorrect if not all its substitution cases are true?

There seem to be only two related ways in which anything near plausibility can be claimed for **C**-logics. The first is by reading subjunctives into the interpretation of deviant theorem-instances, and so in effect altering significance ranges. Then '$\Gamma_0 \supset S\Gamma_0$' would be read 'If Γ_0 were true (which it is not) then Γ_0 is significant'. But **C**-logics which contain only a material implication are not equipped to express subjunctive conditionals satisfactorily. The second possible solution depends on the traditional form/content distinction. Logical truth, we may say, is a matter of form only. Thus such formal theses as *qu*($p \supset p$) and *Barbara* express truths under all substitutions, including the substitution of nonsignificant sentences or even non-words. But such sentences as 'If the Absolute is green, then the Absolute is green' cannot be reckoned to express truths unless either (i) all such sentences as 'The Absolute is green' are admitted as significant, or (ii) the conditional in the sentence is treated as non-classical (in the style of $\mathbf{S}_6$ below), or (iii) a subjunctive interpretation is again introduced. But none of these courses is really viable. To adopt (i) would be to abandon the quest for a significance theory; a three-valued re-interpretation of sort (ii) is impossible if the classical significance matrices are to be retained; and (iii) takes us back to the objection above. (A fuller discussion of these points follows the development of $\mathbf{S}_6$.)

An obvious exit is not to treat seriously the claim that *qu*($p \supset p$), for example, expresses a truth under all substitutions. To keep the intended indicative reading of sentences it is very tempting to take a further step, namely to try to separate out the logical truths from the class of tautologies of **C**-matrix systems. This amounts to following Halldén's suggestion that such theorems of **C**-logics as *qu*($p \supset p$) only represent assertions for replacements which result in significant sentences. Let us call those substitutions which result in nonsignificant theorem-instances, *degenerate*; and the sentences which result, degenerate sentences. Then the theorems which do assert something for all substitutions, *i.e.* which always significantly express truths, can be isolated if the theorem-schemata can be decomposed into non-degenerate and degenerate sections: a maximal section for which no permitted substitutions of constants results in degenerate sentences, and a remainder section which results in degenerate sentences for certain substitutions. In this event, a **C**-logic can be separated into two parts: a good part which admits only theorems which do not give rise to degenerate sentences for any substitutions, and a bad part which contains the remaining theorems. In terms of this, a good thesis of $\mathbf{C}_1$ which has no degenerate substitution instances is *qu*($Sp \equiv S{\sim}p$); a bad thesis, on the other hand, is *qu*($p \supset Sp$). Clearly, in the light of **S**-logics, such a separation can be achieved by introducing two sorts of variables.

These general criticisms of **C**-logics apply *a fortiori* to $\mathbf{C}_1$; and since $\mathbf{C}_1$ is in fact Halldén's logic **C**, they apply to Halldén's system. They are not, however, criticisms of **S**-logics since they turn on the fact that n is a designated value, and this is not the case in **S**-logics. Thus,

$$p \supset Sp$$

is not a thesis of $\mathbf{S}_1$; but only,

$$r \supset Sr$$

where *qu*(r) is a restricted variable. Similarly, *qu*($p \supset p$) is not a thesis, but only *qu*($r \supset r$). **S**-logics thus present a major advance over **C**-logics; all the theorems express logical truths, and there are, therefore, no degenerate substitution instances. Moreover it is just the degenerate cases which are eliminated, for all the non-degenerate theorems of $\mathbf{C}_1$ are theorems of $\mathbf{S}_1$.

As a particular example of implausible theorems of $\mathbf{C}_1$ which have degenerate cases, we may consider the *reductio* variants. For although,

$$((p \supset q) \;\&\; {\sim}Sq) \supset {\sim}Sp$$

and

$$((p \supset q \;\&\; {\sim}q) \;\&\; ({\sim}p \supset q \;\&\; {\sim}q)) \supset {\sim}Sp$$

are theses which appear *prima facie* plausible, we have also the quite implausible,

$$((p \supset q) \;\&\; {\sim}Sq) \supset Sp$$

and

$$((p \supset q \;\&\; {\sim}q) \;\&\; ({\sim}p \supset q \;\&\; {\sim}q)) \supset Sp.$$

In consequence, therefore,

$$((p \supset q) \;\&\; {\sim}Sq) \supset Sp \;\&\; {\sim}Sp$$
$$((p \supset q \;\&\; {\sim}q) \;\&\; ({\sim}p \supset q \;\&\; {\sim}q)) \supset Sp \;\&\; {\sim}Sp;$$

which express the fact that the *reductio* variants are worthless. They do not justify conclusions of the sort: *qu*(p) is nonsignificant.

None of these are theorems of S_1, however, and it is for reasons such as this that our main interests is in S-logics. We develop C-logics alongside S-logics partly for the historical reason that they have been seriously considered, by Halldén and others, as at least partially adequate significance logics, and partly because they have applications in existence-free logic and in intensional logic (see Chapter 7).

5.10 Stage 2: the Logics S_2 and C_2

Even 1-logics are seriously inadequate. For the operators '*T*' and '*F*' which were introduced in §2.12, cannot be defined within them (we show this in detail below). In consequence, it is not possible to formulate many claims commonly made in discussions of nonsignificance (see, *e.g.*, Russell [2], p. 72; Halldén [1], p. 33, p. 37; and Ryle [2]). Consider, to illustrate, such claims as:

(a) A sentence is significant if, and only if, it is true or false (cf. §2.16, $\vdash_{CCL}(99)$);
(b) If a sentence is nonsignificant, it is not true (a consequence of $\vdash_{CCL}(99)$);
(c) A sentence is false if, and only if, its negation is true.

Strong arguments can be given for the view that there are senses of 'true' and 'false' for which these claims are true, given that they operate over the reduced range which excludes incomplete sentences used in indicative acts as well as sentences used in non-indicative acts. Take (b), for example: if a sentence is nonsignificant in a given context, it is thereby excluded from the class of sentences which yield statements, and in particular from those which yield true statements. Hence, given the transfer of 'true' from statements to sentences which yield them, a nonsignificant sentence is not true (does not yield a true statement).

Now it is possible to define in both 0- and 1-logics a sense of 'true' and 'false' by means of classical definitions of the sort examined by Tarski. These we shall represent by 'T^c' and 'F^c' where these absorb quotation functions as '*T*' and '*F*' do; *i.e.* we have,

$$T^c\Delta =_{Df} tru^c\, qu(\Delta)$$

and

$$F^c\Delta =_{Df} fal^c\, qu(\Delta) \qquad (cf.\ §2.12).$$

The classical definitions are then,

$$T^c\Delta =_{Df} \Delta$$
$$F^c\Delta =_{Df} \sim\Delta;$$

and have thesis equivalents:

$$\text{(i) } T^c p \equiv p$$
$$\text{(ii) } F^c p \equiv \sim p.$$

It should be noticed that the laws in this form are not open to Tarski's [1] objections since 'T^c' and 'F^c' absorb quotation functions and not standard quote marks suitable only for use in conjunction with constants.

Suppose, now, however, we try to express (a)–(c) in 1-logics using 'T^c' and 'F^c' along with 'S'. We then have,

$$\text{(a}'\text{)}\ Sp \equiv T^c p \vee F^c p$$
$$\text{(b}'\text{)}\ {\sim} Sp \supset {\sim} T^c p$$
$$\text{(c}'\text{)}\ T^c{\sim}p \equiv F^c p.$$

But none of these are tautologies in **S**-logics, since by the equivalences (i) and (ii), all take the value n when $qu(p)$ takes the value n. For the same reason, though they are all tautologies of **C**-logics, they are not general truths from which we can detach with impunity since the equivalences from which they derive have degenerate instances. Hence, if we wish to construct logics in which (a)–(c) appear as logical truths, we must introduce different senses of 'true' and 'false', in fact those represented by 'T' and 'F', such that,

$$\text{(a}''\text{)}\ Sp \equiv Tp \vee Fp$$
$$\text{(b}''\text{)}\ {\sim} Sp \supset {\sim} Tp$$
$$\text{(c}''\text{)}\ T{\sim}p \equiv Fp$$

are demonstrable in **S**-logics and have no degenerate instances in **C**-logics.

Now given (c″), we may eliminate 'F' in (a'') to obtain,

$$\text{(a}'''\text{)}\ Sp \equiv Tp \vee T{\sim}p;$$

and since $qu(Sp)$ is two-valued, and the connectives are classical significance connectives, neither $qu(Tp)$ nor $qu(T{\sim}p)$ can take the value n. Hence $qu(Tp)$ must be two-valued; and by a similar argument, so must $qu(Fp)$. It follows that, if 'T' and 'F' are to be matrix-accessible over t, f and n, and if (a″) is to hold, their matrices must either coincide with that for 'S' (which would destroy the intended interpretation) or else be one of the following:

p	
t	t
f	f
n	f

p	
t	f
f	t
n	f

But since 'T' and 'F' are required to be, respectively, truth and falsity operators, if $qu(p)$ is true, $qu(Tp)$ must be true and $qu(Fp)$ false. For to say of a true sentence that it is true, is true; while to say of a true sentence that it is false, is false. Again, if $qu(p)$ is false, then $qu(Tp)$ must be false and $qu(Fp)$ true. For to say of a false sentence that it is true, is false; while to say of a false sentence that it is false, is true. These conditions are satisfied by the classical definitions in the form,

$$Tr \equiv r \quad \text{and} \quad Fr \equiv r,$$

and they imply that '*T*' has the matrix on the left and '*F*' the one on the right. That is,

p	Tp	p	Fp
t	t	t	f
f	f	f	t
n	f	n	f

We therefore adopt these matrices for '*T*' and '*F*'; which is to say that we adopt the general condition:

Condition VII (i) to say of a true sentence that it is true, is true; while to say of a true sentence that it is false, is false;
(ii) to say of a false sentence that it is true, is false; while to say of a false sentence that it is false, is true;
(iii) to say of a nonsignificant sentence that it is true, is false; while to say of a nonsignificant sentence that it is false, is false.

We consider that these matrices determine an important sense of 'true' and 'false'. They coincide with classical definitions for significant values ((i) and (ii)); and for all values, they justify assertions such as (a), (b) and (c). They give us, that is to say, those senses of 'true' and 'false' which were used in Part I.

In further support of (iii), we may note the following:

For all sentences under consideration, the values t, f, n are comparable and exhaustive. Consequently, the value n coincides with the value, $\bar{t}$ and $\bar{f}$ (not true and not false). Now since, as we have claimed throughout, it is always significant to say of a nonsignificant sentence that it is nonsignificant, the ascription 'not true and not false' is also significant of nonsignificant sentences. Hence the component descriptions, 'not true' and 'not false' must each be significant of nonsignificant sentences. But these negative descriptions cannot be significant if the positive descriptions 'true' and 'false' are not themselves significant of nonsignificant sentences.

This defence depends on an appeal to features of the classical significance matrices for '&' and '$\sim$', but this is consistent with the development so far. It depends, too, on a connexion between values and operators or descriptions, which can be made explicit given that,

there is a valuation function v such that,
$v(\Delta) = t$, $v(\Delta) = f$, $v(\Delta) = n$
are always significant;

(this of course is the valuation function we have been using throughout). For the assumed connexion is then the one already introduced in §2.15 (D10), *viz.*:

$$(v(\Delta) = t) \equiv T\Delta; (v(\Delta) = f) \equiv F\Delta; (v(\Delta) = n) \equiv \sim S\Delta.$$

This connexion is now of fundamental importance since it leads directly to semantical completeness proofs for axiomatisations of 2-logics.

The operators 'T', 'F' and '$\sim S$' are the Rosser-Turquette [1] J-functors for three-valued logics under a re-labelling and ordering of the values. That is, 'T', 'F and '$\sim S$' are characteristic functors for the values t, f and n and so enable us to reflect valuation rules by theses. It is these thesis-equivalents of valuation rules which provide the basis of completeness proofs for corresponding axiomatic systems since the axioms may be "read off" from the valuation rules (we indicate below how this is done in the case of $\mathbf{S}_2$).

The connective 'T' not only functions as a truth predicate, but also serves as a mapping operator which maps nonsignificance into falsity. This mapping property proves valuable not only for reducing 3-valued systems and 3-valued theses to 2-valued images, and for completeness proofs, but also for extending 2-valued systems and theorems to 3-valued analogues. And in general, the three connectives 'T', 'F' and 'S' may be described as (3:2) mapping operators since they each map three values on to two. Since the two values concerned are t and f, we shall say that the connectives are unlimited. A connective is *unlimited* if its matrix values do not include the value n for any values of its components, otherwise *limited.* In terms of this, the classical significance connectives are limited. Hence, too, 'T', 'F' and 'S' are non-classical.

The idea of an unlimited connective enables us to define a notion which is of particular importance in **C**-logics. Thus, a variable P is said to be *covered* in a wff Δ if, and only if, P occurs in Δ and every occurrence of P in Δ is within the scope of an unlimited connective; otherwise P is *uncovered* in Δ (see Halldén [1], p. 46).

The use of unlimited connectives thus enables us to reach an alternative solution to the problem of obtaining theorems in **S**-logics without at any stage being committed to the assertion of nonsense. For if we restrict attention to covered variables, we may develop a logic in which not only the theses but all components of theses are 2-valued, which means, in terms of the interpretation, that we restrict what we say to statements about (*i.e.* to the use of sentences mentioning) 3-valued sentences. Thus, by using $qu(Tp)$ and $qu(Sp)$, for example, as in $qu(Tp \supset Sp)$, we can effectively express the truth which the non-thesis $qu(p \supset Sp)$ purports to express; or, so far as **C**-logics are concerned, we can express explicitly the non-degenerate cases of degenerate theorems. This is the intuitive basis on which context logic was developed. But this resolution of the problem, as to how to derive theses in a logic in which only t is designated, by mapping all 3-valued sentences on to a 2-valued plane, only succeeds by flouting a general principle of component homogeneity (§5.3): that a sentence is significant if, and only if, each of its components is significant. That is, it succeeds only by the admission of non-classical connectives. As it becomes clear later, however, this mapping resolution is not in general necessary; nor does it allow one to say significantly all that it should be possible to say.

Because 'T' is non-classical, it cannot be defined in terms of classical significance connectives alone. But we now show that it cannot be defined in terms of these connectives together with 'S' as primitive. That is,

Theorem (8): 'T' cannot be defined in 1-logics.

For consider any proposed definition of the form:

$$Tp =_{\mathrm{Df}} \Delta(p)$$

where Δ is some function just of $qu(p)$, containing only '&', '$\vee$', '$\sim$' and 'S'. Now each occurrence of $qu(p)$ must be covered in Δ, otherwise $\Delta(p)$ has value n when $qu(p)$ takes value n. Reduce $\Delta(p)$ by equivalences to conjunctive normal form (for details see §6.2). Then since each variable occurrence is covered, it must be of the form, $qu(Sp)$ or $qu(SSp)$. Therefore,

$$Tp \equiv \Delta(Sp, SSp)$$

where Δ is composed just of '&', '$\vee$', and '$\sim$'. Hence the following valuation table results:

p	$Tp \equiv$	$\Delta(Sp,$	$SSp)$
t	t	t	t
f	f	t	t
n	f	f	t

But there is now no way of producing a matrix from $\Delta(Sp, SSp)$ in which the first line differs from the second. Hence $qu(Tp)$ is not definable in terms of '&', '$\vee$', '$\sim$' and 'S'.

It follows that 0- and 1-logics are not sufficiently rich to permit completeness and related proofs which rely on the use of operators 'T' and 'F'. In particular, then, Rosser-Turquette techniques for dealing with functionally complete many-valued logics—techniques which can be extended to classes of many-valued logics which are not functionally complete but which permit definitions of the J-functors—cannot be extended to 0- and 1-logics. This will become evident later when we discuss the completeness of axiomatised systems.

Given 'T' as primitive, however, together with classical connectives, then 'F', 'S' and '$\simeq$' (strong extensional equivalence, §2.12) can all be defined (see the morphology of 2-logics below). Hence 2-logics, which contain 'T', include 1-logics, since 'S' can be defined; but 1-logics do not include 2-logics since 'T' cannot be defined.

The Logic $\mathbf{S}_2$

(a) *Morphology*

Add to the morphology of $\mathbf{S}_0$, with $\sim$ and $\supset$ taken as primitives:

the improper primitive symbol: T;
the formation rule: If Δ is a wff so is $T\Delta$;
the definitions: $F\Delta =_{\mathrm{Df}} T{\sim}\Delta$; $S\Delta =_{\mathrm{Df}} T\Delta \vee F\Delta$;
$\Delta \simeq \Gamma =_{\mathrm{Df}} (T\Delta \& T\Gamma) \vee (F\Delta \& F\Gamma) \vee ({\sim}S\Delta \& {\sim}S\Gamma)$.

Thus 'T', 'F', 'S' and '$\simeq$' are connectives over S-restricted variables as well as S-unrestricted variables since Δ and Γ are any wffs.

(b) *Valuation Rules*

It is only necessary to add to the valuation rules of $\mathbf{S}_0$ the following:

$$\text{If } v(\Delta) = t, \text{ then } v(T\Delta) = t;$$
$$\text{if } v(\Delta) = f, \text{ then } v(T\Delta) = f;$$
$$\text{if } v(\Delta) = n, \text{ then } v(T\Delta) = f.$$

However, we here give the primitive valuation rules in detail and show the corresponding thesis-equivalents which can be "read off" them, applying the connections: $(v(\Delta) = t) \equiv T\Delta$, $(v(\Delta) = f) \equiv F\Delta$ and $(v(\Delta) = n) \equiv {\sim}S\Delta$.

Valuation Rule	Thesis Equivalent
(1)	
$v(r) = t$ or $v(r) = f$.	$Tr \vee Fr$
(2)	
$v(p) = t$ or $v(p) = f$ or $v(p) = n$.	$Tp \vee Fp \vee {\sim}Sp$
(3)	
If $v(\Delta) = t$, then $v({\sim}\Delta) = f$;	$Tp \supset F{\sim}p$; $Tr \supset F{\sim}r$
if $v(\Delta) = f$, then $v({\sim}\Delta) = t$;	$Fp \supset T{\sim}p$; $Fr \supset T{\sim}r$
if $v(\Delta) = n$, then $v({\sim}\Delta) = n$.	${\sim}Sp \supset {\sim}S{\sim}p$
(4)	
If $v(\Delta) = t$ and $v(\Gamma) = t$, then $v(\Delta \vee \Gamma) = t$;	$Tp \& Tq \supset T(p \vee q)$; $Tr \& Ts \supset T(r \vee s)$
if $v(\Delta) = t$ and $v(\Gamma) = f$, then $v(\Delta \vee \Gamma) = t$;	$Tp \& Fq \supset T(p \vee q)$; $Tr \& Fs \supset T(r \vee s)$
if $v(\Delta) = f$ and $v(\Gamma) = t$, then $v(\Delta \vee \Gamma) = t$;	$Fp \& Tq \supset T(p \vee q)$; $Fr \& Ts \supset T(r \vee s)$
if $v(\Delta) = f$ and $v(\Gamma) = f$, then $v(\Delta \vee \Gamma) = f$;	$Fp \& Fq \supset F(p \vee q)$; $Fr \& Fs \supset F(r \vee s)$
if $v(\Delta) = t$ and $v(\Gamma) = n$, then $v(\Delta \vee \Gamma) = n$;	$Tp \& {\sim}Sq \supset {\sim}S(p \vee q)$
if $v(\Delta) = f$ and $v(\Gamma) = n$, then $v(\Delta \vee \Gamma) = n$;	$Fp \& {\sim}Sq \supset {\sim}S(p \vee q)$
if $v(\Delta) = n$ and $v(\Gamma) = t$, then $v(\Delta \vee \Gamma) = n$;	${\sim}Sp \& Tq \supset {\sim}S(p \vee q)$
if $v(\Delta) = n$ and $v(\Gamma) = f$, then $v(\Delta \vee \Gamma) = n$;	${\sim}Sp \& Fq \supset {\sim}S(p \vee q)$
if $v(\Delta) = n$ and $v(\Gamma) = n$, then $v(\Delta \vee \Gamma) = n$.	${\sim}Sp \& {\sim}Sq \supset {\sim}S(p \vee q)$
(5)	
If $v(\Delta) = t$, then $v(T\Delta) = t$;	$Tp \supset TTp$; $Tr \supset TTr$
if $v(\Delta) = f$, then $v(T\Delta) = f$;	$Fp \supset FTp$; $Fr \supset FTr$
if $v(\Delta) = n$, then $v(T\Delta) = f$.	${\sim}Sp \supset FTp$.

The valuation rules for 'F', 'S' and '$\simeq$' follow by the definitions, with the matrices for 'F' and 'S' as already given. Strong extensional equivalence is thus an unlimited connective with matrix,

$\simeq$	t	f	n
t	t	f	f
f	f	t	f
n	f	f	t

Restricted as it is here to a relation between three-valued sentences over the set $\{t, f, n\}$, as opposed to its more general use in Chapter 2 to relate four-valued sentences, we may say that the relation holds between two sentences if, and only if, they have the same significance value.

(c) *Tautologies*

A wff of $\mathbf{S}_2$ is a tautology if, and only if, it takes the value t for all assignments of values to its variables.

Hence, all $\mathbf{S}_0$- and $\mathbf{S}_1$-tautologies are $\mathbf{S}_2$-tautologies.

(d) *Valuation Tables*

As for $\mathbf{S}_1$ with suitable amendments of 'T' for 'S'.

(e) *Formal Theorems*

A wff of $\mathbf{S}_2$ is a theorem if, and only if, it is a $\mathbf{S}_2$-tautology.

We here give a list of the more interesting theorems even though some have already been listed among the theorems of $\mathbf{S}_0$ and $\mathbf{S}_1$.

$\mathbf{S}_0$-theorems: $r \vee \sim r$; $r \supset r \vee s$; $r \equiv r$; and in general all **RSL** theorems.
$\mathbf{S}_1$-theorems: Sr; $Sr \equiv (r \supset r)$; $r \supset Sr$;
SSp; $Sp \equiv S{\sim}p$; $S(p \vee q) \equiv Sp \,\&\, Sq$; $S(p \,\&\, q) \equiv Sp \,\&\, Sq$; *etc.*

$\mathbf{S}_2$-theorems: $p \simeq p$; $\sim p \vee \sim q \simeq \sim(p \,\&\, q)$; $p \supset q \simeq \sim p \vee q$:
TSr; $Sr \equiv Tr \vee Fr$; $Tr \vee Fr \equiv (r \supset r)$; $Tr \supset Sr$; $Fr \supset Sr$;
$TSSp$; $Sp \equiv Tp \vee Fp$; $\sim Sp \equiv \sim Tp \,\&\, \sim Fp$; $Tp \supset Sp$; $Fp \supset Sp$;
$Sp \simeq Tp \vee Fp$; $\sim Sp \simeq \sim Tp \,\&\, \sim Fp$; $Fp \simeq T{\sim}p$;
STp; SFp; $TTp \equiv Tp$; $TTp \simeq Tp$; $TFp \equiv Fp$;
$TFp \simeq Fp$; $TSp \equiv Sp$; $TSp \simeq Sp$; $Tp \,\&\, Tq \equiv T(p \,\&\, q)$;
$Tp \,\&\, Tq \simeq T(p \,\&\, q)$; $T(p \vee q) \supset Tp \vee Tq$;
$(p \simeq \sim p) \supset \sim Sp$; $Sp \equiv \sim(p \simeq \sim p)$; $Sp \simeq \sim(p \simeq \sim p)$;
$((Tp \supset Fp) \,\&\, (Fp \supset Tp)) \supset \sim Sp$;
$((Tp \supset \sim Tp) \,\&\, (Fp \supset \sim Fp)) \supset \sim Sp$;
$((Tp \supset Tq) \,\&\, \sim Sq) \supset \sim Tp$; $((Tp \supset Tq) \,\&\, \sim Sq) \supset (\sim Sp \vee Fp)$;

together with those listed as thesis-equivalents of valuation rules.

Theorems such as $qu(p \simeq p)$ and $qu(\sim p \vee \sim q \simeq \sim(p \& q))$ are two-valued equivalence laws even though the variables are *S*-unrestricted. This is in marked contrast to laws such as $qu(r \equiv r)$ and $qu[(\sim r \vee \sim s) \equiv \sim(r \,\&\, s)]$ which hold only in terms of *S*-restricted variables, or those **C**-laws such as $qu(p \equiv p)$ and $qu[(\sim p \vee \sim q) \equiv \sim(p \,\&\, q)]$ which take the value *n* for some values of the variables and therefore have degenerate cases. By means of '$\simeq$', therefore, we are able to obtain analogues of certain two-valued **RSL** equivalence laws even though they are expressed entirely in terms of three-valued variables and *t* is taken as the only designated value. In the axiomatic theory such laws permit the replacement of three-valued formulae by their equivalents. It is not however the case that every **RSL** equivalence law has an analogue in terms of *S*-unrestricted variables and the connective '$\simeq$' in place of '$\equiv$'. Thus, for example, $qu(p \vee \sim p \simeq q \vee \sim q)$ is not a theorem of either **C**- or **S**-logics, though $qu(r \vee \sim r \equiv s \vee \sim s)$ is a **RSL** theorem. The reason for this is that the two sides of the equivalence are independent. In general, **RSL** equivalence laws have an analogue in terms of *S*-unrestricted variables and '$\simeq$' provided the same variables occur on both sides of the equivalence. Thus '$\simeq$' is a stronger relation than '$\equiv$'. It singles out those **RSL** equivalence laws which are in some sense synonomy laws.

The thesis $qu(T(p \vee q) \supset Tp \vee Tq)$ shows that '*T*' cannot be interpreted as '$\Box$' ('it is necessary that ...'). For while it is a law, its converse, $qu(Tp \vee Tq \supset T(p \vee q))$, is not. In modal logics, however, the reverse is the case. Thus, $qu(\Box p \vee \Box q \supset \Box(p \vee q))$ is a thesis, but not the converse $qu(\Box(p \vee q) \supset \Box p \vee \Box q)$.

The last few theorems listed above express certain types of *reductio* principle and enable us to evaluate the paradox arguments.

Thus, suppose we have a sentence 'p_0' for which it can be shown that 'p_0' implies '$\sim p_0$', and '$\sim p_0$' implies 'p_0'. Taking 'implies' here in the sense of 'materially implies', then this establishes: $p_0 \equiv \sim p_0$; and by **RSL** this is a contradiction. Now it might be claimed that any sentence 'p_0' which satisfies $qu(p \equiv \sim p)$ is nonsignificant. That is: $(p \equiv \sim p) \supset \sim Sp$. As we have seen, however, though this is a principle of $\mathbf{C}_1$, so is: $(p \equiv \sim p) \supset Sp$. Hence, we have $(p \equiv \sim p) \supset (Sp \,\&\! \sim Sp)$; and so far from avoiding the contradiction, we merely re-affirm it. But we now see that $(p \simeq \sim p) \supset \sim Sp$; and moreover it is not the case that $(p \simeq \sim p) \supset Sp$, *i.e.* the counter law is not a thesis. Hence the contradiction is avoided, and the conclusion that 'p_0' is nonsignificant follows, if 'p_0' is equivalent to '$\sim p_0$' in the strong sense. If, however, we are to conclude from the fact that 'p_0' implies '$\sim p_0$', and '$\sim p_0$' implies 'p_0', that 'p_0' is equivalent to '$\sim p_0$' in the strong sense, then the sense of 'implies' cannot be 'materially implies'. For it is not the case that $((p \supset \sim p) \,\&\, (\sim p \supset p)) \supset (p \simeq \sim p)$, in **S**-logics. This is a thesis of $\mathbf{C}_2$ (below), but so also is its mate, that $((p \supset \sim p) \,\&\, (\sim p \supset p)) \supset \sim(p \simeq p)$; so the conclusion that $p \simeq \sim p$ is valueless if we take

C-logics as a basis. Now it is possible to define a non-classical connective '$\sim\rangle$' such that $((p \sim\rangle \sim p) \,\&\, (\sim p \sim\rangle p)) \supset (p \simeq \sim p)$ is a thesis of **S**-logics while $((p \sim\rangle \sim p) \,\&\, (\sim p \sim\rangle p)) \supset \sim(p \simeq \sim p)$ is not. Thus, if '$\sim\rangle$' is defined by the following matrix:

$\sim\rangle$	t	f	n
t	t	f	f
f	t	t	t
n	t	f	t

then in general we have,

$$((p \sim\rangle q) \,\&\, (q \sim\rangle p)) \supset (p \simeq q);$$

and indeed there are other matrices which yield the same result. Whether this or any other of the possible matrices capture an important sense of implication, however, is not easily determined. We wish simply to point out here that any such connective must be non-classical, though it may, as in the above case, "contain" the two-valued classical '$\supset$' and have affinities to the strong regular connective '$\dot{\supset}$'. Alternatively, of course, it is possible to take the classical three-valued '$\supset$' together with a non-classical definition of '&', but there seems little case for that.

It cannot be claimed, therefore, that paradox arguments of the form, 'p_0' implies '$\sim p_0$', and '$\sim p_0$' implies 'p_0', yield the conclusion that 'p_0' is nonsignificant unless a special sense of 'implies' is justified. If, however, the argument is weakened so that it takes the form:

the assumption that 'p_0' is true implies that 'p_0' is false, and
the assumption that 'p_0' is false implies that 'p_0' is true;

then the conclusion that 'p_0' is nonsignificant does follow by the principle,

$$((Tp \supset Fp) \,\&\, (Fp \supset Tp)) \supset \sim Sp.$$

Similarly, the even weaker form,

the assumption that 'p_0' is true implies that 'p_0' is not true, and
the assumption that 'p_0' is false implies that 'p_0' is not false,

yields the conclusion that 'p_0' is nonsignificant.

Consider now *reductio* arguments of the sort suggested by Ryle:

'p_0' implies 'q_0' and 'q_0' is nonsignificant; therefore 'p_0' is nonsignificant.

These again, as we have seen, are not justified by appeal to principles such as,

$$((p \supset q) \,\&\, \sim Sq) \supset \sim Sp;$$

for although there is an appropriate theorem of $\mathbf{C}_1$, the conclusion is valueless since Sp follows from the same premisses. And even if we put the argument in the form:

the assumption that 'p_0' is true implies that 'q_0' is true, and 'q_0' is nonsignificant,

the only valid conclusion is that 'p_0' is not true (false or nonsignificant), not that 'p_0' is nonsignificant. For,

$$((Tp \supset Tq) \;\&\; \sim Sq) \supset \sim Tp$$

is a thesis of $\mathbf{S}_2$, but

$$((Tp \supset Tq) \;\&\; \sim Sq) \supset \sim Sp$$

is not.

The Logic $\mathbf{C}_2$

$\mathbf{C}_2$-logics are motivated by a combination of those motives which led Halldén to $\mathbf{C}_1$ and those which led to the introduction of 'T'.

(a) *Morphology*

Add to the morphology of $\mathbf{C}_0$, the new additional morphology of $\mathbf{S}_2$.

(b) *Valuation Rules*

Add to the valuation rules of $\mathbf{C}_0$, the valuation rules for '$T\Delta$' as given in $\mathbf{S}_2$.

(c) *Tautologies*

A wff of $\mathbf{C}_2$ is a tautology if, and only if, it takes the value t or the value n for all assignments of values to its variables.

Hence, all $\mathbf{C}_0$- and $\mathbf{C}_1$-tautologies are $\mathbf{C}_2$-tautologies.

(d) *Valuation tables*

As for $\mathbf{C}_1$-formulae with 'T' for 'S'.

(e) *Formal Theorems*

A wff of $\mathbf{C}_2$ is a thesis if, and only if, it is a $\mathbf{C}_2$-tautology.

Among the theorems are all those of $\mathbf{C}_0$ and $\mathbf{C}_1$ together with those listed as theorems of $\mathbf{S}_2$ which do not contain restricted variables. In addition, however, we have a number of theorems which take the value n for some values of the variables and hence have degenerate instances. Such are,

$Tp \equiv p$; $Fp \equiv \sim p$; $((p \supset \sim p) \;\&\; (\sim p \supset p)) \supset (p \simeq \sim p)$;
$((p \supset \sim p) \;\&\; (\sim p \supset p)) \supset \sim (p \simeq \sim p)$.

It should be noted that $qu(Tp \simeq p)$ and $qu(Fp \simeq \sim p)$ are not theses of either **C**- or **S**-logics.

To illustrate the special difficulties of axiomatising **C**-logics, we now show that either one of the two theses,

$$Tp \equiv p; \; p \supset Sp \text{ (also a } \mathbf{C}_1\text{-theorem);}$$

would lead to the conclusion,

Sp (which expresses the fact that every sentence is significant),

unless the rules for the replacement of equivalents and modus ponens are modified appropriately.

Consider first the following argument:

1. $Tp \equiv p$:
2. $T{\sim}p \equiv {\sim}p$: uniform substitution
3. $Fp \equiv {\sim}p$: definition of 'F'
4. $p \vee {\sim}p$:
5. $Tp \vee Fp$: replacing $qu(p)$ in 4 by its equivalent $qu(Tp)$ from 1, and $qu({\sim}p)$ by its equivalent $qu(Fp)$ from 3
6. Sp : definition of 'S'.

Here, 1, 2, 3 and 4 are $\mathbf{C}_2$-tautologies, but 5 and 6 are not. It follows that if we wish to axiomatise $\mathbf{C}_2$, the equivalence-replacement rule used at step 5 must be modified. This difficulty only arises in the case of the equivalence '$\equiv$'; a replacement rule based on '$\simeq$' is admissible, as we shall see. (The above "proof" cannot be formulated using '$\simeq$' since, as we have seen, '$Tp \simeq p$' is not a thesis and the argument cannot begin).

Next consider the following argument:

1. $p \supset Sp$:
2. ${\sim}Sp \supset {\sim}p$: using the strong equivalence law, $qu(p \supset q \simeq {\sim}q \supset {\sim}p)$
3. ${\sim}S{\sim}p \supset {\sim}{\sim}p$: uniform substitution
4. ${\sim}S{\sim}p \supset p$: using $qu({\sim}{\sim}p \simeq p)$
5. ${\sim}S{\sim}p \supset Sp$: from 1 and 4 using $qu((p \supset q \;\&\; q \supset p') \supset (p \supset p'))$ and detachment
6. $S{\sim}p \vee Sp$: using $qu({\sim}p \supset q \simeq p \vee q)$
7. $Sp \vee Sp$: using $qu(S{\sim}p \simeq Sp)$
8. Sp : using $qu(p \vee p \simeq p)$.

Here, every step in the proof is made by reference to a formula which is a $\mathbf{C}_2$ thesis, and besides, all equivalence replacements are based on '$\simeq$' and are acceptable. It follows that either uniform substitution or detachment has to be modified, and it is clear that it has to be detachment since 1, 2, 3 and 4 are $\mathbf{C}_2$-theses, while 5, which arises by detachment, is not.

None of these difficulties arise if we restrict ourselves entirely to the matrix method. For then, what is or is not a theorem is determined simply by constructing a matrix. The above "arguments", however, serve as a reminder that standard proof techniques cannot be applied indiscriminately to known theses without serious risk of interpretational inconsistency.

C-*logics Consistency Theorem: Part* 2

Matrix $\mathbf{C}_2$-logics are simply consistent with respect to all primitive and definable monadic connectives κ which are minimally interpretable as negations.

Proof: We add one further case to the Part 1 theorem (§5.9):

(iv) Δ is of the form $T\Delta_1$. By induction hypothesis Δ_1 does not take value *n* if all the variables in it are assigned value *t*. By the matrix for '*T*', therefore, neither does Δ.

Result by III of §5.6.

Since '*F*' and '$\neg$' are now definable ($F\Delta$ for $T{\sim}\Delta$; and $\neg\Delta$ for ${\sim}T\Delta$), the theorem establishes the simple consistency of $\mathbf{C}_2$-logics with respect to '$\sim$', '$\neg$' and '*F*'.

5.11 Stage 3: The Logics $\mathbf{S}_3$ and $\mathbf{C}_3$

A matrix logic is said to be functionally complete if, and only if, given any matrix over the set of values, it is possible to define, in terms of the primitive connectives, a new connective which has this matrix.

More precisely: a matrix logic **L** has a functionally complete set of connectives over its set of values *V* if, and only if, for any *m*-place operation $\phi: V^m \rightarrow V$, a connective Φ is definable in terms of the connectives of **L** such that the valuations of Φ over V^m are given by ϕ (where *m* is a positive integer). Otherwise, **L** is functionally incomplete.

For example, a three-valued matrix logic **L** with values $\{t,f,n\}$ is functionally incomplete, or has a functionally incomplete set of connectives, if for some *m*-place operation ϕ from $V^m = V \times V \times \ldots \times V$ (*m* times) to *V*, no *m*-place connective Φ is definable in **L** such that the valuations of Φ over V^m are given by ϕ.

Theorem (9): The logics $\mathbf{S}_2$ and $\mathbf{C}_2$ do not have a functionally complete set of connectives.

For the three-valued connective *Q* with matrix,

Γ	*t*	*f*	*n*
$Q\Gamma$	*t*	*n*	*f*

is not definable given just the connectives of 2-logics.

Proof: It is enough to show that *Q* is not definable in terms of '$\supset$', '$\sim$', and '*T*'; *i.e.* there is no wff Δ containing just these connectives such that: $Qp \simeq \Delta(p)$.

Now, when P has value f, QP, has value n, but $\Delta(P)$ can never lead out of the class of values $\{t,f\}$. This follows by induction over the number m of connectives in $\Delta(P)$.

For if $m = 0$, then $\Delta(P)$ is P which has value f when P has value f.

If $m > 0$, there are three cases:

(a) $\Delta(P)$ is of the form $T\Delta_1(P)$. By induction hypothesis, $\Delta_1(P)$ has value t or value f. Therefore by the matrix for 'T', so does $\Delta(P)$.

(b) $\Delta(P)$ is of the form $\sim\Delta_1(P)$. Since $\Delta_1(P)$ has value $\bar{n}(t$ or $f)$, by hypothesis, $\Delta(P)$ also has value $\bar{n}$.

(c) $\Delta(P)$ is of the form $\Delta_1(P) \supset \Delta_2(P)$. Since both $\Delta_1(P)$ and $\Delta_2(P)$ have values from $\{t,f\}$, by induction hypothesis, $\Delta(P)$ has value $\bar{n}$, by the matrix for '$\supset$'.

Corollary: 'Q' is not definable in terms of '$\supset$', '$\sim$' and 'S'. Hence $\mathbf{S}_1$ and $\mathbf{C}_1$ are not functionally complete.

The merits of functionally complete matrix logics will concern us in more detail later (§5.13). Meanwhile, however, the problem arises as to whether more connectives than those definable in 2-logics should be available in any adequate sentential significance logic. That is, we may raise the limited question of whether there are intuitively desirable significance connectives which should be added to 2-logics even though such addition may not entail functional completeness. Special connectives may be desirable for various reasons. They may facilitate the formalisation and assessment of commonly made significance claims and so be desirable at the level of interpretation. Or they may introduce a formal simplicity to the logic—for example, by making it possible to define existing primitives in terms of them.

A potentially valuable connective which satisfies both these kinds of conditions is '$\rightharpoonup$' with matrix,

$\rightharpoonup$	t	f	n
t	t	f	n
f	t	t	t
n	t	t	t.

It will be seen that it is a three-valued extension of the two-valued '$\supset$' and its interpretation is approximately—but only approximately—given by the ordinary language connective 'provided that ... then ...'. Before looking at its advantages, however, we first show that it is not definable in 2-logics.

Theorem (10): '$\rightharpoonup$' is not definable in $\mathbf{S}_2$ or $\mathbf{C}_2$.

Proof: It suffices to show that there is no function Δ containing connectives '$\supset$' '$\sim$' and 'T' such that:

$$(p \rightharpoonup q) \simeq \Delta(p,q)$$

The result follows from the following thesis:

(i) There is no function Δ of P and P' which has value n when P has value t and P' has value n, and values from $\{t,f\}$ when P has value n or P has value f and P' value n.

We prove this by induction over the number m of occurrences of '$\supset$', '$\sim$', and 'T' in Δ.

For if $m = 1$, then $\Delta(P, P')$ is $P \supset P'$ and (i) follows from the matrix for '$\supset$'. If Δ were to contain no connectives, then $\Delta(P, P')$ would coincide with either P or P', and $P \rightharpoonup P'$ would not then be well-defined.

If $m > 1$, there are three cases:

(a) $\Delta(P, P')$ is $T\Delta_1(P, P')$. Since $\Delta(P, P')$ then never has value n, (i) is established.
(b) $\Delta(P,P')$ is $\sim\Delta_1(P,P')$. If $\Delta_1(t,n)$ does not have value n, then $\sim\Delta_1(t,n)$ does not have value n. If, however, $\Delta_1(t,n)$ has value n, then, applying (i) as induction hypothesis, at least one Γ, of $\Delta_1(n, P')$ for $P' = t, f, n$, or $\Delta_1(f,n)$ has value n. Then $\sim\Gamma$ has value n and (i) follows for Δ.
(c) $\Delta(P, P')$ is $(\Delta_1(P, P') \supset \Delta_2(P, P'))$. Suppose $v(\Delta(t, n)) = n$. By the matrix for '$\supset$', either $v(\Delta_1(t, n)) = n$ or $v(\Delta_2(t, n)) = n$; so by the induction hypothesis, either $v(\Delta_1(f, n)) = n$ or $v(\Delta_1(n, P')) = n$ for some value of P' or $v(\Delta_2(f, n)) = n$ or $v(\Delta_2(n, P')) = n$ for some value of P'. Hence by the matrix for '$\supset$', either $v(\Delta(f, n)) = n$ or for some value of P' $v(\Delta(n, P')) = n$. Thus (i) follows for Δ.

Hence '$\rightharpoonup$' is not definable in 2-logics. But the systems obtained by taking it as an extra primitive, which we shall call 3-logics, are nevertheless not functionally complete. Thus,

Theorem (11):
(i) 3-logics are functionally incomplete since the connective 'Q' is not definable in terms of the connective set $\{\sim, \supset, T, \rightharpoonup\}$, or,
(ii) in terms of what proves to be the same, the independent set of connectives $\{\sim, \&, \rightharpoonup\}$.

Proof of (i) extends the proof that 'Q' is not definable in terms of the connectives of 2-logics (*Theorem* (9) above); there is one extra case to be considered:

(d) $\Delta(P)$ is of the form $\Delta_1(P) \rightharpoonup \Delta_2(P)$. Since $\Delta_1(P)$ and $\Delta_2(P)$ have values from $\{t,f\}$, so does $\Delta(P)$ by the matrix for '$\rightharpoonup$'.

As to (ii), '$\supset$' and 'T' are definable thus:

$$\Delta \supset \Gamma =_{\mathrm{Df}} \sim(\Delta \& \sim\Gamma); \; T\Delta =_{\mathrm{Df}} \sim(\Delta \rightharpoonup \sim\Delta).$$

That the connectives $\{\sim, \&, \rightharpoonup\}$ are independent has been shown by Brady [3] as follows: First, '$\sim$' is not definable in terms of '$\&$' and '$\rightharpoonup$' only, since any connective $\Psi(P)$ defined just in terms of these connectives takes value t when P has value t, but '$\sim$' does not. Second, '$\rightharpoonup$' is not definable in terms of '$\sim$' and '&' since any connective $\Psi(P, P')$ so definable takes value n when both P and P' have value n. Lastly, '&' is not definable in terms of '$\sim$' and '$\rightharpoonup$', since for any connective $\Psi(P, P')$ so definable it is not the case that if P has value n and P' has value $\bar{n}$ or P' has value n and P has value $\bar{n}$, then $\Psi(P, P')$ has value n.

But though the connective sets of 3-logics are not functionally complete they are significance complete—in the sense that they contain all matrices that do not turn sense into nonsense—and this may be the most that is desirable (see §5.13). More precisely, an operation $\phi: V^m \rightarrow V$ *preserves significance* iff it maps no element of V^m containing component t or f into n. So, for example, Q does not preserve significance because it maps f to n. A matrix significance logic **L**, with values $V = \{t, f, n\}$ has a *significance complete* set of connectives iff for any m-place operation $\phi: V^m \rightarrow V$ which preserves significance, a connective Φ is definable in terms of the connectives of **L** such that the valuations of Φ over V^m are given by ϕ. These ideas and the following *Theorem* (12) which shows that 3-logics have significance complete connective sets are due in essentials to Brady.

Let $v, v_1, v_2, ..., v_m, ...$ be variables which range over elements of V.

Theorem (12): If $\phi(v_1, ..., v_m)$ is an m-place operation from V^m to V which preserves significance then there is an m-place connective $\Phi(\Delta_1, ..., \Delta_m)$ definable in terms of the connective set $3\mathbf{L} = \{\sim, \&, \rightharpoonup\}$.

Proof is by induction over m.

The induction basis, where $m = 1$, is established by an exhaustion of cases:

Where $t\Delta =_{\text{Df}} \Delta \rightharpoonup \Delta$; $L\Delta =_{\text{Df}} \sim(\Delta \,\&\, \sim\Delta)$
$T\Delta =_{\text{Df}} \sim(\Delta \rightharpoonup \sim\Delta)$; $F\Delta =_{\text{Df}} T\sim\Delta$
$S\Delta =_{\text{Df}} \sim(\sim T\Delta \,\&\, \sim F\Delta)$; $Id\Delta =_{\text{Df}} \Delta$,

the following are all the 1-place significance-preserving connectives:

$t, S, L, \neg T, F, \sim, \neg F, T, Id, \neg S, \neg t, \sim L.$

The induction step assumes the theorem for all m-place significance-preserving operations and extends the proof to $m+1$-place operations. Let $\phi(v_1, ..., v_{m+1})$ be a significance-preserving $m+1$-place operation. Consider each $\phi(v_1, ..., v_m, v) = \phi(\underline{v}, v)$ for $v = t, f, n$ in turn. These provide three significance-preserving operations:

$\phi_t(\underline{v}) = \phi(\underline{v}, t)$, $\phi_f(\underline{v}) = \phi(\underline{v}, f)$ and $\phi_n(\underline{v}) = \phi(\underline{v}, n)$, of m-places.

Accordingly by induction hypothesis, there are m-place connectives,

$\Phi_t(\Delta_1, ..., \Delta_m)$, $\Phi_f(\Delta_1, ..., \Delta_m)$, and $\Phi_n(\Delta_1, ..., \Delta_m)$ definable in terms of 3L.

We show that a connective Φ corresponding to ϕ in the way required in the theorem may be defined as follows:

$$\Phi(\Delta_1, \ldots, \Delta_{m+1}) =_{\text{Df}} \Phi_t(\Delta_1, \ldots, \Delta_m) \cap T\Delta_{m+1}$$
$$\cup . \Phi_f(\Delta_1, \ldots, \Delta_m) \cap F\Delta_{m+1} \cup . \Phi_n(\Delta_1, \ldots, \Delta_m) \cap \sim S\Delta_{m+1}$$

—where the strong regular connectives '$\cap$' and '$\cup$', defined,

$$\Delta \cap \Gamma =_{\text{Df}} \sim [\sim(F\Delta \,\&\, \sim S\Gamma) \,\&\, \sim(\sim S\Delta \,\&\, F\Gamma) \rightharpoonup . \sim(\Delta \,\&\, \Gamma)]$$
$$\Delta \cup \Gamma =_{\text{Df}} \sim(\sim\Delta \cap \sim\Gamma),$$

have as before these matrices:

$\cap$	t	f	n
t	t	f	n
f	f	f	f
n	n	f	n

$\cup$	t	f	n
t	t	t	t
f	t	f	n
n	t	n	n

Now let ϕ^1 be the operation corresponding to Φ. Then, where operations $\cap$ and $\cup$ correspond to connectives '$\cap$' and '$\cup$',

$$\begin{aligned}\phi^1(\underset{\sim}{v}, t) &= \cup\,(\cap\,(\phi_t(\underset{\sim}{v}), t), \cap\,(\phi_f(\underset{\sim}{v}), f), \cap\,(\phi_n(\underset{\sim}{v}), f))\\ &= \cup\,(\phi_t(\underset{\sim}{v}), f, f)\\ &= \phi_t(\underset{\sim}{v})\\ &= \phi(\underset{\sim}{v}, t).\end{aligned}$$

Similarly $\phi^1(\underset{\sim}{v}, f) = \phi(\underset{\sim}{v}, f)$ and $\phi^1(\underset{\sim}{v}, n) = \phi(\underset{\sim}{v}, n)$, whence by the extensionality of operations, $\phi^1(\underset{\sim}{v}, v_{m+1}) = \phi(\underset{\sim}{v}, v_{m+1})$ as required.

Consider now the case for introducing '$\rightharpoonup$' as an additional primitive. The main arguments for its introduction like the main arguments for the introduction of certain further connectives which we shall consider in subsequent sections are, as we have suggested, of two kinds:

(a) the need for the connective at the level of interpretation, so that various kinds of significance claims can be formalised and assessed;
(b) the technical usefulness of the connective.

Consider (a): In a sentential logic adequate to formalise commonly made significance claims, it should be possible significantly to make claims of the sort that a given formula holds provided it takes significant values. Thus, for example, we may wish to assert:

(i) If Sp then $p \supset p$.

We can of course express such a claim indirectly in **S**-logics merely by affirming,

(ii) $r \supset r$;

given that *qu*(*r*) is a restricted variable; but we cannot state the claim explicitly, nor can we explicitly express the condition that *qu*(*r*) is a restricted variable. If, for example, we represent (i) by using a material 'if ... then ...', we have,

(iii) $Sp \supset (p \supset p)$,

which is not a thesis of **S**-logics since it takes value *n* when *qu*(*p*) takes value *n*. It is a thesis of **C**-logics, but it has degenerate instances and therefore does not always express a significant claim, contrary to what was intended. Using '$\rightharpoondown$', however, (i) can be symbolised as,

(iv) $Sp \rightharpoondown (p \supset p)$.

In this form it is a thesis of **S**-logics since it takes the value *t* for all values of the variables.

Similarly, that $qu(p \supset p \vee q)$ holds provided both *qu*(*p*) and *qu*(*q*) are significant, can be expressed by,

(v) $Sp \;\&\; Sq \rightharpoondown (p \supset p \vee q)$.

If we admit '$\rightharpoondown$' as a new primitive connective, therefore, we are able to refute Russell's contention (Russell [2], p. 72) that "we can not make ranges of significance explicit, since the attempt to do so only gives rise to a new proposition in which the same range of significance is implicit". For suppose we wish to say that *qu*(if *x* is a man, then *x* is mortal) is significant provided *qu*(*x*) is restricted in range to a class *K*, say, which might be the class of animate objects or even animate and inanimate objects. Then we may make the range of significance explicit as a sufficient condition by affirming,

provided it is true that *x* is a *K*, then, if *x* is a man, *x* is mortal;

and this we may symbolise by,

$$T(x \in K) \rightharpoondown (x \text{ is a man} \supset x \text{ is mortal}),$$

which is significant for all values of *qu*(*x*). For if the value chosen for *qu*(*x*) is such that *qu*(*x* is a man $\supset$ *x* is mortal) is significant, then *x* belongs to *K* and hence $qu(x \in K)$ is significant, in fact true. If, on the other hand, the value chosen for *qu*(*x*) is such that *qu*(*x* is a man $\supset$ *x* is mortal) is nonsignificant (*e.g.* the value chosen for *qu*(*x*) might be the name of a rainbow), then $T(x \in K)$ is false and the whole formula is significant since $v(f \rightharpoondown n) = t$.

This solution to the problem of the significant formulation of significance conditions is by no means unique, however. Many other three-valued "implications" serve just as well. Indeed a complete three-valued connective is not necessary. The following non-iterable connective '$/\!\!-$' with matrix,

↽	*t*	*f*	*n*
t	*t*	*f*	*n*
f	*t*	*t*	*t*

would serve, since in the formulation of significance conditions the antecedent is always significant. So, too, would '∖–', defined by :Γ ∖– $\Delta =_{Df} T\Gamma \supset T\Delta$. Or at less formal cost, provided bound restricted variables are not wanted, a rule formulation of (i), *viz.*,

(vi) $\vdash Sp \twoheadrightarrow \vdash (p \supset p)$, where '$\twoheadrightarrow$' reads 'yields,'

could be used. Rules such as this are derivable even in 1-logics. Notice that it is not at all easy to establish or refute the claim that the 'provided' in significance conditions formulates an inference rule rather than, what it seems to formulate, an implication of some sort. What can be argued with some plausibility is that if (vi) were replaced by an implicational formulation of the matrix type, the implication adopted would be '⇀'. For a rule formulation expresses the fact that an inference can be made preserving truth (or provability) provided the antecedent premiss is true. If an implication is to have a similar effect, therefore, it must discount all cases where the antecedent is not true. These cases can be discounted by assigning the designated value *t* to all instances of the implication where the antecedent does not have value *t* (the cases would not be discounted by assigning them value *n* except in **C**-logics). Thus the appropriate implication would be obtained from '$\supset$' by replacing values of the matrix where the antecedent does not have value *t*, by the value *t*. The resulting rule-representing connective is then '⇀'.

(b) Perhaps the main arguments in favour of introducing '⇀', however, derive from its technical value. For, as we show in §6.5, it enables us to reduce the variables of **S**-logics from two sorts to one. This is a direct consequence of being able to formulate explicitly the significance conditions on various formulae. And it has a connected advantage in quantified significance logics, since it enables us to formulate restricted variable logics in place of clumsier many-sorted systems which would otherwise be necessary. This will be shown in detail in §7.6 *ff*. The implication '⇀' does not, however, have a unique position in the reduction of sorts, since other connectives will also serve the purpose. Where '⇀' does have a unique role is in the theory of restricted quantification. Suppose we want to restrict the variable X of a universally quantified claim, that for every X, F(X), by the condition that X satisfies G, *i.e.* G(X); *e.g.* we may want to require the X is an integer or that X be a person. Under weak assumptions (discussed in §7.9, §6) the restricted variable claim can be written in the form (*U*X) G(X)/F(X), where '/' is a two-place connective; the claim is read 'for every X such that G(X), F(X)' or 'for every X where X is restricted by G, F(X)'. We expect (as in §3.5 and

§7.7), the universal quantifier 'U' to satisfy the general semantical rule: $(UX)A(X)$ has value t iff $A(X)$ has value t for every assignment to X, and $(UX)A(X)$ has value n iff $A(X)$ has value n for some assignment to X. Further we impose the obvious semantical requirement on the restricted variable claim $(UX)G(X)/F(X)$, *viz.* that it has value t iff F(X) has value t for every assignment to X such the G(X) has value t, and that it has value n iff F(X) has value n for some assignment to X for which G(X) has value t. Finally we suppose that '/' is matrix accessible, and to ensure that it is we make the usual but important assumption that G(X) does have the value t for some assignment to X. Then what matrices, if any, will satisfy '/'? Just one: that for '$\rightharpoondown$'. It follows from a comparison of the semantical rules that for an assignment to X such that G(X) has value t the value of G(X)/F(X) is the same as the value of F(X). Thus, since '/' is matrix-accessible, $v(t/\Delta) = v(\Delta)$. It remains to consider $v(\bar{t}/\Delta)$. Suppose for a given assignment X_0 to X, $G(X_0)$ has value $\bar{t}$, *i.e.* f or n. Then whatever the value of $F(X_0)$, the value of $G(X_0)/F(X_0)$ must, since '/' is matrix-accessible, be t. For, by the semantical rules, when the value of $G(X_0)$ is t the value v of $G(X_0)/F(X_0)$ is discounted in assessing the value of $(UX)G(X)/F(X)$. Now v can only be generally discounted in a conjunctive assessment of the sort demanded by the rule for 'U' if it has the value t; for $(v \;\&\; v_1) \simeq v_1$ iff $v = t$. Accordingly the connective '$\rightharpoondown$' has a unique and essential role in the *extensional* theory of restricted quantification.

Furthermore, given '$\rightharpoondown$' as a primitive, it is possible to define 'T' in terms of it (though as we have seen, the converse is not true) and this means that 2-logics are contained in systems which contain '$\rightharpoondown$'. One suitable definition is,

$$T\Delta =_{\mathrm{Df}} {\sim}(\Delta \rightharpoondown {\sim}(\Delta \rightharpoondown \Delta)),$$

as a valuation table for each will show; another is

$$T\Delta =_{\mathrm{Df}} {\sim}(\Delta \rightharpoondown {\sim}\Delta).$$

These advantages, both interpretational and formal are only obtained however at the expense of sacrificing the principle of component homogeneity. Thus '$\rightharpoondown$' is non-classical since it is not the case that formulae of the sort $P \rightharpoondown P'$ take the value n whenever one of the components takes the value n. To some extent, of course, we have already sacrificed this principle by admitting the connectives 'T' 'F', 'S' and '$\simeq$'. This might be thought to be justified, however, on the grounds that they absorb quotation functions. For it might be accepted that the principle of component homogeneity does not (need not) hold for quoted components. Thus, it might be said, we have not yet sacrificed the principle in the form; a compound sentence is significant if, and only if, each of its non-quoted components is significant. But while it is true that 'T', 'F' and 'S' do not violate this version of the principle, it is less clear that '$\simeq$' does not. Certainly '$\simeq$' is defined in terms of 'T', 'F' and 'S' and from that point of view might be said to absorb quotation functions, as they do; and it can be given a mentioning-interpretation, namely 'has the same significance value as'. But just as '$\equiv$' has a mentioning interpretation, since it expresses the fact that two sentences have the same truth-value, and also a use-interpretation, namely 'if, and only if', so '$\simeq$', too, has a similar use-interpretation. It is at least arguable, therefore, whether '$\simeq$' can be said to absorb quotation. Rather, it abbreviates the use-phrase 'if, and only if' and thereby expresses, but does not state, that the sentences so related have the same significance value. And if this is so, then it violates the principle of component homogeneity even in its revised form.

There is a similar inconclusiveness about the interpretation of '$\rightharpoondown$'. For we wish it to approximate to the ordinary language use-connective 'provided that ... then ...' which does not absorb quotation; and yet 'T', which can be defined in terms of it, does absorb quotation. But of course, even if we take a classical definition of truth, *viz.*:

$$T^{c}\Delta =_{\mathrm{Df}} \Delta$$

then all that is being said is that a metasentence, $T^{c}\Delta$, is equivalent to an object sentence Δ in the sense that they always have the same truth (significance) value, and this does not imply that because quotes occur in $T^{c}\Delta$ so they must also occur in its equivalent, Δ.

What matters, therefore, is not the occurrence of quotation on one side of the definition, but the interpretation which can be put on the connective which is being defined, or in terms of which the definition is made. And from this point of view, since both '$\simeq$' and '$\rightharpoondown$' can be given a satisfactory use-interpretation, it is reasonable to conclude that they do not absorb quotation and that they do, therefore, violate the principle of competent homogeneity.

Now although component homogeneity has often been assumed in the literature which bears on significance theory, and was certainly assumed by Russell, the grounds for its adoption are unfortunately left obscure (see, for example Russell [2] and Tarski [1], p. 162). The grounds for its adoption in the case of the classical significance connectives, we have already discussed, and in general we accept it for this limited set of connectives. But this is not yet to say that there could not be, or ought not to be, any other connective which violates it and yet has a satisfactory interpretation.

The real issue is whether there are sentences of ordinary discourse which are significant when they have nonsignificant components. And in particular, from the present point of view, whether sentences such as,

The number seven lives next door if, and only if, the moon is carnivorous;
If the earth is flat, then the moon is carnivorous;

are significant given a context in which 'The earth is flat' is significant and both 'The number seven lives next door' and 'The moon is carnivorous' are nonsignificant.

A strong case can be made for saying that they are significant. For in the case of the first, we might suppose a context in which it is granted that 'The moon is carnivorous' is nonsignificant; then, by affirming 'The number seven lives next door if, and only if, the moon is carnivorous' we emphasise the nonsignificance of 'The number seven lives next door'. This amounts to pressing the claim that 'The number seven lives next door' is nonsignificant by arguing implicitly that it is as absurd to believe it as it is to believe that the moon is carnivorous. But such an appeal cannot be made unless 'The number seven lives next door if, and only if, the moon is carnivorous' is not only significant but true. Such uses depend on the thesis:

$$\sim Sp \equiv (p \simeq \Gamma_0)$$

where Γ_0 is constant sentence with value *n*. Here, '$\equiv$' would be inadequate as a formalisation of 'if, and only if' on the right.

The case for the second is similar. Sentences such as 'If the earth is flat then the moon is carnivorous' could be used with the intention of asserting truths to the effect that the statement which the antecedent could be used to make is categorically false. In this respect they are analogous to sentences like,

If God exists, then pigs fly,

which are used to emphasise the falsity of the antecedent and succeed because the consequent is acknowledged to be false. They depend on the two-valued thesis,

$$F^c r \equiv (r \supset \Delta_0),$$

where Δ_0 is a constant sentence with value *f*. The three-valued extension of this, exemplified by 'If the earth is flat then the moon is carnivorous' depends on the similar thesis,

$$Fr \equiv T(r \rightarrow \Gamma_0),$$

where, as before, Γ_0 is a constant with value *n* and *qu*(*r*) is truth-valued. Here '$\supset$' is inadequate as a representation of the connective 'if ... then ...' on the right. Thus we are able to express the special ways of emphasising the falsity of a given sentence *P* by affirmations of the sort 'If *P* then anything goes', where the "anything goes" may on occasion be expressed by a nonsignificant sentence.

It is however far from clear that examples such as these should carry much weight. For, it may be argued, dictionaries explicitly list such expressions as abnormal. And since even the best formalised theory should not be expected to take account of all abnormalities and vagaries of discourse, there is no strong case for accepting that they and other similar cases, such as zeugmas, count against a theory which does not explicitly take account of them. A logic which fails to provide means of formalising such abnormal expressions is not thereby inadequate. But against this, it seems pointless to leave such sentences out of account when the means of formalising them is achieved so simply, and when the only sacrifice which has to be made is the violation of the principle of component homogeneity in some cases. Why, after all, should the principle be inviolate with respect to a formal system when there appear to be cases of sentence-use in ordinary language which do not satisfy it?

We do not regard these arguments as conclusive. Rather, it seems to us, the problem of component homogeneity is a characteristic conflict issue calling for a decision rather than a conclusion. Affirmative and negative answers can be supported by inconclusive arguments. The resolution we favour, as the bias in the discussion may indicate, is the introduction of new connectives such as '$\rightharpoonup$' as technical and theoretical adjuncts, and hence the rejection of the component homogeneity principle. Thus we favour stronger logics than 2-logics. Whether or not this decision is accepted, however, it is of considerable interest to examine systems which go beyond 2-logics. To invest wholeheartedly in the principle of component homogeneity would be to restrict significance connectives to classical ones, or at best to these together with 'T', 'F' and 'S'. It does of course remain open that '$\simeq$' and '$\rightharpoonup$' absorb quotation so that they can be admitted and the principle retained in its revised form, but we find this move unnecessary and, in the case of other non-classical connectives, such as 'Q', implausible.

The Logic $\mathbf{S}_3$

(a) *Morphology*

Add to the morphology of $\mathbf{S}_0$,
the primitive improper symbol: $\rightharpoonup$;
the formation rule: if Δ and Γ are wff, so is $(\Delta \rightharpoonup \Gamma)$;
the definitions: $T\Delta =_{\mathrm{Df}} \sim(\Delta \rightharpoonup \sim(\Delta \rightharpoonup \Delta))$

$$F\Delta =_{\mathrm{Df}} T{\sim}\Delta$$
$$S\Delta =_{\mathrm{Df}} T\Delta \vee F\Delta$$
$$\Delta \simeq \Gamma =_{\mathrm{Df}} (T\Delta \mathbin{\&} T\Gamma) \vee (F\Delta \mathbin{\&} F\Gamma) \vee ({\sim}S\Delta \mathbin{\&} {\sim}S\Gamma).$$

Hence, since 'T' can be defined, $\mathbf{S}_3$ contains $\mathbf{S}_2$; and as we know, $\mathbf{S}_2$ contains $\mathbf{S}_1$ which contains $\mathbf{S}_0$.

(b) *Valuation Rules*

Add to the valuation rules of $\mathbf{S}_0$.
If $v(\Delta) = t$, then $v(\Delta \rightharpoonup \Gamma) = v(\Gamma)$;
if $v(\Delta) \neq t$, then $v(\Delta \rightharpoonup \Gamma) = t$.

This determines the matrix for '$\rightharpoonup$'. The valuation rules determining the appropriate matrices for 'T', 'F', 'S' and '$\simeq$' follow by the definitions.

(c) *Tautologies*

A wff of $\mathbf{S}_3$ is a tautology if, and only if, it takes the value *t* for all assignments of values to its variables.

Hence, all $\mathbf{S}_0$-, $\mathbf{S}_1$- and $\mathbf{S}_2$-tautologies are $\mathbf{S}_3$-tautologies.

(d) *Valuation tables*

As for $\mathbf{S}_2$ wff.

(e) *Formal Theorems*

A wff of $\mathbf{S}_3$ is a thesis if, and only if, it is a $\mathbf{S}_3$-tautology.

In addition to the theorems of $\mathbf{S}_2$ we have:

$$(p \supset q) \rightharpoonup (p \rightharpoonup q);\ (p \,\&\, (p \rightharpoonup q)) \rightharpoonup q;\ \sim p \rightharpoonup (p \rightharpoonup q);\ q \rightharpoonup (p \rightharpoonup q);\ p \rightharpoonup p;$$
$$Tp \simeq \sim (p \rightharpoonup \sim (p \rightharpoonup p));\ Sp \rightharpoonup (p \supset p);\ Sp \,\&\, Sq \rightharpoonup (p \supset p \vee q);\ etc.$$

Since some of these theorems contain *S*-unrestricted uncovered variables, we have a further technique for overcoming the problem of obtaining sentential theorems in significance logics which take only *t* as the designated value. The theorem, $qu((p \,\&\, (p \rightharpoonup q)) \rightharpoonup q)$ shows that modus ponens holds for '$\rightharpoonup$'; but $qu(\sim p \rightharpoonup (p \rightharpoonup q))$ and $qu(q \rightharpoonup (p \rightharpoonup q))$ show that '$\rightharpoonup$' has no advantages over '$\supset$' from the point of view of the paradoxes of material implication. The last two theorems listed, and others like them, justify our claim that the use of '$\rightharpoonup$' enables us to state significance conditions explicitly and without incurring nonsignificance.

The Logic $\mathbf{C}_3$

(a) *Morphology*

Add to the morphology of $\mathbf{C}_0$, the special morphology of $\mathbf{S}_3$.

Hence, $\mathbf{C}_3$ contains $\mathbf{C}_2$ contains $\mathbf{C}_1$ contains $\mathbf{C}_0$.

(b) *Valuation Rules*

Add to the valuation rules of $\mathbf{C}_0$, the new valuation rule of $\mathbf{S}_3$.

(c) *Tautologies*

A wff of $\mathbf{C}_3$ is a tautology if, and only if it takes the value *t* or the value *n* for all assignments of values to its variables.

Hence, all $\mathbf{C}_0$-, $\mathbf{C}_1$- and $\mathbf{C}_2$-tautologies are $\mathbf{C}_3$-tautologies.

(d) *Valuation tables*

As for $\mathbf{C}_2$ wff.

(e) *Formal Theorems*

A wff of $\mathbf{C}_3$ is a theorem if, and only if, it is a $\mathbf{C}_3$-tautology.

In addition to the theorems of $\mathbf{C}_2$, we have all those listed as $\mathbf{S}_3$ theorems together with certain theorems such as,

$$(p \rightharpoonup q) \equiv (\sim p \vee q);\ (p \rightharpoonup q) \equiv (p \supset q);$$

which have degenerate instances since they take the value *n* for some values of the variables.

C-*logics Consistency Theorem: Part* 3

Matrix $\mathbf{C}_3$-logics are simply consistent with respect to all primitive and definable monadic connectives κ which are minimally interpretable as negations.

Proof: We add one case to the Part 2 theorem (§5.10):

(v) Δ is of the form $\Delta_1 \rightharpoonup \Delta_2$. By the matrix for '$\rightharpoonup$', Δ can only take the value *n* for all values of the variables in it if Δ_2 takes the value *n* for all values of the variables. By the induction hypothesis, however, Δ_2 does not take value *n* when all variables in it are assigned value *t*.

Result by III of §5.6.

As in Part 2, the theorem establishes the simple consistency of matrix $\mathbf{C}_3$-logics with respect to '$\sim$', '$\neg$' and '*F*'.

5.12 Stage 4: Alternative Formulations of $\mathbf{S}_3$ and $\mathbf{C}_3$, and the 4-Logics

As we have already remarked (in §5.4) Åqvist [1] develops a logic, designed to be a significance logic, which has among its primitives a strong regular connective. Åqvist's logic **Å** has a set of primitive connectives $\{\cup, \sim, T\}$; and in terms of these, other strong regular connectives are definable. The matrix logic based on these connectives proves to be the same as a logic used by Kleene [1] and [2]. Actually two sentential systems can be discerned in Kleene [2]; the first $\mathbf{K}_1$, with primitives $\{\sim, \cup\}$ after reduction, is the system the interpretation of which Kleene discusses and the logic for which he argues, but the second $\mathbf{K}_2$, with primitives $\{\sim, \cup, \simeq\}$ after reduction, is the sentential system in which he states needed theorems and which he uses in his subsequent theory. The remainder of Kleene's three-valued connectives may be defined in the usual way:

$\Delta \supset_K \Gamma$ as $\sim\Delta \cup \Gamma$, $\Delta \cap \Gamma$ as $\sim(\sim \Delta \cup \sim\Gamma)$, and $\Delta \equiv_K \Gamma$ as $((\Delta \supset_K \Gamma) \cap (\Gamma \supset_K \Delta))$.

Both Åqvist's **Å** and Kleene's $\mathbf{K}_2$ prove equivalent in turn to an earlier and more famous system, Łukasiewicz's three-valued logic **L** (discussed at length in Prior [2]). **L** has the advantage over **Å** and $\mathbf{K}_2$ of a smaller set of primitives, since the set $\{\sim, \supset_L\}$ suffices, and of a simple axiomatisation. All the logics, **Å**, $\mathbf{K}_1$, $\mathbf{K}_2$ and **L** are **S**-logics (since *t* is the only designated value), but they are all constructed using *only* *S*-unrestricted variables

$$p, q, p', q', \ldots$$

and they are all complete over a three-valued interpretation with values t, f and n (after relettering), Åqvist's system and Lukasiewicz's systems demonstrably so and Kleene's systems automatically since they are matrix systems. Theoremhood can be defined for each system using the matrix approach, just as for $\mathbf{S}_3$ for instance.

Theorem (13):

(i) The connective sets of **Å**, $\mathbf{K}_2$ and **L** are interdefinable.

(ii) **Å**, $\mathbf{K}_2$ and **L** are deductively equivalent, *i.e.* every theorem of **Å** is a theorem of $\mathbf{K}_2$ and conversely, and every theorem of **Å** is a theorem of **L** and conversely.

Proof: (i) The connectives of **Å** can be defined in terms of those of $\mathbf{K}_2$ as follows:

$$\Delta \supset_A \Gamma =_{Df} \Delta \cap \Gamma \equiv_K \Delta$$
$$\Delta \vee \Gamma =_{Df} (\sim\Delta \supset_A \Gamma) \cap (\sim\Gamma \supset_A \Delta)$$
$$\Delta \,\&\, \Gamma =_{Df} \sim(\sim\Delta \vee \sim\Gamma)$$
$$T\Delta =_{Df} \Delta \simeq (\Delta \simeq \Delta)$$

Conversely the connectives of $\mathbf{K}_2$ can be defined in terms of those of **Å** simply by using the definition of '$\simeq$' already given in $\mathbf{S}_2$ together with the definition: $\Delta \vee \Gamma =_{Df} (\Delta \cap \Gamma) \equiv_K (\Delta \equiv_K \Gamma)$. (See the morphology of $\mathbf{S}_3$-1 below which takes the same primitives as **Å**).

The connectives of **Å** can be defined in terms of those of **L** thus:

$$\Delta \cup \Gamma =_{Df} (\Delta \supset_L \Gamma) \supset_L \Gamma; \; T\Delta =_{Df} \sim(\Delta \supset_L \sim\Delta).$$

The implication '$\supset_L$' has the matrix

$\supset_L$	t	f	n
t	t	f	n
f	t	t	t
n	t	n	t

Conversely, since the connective set of **Å** is significance complete (by *Theorem* 12 and the definitions given below in formulating $\mathbf{S}_3$-1), '$\supset_L$' is definable in **Å**.

(ii) in virtue of (i) the valuation tables for the connectives of **Å**, $\mathbf{K}_2$ and **L** may be interdefined. Therefore by an inductive proof over the number of connectives in a given tautology (in primitive form), any tautology of one system is a tautology of the other, and similarly for the converse. The result then follows since both **Å** and **L** are complete with respect to their matrices and since tautologousness and theoremhood coincide in $\mathbf{K}_2$.

Corollary: **L** and its equivalents are significance complete.

Theorem (14): $\mathbf{K}_1$ is a proper subsystem of $\mathbf{K}_2$.

Proof: Since $\mathbf{K}_1$ is a subsystem of $\mathbf{K}_2$ it suffices to show (I) 'T', and hence '$\simeq$', is not definable in $\mathbf{K}_1$. For then the result follows since $\mathbf{K}_2$ has theorems containing '$\simeq$' which are not derivable (using definitional substitution) in $\mathbf{K}_1$. (I) is immediate, upon considering the value of Tp when $v(p) = n$. For by induction any function Δ(P) of P containing just the connectives of $\mathbf{K}_1$ has value n whereas $v(Tp) = f$.

Kleene's $\mathbf{K}_1$ is scarcely a satisfactory system. For:

Theorem (15): $\mathbf{K}_1$ has no theorems.

Proof: Every theorem must contain some variable, since every theorem is a wff. Suppose wff Δ contains just variables $P_1, ..., P_n$. As in the previous theorem, $v(\Delta) = n$ where $v(P_1) = v(P_2) ... = v(P_n) = n$. Alternatively the previous metatheorem follows from this one, since $\mathbf{K}_2$ has theorems, *e.g.* $p \simeq p$.

Corollary: The implication '$\supset_L$' of **L** is not definable in terms of $\mathbf{K}_1$.

As before, this difficulty, that there are no theorems, can be avoided if $\mathbf{K}_1$ is replaced by a two-sorted system with the same primitive connectives. It is convenient moreover, for comparison purposes, to consider first not **Å** or $\mathbf{K}_2$ but a logic $\mathbf{S}_3$-1 obtained by converting **Å** into a two-sorted system. This has the added advantage of simplifying axiomatic formulations of the logic, at no great cost in extras. For we shall show later how to eliminate the *S*-restricted variables added to **Å** to obtain $\mathbf{S}_3$-1. An elimination can be made because '$\rightharpoonup$' is definable in $\mathbf{S}_3$-1 thus:

$$\Delta \rightharpoonup \Gamma =_{\mathrm{Df}} T\Delta \supset_A \Gamma,$$

where '$\supset_A$' is defined as in **Å**. Conversely, *S*-restricted variables can be introduced into $\mathbf{K}_2$ or **Å** using '$\rightharpoonup$', and doing this facilitates axiomatisation of these systems. Åqvist's axiomatisation of **Å** can be much improved upon in this way. To make the axiomatisation neat and to facilitate quantificational extensions we introduce also a one-sorted formulation of $\mathbf{S}_3$, called $\mathbf{S}_4$, with connective set $\{\sim, \&, \rightharpoonup\}$.

The Logic $\mathbf{S}_3$-1

(a) *Morphology*

Improper primitive symbols: $\cup$, $\sim$, T
Variables as for $\mathbf{S}_0$

Formation rules are the usual with 1-place connectives '$\sim$', and 'T' and 2-place connective '$\cup$'.

Definitions:

$\Delta \cap \Gamma =_{\mathrm{Df}} \sim(\sim\Delta \cup \sim\Gamma)$
$\Delta \supset_K \Gamma =_{\mathrm{Df}} \sim\Delta \cup \Gamma$
$\Delta \equiv_K \Gamma =_{\mathrm{Df}} (\Delta \supset_K \Gamma) \cap (\Gamma \supset_K \Delta)$
$\Delta \supset_A \Gamma =_{\mathrm{Df}} \Delta \cap \Gamma \equiv_K \Delta$
$\Delta \vee \Gamma =_{\mathrm{Df}} (\Delta \cap \Gamma) \equiv_K (\Delta \equiv_K \Gamma)$
$\Delta \,\&\, \Gamma =_{\mathrm{Df}} \sim(\sim\Delta \vee \sim\Gamma)$
$\Delta \supset \Gamma =_{\mathrm{Df}} \sim\Delta \vee \Gamma$
$\Delta \equiv \Gamma =_{\mathrm{Df}} (\Delta \supset \Gamma) \,\&\, (\Gamma \supset \Delta)$
$F\Delta =_{\mathrm{Df}} T\sim\Delta$
$S\Delta =_{\mathrm{Df}} T\Delta \vee F\Delta$
$\Delta \simeq \Gamma =_{\mathrm{Df}} (T\Delta \,\&\, T\Gamma) \vee (F\Delta \,\&\, F\Gamma) \vee (\sim S\Delta \,\&\, \sim S\Gamma)$
$\Delta \rightharpoonup \Gamma =_{\mathrm{Df}} T\Delta \supset_A \Gamma$

(b) *Valuation Rules*

Add to the valuation rules already given for '$\sim$' and 'T' and their bases the following rules for '$\cup$':

For any wffs Δ and Γ,
if $v(\Delta) = t$ or $v(\Gamma) = t$ then $v(\Delta \cup \Gamma) = t$;
if $v(\Delta) = f = v(\Gamma)$ then $v(\Delta \cup \Gamma) = f$; and
if $v(\Delta) = n$ and $v(\Gamma) \neq t$ or $v(\Gamma) = n$ and $v(\Delta) \neq t$ then $v(\Delta \cup \Gamma) = n$.

(c) *Tautologies*
(d) *Valuation Tables* } These are explained as in $\mathbf{S}_3$.
(e) *Formal Theorems*

Examples of theorems are as follows:

$$Tp \vee Tq \supset T(p \cup q); \quad Fp \,\&\, Fq \supset F(p \cup q); \quad S(p \cup q) \supset Sp \vee Tq.$$

But in fact $\mathbf{S}_3$-1 is not a new system at all.

Theorem (16): The theorem sets of $\mathbf{S}_3$ and $\mathbf{S}_3$-1 coincide.

Proof: Since '$\rightharpoonup$' is definable in $\mathbf{S}_3$-1, $\mathbf{S}_3$-1 contains $\mathbf{S}_3$. Conversely since '$\cup$' can be defined in $\mathbf{S}_3$ as follows,

$$\Delta \cup \Gamma =_{\mathrm{Df}} \sim(T\Delta \,\&\, \sim S\Gamma \,\vee .\, \sim S\Delta \,\&\, T\Gamma) \rightharpoonup .\, \Delta \vee \Gamma,$$

every theorem of $\mathbf{S}_3$-1 is a theorem of $\mathbf{S}_3$ (allowing of course for definitional substitution).

The semantical systems so far presented are scarcely formalised. We shall now indicate how to construct more formal versions of the matrix systems. This further formalisation can occur in the formulation of the morphology or in the formulation of the valuation rules or in both. We shall concentrate on cases where the valuation rules and valuation-assessment techniques are formalised. We illustrate this further formalisation, which can be carried out for each semantical system treated, by giving a more elaborate formalisation of S_3-1. Consider first a system S_3-2 which has the same morphology as S_3-1 but in which the valuation rules are amended by replacing ordinary language connectives by logical connectives—in the case we consider, we replace 'not', 'and' and 'or' by the logical connectives '$\sim$', '&' and '$\vee$' partly defined in the system, and 'if ... then ...' by the modal implication connective '$\strictif$'. S_3-2 has the following valuation rules (*cf.* the detailed valuation rules containing ordinary-language connectives given for S_2):

(2′) For every unrestricted variable $qu(q)$,
$(v(q)=t)\vee(v(q)=f)\vee(v(q)=n);\ \ \sim(v(q)=t\ \&\ v(q)=f);$
$\sim(v(q)=t\ \&\ v(q)=n);\ \ \sim(v(q)=f\ \&\ v(q)=n).$

(3′) For every restricted variable $qu(r)$,
$(v(r)=t)\vee(v(r)=f);\ \ \sim(v(r)=t\ \&\ v(r)=f).$

(4′) For every wff Δ,
$v(\Delta)=t\strictif v(\sim\Delta)=f;\ \ v(\Delta)=f\strictif v(\sim\Delta)=t;\ \ v(\Delta)=n\strictif v(\sim\Delta)=n.$

(5′) For every wff Δ and Γ,
$v(\Delta)=t\ \&\ v(\Gamma)=t\strictif v(\Delta\ \&\ \Gamma)=t;$
$((v(\Delta)=f\ \&\ v(\Gamma)\neq n)\vee(v(\Delta)\neq n\ \&\ v(\Gamma)=f))\strictif v(\Delta\ \&\ \Gamma)=f;$
$((v(\Delta)=n)\vee(v(\Gamma)=n))\strictif v(\Delta\ \&\ \Gamma)=n.$

(6′) For every wff Δ,
$v(\Delta)=t\strictif v(T\Delta)=t;\ \ v(\Delta)=f\strictif v(T\Delta)=f;\ \ v(\Delta)=n\strictif v(T\Delta)=f.$

(7′) For every wff Δ and Γ,
$((v(\Delta)=t)\vee(v(\Gamma)=t))\strictif v(\Delta\cup\Gamma)=t;$
$(v(\Delta)=f\ \&\ v(\Gamma)=f)\strictif v(\Delta\cup\Gamma)=f;$
$((v(\Delta)=n\ \&\ v(\Gamma)\neq t)\vee(v(\Gamma)=n\ \&\ v(\Delta)\neq t))\strictif v(\Delta\cup\Gamma)=n.$

When the valuation rules are stated thus formally it is impossible to apply them to assess whether some given wff is a theorem or not, without some statement of the logic which the connectives '&', '$\vee$', '$\sim$' and '$\strictif$' satisfy. Anyone who thinks that this logic need not be developed and that the valuation rules can be applied as they stand, should try to demonstrate that even such simple wff as $qu(\sim(r\ \&\sim r))$ or $qu(T\sim p\supset\sim Tp)$ are tautologies, where 'tautology of S_3-2' is defined in the same way as 'tautology of S_3-1'. However the adjoined logic which the connectives satisfy is far from uniquely determined. Any $\{\sim,\ \&,\ \vee,\ \strictif\}$ logic between Lewis's system S2 (reformulated with these primitives) and classical sentential logic will do; in fact any logic with the same first-degree theorems (*i.e.* theorems which contain no implication '$\strictif$' nested within the scope of another implication) as classical sentential logic will suffice. A significance logic like S_1 with added postulates as,

$$S((v(\Delta) = t) \vee (v(\Delta) = f) \vee (v(\Delta) = n)),$$

will also suffice. On the other hand weaker logics which do not contain appropriate forms of disjunctive syllogism are inadequate as adjoined logics for *this* formulation of the semantics (contrast the semantics developed for S_6 in §5.14), since it is often necessary to apply elimination arguments, *e.g.* of the form

$$[((v(\Delta) = n) \vee (v(\Delta) = t)) \ \& \ v(\Delta) \neq t] \prec v(\Delta) = n.$$

From the point of view of completeness proofs of the corresponding axiomatic systems the simplest logic to adopt for the formalisation of the semantics is classical sentential logic. A significance logic is not needed since

$$S((v(\Delta) = t) \ \& \ (v(\Delta) = f) \ \& \ (v(\Delta) = n)) \text{ holds.}$$

When a non-classical logic is adopted, theoremhood in the semantical system, in the sense of theses provable in the semantics, outstrips theoremhood in the corresponding syntactical systems: in this sense the semantics of significance logic so formulated is richer than the syntax. But the semantics is richer than it need be when it uses a non-classical logic; and the syntax can be extended to embrace more of the semantics than might be suspected. First, and inessentially, the values t, f and n can be defined in S_3^+ systems, which extend S_3 systems by adding the constant sentence 'q_0' with value n, thus:

$$t =_{\mathrm{Df}} (p \rightarrow p)$$
$$f =_{\mathrm{Df}} Ft$$
$$n =_{\mathrm{Df}} q_0.$$

A deficiency of systems up to and including S_3 is that they are not strongly complete; there is no guarantee that any sentence is nonsignificant. In S_5 this defect is remedied; for there n may be defined by,

$$n =_{\mathrm{Df}} Dt.$$

Secondly, even in S_2, valuation functions may be defined:

$$(v(\Delta) = t) =_{\mathrm{Df}} T\Delta; \ (v(\Delta) = f) =_{\mathrm{Df}} F\Delta; \ (v(\Delta) = n) =_{\mathrm{Df}} \sim S\Delta.$$

Thirdly, then, *the semantical valuation rules of* S_3-2, *formulated using classical sentential logic*, may be *proved in the syntactical system*. This will be established when syntactical formulations of S_3 systems are investigated; it offers a very simple way of proving completeness of syntactical systems relative to their intended semantics.

When 'tautology' is defined in more formal detail, enrichment of the semantics is inevitable. For either quantification is needed—as in the usual definition of 'tautology' in terms of *all* values—or modalisation of the logic used in the semantics is needed—as is done in the case of $\mathbf{S}_6$ and could be done here—or provability in the semantics has to be introduced. This last method is a metalogical variation of the second.

In $\mathbf{S}_3$-3 sentential quantifiers and formal predicates are added to the logic used in the valuation postulates (v-rules), and this definition of 'tautology' is adopted, where '*SUV*' and '*SRV*' abbreviate, respectively '*S*-unrestricted variable' and '*S*-restricted variable':

$$\text{Tautology}\ (\Delta) =_{\text{Df}}$$
$$(Uqu(q))\ \{[SUVqu(q) \prec . ((v(q) = t \prec v(\Delta) = t)\ \&\ (v(q) = f \prec v(\Delta) = t)\ \&\ (v(q) = n \prec v(\Delta) = t))]$$
$$\&\ [SRVqu(q) \prec ((v(q) = t \prec v(\Delta) = t)\ \&\ (v(q) = f \prec v(\Delta) = t))]\}.$$

It is unnecessary to restrict the variables to those that occur in Δ, but all variables that occur in Δ must be included.

To be formally explicit some formalisation of the morphology is needed in $\mathbf{S}_3$-3 so that '*SUV*' and '*SRV*' appear as formally specified predicates. How this is done is apparent from the treatment of $\mathbf{S}_6$.

$\mathbf{S}_3$-3 is still not as highly formalised as it might be, since the theory of valuation functions and identity used in the theory is not formalised (though this can be done along the lines of §2.15). Actually identity, though it is easily formalised in the usual way, can be eliminated as far as appearances go, by rewriting '$v(\Delta) = m$' say as '$v(\Delta, m)$', where a valuation predicate replaces the function. Then the semantics uses a constant predicate 'v' which can be introduced as a formal primitive: its relevant role has already been recursively specified in the valuation postulates. Along with predicate 'v', constants 't', 'f' and 'n' should be introduced as primitives.

Then the morphology will have such rules as

(a) Value (t), Value (f), Value (n).

(b) Value (δ) & Wff(Δ) $\prec$. Wff($v(\Delta, \delta)$).

Thus requirements earlier included in the valuation rules now reappear in the morphology.

Finally the semantics can be generalised. Variables distinct from sentential variables, having as substitution values t, f and n would be an initial step in this direction.

The Logic $\mathbf{C}_3$-1

The morphology of the logic is the same as that of $\mathbf{S}_3$-1 except that the variables are those of $\mathbf{C}_0$. The valuation rules and tables are included in those for $\mathbf{S}_3$-1, but tautologies and formal theorems are explained as in $\mathbf{C}_3$, *i.e.* both t and n are designated values.

$\mathbf{C}_3$-1 is deductively equivalent to $\mathbf{C}_3$. Hence $\mathbf{C}_3$-1 is simply consistent with respect to '$\sim$', '$\neg$' and '*F*'.

The 4-Logics

4-logics are one-sorted representations of 3-logics in which *t* is the only designated value; *i.e.*, they are **S**-logics which nevertheless contain only *S*-unrestricted variables, *p*, *q*, ... *etc.*

The satisfactory representation of **S**-logics as one-sorted systems does not depend on their containing connectives which can only be defined in a functionally complete system; it only depends on their containing suitable non-classical connectives. But weaker **S**-systems, like $\mathbf{S}_1$, which can be formulated in terms of unrestricted variables only, are at a serious disadvantage when the systems are enlarged by new operators (see §7.6). 4-logics on the other hand are technically advantageous, *e.g.* in the development of quantified and intensional significance logics; and they vie with 6-logics as the most satisfactory sentential bases for significance developments.

Corresponding to each 3-system there is a 4-system which differs only in that it is one-sorted. Thus the one-sorted systems **Å**, based on the connective set $\{\cup, \sim, T\}$, $\mathbf{K}_2$, based on $\{\sim, \cup, \simeq\}$, and **L**, based on $\{\sim, \supset_L\}$, are all 4-logics. They, of course, by no means exhaust bases for 4-logics. The set $\{\supset_K, \sim, \neg\}$, for example, provides yet another base; and so, too, does $\{\sim, \&, \rightharpoondown\}$. The logic based on this last set we call $\mathbf{S}_4$ and we specify it in detail below.

The Logic $\mathbf{S}_4$

(a) *Morphology*

Improper primitive symbols: $\sim$, &, $\rightharpoondown$
Variables as for $\mathbf{C}_0$

(b) *Valuation Rules* for the primitive connectives are as already given in previous systems. Further connectives are defined as in *Theorem* (12).

An important connective definable in 3- and 4-logics, but not in 2-logics is the *Brady disjunction* 'V' (Brady [3]), which has the following matrix:

V	*t*	*f*	*n*
t	*t*	*t*	*t*
f	*t*	*f*	*f*
n	*t*	*f*	*n*

That 'V' is not definable in 2-logics follows from the following hypothesis which may be proved by induction as in *Theorem* (10):

There is no function $\Delta(P, Q)$ of P and Q which has value n when P and Q both have value n, and value $\bar{n}$ otherwise.

(Proof of the induction step for '$\sim$' uses the fact that only significance preserving connectives are definable in 2-logics.) 'V' is, however, definable in 3-logics as follows:

$$\Delta + \Gamma =_{\mathrm{Df}} (\Delta \rightharpoonup \Gamma) \rightharpoonup \Gamma$$
$$\Delta \mathrm{X} \Gamma =_{\mathrm{Df}} \sim(\sim\Delta + \sim\Gamma)$$
$$\Delta \mathrm{V} \Gamma =_{\mathrm{Df}} (\Delta + \Gamma)\, \mathrm{X}\, (\Gamma + \Delta).$$

Or: $\Delta \oplus \Gamma =_{\mathrm{Df}} \sim((\Delta \rightharpoonup \Gamma) \rightharpoonup \Gamma)$, *i.e.* $\sim(\Delta + \Gamma)$
$$\Delta \mathrm{V} \Gamma =_{\mathrm{Df}} (\Delta \oplus \Gamma) \oplus (\Gamma \oplus \Delta).$$

Or less elegantly,
$$\Delta \mathrm{A}^* \Gamma =_{\mathrm{Df}} (\Delta \vee \sim\Delta)\, \mathrm{U}\, (\Gamma \vee \sim\Gamma)$$
$$\Delta \mathrm{V} \Gamma =_{\mathrm{Df}} (\Delta \mathrm{A}^* \Gamma)\ \&\ (T\Delta \vee T\Gamma).$$

Although 'V' cannot be explicitly defined in $\mathbf{S}_2$ it can be defined by cases using connectives of $\mathbf{S}_2$ thus:

$$\Delta \mathrm{V} \Gamma \simeq .\ \Delta \vee \Gamma, \text{ where } S\Delta \equiv S\Gamma, \text{ and}$$
$$\Delta \mathrm{V} \Gamma \simeq .\ T\Delta \vee T\Gamma, \text{ where } S\Delta \not\equiv S\Gamma.$$

'V' cannot be added to $\mathbf{S}_2$ in place of 'U' to provide a significance complete logic. Consequently the connective set $\{\sim, \&, T, \mathrm{V}\}$ provides an intermediate logic $\mathbf{S}_{2\mathrm{I}}$ between $\mathbf{S}_2$ and $\mathbf{S}_3$. To show this it remains to show that 'U' (or '$\rightharpoonup$') is not definable in $\mathbf{S}_{2\mathrm{I}}$; and this follows since significant values t and f both behave the same with respect to n for all connectives of $\mathbf{S}_{2\mathrm{I}}$ whereas in the case of both 'U' and '$\rightharpoonup$', t combined with n gives a different result from f combined with n. The point is made more perspicuously by using the *s–n* sub-logic (Brady [3], p. 343 ff.), where value s stands in for significant values t and f. Then every one- and two-place connective of $\mathbf{S}_{2\mathrm{I}}$ is among the following (1)–(7):

	(1)	(2)	(8)
s	*s*	*s*	*n*
n	*s*	*n*	*n*

first variable	second variable	(3)	(4)	(5)	(6)	(7)	(9)
s	*s*	*s*	*s*	*s*	*s*	*s*	*n*
s	*n*	*s*	*s*	*s*	*n*	*n*	*n*
n	*s*	*s*	*s*	*n*	*s*	*n*	*n*
n	*n*	*s*	*n*	*n*	*n*	*n*	*n*

To elaborate:—A connective is *direct* iff its operation ϕ on values s and n only maps to n if the operation ϕ with s-values replaced in one or more places by n, also maps to n. A connective is *quite direct* if it is direct and significance-preserving. Connectives determined by (8) and (9) are direct, but not quite direct. The main results are these:—

(I) A connective is quite direct iff it is definable in S_{2I};

(II) A connective is direct iff it is definable in terms of the connectives of S_{2I} together with the Slupecki connective '$\dot{T}$', *i.e.* (8), which has the following vector:

Γ	t	f	n
$\dot{T}\,\Gamma$	n	n	n

That every direct connective is definable in terms of the set 3A $=\{\sim, \&, T$, V, $\dot{T}\}$ is shown, in essentials, in Brady [3], p. 346 ff.; and the argument is readily modified to show that every quite direct connective is definable in S_{2I}. The converse halves are established by an inductive argument over the number of connectives in any putative definition.

The results furnish an alternative proof that S_{2I} is properly included in S_3; for the addition of the Slupecki '$\dot{T}$' to S_3 renders the logic functionally complete (see §5.14, *Theorem* (20)) but its addition to S_{2I} does not, by (II). Result (II) thus reveals a further intermediate system S_{3A}, with connective set 3A which includes S_2 and is included in S_5, but which neither includes nor is included in S_3. Therefore, as expected, the class of S-logics is only partially ordered. In fact appeal to the s–n sub-logic reveals yet another system which deviates from the linearly-ordered S-systems on which we concentrate. The inversion connective 'K+' (of §7.14) can be represented in s–n form:

$K+\Gamma$	s	n
	n	s

but 'K+' is not a direct connective. Hence the class of connectives which can be represented in s–n form provides a further system, between S_{3A} and S_5 which neither includes nor is included in S_3.

Since S_3 properly includes S_{2I}, arguments based on the need for 'V' in significance logics do not take one as far as arguments built on the need for '⇀'. However the need for 'V' does lead one beyond system S_2; and connective 'V' is needed in at least the following places: in the theory of quantification where the particular quantifier 'q' is expanded over finite domains (see §7.7 *ff.*); in the unification of heterogeneous relations (see §5.13); and in the compounding of categories (see §7.14). Brady ([3], p. 13), goes further in defence of his connective and contends that 'V' has applications in ordinary discourse; for example, he argues that the way we ordinarily assign values for the results of substitution in the frame *qu*(*x* is a holiday or *x* likes cheese) reveals that the 'or' used is not the classical disjunction '$\vee$' but 'V'. But this is not clear: it is not even clear that the connective used is matrix-accessible over set $\{t,f,n\}$.

The chief reason for singling 'V' out for attention is that it corresponds to the particular quantifier '$\mathcal{q}$' (of Chapter 7)—precisely as '&' corresponds to the universal quantifier '*U*'—by giving its expansion over finite domains. Under the ordering $t<f<n$ of significance values, '&' and 'V' are in fact related as maximal and minimal operators, *i.e.* $\&(v_1, v_2) = \max(v_1, v_2)$, and $V(v_1, v_2) = \min(v_1, v_2)$ for all values v_1 and v_2. The quantificational behaviour of '*U*' and '$\mathcal{q}$' may be seen as the extrapolation of these functions. If the ordering of $\{t,f,n\}$ is changed, then different maximal and minimal connectives result. For example, under the ordering $t<n<f$, the strong normal connectives '$\cap$' and '$\cup$' emerge as the maximal and minimal connectives respectively.

(c) *Tautologies:* A wff of $\mathbf{S}_4$ is a tautology iff it takes the value *t* for all assignments of values to its variables.

The Logic $\mathbf{C}_4$

$\mathbf{C}_4$ has the morphology of $\mathbf{S}_4$ but differs in its theorem set since both *t* and *n* are designated values. $\mathbf{C}_4$ is deductively equivalent to $\mathbf{C}_3$. Hence it is simply consistent with respect to '$\sim$', '$\neg$' and '*F*'.

5.13 Stage 5: The Logics $\mathbf{S}_5$ and $\mathbf{C}_5$

The involutory connective '*Q*' with matrix

Γ	t	f	n
$Q\Gamma$	t	n	f

is not, by *Theorem* (11), definable in any of the logics formulated so far. Logics like $\mathbf{S}_4$, $\mathbf{\AA}$ and $\mathbf{K}_2$ do not have functionally complete sets of connectives. For similar reasons the related connective '*D*' with matrix

Γ	t	f	n
$D\Gamma$	n	f	t

is also not definable in these logics. In fact '*Q*', '*D*' and other connectives like them which convert sense into nonsense may be discussed together. For in any system containing '$\sim$', '*Q*' and '*D*' are interdefinable, as follows:

$$Q\Delta =_{\mathrm{Df}} \sim D \sim \Delta; \quad D\Delta =_{\mathrm{Df}} \sim Q \sim \Delta;$$

and the systems obtained by adding either 'Q' or 'D' to 3 logics are functionally complete and so contain all like connectives.

The big question is: are such connectives desirable in a significance logic? Two problems may be disentangled here:

Should a significance logic contain (as primitives) connectives which do not have a good interpretation in a natural language? Call this the *first interpretational problem.*

Should a significance logic contain at all connectives which behave so outrageously as to convert not just nonsense into sense (as 'S', 'T' and 'F' do) but sense into nonsense (as 'Q' and 'D' do)? Call this the *second interpretational problem.*

We shall call connectives like 'D' and 'Q' which convert just one of t and f to n *irregular* connectives. Irregular connectives are not significance-preserving. There is however nothing extraordinary about connectives which are not significance-preserving as such. Consider, for example, the operator 'the lazy number thinks ...' where the dots are to be replaced by any that-clause formed from a significant sentence. Or consider 'Φ_1' which absorbs a quotation function over sentences and is interpreted as '... is thinking' where the dots are to be replaced by a quoted sentence: '$\Phi_1\Delta$' then reads 'the sentence $qu(\Delta)$ is thinking'; and this is nonsignificant for all values of Δ. What is extraordinary about the *really* irregular connectives 'D' and 'Q' is their selectivity. For they convert only some values into nonsignificance, and in fact they cannot be defined in terms of connectives such as,

Γ	$\Phi_1\Gamma$	$\Phi_2\Gamma$	$\Phi_3\Gamma$
t	n	n	n
f	n	n	n
n	n	t	f

which convert both the significant values, at least, into n. Accordingly we could design systems which genuinely extend 4-systems but which are not functionally complete.

Certain natural language connectives, however, do, according to some views, act selectively. That is, in such a way that they yield nonsense when applied to some significant sentences and sense when applied to others. It might be claimed, for example, that 'it is calculated (that ...)' is such a connective on the grounds that it makes no sense to speak of calculating the beauty of a given picture, the virtues of a saint, and so on, while it does make sense to speak of calculating the amount of a debt, the value of a function or the size of a room. At least there seems to be no *a priori* reason why such selective connectives should not occur in a natural language. (If none do, it would be interesting to learn why not.) And if this is so, there seems to be no reason why such connectives should not be introduced as part of a formal theory. Admittedly, the sort of connectives indicated, which convert some cases of *t* and *f* to *n*, and others to *t* or *f*, also convert *n* to *n*. Thus, if *qu*(*p*) is nonsignificant, then 'it is calculated that *p*' is also nonsignificant: *e.g.* 'it is calculated that the number seven likes soup' is nonsignificant given that 'the number seven likes soup' is nonsignificant. So these connectives are not exactly like irregular connectives. If we suffix them by '*T*', however, then connectives more like the irregular ones are obtained. But even so, there is still an important difference, for 'it is calculated that *Tp*' is not matrix-accessible. It is not the case that *every* false sentence is converted to a non-significant sentence, for example.

Many matrix-accessible connectives raise the first interpretational problem —which suggests, perhaps, that not many connectives of natural language are matrix-accessible. Even '⊃' raises the first interpretational problem when it is taken as primitive. So it could be said that even in 0-logics there are connectives which lack a good intuitive interpretation, and we might simply claim that the introduction of '*Q*' does not raise any new difficulty. It is commonly claimed in defence of '⊃' that a definitional interpretation and explanation of it can be given in terms of truth values, but similarly we can give an explanation of '*Q*' in terms of values. In the one case we rely on that "well-understood" notion of truth; in the other, on truth and nonsignificance. And it might with reason be claimed that nonsignificance is more readily assessed than truth.

Consider the case against connectives which lack a good interpretation. This turns on the point that significance logics are of interest, indeed are *significance* logics, just because of their interpretation; that significance logics are strictly semantical systems. Connectives like '*Q*' have an interpretation of a sort over the set $\{t,f,n\}$, it is admitted, but if they lack a good intuitive interpretation in terms of a natural language, what role can they have in the formalisation of familiar significance claims and arguments?

Against this kind of attack it can be retorted: even in classical logic any number of connectives which raise interpretational problems can be defined. More important, the further connectives have the same sort of role in significance theory as theoretical expressions have in mathematical or physical theories. Of course there are differences. These allegedly theoretical expressions of significance theory are syncategorematic expressions, whereas typical theoretical expressions of physics are non-syncategorematic expressions which putatively denote items or properties. Moreover the satisfactory appearance of theoretical expressions in physics has to be supported by the adequacy of the theory. And as much, at least, must be true of the theoretical expressions of significance theory. That is, it has to be shown that there is an undeniable theoretical advantage to be gained from their introduction. This, in fact, is the case; and that it is constitutes the main argument for their admission.

We saw in §5.11 that although ‘$\rightharpoondown$’ does have an interpretation of a sort, its real value lies in its formal advantages. And this is so with ‘Q’. The difference between ‘$\rightharpoondown$’ and ‘Q’ is that the latter cannot be given an interpretation, though artificial renderings such as involutory truth can be devised (for $\mathbf{S}_5$, not for $\mathbf{C}_5$). But the formal advantages gained by having ‘Q’ available are quite substantial. For example, using it theories often thought not to be fully formalisable can be formalised. In particular given ‘Q’:

(i) Connections between homogeneous and heterogeneous relations can be expressed (see Goddard [2]).

A full discussion of relations will be given in Part III, but we note here that just as one-place predicates have a significance range associated with them, so m-place relations have m significance ranges associated with them. Thus, we saw in §3.7–§3.8 that the condition for a one-place sentence of the form $qu(fx)$ to be significant is that the subject-term $qu(x)$ be about an item which belongs to the significance range $SD‘f$, *i.e.* $TD‘f \cup FD‘f$, where $TD‘f$ is the truth domain of $qu(f)$, $FD‘f$ is the falsity domain of $qu(f)$, and ‘$\cup$’ indicates set union. Similarly a necessary condition for an m-place relational sentence $qu(f(x,y,z,...))$ to be significant is that x belongs to the s-range $SD_1‘f$, y to the s-range $SD_2‘f$ *etc.*, where $SD_1‘f$, $SD_2‘f$ *etc.* are the significance ranges associated with the 1st, 2nd, ..., mth place of $qu(f)$. In particular, the condition for a two-place relational sentence of the form $qu(f(x,y))$ to be significant is that x belongs to $SD_1‘f$, which we call the significant domain, and y belongs to $SD_2‘f$, which we call the significant converse domain. For example, the condition for $qu(x$ admires $y)$ to be significant is that x belongs to a certain range $SD_1‘$ A (where ‘A’ is ‘admires’) which we may take to be the class of people, and that y belongs to the range $SD_2‘$ A, which we may take, again speaking in the material mode, to be the class of material objects, including people. Here, the significant domain is a subclass of the significant converse domain, though this is not the case with all relations; various different connexions may hold between the ranges, depending on the nature of the relation.

Now it is necessary to distinguish relations according as they satisfy or fail to satisfy the further condition that an arbitrary choice of terms from their significant domains and converse domains guarantees the significance of the relational sentence. Relations which satisfy this condition are said to be homogeneous; those which fail to satisfy it, heterogeneous. ‘Admires’ is an example of a homogeneous relation. For it we choose any term, *i.e.* the name of any person x_0, from the significant domain $SD_1‘$ A and the name of any item y_0 from the significant converse domain $SD_2‘$ A, then ‘x_0 admires y_0’ is significant since it is always significant, though it may be false, to say of any person that he admires any object. An example of a heterogeneous relation is '∈' under type theory. For according to the theory, $qu(x \in y)$ is significant if, and only if, x is of exactly one type less than the type of y. Speaking in terms of items rather than subject terms, because it is simpler (though, strictly, incorrect), this condition cannot be satisfied if y is an individual (*i.e.* of type 0) since no item can have a type less than 0 and the domain place cannot then be filled significantly, but it can be satisfied if x is an individual, a type-1 class, a type-2 class, a type-3 class, *etc.* Hence the significant domain, SD_1'∈, consists of all items (individuals, classes of individuals, classes of classes of individuals, *etc.*), while the significant converse domain SD_2'∈, consists of all these items excluding the individuals. With such ranges, however, it is not the case that an arbitrary choice of values from SD_1'∈ for the domain place, and an arbitrary choice of values from SD_2'∈ for the converse domain place, will result in a significant sentence. For we may choose, say, a type-1 class from SD_1'∈ and a type-3 class from SD_2'∈ which will yield a nonsignificant sentence of the form $qu(x_1 \in y_3)$.

Some heterogeneous relations are so because they have more than one sense. For example, 'is next to' might be used with the sense 'is spatially next to' or 'is temporally next to'. If we take it as one relation, therefore, both the significant domain and the significant converse domain will consist of temporal events and physical objects. In this case, an arbitrary choice of values from the significance ranges will not guarantee a significant sentence. For one may choose, say, a day of the week from the significant domain and a material object from the significant converse domain and obtain non-significant sentences of the sort 'Saturday is next to the table'. In a like manner, one-place predicates, such as 'is a cardinal', which have more than one sense may be regarded as having heterogeneous significance ranges (in this case consisting of prelates and numbers) which do not guarantee significance for an arbitrary choice of value in a given context. But just as heterogeneous, or ambiguous, one-place predicates can be resolved into univocal components, say 'cardinal (1)' and 'cardinal (2)', each associated with a homogeneous significance range, so heterogeneous relations can be resolved into univocal or homogeneous components. The predicate 'is next to', for example, can be resolved into the two homogeneous predicates 'is temporally next to' and 'is spatially next to'. But this resolution cannot be expressed by regarding the original relation as the union-relation of its homogeneous components. For, if $qu(f(x,y))$ is the original predicate, and $qu(f^1(x,y))$ and $qu(f^2(x,y))$ are its homogeneous components, then for given values of x and y, say x_0 and y_0, we may have $qu(f(x_0,y_0))$ significant even though one of $qu(f^1(x_0,y_0))$ and $qu(f^2(x_0,y_0))$ is not significant. But if we take the definition of the union relation in terms of the classical significance connective '∨', the nonsignificance of either one of the components will entail the nonsignificance of the original relation $qu(f(x_0,y_0))$. Using 'Q' however, a wide heterogeneous union can be formed:

$$f(x,y) \simeq (Tf^1(x,y) \vee Tf^2(x,y)) \,\&\, Q(Sf^1(x,y) \vee Sf^2(x,y)).$$

The resultant expression $qu(f(x,y))$ is nonsignificant just in case both of its homogeneous components are nonsignificant. The union is a *wide* heterogeneous union because it permits the composition of relations which would not normally be reckoned to be homogeneous components of *one* heterogeneous relation.

Though this argument illustrates well the type of formalistic argument we think there is for introducing irregular connectives, it is inconclusive. For though it provides a case for going beyond 2-systems to 4-systems, it does not present an argument for 6-systems as distinct from 4-systems. Wide heterogeneous unions can be explicity defined at once in 4-logics, using the Brady disjunction'V', as $f^1(x,y) \mathrm{V} f^2(x,y)$. Indeed $p \mathrm{V} q \simeq .\ (Tp \vee Tq) \,\&\, Q(Sp \vee Sq)$.

(ii) A theory of descriptions, often identified with the Hilbert-Bernays' theory as originally motivated, can be formalised. According to this theory, a description is not significant if the uniqueness condition is not satisfied. Using 'Q', the required description operator can be defined as follows:

$$(\iota x)(fx, gx) \simeq (Ex)(fx \,\&\, (y)(Tfx \supset T(x=y)) \,\&\, gx) \,\&\, Q((Ex)fx),$$

where the existential quantifier 'E' is defined in such a way that '$(Ex)fx$' is significant if, for *some* individual x_0, but not necessarily for all substitutions for 'x' 'fx_0' is significant. (For the exact theory of the quantifier 'E' see §7.9). Given this reformulation of the Hilbert-Bernays' theory, and given the distinction between significance and well-formation, Carnap's objections to description theories of this type are avoided (see Carnap [2] pp. 33–34).

(iii) That interpretation of traditional logic according to which sentences in the traditional forms are nonsignificant if terms occurring in them are empty, can be formalised. For example, qu(All f are g) can be approximated classically by,

$$(x)(fx \supset gx) \,\&\, Q[T(Ex)fx \,\&\, T(Ex)gx].$$

Thus if the term represented by $qu(f)$, say, is empty then qu(All f are g) is nonsignificant. A still better approximation to the relevant interpretation of traditional logic can be obtained by replacing '$\supset$' by some other implication, for example by a connexive implication (such as Angell's [1] '$\rightarrow$') or a conditional assertion relation (as in §7.9, §6).

(iv) The introduction of irregular connectives makes it possible to construct simple functionally complete systems. (For details see §5.14 below). Conversely, a functionally complete system will automatically supply irregular connectives. But given a significance-complete logic it is imperative to advance to a functionally complete logic. For in no merely significance-complete logic is it possible to define a constant nonsignificant sentence. But some sentences are nonsignificant—a main thesis of this book. Let 'n_0' be such a nonsignificant sentence. It is a straightforward matter to define the Slupecki '$\dot{T}$' connective, *e.g.* as follows:

$$\dot{T}\Delta =_{\text{Df}} T\Delta \,\&\, n_0.$$

Then $\dot{T}\Delta$ is nonsignificant whatever value Δ has since $\sim Sn_0$. But '$\dot{T}$' together with significance-complete connectives provides a functionally complete connective set (see *Theorem* (20) ff). In other words, features of both systems S_3 and S_{3A} are needed; but putting these features together results in system S_5.

(v) The full theory of restricted quantification requires not only the connective '$\rightharpoonup$' but, for the theory of the particular quantifier 'q', the connective '1' with matrix:

1	*t*	*f*	*n*
t	*t*	*f*	*n*
f	*n*	*n*	*n*
n	*n*	*n*	*n*

The argument which shows that '1', so characterised, is uniquely determined by the theory of restricted particular quantification, is like that already advanced in the case of universal quantification (in §5.11). The quantifier 'q' satisfies the semantical rule (see §3.5 and §7.7): (qX)*A*(X) takes value *t* iff *A*(X) has value *t* for some assignment to X, and takes value *n* iff *A*(X) has value *n* for every assignment to X. Likewise then the restricted variable form (*b*X)G(X) 1 F(X) takes value *t* iff F(X) has value *t* for some assignment to X for which G(X) has value *t*, and takes value *n* iff F(X) has the value *n* for every assignment to X for which G(X) has value *t*. Assuming that G(X) takes value *t* for some assignment to X, the matrix for '1' is thereby uniquely determined. For if, for an assignment to X, G(X) has value *t* then G(X) 1 F(X) takes the same value as F(X); whence $v(t\ 1\ \Delta) = v(\Delta)$. But if, for an assignment to X, G(X) has value $\bar{t}$ then the value v of $\bar{t}\ 1$ F(X) is discounted in the evaluation of (qX)G(X) 1 F(X). In view then of the disjunctive evaluation of 'q' and the assumed matrix accessibility of '1', $v(\bar{t}\ 1\ \Delta) = n$; for $vVv_1 \simeq v_1$ for every value v_1 if $v = n$. In terms of the connectives '$\rightharpoonup$' and '1' needed for the theory of restricted quantification, however, using just classical negation, all irregular connectives can be defined; for the set $\{\sim, \rightharpoonup, 1\}$ provides a functionally complete set of connectives for S-logics (see §5.14, *Theorem* (20) ff).

A major argument in favour of introducing irregular connectives therefore, and in particular 'Q', is that it enables us to formalise, especially in predicate logic, a number of theories which otherwise could not be readily represented or even represented at all. And this we regard as an important advantage. For the decision as to whether or not such theories should be adopted will depend to some extent on an examination of their consequences; but if they cannot be formalised, their consequences cannot be examined. (Compare the advantages of Platonistic languages over nominalistic ones.) Nevertheless the introduction of irregular connectives is not wholly advantageous. In particular, their presence causes extreme difficulties in the formalisation of category theories, as we shall see in Volume 2 (see also §7.14). What is not clear, however, is whether such problems are avoided by adopting weaker significance logics in which irregular connectives are not definable. Certainly they may be avoided formally, but the problems then reappear at the level of interpretation—for example, why may it not be the case that a sentential variable takes as a value a sentence which contains an irregular connective? It seems to us that by adopting a weaker formalism, the problems are not solved but shelved. In order to bring these problems into the open and to make a satisfactory resolution possible, a wider formalism which includes irregular connectives is necessary.

The Logic S_5

(a) *Morphology*

Add to the morphology of S_2 (which contains '$\sim$', '$\supset$', 'T' as primitive):

the improper symbol: Q

the formation rule: if Δ is a wff, so is $Q\Delta$

the definitions: $F\Delta =_{\mathrm{Df}} T{\sim}\Delta$

$$S\Delta =_{\mathrm{Df}} T\Delta \vee F\Delta$$

$$\Delta \simeq \Gamma =_{\mathrm{Df}} (T\Delta \;\&\; T\Gamma) \vee (F\Delta \;\&\; F\Gamma) \vee ({\sim}S\Delta \;\&\; {\sim}S\Gamma)$$

$$\Delta \rightharpoondown \Gamma =_{\mathrm{Df}} T\Delta \supset Q(\Delta \supset Q\Gamma)$$

$$D\Delta =_{\mathrm{Df}} {\sim}Q{\sim}\Delta$$

$$\Delta \mathrm{U} \Gamma =_{\mathrm{Df}} (T\Delta \vee T\Gamma) \vee D(({\sim}S\Delta \;\&\; {\sim}T\Gamma) \vee ({\sim}S\Gamma \;\&\; {\sim}T\Delta)).$$

Hence S_5 contains S_3 contains S_2 contains S_1 contains S_0.

An alternative basis is provided by adding 'Q' to the morphology of S_1, *i.e.* by taking '$\sim$', '$\supset$', 'S' and 'Q' as primitive, for then 'T' can be defined as follows:—

$$T\Delta =_{\mathrm{Df}} {\sim}SQ{\sim}\Delta.$$

If, however, we simply add 'Q' to S_0 then 'S' is not definable. For any connective definable in terms of '$\sim$', '$\vee$' and 'Q' must contain at least one n in the main column of its matrix. Hence in such a logic, none of 'S', 'T', 'F', '$\simeq$' are definable; neither, therefore, is '$\rightharpoondown$' since 'T' enters into its definition (as above).

(b) *Valuation Rules*

Add to the valuation rules of $\mathbf{S}_2$ the rule,

If $v(\Delta) = t$, then $v(Q\Delta) = t$;
if $v(\Delta) = f$, then $v(Q\Delta) = n$;
if $v(\Delta) = n$, then $v(Q\Delta) = f$.

Valuation rules for the other connectives follow by the definitions.

(c) *Tautologies*

A wff of $\mathbf{S}_5$ is a tautology, if, and only if, it takes the value t for all assignments of values to its variables.

(d) *Valuation Tables*

As for $\mathbf{S}_2$.

(e) *Formal Theorems*

A wff of $\mathbf{S}_5$ is a theorem if, and only if, it is a $\mathbf{S}_5$-tautology. In addition to the theorems of $\mathbf{S}_0$ through $\mathbf{S}_3$, we have:

$r \simeq QQr$; $p \simeq QQp$; $Tp \simeq TQQp$; $Fp \simeq FQQp$; $Tp \simeq TQp$; $Tp \simeq {\sim} SQ{\sim} p$;
${\sim} Sp \simeq FQp$; $(Tp \supset Tq) \supset SQ(p \supset q)$; $(p \rightharpoonup q) \simeq (Tp \supset Q(p \supset Qq))$;
$TQp \simeq TQQp$.

Thus 'Q' operates in some ways analogously to '$\sim$' since 'QQ' cancels out. In view of $qu(TQp \simeq TQQp)$, however, it cannot be interpreted as a "normal" negation.

The Logic $\mathbf{C}_5$

(a) *Morphology*

Add to the morphology of $\mathbf{C}_2$ the new morphology of $\mathbf{S}_5$.

(b) *Valuation Rules*

As for $\mathbf{S}_5$.

(c) *Tautologies*

A wff of $\mathbf{C}_5$ is a tautology if, and only if, it takes the value t or the value n for all assignments of values to its variables.

(d) *Valuation Tables*

As for $\mathbf{C}_2$.

(e) *Formal Theorems*

A wff of $\mathbf{C}_5$ is a theorem if, and only if, it is a $\mathbf{C}_5$-tautology.
$\mathbf{C}_5$ properly includes all preceding $\mathbf{C}$-logics.

C-*logics Consistency Theorem*: *Part* 5

Matrix $\mathbf{C}_5$-logics are simply inconsistent with respect to both '$\sim$' and '$\neg$', and simply consistent with respect to 'F'.

Proof: The simple consistency with respect to 'F' follows immediately from III of §5.6.

The simple inconsistency with respect to both '$\sim$' and '$\neg$' follows from the fact that using 'Q' it is possible to construct a wff Δ which takes the value n for all values of the variables in it. For let Δ be $Q\Delta_1$, where Δ_1 takes the value f for all values of the variables in it; then by the matrix for 'Q', $Q\Delta_1$ takes the value n for all values of the variables. It has only to be shown, therefore, that there is a Δ_1 which takes the value f for all values of the variables in it. Such a Δ_1 is $qu(\sim(p \simeq QQp))$. In fact, the negation, with respect to '$\sim$', of any $\mathbf{C}_5$-wff which happens to be a $\mathbf{S}_5$-theorem will have value f for all values of the variables in it.

Taking such a Δ_1, therefore, $Q\Delta_1$ has value n always. Hence in $\mathbf{C}_5$-logics, $Q\Delta_1$ is a thesis and so is $\sim Q\Delta_1$, since both take the designated value n for all values of the variables. Again, both $Q\Delta_1$ and $\neg Q\Delta_1$ are theses, since $Q\Delta_1$ always takes the value n and $\neg Q\Delta_1$ always takes the value t. In the case of 'F', though $Q\Delta_1$ is a thesis, $FQ\Delta_1$ takes the value f and is not a thesis.

As we show later, *Theorem* (20) of §5.14, 5-logics are functionally complete. Hence the simple inconsistency of $\mathbf{C}_5$ with respect to both '$\sim$' and '$\neg$' in fact follows immediately from IV of §5.6. We recall, however (VII of §5.6), that $\mathbf{C}_5$ is not absolutely inconsistent and not Post-inconsistent. Formally speaking, therefore, it is not a trivial logic. It is, however, trivial from the interpretational point of view. But even this triviality could be used to advantage. For since the only formulae which give rise to the difficulty take the value n for all values of the variables in them, the simple inconsistency could be removed by dividing the theses of $\mathbf{C}_5$ into normal and abnormal. The normal theses are those which take the value t always (in fact the theses of $\mathbf{S}_5$) together with those like,

$$p \equiv Qp; \quad Qp \supset Tp; \quad Fp \supset Qp;$$

which are t for some values and n for others. Abnormal theses take the value n always. There is point in making such a division in terms of the interpretation. For while theses like $qu(p \equiv Qp)$ have degenerate instances for some values of the variables, they have non-degenerate instances for other values (those for which they take the value t). Abnormal theses, on the other hand, have only degenerate instances and are valueless in all applications. Looked at from this point of view, we could define a logic $\mathbf{C}^*_5$ whose theses are just the normal theses of $\mathbf{C}_5$. Such a logic would be midway between $\mathbf{S}_5$ and $\mathbf{C}_5$. It would be functionally complete and nevertheless simply consistent with respect to all negation connectives. This does not refute IV of §5.6, however, since $\mathbf{C}^*_5$ does not satisfy the required condition that the logic have at least two designated values, for now n is only partially designated. In fact theses of $\mathbf{C}^*_5$ would have to be defined as those wff which take only values from the set $\{t,n\}$ provided they do not also take only values from the set $\{n\}$. But this proviso means that $\mathbf{C}^*_5$ cannot be classified as a logic with two designated values in the unrestricted sense in which this is required for IV of §5.6.

5.14 Stage 6: The Logics $\mathbf{S}_6$ and $\mathbf{C}_6$

Except for 4-logics, the **S**-logics so far developed are not completely smooth-running owing to the occurrence of two sorts of variables. In the syntactical systems corresponding to them, we shall see that proofs are complicated to a slight degree because those theorems which are analogues of **RSL** theses are conditional on the significance of components. Attempts to rectify this particular fault by adopting the corresponding **C**-logics as a basis result in awkward qualifications of the modus ponens rule. These blemishes, if such they are, can however be removed by adopting explicitly 3-valued one-sorted systems.

We shall now look at explicitly three-valued one-sorted **S**-systems which take as primitives at least a pair of connectives '$>$' and '$\neg$' (replacing respectively, '$\supset$' and '$\sim$') and which are such that all the classical **RSL** laws are theses even though all the variables are 3-valued and t is taken as the only designated value.

The object, then, is to construct a "classical" significance logic, $\mathbf{S}_6$. To do this, the connective set $\{\supset, \sim\}$ is replaced by a new set $\{>, \neg\}$, and classical implication-negation logic is assumed to hold for "implication" $>$ and "negation" $\neg$. Thus the choice of basic matrices is narrowed by requiring instead of *Condition* IV of §5.3,

Condition IV* All implication-negation theorems of **RSL** hold for 3-valued implication $>$ and negation $\neg$.

Theorem (17): *Condition* IV* is satisfied if, and only if, '$>$' and '$\neg$' satisfy the matrices:

(T_1)

$>$	t	f	n
t	t	$\bar{t}$	$\bar{t}$
f	t	t	t
n	t	t	t

$\neg$	
t	$\bar{t}$
f	t
n	t

where, as before, '$\bar{t}$' symbolises 'not true', *i.e.* 'false or nonsignificant'.

Proof: The sufficiency of this condition for IV* can be established by showing that some postulate set for implication-negation **RSL** comes out valid, with designated value t, under these matrices. For such a demonstration see Prior [2] (p. 239).

That the condition is necessary can be shown thus:

In order that modus ponens is satisfied and $qu(p>p)$ valid, matrices are determined as:

$>$	t	f	n
t	t	$\bar{t}$	$\bar{t}$
f	d	t	b
n	c	a	t

$\neg$	
t	g
f	h
n	e

The places a, b, c and d are determined by considering value-assignments in $qu(p>(q>p))$.

Choosing $\left.\begin{array}{l} v(p)=n \\ v(q)=t \end{array}\right\}$ requires $a=t$

$\left.\begin{array}{l} v(p)=n \\ v(q)=f \end{array}\right\}$ requires $b=t$

$\left.\begin{array}{l} v(p)=n \\ v(q)=f \end{array}\right\}$ requires $c=t$

$\left.\begin{array}{l} v(p)=f \\ v(q)=f \end{array}\right\}$ requires $d=t$.

Places e and h are determined by considering assignments in $qu((\neg p>p)>p)$.

Choosing $v(p)=n$ requires $e=t$
$v(p)=f$ requires $h=t$.

Place g is settled by $(p>(\neg p>q))$.

Choosing $\left.\begin{array}{l} v(p)=t \\ v(q)=f \end{array}\right\}$ requires $v(\neg p)\neq t$, *i.e.* $v(\neg p)=\bar{t}$; hence $g=\bar{t}$.

It thus follows that '$>$' and '$\neg$' must satisfy the matrices (T_1).

These matrices, however, are not limited sufficiently to enable us to formulate significance conditions, *e.g.* to decide whether $Sp>S(p>q)$.

Since the significance logic is intended to be "classical", the classical **RSL** logic should hold when the sentences are significant. Consequently, choice among 3-valued connectives is limited by *Condition* III of §5.2. To meet both IV* and III, as well as I and II of §5.1, prospective matrices for '$>$' and '$\neg$' are restricted to forms:

(T_2)

$>$	t	f	n
t	t	f	$\bar{t}$
f	t	t	t
n	t	t	t

$\neg$	
t	f
f	t
n	t

in which there is a choice (of f or n) in only one place, namely that associated with $v\,(t>n)$ which has the indeterminate value $\bar{t}$.

The matrices (T_2) show that it follows from III and IV* that when in matrices for '$>$' and '$\neg$' the value n is systematically replaced by value f, the resulting values are those of classical **RSL** implication-negation logic. Roughly, **RSL** holds when nonsignificant sentences are systematically replaced by sentences which yield false statements.

Two possibilities remain. Either (i) $v\,(t>n)=f$ or (ii) $v\,(t>n)=n$. If $v\,(t>n)=f$, then the resulting logic has as theses, $qu(S(p>q))$ and $qu(S\neg p)$, and all wff containing a connective are significant. The values f and n can be identified since they have exactly the same logical role. Thus any logic constructed in terms of choice (i) will be built just on **RSL**. The interesting choice is to take $v\,(t>n)=n$. This is the best choice too from the point of view of minimising on connectives both for the logic using only $\{>, \neg\}$ as primitive and for logics built on it by adding further connectives. For '$>$(i)' is definable in terms of '$>$(ii)' and '$\neg$', but the converse is impossible. First,

$$\Delta>(\text{i})\Gamma =_{\text{Df}} \Delta>(\text{ii})\neg\neg\Gamma;$$

the converse is impossible, however, because no 2-place function of '$>$(i)' and '$\neg$' will contain the value n. Choice (ii) is further vindicated when restricted variables are introduced. Thus the matrix selected for '$>$' is such that $> \; = \; >(\text{ii}) \; = \; \rightharpoonup$.

Obviously the primitive connectives '$\rightharpoonup$' and '$\neg$' do not constitute a functionally complete set for 3-valued logic. Even these together with 'Q' do not form a functionally complete set. Thus,

Theorem (18): Neither '$\sim$' nor 'D' is definable in terms of '$\rightharpoonup$', '$\neg$' and 'Q', where 'D' is the involution connective with matrix

Γ	t	f	n
$D\Gamma$	n	f	t

Proof: It suffices to show that 'D' is not so definable. For if '$\sim$' were definable, 'D' could be contextually defined by: $D\Delta =_{\text{Df}} \sim Q \sim \Delta$ (see §5.13).

That 'D' is not definable follows from the thesis:

(J): No unary connectives ρ and τ with columns of the form:

Γ	$\rho(\Gamma)$	$\tau(\Gamma)$
t	n	f
f	f	n
n		

are definable by a wff $\Delta(P)$ containing just '$\neg$', '$\rightharpoonup$' and 'Q'.

(J) is established by induction over the number of connectives in Δ. If Δ contains zero connectives, the result is immediate from the matrix for P. Otherwise, suppose that the first function Δ for which (J) is refuted has $k+1$ connectives. The following cases are exhaustive:

(a) $\Delta(P)$ cannot be $\neg\Delta_1(P)$, for such an expression never has value n.

(b) $\Delta(P)$ is of the form: $\Delta_1(P) \rightharpoonup \Delta_2(P)$

$$\begin{array}{ccc} t & n & n \\ \cdot & \cdot & \cdot \\ t & f & f \end{array}$$

with value assignments $\{n, \cdot, f\}$ as indicated. It follows from the matrix for '$\rightharpoonup$' that for $\Delta(P)$ to have the requisite column, $\Delta_2(P)$ must also have the column for 'ρ', so violating the inductive hypothesis.

(c) $\Delta(P)$ is of the form :$\Delta_1(P) \rightharpoonup \Delta_2(P)$

$$\begin{array}{ccc} t & f & f \\ \cdot & \cdot & \cdot \\ t & n & n \end{array}$$

with value assignments $\{f, \cdot, n\}$ as indicated. Then, much as in (b) $\Delta_2(P)$ must have the column for 'τ', so violating the inductive hypothesis.

(d) $\Delta(P)$ is of the form: $Q\Delta_1(P)$

$$\begin{array}{cc} n & f \\ f & n \\ \cdot & \cdot \end{array}$$

with value assignments as indicated. Then $\Delta_1(P)$ has the column for 'ρ', violating the inductive hypothesis.

(e) Similarly, if $\Delta(P)$ is of the form: $Q\Delta_1(P)$

$$\begin{array}{cc} f & n \\ n & f \\ \cdot & \cdot \end{array}$$

the inductive hypothesis is violated.

This completes the proof that 'D' is not definable in terms of '$\rightharpoonup$', '$\neg$' and 'Q'.

We now show, however, that the connective set $\{\rightharpoonup, \neg, Q\}$ can be completed by replacing 'Q' by 'D'. That the 3-valued logic with connective set $\{\rightharpoonup, \neg, D\}$ is functionally complete follows from the following theorem—the terminology is that of *Theorem* (12), §5.11.

Theorem (19) (Słupecki): If $\phi(v_1, v_2, ..., v_m)$ is an m-place operation from V^m to V then there is an m-place connective $\Phi(\Delta_1, ..., \Delta_m)$ definable in terms of the connective set $6L = \{\rightharpoonup, \neg, D\}$.

Proof parallels proof of *Theorem* (12). The induction basis can most readily be established by an exhaustion of cases.

Where $t\Delta =_{Df} \Delta \rightharpoonup \Delta$; $T\Delta =_{Df} \neg\neg\Delta$;

$$F\Delta =_{Df} \neg(\neg\Delta \rightharpoonup \neg D\Delta);\ S\Delta =_{Df} \neg D\Delta;$$
$$Id\Delta =_{Df} \Delta;\ {\sim}\Delta =_{Df} (\neg\Delta \rightharpoonup D\Delta) \rightharpoonup D\neg\Delta;$$
$$Q\Delta =_{Df} {\sim} D {\sim} \Delta;$$

the following are all vectors:

t, S, QS, $\neg T$, F, $\sim$, $Q\neg T$, $\sim D$, QF, $\neg F$, T, Id, $\neg S$, $\neg t$, $D\neg S$, D, DT, $D\neg F$, $Q\neg F$, Q, QT, $\sim Q$, DF, $D\neg T$, $Q\neg S$, DS, Dt.

The induction step is exactly the same as that in *Theorem* (12). It remains to define the strong regular connectives '$\cap$' and '$\cup$', as follows:

$$\Delta \vee_1 \Gamma =_{Df} (\Delta \rightharpoonup \Gamma) \rightharpoonup \Gamma;\ \Delta \,\&_1\, \Gamma =_{Df} \neg(\neg\Delta \vee_1 \neg\Gamma)$$
$$\Delta + \Gamma =_{Df} {\sim}({\sim}\Delta \vee_1 {\sim} Q(\Delta \,\&_1\, \Gamma))$$
$$\Delta \supset_L \Gamma =_{Df} (\Delta \rightharpoonup \Gamma) + ({\sim}\Gamma \rightharpoonup {\sim}\Delta)$$
$$\Delta \cup \Gamma =_{Df} (\Delta \supset_L \Gamma) \supset_L \Gamma;\ \Delta \cap \Gamma =_{Df} {\sim}({\sim}\Delta \cup {\sim}\Gamma)$$

Alternatively the theorem follows as a special case of the functional completeness of a certain class of many-valued logics. For the connective set $\{\sim, \supset_L, \dot{T}\}$ provides a functionally complete basis for 3-valued logic (see Rosser and Turquette [1], pp. 24–5). So as $\{\dot{T}(\simeq Dt), \sim, \supset_L\}$ can be defined in terms of $\{\rightharpoonup, \neg, D\}$ the latter also constitutes a functionally complete set. Conversely, that $\{\sim, \supset_L, \dot{T}\}$ constitutes a functionally complete set follows from the fact that 6L can be defined in terms of it.

Theorem (20): The set $\{\supset, \sim, T, Q\}$ of primitive connectives of 5-logics provide a functionally complete set over $\{t, f, n\}$.

Proof: The functionally complete set of connectives '$\rightharpoonup$', '$\neg$, and 'D' with appropriate matrices can be defined in terms of the primitives of 5-logics. Thus:

$$D\Delta =_{Df} {\sim} Q {\sim} \Delta;\ \neg\Delta =_{Df} {\sim} T\Delta$$
$$\Delta \rightharpoonup \Gamma =_{Df} T\Delta \supset Q(T\Delta \supset Q\Gamma).$$

Alternative functionally complete sets for 3-logics are provided by the sets $\{\sim, \rightharpoonup, T_n\}$ (Brady [3]); $\{f, \bar{\ }, D, \vee\}$ (Rockingham Gill [1]); $\{\sim, \rightharpoonup, 1\}$; $\{\sim, \rightharpoonup, \bar{\ }\}$; $\{\sim, \supset_L, \bar{\ }\}$ *etc.*, where Brady's 'T_n' has matrix

	t	f	n
T_n	t	n	n

and Rockingham Gill's '$\bar{\ }$' has matrix

	t	f	n
$-$	n	t	t .

Thus $T_n\Delta \simeq \bar{\bar{\Delta}} \simeq QT\Delta$.

The Logic $\mathbf{S_6}$

(a) *Morphology*

Improper primitive symbols: $\rightharpoonup$, $\urcorner$, D, (,).

Variables: (1) p Var, q Var, where 'q Var' abbreviates ' 'q' is a variable', and 'δ Var' abbreviates '$qu(\delta)$ is a variable';
(2) δ Var $\Rightarrow$ δ' Var;

where '$\Rightarrow$' reads 'entails'.

Formation rules (1) δ Var $\Rightarrow$ δ Wff
(2) $\urcorner\delta$Wff $\Leftrightarrow$ δWff
$D\delta$Wff $\Leftrightarrow$ δWff
$(\delta \rightharpoonup \gamma)$Wff $\Leftrightarrow$ δWff & γWff,

where '$\Leftrightarrow$' reads 'co-entails'.

(b) *Valuation Rules*

(1)(a) $v(\delta) = n =_{\mathrm{Df}} v(\delta) \neq t \;\&\; v(\delta) \neq f$.
Hence $v(\delta) = n \Leftrightarrow . \, v(\delta) \neq t \;\&\; v(\delta) \neq f$
and $v(\delta) = n \vee v(\delta) = t \vee v(\delta) = f$.

Thus the definition reflects the three-valued assumption of the semantics.

(b) [δ Var $\Rightarrow$.]
$v(\delta) \neq t \Leftrightarrow . \, v(\delta) = f \vee v(\delta) = n$.
(c) [δ Var $\Rightarrow$.]
$v(\delta) \neq f \Leftrightarrow . \, v(\delta) = t \vee v(\delta) = n$.

In virtue of the rules introduced below, the condition, δ Var, can be deleted; for the consequents of (b) and (c) demonstrably hold for all Wff. Omission of the condition δ Var has the large advantage (given that the condition δ Wff is not required, because Wff are notationally distinguished or because the rules are stated for *all* formulae) that only first degree entailments are needed in the semantics. For, thanks to Belnap's work [1], this part of the theory of entailment is largely understood.

(d) [δ Var $\Rightarrow$.] $v(\delta) = t \Rightarrow . \, v(\delta) \neq f$.
(e) [δ Var $\Rightarrow$.] $v(\delta) = t \Rightarrow . \, v(\delta) \neq n$.
(f) [δ Var $\Rightarrow$.] $v(\delta) = f \Rightarrow . \, v(\delta) \neq n$.

By strengthening the initial valuation clauses (a)–(c), in a quite correct way, the semantics can be developed using a reasonable paradox-free implication in place of a strict implication. Disjunctive Syllogism is no longer needed to guarantee implications like

$$(v(\delta) = t \vee v(\delta) = f) \ \& \ v(\delta) \neq t \Rightarrow . \ v(\delta) = f.$$

For, by (a) and the Rule of Factor, the antecedent implies $v(\delta) \neq n$ & $v(\delta) \neq t$, which in turn implies $v(\delta) = f$ by (c). Similarly other uses of Disjunctive Syllogism can be replaced by entailmentally valid arguments.

(2) [δ Wff $\Rightarrow$.]
$v(\delta) = t \Rightarrow . \ v(\neg\delta) = f$
$v(\delta) = f \Rightarrow . \ v(\neg\delta) = t$
$v(\delta) = n \Rightarrow . \ v(\neg\delta) = t.$

(3) [δ Wff $\Rightarrow$.]
$v(\delta) = t \Rightarrow . \ v(D\delta) = n$
$v(\delta) = f \Rightarrow . \ v(D\delta) = f$
$v(\delta) = n \Rightarrow . \ v(D\delta) = t.$

(4) [δ Wff & γ Wff $\Rightarrow$.]
$v(\delta \rightharpoonup \gamma) = t \ \& \ v(\delta) = t \Rightarrow . \ v(\gamma) = t$ (modus ponens requirement)
$v(\delta) \neq t \Rightarrow . \ v(\delta \rightharpoonup \gamma) = t$
$v(\delta) = t \ \& \ v(\gamma) = t \Rightarrow . \ v(\delta \rightharpoonup \gamma) = t$
$v(\delta) = t \ \& \ v(\gamma) = n \Rightarrow . \ v(\delta \rightharpoonup \gamma) = n$
$v(\delta) = t \ \& \ v(\gamma) = f \Rightarrow . \ v(\delta \rightharpoonup \gamma) = f.$

A simple replacement which entails the last three conditions, given that iterated implications satisfying importation are admitted, and that the identity principle,

$$(v(\delta) = v(\gamma)) \Rightarrow (\Gamma[v(\delta)] \Rightarrow \Gamma[v(\gamma)]),$$

is conceded is:

$$(v(\delta) = t) \Rightarrow (v(\delta \rightharpoonup \gamma) = v(\gamma)).$$

But it might be doubted whether the condensation merits the extra logical apparatus. Without this we can simply require that the entailment '$\Rightarrow$' be defined for first degree wff only and that it satisfy the same first degree principles as **E** (and as a variety of other systems that have the same stable first degree fragment as **E**). For the identity principle though convenient is not essential for the assessment of whether a given wff is a tautology.

(c) *Tautologies*

The usual account of *tautology* for sentential systems is highly Platonistic: for it demands ordinary quantification over all values or all atomic propositions of assessable formulae. In fact it is far from clear that all these values exist. The question arises as to whether 'tautology' can be defined without such quantification. A tentative definition with some initial appeal is the following:

$$\delta \text{ Taut} =_{\text{Df}} v(\delta) = t.$$

Within the semantical system elaborated it is not necessary to require that $v(\delta) = t$ for *all* values of the variables of $qu(\delta)$. For $v(\delta) = t$ is demonstrable *if* (but not only if) it is true for all values of the variables of $qu(\delta)$. To illustrate we show how $(p \rightharpoonup p)$Taut follows. Since

$$v(p) = t \Rightarrow . \ v(p) = t \ \& \ v(p) = t$$

and $\quad v(p) = t \ \& \ v(p) = t \Rightarrow . \ v(p \rightharpoonup p) = t \qquad$ by (4)

(i) $v(p) = t \Rightarrow . \ v(p \rightharpoonup p) = t \qquad$ by Rule Syllogism

Since $\quad v(p) = f \Rightarrow . \ v(p) \neq t \qquad$ by (1) (d)

and $\quad v(p) \neq t \Rightarrow . \ v(p \rightharpoonup p) = t \qquad$ by (4)

(ii) $v(p) = f \Rightarrow . \ v(p \rightharpoonup p) = t \qquad$ by Rule Syllogism

Similarly,

(iii) $v(p) = n \Rightarrow . \ v(p \rightharpoonup p) = t$

Combining (i), (ii) and (iii)

(iv) $v(p) = t \vee v(p) = f \vee v(p) = n \Rightarrow . \ v(p \rightharpoonup p) = t.$

From (iv) by (1) (a) and detachment, $v(p \rightharpoonup p) = t$, as required.

But though the definition enables us to establish that any particular tautology is indeed a tautology, and is adequate for consistency and completeness proofs in the form 'Δ is a theorem of the corresponding axiomatic system iff it is a tautology', it prevents us proving that a sentential variable $qu(p)$ is not a tautology and thereby excludes proof of Post-consistency, according to which a sentential variable alone is not a theorem. This difficulty can be avoided, without recourse to quantification, and at less cost, by modalisation. For the trouble is that we have not guaranteed that every sentential variable *may* take each of its values. To ensure that a variable may assume each of its values, we add to the valuation rules under (1) the further clauses:

(g) $\delta \text{ Var} \Rightarrow . \ \Diamond(v(\delta) = t)$
(h) $\delta \text{ Var} \Rightarrow . \ \Diamond(v(\delta) = f)$
(i) $\delta \text{ Var} \Rightarrow . \ \Diamond(v(\delta) = n)$

where '$\Diamond$' (logical possibility) is defined in terms of logical necessity, symbolised '$\Box$' as follows:

$$\Diamond\Delta =_{\text{Df}} \sim\Box\sim\Delta.$$

The requirement that a variable may take each of its values is included in the usual informal discussion of semantics but it is not reflected in the formalism. To this extent the usual sentential semantics are incompletely formalised.

Given (g)–(i), we amend the definition of *tautology* to:

$$\delta \text{ Taut} =_{\text{Df}} \Box(v(\delta) = t).$$

The satisfactoriness of this definition can also be argued for on the grounds that we require not just that $v(\delta) = t$ be true for $qu(\delta)$ to be a tautology, but that δ Taut if and only if it is *demonstrable* that $v(\delta) = t$.

Only a weak modalisation is needed. Indeed it seems that only a few first degree principles are required such as: $\Box\Delta \Rightarrow \Delta$; $\Box\Delta \;\&\; \Box\Gamma \Rightarrow \Box(\Delta \;\&\; \Gamma)$; and rules,

$$\vdash(\Delta \Rightarrow \Gamma) \twoheadrightarrow \vdash\Box(\Delta \Rightarrow \Gamma)$$
$$\vdash\Box(\Delta \Rightarrow \Gamma) \twoheadrightarrow \vdash(\Box\Delta \Rightarrow \Box\Gamma)$$

where '$\twoheadrightarrow$' reads 'yields'.

The proof that $(p \rightharpoonup p)$ Taut is modified as follows:

(v) $\Box[v(p) = t \vee v(p) = f \vee v(p) = n] \Rightarrow \Box(v(p \rightharpoonup p) = t)$, from (iv) above.
(vi) $(v(\delta) \neq t \;\&\; v(\delta) \neq f \Rightarrow v(\delta) = n)$, from (1) (a).
Also (vi) $\Rightarrow$ (vii), so $\Box$(vi) $\Rightarrow$ $\Box$(vii), where
(vii) is: $v(\delta) = t \vee v(\delta) = f \vee v(\delta) = n$.

Since $\Box$(vi), $\Box$(vii). Finally, detaching from (v), $\Box(v(p \rightharpoonup p) = t)$. Therefore $(p \rightharpoonup p)$ Taut.

It is easy now to prove $\sim(p \text{ Taut})$. For, by (1) (h) and p Var, $\Diamond(v(p) = f)$ follows. Hence by $v(p) = f \Rightarrow v(p) \neq t$, $\Diamond(v(p) \neq t)$ follows. Hence, $\sim\Box(v(p) = t)$; and therefore $\sim(p \text{ Taut})$.

We have explained, then, how the semantics of significance logic, and *ipso facto* of classical sentential logic, can be formalised and investigated using a paradox-free entailment, and without introducing quantification logic.

(d) *Valuation Tables*

As for $\mathbf{C_2}$.
(Since $\mathbf{S_6}$ does not contain restricted variables as primitives, each matrix is a linear array of 3^m rows, where m is the number of distinct variables in the formula. Hence it is a **C**-type matrix rather than an **S**-type.)

(e) *Formal Theorems*

A wff of $\mathbf{S_6}$ is a theorem if, and only if, it is an $\mathbf{S_6}$-tautology.
Among the theorems are the analogues of the **RSL** axioms:

$$p \rightharpoonup (q \rightharpoonup p);\; (p \rightharpoonup (q \rightharpoonup p')) \rightharpoonup ((p \rightharpoonup q) \rightharpoonup (p \rightharpoonup p'));\; (p \rightharpoonup q) \rightharpoonup (\neg q \rightharpoonup \neg p);$$

together with such theses as,

$$S(p \rightharpoonup p);\; Sq \rightharpoonup S(p \rightharpoonup p);\; Tp \rightharpoonup p;\; p \rightharpoonup Tp;\; Tp \rightharpoonup Sp;\; Tp \rightharpoonup F\neg p;\; Tp \rightharpoonup \neg SDp;\; Fp \rightharpoonup T\neg p.$$

Since $\mathbf{S}_6$ is functionally complete, all preceding **S**-systems are contained in it so far as theorems containing only unrestricted variables are concerned. Theorems of preceding **S**-systems which contain restricted variables can be represented by theorems of $\mathbf{S}_6$. This notion of representation is explained in detail for one of the axiomatic formulations of $\mathbf{S}_3$, *viz.* the axiomatic system ${}_2\mathbf{S}_3$ in §6.4.

The Logic $\mathbf{C}_6$

The logic $\mathbf{C}_6$ is similar in every respect to $\mathbf{S}_6$ except that both *t* and *n* are taken as designated values instead of *t* only.

$\mathbf{C}_6$ properly includes all preceding **C**-systems.

(a) *Morphology*
(b) *Valuation Rules*
} As for $\mathbf{S}_6$.

(c) *Tautologies:* A wff of $\mathbf{C}_6$ is a $\mathbf{C}_6$-tautology iff it never takes the value *f* for any assignment to its variables.

Theorem (21): Δ is a $\mathbf{C}_6$-tautology iff (a strict equivalence) $\Box(v(\Delta) \neq f)$; *i.e.* iff $\Box(v(\Delta) = t \vee v(\Delta) = n)$.

(d) *Valuation Tables*: As for $\mathbf{S}_6$.
(e) *Formal Theorems*: A wff of $\mathbf{C}_6$ is a theorem iff it is a $\mathbf{C}_6$-tautology.

C-*logics Consistency Theorem: Part 6*

Matrix $\mathbf{C}_6$-logics are simply inconsistent with respect to '$\sim$' and '$\neg$' and simply consistent with respect to '*F*'.

Proof: The simple inconsistency with respect to '$\sim$' and '$\neg$' follows from the functional completeness and IV of §5.6.

The simple consistency with respect to '*F*' follows from III of §5.6.

As an example of the inconsistency we have $qu(D(p \rightarrow p))$ which takes the value *n* always. Hence,

$$D(p \rightarrow p); \ \sim D(p \rightarrow p); \ \neg D(p \rightarrow p);$$

are all $\mathbf{C}_6$-theses.

As in the case of 5-logics, it would be possible to define a system $\mathbf{C}^*_6$ which has as theses the normal theses of $\mathbf{C}_6$.

In anticipation of the problem of axiomatising $\mathbf{C}_6$, we now introduce an equivalent one-sorted system $\mathbf{D}_6$. $\mathbf{D}_6$ has the advantage that direct axiomatisations of it eliminate the modification on modus ponens required in usual **C**-logics (see §5.6 *fin*).

$\mathbf{D}_6$ has as primitives the connectives '*F*', '*Q*' and '$\rightharpoondown$'. The implicational connective '$\rightharpoondown$' has the following matrix:

$\rightharpoondown$	t	f	n
t	t	f	n
f	t	t	t
n	n	f	t

In view of the details given in previous sections, this information suffices to write out the morphology, valuation rules and specifications of tautologies and theorems of $\mathbf{D}_6$, given that both t and n are designated.

Theorem (22): The set of connectives 'F', 'Q', '$\rightharpoondown$' provides a functionally complete set for three-valued logics.

Proof: It suffices to show that the primitive connectives of $\mathbf{S}_6$ can be defined. For these provide a functionally complete set.

$$D\Delta =_{\mathrm{Df}} QF(\Delta \rightharpoondown \Delta) \rightharpoondown \Delta$$
$$T\Delta =_{\mathrm{Df}} FQD\Delta$$
$$\Delta \rightharpoonup \Gamma =_{\mathrm{Df}} T\Delta \rightharpoondown \Gamma$$
$$\sim\Delta =_{\mathrm{Df}} DQD\Delta$$
$$\neg\Delta =_{\mathrm{Df}} \sim T\Delta.$$

CHAPTER 6

Sentential Significance Logics II The Axiomatic Approach

THE logics so far presented only in a semantical way may be given axiomatic, or so-called syntactical, formulations. In addition, sentential significance systems not amenable to a finite matrix approach can be presented axiomatically. The axiomatic approach to sentential logics is in fact very much more flexible and comprehensive than the finite matrix approach. It does however suffer from the disadvantage that the intuitive appeal of elegant postulates is sometimes far from evident. But this difficulty can be surmounted in many cases by replacing the problematic postulates by more evident equivalent postulates or postulate sets.

By adopting the axiomatic approach many of the objections lodged against significance logics based on the matrix approach can be overcome. For example, much more satisfactory implicational, modal and other non-extensional relations can be introduced in terms of the postulational technique. Again, some of the problems raised by the matrix approach, for example whether the connective set should be functionally complete or not, solve themselves. For in a system where only infinite matrices would suffice, no denumerable language, such as a standard sentential system, could include and characterise a functionally complete set of infinitary connectives.

As with the matrix approach, two basic systems, an **S** system and a **C** system, are considered at each stage. The numeration of the stages corresponds to that adopted in the previous chapter. Thus at Stage 3, we examine axiomatic formulations of the systems $\mathbf{S}_3$ and $\mathbf{C}_3$. Since, however, it is possible to take different postulational bases (by changing the axiom set or the rule set) which are nevertheless deductively equivalent, *i.e.* have the same set of theses, we consider various formulations at several stages. These different formulations are indicated by numerical subscripts before the system letter. So, for example, ${}_1\mathbf{S}_3$, ${}_2\mathbf{S}_3$, ${}_3\mathbf{S}_3$, indicate respectively the first, second and third postulational basis for the system $\mathbf{S}_3$ (where only one postulational basis is given, we omit the subscript prefix). On the one hand, therefore, we may have several different semantic formulations of a given system, say $\mathbf{S}_3$–1, $\mathbf{S}_3$–2, *etc.*, which have the same set of tautologies and differ only in their formal structure, and on the other hand, several different syntactical formulations of the same basic system, say ${}_1\mathbf{S}_3$, ${}_2\mathbf{S}_3$, ${}_3\mathbf{S}_3$, *etc.*, which, allowing for definitional replacement, have the same set of theses as each other but again differ in their formal structure. In spite of these formal variations, however, there is in an important sense only one basic semantic system, say $\mathbf{S}_3$, characterised by the set of tautologies determined by its matrices; and similarly, only one basic syntactic system, $\mathbf{S}_3$, characterised by a set of theses. The question of completeness, therefore, whether every tautology is a thesis of the postulational system, is not compounded by the multiplication of the formal variants. For each basic **S** or **C** system, there is only one problem of completeness, and similarly only one converse problem of consistency or plausibility—whether every thesis of the syntactic system is a tautology. Of course, that these questions of completeness and plausibility are singular depends on showing that the various semantic formulations are deductively equivalent (as they are), and similarly, that the various axiomatic formulations are deductively equivalent (as we shall see). Then the problems of completeness and plausibility together amount to asking whether the basic semantic system, say $\mathbf{S}_3$, however characterised formally, is deductively equivalent to the basic syntactic system $\mathbf{S}_3$, however characterised formally.

In addition to examining the basic **S** and **C** systems at each stage, however, at some stages we also investigate significance extensions of certain non-classical sentential logics which have not been examined at the corresponding stage of the matrix chapter, since the matrix approach is inappropriate. Thus at stage 0, for example, we examine a system $\mathbf{ES}_0$ which is an $\mathbf{S}_0$-type logic obtained from Anderson and Belnap's entailment system **E** (see Anderson [1]) by interpreting it over a further value *n*.

In general, therefore, Stage *m* in the axiomatic approach considers systems like and including those considered at Stage *m* in the matrix approach; and in the case of the basic systems already described at the semantic level, we do not give the detailed morphology again since it is the same. In general, too, we use the same symbolism, but one minor symbolic difference should be noted: to facilitate the reading of metatheorems we generally use '*A*', '*B*', '*C*', '*D*', *etc.* as metavariables standing for any wff where previously we used 'Δ' and 'Γ'. This is simply to ensure a larger supply of letters and hence a reduction in the numbers of primes which it would otherwise be necessary to use. Since we are here concerned only with many-valued sentential systems, there is no clash with the earlier use of '*A*', '*B*', *etc.* to stand for wffcl.

The substitution notation:

$$\mathsf{S}^P_B A \,|$$

represents the formula which results from *A* by substituting the wff *B* for the variable *P* uniformly for each occurrence of *P* in *A*. (The notation is that of Church [4], p. 192. Our substitution notation and bracketing conventions are all adapted from Church). For the purpose of effective formulations of axiomatic systems and for effective proofs and metaproofs it is desirable to provide as well a recursive analysis of the substitution notation. In the case of $\mathrm{S}^P_B A\,|$ the specification is as follows:

Basis: $\mathrm{S}^P_B P\,| = B$
$\mathrm{S}^P_B P'\,| = P'$, where P' differs from P.

Recursion steps:

1. Where Φ is a 1-place sentence connective (*e.g.* '$\sim$'),
   if $\mathrm{S}^P_B A\,| = A'$ then $\mathrm{S}^P_B \Phi\, A\,| = \Phi\, A'$.
   Hence $\mathrm{S}^P_B \Phi\, A\,| = \Phi\, \mathrm{S}^P_B A\,|$.
2. Where Ψ is a 2-place connective (*e.g.* '$\vee$'),
   if $\mathrm{S}^P_B A\,| = A'$ and $\mathrm{S}^P_B C\,| = C'$,
   then $\mathrm{S}^P_B \Psi(A, C)\,| = \Psi(A', C')$;
   thus $\mathrm{S}^P_B A \vee C\,| = A' \vee C'$.
   Hence $\mathrm{S}^P_B\, A\, \Psi\, C\,| = \mathrm{S}^P_B A\,|\,\Psi\, \mathrm{S}^P_B C\,|$.

We use the assertion sign '$\vdash$', read 'is a thesis', to indicate theorems and the arrow sign '$\twoheadrightarrow$', read 'yields', to represent a derivation rule. Thus a sentential substitution rule will appear in the form,

$$\vdash A \twoheadrightarrow \ \vdash \mathrm{S}^P_B A\,|.$$

Similarly, a material detachment rule will appear in the form,

$$\vdash A, \vdash A \supset B \twoheadrightarrow \ \vdash B.$$

6.1 Stage 0: Axiomatic formulations of 0-systems

Consider first the logic S_0 with morphology as given in §5.8. We compare this with the system presented in *1–*5 of *PM*.

S_0 has two sorts of variables: *S*-restricted and *S*-unrestricted. The *S*-restricted variables,

$$r, s, r', s', etc.,$$

correspond directly to the elementary propositional variables in *1 of *PM*. For elementary propositional variables are significance and indicative restricted. Every elementary sentence is significant in terms of the philosophical presuppositions of each edition of *PM*. However, the range of the *elementary* propositional variables is a subrange of the range of *S*-restricted variables since the latter have in their range sentences which contain quantifying expressions like 'all' and 'some'.

Significance-restricted formulae (*srf*) are characterised by the following recursion rules, which mirror those in *PM* for elementary propositions:

1. An *S*-restricted variable is an srf.
2. If *A* is an srf, then $\sim A$ is an srf, and conversely [*cf.* *1.7 of *PM*].
3. If *A* and *B* are srf, then $A \vee B$ is an srf, and conversely [*cf.* *1.71 of *PM*].

It is clear from the commentary on *1.7–*1.72 (*PM*, p. 97), and on the axiom of identification of real variables (*PM*, p. 109; Russell [2], p. 86), that the intention is to preserve significance over elementary formulae. The restriction to elementary propositions as arguments is sufficient, under the assumptions of type-theory and logical atomism, to guarantee preservation of significance by the recursion rules.

The recursion postulates for elementary propositional functions in *PM* are in fact deficient. The following addition,

> *1.73 If ϕp is an elementary propositional function which takes elementary propositions as arguments, then $\sim\phi p$ is an elementary propositional function,

is essential for the proof of *3.03 (see *PM*, pp. 111, 109). The formalisation and proof of what amounts to *3.03 in S_0 is of interest in this connexion. It is:

$$\text{if } \vdash A \text{ and } \vdash B, \text{ where } A \text{ and } B \text{ are srf, then } \vdash A \;\&\; B.$$

The proof which follows uses the definitions,

$$A \supset B =_{\mathrm{Df}} \sim A \vee B \quad (\text{*1.01 of } PM)$$
$$A \;\&\; B =_{\mathrm{Df}} \sim(\sim A \vee \sim B) \quad (\text{*3.01 of } PM).$$

Now if *A* and *B* are srf, then $\sim A$ and $\sim B$ are srf. Thus $\sim A \vee \sim B$ is an srf, and consequently $\sim(\sim A \vee \sim B)$ is an srf. So $A \;\&\; B$ is an srf.

But we can show in S_0,

$$\vdash r \vee \sim r;$$

and by restricted substitution, since $\sim A \vee \sim B$ is an srf, given that *A* and *B* are,

$$\vdash(\sim A \vee \sim B) \vee \sim(\sim A \vee \sim B).$$

Applying definitions and association we then have,

$$\vdash A \supset (B \supset (A \;\&\; B)).$$

Hence,

$$\vdash A \;\&\; B$$

follows by material detachment given $\vdash A$ and $\vdash B$.

The standard modern, but quite inadequate, fuller formalisation of the propositional logic of *PM* which takes *1.7–*1.72 as formation rules for wff is partly explained by the following metatheorem:

Theorem (1): If a formula A contains only S-restricted variables and improper symbols of $\mathbf{S}_0$, then A is a wff if, and only if, A is an srf.

Proof: By induction over the number of connectives in A.

The modern formalisations of the propositional logic of *PM* are inadequate because they eliminate significance in favour of grammaticalness or well-formation, or, perhaps more accurately, because they conflate what Whitehead and Russell *would* have distinguished: on the one hand, significance; on the other, grammaticalness and well-formation.

Postulates of $\mathbf{S}_0$

Axioms

0.1 (*1.2 of *PM*) $\vdash r \vee r \supset r$
0.2 (*cf.* *1.3 of *PM*) $\vdash r \supset r \vee s$
0.3 (*1.4, *1.6) $\vdash (s \supset s') \supset (r \vee s \supset s' \vee r)$.

Rules of Inference

R0.1 (*1.1, *1.11) $\vdash A, \vdash A \supset B \twoheadrightarrow \vdash B$ (Material Detachment).
R0.2 $\vdash A \twoheadrightarrow \vdash \mathbf{S}^R_B A\,|$, where R is an S-restricted variable and A and B are srfs. (Restricted Substitution).

It should be noticed that although $\mathbf{S}_0$ has wff which are not srf (see the morphology in §5.8), the axioms and rules are expressed entirely in terms of srf. Hence although $\mathbf{S}_0$ does not have the same class of wffs as classical sentential logic, its class of srfs coincides (with re-lettering if necessary) with the class of wffs of classical logic and it has the same class of theorems.

The main features of $\mathbf{S}_0$ are contained in the following metatheorems:

Theorem (2): If $\vdash A$, then A is an srf.

Proof: By induction over proof lengths. Every axiom is an srf and the rules of inference preserve this property.

For R0.1 the following lemma suffices:

Lemma: Any sub-wff of an srf is also an srf. This is proved for all srf of a wff A by induction over the number of connectives in the sub-wff.

The lemma required for R0.2 is as follows:

Lemma: If A is an srf, then the wff obtained by substituting srf B uniformly for an S-restricted variable R of A is also an srf. This is proved by induction over the number of occurrences of connectives in A.

Theorem (3): The system $\mathbf{S}'_0$ obtained from $\mathbf{S}_0$ by taking as sentential variables the *S*-restricted variables of $\mathbf{S}_0$; as wff, the srfs of $\mathbf{S}_0$; and as theorems, the theorems of $\mathbf{S}_0$; coincides formally with classical sentential logic (**RSL**).

Proof: is based on the following lemmata:

Lemma: *A* is an srf of $\mathbf{S}_0$ if, and only if, *A* is a wff of classical sentential logic (where the sentential variables of classical logic are chosen to coincide type-wise with the *S*-restricted variables, and where the connectives '$\sim$' and '$\vee$' are taken as primitive).

Proof: By induction over the number of occurrences of connectives in *A*, the induction using the parallel recursive specifications of $\mathbf{S}_0$-srf and classical-wff.

Lemma: *A* is a theorem of $\mathbf{S}_0$ if, and only if, *A* is a theorem of classical sentential logic.

Proof: Axiomatise classical sentential logic using 0.1–0.3, R0.1 and Uniform Substitution. But in virtue of the previous lemma, the rule of Uniform Substitution for classical logic coincides with the rule of Restricted Substitution for $\mathbf{S}_0$. Hence the theorem sets coincide.

Theorem (4): An srf *A* is an $\mathbf{S}_0$-tautology if, and only if, *A* is a classical tautology.

Proof: As indicated in §5.8.

Theorem (5): *A* is a theorem derivable from the postulates of $\mathbf{S}_0$ if, and only if, *A* is an $\mathbf{S}_0$-tautology.

Proof: If *A* is a theorem derivable from the postulates of $\mathbf{S}_0$, then it is a classical theorem, hence a classical tautology. Conversely, if *A* is an $\mathbf{S}_0$-tautology, then it is a classical tautology, and so a theorem of classical logic. Therefore it is an srf also, and so a theorem of $\mathbf{S}_0$.

Various consistency and completeness results follow as corollaries.

Postulates of $\mathbf{C}_0$

As a postulate set for the logic $\mathbf{C}_0$, we take the axioms:

$$p \vee p \supset p;\ q \supset p \vee q;\ (q \supset q') \supset (q \vee p \supset p \vee q');$$

and the rules of material detachment and uniform unrestricted substitution. Since this postulate set coincides with a postulate set for classical sentential logic, and since every wff of classical sentential logic is a wff of $\mathbf{C}_0$, and *vice versa*, there being no restricted variables in $\mathbf{C}_0$, it follows:

Theorem (6): A wff A is a theorem derivable from the postulate of $\mathbf{C}_0$ if, and only if, A is a theorem of classical sentential logic.

Theorem (7): A is a theorem derivable from the postulates of $\mathbf{C}_0$ if, and only if, A is a $\mathbf{C}_0$-tautology.

Proof: A is a theorem of $\mathbf{C}_0$ if, and only if, A is a classical theorem; hence if, and only if, A is a classical tautology; hence if, and only if, A is a $\mathbf{C}_0$-tautology; by results from §5.8.

Consistency and completeness results for $\mathbf{C}_0$ follow as corollaries from this theorem.

The completeness results on $\mathbf{S}_0$ and $\mathbf{C}_0$ are, for reasons already explained in §5.8, of no great value as *completeness* results. For there is no way of expressing within these systems quite elementary theses about significance. Theses which exemplify instances of the general assumption that a compound sentence is significant if, and only if, its components are, must remain extra-systematic or metatheoretic. To say that $\mathbf{S}_0$ and $\mathbf{C}_0$ are complete is therefore something of a sham. They are technically complete with respect to certain restricted assumptions, but they are in no sense complete theories of significance. The merits of formal "completeness" theorems will be discussed in more detail subsequently.

Any sentential logic can be converted into a Stage 0 significance logic, either **S**-type or **C**-type. Consider, for example, a system with axioms A_1, ..., A_m and rules of uniform substitution, modus ponens and adjunction. Formally speaking it is a $\mathbf{C}_0$-type logic as it stands. To convert it into a significance $\mathbf{C}_0$-logic, we simply re-interpret it over a further value n. To arrive at an $\mathbf{S}_0$-type logic, we introduce a new set of variables in addition to the sorts which already occur—say the S-restricted variables of $\mathbf{S}_0$. We then re-state A_1, ..., A_m using S-restricted variables and introduce a rule of restricted uniform substitution.

To illustrate, we sketch below a significance logic $\mathbf{ES}_0$, an $\mathbf{S}_0$-type logic obtained from Anderson and Belnap's entailment system **E** (Anderson [1]).

The morphology of $\mathbf{ES}_0$ is the same as that of $\mathbf{S}_0$ except that the 2-place connectives '→' and '&' are added to the improper symbols. Definitions of wff and srf are extended to allow for these connectives. The axioms of $\mathbf{ES}_0$ are obtained from those of **E** by replacing axiom-schemata of **E** by axioms formulated using S-restricted variables. For example, instead of,

$$\mathbf{E1}: A \rightarrow . A \rightarrow . B \rightarrow B$$

we now have,

$$\mathbf{ES}_0\mathbf{1}: r \rightarrow . r \rightarrow . s \rightarrow s$$

$\mathbf{ES}_0$ has the following rules:

Modus ponens for '→'; Adjunction; Restricted Uniform Substitution. It could also have the rule of Uniform Substitution, but in $\mathbf{ES_0}$ this rule would be otiose.

Theorem (8): There are no finite characteristic matrices for $\mathbf{ES_0}$.

Proof: If there were, there would be finite characteristic matrices for **E**. But Dugundji's [1] proof that there are no finite characteristic matrices for the Lewis systems S1–S5 also establishes that there are no finite characteristic matrices for **E**.

Non-classical S_0-type logics are of interest in that they provide a basis on which more powerful significance logics with non-extensional connectives can be constructed. Such logics cannot of course be investigated by the matrix method.

6.2 Stage 1: Axiomatic Formulations of 1-systems

Stage 0 logics are both semantically and syntactically weak. They are semantically weak because it is impossible to state within them as theses important significance laws. They are syntactically weak because certain formal conditions cannot be stated formally. An example is provided by the rule of Restricted Substitution, where a metatheoretic qualification is made to the effect that substituted wff are significant. It is an advantage of Stage 1 logics that this restriction can be stated formally and Restricted Substitution takes on a more satisfactory form. Stage 1 logics are also, of course, more satisfactory from the semantic point of view, though they are still far too weak to be of real value as significance theories.

S_1-logics have two substitution rules:

(a) the rule of substitution for *S*-restricted variables is a restricted uniform substitution rule which permits substitution of significant wff only for a variable. The rule is:

$$\vdash A, \vdash SB \twoheadrightarrow \vdash \mathrm{S}^{R}_{B}A\,|,$$

where *A* and *B* are wff and *R* is an *S*-restricted variable.

(b) the rule of substitution for *S*-unrestricted variables, whether covered or not, is simply uniform substitution, *i.e.*,

$$\vdash A \twoheadrightarrow \vdash \mathrm{S}^{P}_{B}A\,|,$$

where *A* and *B* are wff and *P* is an *S*-unrestricted variable.

These substitution rules are not uncontroversial. What is particularly questionable is whether significance is preserved under replacement of any significant sentence by any other. For do not sentence connectives impose limitations on the sentences they can significantly couple? Consider such examples as 'She came home in a flood of tears and a sedan chair'. Should not relevance conditions be taken seriously? Although natural-language analogues or transcriptions of sentential connectives do appear to impose relevance and other restrictions, it has been assumed in the matrices that classical connectives do not. Connectives approximating more closely to natural-language connectives such as 'and' and 'or' could be introduced as sentence relations at a later stage when more and finer features of sentence structure are taken into account. This general problem will be discussed in more detail in Volume 2.

Postulates of ${}_1\mathbf{S}_1$

The morphology is as given in §5.9.

Axioms

1.1 $\vdash r \supset (r' \supset r)$
1.2 $\vdash (r'' \supset (r \supset r')) \supset ((r'' \supset r) \supset (r'' \supset r'))$
1.3 $\vdash (\sim r \supset \sim r') \supset (r' \supset r)$
1.4 $\vdash Sp \supset (Sq \supset S(p \supset q))$
1.5 $\vdash S(p \supset q) \supset (Sp \supset Sq)$
1.6 $\vdash S(p \supset q) \supset Sp$
1.7 $\vdash Sp \supset S{\sim}p$
1.8 $\vdash S{\sim}p \supset Sp$

Rules of Inference

R1.1 $\vdash A \rightarrow \vdash SA$
R1.2 $\vdash A, \vdash SB, \rightarrow \vdash \mathrm{S}^R_B A\,|$, where R is an S-restricted variable
R1.3 $\vdash A \rightarrow \vdash \mathrm{S}^P_B A\,|$, where P is an S-unrestricted variable and B is a wff, *or* both P and B are S-restricted variables.
R1.4 $\vdash A, \vdash A \supset B \rightarrow \vdash B$.

The usual definitions of the classical connectives, as given in the morphology, are adopted, *viz.*:

$A \vee B =_{\mathrm{Df}} \sim A \supset B$; $A \,\&\, B =_{\mathrm{Df}} \sim(A \supset \sim B)$;
$A \equiv B =_{\mathrm{Df}} ((A \supset B) \,\&\, (B \supset A))$.

Results on $\mathbf{S}_1$

1. *Theorem* (9): Classical sentential logic **RSL** holds for any wff B of the form $B(A_1, \ldots, A_n)$, where B contains only classical connectives and each of $A_1, \ldots, A_n$ is provably significant, taking $A_1, \ldots, A_n$ as the components of B.

Proof: From 1.1 it follows by R1.2, $\vdash A \supset (B \supset A)$, provided $\vdash SA$ and $\vdash SB$. Similarly, remaining axioms for a complete sentential logic follow from 1.2 and 1.3 provided the component expressions are significant. Finally the rules of modus ponens and uniform substitution hold without qualification for significant components and significant replacements.

2. *Formal Theorems:*

T1. $\vdash SSp$

Proof: 1. $\vdash S(Sp \supset S{\sim}p)$ from 1.7 by R1.1
2. $\vdash SSp$ from 1 by 1.6, R1.3 and R1.4

T2. $\vdash Sr$

Proof: 1. $\vdash S(r \supset (r' \supset r))$ from 1.1 by R1.1
2. $\vdash Sr$ by 1.6, R1.3 and R1.4

T3. $\vdash Sp \equiv S{\sim}p$

T4. $\vdash S(p \supset q) \equiv Sp \mathbin{\&} Sq$

T5. $\vdash S(p \mathbin{\&} q) \equiv Sp \mathbin{\&} Sq$

T6. $\vdash S(p \vee q) \equiv Sp \mathbin{\&} Sq$

T7. $\vdash S(p \equiv q) \equiv Sp \mathbin{\&} Sq$

T8. $\vdash S{\sim}Sp$

Proof: T1, T3.

T9. $\vdash S{\sim}r$

T10. $\vdash S(r \supset r')$

T11. $\vdash r \supset Sr$

T12. $\vdash SSp \equiv SSSp$

3. The axiom set can be shortened by replacing 1.4–1.8 by the equivalent set,

1.4′$\vdash S(p \supset q) \equiv Sp \mathbin{\&} Sq$

1.5′$\vdash Sp \equiv S{\sim}p$

This equivalence is proved by T3, T4 and *Theorem* (9).

Call the resulting system $_2\mathbf{S}_1$.

Note that 1.4′ cannot be replaced by $qu(S(p \mathbin{\&} q) \equiv Sp \mathbin{\&} Sq)$ unless '$\supset$' is supplanted as primitive 2-place connective by '&'.

4. $\mathbf{S}_1$ can be axiomatised alternatively by omitting R1.1 and adding instead the following axiom schema:

1.9 $\vdash SA$, provided all uncovered variables in A are S-restricted.

Call the resulting system $_3\mathbf{S}_1$.

We recall that a variable P is covered in a wff A iff P occurs in A and every occurrence of P in A is within the scope of an unlimited connective when A is in primitive notation (§5.10). Since 'S' is the only unlimited connective in $\mathbf{S}_1$-logics, a variable is covered iff it is within the scope of 'S'. Note, too, that since all occurrences of unrestricted variables in the axioms of $_1\mathbf{S}_1$ are covered, and since the rules of $_1\mathbf{S}_1$ preserve the cover of unrestricted variables, an unrestricted variable never occurs uncovered in a theorem of $_1\mathbf{S}_1$ or any equivalent system.

We now show that $_1\mathbf{S}_1$ and $_3\mathbf{S}_1$ are deductively equivalent by proving, first, that 1.9 is a metatheorem of $_1\mathbf{S}_1$ and, secondly, that R1.1 is a derived rule of $_3\mathbf{S}_1$.

That 1.9 is a metatheorem of ${}_1S_1$ can be shown by induction on the number of connectives in wff A. If A has no connectives, then it is of the form R. But $\vdash SR$, a result proved, after appropriate substitution in 1.1 using R1.3, like T2. The induction step is proved immediately from T1 (without the induction hypothesis) in the case that A is of the form SB, from T5 if A is of the form $(B_1 \supset B_2)$, and from T3 if A is of the form $\sim B$.

Conversely, that R1.1 is a derived rule of the alternative system ${}_3S_1$, follows by induction on the lengths of proofs of theorems of the system. The induction basis, that for every axiom A, $\vdash SA$, follows at once from 1.9. And if B follows from earlier members of a proof sequence by application of one of the inference rules, then $\vdash SB$ follows from the significance closures (closure by 'S') of the premisses used by application of the same rule. Thus every theorem of ${}_3S_1$ is significant.

The schema 1.9 of ${}_3S_1$ can be equivalently replaced by the axioms:

$\vdash Sr$; $\vdash SSp$.

These postulates follow immediately from 1.9. The converse can be proved, much as above, by induction on the number of connectives in A of 1.9.

Call this system ${}_4S_1$.

5. We recall that a wff A is a tautology of S_1 if, and only if, it has the value t for all assignments of values from the set $\{t,f\}$ for its S-restricted variables and from the set $\{t,f,n\}$ for its S-unrestricted variables under the matrices assigned for classical '$\supset$' and '$\sim$' and for unlimited 'S'.

5.1 Every theorem of ${}_1S_1$ is a tautology of S_1 (The Plausibility Theorem).
5.2 ${}_1S_1$ is simply, absolutely and Post-consistent.
5.3 ${}_1S_1$ is S-consistent; *i.e.* not every significant wff is a theorem.
5.4 ${}_1S_1$ is S-correct; *i.e.* no nonsignificant wff is a theorem.

Proofs of these results are included in proofs of analogous results for S_2 given in §6.3.

Since ${}_2S_1$, ${}_3S_1$ and ${}_4S_1$ are deductively equivalent to ${}_1S_1$, similar results follow trivially for these systems.

6. *Substitutivity of equivalence results*

Substitutivity Theorem 1: If A' is obtained from A by substitution of wff B' for a well-formed part B of A, then: $\vdash SB' \equiv SB \supset . SA' \equiv SA$.

Proof: 1. Special cases:

(i) If the substitution is at zero places then A' is the same as A and the result follows by substitution in $\vdash r' \supset . r \equiv r$.

(ii) If substitution is for the whole of A then $SB' \equiv SB$ is the same as $SA' \equiv SA$.

2. Induction over the number of occurrences of primitive connectives in A is used. If A has zero connectives, the result follows

by 1. If the number of connectives is greater than zero, there are three cases to consider:

(a) A is of the form SC. Then A' is of the form SC', where C' results from C by substitution of B' for B in C. But $\vdash SSC' \equiv SSC$. Therefore, $\vdash SB' \equiv SB \supset . SSC' \equiv SSC$, by **RSL** (in view of *Theorem* (9)), as required.

(b) A is of the form $\sim C$. Then A' is of the form $\sim C'$ where C' results from C by substitution of B' for B. By induction hypothesis, $\vdash SB' \equiv SB \supset . SC' \equiv SC$. By T3 and **RSL** it follows:
$\vdash SB' \equiv SB \supset . S{\sim}C' \equiv S{\sim}C$, as required.

(c) A is of the form $(C_1 \supset C_2)$. Then A' is of the form $(C'_1 \supset C'_2)$, where C'_1 and C'_2 result from C_1 and C_2 respectively by substitution of B' for B at zero or more places. By induction hypothesis:
$\vdash SB' \equiv SB \supset . SC_1 \equiv SC'_1$
$\vdash SB' \equiv SB \supset . SC_2 \equiv SC'_2$.
The result follows by restricted substitution in the **RSL** theorem $\vdash [r \supset . r' \equiv r''] \supset . [r \supset . r''' \equiv r''''] \supset . [r \supset (r' \,\&\, r''' \equiv r'' \,\&\, r'''')]$, then by T4.

Substitutivity Theorem 2: If $\vdash SA$, and if A' is obtained from A by substitution of B' for B at zero or more places in A, and if both $\vdash SB' \equiv SB$ and, $\vdash B' \equiv B$ provided $\vdash SB$, then $\vdash A' \equiv A$.

The clause 'if $\vdash SB' \equiv SB$ and, $\vdash B' \equiv B$ provided $\vdash SB$', is a syntactical way of saying 'if B' demonstrably has the same value as B for all values of the variables', a condition which can be expressed in stronger significance logics as 'if $\vdash B' \simeq B$'.

Proof: 1. Special cases:

(i) If the substitution is at zero places, then A' is the same as A, and $\vdash SA'$. The result is by restricted substitution in $\vdash r \equiv r$.

(ii) If the substitution is for the whole of A then B is A, so $\vdash SB$ and $\vdash B' \equiv B$, *i.e.* $\vdash A' \equiv A$.

2. By induction over the number of connectives in A; if A has zero connectives, the result follows by 1. Otherwise there are three cases:

(a) A is of the form SC. Then A' is of the form SC' where C' results from C. By the previous metatheorem, $\vdash SB' \equiv SB \supset . SC' \equiv SC$. So the case is established.

(b) A is of the form $\sim C$. Then A' is of the form $\sim C'$, where C' results from C. By T3, $\vdash SC$. By induction hypothesis then, $\vdash C' \equiv C$ and $\vdash SC'$. Hence the result follows by restricted substitution in $\vdash r \equiv r' \supset . \sim r' \equiv \sim r$.

(c) A is of the form $(C_1 \supset C_2)$. Then A' is of the form $(C'_1 \supset C'_2)$

where C'_1 and C'_2 result from C_1 and C_2 respectively. $\vdash SC_1$ and $\vdash SC_2$ follow by T4. So by induction hypothesis, $\vdash C'_1 \equiv C_1$ and $\vdash C'_2 \equiv C_2$, and $\vdash SC'_1$ and $\vdash SC'_2$. The result follows by restricted substitution in the **RSL** theorem

$$\vdash r \equiv r' \supset . r'' \equiv r''' \supset (r \supset r'' \equiv r' \supset r''').$$

7. Normal forms may be defined for wff of ${}_1\mathbf{S}_1$. A wff of ${}_1\mathbf{S}_1$ is in conjunctive normal form (CNF) if it is a conjunction of disjunctions (in the sense defined for classical sentential logic) where components of the disjuncts may have the following forms:

$$P, \sim P, SP, SSP, \sim SP, \sim SSP.$$

Every wff A of ${}_1\mathbf{S}_1$ may be put into CNF A', *i.e.* effectively transformed to a strongly equivalent ($\simeq$) wff A' which is in CNF. To be exact, since '$\simeq$' is not definable in ${}_1\mathbf{S}_1$, $\vdash A \equiv A'$ provided $\vdash SA$, and $\vdash SA \equiv SA'$. The reduction of A to CNF A' is effected thus:

(i) Distribute all occurrences of 'S' on to variables or on to the S-covered variables (*i.e.* expressions of the form SP) by application of T3, T4, T5, T6, T7 and T8. Because of T12, more than doubly iterated 'S's need not be considered.

(ii) Convert the expression resulting when (i) is completed to CNF by the procedure for reduction to CNF for classical sentential logic.

That $\vdash SA \equiv SA'$ follows by Substitutivity Theorem 1 from the following results:

T13. $\vdash S(A \supset B) \equiv S(\sim A \vee B)$
T14. $\vdash S\sim\sim A \equiv SA$
T15. $\vdash S\sim(A \vee B) \equiv S(\sim A \,\&\, \sim B)$
T16. $\vdash S\sim(A \,\&\, B) \equiv S(\sim A \vee \sim B)$
T17. $\vdash S(A \vee (B \,\&\, C)) \equiv S(A \vee B \,\&\, A \vee C)$.

Also, since $\vdash A \equiv B \rightarrow \vdash SA \equiv SB$,

T18. $\vdash S(S(A \supset B)) \equiv S(SA \,\&\, SB)$ *etc.*

That $\vdash A \equiv A'$ provided $\vdash SA$, follows by Substitutivity Theorem 2 using the above theorems and such **RSL** results as,

T19. $\vdash r \supset r' \equiv \sim r \vee r'$
T20. $\vdash \sim\sim r \equiv r$
T21. $\vdash \sim(r \vee r') \equiv \sim r \,\&\, \sim r'$
T22. $\vdash \sim(r \,\&\, r') \equiv \sim r \vee \sim r'$ *etc.*

The following CNF metatheorems hold:

CNF1: A wff A in CNF is a tautology of $\mathbf{S}_1$ iff A is a *valid CNF*, *i.e.* each disjunct of a conjunct of A contains either:

(i) an S-restricted variable and its negation; or
(ii) an S-covered variable SP and its negation; or
(iii) SSP for some P; or
(iv) SR for some S-restricted variable R;
and all S-unrestricted variables in A are covered.

Proof: 1. *Sufficiency.* $\vdash SA$ by 1.9. Also if A is of the form $B_1 \,\&\, B_2 \,\&\, \ldots \,\&\, B_n$, then SB_i for each i, by 1.9. For since by **RSL**, $\vdash SP \vee {\sim} SP \vee B$, provided SB; $\vdash R \vee {\sim} R \vee B$, provided SB; $\vdash SR \vee B$, provided SB; and $\vdash SSP \vee B$, provided SB; then $\vdash B_i$ for each i follows from (i)or (ii) or (iii) using commutativity. Since $\vdash B_i, \vdash B_j \longrightarrow \vdash (B_i \,\&\, B_j)$ is a derived rule (again by **RSL**), it follows, using associativity, $\vdash A$. Hence by 5.1, A is a tautology of $\mathbf{S}_1$.

2. *Necessity.* If an S-unrestricted variable P in A is not covered, then A has value n when P has value n (by contraposing T3, T4, *etc.*). Hence A is not a tautology. If, alternatively, some disjunct B_j does not satisfy any of (i), (ii), (iii) and (iv), then B_j is of the form:

$$f \vee SP_1 \vee \ldots \vee SP_{i1} \vee {\sim} SP'_1 \vee \ldots \vee {\sim} SP'_{i2} \vee R_1 \vee R_2 \vee \ldots \vee R_{i3} \vee {\sim} R'_1 \vee \ldots \vee {\sim} R'_{i4}$$

where 'f' represents disjunctions of expressions of the form ${\sim} SR$ and ${\sim} SSP$, and where all variables are distinct. Then B_j has value f for assignments of values n to $P_1 \ldots P_{i1}$, t (or f) to $P'_1 \ldots P'_{i2}$, f to $R_1 \ldots R_{i3}$, and t to $R'_1 \ldots R'_{i4}$. Hence B_j is not a tautology. Hence, also, by **RSL** if SB_i for all i, and by contraposing T5 otherwise, A is not a tautology.

CNF2: If a wff in CNF is a tautology of $\mathbf{S}_1$ then it is a theorem of ${}_1\mathbf{S}_1$.

Proof: If a wff A is a tautology of $\mathbf{S}_1$ then it is a valid CNF by part 2 of *CNF1*. But if A is a valid CNF then that A is a theorem of ${}_1\mathbf{S}_1$ follows as in part 1 of *CNF1*.

CNF3: A wff A is a tautology of $\mathbf{S}_1$ iff its CNF A' is a tautology.

Proof: If A is a tautology then so is SA. If SA is a tautology then, as A never takes value n, so $A \vee {\sim} A$ is a classical tautology. Hence by completeness of classical logic, $\vdash A \vee {\sim} A$, and so by R1.1, $\vdash S(A \vee {\sim} A)$; whence $\vdash SA$ by T6. A' is obtained from A by a finite number of substitutions each of which satisfies the premisses of Substitutivity Theorem 2. Hence by iteration of this theorem, $\vdash A \equiv A'$, since $\vdash SA$. Then by 5.1, $(A \equiv A')$ is a tautology. So by the matrix for '$\equiv$', A is a tautology. Similarly if A' is a tautology.

CNF4: A wff A is a theorem of ${}_1\mathbf{S}_1$ iff its CNF A' is a theorem.

Proof: If $\vdash A$ then $\vdash SA$. Then as in the previous theorem, $\vdash A \equiv A'$. Hence $\vdash S(A \equiv A')$ and $\vdash SA'$ by T7. Therefore $\vdash A'$ by **RSL**. Similarly if $\vdash A'$ then $\vdash A$.

The above results, and those which follow for ${}_1\mathbf{S}_1$, hold also for the equivalent systems ${}_2\mathbf{S}_1$, ${}_3\mathbf{S}_1$, ${}_4\mathbf{S}_1$.

8. *Completeness Theorems for* ${}_1\mathbf{S}_1$

8.1 Every tautology of $\mathbf{S}_1$ is a theorem of ${}_1\mathbf{S}_1$.

Proof: By *CNF2, 3, 4.*

8.2 ${}_1\mathbf{S}_1$ is complete.

8.3 ${}_1\mathbf{S}_1$ is not strongly complete.

Proof: *Sp* can be added as a further axiom. Now *Sp* is not a tautology of $\mathbf{S}_1$, hence not a theorem of ${}_1\mathbf{S}_1$ (by 5.1 and 8.1). But the resulting system ${}_1\mathbf{S}_1 + [Sp]$ is consistent, as can be shown by these matrices:

$\supset$	t	f
t	t	f
f	t	t

$\sim$	
t	f
f	t

S	
t	t
f	t

where *S*-unrestricted variables have values from the set $\{t, f\}$. Therefore ${}_1\mathbf{S}_1$ is not strongly complete.

Strong completeness can be gained however by adding to ${}_1\mathbf{S}_1$ the constant 'q_0' and the axiom,

1.10 $\vdash \sim Sq_0$.

Call the resulting system $\mathbf{S}_1^\dagger$ (*cf.* the system $\mathbf{S}_3^\dagger$ of §5.12). $\mathbf{S}_1^\dagger$ is not deductively equivalent to ${}_1\mathbf{S}_1$ since it contains theorems (*e.g.* $\vdash \sim Sq_0$) which ${}_1\mathbf{S}_1$ does not contain.

8.4 $\mathbf{S}_1^\dagger$ is complete.

Proof: The proof is similar to that given for $\mathbf{S}_1$ except that 'valid CNF' must be re-defined to allow for the possibility that disjuncts are valid because they contain $\sim Sq_0$ or *SSp* for some constant.

8.5 $\mathbf{S}_1^\dagger$ is strongly complete.

Proof: Suppose *A* which is not tautologous is added to $\mathbf{S}_1^\dagger$ as a further axiom. Then $\vdash SA$ by R1.1. Hence $\vdash A'$ where A' is the CNF of *A* by *CNF4*. Since *A* is not a tautology, A' is not a tautology, by *CNF3*. Hence some conjunct, say B_j, of A' is not a valid CNF, by *CNF1*. Also $\vdash SB_j$ and $\vdash B_j$ by **RSL**. Now substitute in B_j, 'Sq_0' for every unnegated *S*-restricted variable, '$\sim Sq_0$' for every negated *S*-restricted variable, and 'q_0' for every *S*-unrestricted variable. Since $\vdash S\sim Sq_0$, $\vdash SSq_0$, the rules permit these substitutions. The resulting wff is of the form $(Sq \vee f)$. For since B_j is not a valid CNF, it does not contain $\sim Sq_0$, SSq_0, *etc.* Since SB_j, all disjuncts of the resulting wff not identical with Sq_0 have value *f*. Now since $\vdash B_j$, $\vdash Sq_0$. But also $\vdash \sim Sq_0$ (axiom). Thus we have a contradiction and hence $\mathbf{S}_1^{\dagger}$ is strongly complete.

The strategy of adding constants and axioms on them can often be exploited to convert completeness into strong completeness. It will work, for example, for sentential systems where a normal form completeness proof can be found.

8.6 The matrices for classical '$\supset$' and '$\sim$' and unlimited '*S*' are characteristic for ${}_1\mathbf{S}_1$.
Proof: Immediately from 5.1 and 8.1.

8.7 There are no theorems containing only classical connectives which also contain uncovered *S*-unrestricted variables.
Proof: For such wffs do not express tautologies.

Further Developments

1. Certain deficiences of $\mathbf{S}_1$ as a working sentential significance logic soon emerge. Some of these, as we have seen, arise from the inexpressibility of important significance connectives such as '*T*' and '$\rightharpoonup$'. Overcoming these leads to the development of progressively more adequate basic logics $\mathbf{S}_2$ through $\mathbf{S}_6$. At the same time, the intuitive ties of these more comprehensive logics become tenuous at the level of interpretation. Thus, S_5 contains '*Q*' which has no immediate intuitive interpretation. It is hoped, however, that the formal development of these systems will reveal the rationale of adopting stronger S-logics as a working basis. There are however other deficiencies which are features of all the systems $\mathbf{S}_1$ through $\mathbf{S}_6$. These arise because each of them contains only matrix-accessible connectives: in particular, therefore, none of them contains a satisfactory implication relation (though '$\rightharpoonup$' of $\mathbf{S}_3$ and stronger systems has many advantages over '$\supset$'). It is possible, however, to construct at each stage other formal systems which go some way towards meeting this particular difficulty since, by means of the axiomatic approach, connectives which are not matrix-accessible can be introduced.

We now illustrate this by showing how to construct significance logics of $\mathbf{S}_1$-strength based on sentential implicational systems which are not amenable to the matrix approach.

We examine a logic $\mathbf{IS}_1$ based on a sentential implicational system **I** (characterised by the axioms I1–I9 below). The implication '$\rightarrow$' is paradox free at the first degree. The technique of construction serves to show how any sentential system can be converted into a significance logic of $\mathbf{S}_1$-strength. A simple extension of the technique enables any sentential system to be converted into a significance logic of higher strength.

The Logic $\mathbf{IS_1}$

The logic has *S*-restricted and *S*-unrestricted variables and these improper symbols:

$$\sim, \&, \rightarrow, S, (\ ,\ ).$$

The formation rules are analogous to those of ${}_1\mathbf{S}_1$: '$\sim$' and 'S' are one-place connectives and '&' and '$\rightarrow$' are 2-place connectives. '$\vee$', '$\supset$' and '$\equiv$' are defined in the usual way. '$\leftrightarrow$' is defined by: $A\leftrightarrow B =_{\mathrm{Df}}(A\rightarrow B)\ \&\ (B\rightarrow A)$.

Postulates

I1. $\vdash(r\rightarrow s) \rightarrow (s \rightarrow s' \rightarrow .\ r \rightarrow s')$
I2. $\vdash(\sim r \rightarrow s) \rightarrow (\sim s \rightarrow r)$
I3. $\vdash r\ \&\ s \rightarrow \sim\sim r$
I4. $\vdash(r\ \&\ (r \rightarrow s)) \rightarrow s$
I5. $\vdash r \rightarrow r\ \&\ r$
I6. $\vdash(r \rightarrow s) \rightarrow (s'\ \&\ r \rightarrow s\ \&\ s')$
I7. $\vdash(r \rightarrow s)\supset \sim(r\ \&\ \sim s)$
I8. $\vdash(r\ \&\ (s \vee s')) \rightarrow ((r\ \&\ s) \vee (r\ \&\ s'))$
I9. $\vdash(r \rightarrow s) \rightarrow \sim(r\ \&\ \sim s)$
S1. $\vdash S{\sim}p \leftrightarrow Sp$
S2. $\vdash S(p \rightarrow q) \leftrightarrow Sp\ \&\ Sq$
S3. $\vdash S(p\ \&\ q) \leftrightarrow Sp\ \&\ Sq$

The rules are the same as those of ${}_1\mathbf{S}_1$ together with modus ponens for '$\rightarrow$'.

Some of the metatheorems which hold are as follows (we state them without proof):

IMT1. $\mathbf{IS_1}$ includes ${}_1\mathbf{S}_1$; *i.e.* every theorem of ${}_1\mathbf{S}_1$ is a theorem of $\mathbf{IS_1}$.
IMT2. $\mathbf{IS_1}$ is consistent.
IMT3. There are no finite characteristic matrices for $\mathbf{IS_1}$. Thus $\mathbf{IS_1}$ is not matrix-accessible.
IMT4. There are no theorems of $\mathbf{IS_1}$ of the form $A\rightarrow(B\rightarrow C)$ where A is a sentential variable of some sort.

Thus the system is constructed simply by introducing two sorts of variable, *S*-restricted and *S*-unrestricted and then writing the axioms of the background sentential system in terms of *S*-restricted variables and the axioms of the imposed significance system in terms of *S*-unrestricted variables. The further essential requirement is that there should be a rule of restricted substitution. This technique can clearly be extended to stronger **S** systems. Thus, in order to construct $\mathbf{IS_2}$ it is necessary only to replace S1–S3 by axioms which characterise the significance part of $\mathbf{S_2}$ systems.

If we replace I4 by
I4′. $r\ \&\ (r\supset s)\rightarrow s,$

then the set I1-I3, I4′, I5–I9 characterises the Lewis modal system S3. The familiar modal connective '□' may be defined: $\Box A =_{\mathrm{Df}} \sim A \rightarrow A$. The resultant significance logic $S3S_1$ contains the theorem $\vdash S\Box A \leftrightarrow SA$. Hence '□' is a classical significance connective since $\Box A$ takes the value n when, and only when A takes the value n.

2. As a second development we now show how to develop postulational bases for four-valued matrix systems of $\mathbf{S}_1$-strength. The four values are $\{t, f, i, n\}$ where 'i' represents 'significant but incomplete'.

The Logic $\mathbf{L}_3\mathbf{S}_1$

We consider a significance version of Lukasiewicz's three-valued logic $\mathbf{L}_3$ extended to four values by the introduction of 'n' and formulated with primitives '$\sim$', '$\supset_L$' and 'S' and two sorts of variables. The S-restricted variables 'r', 's', ..., are now three-valued, ranging over the set $\{t, f, i\}$; the S-unrestricted variables, 'p', 'q', ..., are now four-valued, ranging over the set $\{t, f, i, n\}$. There is still only one designated value, t. The intended valuation rules can be read off the following (characteristic) matrices:

$\supset_L$	t	f	i	n
t	t	f	i	n
f	t	t	t	n
i	t	i	t	n
n	n	n	n	n

$\sim$	
t	f
f	t
i	i
n	n

S	
t	t
f	t
i	t
n	f

Hence '$\supset_L$' and '$\sim$' are classical since $A \supset_L B$ takes the value n if either A or B take the value n; and $\sim A$ takes the value n when, and only when, A takes the value n.

Postulates

The logic has as postulates the rules of inference of ${}_1\mathbf{S}_1$ together with the following axioms (for L1–L4 see Tarski [2], p. 49):

L1. $\vdash r \supset_L (s \supset_L r)$
L2. $\vdash (r \supset_L s) \supset_L ((s \supset_L r') \supset (r \supset_L r'))$
L3. $\vdash ((r \supset_L \sim r) \supset_L r) \supset_L r$
L4. $\vdash (\sim r \supset_L \sim s) \supset_L (s \supset_L r)$
S1. $\vdash S(p \supset_L q) \equiv Sp \;\&\; Sq$
S2. $\vdash S \sim p \equiv Sp$

where these definitions are adopted:

$A \vee B =_{\mathrm{Df}} (A \supset_L B) \supset_L B$;
$A \;\&\; B =_{\mathrm{Df}} \sim(\sim A \vee \sim B)$;
$A \equiv B =_{\mathrm{Df}} ((A \supset_L B) \;\&\; (B \supset_L A))$

Again it is evident how this technique can be extended to convert other three-valued sentential logics into four-valued significance systems. However an alternative technique consists in introducing three sorts of variables: unrestricted, ranging over the set $\{t,f,i,n\}$; *S*-restricted, ranging over the set $\{t,f,i\}$; and *I*-restricted, ranging over the set $\{t,f\}$. We shall illustrate this by developing a three-sorted system $\mathbf{AS_1}$.

The Logic $\mathbf{AS_1}$

The logic is technically of $\mathbf{S_1}$-strength since it contains the significance operator '*S*' but not '*T*'. It does however contain a *T*-operator, 'T^*' similar in some respects to the '*T*' of $\mathbf{S_2}$ and higher logics, but different in that T^*A takes the value *n* when *A* takes the value *n*. There is indeed a sense of 'true' for which this feature holds and the introduction of 'T^*' (and definable associated connectives 'F^*' and 'S^*') will serve to illustrate in passing how to develop systems which contain a truth-connective with this characteristic.

Morphology

Unrestricted variables: $p, q, p', q', \ldots$
S-restricted variables: $u, v, u', v', \ldots$
I-restricted variables: $r, s, r', s', \ldots$
Improper primitive symbols: $\sim$; &; *S*; T^*;

Formation rules:
1. A sentential variable standing alone is a wff.
2. If *A* and *B* are wff, then so are $\sim A$, SA, T^*A and $A \,\&\, B$.
3. Defined equivalents of wff are wff.

Definitions:
$$A \supset B =_{\mathrm{Df}} \sim(A \,\&\, \sim B)$$
$$A \vee B =_{\mathrm{Df}} \sim(\sim A \,\&\, \sim B)$$
$$A \equiv B =_{\mathrm{Df}} ((A \supset B) \,\&\, (B \supset A))$$
$$F^*A =_{\mathrm{Df}} T^* \sim A$$
$$S^*A =_{\mathrm{Df}} T^*A \vee F^*A.$$

Valuation rules

The valuation rules which are intended to provide the semantics of $\mathbf{AS_1}$ may be read off the following matrices:

&	t	f	i	n
t	t	f	i	n
f	f	f	f	n
i	i	f	i	n
n	n	n	n	n

~	
t	f
f	t
i	i
n	n

S	
t	t
f	t
i	t
n	f

T*	
t	t
f	f
i	f
n	n

$\supset$	t	f	i	n
t	t	f	i	n
f	t	t	t	n
i	t	i	i	n
n	n	n	n	n

$\vee$	t	f	i	n
t	t	t	t	n
f	t	f	i	n
i	t	i	i	n
n	n	n	n	n

	F^*
t	f
f	t
i	f
n	n

	S^*
t	t
f	t
i	f
n	n

Note that all the connectives, with the exception of 'S', are classical in the sense that they are nonsignificant when and only when any one of their sentential components is nonsignificant. Note, too, that the '&' and '$\sim$' matrices coincide with those for $\mathbf{L_3S_1}$ and that the matrix for '$\supset$' differs in only one place, namely that the value of $A \supset B$ when $v(A) = v(B) = i$ is i, not t. In the construction of valuation tables, values for the unrestricted variables are chosen from the full set $\{t, f, i, n\}$, for the S-restricted variables from the limited set $\{t, f, i\}$, and for the I-restricted variables from the two-valued set $\{t, f\}$.

Tautologies

A wff of $\mathbf{AS_1}$ is a tautology iff it has the value t for all assignments of values to the variables.

Postulates

The postulates are as follows:

A. Some set of axioms for classical sentential logic (**RSL**) formulated using I-restricted variables.

B.1. $S{\sim}p \equiv Sp$
2. $S(p \,\&\, q) \equiv Sp \,\&\, Sq$
3. $ST^*p \equiv Sp$

C.1. $r \supset T^*r$
2. S^*r
3. S^*T^*u
4. $T^*u \supset u$
5. $T^*{\sim}u \supset T^*{\sim}(u \,\&\, v)$

D.1. $\vdash A \rightarrow \vdash SA$
2. $\vdash A, \vdash A \supset B \rightarrow \vdash B$
3. $\vdash A \rightarrow \mathrm{S}^P_B A|$, where P is an unrestricted variable and B is a wff, or both P and B are S-restricted variables, or both are I-restricted variables.
4. $\vdash A, \vdash SB, \rightarrow \vdash \mathrm{S}^U_B A|$, where U is an S-restricted variable.
5. $\vdash A, \vdash S^*B, \rightarrow \vdash \mathrm{S}^R_B A|$, where R is an I-restricted variable.
6. $\vdash A \equiv B \rightarrow \vdash \mathrm{S}^{R_1 \dots R_n}_{U_1 \dots U_n}(T^*A \equiv T^*B)|$, where A and B do not contain 'S' or 'T^*' or other connectives only definable in terms of these, and where both A and B contain just the same I-restricted

variables $R_1 \ldots R_n$. (Rules similar to this are discussed in detail in Stage 2, §6.3.)

We shall not prove any metatheorems for this system since many of them, *e.g.* completeness and consistency results, can be derived by making minor modifications in similar proofs for $\mathbf{S}_2$ systems which are given in detail in the next section.

The C-*logic* $\mathbf{C}_1$

The first axiomatised sentential significance logic to appear in the literature was Halldén's system **C** in 1949 (Halldén [1]). It is deductively equivalent to the following system $\mathbf{C}_1$, in which the primitives '$\sim$', '&' and '+' of Halldén's **C** are replaced by '$\sim$', '$\supset$' and '*S*' respectively. The morphology of $\mathbf{C}_1$ has already been presented in §5.9; the postulates are as follows:

Postulates

Axioms:

1.1′ $\vdash p \supset (q \supset p)$
1.2′ $\vdash (p' \supset (p \supset q)) \supset ((p' \supset p) \supset (p' \supset q))$
1.3′ $\vdash (p \supset q) \supset (\sim q \supset \sim p)$
1.4′ $\vdash S(p \supset q) \equiv Sp \;\&\; Sq$
1.5′ $\vdash Sp \equiv S{\sim}p$
1.6′ $\vdash p \supset Sp$

Rules:

R1′ uniform substitution
R2′ $\vdash A$, $\vdash A \supset B \longrightarrow \vdash B$, provided no variable is uncovered in A and covered in B (modified modus ponens).

The definition of 'cover' is as before, specialised since '*S*' is the only unlimited connective to: A variable P is covered in a wff A iff P occurs in A and every occurrence of P in A is within the scope of some occurrence of '*S*' when A is written in primitive notation; a variable P is uncovered in A iff P occurs in A and not every occurrence of P in A is within the scope of some occurrence of '*S*'.

By Halldén's results ([1], pp. 48–52), the three-valued matrices for '$\sim$', '&' and '*S*' determined by the valuation rules for $\mathbf{C}_1$ (§5.9), with both t and n taken as thesis-determining values, are characteristic for **C**. The same results show that the three-valued matrices for '$\sim$', '$\supset$' and '*S*' are characteristic for $\mathbf{C}_1$. Hence,

Theorem (10): $\mathbf{C}_1$ is consistent and complete, since a wff A is a theorem of $\mathbf{C}_1$ iff A is a $\mathbf{C}_1$-tautology.

Alternatively, a normal-form completeness proof could be given.

The "good" theses of $\mathbf{C}_1$, *i.e.* those which take only the value t for all assignments of values to the variables and hence have no degenerate substitution instances when the logic is applied, can, as we have seen, be separated from the "bad" theses by matrix evaluations. There are various ways in which axiomatic analogues of this semantic separation may be achieved. We now develop one syntactic analogue of the separation by setting up a formal system which facilitates the separation of $\mathbf{C}_1$ theorems into "good" and "bad". At the same time we gain a comparison of ${}_1\mathbf{S}_1$ with $\mathbf{C}_1$ and Halldén's **C**.

The Logic $\mathbf{HS}_1$

$\mathbf{C}_1$ is first compared with a system $\mathbf{HS}_1$, an extension of ${}_1\mathbf{S}_1$, which is obtained by adding to ${}_1\mathbf{S}_1$ the following H-rule:

(H) $\vdash A \twoheadrightarrow \vdash_{\mathrm{H}} \mathrm{S}^{R_1 \ldots R_n}_{P_1 \ldots P_n} A \,|$, where $R_1 \ldots R_n$ include all the S-restricted variables in A (if any), *provided A* contains no S-uncovered components of the form SB, where B contains an uncovered restricted variable which does not also appear uncovered elsewhere in A.

Call this proviso on (H), proviso (p).

From (H) and T11 of ${}_1\mathbf{S}_1$,

TH.1 $\vdash_{\mathrm{H}} p \supset Sp$

follows, but $\vdash_{\mathrm{H}} Sp$, which would otherwise follow from T2 of ${}_1\mathbf{S}_1$, is excluded by proviso (p).

$\mathbf{HS}_1$ has two classes of theorems, the **H**-theorems, introduced by '$\vdash_{\mathrm{H}}$', and the non-**H**-theorems, which coincide with the theorems of ${}_1\mathbf{S}_1$. Thus ${}_1\mathbf{S}_1$ is a sub-logic of $\mathbf{HS}_1$.

MTH.1 (i) No **H**-theorems of $\mathbf{HS}_1$ contain restricted variables, in virtue of rule (H).
(ii) If A contains no restricted variables, $\vdash_{\mathrm{H}} A$ if $\vdash A$.

MTH.2 If $\vdash A$ is a theorem of $\mathbf{C}_1$, then $\vdash_{\mathrm{H}} A$ is an **H**-theorem of $\mathbf{HS}_1$, *i.e.* the theorems of $\mathbf{C}_1$ are included in the **H**-theorems of $\mathbf{HS}_1$.

Proof: That 1.1′, 1.2′ and 1.3′ are **H**-theorems of $\mathbf{HS}_1$ follows by applying (H) to 1.1, 1.2 and 1.3. Further, that 1.4′ and 1.5′ are **H**-theorems of $\mathbf{HS}_1$ follows by MTH.1 (ii) above. Again, 1.6′ is given by TH. 1 above. It remains to show that derived rules of $\mathbf{HS}_1$ leading from **H**-theorems to **H**-theorems include the rules of $\mathbf{C}_1$.

(a) *Uniform Substitution:* It needs to be shown,

$$\vdash_{\mathrm{H}} A \twoheadrightarrow \vdash_{\mathrm{H}} \mathrm{S}^{P}_{B} A \,|,$$

where B contains no restricted variables.

Now A is $\mathrm{S}^{R_1 \ldots R_n}_{Q_1 \ldots Q_n} A' \,|$, for some n and for some A' which satisfies proviso (p) and for which $\vdash A'$.

If P is Q_j, for some j, choose P' as R_j; otherwise choose P' as P.

Since $\vdash_H A$ is

$\vdash_H \mathsf{S}^{R_1 \dots R_n}_{Q_1 \dots Q_n} A'|$,

$\vdash_H A \longrightarrow \vdash A'$; by choice of A' above;

$\longrightarrow \vdash \mathsf{S}^P_B A'|$; by R 1.2;

$\longrightarrow \vdash_H \mathsf{S}^{R_1 \dots R_n}_{Q_1 \dots Q_n} \mathsf{S}^{P'}_B A'||$; by (H),

since no restricted variables occur in B.

But,

$\mathsf{S}^{R_1 \dots R_n}_{Q_1 \dots Q_n} \mathsf{S}^{P'}_B A'||$ is $\mathsf{S}^P_B \mathsf{S}^{R_1 \dots R_n}_{Q_1 \dots Q_n} A'||$,

since P is Q_i if P' is R_i for some i $(1 \leqslant i \leqslant n)$ and P' otherwise, and since P' is distinct from $Q_1 \dots Q_n$.

Therefore,

$\mathsf{S}^{R_1 \dots R_n}_{Q_1 \dots Q_n} \mathsf{S}^{P'}_B A'||$ is $\mathsf{S}^P_B A|$.

Hence,

$\vdash_H \mathsf{S}^P_B A|$.

(b) *Modified Modus Ponens* (MMP): Since,

$\vdash_H A$, $\vdash_H (A \supset B)$, as premisses,

for some n and some m, with $m \geqslant n$, and for some A' and B',

$\vdash \mathsf{S}^{R_1 \dots R_n}_{Q_1 \dots Q_n} A'|$, $\vdash \mathsf{S}^{R_1 \dots R_m}_{Q_1 \dots Q_m} A' \supset B'|$,

after adjustment (if necessary) of variables in

$\mathsf{S}^{R_1 \dots R_n}_{Q_1 \dots Q_n} A'|$,

where A' and $A' \supset B'$ both satisfy proviso (p).

Hence, by results for 'S',

$\vdash \mathsf{S}^{R_1 \dots R_n}_{Q_1 \dots Q_n} A'|$, $\vdash \mathsf{S}^{R_1 \dots R_n}_{Q_1 \dots Q_n} A'| \supset \mathsf{S}^{R_1 \dots R_m}_{Q_1 \dots Q_m} B'|$

$\longrightarrow \vdash \mathsf{S}^{R_1 \dots R_m}_{Q_1 \dots Q_m} B'|$, by R 1.3 and deleting superfluous substitutions

$\longrightarrow \vdash_H B$,

provided B' satisfies proviso (p).

Now A' and $(A' \supset B')$ satisfy their provisos, therefore B' satisfies proviso (p) if no restricted variable which is uncovered in A' is covered in B'. Consequently, a sufficient condition for modus ponens:

$$\vdash_H A, \vdash_H A \supset B \longrightarrow \vdash_H B$$

is that no variable is uncovered in A and covered in B. Thus MMP holds.

MTH.3 If $\vdash A$ with proviso (p) on A, then $\mathsf{S}^{R_1 \dots R_n}_{Q_1 \dots Q_n} A|$ has value $\bar{f}$, *i.e.* t or n, for all systems of values for its variables.

Proof: Since $\vdash A$, A has value t for all values of its variables (from 5.1 above). Transform A to conjunctive normal form. Then each occurrence of a restricted variable R_i is either not within the scope of 'S' or is within the scope of 'SS'. Proceed by induction over the number m of restricted variables in A.

If $m=0$, then no substitutions are made. Therefore $\mathsf{S}A|$ is A. But A always has value t, therefore $\mathsf{S}A|$ always has value t.

If $m>0$ consider what happens when a given variable R_k is replaced by Q_k. Occurrences of R_k within the scope of 'SS' can be neglected since SSA always has value t irrespective of the value of A. There remain two cases to examine:

(i) No occurrences of R_k are within the scope of 'S'.

(ii) Some occurrence of R_k is within the scope of 'S'. Proviso (p) then requires that some other occurrence of R_k in A is not within the scope of 'S'.

In case (i), replacement of R_k by Q_k only itself affects the lines of the matrix for A for which Q_k has value n. But when Q_k has value n, then A has value n since Q_k only occurs coupled with classical connectives.

In case (ii), whenever Q_k has value n and so SQ_k has value f, Q_k also has an occurrence which is not within the scope of 'S'. Therefore as in case (i), such lines have value n.

In conclusion, then, all lines of $\mathsf{S}A|$ for which variables substituted for restricted variables have values t or f have resultant value t (since A is a tautology under these assignments) and all other lines have value n. That is, $\mathsf{S}A|$ has value $\bar{f}$, as required.

MTH.4 The matrices for '$\supset$', '$\sim$' and 'S' with both t and n as thesis-determining values, are characteristic for $\mathbf{C}_1$ (by Halldén, pp. 48–62).

MTH.5 Every **H**-theorem of $\mathbf{HS}_1$ is a theorem of $\mathbf{C}_1$.

Proof: Every **H**-theorem of $\mathbf{HS}_1$ has value $\bar{f}$, from MTH.3. But all wff with value t or n for unrestricted variables are theorems of $\mathbf{C}_1$; by MTH.4.

MTH.6 A is an **H**-theorem of $\mathbf{HS}_1$ iff A is a theorem of $\mathbf{C}_1$.

Proof: By MTH.2 and MTH.5.

MTH.7 $\mathbf{HS}_1$ is deductively equivalent to the system characterised by the theorem set: {theorems of $_1\mathbf{S}_1$}$\cup${theorems of $\mathbf{C}_1$}, where each theorem of $\mathbf{C}_1$ is **H**-asserted ('$\cup$' indicates the union set}).

Proof: By MTH.6 since by construction $\mathbf{HS}_1$ is characterised by the theorem set: {theorems of $_1\mathbf{S}_1$}$\cup${the **H**-theorems of $\mathbf{HS}_1$}. [Note that the theorem sets of $_1\mathbf{S}_1$ and $\mathbf{C}_1$ are not disjoint. That is, a given A of $\mathbf{HS}_1$ may be such that both $\vdash A$ and $\vdash_{\mathbf{H}} A$. This will be so, for example if A is a theorem of ${}_1\mathbf{S}_1$ which contains no restricted variables, for then rule (H) can be applied trivially with nil substitutions (MTH.1(ii))].

Definition 1: A is a *bad* theorem of $\mathbf{HS}_1$ if $\vdash_{\mathbf{H}} A$ but not $\vdash A$; otherwise a theorem A of $\mathbf{HS}_1$ is *good* (*i.e.* if both $\vdash_{\mathbf{H}} A$ and $\vdash A$ or if $\vdash A$ but not $\vdash_{\mathbf{H}} A$).

Thus, $qu(p \supset Sp)$ is a bad theorem, but $qu(r \supset Sr)$ is good. Again, $qu(Sp \equiv S{\sim}p)$ is a good theorem since both $\vdash Sp \equiv S{\sim}p$ and $\vdash_{\mathbf{H}} Sp \equiv S{\sim}p$.

Not all of the good theorems of $\mathbf{HS}_1$ are **H**-theorems. In particular, if A is a theorem which contains restricted variables, then A is a good theorem of $\mathbf{HS}_1$ (Defn. 1) but is not an **H**-theorem.

Definition 2: A is a *good* **H**-*theorem* of $\mathbf{HS}_1$ iff both $\vdash_{\mathbf{H}} A$ and $\vdash A$.

Accordingly, by applying MTH.6 we may now connect the separation of good and bad theorems of $\mathbf{HS}_1$ with the separation of theses of the matrix system $\mathbf{C}_1$ into good and bad.

MTH.8 A is a good theorem of $\mathbf{C}_1$ iff A is a good **H**-theorem of $\mathbf{HS}_1$.

Therefore, since $\mathbf{C}_1$ is complete,

MTH.9 A is a good tautology of $\mathbf{C}_1$ iff A is a good theorem of $\mathbf{C}_1$.

Thus, the bad theorems of $\mathbf{C}_1$ have degenerate substitution instances but the good theorems do not.

Hence, applying MTH.7, $\mathbf{C}_1$ can be extended to a system ${}_1\mathbf{S}_1 \cup \mathbf{C}_1$ which has a complete good part. Halldén's restriction on *modus ponens* is automatically satisfied in ${}_1\mathbf{S}_1$ and so is satisfied in ${}_1\mathbf{S}_1 \cup \mathbf{C}_1$.

We note finally that the derived rule of ${}_1\mathbf{S}_1$,

$$\vdash A \supset B \twoheadrightarrow \vdash {\sim}SB \supset {\sim}SA$$

which is proved as follows:

$$\begin{aligned} &\vdash A \supset B \twoheadrightarrow \vdash S(A \supset B) && \text{(R1.1)} \\ &\twoheadrightarrow \vdash SA \supset SB && \text{(by 1.5)} \\ &\twoheadrightarrow \vdash {\sim}SB \supset {\sim}SA && \text{(by RSL)} \end{aligned}$$

will licence literal *reductio ad absurdum* arguments in $\mathbf{HS}_1$ (for which it is also a derived rule) but not in ${}_1\mathbf{S}_1$. For in ${}_1\mathbf{S}_1$, the antecedent $A \supset B$ being a theorem, can only take the value t; hence B cannot take the value n and therefore ${\sim}SB$ is never t. Hence we cannot detach to draw the conclusion $\vdash {\sim}SA$. In $\mathbf{HS}_1$, however, $A \supset B$ may take the value n and so may B; hence ${\sim}SB$ is t and we may therefore affirm $\vdash {\sim}SB$, hence $\vdash {\sim}SA$. That such arguments are suspect, however, is evidenced by the fact that the rule has a bad theorem analogue:

TH.2 $\vdash_H (p \supset q) \supset (\sim Sq \supset \sim Sp)$ (Ryle's hypothesis).

Proof: $\vdash (r \supset r') \supset (\sim Sr' \supset \sim Sr)$; by T11, 1.4′, 1.3.
Result by (H).

We do not, however, have $\vdash (p \supset q) \supset (\sim Sq \supset \sim Sp)$ in ${}_1\mathbf{S}_1$ but only $\vdash (r \supset r') \supset (\sim Sr' \supset \sim Sr)$. In the latter case, however, as in the case of the use of the rule in ${}_1\mathbf{S}_1$, the condition $\sim Sr'$ can never be fulfilled and so we can never affirm $\vdash \sim Sr$. (See also the discussion following $\mathbf{C}_1$ in §5.9.)

6.3 Stage 2: Axiomatic Formulations of 2-systems

Stage 1 logics, though an improvement on Stage 0 logics, are still both semantically and syntactically weak. Their semantical weakness has been explained at length in §5.10. Clear examples of the syntactical weakness of 1-logics are provided by *Substitutivity Theorem* 2 (in §6.2) and by *CNF* theorems, where strong equivalence ($\simeq$) connections have to be expressed in a quite roundabout way. Stage 2 logics are much richer, and can even be defended as suitable sentential bases for a significance theory, especially since they contain all the matrix-accessible sentential connectives that appear in major significance theories to date. The shortcomings of stage 2 logics as adequate sentential bases have, however, been discussed in chapter 5.

The Logic ${}_1\mathbf{S}_2$

There are, of course, various ways in which the logic $\mathbf{S}_2$ may be axiomatised. We shall study first a system ${}_1\mathbf{S}_2$ with primitive connective set $\{\sim, \&, T\}$ —it could equally well have primitives $\{\sim, \vee, T\}$ or $\{\sim, \supset, T\}$—and with these postulates:

Axioms

2.1 $\vdash r \supset . r' \supset r$
2.2 $\vdash r'' \supset (r \supset r') \supset . r'' \supset r \supset . r'' \supset r'$
2.3 $\vdash \sim r \supset \sim r' \supset . r' \supset r$
2.4 $\vdash Tr \equiv r$
2.5 $\vdash STp$
2.6 $\vdash T \sim p \supset \sim Tp$
2.7 $\vdash T(p \vee q) \supset . Tp \vee Tq$
2.8 $\vdash T(p \,\&\, q) \equiv Tp \,\&\, Tq$
2.9 $\vdash S(p \,\&\, q) \supset Sp$
2.10 $\vdash Sp \,\&\, Sq \supset . S(p \,\&\, q)$
2.11 $\vdash Sr$

Rules:

R2.1 $\vdash A, \vdash SB \twoheadrightarrow S^R_B A \mid$
R2.2 $\vdash A \twoheadrightarrow \vdash S^P_B A \mid$,

where P is a variable and B a wff, or both P and B are S-restricted variables.

R2.3 $\vdash A, A \supset B \twoheadrightarrow \vdash B$
R2.4 $\vdash A \equiv B \twoheadrightarrow \vdash \mathrm{S}^{R_1 \dots R_n}_{P_1 \dots P_n}(TA \equiv TB)\,|$,

where A and B each contain only classical connectives and just the same restricted variables $R_1, R_2, \dots, R_n$ (E.T. rule).

The restrictions on the Equivalential T-rule, R2.4, are necessary. The rule does not hold if A and B contain different variables. For instance, '$r \equiv . r \,\&\, (r' \vee \sim r')$' and '$r \vee \sim r \equiv . r' \vee \sim r'$' are theorems, but '$Tp \equiv . T(p \,\&\, . q \vee \sim q)$' and '$T(p \vee \sim p) \equiv T(q \vee \sim q)$' are not, since they are not valid under the intended interpretation (see §5.10). Moreover the E.T. rule cannot in general be extended to non-classical connectives such as 'T'. Suppose, for example, A contains 'T'. Then although '$Tr \equiv r$' and '$\sim Tr \equiv \sim r$' hold, what could then be obtained by successive applications of the E.T. rule, '$STp \equiv Sp$', is not a theorem. Stronger T-rules which facilitate the elimination of some of the axioms of $_1\mathrm{S}_2$ are presented below.

Results on $_1\mathrm{S}_2$

Theorem (11): Classical sentential logic **RSL** holds for wff with significant components as atoms where these components are connected by classical connectives.

Proof: As for *Theorem* (9) of S_1, from 2.1, 2.2, 2.3 and R2.1 and R2.3.

Formal Theorems

T1 $\vdash TTp \equiv Tp$. *Proof* uses 2.4, 2.5 and R2.1
T2 $\vdash ST{\sim}p$. *Proof* uses 2.5 and R2.2
T3 $\vdash ST{\sim}{\sim}p$
T4 $\vdash Sp \equiv S{\sim}p$

Proof: $\vdash Tp \vee T{\sim}p \equiv T{\sim}p \vee Tp$, by 2.5, T2, **RSL**.
$\vdash Tp \equiv T{\sim}{\sim}p$, by **RSL**, E.T. rule.
$\vdash Tp \vee T{\sim}p \equiv T{\sim}p \vee T{\sim}{\sim}p$, by 2.5, T2, T3, **RSL**.
The result now follows by the definition of 'S'.

From now on we omit details of the more immediate proofs and content ourselves with sketches in other cases. In almost all cases full proofs are easily constructed.

T5 $\vdash S{\sim}Tp$
T6 $\vdash T{\sim}Tp \equiv {\sim}Tp$
T7 $\vdash Sp \equiv . Tp \vee Fp$
T8 $\vdash Tp \supset {\sim}T{\sim}p$
T9 $\vdash T(p \supset q) \supset . {\sim}T{\sim}p \supset Tq$
T10 $\vdash T(p \supset q) \supset . Tp \supset Tq$
T11 $\vdash S(p \vee q) \supset Sp$

T12 $\vdash (Tp \supset {\sim}T{\sim}q) \supset T(p \supset q) \supset .\ Sp$
T13 $\vdash Sp \,\&\, Sq \supset S(p \supset q)$
T14 $\vdash Sp \,\&\, Sq \supset S(p \vee q)$
T15 $\vdash SSp$

Proof: $\vdash Sr$, *i.e.* $Tr \vee T{\sim}r$; 2.11
$\vdash S{\sim}r$, *i.e.* $T{\sim}r \vee T{\sim}{\sim}r$; T4, R2.3

Hence $\vdash T(r \vee {\sim}r) \vee T{\sim}(r \vee {\sim}r)$, by T14, R2.2

Therefore
$\vdash S(r \vee {\sim}r)$
$\vdash T(r \vee {\sim}r) \equiv (r \vee {\sim}r)$, 2.4, R2.1
$\vdash ST(r \vee {\sim}r)$
$\vdash T(r \vee {\sim}r)$, by **RSL**
$\vdash S(Tp \vee T{\sim}p)$, by T14, 2.5, R2.3
$\vdash T((Tp \vee T{\sim}p) \vee {\sim}(Tp \vee T{\sim}p))$, by R2.1
$\vdash T(Tp \vee T{\sim}p) \vee T{\sim}(Tp \vee T{\sim}p))$, by 2.7, R2.3

T16 $\vdash S(p \supset q) \supset .\ Sp \,\&\, Sq$

Theorem (12): $\vdash SA$, where A contains just classical connectives and S-restricted variables.

Proof: By induction over the number of occurrences of '$\sim$' and '&' in A.
1. $\vdash SR$, where R is some S-restricted variable. For $\vdash SR$ by substitution in 2.11 according to R2.2.
2. In the induction step there are two cases to consider:

(a) A is of the form ${\sim}B$. By induction hypothesis, $\vdash SB$.
Hence, by T4 and T15, $\vdash S{\sim}B$, *i.e.* $\vdash SA$.
(b) A is of the form $(B_1 \,\&\, B_2)$. By hypothesis, $\vdash SB_1, \vdash SB_2$.
Hence, using T17, **RSL** and 2.10, $\vdash S(B_1 \,\&\, B_2)$.

T17 $\vdash S{\sim}Sp$
T18 $\vdash {\sim}Sp \supset {\sim}Tp$; $\vdash Tp \supset Sp$
T19 $\vdash {\sim}Sp \supset {\sim}Fp$; $\vdash Fp \supset Sp$

Analogues of the E.T. rule for the characteristic functors 'F' and 'S' are also forthcoming.

Theorem: (13) $\vdash A \equiv B \longrightarrow \vdash \mathrm{S}^{R_1 \ldots R_n}_{P_1 \ldots P_n} FA \equiv FB|$, where A and B each contain only classical connectives and just the restricted variables $R_1, R_2, \ldots, R_n$.

Theorem (14): $\vdash A \equiv B \longrightarrow \vdash \mathrm{S}^{R_1 \ldots R_n}_{P_1 \ldots P_n} SA \equiv SB|$, with proviso as in *Theorem* (13).

T20 $S(p \equiv q) \equiv .\ Sp \,\&\, Sq$

Theorem (15): $\vdash A \twoheadrightarrow \vdash SA$ (full S-rule).

Proof: By induction over the length of proofs.

1. If A is an axiom then $\vdash SA$. For axioms 2.1, 2.2, 2.3 the result follows by *Theorem* (12). Since $\vdash Sr, \vdash STr$; by T20 $\vdash S(Tr \equiv r)$. This gives the result for 2.4. For 2.5 the result follows from T15; similarly for 2.11 For 2.6, since $\vdash ST{\sim}p$ and $\vdash S{\sim}Tp$, by T13 $\vdash S(T{\sim}p \supset {\sim}Tp)$. The proofs for 2.7–2.10 are similar in principle.

2. The rules preserve the significance property.

R2.1 Given $\vdash A, \vdash SB \twoheadrightarrow \mathsf{S}^R_B A|$, by induction hypothesis $\vdash SA$. Also $\vdash SB$. Hence by R2.1, $\vdash \mathsf{S}^R_B SA|$ as required.

R2.2 Given $\vdash A \twoheadrightarrow \vdash \mathsf{S}^P_B A|$, by induction hypothesis $\vdash SA$. Therefore by R2.2 $\vdash \mathsf{S}^P_B SA|$.

R2.3 By induction hypothesis $\vdash SA, \vdash S(A \supset B)$. Therefore by T13, T16, and **RSL** $\vdash SA \supset SB$. Thus by R2.3, $\vdash SB$.

R1.4 $\vdash STA, \vdash STB$, from 2.5. Therefore, by T20, $\vdash \mathsf{S}^{R_1}_{P_1} \cdots ^{R_n}_{P_n} S(TA \equiv TB)|$ as required.

Theorem (16): $\vdash A \twoheadrightarrow \vdash TA$

Proof: $A \twoheadrightarrow \vdash SA$, *Theorem* (15). Hence $\vdash A \equiv TA$, by 2.4 and R2.1. Also since $\vdash STA$, $\vdash A \supset TA$, by **RSL**. Then by R2.3, $\vdash TA$.

Theorem (17): $\mathbf{S_1}$ is a proper subsystem of $_1\mathbf{S_2}$, *i.e.* every theorem of $\mathbf{S_1}$ is a theorem of $_1\mathbf{S_2}$ but not conversely. *Proof* of the inclusion follows from the fact that the postulates of $_2\mathbf{S_1}$ have been derived for $_1\mathbf{S_2}$ and the completeness of $_2\mathbf{S_1}$ with respect to $\mathbf{S_1}$. But since 'T' is not definable in $\mathbf{S_1}$ no T-theorems are theorems of $\mathbf{S_1}$.

T21 $\vdash {\sim}Tp \supset . {\sim}T{\sim}(p \supset q)$

T22 $\vdash {\sim}T{\sim}q \supset . {\sim}T{\sim}(p \supset q)$

T23 $\vdash Fr \equiv {\sim}r$

T24 $\vdash FTp \equiv {\sim}Tp$

T25 $\vdash Sp \equiv . {\sim}Tp \equiv T{\sim}p$

T26 $\vdash Sp \;\&\; Sq \supset . {\sim}T(p \vee q) \supset T{\sim}p$

Proof uses these theorems: $Sp \;\&\; Sq \supset . S(p \vee q)$; $S(p \vee q) \;\&\; {\sim}T(p \vee q) \supset . T{\sim}(p \vee q)$; $T{\sim}(p \vee q) \supset T{\sim}p$.

T27 $\vdash Sp \;\&\; Sq \supset . Tp \vee Tq \supset . T(p \vee q)$

T28 $\vdash Sp \;\&\; Sq \supset . Tp \supset Tq \supset . T(p \supset q)$

T29 $\vdash Sp \supset T(p \supset p)$; $\vdash Sp \equiv T(p \supset p)$

T30 $\vdash Tp \supset F{\sim}p$; $\vdash Fp \supset T{\sim}p$

T31 $\vdash {\sim}Sp \supset FTp$

T32 $\vdash Fp \supset FTp$

T33 $\vdash Fp \;\&\; Sq \supset F(p \;\&\; q)$; $\vdash Sp \;\&\; Fq \supset F(p \;\&\; q)$

Proof uses these theorems: $Fp \,\&\, Sq \supset . S(p \,\&\, q)$; $Fp \supset \sim T(p \,\&\, q)$; and T25.

T34 $\vdash Tp \vee Fp \vee \sim Sp$
T35 $\vdash Tp \supset Tq \supset . Fp \supset Tq \supset . \sim Sp \supset Tq \supset . Tq$
T36 $\vdash Tp \supset Sq \supset . Fp \supset Sq \supset . \sim Sp \supset Sq \supset . Sq$

We now connect the axiomatic system $_1\mathbf{S}_2$ with the semantic system $\mathbf{S}_2$.

Theorem (18): Every theorem of $_1\mathbf{S}_2$ is an $\mathbf{S}_2$-tautology.

Proof is by induction over proofs in $_1\mathbf{S}_2$.

1. Each of the axioms of $_1\mathbf{S}_2$ is an $\mathbf{S}_2$-tautology, by matrix computation.
2. The rules preserve $\mathbf{S}_2$-tautologousness. Thus:

R2.3 If A and $(A \supset B)$ are tautologies then both have value t. Therefore by the matrix for '$\supset$' B has value t.

R2.4 If $A \equiv B$ is a tautology then all lines of the matrix table for $\mathsf{S}^{R_1 \dots R_n}_{P_1 \dots P_n}(TA \equiv TB)|$ for which $P_1, P_2, \dots, P_n$ have values from $\{t,f\}$ have value t. For if $\mathsf{S}^{R_1 \dots R_n}_{P_1 \dots P_n}A|$ has value t or f then $\mathsf{S}TA|$ has the same value. If, alternatively, one or more of $P_1, P_2, \dots, P_n$ have value n then $\mathsf{S}A|$ and $\mathsf{S}B|$ have value n by matrices for classical connectives. Then both $\mathsf{S}TA|$ and $\mathsf{S}TB|$ have value f. Hence $\mathsf{S}TA| \equiv \mathsf{S}TB|$, and so $\mathsf{S}(TA \equiv TB)|$ has value t, *i.e.* it is a tautology.

In the case of R2.1 and R2.2 the following lemma is needed:

(A) If $P_1, P_2, \dots, P_n, P$ are distinct variables among which are all the variables in A and in B, and if for values $a_1, a_2, \dots, a_n, a$ of $P_1, P_2, \dots, P_n, P$ the value of B is b and the value of $\mathsf{S}^P_B A|$ is c then the value of A, for values $a_1, a_2, \dots, a_n, b$ of $P_1, P_2, \dots, P_n, P$ is c.

Proof is by induction over the number of occurrences of primitive connectives in A.

1. If A is a variable alone then either (i) no substitution is made, in which case $\mathsf{S}^P_B A|$ is A so A has the value c; or (ii) substitution is just for P, *i.e.* A. Then $\mathsf{S}^P_B A|$ is B. Since $b = c$, (A) follows.
2. A is $\sim A_1$. So $\mathsf{S}^P_B A|$ is $\sim \mathsf{S}^P_B A_1|$. By induction hypothesis, if the value of $\mathsf{S}^P_B A_1|$ is c' then the value of A_1 is c'. Therefore if the value of $\mathsf{S}^P_B A|$ is c, $c = f, n, t$ according as $c' = t, n, f$. So the value of A, *i.e.* $\sim A_1$, is f, n, t according as the value of c is f, n, t, *i.e.* the value of A is c.
3. A is TA_1. So $\mathsf{S}^P_B A|$ is $T\,\mathsf{S}^P_B A_1|$. Apply the induction hypothesis as in 2. If the value of $\mathsf{S}^P_B A|$ is c, $c = t, f, f$ according as $c' = t, n, f$, *i.e.* according as A has value t, f, f. Thus the value of A is c.
4. A is $(A_1 \,\&\, A_2)$; so $\mathsf{S}^P_B A|$ is $\mathsf{S}^P_B A_1| \,\&\, \mathsf{S}^P_B A_2|$. By induction hypothesis if the value of $\mathsf{S}^P_B A_1|$ is c', that of A_1 is c', and if the value of $\mathsf{S}^P_B A_2|$

is c″ that of A_2 is c″. Then the same matrix assignments which result in value c for values c' and c'' of $S^P_B A_1|$ and $S^P_B A_2|$ result in value c for $(A_1 \& A_2)$ for values c' and c'' for A_1 and A_2.

It is clear in fact that the induction will work for any matrix-accessible connective.

R2.2 (first part). If $S^P_B A|$ has value $\bar{t}$ for some set of values of the variables then by lemma (A), A has value $\bar{t}$ for some set of values of its variables. Therefore, contraposing, if A is an S_2-tautology $S^P_B A|$ is an S_2-tautology.

R2.1 If A and SB are tautologies then A has value t and B has value $\bar{n}$, *i.e.* t or f. Lemma (A) will still apply provided the variable for which B is substituted has just the values t and f. This point is guaranteed by the requirement that B is substituted for a restricted variable R. Hence if $S^R_B A|$ were to have value $\bar{t}$ then by lemma (A), A would have value $\bar{t}$ for some set of values of its variables, contrary to the assumption that A is a tautology.

R2.2 (second part) is similar to R2.1 as just given. Lemma (A) continues to apply when neither a nor b equals n.

Theorem (19) (Consistency theorems): ${}_1S_2$ is simply, absolutely, and Post-consistent; it is S-consistent and S-correct.

Proof: $\sim(r \supset r)$ is not a theorem, by *Theorem* (18); therefore not both A and $\sim A$ are theorems when SA. And if $\sim SA$ then neither A nor $\sim A$ is a theorem. Also P is not a theorem.

Strong equivalence is defined as before, *viz.*: $A \simeq B =_{Df} TA \,\&\, TB \vee . FA \,\&\, FB \vee . \sim SA \,\&\, \sim SB$. The following theorems which serve in particular to characterise $\simeq$ as an identity relation between sentences—it is an equivalence relation which guarantees intersubstitutivity in all S_2 sentence contexts —may be established alternatively by applying the completeness theorem, *Theorem* (25), proved below.

T37 $\vdash S(p \simeq q)$

T38 $\vdash p \simeq p$; $\vdash p \simeq q \supset . q \simeq p$

$\vdash p \simeq q \,\&\, q \simeq r \supset . p \simeq r$

Thus $\simeq$ is an equivalence relation.

T39 $\vdash p \simeq q \supset . Tp \equiv Tq$; $\vdash p \simeq q \supset . Sp \equiv Sq$.

The converses are not theorems; $Tp \equiv Tq$ does not imply $p \simeq q$. However,

T40 $\vdash (Tp \equiv Tq) \,\&\, (Fp \equiv Fq) \supset . p \simeq q$

T41 $\vdash Sp \equiv . Tp \simeq p$

Theorem (20): $\vdash A, \vdash A \simeq B \twoheadrightarrow \vdash B$

Proof: $\vdash A, \vdash A \simeq B \twoheadrightarrow \vdash TA, \vdash TA \equiv TB$; *Theorem* (16), T39. Then follow in turn TB, by **RSL**; SB, by T18; $TB \equiv B$, by 2.4 and R2.1; and B, by R2.3 and **RSL**.

Theorem (21) (Replacement theorem): $\vdash A \equiv B \twoheadrightarrow \vdash \mathrm{S}^{R_1 \ldots R_n}_{P_1 \ldots P_n} A \simeq B|$, where A and B each contain only classical connectives and have just the variables R_1, R_2, ..., R_n. *Proof* follows from E.T. rules by T40. The replacement theorem in turn implies the E.T. rules.

The full equivalential logic does not hold for '$\simeq$'. For $p \simeq q \simeq . r \simeq q \simeq . p \simeq r$ is not a theorem, since by matrix assessment it is not an S_2-tautology.

T42 $\vdash p \simeq q \supset . \sim p \simeq \sim q$

T43 $\vdash p \simeq q \supset . Tp \simeq Tq$

T44 $\vdash p \simeq q \supset . p \,\&\, r \simeq q \,\&\, r$

T45 $\vdash p \simeq q \supset . r \,\&\, p \simeq r \,\&\, q$

Theorem (22): If B results from A by substitution of B_1 for A_1 at zero or more places in A, then

(1) $\vdash A_1 \simeq B_1 \supset . A \simeq B$

Proof by induction over the number of primitive connectives in A.

1. Special cases:
   (a) Substitution is at zero places in A. Then B is A; hence $\vdash A \simeq B$, and the result follows by **RSL**.
   (b) A_1 coincides with A. Then B_1 is the same as B, and the result follows from $\vdash r \supset r$ by T38 and R2.1.
2. If A contains zero connectives the result follows by 1. If A contains k connectives, with $k > 0$, there are 3 cases:
   (a) A is $\sim A'$. Then unless a special case obtains, B is B' where B' is obtained from A' by substitution of B_1 for A_1. By induction hypothesis $\vdash A_1 \simeq B_1 \supset . A' \simeq B'$. But then (1) follows by T42 and T38.

   Cases (b) and (c) where A is TA' and $A' \,\&\, A''$, respectively, are similar, and follow using T43, and T44 and T45, respectively.

Theorem (23) (Substitutivity of strong equivalence): If B results from A by substitution of B_1 for A_1 at zero or more places in A, then if $\vdash A_1 \simeq B_1$ and $\vdash A$ then $\vdash B$.

Proof applies *Theorem* (20) and *Theorem* (22).

Theorem (24) (Completeness lemma): Let B be a wff of ${}_1\mathrm{S}_2$. Let P_1, P_2, ..., P_n be the distinct variables occurring in B. Let A_i be TP_i, FP_i or $\sim SP_i$ according as the value a_i of P_i is t, f, or n. Let B' be TB, FB or SB according as the value of B for the values a_1, a_2, ..., a_n of P_1, P_2, ..., P_n is t, f, or n.

Then, where 'proof from hypotheses' is defined as in **RSL**,

(1) $A_1, A_2, ..., A_n \vdash B'$.

Note: If P_i is an S-restricted variable R_i then A_i is TR_i or FR_i, *i.e.* R_i or $\sim R_i$.

Proof: Each A_i is significant and B' is significant, by 2.5, T2, T17. Thus (1) is well-defined, and **RSL** holds for each A_i and for B', by *Theorem* (11).

The proof is by induction over the number of occurrences of primitive connectives in B.

1. Basis. If there are no connectives in B then B is either an S-unrestricted variable P_i or an S-restricted variable R_i. Then B' is the same as A_i, whence (1) follows.
2. Induction step. When B contains connectives there are 3 cases to consider.

(a) B is $\sim B_1$. By induction hypothesis,

(2) $A_1, A_2, ..., A_n \vdash B_1'$.

If B_1 has value t (f, n), B has value f (t, n); then B_1' is TB_1 (FB_1, $\sim SB_1$) and B' is $F{\sim}B_1$ ($T{\sim}B_1$, $\sim S{\sim}B_1$). Then (1) follows from (2) using T30 (T30, T4).

(b) B is TB_1. If B_1 has value t (f, n), B has value t (f, f); so B_1' is TB_1 (FB_1, $\sim SB_1$) and B' is TTB_1 (FTB_1, FTB_1). (1) follows from induction hypothesis (2) using T1 (T32, T31).

(c) B is (B_1 & B_2). By induction hypothesis (2)

and

(3) $A_1, A_2, ..., A_n \vdash B_2'$

both hold. If either B_1 or B_2 has value n, then B has value n. Then either (i) B_1' is $\sim SB_1$ or (ii) B_2' is $\sim SB_2$, and B' is $\sim S(B_1 \,\&\, B_2)$. If (i) holds then (1) follows from (2) using 2.9 and **RSL**; likewise if (ii) holds (1) follows from (3). If B_1 and B_2 both have value t so does (B_1 & B_2); then B_1' is TB_1, B_2' is TB_2, and B' is $T(B_1 \,\&\, B_2)$; and (1) follows from (2) and (3) using 2.8 and **RSL**. If B_1 has value f and B_2 has value $\bar{n}$ then B has value f; so B_1' is FB_1, B_2' is SB_2 applying T7, and B' is FB. (1) follows from (2) and (3) by T33. The remaining case where B_1 has value $\bar{n}$ and B_2 has value f is similar.

Theorem (25) (Completeness theorem): If B is a theorem of matrix system $\mathbf{S}_2$ then B is a theorem of ${}_1\mathbf{S}_2$.

Proof: Let $P_1, P_2, ..., P_n$ be the distinct variables of B and define $A_1, A_2, ..., A_n$ as in the lemma. Since B is a tautology of $\mathbf{S}_2$, B' of the lemma is TB.

Therefore by the lemma

(1) $A_1, A_2, ..., A_n \vdash TB$

(1) holds irrespective of whether P_n has value *t*, *n*, or *f*. Therefore it follows from (2) using the deduction theorem (a consequence of **RSL**):

$$A_1, A_2, ..., A_{n-1} \vdash TP \supset TB$$
$$A_1, A_2, ..., A_{n-1} \vdash \sim SP \supset TB$$
$$A_1, A_2, ..., A_{n-1} \vdash FP \supset TB$$

Applying T35 and R2.3, it follows:

$$A_1, A_2, ..., A_{n-1} \vdash TB$$

By $(n-1)$ iterations of this procedure all hypothesis are eliminated. Hence $\vdash TB$. Hence also, by T18, $\vdash SB$. Therefore by R2.1 and 2.4 $\vdash TB \supset B$, and by R2.3, $\vdash B$.

An alternative axiomatisation of $\mathbf{S}_2$ can be read off the completeness theorem (& lemma). Such a system, ${}_2\mathbf{S}_2$, has as axioms then 2.1, 2.2, 2.3, 2.4, 2.5, 2.8, 2.9, 2.11, T1, T2, T4, T7, T17, T18, T30, T32, T33 and as rules R2.1, R2.2 and R2.3. Although ${}_2\mathbf{S}_2$ avoids the E.T. rule it is apparent that ${}_1\mathbf{S}_2$ constitutes quite a simplification over ${}_2\mathbf{S}_2$. ${}_2\mathbf{S}_2$ almost coincides with the system that may be read off the valuation rules for $\mathbf{S}_2$ in the way explained in §5.10. The minor differences are these. First, there are redundancies in the axioms read off the valuation rules. Second, what is really read off the rules are axiom schemata which combine with a sole rule of detachment, not axioms proper with substitution rules. For example, what we read off the rule "For every *S*-restricted variable *R*, $v(R) = t \vee v(R) = f$" is really the schema $TR \vee FR$, *i.e.* *SR*, which results from 2.11 combined with R2.1. To obtain axioms proper with substitution rules the schemata read off the valuation rules have to be analysed; and it is readily seen that these yield axioms proper and two substitution rules, one for *S*-restricted substitution and the other for unrestricted substitution. Now the gap between the axiomatisation read off the valuation rules and the axiomatisation read off the completeness proof can be closed by appealing to axiom schemata in the completeness proof itself and by writing down definitionally redundant schemes. Then the axiomatisation read off the valuation rules coincides with that read off the completeness proof: the axiomatisation is provided by rule R2.3, by schemes corresponding to 2.1 through T33, and by schemes like $FP \supset T{\sim}P$ corresponding to definitions of connectives '*F*' and '*S*'. Finally the coincidence vindicates our claim that complete axiomatisations of certain logics, such as $\mathbf{S}_2$, can be read off the valuation rules.

By strengthening the rule structure of ${}_1\mathbf{S}_2$ a more compact axiomatisation of $\mathbf{S}_2$ can be obtained. For consider the following system ${}_3\mathbf{S}_2$ which has just these axioms:

Some set of axioms for **RSL**, say 2.1–2.3.

2.4′ $\vdash r \supset Tr$; 2.5 $\vdash STp$;
2.6′ $\vdash {\sim}T(p \,\&\, {\sim}p)$

${}_3S_2$ retains the rules R2.1 and R2.3 of ${}_1S_2$, but strengthens the remaining rules as follows:

R2.2′ $\vdash A \longrightarrow \vdash S^P_B A \,|$,

where P is a variable and B is a wff, or P is an S-restricted variable and B is a wff containing only S-restricted variables and classical connectives (*i.e.* connectives definable in terms of '$\sim$' and '&'). The last clause substantially strengthens R2.2, which permits only alphabetic change of S-restricted variables, and could be further strengthened to eliminate 2.5 (see HAT3 below).

R2.4 is replaced by the T-rule:

$$\text{R2.4}': \vdash A_1 \,\&\, A_2 \,\&\, A_3 \,\&\, \ldots \,\&\, A_k \supset . \, B_1 \vee B_2 \vee \ldots \vee B_m$$
$$\longrightarrow \vdash S^{R_1 \ldots R_n}_{P_1 \ldots P_n}(TA_1 \,\&\, TA_2 \,\&\, \ldots \,\&\, TA_k \supset . \, TB_1 \vee TB_2 \vee \ldots \vee TB_m) \,|$$

where $A_1, A_2, \ldots, A_k, B_1, B_2, \ldots, B_m$ contain only classical connectives and all variables of $B_1, \ldots, B_m$ are among (or coincide with) those (*viz.* $R_1, \ldots, R_n$) of $A_1, A_2, \ldots, A_k$.

The T-rule lends itself admirably to formulation in Gentzen sequents notation where it becomes:

$$\frac{A_1, A_2, \ldots, A_k \rightarrow B_1, B_2, \ldots, B_m}{TA_1, TA_2, \ldots, TA_k \rightarrow TB_1, TB_2, \ldots, TB_m}$$

To show that ${}_3S_2$ is deductively equivalent to ${}_1S_2$ it suffices, in view of the completeness of ${}_3S_2$, to derive the postulates of ${}_1S_2$ and to show that ${}_1S_2$ is correct.

Theorem (26) (**RSL**): As for *Theorem* (11) of ${}_1S_2$.
Theorem (27) (Implication T-rule):

$$\vdash A \supset B \longrightarrow \vdash S^{R_1 \ldots R_n}_{P_1 \ldots P_n} TA \supset TB \,|$$

where A and B are wff containing only classical connectives and S-restricted variables $R_1, R_2, \ldots, R_n$ and all B's variables are included in those of A.

Proof: By specialisation of T-rule; $k = m = 1$.

The following formal theorems arise just from **RSL** and the T-rule.

TA1 $\vdash T(p \,\&\, q) \supset Tp$; $\vdash T(p \,\&\, q) \supset Tq$
$\vdash T(p \,\&\, q) \supset . \, T(q \,\&\, p)$
TA2 $\vdash Tp \,\&\, Tq \supset . \, T(p \,\&\, q)$
TA3 $\vdash T(p \vee q) \supset . \, Tp \vee Tq$
TA4 $\vdash T{\sim}{\sim}p \supset Tp$; $\vdash Tp \supset T{\sim}{\sim}p$

Theorem (28): The full classical sentential logic holds for wff in which all unrestricted variables are covered (by 'T'), with respect to T-ed components, where a component is T-ed if it is of the form TA.

Proof: For $\vdash STA$ by 2.5 and R2.2. Alternatively a further clause guaranteeing substitutivity of T-ed wff for S-restricted variables could be added to R2.2 so eliminating the need for 2.5.

Using *Theorem* (28) one may transform previous theorems, for example to obtain

TA5 $\vdash T(p \,\&\, q) \equiv . \, Tp \,\&\, Tq$, and
TA6 $\vdash Sp \equiv S{\sim}p$.
TA7 $\vdash Sp \,\&\, Sq \supset . \, S(p \,\&\, q)$

Proof applies *Theorem* (28) to the following results of the T-rule:

$$\vdash Tp \,\&\, Tq \supset . \, T(p \,\&\, q); \quad \vdash T{\sim}p \,\&\, Tq \supset . \, T{\sim}(p \,\&\, q);$$

$$\vdash Tp \,\&\, T{\sim}q \supset . \, T{\sim}(p \,\&\, q); \quad \vdash T{\sim}p \,\&\, T{\sim}q \supset . \, T{\sim}(p \,\&\, q).$$

TA8 $\vdash S(p \,\&\, q) \supset Sp$

Proof is similar to that of TA7 but uses the following result of the T-rule.

$$\vdash T{\sim}(p \,\&\, q) \supset . \, T({\sim}p \,\&\, q) \vee T(p \,\&\, {\sim}q) \vee T({\sim}p \,\&\, {\sim}q)$$

TA9 $\vdash T{\sim}p \supset {\sim}Tp$; from 2.6′
TA10 $\vdash Tr \equiv r$

Proof: By 2.4′ and R2.2′, $\vdash {\sim}r \supset T{\sim}r$; by TA9 $\vdash T{\sim}r \supset {\sim}Tr$. Then by R2.1 and **RSL** $\vdash {\sim}r \supset {\sim}Tr$, and $\vdash Tr \supset r$.

TA11 $\vdash Sr$
Proof uses R2.2′ to obtain $\vdash r \vee {\sim}r \supset . \, T(r \vee {\sim}r)$ and then applies TA3.

Theorem (29): E.T. rule
Proof: $\vdash A \equiv B \longrightarrow \vdash A \supset B, \vdash B \supset A$, by R2.2′ and **RSL**, where A and B satisfy conditions for application of the E.T. rule

$$\longrightarrow \vdash \mathbf{S}^{R_1 \ldots R_n}_{P_1 \ldots P_n}(TA \supset TB)|, \quad \vdash \mathbf{S}^{R_1 \ldots R_n}_{P_1 \ldots P_n}(TB \supset TA)|,$$

with appropriate provisos, by *Theorem* (27);

$$\longrightarrow \vdash \mathbf{S}^{R_1 \ldots R_n}_{P_1 \ldots P_n}(TA \supset TB)| \quad \& \quad \mathbf{S}^{R_1 \ldots R_n}_{P_1 \ldots P_n}(TB \supset TA)|,$$

with provisos, by **RSL** since $\vdash STA$, $\vdash STB$ by 2.5 and R2.2′;

$$\longrightarrow \vdash \mathbf{S}^{R_1 \ldots R_n}_{P_1 \ldots P_n}(TA \equiv TB)|,$$ provided the variables of A coincide with those of B.

Theorem (30): Every theorem of ${}_1\mathbf{S}_2$ is a theorem of ${}_3\mathbf{S}_2$. For the postulates of ${}_1\mathbf{S}_2$ have been derived.

Theorem (31): Every theorem of ${}_3S_2$ is an S_2-tautology.

Proof: It suffices, given *Theorem* (18), to show that the new rules preserve tautologousness. The case of R2.2′ is as before since by induction when A is composed of S-restricted variables and classical connectives, A has value $\bar{n}$. As to the T-rule, if $A_1 \,\&\, A_2 \,\&\, \ldots\, A_k \supset . B_1 \vee B_2 \vee \ldots \vee B_m$ is a tautology then all lines of the matrix table for $S^{R_1 \ldots R_n}_{P_1 \ldots P_n}\, TA_1 \,\&\, \ldots\, TA_k \supset . TB_1 \vee \ldots TB_m |$ for which $P_1, P_2, \ldots, P_n$ have values from $\{t, f\}$ have value t. For under this restriction TA has the same value as A. Alternatively, one of $P_1, P_2, \ldots, P_n$ (each of which occurs in at least one of $A_1, A_2, \ldots, A_k$) has value n. Suppose P_i which occurs in A_j has value n. Then, as A_j contains only classical connectives, TA_j has value f. Also, by the matrix for 'T', TA has either value t or value f, for any A. Therefore $S^{R_1 \ldots R_n}_{P_1 \ldots P_n}\, TA_1 \,\&\, \ldots\, TA_k \supset . TB_1 \vee \ldots TB_m |$ has value t. Moreover this exhausts all cases since every variable which occurs in $B_1 \ldots B_m$ also occurs in at least one of $A_1 \ldots A_k$, so that if some B_i has value n under substitution then some A_j has the value n under the same substitution.

Theorem (32): Every theorem of ${}_3S_2$ is a theorem of ${}_1S_2$.

Proof: If A is a theorem of ${}_3S_2$ then A is an S_2-tautology and so, by *Theorem* (25), a theorem of ${}_1S_2$.

The C-logic C_2

C_2 strengthens C_1 by supplanting 'S' by 'T'; it has the following postulates.

Axioms

C1 $\vdash p \supset . q \supset p$

C2 $\vdash p' \supset (p \supset q) \supset . p' \supset p \supset . p' \supset q$

C3 $\vdash \sim p \supset \sim q \supset . q \supset p$

C4 $\vdash p \supset Tp$

C5 $\vdash T{\sim}p \supset \sim Tp$

Rules

RC1 (uniform substitution)

RC2 (modified modus ponens: MMP)

$\vdash A,\ \vdash A \supset B \longrightarrow \vdash B$, provided no variable is uncovered in A and covered in B.

RC3 (C.T. rule)

$$\vdash A_1 \,\&\, A_2 \,\&\, \ldots\, A_k \supset . B_1 \vee B_2 \ldots B_m \longrightarrow \vdash TA_1 \,\&\, \ldots\, TA_k \supset . TB_1 \vee \ldots TB_m,$$

where $A_1, \ldots A_k, B_1 \ldots B_m$ contain only classical connectives and all variables of $B_1 \ldots B_m$ are among those of $A_1 \ldots A_k$.

Further classical connectives and '*F*' and '*S*' are defined as before. Since '*T*' is the only unlimited primitive connective of $\mathbf{C}_2$ the definition of *cover* can be restricted accordingly for $\mathbf{C}_2$. So, for example, a variable is covered in *A* iff the variable occurs in *A* and every occurrence of the variable in *A* is within the scope of an occurrence of '*T*'.

Theorem (33): Full classical sentential logic holds (i) for all wff which contain only classical connectives, (ii) for wff the components of which are expressions of the form *TA*, *FA* or *SA* where these components are connected only by classical connectives (**RSL** result).

For the qualification on modus ponens is automatically met. It follows that such theorems as the deduction theorem can be applied when (i) or (ii) is satisfied.

Formal theorems of $\mathbf{C}_2$ leading to those theorems required to establish the completeness of $\mathbf{C}_2$ are listed. Proofs where not similar to those of ${}_3\mathbf{S}_2$ are similar to those of ${}_1\mathbf{S}_2$.

T1 $\vdash T(p \,\&\, q) \supset Tp$
T2 $\vdash Tp \,\&\, Tq \supset T(p \,\&\, q)$
T3 $\vdash T(p \,\&\, q) \equiv .\, Tp \,\&\, Tq$
T4 $\vdash T(p \supset q) \supset .\, Tp \supset Tq$
T5 $\vdash T(p \vee q) \supset .\, Tp \vee Tq$
T6 $\vdash S(p \,\&\, q) \equiv .\, Sp \,\&\, Sq$
T7 $\vdash STp$

Proof: $\vdash T(Tp \vee \sim Tp)$ by *Theorem* (33) and C4. The result follows by MMP and T5.

T8 $\vdash TTp \equiv Tp$
T9 $\vdash Tp \supset F\sim p$
T10 $\vdash Sp \equiv S\sim p$
T11 $\vdash Fp \supset T\sim p$
T12 $\vdash FTp \equiv \sim Tp$
T13 $\vdash Tp \supset Sp$
T14 $\vdash \sim Sp \supset FTp$
T15 $\vdash Fp \supset FTp$
T16 $\vdash Fp \supset Sp$
T17 $Fp \supset .\, Sq \supset F(p \,\&\, q)$
T18 $\vdash Sp \supset .\, Fq \supset F(p \,\&\, q)$
T19 $\vdash Tp \supset q \supset .\, Fp \supset q \supset .\, \sim Sp \supset q \;\supset .\, q$
T20 $\vdash p \supset Sp$
T21 $\vdash Tp \vee \sim Sp \supset .\, p$

By T20, T10, T8 and *Theorem* (33).

T21 though an implausible theorem, simply reflects the semantical fact that *n* as well as *t* is a designated value.

Short proofs of certain postulates and theorems of $\mathbf{C}_2$ could be obtained if a full modus ponens rule were available. For example C5 would follow thus:

$\vdash T{\sim}p \supset {\sim}p$; $\vdash {\sim}p \supset {\sim}Tp$

$\vdash T{\sim}p \supset {\sim}p \supset . {\sim}p \supset {\sim}Tp \supset . T{\sim}p \supset {\sim}Tp$ RSL

$\vdash {\sim}p \supset {\sim}Tp \supset . T{\sim}p \supset {\sim}Tp$ MMP

$\vdash T{\sim}p \supset {\sim}Tp$ by MP

But the last step is inadmissible, and given the intended interpretation could only be applied when the conclusion is known to be—just what is to be established—a theorem. A full modus ponens rule would result in *S*-incorrectness. For *Sp* would then follow from T20 and T6 upon substituting a tautology such as '$p \supset p$' for 'p' in T20.

Theorem (34): Every theorem of $\mathbf{C}_2$ is a $\mathbf{C}_2$-tautology.

Proof: Every axiom of $\mathbf{C}_2$ is a tautology of $\mathbf{C}_2$. The following lemmata are used to show that the rules preserve tautologousness.

(A) If every variable in *A* occurs in *B* and if no variable is uncovered in *A* and covered in *B* then it is not the case that *A* has value *n* and *B* has value *f*.

For (i) if variable *P* is uncovered in *A* then *A* has value *n* when *P* has value *n*. By induction over the number of occurrences of '&' and '$\sim$' in *A*. (ii) If *A* has value *n* for some assignment of values to variables then some variable *P* in *A* is uncovered and assigned value *n*. By induction over the number of connectives in *A*.

Now if *A* has value *n* then by (ii) some variable, *P* say, must be uncovered in *A* and have value *n*. If the hypothesis of lemma (A) is satisfied then *P* occurs uncovered in *B*, therefore by (i) *B* has value *n* for value *n* of *P*. So *B* does not have value *f*. (A) follows by propositional logic.

(B) RC2 can be replaced exactly preserving theoremhood by the following modus ponens rule

RC2′ $\vdash A$, $\vdash A \supset B \longrightarrow \vdash B$ provided all variables in *A* occur in *B* and no variable is uncovered in *A* and covered in *B*. *Proof* is as in Halldén [1], p. 48.

Instead of showing RC2 preserves tautologousness it suffices by (B) to show RC2′ preserves tautologousness. There are two cases:

1. If *A* has value *t* and $(A \supset B)$ has value $\bar{f}$ then *B* has value $\bar{f}$ by the matrix for '$\supset$'.
2. If *A* has value *n*, then since the hypothesis on (A) follows from the requirements of RC2′, *B* cannot have value *f*, by (A). Therefore as required *B* has value $\bar{f}$.

RC1 If $S^P_B A|$ has value f for some set of values of its variables then, by lemma (A) of *Theorem* (18), A has value f for the related set of values of its variables. Therefore if A is a tautology, *i.e.* has value $\bar{f}$, so is $S^P_B A|$.

RC3 If A_1 & ... & $A_k \supset . B_1 \vee \ldots \vee B_m$ has value t then each A_i has value t and each B_i has value $\bar{n}$, so TA_1 & ... $TA_k \supset . TB_1 \vee \ldots TB_m$ has value t. If however A_1 & ... $A_k \supset B_1 \vee \ldots B_m$ has value n then some A_i has value n or some B_j has value n. In the second case, by the proviso on RC3, some A_i has value n; so in any case some A_i has value n. Therefore TA_1 & ... TA_k has value f, and TA_1 & ... $TA_k \supset . TB_1 \vee \ldots TB_m$, since it necessarily has either value t or value f, has value t.

Theorem (35): Every $\mathbf{C}_2$-tautology is a theorem of $\mathbf{C}_2$.

Proof: 1. The completeness lemma *Theorem* (24) holds. The proof is as before, using *Theorem* (33), T3, T6, T7, T8, T9, T10, T11, T14, T15, T17, T18 and T19.

2. Let $P_1, \ldots, P_n$ be distinct variables of tautology B and define $A_1, \ldots, A_n$ as in the lemma. Since B is a tautology B' is either TB or $\sim SB$. Therefore, by the lemma and the deduction theorem (a consequence of *Theorem* (33)),

(1) $A_1, A_2, \ldots, A_n \vdash TB \vee \sim SB$

(Alternatively since B' is a tautology B' is not FB, so by excluded fourth, B' is $TB \vee \sim SB$). As in *Theorem* (25), the premisses $A_n, A_{n-1}, \ldots, A_1$ can be peeled off in turn using *Theorem* (33), T19, RC2. So finally $\vdash TB \vee \sim SB$; whence $\vdash B$ by T21 and MMP.

Theorem (36): $\mathbf{C}_2$ is simply and absolutely consistent.

Theorem (37): $\mathbf{C}_2$ is complete, but not strongly complete.

A rule-simplified formulation of $\mathbf{C}_2$ with just the rules of uniform substitution and modified modus ponens can be devised using the completeness proof, *Theorem* (35). This rule-simplified system has as axioms C1, C2, C3 (or some other basis for **RSL**) and the theorems of $\mathbf{C}_2$ listed in the proof of *Theorem* (35); for these theorems are just those required to prove completeness. This rather uneconomical class of axioms can of course be reduced.

The theorems of $\mathbf{C}_2$ coincide with the **H**-theorems of $\mathbf{HS}_2$, where $\mathbf{HS}_2$ is an extension of ${}_1\mathbf{S}_2$ obtained by adding to ${}_1\mathbf{S}_2$ the following H-rule:

(H′) $\vdash A \longrightarrow \vdash_H S^{R_1 \ldots R_n}_{P_1 \ldots P_n} A|$,

where $R_1, R_2, \ldots, R_n$ are all the restricted variables in A, provided A contains no components of the form TB where B contains an uncovered restricted variable which does not also appear T-uncovered in A.

Call the proviso (p′). It follows, by (H′)

$\vdash_{H} Tp \equiv p$

MH.1 If A is a theorem of $\mathbf{C}_2$ then A is an **H**-theorem of $\mathbf{HS}_2$.

Proof: The proof of MH.1 is similar to that of MTH.2 (of $\mathbf{HS}_1$) except for the following additional case:

C.T. rule. If $\vdash_{H} A_1 \,\&\, \ldots\, A_k \supset . B_1 \vee \ldots B_m$ then, for some m and some $A_1{}' \ldots B_m{}'$,

$\vdash_{H} \mathsf{S}^{R_1 \ldots R_n}_{P_1 \ldots P_n} A_1{}' \,\&\, \ldots\, A_k{}' \supset . B_1{}' \vee \ldots B_m{}' |$, and
$\vdash A_1{}' \,\&\, \ldots\, A_k{}' \supset . B_1{}' \vee \ldots B_m{}'$. Therefore by the T-rule
$\vdash \mathsf{S}^{R_1 \ldots R_n}_{P_1 \ldots P_n} TA'_1 \,\&\, \ldots\, TA_k{}' \supset . TB_1{}' \vee \ldots TB_m{}' |$. Hence
$\vdash TA_1 \,\&\, \ldots\, TA_k \supset . TB_1 \vee \ldots TB_m$, and therefore by MTH.1 (of $\mathbf{HS}_1$)
$\vdash_{H} TA_1 \,\&\, \ldots\, TA_k \supset . TB_1 \vee \ldots \vee TB_m$

MH.2 If $\vdash A$ with proviso (p′) on A, then $\mathsf{S}^{R_1 \ldots R_n}_{P_1 \ldots P_n} A|$ has value $\bar{f}$ for systems of values of its variables.

Proof: Since $\vdash A$, A has value t for values of its variables. If the number n of restricted variables in A is zero then $\mathsf{S}A|$ coincides with A. Since $\mathsf{S}A|$ has value t it has value $\bar{f}$. If $n > 0$ consider what happens to the truth table of A when a given variable R_k is replaced by P_k. Because of proviso (p′) some occurrence of P_k will be T-uncovered. The resulting table for $\mathsf{S}^{R_k}_{P_k} A|$ differs from the table for A only in additional lines where P_k has value n. Since some occurrence of P_k is uncovered in these lines, as in (B) of *Theorem* (34), $\mathsf{S}^{R_k}_{P_k} A|$ has value n for these lines. By iteration then $\mathsf{S}^{R_1 \ldots R_n}_{P_1 \ldots P_n} A|$ has value t or value n for all lines of its table.

MH.3 Every H-theorem of $\mathbf{HS}_2$ is a theorem of $\mathbf{C}_2$.

Proof is similar to MTH5 (of $\mathbf{HS}_1$).

MH.4 A is an H-theorem of $\mathbf{HS}_2$ iff A is a theorem of $\mathbf{C}_2$.

This metatheorem provides one interesting connection between $\mathbf{C}_2$ and $\mathbf{S}_2$, and it provides the basis on which we can claim that the *good* theorems of $\mathbf{C}_2$ are those of $\mathbf{S}_2$.

6.4 Stage 3: Axiomatic Formulations of 3-systems

We examine briefly axiomatic formulations of $\mathbf{S}_3$, $\mathbf{S}_3$-1, and $\mathbf{C}_3$. First we consider a formulation ${}_1\mathbf{S}_3$ with primitive set {∼, &, T, ⥛}, or with set {∼, &, ⥛} and with 'T' defined, which simply builds on that of ${}_3\mathbf{S}_2$; it has just these further axioms:

3a. $\vdash \sim Tp \supset . T(p \rightharpoonup q)$
3b. $\vdash Tp \supset . Tq \supset . T(p \rightharpoonup q)$
3c. $\vdash Tp \supset . S(p \rightharpoonup q) \supset Sq$
3d. $\vdash Tp \supset . Fq \supset . F(p \rightharpoonup q)$.

As for ${}_1\mathbf{S}_2$ it follows:

Theorem (38): A wff B is a theorem of ${}_1\mathbf{S}_3$ iff it is an $\mathbf{S}_3$-tautology.

Corollary: When A and $A \rightharpoonup B$ are theorems of ${}_1\mathbf{S}_3$ so is B.

Axioms 3a–3d of ${}_1\mathbf{S}_3$ are precisely those needed to guarantee the induction step for '$\rightharpoonup$' in the completeness lemma for ${}_1\mathbf{S}_3$. Examination of this induction step also shows that axioms 3b–3d can be equivalently replaced by the single axiom 3b′. $\vdash Tp \supset . p \supset q \simeq p \rightharpoonup q$.

Next we consider an S-logic ${}_2\mathbf{S}_3$, with the same morphology as $\mathbf{S}_3$-1 (of §5.12) and with a postulate set designed to yield the completeness of ${}_2\mathbf{S}_3$ with respect to $\mathbf{S}_3$-1 immediately, given our results on $\mathbf{S}_2$. The logic ${}_2\mathbf{S}_3$ equally provides an axiomatisation of an extension of $\mathbf{S}_3$-1 obtained by adding '&' (or '$\supset$') as a further primitive. Thus ${}_2\mathbf{S}_3$ contains minor redundancies: we take advantage of our knowledge of $\mathbf{S}_2$ at the cost of losing independence of connectives. ${}_2\mathbf{S}_3$ is based on the classical implication '$\supset$' defined, given '&', by: $A \supset B =_{\mathrm{Df}} \sim(A \;\&\sim B)$.

Axiomatisations of $\mathbf{S}_3$ based on alternative implications such as '$\supset_A$' are readily designed.

The Logic ${}_2\mathbf{S}_3$

${}_2\mathbf{S}_3$ has as postulates the postulates of ${}_3\mathbf{S}_2$, together with these axioms;

3e. $Tp \vee Tq \supset T(p \cup q)$
3f. $Fp \;\&\; Fq \supset F(p \cup q)$
3g. $S(p \cup q) \supset Sp \vee Tq$
3h. $S(p \cup q) \supset S(q \cup p)$

Theorem (39): A is a theorem of ${}_2\mathbf{S}_3$ if, and only if, A is a theorem of $\mathbf{S}_3$-1.

Hence ${}_2\mathbf{S}_3$ is simply and absolutely consistent and complete.

The **C**-*logic* ${}_1\mathbf{C}_3$

We saw in §5.13 and §5.14 that the matrix **C**-logics $\mathbf{C}_5$ and $\mathbf{C}_6$ are simply inconsistent but absolutely consistent. This means that, in the axiomatic approach, we must find some way of representing these systems in such a way that the simple inconsistency does not spread to absolute inconsistency. If, however, 'Q' (hence 'D') are admitted as not affecting cover, then the existing

MMP of $\mathbf{C}_2$ and lower systems is inadequate, for the following deduction can be made:

1. $\vdash D(p \rightharpoonup p)$
2. $\vdash \sim D(p \rightharpoonup p)$
3. $\vdash D(p \rightharpoonup p) \;\&\; \sim D(p \rightharpoonup p)$

But also 4. $\vdash (D(p \rightharpoonup p) \;\&\; \sim D(p \rightharpoonup p)) \supset q$

Hence, since in 4 no variable is uncovered in the antecedent and covered in the consequent, the existing restriction on modus ponens does not apply and we may therefore derive,

5. $\vdash q$

Now although matrix-$\mathbf{C}_3$ is simply consistent (§5.11), and we need not at this point concern ourselves with the problem of simple inconsistency which runs to absolute inconsistency, MMP is no longer adequate even at this level since '$\rightharpoonup$' is neither classical nor unlimited. We therefore present a system ${}_1\mathbf{C}_3$ which has a two-tier modus ponens rule: in effect the existing MMP for formulae of $\mathbf{C}_3$ which are also formulae of $\mathbf{C}_2$, together with a special modus ponens rule for formulae containing '$\rightharpoonup$'. And as we noted in §5.6, this two-tier rule is in itself sufficient to stop the spread of simply inconsistencies in the stronger systems. This we show in the discussion of ${}_1\mathbf{C}_5$ (§6.6). In these stronger systems, however, simpler options are available. One of these, switching to a different implication connective '$\rightharpoondown$' for which full modus ponens holds, we illustrate at Stage 6 (§6.7), though a parallel move could be made at earlier stages. It is not possible, however, to make this sort of move by taking the $\mathbf{C}_3$ connective '$\rightharpoonup$', as the main implication connective, along with an unqualified modus ponens rule. For $(p \supset p) \rightharpoonup Sp$ is a theorem of $\mathbf{C}_3$; hence, if detachment for '$\rightharpoonup$' were unrestricted, Sp would be a theorem and $\mathbf{C}_3$ (and later systems) would collapse at the level of interpretation. Another option which is satisfactory is to omit C1–C3 (of $\mathbf{C}_2$) and MMP and add RC4 (below) together with the schema: If A is a substitution instance of a classical tautology, then $\vdash A$.

The postulates of ${}_1\mathbf{C}_3$ are as follows:

The postulates of $\mathbf{C}_2$ (except for MMP) and the axioms 3a–3d of ${}_1\mathbf{S}_3$.

The rules:

RC2 ($\mathbf{C}_2$ Modified modus ponens, or $\mathbf{C}_2$-MMP)

$\vdash A,\ \vdash A \supset B \longrightarrow \vdash B,$

provided: (i) no variable is uncovered in A and covered in B;
(ii) $A \supset B$ is a wff of $\mathbf{C}_2$.

RC4 (*T*-ed MP)

$\vdash TA,\ \ \vdash A \supset B \longrightarrow \vdash B$

Theorem (40): A is a theorem of ${}_1\mathbf{C}_3$ if, and only if, A is a theorem of matrix $\mathbf{C}_3$. ${}_1\mathbf{C}_3$ is therefore simply consistent, absolutely consistent and complete. In the proof of completeness, RC4 is applied repeatedly. Detachment can be made from such theorems as $(Tp \vee \sim Sp) \supset p$ using RC4.

The Logic ${}_2\mathbf{C}_3$

${}_2\mathbf{C}_3$ has as postulates the postulates of ${}_1\mathbf{C}_3$ except that the above axioms 3e–3h replace 3a–3d of ${}_1\mathbf{C}_3$.

Theorem (41): A is a theorem of ${}_2\mathbf{C}_3$ if, and only if, A is a theorem of matrix $\mathbf{C}_3$. ${}_2\mathbf{C}_3$ is therefore simply and absolutely consistent and complete.

6.5 Stage 4: Axiomatic Formulations of 4-systems

Because '$\rightharpoonup$' may occur as a primitive in 3-logics, the two sorts of variables in $\mathbf{S}_3$ can very easily be reduced (value-functionally) to a single sort. The resulting one-sorted formulations of 3-logics are called 4-logics (see §5.12). We provide first an axiomatisation ${}_1\mathbf{S}_4$ of 4-logic $\mathbf{S}_4$, with connective set $\{\sim, \&, \rightharpoonup\}$. The axiomatisation is rule-simplified, *i.e.* detachment (for '$\rightharpoonup$') and substitution are the sole rules, and it leads readily to a completeness result. ${}_1\mathbf{S}_4$ has the following axioms:

1. $\vdash p \rightharpoonup . q \rightharpoonup p$
2. $\vdash p \rightharpoonup (q \rightharpoonup q') \rightharpoonup . p \rightharpoonup q \rightharpoonup . p \rightharpoonup q'$
3. $\vdash p \,\&\, q \rightharpoonup p$
4. $\vdash p \,\&\, q \rightharpoonup q$
5. $\vdash p \rightharpoonup . q \rightharpoonup . p \,\&\, q$
6. $\vdash p \rightharpoonup . \sim p \rightharpoonup q$
7. $\vdash p \rightharpoonup \sim\sim p$
8. $\vdash \sim\sim p \rightharpoonup p$
9. $\vdash p \,\&\, \sim q \rightharpoonup \sim (p \rightharpoonup q)$
10. $\vdash \sim (p \rightharpoonup q) \rightharpoonup \sim (p \,\&\, q)$
11. $\vdash \sim (p \,\&\, q) \rightharpoonup \sim (q \,\&\, p)$
12. $\vdash Tp \rightharpoonup p$
13. $\vdash p \rightharpoonup Tp$
14. $\vdash p \,\&\, \sim (p \,\&\, q) \rightharpoonup . \sim q$
15. $\vdash \sim Sp \rightharpoonup . \sim (p \,\&\, q) \rightharpoonup p \,\&\, q$,

where $TA =_{\mathrm{Df}} \sim (A \rightharpoonup \sim A)$, $FA =_{\mathrm{Df}} \sim (\sim A \rightharpoonup A)$,
and $\sim SA =_{\mathrm{Df}} (A \rightharpoonup \sim A) \,\&\, (\sim A \rightharpoonup A)$.

Axioms 1.–5. may be replaced by any set of axioms for the implication-conjunction fragment of positive logic or for that matter of classical logic since Peirce's law, $((p \rightharpoondown q) \rightharpoondown p) \rightharpoondown p$, is valid. The clumsiness of the remaining axioms 6–15 is due to the fact that, as with Fitch's system, contraposition principles break down. Simpler axiomatisations can be obtained at the cost of rule simplification; we suspect however that there is a much superior rule-simplified axiomatisation that we have not uncovered. We list theorems used in a completeness proof—of the type we have given for 2-logics—for ${}_1\mathbf{S}_4$.

T1 $\vdash \sim Tp \rightharpoondown . p \rightharpoondown q$
T2 $\vdash Tp \rightharpoondown . Tq \rightharpoondown T(p \rightharpoondown q)$
T3 $\vdash Tp \rightharpoondown . Fq \rightharpoondown F(p \rightharpoondown q)$
T4 $\vdash Tp \rightharpoondown . \sim Sq \rightharpoondown \sim S(p \rightharpoondown q)$
T5 $\vdash Tp \rightharpoondown T \sim \sim p$
T6 $\vdash \sim Sp \rightharpoondown \sim S \sim \sim p$
T7 $\vdash Tp \rightharpoondown . Tq \rightharpoondown T(p \,\&\, q)$
T8 $\vdash Fp \rightharpoondown . Tq \rightharpoondown F(p \,\&\, q)$
T9 $\vdash F(p \,\&\, q) \rightharpoondown F(q \,\&\, p)$
T10 $\vdash \sim Sp \rightharpoondown \sim S(p \,\&\, q)$
T11 $\vdash \sim Sq \rightharpoondown \sim S(p \,\&\, q)$
T12 $\vdash p \rightharpoondown Sp$

Theorem (42): B is a theorem of ${}_1\mathbf{S}_4$ iff B is an $\mathbf{S}_4$-tautology.

The proof of completeness reveals, however, that ${}_1\mathbf{S}_4$ has little advantage in conciseness over the rule-simplified system ${}_2\mathbf{S}_4$ that may be read off completeness requirements. ${}_2\mathbf{S}_4$ has as postulates the axioms 1, 2 and the rules of ${}_1\mathbf{S}_4$ and T1–T11.

Although ${}_1\mathbf{S}_4$ and ${}_2\mathbf{S}_4$ are complete as 4-logics, they are not complete as 3-logics. For ${}_1\mathbf{S}_4$, for example, has only one sort of variable whereas $\mathbf{S}_3$ has two sorts. Nevertheless every tautology of $\mathbf{S}_3$ can be *represented* in ${}_1\mathbf{S}_4$.

Theorem (43): The theorems of ${}_1\mathbf{S}_4$ coincide with those tautologies of $\mathbf{S}_3$ which contain just unrestricted variables.

Proof: Every theorem of ${}_1\mathbf{S}_4$ is an $\mathbf{S}_3$-tautology, and by the preceding *Theorem* every $\mathbf{S}_3$-tautology containing only unrestricted variables is a theorem of ${}_1\mathbf{S}_4$.

Theorem (44): The wff

$$(\alpha): \quad A(r_1, \ldots, r_m, p_1, \ldots, p_n)$$

is an $\mathbf{S}_3$-tautology iff its *representative*

$$(\alpha^*): \quad Sq_1 \rightharpoondown . Sq_2 \ldots . Sq_m \rightharpoondown A(q_1, \ldots, q_m, p_1, \ldots, p_n)$$

is a theorem of ${}_1\mathbf{S}_4$, where $r_1, \ldots, r_m$ include all the S-restricted variables and $p_1, \ldots, p_n$ all the S-unrestricted variables in (α), and $A(q_1, \ldots, p_n)$ is obtained from (α) by simultaneous substitution of new distinct unrestricted variables $q_1, \ldots, q_m$ for $r_1, \ldots, r_m$.

Proof: 1. If (α*) is a theorem of $_1\mathbf{S}_4$ then (α*) is an $\mathbf{S}_3$-tautology, by the previous *Theorem*; and thus (α*) is a theorem of $_1\mathbf{S}_3$. Hence (α) is a theorem of $_1\mathbf{S}_3$ by substitution and detachment, since Sr_i is a theorem of $_1\mathbf{S}_3$. Finally then (α) is a tautology of $\mathbf{S}_3$.

2. If (α) is a tautology of $\mathbf{S}_3$ then so also is (α*). For when
$v(q_i) \neq n$ for each i ($i = 1, \ldots, m$),
$v(A(q_1, \ldots, q_m, p_1, \ldots, p_n)) = t$ since (α) is a tautology.
Hence by finite induction $v(\alpha^*) = t$ since $v(B_1 \rightharpoonup B_2) = t$ when $v(B_2) = t$. Also when $v(q_i) = n$ for some i,
then $v(Sq_i \rightharpoonup . Sq_{i+1}. \ldots \rightharpoonup . Sq_m \rightharpoonup A(q_1, \ldots, q_m, p_1, \ldots, p_n)) = t$
since $v(Sq_i) = f$ and $v(B_1 \rightharpoonup B_2) = t$ when $v(B_1) = f$.
Since for all assignments $v(\alpha^*) = t$, (α*) is a theorem of $_1\mathbf{S}_4$.

Corollary: Every 2-sorted theorem (α) of $_1\mathbf{S}_3$ can be *represented* by a deductively equivalent 1-sorted theorem (α*) of $_1\mathbf{S}_4$.

Åqvist's system **Å**, based on the connective set $\{\cup, \sim, T\}$ may be given various one-sorted formulations. One adequate rule-simplified formulation, using the defined *provided that* connective '$\rightharpoonup$' as operative implication and with axioms chosen so as to yield a completeness result immediately, has these axioms:

$p \rightharpoonup (q \rightharpoonup p)$	$p \rightharpoonup (q \rightharpoonup q') \rightharpoonup ((p \rightharpoonup q) \rightharpoonup (p \rightharpoonup q'))$
$(p \rightharpoonup q) \rightharpoonup (\sim Tq \rightharpoonup \sim Tp)$	$Tp \rightharpoonup F\sim p$
$S\sim p \rightharpoonup Sp$	$Tp \rightharpoonup TTp$
$Fp \rightharpoonup FTp$	$\sim Sp \rightharpoonup FTp$
$Tp \rightharpoonup T(p \cup q)$	$T(p \cup q) \rightharpoonup T(q \cup p)$
$Fp \rightharpoonup (Fq \rightharpoonup F(p \cup q))$	$\sim Tp \rightharpoonup (\sim Sq \rightharpoonup \sim S(p \cup q))$
$\sim S(p \cup q) \rightharpoonup \sim S(q \cup p)$	$Tp \rightharpoonup p$

A very economical axiomatisation of 4-S-systems is provided by Wajsberg's axiomatisation of **L** (given in Prior [2]). This axiomatisation, $\mathbf{L}_1$, has as rules detachment for '$\supset_L$' and substitution; and it has these axioms:

L1 $\vdash p \supset_L . q \supset_L p$
L2 $\vdash p \supset_L q \supset_L . q \supset_L p' \supset_L . p \supset_L p'$
L3 $\vdash (p \supset_L \sim p) \supset_L p \supset_L . p$
L4 $\vdash \sim p \supset_L \sim q \supset_L . q \supset_L p$

Strong normal and characteristic connectives are defined thus:

$A \cup B =_{Df} (A \supset_L B) \supset_L B$;	$A \cap B =_{Df} \sim(\sim A \cup \sim B)$
$TA =_{Df} \sim(A \supset_L \sim A)$;	$FA =_{Df} T\sim A$
$SA =_{Df} TA \cup FA$;	$A \rightharpoonup B =_{Df} \sim TA \cup B$

The postulates of $\mathbf{L}_1$ suffice to prove the following theorems, which in turn suffice for a completeness proof of the style given for 2-systems:

$$Tq \supset_L . T(p \supset_L q); \quad Fq \supset_L . T(p \supset_L q); \quad Tp \supset_L . Fq \supset_L F(p \supset_L q);$$
$$Tp \supset_L . \sim Sq \supset_L . \sim S(p \supset_L q); \quad \sim Sp \supset_L . \sim Sq \supset_L . T(p \supset_L q);$$
$$\sim Sp \supset_L . Fq \supset_L . \sim S(p \supset_L q); \quad Tp \supset_L p; \quad Tp \supset_L F{\sim}p; \quad p \supset_L p;$$
$$\sim Sp \supset_L {\sim} S{\sim}p; \quad Tp \supset_L q \supset_L . Fp \supset_L q \supset_L . \sim Sp \supset_L q \supset_L . q$$

Proof of these theorems and completeness is facilitated by the following deduction theorem:

Theorem (45): If $A_1, A_2, \ldots, A_m \vdash B$, *i.e.* there is a proof of B from hypotheses $A_1, \ldots, A_m$ using just the hypotheses, theorems of $\mathbf{L}_1$ and $\supset_L$-detachment, then

$\vdash C_1 \supset_L . C_2 \supset_L \ldots C_n \supset_L B$ where each C_i is one of $A_1, \ldots, A_m$ and occurs at most twice.

Proof uses L2 and the theorems

$$\vdash p \supset_L p; \ \vdash p \supset_L (q \supset_L p') \supset_L . q \supset_L . p \supset_L p';$$
$$\vdash p \supset_L (p \supset_L . p \supset_L q) \supset_L . p \supset_L . p \supset_L q.$$

(see Curry [2]). In view of L1 additional irrelevant elements may also be inserted among the hypotheses.

The deduction theorem for $\mathbf{L}_1$ may be presented in a more perspicuous, and familiar, form if (following Brady [3]) we redefine '$\rightharpoonup$' thus:

$$A \rightharpoonup B =_{Df} A \supset_L (A \supset_L B)$$

Theorem (46): If $A_1, \ldots, A_m \vdash B$ then $A_1, \ldots, A_{m-1} \vdash A_m \rightharpoonup B$.

Since $\vdash_{L_1} p \rightharpoonup . q \rightharpoonup p$ and $\vdash_{L_1} p \rightharpoonup (p' \rightharpoonup q) \rightharpoonup . p \rightharpoonup p' \rightharpoonup . p \rightharpoonup q$, proof of this deduction theorem is standard (for details see *e.g.* Church [4], p. 88).

Theorem (47): (a) A wff B is a theorem of $\mathbf{L}_1$ iff B is an **L**-tautology.
(b) Every theorem of $\mathbf{L}_1$ is a theorem of ${}_1\mathbf{S}_4$ and conversely (using definitions).

Instead of just proving (a) and then deducing (b), part of (b) and the completeness of ${}_1\mathbf{S}_4$ may be used to prove the completeness of $\mathbf{L}_2$, as follows:

Lemma. If B is a theorem of ${}_1\mathbf{S}_4$ then B is a theorem of $\mathbf{L}_1$.
For a proof it suffices to show that each axiom of ${}_1\mathbf{S}_4$ is a theorem of $\mathbf{L}_1$ and for $\rightharpoonup$-detachment that

$((p \supset_L {\sim} p) \supset_L q) \supset_L q \supset_L . \ p \supset_L q$ is a theorem of $\mathbf{L}_1$.

Now if B is an **L**-tautology then B is an $\mathbf{S}_4$-tautology, so B is a theorem of ${}_1\mathbf{S}_4$ and so of $\mathbf{L}_1$.

4-S-systems are especially important for two reasons:

(1) Since they contain the strong normal connectives $\{\sim, \cup, \cap\}$, and since the matrix logic of incompleteness is very like that of these connectives (see §5.4), the systems are important for the study of incompleteness and indeterminacy. In fact **L** was devised by Lukasiewicz precisely to treat logically a case of indeterminacy, that of future contingents (see Prior [2]), and use of **L** in the investigation of intensional indeterminacy and logical paradoxes has been carried further by Prior and others (see Prior [3], IX and X: we return to the matter in the discussion of quantified **L** in §7.8).

(2) Because they are significance complete and because they are strong enough for most sentential significance purposes (especially those of expressibility of significance claims and significance restrictions), they provide a viable and flexible extensional base for significance theory. In fact the real choice of a sentential base for significance theory is the choice between 4-S-systems and 6-S-systems.

One outcome of (1) and (2) is that, for a four-valued logic to provide a satisfactory treatment of nonsignificance, and of incompleteness as a further distinct value, it would need to include 4-systems for both nonsignificance and incompleteness as three-valued sublogics. (For such a system see Brady [3]).

6.6 Stage 5: Axiomatic Formulations of 5-Systems

As at earlier stages there are many complete axiomatisations of 2-sorted matrix system $\mathbf{S}_5$, depending on the primitives selected, the axioms assumed for these connectives and the rule structure imposed. In particular each of the systems studied at stage 3 can be extended to yield a complete axiomatisation of $\mathbf{S}_5$ by adding the axioms on 'Q' that we adopt for system ${}_1\mathbf{S}_5$. For system ${}_1\mathbf{S}_5$ we take as postulates the postulates of ${}_3\mathbf{S}_2$ together with the following axioms on 'Q':

5.7 $\vdash Tp \supset TQp$; 5.8 $\vdash Fp \supset \sim SQp$; 5.9 $\vdash \sim Sp \supset FQp$.

These axioms are precisely those needed to extend the completeness proof given for ${}_1\mathbf{S}_2$ to cover the further case in the induction caused by the addition of 'Q'. For *Theorem* (24) is extended by the case where B is QB_1 for some B_1. By induction hypothesis

(2) $A_1, \ldots, A_n \vdash B_1'$.

If B_1 has value t (f,n), B has value t (n,f); hence B_1' is TB_1 $(FB_1, \sim SB_1)$ and B' is TQB_1 $(\sim SQB_1, FQB_1)$. Thus $A_1, \ldots, A_n \vdash B'$ follows from (2) by 2.7 (2.8, 2.9). As at stage 2 this establishes the completeness theorem.

Theorem (48): If B is a theorem of matrix $\mathbf{S}_5$ then B is a theorem of ${}_1\mathbf{S}_5$.

Likewise consistency is proved as at stage 2 using

Theorem (49): If B is a theorem of ${}_1S_5$ then B is a theorem of matrix S_5.

Substitutively of strong equivalence (as stated in *Theorem* (23)) now follows as a corollary.

To give a fuller picture of ${}_1S_5$ (and of S_5) we state a number of representative formal theorems in an order in which they may readily be proved, without use of metatheoretic results such as *Theorem* (48) if so desired; formal theorems proved for ${}_3S_2$ are of course used.

T1	$\vdash Tp \supset TQQp$	T2	$\vdash Fp \supset FQQp$
T3	$\vdash \sim Sp \supset \sim SQQp$	T4	$\vdash TQQp \supset Tp$
T5	$\vdash FQQp \supset Fp$	T6	$\vdash QQp \simeq p$
T7	$\vdash TQp \supset Tp$	T8	$\vdash TQp \simeq Tp$
T9	$\vdash T\sim Qp \simeq \sim Sp$	T10	$\vdash Tp \simeq \sim SQ\sim p$
T11	$\vdash T(Tp \supset Q(Tp \supset Qq)) \supset . Tp \supset Tq$	T12	$\vdash SQ\sim\sim p \equiv SQp$
T13	$\vdash \sim Sp \supset SQp$	T14	$\vdash Tp \supset Tq \supset . SQ(p \supset q)$
T15	$\vdash Tp \supset Tq \supset . T(Tp \supset Q(Tp \supset Qq))$		
T16	$\vdash Tp \supset Tq \simeq . T(Tp \supset Q(Tp \supset Qq))$		
T17	$\vdash r \equiv SQr$	DR1	$\vdash A \rightarrow \vdash QA$
T18	$\vdash Qp \,\&\, p \simeq . Qp \supset p$	T19	$\vdash T(p \supset q) \supset . SQp \supset SQq$

The implication '$\rightharpoonup$' is here defined:

$$A \rightharpoonup B =_{\mathrm{Df}} TA \supset Q(TA \supset QB)$$

Alternative definitions of '$\rightharpoonup$' are available, in view of the equivalences:

$$\vdash p \rightharpoonup q \simeq . Tp \supset Q(p \supset Qq) \text{ and } \vdash p \rightharpoonup q \simeq . Q(Tp \supset Qq) \vee (Tp \supset Tq).$$

T20 $\vdash S(p \rightharpoonup q) \simeq . \sim Tp \vee Sq$ T21 $\vdash p \rightharpoonup . q \rightharpoonup p$

The proof of T21, and of some of the following theorems, is conveniently divided into two parts, in one of which TA (A is '$p \rightharpoonup . q \rightharpoonup p$' in the case of T21) is proved, and in the other SA is proved. The theorem then follows using $\vdash Sp \supset . Tp \simeq p$.

T22 $\vdash p' \rightharpoonup (p \rightharpoonup q) \rightharpoonup . p' \rightharpoonup p \rightharpoonup . p' \rightharpoonup q$

Unlimited negation '$\urcorner$' is defined: $\urcorner A =_{\mathrm{Df}} \sim TA$; and the irregular connective 'D' is defined $DA =_{\mathrm{Df}} \sim Q \sim A$.

T23 $\vdash S \urcorner p$
T24 $\vdash \urcorner p \rightharpoonup \urcorner q \rightharpoonup . q \rightharpoonup p$
T25 $\vdash \urcorner Dp \simeq Sp$
T26 $\vdash \urcorner D(p \rightharpoonup p)$
T27 $\vdash \urcorner D(p \rightharpoonup q) \rightharpoonup . p \rightharpoonup \urcorner Dq$
T28 $\vdash \urcorner D \urcorner p$

T29 $\vdash p \rightharpoonup DDp$
T30 $\vdash DDp \rightharpoonup p$
T31 $\vdash p \simeq q \simeq . Qp \simeq Qq$
DR2 $\vdash A, \vdash A \rightharpoonup B \twoheadrightarrow \vdash B$.

The Logic $_1\mathbf{C}_5$

A $\mathbf{C}_5$-system is obtained by adding the primitive 'Q' to $\mathbf{C}_2$. Since 'Q' is a limited non-classical connective, addition of 'Q' does not demand revision of the general definition of *cover* for $\mathbf{C}_2$. However, once 'Q' and its postulates are added, the modification on modus ponens in MMP is, as we have shown in discussing $_1\mathbf{C}_3$ (§6.4), altogether too liberal. Accordingly we adopt a formulation of $\mathbf{C}_5$ like that provided for $_1\mathbf{C}_3$ with a two-tier modus ponens rule. (An alternative approach is given in the construction of $\mathbf{D}_6$ (§6.7).)

$_1\mathbf{C}_5$ has as postulates the postulates of $_1\mathbf{C}_3$, except for special axioms 3a–3d, together with the special axioms of $_1\mathbf{S}_5$.

The main features of $_1\mathbf{C}_5$ are summarised in the following metatheorems.

Theorem (50): Every theorem of $_1\mathbf{C}_5$ is a tautology of $\mathbf{C}_5$.
Proof: Essentially as for *Theorem* (34), but with additions to cater for RC4 of $_1\mathbf{C}_3$.

Theorem (51): Every tautology of $\mathbf{C}_5$ is a theorem of $_1\mathbf{C}_5$.
Proof: The completeness lemma *Theorem* (24), holds for $_1\mathbf{S}_5$. Thus the result follows by *Theorem* (35) and *Theorem* (24) using RC4 where RC2 cannot be applied.

Theorem (52): $_1\mathbf{C}_5$ is absolutely consistent.

Theorem (53): $_1\mathbf{C}_5$ is complete.

Consequently $_1\mathbf{C}_5$ is not simply consistent since matrix $\mathbf{C}_5$ is not. It is not, therefore, a satisfactory significance logic.

6.7 Stage 6: Axiomatic Formulations of 6-systems

We turn now to one-sorted axiomatisations of functionally complete 3-valued logics. The main S-system we shall study, a simple and flexible system $_1\mathbf{S}_6$, with primitives $\{\urcorner, \rightharpoonup, D\}$, has these postulates:

Axioms

6.1 $\vdash p \rightharpoonup . q \rightharpoonup p$
6.2 $\vdash p' \rightharpoonup (p \rightharpoonup q) \rightharpoonup . p' \rightharpoonup p \rightharpoonup . p' \rightharpoonup q$
6.3 $\vdash \urcorner p \rightharpoonup \urcorner q \rightharpoonup . q \rightharpoonup p$
6.4 $\vdash S(p \rightharpoonup p)$
6.5 $\vdash Sq \rightharpoonup S(p \rightharpoonup q)$

6.6 $\vdash S(p \rightharpoonup q) \rightharpoonup . p \rightharpoonup Sq$

6.7 $\vdash S \urcorner p$

6.8 $\vdash p \rightharpoonup DDp$

6.9 $\vdash DDp \rightharpoonup p$

Rules

R6.1: Uniform substitution. R6.2 Material detachment for '$\rightharpoonup$'.
The characteristic connectives '*T*', '*F*' and '$\sim S$' are defined through the following definitions:

$$TA =_{\mathrm{Df}} \urcorner \urcorner A; \quad FA =_{\mathrm{Df}} \urcorner (\urcorner A \rightharpoonup DA); \quad SA =_{\mathrm{Df}} \urcorner DA.$$

Thus $\vdash \sim SA \simeq \urcorner \urcorner DA$.

Axioms 6.1–6.3 provide, given R6.1 and R6.2, a complete set of axioms for classical sentential logic, formulated using connectives '$\rightharpoonup$' and '$\urcorner$': they may be replaced by any other complete set of implication-negation axioms for classical logic without affecting the class of theorems of $\mathbf{S}_6$. Though '*D*' appears to lack a natural language intuitive interpretation, as distinct from a matrix one, '$\urcorner D$' is extensionally synonymous with '*S*', and consequently axioms 6.4–6.7 have been made intelligible by writing them in abbreviated form. Finally 6.8 and 6.9 sum up the involutory character of '*D*'; they may be replaced by the axioms: $SDp \rightharpoonup \urcorner p$ and $\urcorner p \rightharpoonup SDp$.

A formal system like ${}_1\mathbf{S}_6$ was first investigated by Slupecki [1]. His axiom set is displayed in Prior [2] (pp. 237–8).

Results on system ${}_1\mathbf{S}_6$

1 Two-valued classical implication-negation sentential calculus, with '$\rightharpoonup$' representing implication and '$\urcorner$' negation, is a subsystem of ${}_1\mathbf{S}_6$. For 6.1, 6.2 and 6.3 provide Lukasiewicz's 1930 axioms for the full sentential calculus. Thus theorems isomorphic with ($\supset$, $\sim$)-theorems of sentential logic are derivable. (Labels for theorems of 2-valued sentential logic, *e.g.* 'Syll', 'Comm', are transferred to their analogues in ${}_1\mathbf{S}_6$. Theorems of ${}_1\mathbf{S}_6$ which are derived just as for classical sentential logic are said to be derived by **RSL**). It also follows that certain metatheorems of classical sentential logic have analogues for ${}_1\mathbf{S}_6$. For example:

2 *Deduction theorems* such as the following:

If $A_1, A_2, \ldots, A_n \vdash B$ then $A_1, \ldots, A_{n-1} \vdash A_n \rightharpoonup B$
are provable for ${}_1\mathbf{S}_6$.

Proofs of these metatheorems are straightforward rewritings of those given in Church [4] for systems $\mathbf{P}_1$ or $\mathbf{P}_2$.

3. A wff is *D-free* if '*D*' does not occur among its primitive connectives. If *A* is a *D*-free wff of ${}_1\mathbf{S}_6$, then *A* is a theorem of ${}_1\mathbf{S}_6$ iff *A* is a tautology of

S_6. Thus the *D*-free segment of ${}_1S_6$ is consistent and complete. Any theorem of ${}_1S_6$ which is *D*-free can be obtained by a *D*-free derivation. This *D*-elimination theorem follows from the completeness of the $(\rightharpoonup, \neg)$ segment of ${}_1S_6$.

4 *Consistency theorems*

Every theorem of ${}_1S_6$ is a tautology of S_6. ${}_1S_6$ is absolutely consistent and Post-consistent. ${}_1S_6$ is *S*-consistent, *i.e.* not every significant wff is a theorem. The wff $\neg(p \rightharpoonup p)$ which has value *f* provides an example. Proofs are similar to those for earlier systems.

5 Formal theorems of ${}_1S_6$ used to establish completeness theorems are listed in an order in which they may easily be derived. Some of the proofs, for instance those of T6 and T8, may be simplified by application of the deduction theorem 2.

T1 $\vdash p \rightharpoonup Sp$
T2 $\vdash Tp \rightharpoonup p$
T3 $\vdash p \rightharpoonup Tp$
T4 $\vdash Tp \rightharpoonup Sp$
T5 $\vdash Tp \rightharpoonup . S(p \rightharpoonup q) \rightharpoonup Sq$
T6 $\vdash Tp \rightharpoonup F\neg p$
T7 $\vdash Tp \rightharpoonup \neg SDp$
T8 $\vdash Tp \rightharpoonup q \rightharpoonup . Fp \rightharpoonup q \rightharpoonup . \neg Sp \rightleftharpoons q \rightharpoonup . q$
T9 $\vdash Tp \rightharpoonup T(q \rightharpoonup p)$
T10 $\vdash \neg Sp \rightharpoonup T(p \rightharpoonup q)$
T11 $\vdash Fp \rightharpoonup T(p \rightharpoonup q)$
T12 $\vdash Tp \rightharpoonup . \neg Sq \rightharpoonup \neg S(p \rightharpoonup q)$
T13 $\vdash Tp \rightharpoonup . Fq \rightharpoonup F(p \rightharpoonup q)$
T14 $\vdash \neg Sp \rightharpoonup T\neg p$
T15 $\vdash Fp \rightharpoonup T\neg p$
T16 $\vdash \neg Sp \rightharpoonup TDp$
T17 $\vdash Fp \rightharpoonup FDp$

In the case of the second completeness proof we provide, further theorems concerning the connectives '$\&_1$' and '$\vee_1$' are wanted. The connectives concerned are defined thus:

$$A \,\&_1\, B =_{\mathrm{Df}} \neg(A \rightharpoonup \neg B); \quad A \vee_1 B =_{\mathrm{Df}} \neg A \rightharpoonup B.$$

But though the syntactical definitions may be familiar, the significance features of the connectives defined will not be. For example $A \vee_1 B$ has value *n* when $v(A) = f$ and $v(B) = n$, but has value *f* when $v(A) = n$ and $v(B) = f$; thus $\vee_1$ is not significance symmetric.

T18 $\vdash \neg Sp \rightharpoonup . \neg Tp \,\&_1 \neg Fp$
T19 $\vdash \neg Tp \,\&_1 \neg Fp \rightharpoonup . \neg Sp$
T20 $\vdash \neg Tp \rightharpoonup . Fp \vee_1 \neg Sp$

T21 $\vdash Fp \vee_1 \neg Sp \rightharpoonup . \neg Tp$

T22 $\vdash \neg Fp \rightharpoonup . Tp \vee_1 \neg Sp$

T23 $\vdash Tp \vee_1 \neg Sp \rightharpoonup . \neg Fp$

T24 $\vdash T(p \rightharpoonup q) \,\&_1\, Tp \rightharpoonup . Tq$

T25 $\vdash Tp \,\&_1\, Tq \rightharpoonup . T(p \rightharpoonup q)$

T26 $\vdash Tp \,\&_1\, \neg Sq \rightharpoonup . \neg S(p \rightharpoonup q)$

T27 $\vdash Tp \,\&_1\, Fq \rightharpoonup . F(p \rightharpoonup q)$

6 *Completeness theorems: First method*

6.1 *Completeness lemma:* If B is a wff of ${}_1\mathbf{S}_6$, and $P_1, P_2, ..., P_n$ are the distinct variables occurring in B, and if A_i is TP_i, FP_i or $\neg SP_i$ according as the value v_i of P_i is t, f or n, and B' is TB, FB or $\neg SB$ according as the value of B for values $v_1, ..., v_n$ of its variables is t, f or n; then

$$(\alpha) \quad \vdash A_1 \rightharpoonup . A_2 ... \rightharpoonup . A_n \rightharpoonup B'.$$

Proof: The proof is by induction over the number of occurrences of '$\rightharpoonup$', '$\neg$', 'D' in B. If zero connectives occur in B then B is P_1; and the result follows at once using $p \rightharpoonup p$.

If there are occurrences of the primitive connectives in B then there are three cases to consider.

Case 1 B is $B_1 \rightharpoonup B_2$. By hypothesis of induction

$$(\beta) \quad \vdash A_1 \rightharpoonup . A_2 ... \rightharpoonup . A_n \rightharpoonup B_1'$$
$$(\gamma) \quad \vdash A_1 \rightharpoonup . A_2 ... \rightharpoonup . A_n \rightharpoonup B_2'$$

If $v(B_2) = t$, $v(B) = t$. So B_2' is TB_2 and B' is $T(B_1 \rightharpoonup B_2)$. Using $TB_2 \rightharpoonup T(B_1 \rightharpoonup B_2)$, a substitution-instance of T9, (α) follows from (γ) applying Implicative Syllogism. If $v(B_1) = n\ (f)$, $v(B) = t$. So B_1' is $\neg SB_1$ (FB_1) and B' is $T(B_1 \rightharpoonup B_2)$. Then ($\alpha$) follows from ($\gamma$) using, from T10, $\neg SB_1 \rightharpoonup T(B_1 \rightharpoonup B_2)$ (from T11, $FB_1 \rightharpoonup T(B_1 \rightharpoonup B_2)$) and Implicative Syllogism. If $v(B_1) = t$ and $v(B_2) = n$, $v(B) = n$. So B_1' is TB_1, B_2' is $\neg SB_2$, and B' is $\neg SB$. Then (α) follows from (β) and (γ) using, from T12, $TB_1 \rightharpoonup . \neg SB_2 \rightharpoonup \neg S(B_1 \rightharpoonup B_2)$, applying Implicative Syllogism and Self-distribution axiom 6.2. If $v(B_1) = t$ and $v(B_2) = f$, then $v(B) = f$. Thus B_1' is TB_1, B_2' is FB_2, and B' is FB. Then, as before, (α) follows from (β) and (γ) using $TB_1 \rightharpoonup . FB_2 \rightharpoonup F(B_1 \rightharpoonup B_2)$, a substitution-instance of T13.

Case 2 B is $\neg B_1$; so (β) holds.

If $v(B_1) = t\ (n, f)$ then $v(B) = f\ (t, t)$. Then B_1' is TB_1 ($\neg SB_1$, FB_1) and B' is $F\neg B_1$ ($T\neg B_1$, $T\neg B_1$). So (α) follows from (β) using T6 (T14, T15), and Implicative Syllogism.

Case 3 B is DB_1; so (β) holds.

If $v(B_1) = t\ (n, f)$, then $v(B) = n\ (t, f)$. Thus B_1' is TB_1 ($\neg SB_1$, FB_1) and B'

is $\neg SDB_1$ (TDB_1, FDB_1). Then (α) follows from (β) using $TB_1 \rightharpoonup \neg SDB_1$ (T7) ($\neg SB_1 \rightharpoonup TDB_1$ (T16), $FB_1 \rightharpoonup FDB_1$ (T17)).

6.2 *Completeness theorem:* If B is a tautology of S_6 then B is a theorem of ${}_1S_6$.

Proof: Let P_1, ..., P_n be the distinct variables of B and define A_1, ..., A_n as in the lemma. Since B is a tautology, B' is TB. Therefore by the lemma

$$A_1 \rightharpoonup . A_2 \rightharpoonup \dots \rightharpoonup . A_n \rightharpoonup TB.$$

Since this holds both when $v(P_i) = t$, $v(P_i) = n$ and $v(P_i) = f$, it follows, taking $i = 1$.

$$\vdash TP_1 \rightharpoonup . A_2 \dots \rightharpoonup TB$$
$$\vdash FP_1 \rightharpoonup . A_2 \dots \rightharpoonup TB$$
$$\vdash \neg SP_1 \rightharpoonup . A_2 \dots \rightharpoonup TB$$

Since by T8, $TP_1 \rightharpoonup C \rightharpoonup . FP_1 \rightharpoonup C \rightharpoonup . \neg SP_1 \rightharpoonup C \rightharpoonup . C$, by detachment, $A_2 . A_3 \rightharpoonup \dots \rightharpoonup TB$.

Reiterating these steps n times, $\vdash TB$, and hence by T2, $\vdash B$.

7 *Consistency and Completeness Theorems: Second Method*

7.1 *Consistency Theorem:* If $\vdash B$ in ${}_1S_6$ then B Taut in S_6.

Proof is by induction over the length of B's proof in ${}_1S_6$. As in earlier proofs it is shown that each axiom of ${}_1S_6$ is an S_6-tautology, and that the rules preserve the property of S_6-tautologousness. We illustrate the proof that the axioms are tautologies by considering axiom 6.4. We have already shown, in discussing S_6, $(p \rightharpoonup p)$ Taut. Now $v(p \rightharpoonup p) = t \Rightarrow .v(D(p \rightharpoonup p)) = n$, and $v(D(p \rightharpoonup p)) = n \Rightarrow v(\neg D(p \rightharpoonup p)) = t$. Hence $\Box\ [v(p \rightharpoonup p) = t] \Rightarrow \Box [v(\neg D(p \rightharpoonup p)) = t]$, and detaching $(\neg D(p \rightharpoonup p))$ Taut. That R6.2 preserves tautologousness follows from this consequence of the modus ponens requirement:

$$\Box(v(A \rightharpoonup B) = t)\ \&\ \Box(v(A) = t) \Rightarrow . \Box(v(B) = t).$$

For R6.1 the proof formalises that given for ${}_1S_2$ in *Theorem* (18).

7.2 *Completeness Theorem:* If B Taut in S_6 then $\vdash B$ in ${}_1S_6$.

Proof: Since B Taut, $v(B) = t$ in S_6. The proof that $v(B) = t$ in S_6 yields a proof of B in ${}_1S_6$ upon making the following replacements:

(i) Eliminate all occurrences of 'v' using the coentailments: $v(B) = t \Leftrightarrow TB$; $v(B) = f \Leftrightarrow FB$; $v(B) = n \Leftrightarrow \neg SB$. Replace '$\Leftrightarrow$' using $A \Leftrightarrow B \Leftrightarrow . A \Rightarrow B\ \&\ B \Rightarrow A$.

(ii) Replace '$\sim$' by '$\neg$', '$\Rightarrow$' by '$\rightharpoonup$', '&' by '$\&_1$', and '$\vee$' by '$\vee_1$' throughout.

In this way the proof that $v(B) = t$ in S_6 is transformed into a proof of TB in ${}_1S_6$. For each valuation rule is replaced by a theorem of ${}_1S_6$, and the theorems on entailment '$\Rightarrow$' are a proper subset of those for '$\rightharpoonup$', over significant wff. Hence $\vdash TB$ in ${}_1S_6$ and so $\vdash B$ in ${}_1S_6$.

This style of completeness proof works for a large class of many-valued logics, and reveals how *a complete axiomatisation of any 3-valued significance logic which includes* S_2, for example, *can be read off the formalised semantics of the system* using the relations:

$$v(B) = t \equiv . TB;\ v(B) = f \equiv . FB;\ v(B) = n \equiv . \sim SB \text{ (or } \neg SB).$$

Yet a third completeness proof for ${}_1S_6$, by a maximal consistent set construction, is a special case of the proof given in chapter 7 for quantificational developments of ${}_1S_6$ and for modal extensions of S_6.

8 Several metatheorems established for ${}_1S_2$ extend to the stronger S-systems we have studied; for example,

Equivalence theorem: If B results from A by substitution of B_1 for A_1 at one or more places in A, and if $\vdash A_1 \simeq B_1$ then $\vdash A \simeq B$.

A proof using completeness results parallels that in Church [4] (p. 101).

Substitutivity of equivalence: If B results from A by substitution of B_1 for A_1 at one or more places in A, and if $\vdash (A_1 \simeq B_1)$ and $\vdash A$ then $\vdash B$.

Proof: By the equivalence theorem, $\vdash (A \simeq B)$. From $\vdash p \simeq q \rightharpoonup . p \rightharpoonup q$, by modus ponens, $\vdash (A \rightharpoonup B)$. Then by modus ponens, $\vdash B$.

Replacement theorem: If $\vdash (A \equiv B)$ in the classical sentential calculus and every variable which occurs in A occurs in B and conversely then $\vdash (A \simeq B)$.

Proof: If $\vdash (A \equiv B)$ in classical logic then A and B contain only classical connectives. Thus if one variable in A has value n then A has value n; since by hypothesis that variable occurs in B it too has value n. When no variable in A has value n no variable in B has value n and conversely. Then both A and B have value t or both have value f. Therefore in any case (for any value assignments) A and B either both have value n, or both have value t or both have value f. Hence $(A \simeq B)$ is a tautology, and therefore $\vdash (A \simeq B)$.

Since the relation $Tp \simeq p$ holds if and only if Sp, the following definition can be adopted:

$$A \text{ is 2-valued} =_{\mathrm{Df}} TA \simeq A.$$

It follows that classical logic is a special case of S-logics. For classical $\{\&, \vee, \sim\}$-logic holds for 2-valued wff.

9 *Paradox resolution in S-logics*

In S-logics including $\mathbf{S}_2$ various paradox-resolving theorems can be proved, in particular:

$\vdash p \simeq \sim p \supset . \sim Sp$; $\vdash p \simeq \sim p \supset . \sim S \sim p$
$\vdash p \simeq \sim p \simeq . \sim Sp$; $\vdash p \simeq \sim p \simeq . \sim S \sim p$
$$\left.\begin{array}{l}\vdash p \equiv \sim p \rightharpoonup . \sim Sp \\ \vdash p \equiv \sim p \rightharpoonup . \sim S \sim p\end{array}\right\} \text{but the converses fail.}$$
$\vdash p \simeq q \supset . Sp \simeq Sq$

It is easy to see how, applying certain of these principles, Russell was able to conclude that it is not significant to say that the class of all classes not members of themselves is (or is not) a member of itself; and were he working in the context of such a significance logic his reasoning would be correct. It would be a mistake however to think that the *weaker* of these theorems show very much, or support a significance resolution of paradoxes; for also $\vdash p \equiv \sim p \rightharpoonup . q$.

S-logics do, moreover, have a real importance for the consistent resolution of logical and semantical paradoxes, especially since one of the few philosophically viable resolutions of these paradoxes is a significance resolution according to which paradox-generating sentences are not significant. And even at the sentential stage some results relevant to the consistent resolution of semantical paradoxes can be obtained. In particular, any form of self-reference reducible to the form

$$c\colon\ g(c)$$
i.e. in $\mathbf{S}_6$ $\quad c \simeq g(c)$

where 'g' is a predicate and 'c' a constant, can be consistently admitted into an S-logic provided

(α) predicate 'g' is definable using just classical connectives.

This includes not only simple cases of the liar paradox, formulated with *classical* truth and falsity (see §5.10), of the form

c: *c* is false

but also chain self-reference as is illustrated by the pair

c: *d* is false
d: *c* is true

for this is reducible by substitution to

c: (*c* is true) is false.

Self-reference theorem 1. The system obtained from $\mathbf{S}_6$ or any subsystem of $\mathbf{S}_6$ by adding the constant *c* and the axiom

$$(1) \quad c \simeq g(c)$$

whatever '*g*' so long as proviso (α) is met, is consistent.

Proof: Since '*f*' is definable in $\mathbf{S}_6$ using only classical connectives:

$$g(c) \simeq . A \,\&\, c \vee B \,\&\, {\sim} c$$

where *A* and *B* are independent of *c* and take values *t*, *f*, *n*. This is an immediate extension of Boole's law of development (see Church [4], p. 155 and Lewis [1], p. 133). Thus, by (1)

$$c \simeq . A \,\&\, c \vee B \,\&\, {\sim} c.$$

The theorem then follows since the relation is always satisfiable by assigning *c* value *n*. Provided neither of *A* and *B* have value *n* and *B* does not have value *t* when *A* has value *f*, the relation is also satisfiable for other values of *c*.

This self-reference theorem can be extended to admit predicates definable using other classes of connectives. For example it can be extended to predicates definable using just clean connectives, where a connective is *clean* if it is definable in terms of classical connectives and '*S*'. All connectives of $\mathbf{S}_1$ are clean, and all classical connectives are clean. But not all connectives are clean, *e.g.* '¬' and '*D*' are not clean.

Self-reference theorem 2. The system obtained from any system included in $\mathbf{S}_6$ by adding the constant '*c*' and axiom (1) is consistent provided '*g*' is definable using only clean connectives.

Proof: Consider any such function of *c*. Since all connectives occurring in its expression are clean it can be transformed preserving '&', '∨', '∼', '*S*'. Now if '*S*' covers a compound expression, use compounding theorems for '*S*' to distribute '*S*' into elementary components. If any expression of the form *SSp* appears replace it by *t*, and replace $\sim(t)$ by *f*, $(t \vee q)$ by *t*, $(t \,\&\, q)$ by *q*, $(f \vee q)$ by *q*, and $(f \,\&\, q)$ by *f*. In this way any function of *c* can be transformed into an expression with the following basic expressions: *c*, or $\sim c$, or $(c \,\&\, {\sim} c)$, or *Sc*, or $\sim Sc$, or $Sc \,\&\, {\sim} Sc$, or expressions independent of *c*. Thus after applying distributive laws to classical connectives, the general form of a function in *c*, after coefficients have been collected, can be expressed:

$$g(c) \simeq . A' \,\&\, c \vee B' \,\&\, {\sim} c \vee C' \,\&\, (c \,\&\, {\sim} c) \vee E' \,\&\, Sc \\ \vee G' \,\&\, {\sim} Sc \vee H' \,\&\, (Sc \,\&\, {\sim} Sc) \vee J'$$

where the coefficients are independent of c. Now $(Sc \;\&\; {\sim}Sc)$ always has value f, so unless H' has value n the compound can be discarded (otherwise if J' has value n, then c, where $c \simeq g(c)$, has value n). Unless $(c \;\&\; {\sim}c)$ has value n, $(c \;\&\; {\sim}c)$ has value f; so that unless C' has value n, the component $C' \;\&\; (c \;\&\; {\sim}c)$ can be omitted. (If either C' or $(c \;\&\; {\sim}c)$ has value n, then c can be assigned value n.) Finally since

$$J' \simeq .\, J' \;\&\; Sc \vee H' \;\&\; {\sim}Sc,$$

the theorem is established if the relation

$$c \simeq g(c) \simeq (A \;\&\; c \vee B \;\&\; {\sim}c \vee C_1 \;\&\; {\sim}Sc \vee E \;\&\; Sc),$$

with A, B, C_1, E independent of c, is satisfiable, and hence consistent with ${}_1\mathbf{S}_6$.

The following cases are exhaustive:

(i) If any of A, B, C_1 or E has value n, then c can have value n. So for remaining cases it is supposed that none of A, B, C_1 and E have value n.

(ii) If C_1 and E both have value f, then

$$c \simeq g(c) \simeq A \;\&\; c \vee B \;\&\; {\sim}c.$$

By self-reference theorem 1, this relation is satisfiable.

(iii) If C_1 and E both have value t, or C_1 has value f and E has value t, then c has value t.

(iv) If C_1 has value t and E has value f, then if B has value f, c can have value f, and if B has value t, c can have value n.

These self-reference theorems suggest ways of consistently extending ${}_1\mathbf{S}_6$ to a protothetic without types; namely, by qualifying the function abstraction axiom scheme:

$$(P\lambda g)(U\S p)(\S p \; i \; \lambda g \simeq A(\S p)),$$

by which new functions are introduced, with the proviso that wff A (which does not contain 'λg' free) contains only classical, or clean, connectives. Such a type-free protothetic is indeed consistent, but has severe limitations since connectives such as 'F' and '$\neg$' cannot occur in A.

Consistent significance class logics can similarly be obtained by limiting the class of connectives and quantifiers that may appear on the right of the class abstraction scheme:

$$(Py)(Ux)(x \in y \simeq A(x)),$$

where y is not free in A. An important result here is Brady's [1] theorem according to which the class logic (with the axiom of extensionality), where $A(x)$ contains only strong regular connectives and the universal quantifier corresponding to the conjunction '$\cap$', is consistent (relative to a fragment of Zermelo-Fraenkel set logic).

10 *Some Subsystems of* ${}_1\mathbf{S}_6$ *based on* $\{\rightharpoonup, \neg\}$

The 3-valued system $\mathbf{S}_{6\cdot 0}$ with primitive connectives '$\rightharpoonup$' and '$\neg$' and postulates 6.1, 6.2, 6.3, R6.1 and R6.2 coincides syntactically with classical sentential logic. From this possibility of reinterpretation of $\mathbf{S}_{6\cdot 0}$ it can be seen that the connective 'S' is not definable in $\mathbf{S}_{6\cdot 0}$. For since the logic would have to bring out as a theorem $S\neg p$ and '$\neg$' is under 2-valued reinterpretation an involution operator, 'S' would have to convert both values, t and f, of the 2-valued interpretation to t. Then, however, Sp would be a theorem, contradicting the intended significance interpretation. Likewise 'F' is not definable in $\mathbf{S}_{6\cdot 0}$ though 'T' is, using $Tp \simeq \neg\neg p$. For 'F' and 'S' are interdefinable with respect to 3-valued connective set $\{\rightharpoonup, \neg\}$, thus:

$$FA =_{\mathrm{Df}} \neg(SA \rightharpoonup TA)$$
$$SA =_{\mathrm{Df}} \neg TA \rightharpoonup FA.$$

A second subsystem $\mathbf{S}_{6\cdot 1}$ has then as primitives the set $\{\neg, \rightharpoonup, S\}$—or the set $\{\neg, \rightharpoonup, F\}$. A complete axiomatisation of $\mathbf{S}_{6\cdot 1}$ may be read off the valuation rules for its connectives, in the way we have explained. It is then easy to show that $\mathbf{S}_{6\cdot 1}$ may be *completely* axiomatised by adding to the postulates of $\mathbf{S}_{6\cdot 0}$ these axioms:

$$\vdash S(p \rightharpoonup p);\quad \vdash S(p \rightharpoonup q) \rightharpoonup .\, p \rightharpoonup Sq;\quad \vdash Sq \rightharpoonup S(p \rightharpoonup q);\quad \vdash S\neg p;\ \vdash SSp.$$

In other words $\mathbf{S}_{6\cdot 1}$ is obtained from ${}_1\mathbf{S}_6$ by omitting 6.8 and 6.9 and adding instead $\vdash SSp$. However $\mathbf{S}_{6\cdot 1}$ lacks classical connectives.

Theorem (54): '$\sim$' is not definable in terms of $\{\rightharpoonup, \neg, S\}$.

Proof: Induction over the number of connectives in the definiens establishes the following sufficient hypothesis: there is no one-place function $\Phi(P)$ of P such that $\Phi(P)$ has value n when P has n and value f when P has value t. Since the induction basis is immediate and the induction step is immediate in the case of unlimited connectives '$\neg$' and 'S', it remains to consider the case where $\Phi(P)$ is of the form $\Phi_1(P) \rightharpoonup \Phi_2(P)$. Then because of the matrix for '$\rightharpoonup$', $\Phi_2(P)$ as well as $\Phi(P)$ must violate the hypothesis, contradicting the induction hypothesis.

Our next subsystem $\mathbf{S}_{6\cdot 2}$ has then connective set $\{\rightharpoonup, \neg, \sim\}$. The connectives '$F$' and '$S$' can be defined: $FA =_{\mathrm{Df}} T\sim A$, and $SA =_{\mathrm{Df}} \neg A \rightharpoonup \neg\neg\sim A$. That significance is definable in a two-negation system using just an implication—'$\supset$' for instance will do as well as '$\rightharpoonup$'—reveals that those who try to eliminate nonsignificance in favour of falsity (see Introduction §2) but who, in order to account for the facts, admit two suitable negations, are thereby committed to a significance theory. A complete axiomatisation of $\mathbf{S}_{6\cdot 2}$ can once again be read off the valuation rules for its connectives, and this axiomatisation too can be replaced by more economical ones.

'Q' of course is not definable in terms of $\{\rightharpoonup, \neg, \sim\}$. For no way of converting t or f to n is provided by $\{\rightharpoonup, \neg, \sim\}$. Further subsystems can accordingly be constructed starting with the system $\mathbf{S}_{6\cdot 3}$ with primitive set $\{\rightharpoonup, \neg, Q\}$ and adding 'F'. The complete system obtained by adding '$\sim$' to either $\mathbf{S}_{6\cdot 3}$ or to system $\mathbf{S}_{6\cdot 4}$ based on $\{\rightharpoonup, \neg, Q, F\}$ coincides with ${}_1\mathbf{S}_6$ itself. $\mathbf{S}_{6\cdot 3}$ and $\mathbf{S}_{6\cdot 4}$ are however distinct systems in view of the following results.

Theorem (55): 'F' is not definable in terms of $\{\rightharpoonup, \neg, Q\}$.
Proof establishes by induction the following hypothesis: there is no function $\Phi(P)$ of P which converts just one of f and n to t.

Theorem (56): '$\sim$' is not definable in terms of $\{\rightharpoonup, \neg, Q, F\}$.
Proof consists in establishing by induction the hypothesis: there is no function $\Phi(P)$ of P such that when P has values $\langle t,.,n\rangle$, $\Phi(P)$ has either values $\langle f,.,n\rangle$ or $\langle n,.,f\rangle$.

A complete axiomatisation of $\mathbf{S}_{6\cdot 4}$ may be read off its semantics.

The subsystems of $\mathbf{S}_6$ we have noticed are by no means exhaustive. For example, another system based on connectives we have been considering is that built on set $\{\rightharpoonup, Q, \sim\}$.

Certain subsystems of $\mathbf{S}_6$ which include $\{\rightharpoonup, \neg, S\}$ are of particular interest as intermediate sentential significance logics. For they are one-sorted, they are easily provided with complete axiomatisations, they have easy quantificational extensions, and yet they can avoid certain connectives which are difficult to interpret and control, such as 'Q'.

11 *The equivalence of* ${}_1\mathbf{S}_5$ *and* ${}_1\mathbf{S}_6$

First, ${}_1\mathbf{S}_5$ deductively includes ${}_1\mathbf{S}_6$ but not conversely, *i.e.*

Theorem (57): Every theorem of ${}_1\mathbf{S}_6$ is a theorem of ${}_1\mathbf{S}_5$, but not conversely.
For the postulates of ${}_1\mathbf{S}_6$ were derived in ${}_1\mathbf{S}_5$, but restricted variable theorems of ${}_1\mathbf{S}_5$ are not found in ${}_1\mathbf{S}_6$.

To show that ${}_1\mathbf{S}_5$ is, in a strong sense, included in ${}_1\mathbf{S}_6$ some device is needed for *representing* S-restricted variables in ${}_1\mathbf{S}_6$. We say that a wff

$$(\alpha) \quad A(R_1, \ldots, R_n)$$

of ${}_1\mathbf{S}_5$ is *represented* in ${}_1\mathbf{S}_6$ by the wff

$$(\alpha^*) \quad SP_1 \rightharpoonup . \; SP_2 \ldots . \; SP_n \rightharpoonup A(P_1, \ldots, P_n),$$

where $A(P_1, ..., P_n)$ is obtained from $A(R_1, ..., R_n)$ by simultaneous substitution of $P_1, ..., P_n$ for $R_1, ..., R_n$, where $R_1, ..., R_n$ are all the distinct S-restricted variables, if any, of $A(R_1, ..., R_n)$ and $P_1, ..., P_n$ are the first distinct S-unrestricted variables in alphabetical order that do not occur in $A(R_1, ..., R_n)$. (α^*) is the ${}_1\mathrm{S}_6$ *representative* of (α).

Theorem (58): Where B is a wff of ${}_1\mathrm{S}_5$, B is a theorem of ${}_1\mathrm{S}_5$ iff its representative is a theorem of ${}_1\mathrm{S}_6$, after definitional substitution for connectives.

Proof: The primitive connectives of ${}_1\mathrm{S}_5$ can be defined in terms of those of ${}_1\mathrm{S}_6$ since ${}_1\mathrm{S}_6$ has a functionally complete set. Specifically, we may define the connectives as follows (*cf. Theorem* (19) of §5.14):

$$\sim A =_{\mathrm{Df}} (\neg A \rightharpoonup DA) \rightharpoonup D\neg A;\ QA =_{\mathrm{Df}} D{\sim}DA;$$
$$TA =_{\mathrm{Df}} \neg\neg A;\ A \supset B =_{\mathrm{Df}} (A * B) + QSA,$$

where $A \vee_1 B =_{\mathrm{Df}} (A \rightharpoonup B) \rightharpoonup B$; $A \,\&_1 B =_{\mathrm{Df}} \neg(\neg A \vee_1 \neg B)$; $A + B =_{\mathrm{Df}} {\sim}({\sim}A \vee_1 {\sim}Q(A \,\&_1 B))$; $A \supset_{\mathrm{L}} B =_{\mathrm{Df}} (A \rightharpoonup B) + ({\sim}B \rightharpoonup {\sim}A)$; $A * B =_{\mathrm{Df}} [{\sim}({\sim}A \,\&_1 {\sim}B) \rightharpoonup . (\neg{\sim}(A \rightharpoonup B)) + (A \vee_1 B)] + (A \supset_{\mathrm{L}} B)$. Now if B is a wff of ${}_1\mathrm{S}_5$ with representative B^* in ${}_1\mathrm{S}_6$ and B^* is a theorem of ${}_1\mathrm{S}_6$, then B^* is a theorem of ${}_1\mathrm{S}_5$ by *Theorem* (57). By substitution then, where $R_1, ...,$ R_n were the variables supplanted by $P_1, ..., P_n$ in obtaining B^* from B, $\mathrm{S}^{P_1 \ldots P_n}_{R_1 \ldots R_n} B^* |$ is also a theorem. Since however $\vdash SR_i$ for each i, $1 \leq i \leq n$, by $\rightharpoonup$-detachment B is a theorem of ${}_1\mathrm{S}_5$. Conversely if B is a theorem of ${}_1\mathrm{S}_5$ then we show by induction over the proof of B that its representative B^* is a theorem of ${}_1\mathrm{S}_6$. Since $Sp \rightharpoonup . Sq \rightharpoonup . p \supset . q \supset p$, $Sp \rightharpoonup . Sq \rightharpoonup . Sp' \rightharpoonup . p \supset (q \supset p') \supset . p \supset q \supset . p \supset p'$ and $Sp \rightharpoonup . Sq \rightharpoonup . {\sim}p \supset {\sim}q \supset . q \supset p$ are theorems of ${}_1\mathrm{S}_6$, since they are S_6-tautologies, the first three axioms of ${}_1\mathrm{S}_5$ are represented by theorems in ${}_1\mathrm{S}_6$, and similarly for the remaining axioms of ${}_1\mathrm{S}_5$.

Since uniform substitution is a rule of ${}_1\mathrm{S}_6$ it remains to show that restricted substitution, modus ponens and T- rules can be represented as derived rules of ${}_1\mathrm{S}_6$. R2.1 is immediate for zero substitution. In the case of non-zero substitution R2.1 is represented in ${}_1\mathrm{S}_6$ by:

(1) $\vdash SP_1 \rightharpoonup \ldots \rightharpoonup . SP_n \rightharpoonup A, \vdash SP'_1 \rightharpoonup \ldots \rightharpoonup . SP'_m \rightharpoonup SB \twoheadrightarrow$
$\vdash SP'_1 \rightharpoonup \ldots \rightharpoonup . SP_n \rightharpoonup \mathrm{S}^{P_i}_B A |$,

where A has n restricted variables upon one of which (replaced by P_i) restricted substitution is made, and where B has m restricted variables. Because commutation holds in ${}_1\mathrm{S}_6$ the order in which significance qualifications occur is immaterial. (1) is a derivable rule of ${}_1\mathrm{S}_6$: for

$\vdash SP_1 \rightharpoonup \ldots \rightharpoonup SP_n \rightharpoonup . SP_i \rightharpoonup A \twoheadrightarrow \vdash \mathrm{S}^{P_i}_B (SP_1 \rightharpoonup \ldots \rightharpoonup . SP_i \rightharpoonup A) |$; R6.1

$\twoheadrightarrow \vdash SP_1 \rightharpoonup \ldots \rightharpoonup S^{P_i}_B(SP_i \rightharpoonup A)|$, by rules for S since variables are distinct

$\twoheadrightarrow \vdash SB \rightharpoonup . SP_1 \rightharpoonup \ldots \rightharpoonup . SP_n \rightharpoonup S^{P_i}_B A|$, by substitution and commutation. (1) now follows by **RSL**.

The proof that the second part of R2.2 can be represented by a derived rule of ${}_1\mathbf{S}_6$ is just a special case of the argument for R2.1. That the modus ponens rule R2.3 is a derived rule of ${}_1\mathbf{S}_6$ follows from the matrix for '$\supset$' and the completeness theorem. The extension required for representability $\vdash SP_1 \ldots \rightharpoonup . SP \rightharpoonup A, \vdash SP'_1 \rightharpoonup \ldots \rightharpoonup . SP \rightharpoonup (A \supset B) \rightarrow \vdash SP'_1 \ldots \rightharpoonup . SP' \rightharpoonup B$ follows then by modus ponens for '$\rightharpoonup$' and the deduction theorem. The T-rule is represented in ${}_1\mathbf{S}_6$ by a rule of the form

(1) $\vdash SP_1 \ldots \rightharpoonup . SP_n \rightharpoonup . A_1 \& \ldots \& A_k \supset . B_1 \vee \ldots \vee B_m$
$\twoheadrightarrow \vdash S^{R_1}_{P_1} \ldots ^{R_n}_{P_n} TA_1 \& \ldots TA_k \supset . TB_1 \vee TB_2 \vee \ldots TB_m|$,

where the n restricted variables of $A_1 \ldots, A_k, B_1 \ldots, B_m$ are replaced by $P_1 \ldots, P_n$.

It suffices to show

(2) $\vdash SP_1 \rightharpoonup \ldots \rightharpoonup . SP_n \rightharpoonup . A_1 \& \ldots \& A_k \supset . B_1 \vee \ldots B_m \rightharpoonup .$
$TA_1 \& \ldots TA_k \supset . TB_1 \vee \ldots TB_m$

since (1) then follows from (2) by R6.2 and applications of R6.1. But (2) is a theorem of ${}_1\mathbf{S}_6$. For if SP_i is true for all i, $1 \leqslant i \leqslant n$, then SA_j and SB_l for all j and l, $1 \leqslant j \leqslant k$, $1 \leqslant l \leqslant m$, by the proviso on the T-rule. Since $SB \rightharpoonup . TB \simeq B$, the value of the antecedent of (2) is the same as the value of the consequent, and (2) follows from $(B \rightharpoonup B)$. If however for some i, SP_i is false the antecedent of (2) has value t. Also either some A_j or some B_i and some A_j have value n by the T-rule proviso. Therefore either TA_j has value f or both TA_j and TB_i have value f. In either case the consequent of (2) has value t, so (2) has value t. Thus (2) is a tautology and so a theorem.

Corollary: If B is a theorem of $\mathbf{S}_1$, $\mathbf{S}_2$ or $\mathbf{S}_3$ then its representative is a theorem of ${}_1\mathbf{S}_6$.

Thus unlike the step from $\mathbf{S}_3$ down to $\mathbf{S}_2$ say, there is no loss in expressive power in adopting $\mathbf{S}_6$ instead of $\mathbf{S}_5$ as a sentential basis on which to build a significance theory—and S-restricted variables can be added to $\mathbf{S}_6$ as meta-theoretically dispensable adjuncts.

12 *Connections between* **C**-*logics and* ${}_1\mathbf{S}_6$

$\mathbf{C}_1$ is syntactically a subsystem of ${}_1\mathbf{S}_6$. For consider the mapping $\mathcal{M}$ which maps '$\supset$' to '$\rightharpoonup$', '$\sim$' to '$\urcorner$' and 'S' to 'V', where $VB =_{\mathrm{Df}} B \rightharpoonup B$. Using completeness proofs it can be shown that every theorem of $\mathcal{M}(\mathbf{C}_1)$ is a theorem of ${}_1\mathbf{S}_6$; *e.g.* $\mathcal{M}(p \supset Sp)$ *i.e.* $p \rightharpoonup Vp$, is an $\mathbf{S}_6$-tautology. But such a mapping, though less trivial than the mapping of one denumerable set of theorems into another which holds between all logistic systems, is without much interest since it does not preserve even approximately the intended interpretation of 'S'. Indeed no 1-1 mapping does.

Theorem (59) : There is no 1–1 mapping of $\mathbf{C}_1$ into $\mathbf{S}_6$ which preserves the matrix for 'S'.

Proof: To preserve the matrix for 'S' and to keep analogues of **RSL** theorems the mapping $\mathscr{M}$ must map '$\supset$' to '$\rightharpoonup$', '$\sim$' to '$\urcorner$' and 'S' to 'S', where '$\urcorner$' and 'S' have the valuations assigned for ${}_1\mathbf{S}_6$ and '$\rightharpoonup$' has matrix

$\rightharpoonup$	t	t	t
t	t	$\bar{t}$	$\bar{t}$
f	t	t	t
n	t	t	t

by *Theorem* (17) of §5.14.

But then $\mathscr{M}(S{\sim}p \equiv Sp)$ is not a theorem.

The reason why simple mapping connections between $\mathbf{C}_1$ and ${}_1\mathbf{S}_6$, or more generally between **C**-logics and **S**-logics, are precluded can be traced back to the difference in number of designated values. **S**-logics cannot consistently be reinterpreted (without change of postulates) as logics for which both t and n are designated values. Consider the disastrous consequences if ${}_1\mathbf{S}_6$ were so reinterpreted. $(p \rightharpoonup Dp)$ would be valid under the interpretation. Consequently, by ${}_1\mathbf{S}_6$, $(\urcorner D \urcorner p \rightharpoonup p)$, and so p itself, would be valid. ${}_1\mathbf{S}_6$ would be absolutely inconsistent under this "interpretation". What can be done however is what we have already done, namely to separate out the good theorems of certain **C**-logics and show that the resulting systems are **S**-logics.

Weaker **C**-logics can moreover be related to a rule-simplified system $\mathbf{D}_6$, with connective set $\{\rightharpoondown, F, Q\}$. $\mathbf{D}_6$ provides a complete axiomatisation of $\mathbf{C}_6$, where the primitive connectives of $\mathbf{C}_6$ are defined as in *Theorem* (22) of §5.14. $\mathbf{D}_6$ facilitates quantificational extensions of **C**-logics; for as it has an unmodified modus ponens rule it avoids the difficulties Halldén [1] finds in extending his modified modus ponens rule to quantified logic. $\mathbf{D}_6$ has these postulates:

Axioms of $\mathbf{D}_6$

D6.1 $\vdash p \rightharpoondown . q \rightharpoondown p$

D6.2 $\vdash p' \rightharpoondown (p \rightharpoondown q) \rightharpoondown . p' \rightharpoondown p \rightharpoondown . p' \rightharpoondown q$

D6.3 $\vdash Fq \rightharpoondown Fp \rightharpoondown . p \rightharpoondown q$

D6.4 $\vdash QFp$

D6.5 $\vdash Q(p \rightharpoondown Fp)$

D6.6 $\vdash p \rightharpoondown QQp$

D6.7 $\vdash QQp \rightharpoondown p$

D6.8 $\vdash Qp \rightharpoondown . Qq \rightharpoondown Q(p \rightharpoondown q)$

D6.9 $\vdash (Q(p \rightarrow q) \rightarrow p) \rightarrow p$
D6.10 $\vdash p \rightarrow . Qp \rightarrow . Q(p \rightarrow q) \rightarrow Qq$
D6.11 $\vdash p \rightarrow . Qp \rightarrow . Q(q \rightarrow p) \rightarrow Qq$
D6.12 $\vdash FQp \rightarrow . Q(p \rightarrow q) \rightarrow F(p \rightarrow q) \rightarrow Qq$

As transformation rules $_1\mathbf{D}_6$ has uniform substitution and modus ponens for '$\rightarrow$'.

Formal development of $\mathbf{D}_6$

1 Classical sentential logic holds for wff containing just '$\rightarrow$' (for classical implication) and 'F' (for classical negation) and for substitution-instances of these wff.

2 Theorems required for a completeness proof are listed in an order in which they are readily derivable.

T1 $\vdash (QFp \rightarrow p) \rightarrow p$
T2 $\vdash Fp \rightarrow . p \rightarrow FQp$
T3 $\vdash FQp \rightarrow . Qp \rightarrow Fp$
T4 $\vdash FQ(p \rightarrow q) \rightarrow . Qq \rightarrow FQp$
T5 $\vdash Fp \rightarrow Qp$
T6 $\vdash FQ(p \rightarrow q) \rightarrow . Qp \rightarrow FQq$
T7 $\vdash FQ(p \rightarrow q) \rightarrow . FFQp \rightarrow FQq$
T8 $\vdash Fp \rightarrow . p \rightarrow q$

The characteristic connectives are defined as follows:

$$TA =_{\mathrm{Df}} F(A \rightarrow FQA); \quad SA =_{\mathrm{Df}} FFQA$$

T9 $\vdash Tp \rightarrow p$
T10 $\vdash Fp \rightarrow . FQ(p \rightarrow q) \rightarrow p$
T11 $\vdash Sp \rightarrow . p \rightarrow Tp$
T12 $\vdash Tp \rightarrow FFp$
T13 $\vdash FQp \rightarrow FFp$
T14 $\vdash FQp \rightarrow p$
T15 $\vdash Fp \rightarrow TFp$
T16 $\vdash Tp \rightarrow TQp$
T17 $\vdash Fp \rightarrow T(p \rightarrow q)$
T18 $\vdash Fp \rightarrow FQQp$
T19 $\vdash Tp \rightarrow . Tq \rightarrow T(p \rightarrow q)$
T20 $\vdash Tp \rightarrow q \rightarrow . Fp \rightarrow q \rightarrow . FQp \rightarrow q \rightarrow . q$
T21 $\vdash FQp \rightarrow . Fq \rightarrow F(p \rightarrow q)$
T22 $\vdash Tp \rightarrow . Fq \rightarrow F(p \rightarrow q)$
T23 $\vdash Tp \rightarrow . FQq \rightarrow FQ(p \rightarrow q)$
T24 $\vdash FQp \rightarrow . Tq \rightarrow FQ(p \rightarrow q)$
T25 $\vdash FQp \rightarrow . FQq \rightarrow T(p \rightarrow q)$

3 Every theorem of $\mathbf{D}_6$ is a theorem of $\mathbf{C}_6$; hence $\mathbf{D}_6$ is absolutely consistent.

4 Every theorem of $\mathbf{C}_6$ is a theorem of $\mathbf{D}_6$; hence $\mathbf{D}_6$ is complete. For
(a) Let B be a wff of $\mathbf{D}_6$. Let $P_1, P_2, ..., P_n$ be the distinct variables occurring in B. Let A_i be TP_i, FP_i or FQP_i according as the value a_i of P_i is t, f or n. Let B' be TB, FB or FQB according as the value B for values $a_1, ..., a_n$ of $P_1, ..., P_n$ is t, f or n.
Then $A_1, A_2, ..., A_n \vdash B'$.
(b) If B is a $\mathbf{C}_6$-tautology then B is a theorem of $\mathbf{D}_6$.

Using 3 and 4 it follows

5 $\mathbf{C}_5$, $\mathbf{C}_6$ and $\mathbf{D}_6$ are deductively equivalent, and $\mathbf{C}_1$, $\mathbf{C}_2$ and $\mathbf{C}_3$ are sub-logics of $\mathbf{D}_6$. Thus **C**-logics can be related to an explicitly 3-valued logic $\mathbf{D}_6$, with unrestricted rules. Much as the qualifications of $\mathbf{S}_5$ on substitution were eliminated by changing the connective set, so qualifications on modus ponens of lower **C**-logics can be eliminated by the same strategy.

CHAPTER 7

Intensional and Quantified Significance Logics

7.1 EXTENDING SENTENTIAL SIGNIFICANCE LOGICS

If we are to make good our claims that significance logics are as viable, formally and interpretationally, as classical two-valued logics, and that they provide a better basis for philosophical investigation, we can hardly rest content with the extensional sentential systems we have so far developed. For there are severe limits on what we can express and formalise within them. Many of the important philosophical problems associated with significance claims cannot be examined at this level. Thus the issue of the non-contingency of significance claims only appears properly when the logics are modalised, and the matter of people believing absurdities when the logics are intensionalised. And major issues such as those that centre on categories and significance ranges only emerge formally when subject-predicate analysis is undertaken. The development of a logic of sentence parts is, of course, necessary also for the standard reasons that the validity of arguments such as 'Horses are animals and Dobbin is a horse, therefore Dobbin is an animal', and the special relations which hold between sentences like 'Tom is happy', 'Someone is happy' and 'Everyone is happy', cannot be assessed in terms of logics where the only relations considered are those which hold between whole sentences.

Development beyond the sentential level, however, raises several new problems. Indeed it is sometimes thought that quantified significance logics cannot be constructed at all, or at all satisfactorily (*cf.* Strawson [2], ch. 8). Even Halldén [1] asserts "At the present it is impossible to develop a connected and comprehensive theory for this field (*i.e.* quantificational logic of significance and nonsense)". If, however, significance theory is to be in genuine competition with other logical systems, it is crucial that such logics be developed.

Most important, from our point of view, a logic of sentence parts is necessary as a preliminary to the formal elucidation of factor theories of significance, which we investigate in Part III, since in terms of these, the significance or nonsignificance of any given sentence depends on the couplings of sentential components. Such theories are exemplified by Russell's type-theory and Ryle's theory of categories. By contrast, unit theories of significance, exemplified by the logical positivists, do not presuppose an analysis of sentential components as a preliminary to the *significance* claims which are made, though presumably some sort of predicate logic has to be adopted for the ordinary analysis of arguments. What seems to be an outcome of the positivist position is that once the significance or nonsignificance of a sentence has been determined by means of the criterion of verification, and the consequences of this determination examined by means of a preferred sentential logic selected from those developed in Chapters 5 and 6, the nonsignificant sentences are eliminated and the analysis of arguments at both the predicate and sentential levels is then carried out in the two-valued classical logic. If, for example, it is shown that a given sentence 'p_0' is nonsignificant, and principles such as,

$$\sim Sp \supset \sim S(p \,\&\, q); \quad \sim Sp \supset \sim S(p \vee q): \quad \sim Sp \supset \sim S(p \supset q);$$

which characterize classical connectives, are adopted, then all sentences of the form,

$$p_0 \,\&\, q; \quad p_0 \vee q; \quad p_0 \supset q; \quad etc.,$$

are nonsignificant and can be eliminated. Sentences which remain are significant, and classical two-valued logic applies.

This use of significance principles as principles of elimination is not of course restricted to unit theories, but except in the case of unit theories the technique cannot be applied before the development of a three-valued predicate theory and subsequent range theory, since the determination of whether or not a given sentence is nonsignificant cannot, in the case of factor theories, be made at the level of whole sentences.

It is however arguable whether even the positivists can carry out their programme successfully in terms of a significance theory restricted to sentential principles. For it is quite unclear how the criterion of verification is meant to apply to quantified sentences. Is it the case, for example, that a universal sentence of the form,

$$\text{for all } x, \; f_0x \supset g_0x,$$

is not in principle verifiable (hence nonsignificant) if there is one and only one value of *qu*(x), say 'x_0', such that,

$$f_0x_0 \supset g_0x_0,$$

is nonsignificant? This is only a particular aspect of the general problem concerning the verifiability of universal sentences, but it does illustrate the difficulty of applying crude unit theories. Since some units are quantified sentences, some rules have to be given for determining their significance value; and once this is done, a two-valued predicate logic is quite inadequate.

The mere adoption of a three-valued predicate logic, however, does not of itself imply a factor theory since we may still say, for example, that the significance of a sentence such as 'f_0x_0' is to be determined as a whole (in terms of its verifiability, *e.g.*) rather than, as in the case of a factor theory, by examining whether or not there is a clash or incorrect coupling between 'f_0' and 'x_0'. And on such a basis it is possible to give rules in terms of which the significance value of quantified sentences can be determined. As we shall see, however, many of the problems which arise in the actual development seem to demand some sort of factor solution; and from this point of view unit theories appear over-simple. It is in general the case, however, that although the predicate logics developed in this chapter stand as a necessary preliminary to the elucidation of factor theories, they do not necessarily presuppose a factor theory.

One of the first problems we have to face concerns the sentential basis of the modal and quantification theory. For any such logic, of however many values, can only be built on a sub-logic of sentential principles. In the case of two-valued classical logic this is not a problem since there is only one full extensional sentential basis, the classical **RSL**, which can be chosen. Of course if this is supplanted by some non-classical implicational or entailment system then the quantification logic has to be erected on a different basis. But in the case of a three- or higher-valued logic there is an embarrassment of choice. And in particular, for three-valued significance quantification theory, we may take as a basis any one of the sentential systems developed in Chapters 5 and 6. We could of course build separate quantification theories on each sentential logic we have considered, but this is pointless in view of some of the results which have been reached. In particular, it now seems to us unnecessary to develop **C**-logics further. For if we are to avoid simple inconsistency with respect to '$\sim$' and '$\urcorner$', we must take $\mathbf{C}_4$ or a weaker system as the basis; and if we are to avoid systems which contain counter-intuitive principles, we cannot take a stronger system than $\mathbf{C}_0$ (*cf.* remarks in §5.8 ff.); but $\mathbf{C}_0$ is far too weak to be of any value as a significance logic.

We therefore confine attention to quantification and modal systems based on **S**-logics; and especially to the systems based on $\mathbf{S}_6$. For although $\mathbf{S}_6$ contains some unfamiliar connectives, such as 'Q' and 'D', it has the advantage that, being functionally complete, any required sentential connective can be defined. Again, since it is one-sorted, it is simpler to manipulate than the weaker two-sorted **S**-logics; and in any case, by reducing the number of primitive connectives, logics based on the weaker systems $\mathbf{S}_2$, $\mathbf{S}_3$, $\mathbf{S}_4$ and $\mathbf{S}_5$ can be obtained (see §7.8).

The three-valued quantified and intensionalised significance logics which we develop in this chapter provide a framework which is adequate for the expression of a non-negligible part of discourse, but even the most comprehensive system we examine still falls far short of a fully-developed version of the logic of semi-wffs foreshadowed in Chapter 3. There are several reasons for this. First, the basic restrictions of the sentential systems of Chapters 5 and 6 are carried over to quantificational and modal extensions. Thus, the role of context is neglected (though some indication of the way context can be treated is given in §7.2, §8 and §7.11, §4), since the theses of each system are asserted parametrically with respect to a pre-determined context *c*; and incomplete sentences are excluded or re-analysed as three-valued. But the most important reason is that one of our main concerns here is to produce a viable and detailed semantic theory for each system we develop. A full semantic theory for the logic of the semi-wffs of Chapter 3, however, would be extremely complex—compare, for example, the problems of developing standard model theory to treat of the large variety of subject terms (clauses, quotation expressions, attributives, descriptions, *etc.*) which are allowed as values of the subject variables in semi-wffs of the sort $F(X_1, ..., X_n)$. We do make some progress towards the solution of these problems, but it is necessary to proceed in a step-by-step way and even then we do not fully achieve our goal. The second-order theory which we develop, for example, which we call second-stage logic, cannot contain both clauses and sentences (as Chapter 3 logic does) and it can only contain an analogue of sentential quantification if it is interpreted as a clause logic. This halting progress is, however, a valuable means to an end. It provides the semantic basis for sections of the full logic of semi-wffs, embedded in **CL**, which we examine in greater detail in volume 2.

A different problem, which we leave on one side here but take up in the next volume, is the question of how actual sentences of the background language are to be systematically transformed into formulae of the logics we develop. A modified transformational grammar appears to provide the key to this problem.

Finally we note that many other interpretations, apart from the intended significance interpretation, can be given to the systems in this chapter. In particular, they can be interpreted assertorically instead of sententially, for we do not escape entirely the double interpretation discussed in §1.1 since significance assignments may be mapped back into assertions.

We begin with intensional logics and then move to quantification theory, but we vary the approach of Chapters 5 and 6, treating each system both semantically and axiomatically before we advance to the next. It would, however, be a simple matter to reorder our treatment, and develop the generalization of the matrix approach, model theory, for all the systems first before turning to the postulational theory. We vary, too, the metasymbolism of the earlier chapters. The reason for this is that we now need to make distinctions different from those made previously. As in Chapter 6, all the wffs we now consider are of the same kind (*i.e.* all are semi-wffs from the point of view of context logic) and since wffcl do not occur at all, we are able to use *A*, *B*, *etc.* to stand for wffs. This frees Φ, Ψ, Γ, Δ, *etc.* for other uses so that, for example, we often use Φ to stand for any one-place connective and Γ and Δ to stand for sets of wffs. Because of the complexity of some of the semantic theory it is necessary sometimes to use symbols of the same type to do different jobs, and we do not, therefore, adopt general conventions for the use of metasymbols. The particular use of a symbol is always explained when it is introduced, however (by leaders of the sort 'where Φ is a one-place connective ...'), and provided it is remembered that the role of the symbols is determined in this way, and not by the earlier general conventions, there should be no confusion with the metasymbolism of the previous chapters.

7.2 Intensional Superstructures

§1 *Inadequacies of extensional significance logics: non-value-functional connectives*

Modal connectives such as 'It is necessary (that)', 'Probably' and 'Demonstrably'; epistemic connectives, like 'It is known (that)', 'A rational person would believe (that)' and 'Rover smells (that)'; and indeed most sentence connectives of natural language, are not matrix-accessible over a finite set of values. (In certain cases this claim can be formally established: see below.) Sentence predicates which are beyond the scope of the sentential significance logics we have studied we can group together as non-value-functional. Intuitively, a one-place connective Φ is value-functional (3-functional in the cases we consider) if, and only if, strongly equivalent sentences can be intersubstituted under Φ without changing values. This intuitive notion, however, appears to involve quantification over sentences since it would have to be expressed formally in some such way as the following:

$$Val\ (\Phi) \text{ iff } (Up)(Uq)(p \simeq q \rightharpoonup . \Phi p \simeq \Phi q);$$

and this is unacceptable. It is nevertheless possible to give a satisfactory formal definition in terms of an acceptable analogue of sentential quantification if we are rather more precise about the nature of a connective.

There are many cases of apparent sentential quantification which appear to be unexceptionable. The above "definition" provides an example. So, too, do formulae such as '$(Up)Sp$', '$(Up)\Box p$' and '$(Up)B_0 p$', where 'B_0' is, say, 'John believes'. In these latter cases it is easy to see why the quantification appears to be, and in fact is, innocuous. It is simply that, in each case, there is a genuine subject which is open to quantification since it is associated with a predicate. Thus, the quantification is not over sentences at all, but over restricted subjects such as quotation expressions and that-clauses. Writing these three formulae in a form in which the quantification is correct, we have:

> *(Uqu(p))(sig qu(p))*; *(U§p)(§p is necessary)* [or, on a different account of modal properties: *(Uqu(p))(qu(p) is necessary)*]; *(U§p) (John believes §p)*.

Thus, we can often expand a formula of the sort Φp, where Φ is a connective, to obtain what we may call its s-p (subject-predicate) transform $\Phi'\tau p$, where τ is a descriptor (usually '§' or '*qu*')—and hence 'τp' is a subject, and Φ' is a predicate. The necessary and sufficient condition for such a transformation is that Φ and $\Phi'\tau$ be related by an appropriate absorption principle. Conversely, given $\Phi'\tau p$, and an absorption principle, we may obtain its c-transform (connective-transform), Φp. Now whereas 'p' in Φp is not open to quantification, 'τp' in $\Phi'\tau p$ is: *i.e.* '$(U\tau p)\ \Phi'\tau p$' is a wff, but '$(Up)\Phi p$' is not. We may, however, use the latter to *abbreviate* the former. This explains why some cases of apparent sentential quantification appear to be unexceptionable.

There is nothing in this account to exclude ordinary extensional connectives such as '$\sim$' provided there is an appropriate absorption principle. Indeed we may regard '$\sim p$' as a c-transform of '$\sim'\tau p$' where τ is '§' and '$\sim'$§' is related to '$\sim$' by an absorption principle; clearly, '$\sim'$§p' may be interpretated as 'that-p is not the case' (*i.e.* 'it is not the case that-p') where '§p' is the subject and '$\sim'$' is the predicate 'is not the case'.

It is important to notice that although Φ and $\Phi'\tau$ are related by an absorption principle, this is not to say that the connective Φ *is* a predicate plus a descriptor, nor yet that it is a predicate which absorbs a descriptor. An expression such as 'It is necessary' or 'It is not the case' is neither a connective nor, in terms of traditional grammar, a predicate. Such expressions only have a use in complexes of the sort 'it is necessary that-p' which are idiomatic variants of the correct form 'that-p is necessary', where the 'that' is part of the subject expression and not part of the predicate. That is why we usually write the word 'that' in parentheses in the connective, *i.e.* as 'it is necessary (that)'. Since, however, there is always an elementary transformation which turns expressions like 'it is necessary' into predicates —namely removal to the end of the sentence and deletion of 'it'—they may perhaps be called predicate transforms. So a connective like 'it is necessary that', though it can be regarded as being grammatically composite in that it can always be transformed into a predicate plus a descriptor, is not itself a predicate plus a descriptor. This becomes obvious if the connective is read as 'necessarily' instead of as 'it is necessary that'.

The treatment of two (or more)-place connectives is similar. The s-p transform of $\Phi(p, q)$ is $\Phi'(\tau p, \tau q)$, where Φ is related to $\Phi'(\tau\ ...,\ \tau\ ...)$ by an absorption principle. For example, let $\Phi(p, q)$ be '$\simeq(p, q)$', *i.e.* '$p \simeq q$', then the s-p transform $\Phi'(\tau p, \tau q)$ is '$\simeq'(\tau p, \tau q)$', *i.e.* '$\tau p \simeq' \tau q$', provided the connective '$\simeq$' is related by an absorption principle to the predicate form '$\simeq'(\tau\ ...,\ \tau\ ...)$'—as in fact it is (§2.12), where '$\simeq'$' is the predicate '*stee*' ('has the same significance value as') and τ is '*qu*'; hence $\Phi'(\tau p, \tau q)$ is '*stee* (*qu*(p), *qu*(q))', *i.e.* '*qu*(p) *stee* *qu*(q)'. Similarly, the s-p transform of '$\supset(p, q)$' is '$\supset'$(*qu*(p), *qu*(q))' or '$\supset'$(§p, §q)', where in either case '$\supset'$' is the predicate 'materially implies'. This connexion between connective formulae and their s-p transforms explains the ready shift from a use-interpretation of sentential logic to a mentioning-interpretation.

As the example shows, there may be more than one kind of s-p transform of a given connective formulae. Thus, the s-p transform of '$\Box p$' may be taken to be either '*qu*(p) is necessary' or '§p is necessary'; similarly, the s-p transform of '$\supset(p, q)$' may be taken as 'materially implies (*qu*(p), *qu*(q))' or as 'materially implies (§p, §q)'. Which transform is preferred will depend partly on the context and partly on the analysis which is given of the predicates involved. We have, for example, so far taken the s-p transform of '$\simeq(p, q)$' to be 'has the same significance value as (*qu*(p), *qu*(q))', but it is possible to affirm a different absorption principle and to take it as 'is strongly equivalent to (§p, §q)'. What is essential, however, is that the same descriptor should be applied uniformly to each variable throughout a formula when the s-p transform is constructed. Thus, neither 'materially implies (§p, *qu*(q))' nor 'materially implies (*qu*(p), §q)' is acceptable. This condition ensures that there is no ambiguity in the predicate and no confusion of levels in the s-p transform. It is not, however, necessary to convert all connectives in a formula in order to obtain an adequate s-p transform. For example, we may move from '$(p\supset q)\supset r$' to the partial s-p transforms, '§$(p\supset q)$ materially implies §r' or '(§p materially implies §q)$\supset r$', or to the full transform '§(§p materially implies §q) materially implies §r'. It is necessary to ensure, however, when constructing s-p transforms, that subject terms are connected by predicates and sentential terms by connectives, otherwise the resultant expression will not be well formed. For example, '(p materially implies q)$\supset$§r' is not an acceptable transform of '$(p\supset q)\supset r$'.

We shall return to most of these points in §7.12 where we face the problem of providing an acceptable analogue of extended sentential logic. There we give an alternative to the path sketched here since we wish to develop an analogue which has the same syntactical structure as extended logic—essentially a logic which has only one set of connectives. The use of s-p transforms, however, requires the use of three different sets of "connectives"—the ordinary sentential connectives, descriptors, and mentioning-connectives (predicates)—and therefore gives a much more complicated system. Not only this, but the quantification theory over s-p transforms is not standard. Consider, for example, a thesis like '$Sp \simeq$ *sig qu*(p)' (the formal expression of an absorption principle), in which '*qu*(p)', being a subject term, is open to quantification while 'p', being a sentential parameter, is not. Using standard principles of quantification we can then pass to '(U*qu*(p))($Sp \simeq$ *sig qu*(p))', hence '$Sp \simeq$ (U*qu*(p)) *sig qu*(p)' since '*qu*(p)' does not appear free in 'Sp'. But this last is clearly unacceptable.

These problems need not concern us here since we only wish at this stage to give an acceptable definition of value-functional connectives, and this we are now in a position to do. If we consider the formula in the scope of the quantifiers in the earlier intuitive definition, *viz.*:

$$p \simeq q \rightharpoonup. \Phi p \simeq \Phi q,$$

we may introduce quantifiable variables in place of the parameters by writing 'τp' for 'p' and 'τq' for 'q' throughout provided we also introduce simultaneously appropriate associated predicates in place of the connectives. This gives the partial s-p transform:

$$(\tau p \simeq' \tau q) \rightharpoonup. \Phi' \tau p \simeq \Phi' \tau q.$$

Notice that it would be incorrect to change the second occurrence of '$\simeq$' unless appropriate subject terms are introduced, as in the following further partial s-p transform:

$$(\tau p \simeq' \tau q) \rightharpoonup. \pi \Phi' \tau p \simeq' \pi \Phi' \tau q.$$

Here π need not be the same descriptor as τ, though it may be. If, for example, τ is '*qu*' and π is '§', then the first occurrence of '$\simeq'$' reads 'has the same significance value as' while the second reads 'is strongly equivalent to'. For our purposes, however, it is unnecessary to take this second step and we therefore define value-functional connectives as follows:

Where Φ is a one-place connective which is related by an absorption principle to $\Phi'\tau$, and '$\simeq(\ ,\ )$' is related by an absorption principle to '$\simeq'(\tau \ldots, \tau \ldots)$',

$$Val(\Phi) \text{ iff } (U\tau p)(U\tau q)(\tau p \simeq' \tau q \rightharpoonup. \Phi' \tau p \simeq \Phi' \tau q).$$

The original "definition" may now be regarded as an abbreviation of this.

An *n*-place sentence connective is value-functional in the *i*-th place iff strongly equivalent sentences can be intersubstituted in the *i*-th place preserving truth and significance values, and is fully value-functional iff it is value-functional in every place.

To illustrate: the connective 'It is logically necessary (that)' is not value-functional, because, although 'Aristotle = Aristotle $\simeq$. Aristotle does not exist' is true, it is logically necessary that Aristotle = Aristotle and not logically necessary that Aristotle does not exist. It follows however from the substitutivity theorem for $\mathbf{S}_6$ that all one-place connectives of $\mathbf{S}_6$ are value-functional and that all *n*-place connectives are fully value-functional. Accordingly most natural-language sentence connectives are excluded from consideration in $\mathbf{S}_6$ and its sub-systems; and such logics are hardly adequate for the formalisation of sentential discourse. The matter is not rectified by a finite increase in the number of values, by switching to a many-valued logic with more values than three, since many non-value-functional connectives of interest are not matrix-accessible. In fact the matrix approach breaks down completely when significance logics are enriched either by intensional connectives such as modal operators, or by quantifiers. Matrices can however be extended to models (roughly, each line of the matrix valuation taken for a wff is replaced by a model for the wff), and using models, semantics can be provided for the richer systems. We shall provide semantics for a class of intensional significance logics, through the theory of situations and set-ups, and we shall indicate how, by relaxing and varying the conditions on situations, semantics can be provided for a much more extensive class of intensional logics, and for some notions usually counted as metalinguistic.

The inadequacies of sentential significance logic can be located with more precision. It is not the addition of the subject-predicate notation (of chapter 3), to $\mathbf{S}_6$ say, that causes a breakdown, as the following typical separation results make plain:—

Theorem (1):

(1) Where A is a quantifier-free wff, *i.e.* a wff containing no quantifiers, of system $\mathbf{QS}_6$ (of §7.7) then A is a theorem of $\mathbf{QS}_6$ iff A is a substitution instance of a theorem of $\mathbf{S}_6$.

(2) Where A is a wff of $\xi\mathbf{S}_6$ (of §7.11) containing no occurrences of the choice descriptor 'ξ', then A is a theorem of $\xi\mathbf{S}_6$ iff A is a substitution instance of a theorem of $\mathbf{S}_6$.

(3) Where A is a wff of $\mathbf{IS}_6$ (of this section) containing no connective 'Φ_i' for any i, then A is a theorem of $\mathbf{IS}_6$ iff A is a substitution instance of a theorem of $\mathbf{S}_6$.

(These results follow using completeness theorems, established below, for $\mathbf{QS}_6$, $\xi\mathbf{S}_6$ and $\mathbf{IS}_6$. A substitution instance of a wff A of $\mathbf{S}_6$ is the result of a *uniform* substitution on the sentential parameters of A.) In fact the semantics provided for $\mathbf{S}_6$ can be taken over intact for notational enlargements of $\mathbf{S}_6$ which preserve substitution, *e.g.* for enlargements by the introduction of subject-predicate notation and by the addition of such operators as '§' (in advance of proper axiomatisation of 'that'). Complex wff such as '$f§(p\,\&q)$' and '$g(x,y,§p)$' are simply assigned values *en bloc*. Some enlargements so obtained are already intensional logics, since for example $v(f§(Tp \vee \neg p))$ may differ from $v(f§(Tq \vee \neg q))$ even though $\vdash Tp \vee \neg p \simeq . Tq \vee \neg q$; but they are uninteresting and unhelpful as intensional logics since the complex wff are given no analysis.

It is the introduction, first, of postulate-governed operators, and, second, of predicate constants with a determinate logical behaviour, that raises the new logical problems. In the light of the separation theorems the quantified significance logic $\mathbf{QS}_6$ can be seen, in terms of $\mathbf{S}_6$, as the theory of quantifiers, and $\xi\mathbf{S}_6$ as the theory of the selection operator 'any'. In much the same way the inadequacy of $\mathbf{S}_6$ as an intensional logic can be ascribed, in part, to its failure to provide a theory of subject-forming operators such as '§'—especially since the logical differences created by intensional predicates can be largely isolated in operators like '§' (see §3.5 and §7.13). But the treatment of operators like '§' is only a part of intensional logic; the other part is the analysis of constant assertional predicates of logical and philosophical importance, in particular of such predicates as 'is logically necessary', 'entails', 'is probable', 'is possible' and 'is believed by x_0' (*e.g.* by the rational man). Similarly the treatment of quantifiers is only a part of the standard extensional logic and theory of "individuals"; equally important is the analysis of constant predicates such as 'is identical with', 'exists' and 'is part of'.

As in the standard logic it helps to break the development down into stages. We begin with the theory of operators such as '§' and with the general theory of predicates; and later we consider conditions which distinguish special predicate constants. We add, then, to sentential significance logics the descriptors '§' and '*qu*': where A is a wff §A is a (assertoric) subject or, as we shall say, an *assertional*, and $qu(A)$ is a (quotational) subject. In order to form sentences with these subjects, we shall have to add also a set of predicates. So we take over the notation and rules for predicates of §3.5, under the simplifying assumptions of chapter 5: thus where τ is a descriptor, τA_1, ..., τA_m are m subjects (assertoric or quotational), and F^m is an m-place predicate term, $F^m(\tau A_1, ..., \tau A_m)$ is a wff. If, for example, our predicate constants include the one-place predicates 'f_0' read 'is necessary' or 'it is necessary' and 'g_0' read 'is true' [*i.e.* 'g_0' is the predicate '*tru*'], then the wff 'f_0(§p)' reads 'that-p is necessary' and '$g_0qu(p)$' reads 'quotes-p is true'. The connectives '$\Box$' ('it is necessary that') and 'T' are then obtained via the definitions: $\Box A =_{\text{Df}} f_0§A$, and $TA =_{\text{Df}} g_0qu(A)$. Indeed the connectives of sentential significance logic are themselves but predicate constants coupled with descriptors. (Hence the adequacy of the definition of *Val* above).

Where the underlying sentential logic is $\mathbf{S_6}$, we call the system which results when descriptors and predicate terms are added, $\mathbf{LS_6}$. We introduce the new significance symbol '$\mathscr{S}$', defined by: $\mathscr{S}A =_{\text{Df}} \Box SA$, because, in logics like $\mathbf{LS_6}$ where logical necessity can be expressed, we can distinguish the rival significance notions discussed in §4.8, namely non-contingent significance, now represented by '$\mathscr{S}$', and significance which may be a contingent matter represented, as previously, by 'S', and defined as in $\mathbf{S_6}$ by: $SA =_{\text{Df}} \urcorner DA$.

The formation rules of $\mathbf{LS_6}$ permit the iteration of connectives, *i.e.* of predicates concatenated with descriptors, though this could be simply prevented. However, the usual objection to iterated intensional connectives is that the results do not make sense; in which case the formation rules should stand and a principle added to the effect: $\sim SA$, where A contains iterated connectives. Since there are many counterexamples to such a principle, however, we shall not adopt it.

How are wff of the sort $F^m(\tau A_1, ..., \tau A_n)$ to be evaluated? The standard recipe for subject-predicate sentences of quantification logic (given in §3.7 and developed in §7.7) can be applied, and with the usual justification, *viz.*

$v(\mathrm{F}^m(\tau A_1, ..., \tau A_m)) = t$ iff $T(\langle \mathrm{I}(\tau A_1), ..., \mathrm{I}(\tau A_m)\rangle\, i\, \mathrm{I}(\mathrm{F}^m))$,
$v(\mathrm{F}^m(\tau A_1, ..., \tau A_m)) = n$ iff $\sim S(\langle \mathrm{I}(\tau A_1), ..., \mathrm{I}(\tau A_m)\rangle\, i\, \mathrm{I}(\mathrm{F}^m))$,

where I is an interpretation function which assigns to every subject τA an element of non-null domain d and $\mathrm{I}(\mathrm{F}^m)$ is an m-place relation on d. Validity for $\mathbf{LS}_6$ can be defined as for quantification theory; *i.e.* A is $\mathbf{LS}_6$-valid iff $v(A) = t$ for every interpretation I over each choice of domain d. (For fuller details see §7.7).

Theorem (2): B is $\mathbf{LS}_6$ valid iff B is a substitution instance of a theorem of $\mathbf{S}_6$; and $\mathbf{LS}_6$ may be axiomatised using just the axioms of $\mathbf{S}_6$ and the rules of $\rightharpoondown$-detachment and uniform substitution.

The semantics given for $\mathbf{LS}_6$ is however uninformative, and, as already explained, breaks down for extensions of $\mathbf{LS}_6$. In fact introduction of domain d and function I is, in the case of $\mathbf{LS}_6$, a logical frill; for values could simply be assigned *en bloc* to wff of the form $\mathrm{F}^m(\tau A_1, ..., \tau A_m)$. For a less trivial semantics which does admit extensions the problem is to give an account of the contracted item $\mathrm{I}(\tau A)$ since the interpretation $\mathrm{I}(\mathrm{F}^m)$ of F^m as an m-place relation on set $\mathrm{d} = \{\mathrm{I}(\tau A) : A \text{ is a sentence}\}$ can be retained. In the case of $\mathrm{I}(\S A)$, where A is a sentence, we already know the answer: 'that-A' is s-about the assertion, or statement, I(that-A). Thus the domain d is the class of assertions yielded by sentences of the system. The trouble is that this answer is not, on its own, sufficiently illuminating: it does not provide a tight enough connection between the evaluation of a sentence A and the assertion yielded by *it*, and accordingly it neglects characteristic iteration features of sentential connectives. For a more illuminating account we shall have to specify the assertion yielded by sentence A in terms of the values assigned to A. This cannot be done extensionally, but it can be done through the theory of situations. For, given an appropriate account of situations, the assertion $\S A$ is determined by the class of situations H such that A holds in H and the class of situations H′ such that A significantly holds in H′; formally $\S A = \phi\langle\{\mathrm{H} : v(A, \mathrm{H}) = t\}, \{\mathrm{H}' : v(A, \mathrm{H}') = n\}\rangle$, where ϕ is a one-one function, the *giving* function, whose inverse is an interpretation function. This account takes up, in a significance way, the standard position according to which an assertion, or statement, is "given by" the circumstances or situations in which it holds and the circumstances in which it does not hold; and it has as a consequence the familiar thesis that two assertions are identical iff they hold in the same situations and do not hold in the same situations, *i.e.*

$\S A = \S B$ iff
$\{\mathrm{H} : v(A, \mathrm{H}) = t\} = \{\mathrm{H} : v(B, \mathrm{H}) = t\}\ \&.\ \{\mathrm{H}' : v(A\ \mathrm{H}') = n\} = \{\mathrm{H}' : v(B, \mathrm{H}') = n\}$
i.e., iff $(U\mathrm{H})(v(A, \mathrm{H}) = v(B, \mathrm{H}))$.

The account of assertion proposed is not incompatible with, but indeed, given the identity account, a consequence of, the characterisation of statements suggested in §3.6 in terms of contractions; and moreover the derivation makes the giving function explicit. For

§$A = (\iota B)(A = B)$, for the relation of assertional identity,
$= (\iota B)(U\text{H})(v(A,\text{H}) = v(B,\text{H}))$, by the account of identity,
$= \{\text{H} : v(A,\text{H}) = t\} \ldots \{\text{H}' : v(A,\text{H}') = n\} \ldots$ expanding as on the previous page but neglecting components involving B since it is bound,
$= \phi\langle\{\text{H} : v(A,\text{H}) = t\}, \{\text{H}' : v(A,\text{H}') = n\}\rangle$, defining '$\phi$'.

But for an adequate theory of assertions, marking out correctly the extent of the class of situation is quite crucial: otherwise too many or, more likely, too few assertions, will result.

§2 *Set-up and situational semantics*

A set-up is a class of sentences or wff, or a sequence of such classes; it is typically governed by a set of rules which distinguish the sort of set-up it is. For example each set-up H which classical (two-valued) modal logic introduces satisfies these conditions:

$(A \& B)$ is in H iff A is in H and B is in H (normal &-rule);
$(A \vee B)$ is in H iff A is in H or B is in H (normal $\vee$-rule);
$\sim A$ is in H iff A is not in H (classical $\sim$-rule);

where A is in H iff A holds in H, which we write: $v(A,\text{H}) = t$. A set-up satisfying the above conditions is variously called *a state (-of-affairs)*, or *a model set* or *an index*. The explanation of a set-up as a sequence of classes of wff (and so, in *the* special case, as a class of wff) is inessential for the formal semantics we shall give; the rule-governed parameter H can be interpreted in other ways, *e.g.* though this is misleading, as a world or as a point of reference.

In contrast to model sets, each set-up H used in the semantical theory of entailment satisfies, as well as the normal &- and $\vee$-rules, the following negation rule:

$\sim A$ is in H iff A is not in H* (normal $\sim$-rule), where the image set-up H*, under one-one function *, is such that H** = H. Such a set-up we call a *situation* (for reasons developed in R. and V. Routley [2]) and its members can be treated as statements; coincidence over such set-ups serves to define statemental identity. The actual or base situation G, in the special case the class of true statements or what is the case, is of course classical, so G = G* since G is both consistent and complete, *i.e.* $\sim A$ is in G iff A is not in G. Extensional value assignments (of the form $v(A)$) are always value assignments with respect to situation G (thus $v(A) = v(A,\text{G})$); they never take us beyond G to some other situation (see also §7.13). In the model theory for intensional systems we relativise the valuation function v to situations, replacing $v(A)$ by $v(A, \mathrm{H})$. At the same time, since these situations in general differ from G, and may not be realisable at all, we change the interpretation of v, and regard v as a characteristic function showing, in case $v(A, \mathrm{H})$, whether A holds in H or otherwise. Thus where A is an indicative sentence, $v(A, \mathrm{H}) = t$ simply asserts that A holds in H or is indeed in H, and $v(A, \mathrm{H}) = n$ that A does not significantly hold in H. The relation 'holds in', though analogous to 'true in' (see Carnap [8], p. 78) differs from the truth predicate. To regard it as a truth predicate suggests that H's are in some way just alternatives to G (the actual world) whereas they may be seriously incomplete and even inconsistent; it suggests, too, that H's somehow exist. The theory of set-ups also allows for the analysis of non-indictives such as orders and questions: in the case of a non-indicative sentence A, however, 'A is in H' cannot, without further ado, be read as 'A holds in H'.

An important difference between the descriptors '§' and '*qu*' lies in the different classes of set-ups in terms of which they are assessed. The assertional descriptor '§' is interpreted, in the two-valued case, over the class of situations, or, under the unsatisfactory account classical modal logic offers, over states-of-affairs. The quotational descriptor '*qu*' is, however, assessed over all set-ups, including what we call *Rafferty* set-ups, because these are governed only by Rafferty's rules—in effect by no rules. Whereas situations determine assertional identity, set-ups in general determine only type identity of sentences. For this reason the class of all set-ups is important for the analysis of quotation and syntactical matters commonly relegated to metatheory.

The evaluation of even basic connectives may be queer over set-ups which are not situations, without being ruleless. An important example is Montgomery's &-rule for connexive logics, under which conjunction is evaluated like equivalence for a class of set-ups Q; *i.e.* according to this rule if H is a state then the normal &-rule holds, but if $\mathrm{H} \in \mathrm{Q}$ then

$A \& B$ is in H when, and only when, A is in H iff B is in H.

Set-ups governed by queer rules are of interest for the semantical analysis of non-standard logics, and set-ups governed by very limited rules are of interest for the semantical analysis of connectives of ultra-intensional order such as 'It is said (that)', 'It is trivial (that)', 'It caught our attention (that)' and 'It is illuminating (that)'.

Set-up rules are not restricted to rules for connectives; they may also include rules for compounding, rules for subject-predicate evaluation, and transformational rules—such as those for converting predicates + *that* to connectives, or for converting active to passive, *e.g.*

Passive 'A is in H iff A is in H.

The enlarged significance logics we have introduced do not, of course, admit the full range of subject factors of the logic of §3.5 but only subjects formed from wff by '§' and '*qu*' (and similar descriptors). Because of the limited class of subjects so far admitted many predicates, especially characteristic individual predicates such as 'is blue' and 'is hard', are always non-significant however coupled with subjects, *i.e.* such a predicate F satisfies the condition $\sim S\text{F}(\tau A)$. Remaining predicates divide into those which sometimes apply significantly, and those which are always significant. Now predicates also divide into value-functional predicates, which can be treated, as before, through connectives of sentential logic, and non-value-functional predicates. We shall make the simplifying assumption that all *primitive non-value-functional* predicates of the logic are *classical significance connectives*, *i.e.* they satisfy the postulates:

J1. $S\text{F}^m(\tau A_1, ..., \tau A_m) \supset .SA_1 \;\&\; ... \;\&\; SA_m$; and
J2. $SA_1 \;\&\; ... \;\&\; SA_m \supset .S\text{F}^m(\tau A_1, ..., \tau A_m)$.

Thus, for example, taking 'τ' to be '§' and 'F^m' to be 'is necessary', where $m=1$, we have $\vdash S\Box A \equiv SA$. Since the postulates are restricted in application to non-value-functional predicates, we cannot of course take, say 'τ' to be '*qu*' and 'F^m' to be '*tru*', to obtain the falsehood $STA \equiv SA$, since '*tru*' is value-functional.

It might perhaps be thought that the thesis $\vdash S\Box A \equiv SA$ is unacceptable since there is a necessity connective which, like '*tru*', is not a classical significance connective. Thus, we might argue: if A is nonsignificant, SA is false; but again, if A is nonsignificant, $\Box A$, like TA, is also false, and hence $S\Box A$ is true. In this case, $S\Box A$ is not equivalent to SA. But the postulates do not imply that *all* non-value-functional predicates are classical; we have only made this assumption for *primitive* non-value-functional predicates, and in fact many, and typical, non-classical assertoric predicates can be defined in terms of primitive classical predicates using also non-classical connectives. For example, the two-valued necessity connective used in the above argument, which we may write as '$\Box^+$', which corresponds to the unlimited truth connective 'T' and is always significant, can be defined using 'T' itself as, '$T\Box$'. In fact, the postulates are not nearly as restrictive as might at first appear; all predicates that are always significant can be defined in terms of related predicates which do satisfy J1 and J2, and likewise predicates which are never significant.

We call the system which results from $\mathbf{S_6}$ by addition of J1 and J2, $\mathbf{KS_6}$. $\mathbf{KS_6}$-validity may be defined just as $\mathbf{LS_6}$-validity except that we change the valuation rule in case $v(\text{F}^m(\tau A_1, ..., \tau A_m)) = n$ to:

$$v(\text{F}^m(\tau A_1, ..., \tau A_m)) = n \text{ iff } v(A_1) = n \text{ or } v(A_2) = n \text{ or } ... \; v(A_m) = n.$$

The postulates of $\mathbf{KS_6}$ are those of $\mathbf{LS_6}$ together with J1 and J2. Then B is $\mathbf{KS_6}$-valid iff B is a theorem of $\mathbf{KS_6}$. But the semantics of $\mathbf{KS_6}$, like that of $\mathbf{LS_6}$, precludes extensions we shall want to make; and in particular does not distinguish wff involving '§' from wff containing '*qu*'. For a less superficial analysis we appeal to set-up models.

The postulates J1 and J2 reduce the problem of assessing whether or not $F^m(\tau A_1, ..., \tau A_m)$ is in, or significantly in, a situation H, *i.e.* the problem of determining the value v of the holding function $v(F^m(\tau A_1, ..., \tau A_m), H)$, to the two-valued problem of determining precisely when $v(F^m(\tau A_1, ..., \tau A_m), H) = t$. For in view of J1 and J2, we can require:

$$v(F^m(\tau A_1, ..., \tau A_m), H) = n \text{ iff } v(A_1, H) = n \text{ or ... or } v(A_m, H) = n;$$

i.e. the assessment can be made without going beyond H, and independently of the predicate F^m. Furthermore for two-valued problems of evaluating $v(F^m(\tau A_1, ..., \tau A_m), H)$, the characterisation of the assertion $\S A$, given above, can be simplified to,

$\S A = \theta(\{H \in K_1 : v(A, H) = t\})$ where θ is a one-one function and K_1 is the class of situations. Similarly,

$qu(A) = \theta'(\{H \in K : v(A, H) = t\})$ where K is the class of all set-ups including Rafferty set-ups.

To arrive at a rule for assessing $v(F^m(\tau A_1, ..., \tau A_m), H) = t$, *i.e.* whether $F^m(\tau A_1, ..., \tau A_m)$ holds in H, we apply the quantificational recipe only we relativise it to set-up H and we set $I(\tau A) = \theta^{-1}(\tau A) = \{H \in K_\tau : v(A, H) = t\}$, where θ^{-1} is the inverse of function θ and K_τ is the class of set-ups correct for the corresponding descriptor. Then the rule can be put in the form:

$$v(F^m(\tau A_1, ..., \tau A_m), H) = t \text{ iff}$$

$$\langle\{H \in K_\tau : v(A_1, H) = t\}, ..., \{H \in K_\tau : v(A_m, H) = t\}\rangle \; i \; I(F^m, H),$$

where $I(F^m, H)$ is an m-place relation on classes of set-ups. For the relativisation of assessment can be isolated in the interpretation of F^m. The rule may be rewritten, without loss of generality:

$$v(F^m(\tau A_1, ..., \tau A_m), H) = t \text{ iff } \langle\{H' \in K_\tau : v(A_1, H') = t\}, ...\rangle \mathbf{R}_{F^m} H,$$

where $\mathbf{R}_{F^m}$ is a relation between ordered classes of set-ups and a set-up. The same rule (given in essentials in Montague [1], p. 109) may be reached without the detour through application of the quantificational recipe. Consider, to illustrate, the rule for connective '□':

$v(\Box A, H) = t$ iff $v(f_0 \S A, H) = t$,
i.e. iff $\S A$ has the property λf_0 holds in H,
i.e. iff for H, $\theta\{H' \in K_1 : v(A, H') = t\} \; i \; \lambda f_0$,
i.e. (definitionally) $\{H' \in K_1 : v(A, H') = t\} \mathbf{R}_{f_0} H$.

The problem, as to precisely what relation $\mathbf{R}_{f_0}$ or $\mathbf{R}_{F^m}$ is, may be shelved, to begin with, by allowing the relation to be simply laid down in the modelling, as follows:—

A *model* $\mathscr{M}$ (more precisely an $\mathbf{MS}_6$-model) is a structure $\mathscr{M} = \langle G, K, K_2, I \rangle$ where K is a set (of set-ups), $G \in K_2$, $K_2 \subseteq K$ and I is an interpretation function which assigns to every m-place predicate F^m an $(m+1)$-place relation $\mathbf{R}_{F^m}$ in $(\mathscr{P}K)^m \times K$ and to every element of $At \times K$ one of the elements of V $= \{t, f, n\}$, where At is the class of atomic wff. Here $\mathscr{P}K$ is the power set of K, and $\times$ is the Cartesian product of sets. Thus $I(P, H) = t$ or $= f$ or $= n$ for every atomic wff P and every $H \in K$; and so I informs us whether an atomic wff holds, or holds significantly, in each set-up. In order to determine validity, the notion that generalises tautologousness as models generalise matrices, we define a valuation function v which informs us for each wff whether it holds, or holds significantly, in a given set-up. Function v, from wff and set-ups to V, is defined recursively as follows, for every $H \in K$:—

$$v(P, H) = I(P, H), \text{ for } P \in At.$$

For every $H \in K_2$ (*i.e.* for every state H), and every wff A,

$$v(\neg A, H) = t \text{ iff } v(A, H) \neq t;\ v(\neg A, H) \neq n;$$
$$v(A \rightharpoonup B, H) = t \text{ iff } v(A, H) \neq t \text{ or } v(B, H) = t;$$
$$v(A \rightharpoonup B, H) = n \text{ iff } v(A, H) = t \text{ and } v(B, H) = n;$$
$$v(DA, H) = t \text{ iff } v(A, H) = n;\ v(DA, H) = n \text{ iff } v(A, H) = t.$$

For every $H \in K$, every m-place predicate F^m and every wff $A_1, \ldots, A_m$,

$$v(F^m(\tau A_1, \ldots, \tau A_m), H) = t \text{ iff } \langle \{H \in K_\tau : v(A_1, H) = t\}, \ldots \rangle\, \mathbf{R}_{F^m} H,$$
$$v(F^m(\tau A_1, \ldots, \tau A_m), H) = n \text{ iff } v(A_1, H) = n \text{ or } \ldots\ v(A_m, H) = n.$$

K_τ is either K or K_2 according as the operator τ, concatenated with the distinguishing wff A_i, is '*qu*' or '§'. It remains to say how $\neg A$, $A \rightharpoonup B$ and DA are evaluated over $H \in (K - K_2)$. Since the set-ups concerned are Rafferty set-ups we can simply assign values arbitrarily to $v(\neg A, H)$, *etc.*, for $H \in (K - K_2)$, and we can let the interpretation function I make the choice of values. Thus we require of model $\mathscr{M}$ that $I(\neg A, H)$ is one of set $V = \{t, f, n\}$ for each wff $\neg A$ and each $H \in (K - K_2)$; and similarly for $I(A \rightharpoonup B, H)$ and $I(DA, H)$ for $H \in (K - K_2)$. Then our rule is: for every $H \in (K - K_2)$,

$$v(\neg A, H) = I(\neg A, H);\ v(A \rightharpoonup B, H) = I(A \rightharpoonup B, H); \text{ and}$$
$$v(DA, H) = I(DA, H).$$

The rules for extensional connectives '$\neg$' '$\rightharpoonup$' and 'D' in states, or G-like situations, are just the evaluation rules with respect to G restated for arbitrary state H. The value-functionality of a connective Φ definable in $\mathbf{S}_6$ is shown in the feature that $\Phi(A_1, \ldots, A_m)$ can be evaluated in state H just in terms of the evaluations of $A_1, \ldots, A_m$ in H, *i.e.* without going beyond state H.

Since truth is, of course, holding in the actual case G, we define: A wff B *is true* in model $\mathscr{M}$ iff $v(B,\mathrm{G}) = t$. Validity is truth in every model, thereby generalising tautologousness as truth in every line of a valuation table. Thus B is *valid* iff B is true in every model $\mathscr{M}$; or, more precisely, B is $\mathbf{MS}_6$-valid iff B is true in every $\mathbf{MS}_6$-model.

$\mathbf{MS}_6$ may be axiomatised as follows: it has axiom-schemata corresponding to the axioms of $\mathbf{S}_6$; J1 and J2; the rule of $\rightharpoonup$-detachment (R6.2); and the following substitutivity of equivalents rule:

if $\vdash A_i \rightharpoonup B_i$ and $\vdash B_i \rightharpoonup A_i$ then $\vdash \mathrm{F}^m(\tau A_1, ..., \S A_i, ..., \tau A_m) \rightharpoonup \mathrm{F}^m(\tau A_1, ..., \S B_i, ..., \tau A_m)$.

It then follows using J1 and J2:

if $\vdash A_i \simeq B_i$ then $\vdash \mathrm{F}^m(\tau A_1, ..., \S A_i, ..., \tau A_m) \simeq \mathrm{F}^m(\tau A_1, ..., \S B_i, ..., \tau A_m)$.

Using this rule part (1) of *Theorem* (3) of our results on $\mathbf{MS}_6$ follows, by induction over the number of predicates and connectives in A:

Theorem (3):

(1) If $\vdash A_i \simeq B_i$ then $\vdash A \simeq B$, where B results from A by replacing one or more occurrences of A_i in A by B_i provided only that no occurrence of A_i that is replaced is in the scope of descriptor '*qu*'.

(2) B is a theorem of $\mathbf{MS}_6$ iff B is $\mathbf{MS}_6$-valid.

(3) $\mathbf{MS}_6$ is decidable.

Proof of (2) is provided after appropriate formal preliminaries have been developed; and proof of (3) will follow using the filtration method of §6.

In view of (1), $\mathbf{MS}_6$ improves on $\mathbf{KS}_6$ by separating the roles of descriptors '§' and '*qu*'. But $\mathbf{MS}_6$ scarcely provides an adequate theory of assertions; for it effectively identifies all $\mathbf{S}_6$-tautologies, and also identifies all $\mathbf{S}_6$-contradictions. For example, for every $qu(f), f(\S(Tp \vee \neg p)) \simeq f(\S(Tq \vee \neg q))$; whence $\S(Tp \vee \neg p) = \S(Tq \vee \neg q)$. $\mathbf{MS}_6$ provides at best a theory of *strict* assertions, not of statements. A better theory emerges if we replace model-sets, *i.e.* elements of K_2, by propositional-sets. We replace the connective '$\neg$' by the connectives '$\sim$' and 'T'; and we add to $\mathbf{MS}_6$-model $\mathscr{M}$ a new set K_1 (of propositional-sets), with $\mathrm{K}_2 \subseteq \mathrm{K}_1 \subseteq \mathrm{K}$, and a new operation * such that for $\mathrm{H} \in \mathrm{K}_1, \mathrm{H}^{**} = \mathrm{H}$, and for $\mathrm{H} \in \mathrm{K}_2, \mathrm{H}^* = \mathrm{H}$.

The new evaluation rules are as follows: For $\mathrm{H} \in \mathrm{K}_1$, $v(\sim A, \mathrm{H}) = t$ iff $v(A, \mathrm{H}^*) = f$; $v(\sim A, \mathrm{H}) = n$ iff $v(A, \mathrm{H}) = n$; $v(TA, \mathrm{H}) = t$ iff $v(A, \mathrm{H}) = t$; $v(TA, \mathrm{H}) \neq n$. Connectives 'D' and '$\rightharpoonup$' are assessed as before. For $\mathrm{H} \in (\mathrm{K} - \mathrm{K}_1)$ (not $\mathrm{K} - \mathrm{K}_2$) values of compounds are assigned (arbitrarily) by I. Finally $\S A_i$ is assessed over $\{\mathrm{H} \in \mathrm{K}_1 : v(A, \mathrm{H}) = t\}$; *e.g.* $v(\mathrm{F}(\S A), \mathrm{H}) = t$ iff $\{\mathrm{H}' \in \mathrm{K}_1 : v(A, \mathrm{H}') = t\}\ \mathbf{R}_\mathrm{F}\mathrm{H}$. This semantics, for system $\mathbf{NS}_6$, does not validate (1); but it nevertheless separates descriptors properly. For example, while $f(\S\sim\sim p) \simeq f(\S p)$ is valid, $f(qu(\sim\sim p)) \simeq f(qu(p))$ of course is not.

Given these semantics for $\mathbf{MS}_6$ and $\mathbf{NS}_6$ it is not difficult to see how to generalise the theory in various directions. First, the conditions imposed by postulates J1 and J2 can be removed simply by replacing the valuation rule for $v(F^m(\tau A_1, ..., \tau A_m), H)$ by the following rules:

$v(F^m(\tau A_1, ..., \tau A_m), H) = t$ iff
$T\langle\langle\{H \in K_\tau : v(A_1,H) = t\}, \{H \in K_\tau : v(A_1,H) = n\}\rangle, ...\rangle\ \mathbf{R}_{F^m}H$; and
$v(F^m(\tau A_1, ... \tau A_m), H) = n$ iff
$\sim S\langle\langle\{H \in K_\tau : v(A_1,H) = t\}, \{H \in K_\tau : v(A_1,H) = n\}\rangle, ...\rangle\ \mathbf{R}_{F^m}H$ (or still more generally: iff $\langle\langle\{H \in K_\tau : v(A_1,H) = t\}, \{H \in K_\tau : v(A_1,H) = n\}\rangle, ...\rangle\ \mathbf{R}'_{F^m}H$, where $\mathbf{R}'_{F^m}$ is a relation distinct from $\mathbf{R}_{F^m}$.)

Secondly, the resulting intensional theory can be extended from three to many values by using the Rosser-Turquette J-functors. Thirdly, the range of behaviour of each connective can be modelled in more detail by associating, with each connective, classes of set-ups, each group of which measures its different relevant logical behaviour; world, situational, quotational, *etc.*

§3 *Accounting for specific properties of intensional connectives*

$\mathbf{MS}_6$ and $\mathbf{NS}_6$, though they control the behaviour of descriptors '§' and '*qu*', impose no conditions on predicates. For example they do not give enough information to determine recognised properties of '$\Box$'. However, the semantics for $\mathbf{MS}_6$ and $\mathbf{NS}_6$ improve vastly on that for $\mathbf{KS}_6$ in that very often we can reflect semantically, by conditions imposed on relation $\mathbf{R}_{F^m}$, or by a more detailed analysis of relation $\mathbf{R}_{F^m}$, axiomatic conditions imposed on predicate F^m. The simplest condition corresponds to that required by the one-place identity predicate 'f_1' which satisfies the postulate $f_1(\S A) \simeq A$. Then $\{H \in K_2 : v(A,H) = t\}\ \mathbf{R}_{f_1}H$ iff $H \in \{H \in K_2 : v(A,H) = t\}$; *i.e.* $v(A,H) = t$ for $H \in K_2$. But this simplest case can of course be handled in terms of $\mathbf{KS}_6$. An example that cannot be so dealt with in $\mathbf{KS}_6$ is provided by a significance version of Lemmon's S0.5 which imposes on predicate 'f_0' the following postulates: $\Box A \rightharpoonup A$; $\Box(A \rightharpoonup B) \rightharpoonup . \Box A \rightharpoonup \Box B$; and $\Box A$, where A is an $\mathbf{S}_6$-tautology. (For axiomatic and semantical formulations of S0.5 itself, see Hughes and Cresswell [1].) Then the corresponding modelling condition on $\mathbf{R}_{f_0}$ is: $\{H' \in K_2 : v(A,H') = t\}\ \mathbf{R}_{f_0}G$ iff $\{H' \in K_2 : v(A,H') = t\} = K_2$; and for $H \neq G$, $\{H' \in K_2 : v(A,H') = t\}\ \mathbf{R}_{f_0}H$ is assigned arbitrarily; *i.e.* $\mathbf{R}_{f_0}$ is constrained by a condition only in case $H = G$. Hence, where $H \neq G$, $v(\Box A,H) = t$ is assigned (arbitrarily) by the model, and where $H = G$, $v(\Box A,G) = t$ iff, for every $H' \in K_2$, $v(A,H') = t$.

By extending the constraint on $\mathbf{R}_{f_0}$ to every $H \in K_2$ a modelling for system $S5S_6$, a significance version of Lewis' system S5, results. In other words $\mathbf{R}_{f_0}$ is constrained by the condition: $\{H' \in K_2 : v(A,H') = t\}\ \mathbf{R}_{f_0}H$ iff $\{H' \in K_2 : v(A,H') = t\} = K_2$, for every $H \in K_2$; whence $v(\Box A,H) = t$ iff, for every $H \in K_2$, $v(A,H') = t$. That is, $\Box A$ holds in state-of-affairs H iff A holds in every state-of-affairs—the positivist adaption of Leibniz's prescription for logical necessity (see, *e.g.*, Carnap [2]).

The *mixed method* so far followed in stating further modelling conditions entangles the valuation function with conditions on the relation $\mathbf{R}_{Fm}$; thus it seriously reduces the effectiveness of specification of the model. The *pure method* avoids these problems; according to it, the condition for the identity predicate is $\alpha\mathbf{R}_{f_1}$H iff H$\in\alpha$, for significance S0.5 $\alpha\mathbf{R}_{f_0}$G iff $\alpha = K_2$, and for $S5S_6$ $\alpha\mathbf{R}_{f_0}$H iff $\alpha = K_2$ for H$\in K_2$, where $\alpha\in\mathscr{P}K_2$, *i.e.* α is a set of states. In short, the pure method works by enlarging the domain of the relation.

In case 'f_0' is the only intensional predicate of $\mathbf{MS}_6$ and '§' the only descriptor, K can be omitted from the modelling, and since $\mathbf{R}_{f_0}$ vanishes under analysis, interpretation function I can be replaced by valuation function v. Thus, after simplification, an $S5S_6$-model $\mathscr{M}$—for a system obtained from $\mathbf{S}_6$ by the addition of the one-place logical necessity connective '□'—is a structure $\mathscr{M} = \langle G, K_2, v\rangle$, where K_2 is a set to which G belongs, and v is a holding function which assigns, for every H$\in K_2$ and $P\in At$, exactly one element of $V = \{t, f, n\}$ to $v(P, H)$. The recursive extension of v from atomic wff to all wff of $S5S_6$ has already been given: that is to say the connectives '$\rightharpoondown$', '$\neg$', 'D' are evaluated as before over K_2, and '□' is evaluated thus, for H$\in K_2$:

$$v(\Box A, H) = t \text{ iff } (UH'\in K_2)v(A, H') = t;\ v(\Box A, H) = n \text{ iff } v(A, H) = n.$$

It will follow, as a special case of subsequent theorems, that $S5S_6$ is exactly axiomatised by the addition to $\mathbf{S}_6$ of the following postulates:

R□. $\vdash A \rightarrow \vdash \Box A$ (necessitation)
□1. $\vdash \Box p \rightharpoondown p$ (axiom of necessity)
□2. $\vdash \Box(p \rightharpoondown q) \rightharpoondown . \Box p \rightharpoondown \Box q$ (axiom of distribution)
□3. $\vdash \neg \Box p \rightharpoondown \Box \neg \Box p$ (S5 axiom).
□4. $\vdash S\Box p \rightharpoondown Sp$
□5. $\vdash Sp \rightharpoondown S\Box p$.

The two-valued ($\sim$, $\supset$) analogues of □1–□3 together with R□—a satisfactory rule only provided that the system contains no contingent theses—provide Gödel's axiomatisation of Lewis's system S5. □4 and □5 state the (weak) significance conditions for classical '□' and correspond to J1 and J2.

In two-valued modal systems a single logical possibility connective '◇', with expected S5 properties, is definable in terms of '□' as: $\Diamond A =_{Df} \sim \Box \sim A$. In significance modal logics however, the expected S5 properties do not follow using such a definition; *e.g.* $\Box(p \rightharpoondown q) \rightharpoondown . \Diamond p \rightharpoondown \Diamond q$ is not derivable (for it is not valid). Although the expected S5 properties do follow using the definition $\Diamond A =_{Df} \neg \Box \neg A$, the desired analogue of □4, viz. $S\Diamond p \rightharpoondown Sp$, breaks down since under this definition $S\Diamond A$ is a theorem, whatever A. Though '◇' has other definable analogues such as '$D\Box D$' none of these quite suffice; to obtain a satisfactory logical possibility connective the only course is to introduce '◇' as well as '□' as a primitive connective and to axiomatise it separately. Thus even in the case of logical modalities we are forced to consider multiply modal logics, *i.e.* logics with several independent modal connectives.

Conditions so far imposed on relation $\mathbf{R}_F$ have, in effect, eliminated the relation. In the case of identity predicates also the relation vanishes under constraint. In terms of $\mathbf{NS}_6$ semantics, for example, we can distinguish three identity determinants: identity of type sentences, which is assessed in terms of coincidence over K; identity of assertions proper, assessed over K_1; and identity of strict assertions, assessed over K_2. For assertional identity '=' the relation $\mathbf{R}_=$ is constrained by the condition (over K_1):

$$\langle \alpha_1, \alpha_2 \rangle \mathbf{R}_= G \quad \text{iff} \quad \alpha_1 = \alpha_2, \text{ for } \alpha_1, \alpha_2 \in \mathscr{P}K_1.$$

In general, however, constraints on relation $\mathbf{R}_F$ are not quite so simple. Though $\mathbf{R}_F$ cannot always be eliminated it can very often be replaced by a simpler relation R_F between set-ups: then $(m+1)$-place relation $\mathbf{R}_{F^m}$ in $(\mathscr{P}K)^m \times K$ is traded for an $(m+1)$-place relation R_{F^m} on K^{m+1}. The $S5S_6$ requirement can be re-expressed, for $H \in K_2$, as follows:

$$\alpha \mathbf{R}_{f0} H \text{ iff } K_2 \subseteq \alpha, \text{ for } \alpha \in \mathscr{P}K_2,$$

or under the mixed method:

$$\{H' \in K_2 : v(A, H') = t\} \, \mathbf{R}_{f_0} H \text{ iff } K_2 \subseteq \{H' \in K_2 : v(A, H') = t\}.$$

The "normal" modal logics result upon replacing K_2 on the right by its subset: $\{H' \in K_2 : HR_{f_0}H'\}$; whence the usual rule of '$\Box$': $v(\Box A, H) = t$ iff, for every $H' \in K_2$, if $HR_{f_0}H'$ then $v(A, H') = t$; for $H \in K_2$. The customary array of modal and tense logics emerge, only as significance logics, by imposing further conditions on relation R_{f_0}. (For details, which are readily transposed to significance modal and tense logics, see Montague [1].) It is of course not necessary to replace $\mathbf{R}_{f_0}$: it suffices to require that for each $H \in K_2$ $\mathbf{R}_{f_0}$ satisfies the condition that for some wff C (C_H let us say), $\{H' \in K_2 : v(A, H') = t\} \, \mathbf{R}_{f_0} H$ when and only when $\{H' \in K_2 : v(C, H') = t\} \subseteq \{H' \in K_2 : v(A, H') = t\}$. But under these conditions R_{f_0} can be introduced, as above, upon defining $HR_{f_0}H'$ iff $v(C_H, H') = t$; whence $\{H' \in K_2 : HR_{f_0}H'\} = \{H' \in K_2 : v(C_H, H') = t\}$, and so $\{H' \in K_2 : HR_{f_0}H'\} \, \mathbf{R}_{f_0} H$. It follows at once, using the condition and definitional connections, that $v(\Box A, H) = t$ iff $\{H' \in K_2 : HR_{f_0}H'\} \subseteq \{H' \in K_2 : v(A, H') = t\}$ for $H \in K_2$. For the "non-normal" modal logics, which we shall consider in detail shortly, the general requirement on R_{f_0} further restricts H to a subset K_3 (of modally normal states-of-affairs) of K_2, as follows: $\alpha \mathbf{R}_{f_0} H$ iff both $\{H' \in K_2 : HR_{f_0}H'\} \subseteq \alpha$ and $H \in K_3$, for $\alpha \in \mathscr{P}K_2$. As a final example we give the mixed conditions on relation $\mathbf{R}\Rightarrow$ required for significance entailment logics built on system $\mathbf{NS}_6$: $\langle \{H' \in K_1 : v(A, H') = t\}, \{H' \in K_1 : v(B, H') = t\} \rangle \, \mathbf{R}\Rightarrow H$ iff for every $H' \in K_1$ and $H' \in K_1$ if $R\Rightarrow HH'H''$ and $v(A, H') = t$ then $v(B, H'') = t$; for $H \in K_1$. For basic entailment we require further that if $R\Rightarrow G H'H''$ then $H' \subseteq H''$; an array of entailment systems results on imposing further conditions on $R\Rightarrow$. (See Routley and Meyer [1].)

Formally pleasing though the synthesis of intensional logics is, it is not always illuminating—especially in view of the complex conditions required to derive such evaluation rules as the rule for connective '□' in "normal" modal logics. However such an objection to the general semantics is likely to devolve to an objection to the usual rule for '□' itself: that the fundamental relation R_{f_0} is unexplained, and that it is not explained through such metaphors as accessibility between worlds. Even if an accessibility account of inter-world relations can be made out, the account does not extend satisfactorily to relations between set-ups which are not world-like, especially such set-ups as Rafferty set-ups.

For a better account we first distinguish, from among one-place connectives, &-distributive connectives and ∨-distributive connectives. Connective Φ is *&-distributive* iff, for every wff A and B, $\Phi(A \& B) \simeq . \Phi A \& \Phi B$; and is *∨-distributive* iff, for every A and B, $\Phi(A \vee B) \simeq . \Phi A \vee \Phi B$. The necessity-type connectives, symbolised '□', for which we have given semantics, are &-distributive, whereas possibility-type connectives, symbolised '◇', are ∨-distributive. Whereas the normal evaluation rule for an &-distributive connective Φ over situations is that already given, *viz.*

$$v(\Phi A, \mathrm{H}) = t \text{ iff } (U\mathrm{H}')(\mathrm{HR}_\Phi \mathrm{H}' \supset v(A, \mathrm{H}') = t),$$

the normal rule for a ∨-distributive connective Ψ over situations is:

$$v(\Psi A, \mathrm{H}) = t \text{ iff } (P\mathrm{H}')(\mathrm{HR}_\Psi \mathrm{H}' \ \&\ v(A, \mathrm{H}') = t).$$

We fixed the connexion between $\mathbf{R}_{f_0}$ and R_{f_0} so that the &-distributive rule resulted: that some other rule did not result reveals both how contrived the connexion is, and that the &-distributive rule must be explained independently of the rule using $\mathbf{R}_{f_0}$.

Consider then an &-distributive connective Φ of modal or propositional order, *e.g.* '□' or '*Be*' (read 'x_0 believes (that)'). As a propositional function determined through its values, what Φ really does (as V. Routley has pointed out to us) is to pick out a set-up $\mathrm{H}' = \{A : \Phi A \text{ is true}\}$; *e.g.* '□' picks out the class of necessary truths, and '*Be*' selects x_0's belief set. Generalising, what Φ does in arbitrary situation H is to pick out another set-up $\mathrm{H}' = \{A : v(A, \mathrm{H}) = t\}$. If we represent the relation between H and H′ so determined by Φ as R'_Φ then the evaluation rule:

$v(\Phi A, \mathrm{H}) = t$ iff $v(A, \mathrm{H}') = t$ where $\mathrm{HR}'_\Phi \mathrm{H}'$, follows. Though the connexion already gives an intelligible semantical rule for assessing Φ, the set-up H′ so introduced may not be logically tractable. Though H′ will be &-normal when H is a situation since Φ is &-distributive, it will not in general be ∨-normal. What we can do when H is a situation is to *complete* H′ by adding further wff to it, in the hope of eventually obtaining a situation. H′ will not have a unique completion but a set $\{\mathrm{H}_1, \mathrm{H}_2, ..., \mathrm{H}_i, ...\}$ of alternative completions.

We complete a set by repeated application of the steps of disjunctive and conjunctive extension. Where H is a set-up which contains $(B \vee C)$ but contains neither B nor C, we form two disjunctive extensions $H \cup \{B\}$ and $H \cup \{C\}$ of H; we then supplement these new sets to ensure that they are &-normal. By organising the extensions according to Lindenbaum's method (see Lemma 2 below) we can arrive finally at the completions of H′ by taking the unions of each extension branch of the tree of alternative extensions of H′. Given simple axiomatic conditions on Φ these completions will be situations. The method is indicated roughly by considering how we might complete x_0's incomplete belief set by resolving each alternative that his belief set leaves open. Where $H'R''_{\Phi}H_i$, *i.e.* H_i is a Φ-completion of H′, then if $v(A, H') = t, v(A, H_i) = t$ since $H' \subseteq H_i$; and H′ is the intersection of its alternative Φ-completions. Since, then, $v(\Phi A, H) = t$ iff $v(A, H') = t$ where $HR_{\Phi}H'$, $v(\Phi A, H) = t$ iff $v(A, H_i) = t$ for every H_i which is a Φ-completion of H′, *i.e.* $v(\Phi A, H) = t$ iff $v(A, H_i) = t$ for every H_i such that $HR_{\Phi}H_i$ where R_{Φ} is the relative product of selection and completion relations. That is, $HR_{\Phi}H_i$ iff $HR'_{\Phi}|R''_{\Phi}H_i$ *i.e.* $(PH')(HR'_{\Phi}H' \;\&\; H'R''_{\Phi}H_i)$. The account thus explains both why we adopt the normal evaluation rule for &-distributive Φ, and the role of relation R_{Φ}; and it also explains why we can, though less revealingly, replace R_{Φ} by a relation $\mathbf{R}_{\Phi}$ between situation H and a set of situations, namely the alternative completions of H′. Finally, the account suggests what will be our canonical characterisation of R_{Φ}, as the relation $\hat{H}\hat{H}_i(UA)(v(\Phi A, H) = t \supset v(A, H_i) = t)$. One half of the illuminating connexion, $HR_{\Phi}H_i$ iff $(UA)(v(\Phi A, H) = t \supset v(A, H_i) = t)$ is immediate; for the converse we await a completeness result (*Theorem* (6) below for modal logics.)

§4 *Multiply modal significance logics*

For the formal axiomatic development of intensional significance logics we begin by enriching S_6—or S_4 since most of our claims and results could as well be based on it—with further intensional connectives. For simplicity we now restrict discussion to one-place &-distributive sentence connectives and thereby to the class of connectives definable in terms of these one-place connectives together with S_6 connectives. Thus we shall deal with significance versions of two-place connectives such as strict implication but not with such irreducibly two-place connectives as entailment, and identity of logical content.

Consider then an extension $\mathbf{IS}_6$ of S_6 which adds to the morphology of S_6 m one-place connective constants, $\Phi_1, \ldots, \Phi_m$. To ensure that these connectives are indeed sentence connectives we introduce for each i, the formation rule:

If A is a wff then $\Phi_i A$ is a wff (of the relevant system).

But this step of simply introducing unexplained symbolism raises more questions than it answers. In particular the questions arise: What axioms do these connectives satisfy? What guarantees that these connectives are intensional?

There is at present no substantial agreement as to precisely which axioms even the most logically familiar English intensional connectives satisfy. Though, for example, we should defend the claim that the connective 'It is logically necessary (that ...)' satisfies the set of (Lewis S5) axioms given for $S5S_6$, well-known objections have been made to this claim (for our defence against some of these objections see Routley [5]). The main disagreement over the sentential formalisation of logical necessity turns on whether $\Box 3$ should be accepted or weakened to an S4 axiom $\Box 3'. \vdash \Box p \rightarrow \Box\Box p$. A related controversy affects the axiomatisation of (absolute) modal connective '*Pr*' read 'It is probable (that)' and its two-place mate, namely as to whether the principle $Prp \rightarrow \Box Prp$ holds.

The position with regard to more highly intensional connectives is much worse; which axioms belief and rational belief notions, for example, satisfy is at present opaque, and proposals on the matter are likely to be highly controversial. Hintikka (in [1] and [2]) has suggested that, given proper qualifications, belief satisfies an S4° modal structure and knowledge an S4 modal structure. But the paradoxes of material and strict implication suffice to destroy this well-hedged claim. For it is simply not true, and not defensible, that for any necessary truth A, or even any theorem A of S4° belief logic, that person x_0 believes A, or for that matter can be brought to believe A. Part of the trouble with Hintikka's suggestion can undoubtedly be traced back to the comparative strength of "normal" modal logics about S4. So it may be thought that all the troubles can be avoided by retreating to much weaker Lewis or Lemmon modal systems, systems which do not contain such epistemically damaging principles as the rule of necessitation (R$\Box$: doxastically, if A is a theorem then that x_0 believes A is a theorem) and the transitivity principle ($\Box 3'$: doxastically, if x_0 believes A then x_0 believes that he believes that A.) It is already apparent, however, without further clarification of the epistemic notions concerned, that even the minimal modal systems, C1 and S0.5°, contain principles which render them quite unsuitable for application as epistemic logics. For all these logics contain as a derived or primitive rule

RD. $\vdash A \rightarrow B \twoheadrightarrow \vdash \Phi A \rightarrow \Phi B$, where $A \rightarrow B$ is a substitution instance of a (S_6-) tautology.

Here Φ is a parameter standing in for relevant epistemic connectives, such as 'x_0 believes (that ...)', 'y_0 knows (that ...)'. (In the case of systems S0.5° and S1, RD follows from the rule of tautology, $\vdash \Phi A$, where A is a tautology, together with distribution, $\Box 2$). Because of the paradoxes of implication a false statement implies an arbitrarily selected assertion, even a nonsignificant one, 'q_0' say; hence, by RD, if Quine believes some false statement then he believes q_0! Likewise if he believes (or knows) any truth, he believes (knows) any other truth! Modal logics accordingly break down under such epistemic interpretations as soon as they are applied. Even if the assertion of contingent truths—surely permissible in an application—is ruled out in an effort to refloat the logic, difficulties arise from applications within logic; for it is not true in general of any arbitrarily chosen logical truth that even Quine knows and believes it.

Similar problems confront *modal* analyses of such notions as obligation, logical consequence, cause and subjunctive conditional (for such analyses see, *e.g.* Montague [1]). For example, Montague's proposed analysis of the future conditional: 'If it should be the case that A, it would (at the same time) be the case that B' as '$\sim \Diamond F(A \,\&\! \sim B)$', where 'F' is Prior's futurity connective and '$\Diamond$' is a temporal possibility connective, succumbs at once to paradoxes.

It would be a mistake to jump from the inadequacy of modal logics as epistemic or deontic logics, and the inadequacy of modal analyses of conditional notions, to scepticism about the possibility of epistemic logic and about the possibility of analyses of conditionals. For many logical principles which fail to hold for the relevant notions, *e.g.* the sceptic's principle $\Phi(A \rightharpoondown B) \rightharpoondown . A \rightharpoondown \Phi B$, are widely recognised as failing to hold; and many logical principles which do hold for the relevant notions are uncontroversial. For example, we should have as theses: x *believes* $(p \,\&\, q) \rightharpoondown x$ *believes* p; x *knows* $p \rightharpoondown x$ *believes* p; x *knows* $p \rightharpoondown . p$, and, where 'z' is a significant subject, $Sp \rightharpoondown S$ (z *believes* p). Thus there is room for epistemic logics which marshall these sets of principles; and indeed there is a serious need for the logics in order to investigate the consequences of common epistemological assumptions. However the main problems concerning the formulation of adequate epistemic logics—many of which would be resolved by finding suitable semantical analyses for rival syntactical accounts—are not problems peculiar to significance logic: the main problems arise at the two-valued stage. Indeed once the problems of setting up satisfactory epistemic logics, based on classical sentential logic for significant sentences, are solved, there should be little difficulty in extending the solutions to significance epistemic logics, along similar lines to those we shall use to extend standard modal logics to significance modal logics. Moreover we believe that semantical analyses of some at least of the important non-modal intensional notions can be provided within the framework of the theory of situations (see semantical $\mathbf{NS}_6$ above).

The multiply modal system $\mathbf{IS}_6$ that we study combines significance versions of analyses that have been claimed, quite exorbitantly, to capture at the sentential level, not only necessity, entailment and provability and epistemic notions such as knowledge, belief and verification, but also such diverse notions as obligation, permissibility and imperative notions, judgement, decision and cause, pastness and futurity under various accounts of time, and even referential opacity generally (for details see *e.g.* Montague [1] and Rennie [1]). Apart from our claim concerning logical modalities, however, we risk only the much weaker claim that these analyses constitute important steps in the direction of obtaining satisfactory analyses.

We shall suppose that our one-place sentence connectives divide into several classes, and we shall treat in detail two classes of these connectives: the L-connectives $\Phi_1, ..., \Phi_g$ and the F-connectives $\Phi_j, ..., \Phi_m$, where $j \geqslant g+1$. We suppose that each L-connective satisfies at least the postulates of a significance version of Lemmon's system C2, *viz.* for each i, $1 \leqslant i \leqslant g$,

D. $\vdash \Phi_i(p \rightharpoonup q) \rightharpoonup . \Phi_i p \rightharpoonup \Phi_i q$ (Distribution)
RD. $\vdash A \rightharpoonup B \longrightarrow \vdash \Phi_i A \rightharpoonup \Phi_i B$ (Rule Distribution)

and that each F-connective satisfies at least the postulates of a significance version of Fey's system S2°, *viz.* for each i, $j \leqslant i \leqslant m$, the following rules obtain:

RB. $\vdash \Phi_i(A \rightharpoonup B) \longrightarrow \vdash \Phi_i(\Phi_i A \rightharpoonup \Phi_i B)$ (Becker Rule)
RT. $\vdash \Phi_i A$, provided A is an S_6-tautology or axiom D.
RS. $\vdash \Phi_i A \longrightarrow \vdash A$.

Finally each L-connective and each F-connective satisfies

J1. $\vdash S\Phi_i p \rightharpoonup Sp$; and
J2. $\vdash Sp \rightharpoonup S\Phi_i p$.

J1 and J2 facilitate the conversion of axiomatisations of modal connectives on the basis of two-valued logic into significance axiomatisations. For given an axiomatisation of connective, Φ_i say, in terms of axioms written using the connective set $\{\sim, \supset, \Phi_1, ..., \Phi_m\}$ *a* significance axiomatisation can be simply obtained by rewriting the postulates in terms of the set $\{\neg, \rightharpoonup, \Phi_1, ..., \Phi_m\}$ and adding J1 and J2 for each Φ_i; then a significance analysis of the notion results. In this way significance axiomatisations, and semantics, can be obtained for classes of connectives corresponding to the connectives of weak modal logics, such as Cl, S0.5 and Sl.

Theorem (4): There are no finite characteristic matrices for the F-connectives and L-connectives of $\mathbf{IS}_6$.

Proof applies the standard proof for modal logics (given in Dugundji [1]); for if there were such matrices, so would there be in the two-valued case.

A connective may satisfy a number of conditions beyond those minimal conditions required of F- and L-connectives; it may for example satisfy any consistent selection from the following sample postulates, given for Φ_i.

1. $\Phi_i p \rightharpoonup p$
2. $\Phi_i p \rightharpoonup . \neg \Phi_i \neg p$
3. $p \rightharpoonup \Phi_i \neg \Phi_i \neg p$
4. $\Phi_i(p \rightharpoonup q) \rightharpoonup \Phi_i(\Phi_i p \rightharpoonup \Phi_i q)$
5. $\Phi_i t \rightharpoonup \Phi_i \Phi_i t$
6. $\Phi_i p \rightharpoonup \Phi_i \Phi_i p$
7. $\neg \Phi_i p \rightharpoonup \Phi_i \neg \Phi_i p$
8. $\neg \Phi_i \Phi_i p$
9. $\Phi_i p \rightharpoonup D\Phi_i Dp$
10. $D\Phi_i Dp \rightharpoonup \Phi_i D\Phi_i Dp$
11. $\Phi_i p \rightharpoonup . \sim \Phi_i \sim p$
12. $\sim \Phi_i p \rightharpoonup \sim \Phi_i \sim p$

Also where many connectives are considered at once, a set of connectives may satisfy axioms which interrelate the notions. Simple examples already noticed are $\Box p \rightharpoonup \Diamond p$ and *x knows* $p \rightharpoonup$ *x believes p*. Typical axioms which can be added include

13. $\Phi_{i_1} \ldots \Phi_{i_k} p \rightharpoonup p$ 14. $\Phi_{i_1} \ldots \Phi_{i_k} p \rightharpoonup \Phi_{s_1} \ldots \Phi_{s_m} p$.

(For a treatment of multiple conditions for "normal" modal connectives in the two-valued case, see Rennie [1]). Where Φ_i is an F-connective, Φ_i may also satisfy $\Phi_i \gamma$ where γ is any one of postulates 1–14.

A semantical analysis of truth and validity for $\mathbf{IS}_6$ can be provided, along the lines which generalise on standard semantics for modal logics discussed above. An $\mathbf{IS}_6$-*model* $\mathcal{M}$ is a structure $\mathcal{M} = \langle G, K, N_1, \ldots, N_g, N_j, \ldots, N_m, R_1, \ldots, R_g, R_j, \ldots R_m, v\rangle$, where $K = K_2$ is a set (of set ups, states-of-affairs, or contexts), G (the actual set up, the current context) is an element of K, each N_i is a subset of K such that, for $j \leqslant i \leqslant m$, $G \in N_i$, each R_i is a two place relation on K, and v is a valuation function from initial wff and elements of K to $V = \{t, f, n\}$. A wff is *initial* iff it is atomic or is of the form $\Phi_i A$ where $g < i < j$. Of course if axiomatic conditions were imposed on these Φ_i, for $g < i < j$, then values could not be arbitrarily assigned to $\Phi_i A$. According as further axiomatic conditions are imposed on a given Φ_i, for example from among 1–14, so further model conditions are imposed, conditions which usually affect the corresponding "accessibility" relation R_i. So, for example, with $1 \leqslant i \leqslant g$, when Φ_i satisfies axiom 1, the model conforms to the condition that R_i is reflexive; when Φ_i satisfies axiom 4, then the modelling condition is that R_i is transitive; when $\Phi_{i_1}, \ldots, \Phi_{i_k}$, with $1 \leqslant i_j \leqslant g$, satisfy 13, the iterated relative product $((\ldots R_{i_1} \ldots | (R_{i_{k-1}} | R_{i_k}) \ldots)$ is reflexive.

By virtue of the specification of v in a given model, 3-valued holding conditions are specified for all initial wff. Relative to the given model holding conditions are specified recursively for every wff with respect to every set-up or context H in K thus (*i.e.* as for $\mathbf{MS}_6$ for $H \in K_2$):

1. $v(\neg A, H) = t$ iff $v(A, H) \neq t$
   $v(\neg A, H) = f$ iff $v(A, H) = t$

2. $v(A \rightharpoonup B, H) = t$ iff $v(A, H) \neq t \vee v(B, H) = t$
   $v(A \rightharpoonup B, H) = f$ iff $v(A, H) = t \ \& \ v(B, H) = f$
   $v(A \rightharpoonup B, H) = n$ iff $v(A, H) = t \ \& \ v(B, H) = n$

3. $v(DA, H) = t$ iff $v(A, H) = n$
   $v(DA, H) = f$ iff $v(A, H) = f$
   $v(DA, H) = n$ iff $v(A, H) = t$.

These valuations are of course precisely those specified for S_6 relativised to each $H \in K$. Only the rules for Φ_i are new:

4. $v(\Phi_i A, \mathrm{H}) = t$ iff $(U\mathrm{H}')(\mathrm{HR}_i\mathrm{H}' \supset v(A, \mathrm{H}') = t)$ & $\mathrm{H} \in \mathrm{N}_i$
$v(\Phi_i A, \mathrm{H}) = n$ iff $v(A, \mathrm{H}) = n$
$v(\Phi_i A, \mathrm{H}) = f$ otherwise.

This specification is made for each i such that $1 \leqslant i \leqslant g$ and $j \leqslant i \leqslant m$. In the case of $g < i < j$, $v(\Phi_i A, \mathrm{H})$ is specified by the model—effectively $\Phi_i A$ is treated as a new atomic wff.

A wff B is *true in* $\mathbf{IS}_6$-model $\mathcal{M}$ iff $v(B, \mathrm{G}) = t$, *false in* $\mathcal{M}$ iff $v(B, \mathrm{G}) = f$, *nonsignificant in* $\mathcal{M}$ iff $v(B, \mathrm{G}) = n$. B is $\mathbf{IS}_6$-*valid* iff B is true in every $\mathbf{IS}_6$-model. $\mathcal{M}$ *satisfies* B(set S) iff B(every $B \in$ S) is true in $\mathcal{M}$; $\mathcal{M}$ is an $\mathbf{IS}_6$-*counter-model* to B iff B is not true in $\mathcal{M}$.

These analyses of truth and validity simply extend analogous notions defined for $\mathbf{S}_6$. There, however, a model was simply an assignment of specific values to each variable—those written at the start of a single line of a valuation table together with arbitrary assignments for remaining variables not occurring in the table—and not distinctive enough for separate treatment. But as before validity is defined as truth in every model, *i.e.* truth for every assignment of values to atomic wff in the case of $\mathbf{S}_6$.

The structure $\mathcal{M}$ can be expressed in the more manageable form $\langle \mathrm{G}, \mathrm{K}, v \rangle$ by assigning v a larger task, effectively by letting $\mathrm{R}_i = v(\Phi_i)$ and $\mathrm{N}_i = v(\mathrm{K}, \Phi_i)$.

§5 *Soundness and Completeness Theorems*

We show now that $\mathbf{IS}_6$-validity is tantamount, as it was designed to be, to $\mathbf{IS}_6$-theoremhood.

Theorem (5): If B is a theorem of $\mathbf{IS}_6$ then B is $\mathbf{IS}_6$-valid.

Proof is by induction over the length of the proof of B. We give some illustration of details. Suppose, to show by reductio that axiom D is valid, that for some H in some model and for some i:

$$1 \leqslant i \leqslant g \text{ or } j \leqslant i \leqslant m \text{ and } v(\Phi_i(p \rightharpoonup q) \rightharpoonup .\Phi_i p \rightharpoonup \Phi_i q, \mathrm{H}) \neq t.$$

In case $1 \leqslant i \leqslant g$, H is the base set-up G of the model, and in case $j \leqslant i \leqslant m$, H is either G or else, to deal with □D obtained from D by RT, H is such that GRH. Applying the rule for '$\rightharpoonup$':

$$v(\Phi_i(p \rightharpoonup q), \mathrm{H}) = t \ \& \ v(\Phi_i p, \mathrm{H}) = t \ \& \ v(\Phi_i q, \mathrm{H}) \neq t.$$

Hence

$$(U\mathrm{H}')(\mathrm{HR}_i\mathrm{H}' \supset v(p \rightharpoonup q, \mathrm{H}) = t) \& (U\mathrm{H}')(\mathrm{HR}_i\mathrm{H}' \supset v(p, \mathrm{H}) = t) \& \mathrm{H} \in \mathrm{N}_i.$$

But also $\mathrm{H} \notin \mathrm{N}_i \vee (P\mathrm{H}')(\mathrm{HR}_i\mathrm{H}' \ \& \ v(q, \mathrm{H}') \neq t)$. Hence for some set-up, H_1 say, $\mathrm{HR}_i\mathrm{H}_1$ & $v(q, \mathrm{H}_1) \neq t$; and also $v(p \rightharpoonup q, \mathrm{H}_1) = t$ & $v(p, \mathrm{H}_1) = t$. Thus $v(q, \mathrm{H}_1) = t$ and $v(q, \mathrm{H}_1) \neq t$, a contradiction. Suppose next $(S\Phi_i p \rightharpoonup Sp, \mathrm{H}) \neq t$ for some H in some model for relevant i; then $v(S\Phi_i p, \mathrm{H}) = t$ & $v(Sp, \mathrm{H}) \neq t$; so $v(\Phi_i p, \mathrm{H}) \neq n$ & $v(p, \mathrm{H}) = n$, which yields a contradiction; thus J1 is verified.

To show that the rule RD of distribution preserves validity requires a further detail. Suppose, again using reductio, that for some model $v(\Phi_i A \rightharpoonup \Phi_i B, \mathrm{H}) \neq t$. Then, as with axiom D, it follows that for some H_1 such that $\mathrm{HR}_i\mathrm{H}_1$, $v(A, \mathrm{H}_1) = t$ & $v(B, \mathrm{H}_1) \neq t$, so that $v(A \rightharpoonup B, \mathrm{H}_1) \neq t$. But we have here a countermodel with base H_1 to $A \rightharpoonup B$, contradicting the premiss that it is valid.

The completeness theorem wants further preliminaries. A set ∇ of wff of $\mathbf{IS}_6$ is $\urcorner$*-inconsistent* iff for some wff E and some $A_1, ..., A_n \in \nabla$, $A_1, ..., A_n \vdash E$ and $A_1, ..., A_n \vdash \urcorner E$, where a proof, $A_1, ..., A_n \vdash B$, of B from hypothesis $A_1, ..., A_n$ is defined as for $\mathbf{S}_6$ (see also §7.7): theorems of $\mathbf{IS}_6$ may of course occur in the proof sequence. Otherwise ∇ is $\urcorner$*-consistent*. A set ∇ of wff of $\mathbf{IS}_6$ is *maximal* $\urcorner$*-consistent* iff ∇ is $\urcorner$-consistent and every wff C $\urcorner$-consistent with ∇ belongs to ∇, where C is $\urcorner$*-consistent with* Γ iff $\Gamma \cup \{C\}$ is $\urcorner$-consistent.

Lemma: Every $\urcorner$-consistent set ∇ of wff of $\mathbf{IS}_6$ has a maximal $\urcorner$-consistent extension ∇^+, with $\nabla \subseteq \nabla^+$.

Proof: Using some standard method, *e.g.* Gödel numbering, enumerate the wff of $\mathbf{IS}_6$ in sequence $B_0, ..., B_j, ...$ so that C precedes $\urcorner C$ and DC, and C and E precede $C \rightharpoonup E$.

Define:
$$\nabla_0 = \nabla$$
$$\nabla_{j+1} = \nabla_j, \text{ if } B_j \text{ is } \urcorner\text{-inconsistent with } \nabla_j$$
$$\nabla_{j+1} = \nabla_j \cup \{B_j\}, \text{ if } B_j \text{ is } \urcorner\text{-consistent with } \nabla_j$$
$$\nabla^+ = \bigcup_{j<\omega} \nabla_j$$

Then each ∇_j is $\urcorner$-consistent; for ∇_0 is $\urcorner$-consistent since ∇ is, and where ∇_k is $\urcorner$-consistent ∇_{k+1} is, by construction. Accordingly ∇^+ is $\urcorner$-consistent, since if it were not some ∇_k would be $\urcorner$-inconsistent as a proof of inconsistency uses only finitely-many wff. Also ∇^+ is maximal. Suppose some wff, B_k say, is $\urcorner$-consistent with ∇^+. Then, by construction of ∇^+, B_k is $\urcorner$-consistent with ∇_k. But then $B_k \in \nabla_{k+1}$, and so $B_k \in \nabla^+$.

Lemma: Where ∇ is a maximal $\urcorner$-consistent set of $\mathbf{IS}_6$,

(i) Every theorem of $\mathbf{IS}_6$ belongs to ∇.

(ii) ∇ is closed under $\rightharpoonup$-detachment, *i.e.* if $A \in \nabla$ and $A \rightharpoonup B \in \nabla$ then $B \in \nabla$.

(iii) $TA \in \nabla$ iff $A \in \nabla$.

(iv) $A \notin \nabla$ iff $\urcorner A \in \nabla$.

Proofs: (i) Suppose some theorem $A \notin \nabla$. Then $\nabla \cup \{A\}$ is $\urcorner$-inconsistent. But then ∇ is $\urcorner$-inconsistent.

(ii) Suppose $A \in \nabla$ and $A \rightharpoonup B \in \nabla$ but $B \notin \nabla$. Then for some E, and $C_1, ..., C_n \in \nabla$, $C_1, ..., C_n, B \vdash E$ and $C_1, ..., C_n, B \vdash \urcorner E$, whence $C_1, ..., C_n, A, A \rightharpoonup B \vdash \urcorner B$. But also $C_1, ..., C_n, A, A \rightharpoonup B \vdash B$, contradicting the $\urcorner$-consistency of ∇.

(iii) follows from (i) and (ii) using $Tp \rightharpoonup p$ and $p \rightharpoonup Tp$.

(iv) If $\neg A \in \nabla$ then $A \notin \nabla$ for if it did $\neg$-consistency would ensue. If $A \in \nabla$ then $\nabla \cup \{A\}$ is $\neg$-inconsistent. Hence for $C_1, ..., C_n \in \nabla$, $C_1, ..., C_n \vdash \neg A$. By the deduction theorem $\vdash C_1 \rightharpoonup . C_2 ... \rightharpoonup . C_n \rightharpoonup \neg A$, and by (i) and (ii) $\neg A \in \nabla$.

A *mac* set (of **L**) is a maximal $\neg$-consistent set of $\mathbf{IS}_6$ (of system **L** which includes $\mathbf{IS}_6$ and is closed under its rules), and a *con* set (of **L**) is a $\neg$-consistent set of $\mathbf{IS}_6$ (of **L**).

Where system $\mathbf{IS}_6$ is obtained by requiring that F-connectives $\Phi_{i_1}, ..., \Phi_{i_n}$ satisfy the Φ-axioms $\Phi_{i_1}A_1, ..., \Phi_{i_p}A_p$, the *filter system* $\mathbf{IS}_6$* of $\mathbf{IS}_6$ is the system which differs from $\mathbf{IS}_6$ just by replacing $\Phi_j, ..., \Phi_m$ by L-connectives $\Phi_j, ..., \Phi_m$ which satisfy the additional axioms $\Phi_{i_1}p \rightharpoonup A_1, ..., \Phi_{i_p}p \rightharpoonup A_p$. (As to how to define the filter system in other cases see Routley [6]).

Lemma: Where ∇ is a mac set which contains both wff $\neg\Phi_i A$ and $\Phi_i B$ for some A and B, where connective Φ_i satisfies axiom D and rule RD, and where ∇^1 is the set whose elements are $\neg A$ and every wff C such that $\Phi_i C$ is in ∇, ∇^1 is a con set.

Proof: Suppose ∇^1 is not a con set. Then for some $C_1, ..., C_k$ with $\Phi_i C_1, ..., \Phi_i C_k$ in ∇ and $k \geqslant 1$, and for some E,

$$C_1, ..., C_k, \neg A \vdash E \text{ and } C_1, ..., C_k, \neg A \vdash \neg E.$$

By the deduction theorem for $\mathbf{S}_6$, $C_1, ..., C_k \vdash A$, whence $\vdash C_1 \rightharpoonup . C_2 .... C_k \rightharpoonup A$. Then by RD, $\vdash \Phi_i C_1 \rightharpoonup \Phi_i(C_2 ... \rightharpoonup A)$, and by repeated uses of axiom D, $\vdash \Phi_i C_1 \rightharpoonup . \Phi_i C_2 .... \Phi_i C_k \rightharpoonup \Phi_i A$.

Then by the previous lemma, since ∇ is a mac set, $\Phi_i A \in \nabla$, contradicting the $\neg$-consistency of ∇.

Theorem (6): If Γ is a con set of $\mathbf{IS}_6$ then there is an $\mathbf{IS}_6$-model

$$\mathscr{M} = \langle G, K, N_1, ..., N_g, N_j, ..., N_m, R_1, ..., R_g, R_j, ..., R_m, v \rangle$$

which satisfies Γ.

Proof: Define a canonical $\mathbf{IS}_6$-model $\mathscr{M}$ as follows:

(i) $G = \Gamma^+$ where Γ^+ is the $\mathbf{IS}_6$ maximal extension of Γ provided by a preceding lemma.

(ii) K is the set of mac sets of $\mathbf{IS}_6$*.

(iii) N_i is that subset of K to which some wff of the form $\Phi_i B$ belongs, for each i, $1 \leqslant i \leqslant g$ or $j \leqslant i \leqslant m$.

(iv) $H_1 R_i H_2$ obtains iff for every wff B, $\Phi_i B \in H_1$ materially implies $B \in H_2$, for each relevant i and of course for $H_1, H_2 \in K$.

(v) $v(P, H) = t$ iff $TP \in H$
$v(P, H) = f$ iff $FP \in H$

$v(P, \mathrm{H}) = n$ iff $\neg SP \in \mathrm{H}$;
for every $\mathrm{H} \in \mathrm{K}$ and every P which is either atomic or of the form $\Phi_i B$ with $g < i < j$.

Then:

(α). $\mathscr{M}$ is an $\mathbf{IS}_6$-model. The following points complete the details. $\mathrm{G} \in \mathrm{K}$, for if G is a mac set of $\mathbf{IS}_6$ then G is a mac set of $\mathbf{IS}_6{}^*$ since every theorem of $\mathbf{IS}_6{}^*$ is a theorem of $\mathbf{IS}_6$. $\mathrm{G} \in \mathrm{N}_i$ for each i such that $j \leqslant i \leqslant n$, since $\vdash \Phi_i(p \rightharpoonup p)$ by RT.

(β). For every wff C and every $\mathrm{H} \in \mathrm{K}$,
$v(C, \mathrm{H}) = t$ iff $TC \in \mathrm{H}$,
$v(C, \mathrm{H}) = f$ iff $FC \in \mathrm{H}$, and
$v(C, \mathrm{H}) = n$ iff $\neg SC \in \mathrm{H}$.

Proof of (β) is by induction over the number of connectives in A, on the basis of (v). The induction steps are these:

Where C is DA for some A,
$v(DA, \mathrm{H}) = t$ iff $v(A, \mathrm{H}) = n$, by a valuation rule
iff $\neg SA \in \mathrm{H}$, by induction hypothesis
iff $TDA \in \mathrm{H}$, since $\neg Sp \simeq TDp$.
$v(DA, \mathrm{H}) = f$ iff $v(A, \mathrm{H}) = f$; *i.e.* iff $FA \in \mathrm{H}$
iff $FDA \in \mathrm{H}$, since $Fp \rightharpoonup FDp$ and $FDp \rightharpoonup Fp$ and H is a mac set.
$v(DA, \mathrm{H}) = n$ iff $v(A, \mathrm{H}) = t$; *i.e.* iff $TA \in \mathrm{H}$
iff $\neg SDA \in \mathrm{H}$, since $Tp \simeq \neg SDp$.

The steps for '$\neg$' and '$\rightharpoonup$' are similar; they simply use the appropriate valuation rules, the induction hypothesis, theorems of $\mathbf{S}_6$ and properties of mac sets, and details are similar to those provided in the quantificational case in §7.7.

Where C is $\Phi_i A$, the result follows from (v) where $g < i < j$.
For the remaining cases we prove

(1) if $T\Phi_i A \in \mathrm{H}$ then $v(\Phi_i A, \mathrm{H}) = t$. For first if $T\Phi_i A \in \mathrm{H}$ then $\Phi_i A \in \mathrm{H}$ whence $\mathrm{H} \in \mathrm{N}_i$. Also if $\Phi_i A \in \mathrm{H}$ then $(U\mathrm{H}')((UB)(\Phi_i B \in \mathrm{H} \supset B \in \mathrm{H}') \supset . A \in \mathrm{H}')$, whence applying the induction hypothesis and (iv), $(U\mathrm{H}')(\mathrm{HR}_i\mathrm{H}' \supset v(A, \mathrm{H}') = t)$.

(2) if $v(\Phi_i A, \mathrm{H}) = t$ then $T\Phi_i A \in \mathrm{H}$. Suppose $T\Phi_i A \notin \mathrm{H}$. Then $\neg \Phi_i A \in \mathrm{H}$, and since $v(\Phi_i A, \mathrm{H}) = t$, $\Phi_i E \in \mathrm{H}$ for some E since $\mathrm{H} \in \mathrm{N}_i$, and
$(U\mathrm{H}')((UB)(\Phi_i B \in \mathrm{H} \supset B \in \mathrm{H}') \supset A \in \mathrm{H}')$. Hence by a lemma there is a con set Γ such that $\neg A \in \Gamma$ and
$(UB)(\Phi_i B \in \mathrm{H} \supset B \in \Gamma)$. Let H_1 be the maximal extension of Γ. Then $\neg A \in \mathrm{H}_1$; but also since $B \in \Gamma \supset B \in \mathrm{H}_1$, $A \in \mathrm{H}_1$, contradicting the $\neg$ - consistency of H_1.

(3) $\neg S\Phi_i A \in \mathrm{H}$ iff $v(\Phi_i A, \mathrm{H}) = n$;
for $\neg S\Phi_i A \in \mathrm{H}$ iff $\neg SA \in \mathrm{H}$; *i.e.* iff $v(A, \mathrm{H}) = n$; *i.e.* iff $v(\Phi_i A, \mathrm{H}) = n$.

(γ). $\mathscr{M}$ satisfies Γ. For any A, if $A \in \Gamma$ then $A \in \mathrm{G}$; so by (β), $v(A, \mathrm{G}) = t$.

(δ). When connectives $\Phi_{i(1)}, ..., \Phi_{i(k)}$ satisfy axiom Aα corresponding modelling condition $\mathscr{M}\alpha$ is met. No technique general enough to handle even the examples we have provided is at present known, so a case by case treatment is inevitable. We content ourselves with some simple cases; the treatment of more complicated cases in Rennie [1] and Routley [6] is readily adaptable.

When $\Phi_i p \rightharpoonup p$ is an axiom for relevant i then R_i is reflexive (on N_i). For, given the axiom, $\Phi_i B \in H \supset B \in H$. Hence HR_iH for every $H(\in N_i)$.

When the rule of necessitation, $\vdash A \twoheadrightarrow \vdash \Phi_i A$, holds for Φ_i then $N_i = K$. For given the rule $\Phi_i(p \rightharpoonup p)$ is a theorem, whence $\Phi_i(p \rightharpoonup p) \in H$ for every $H \in K$, since H is a mac set. Hence $H \in N_i$ iff $H \in K$.

Theorem (7): If B is $\mathbf{IS}_6$-valid then B is a theorem of $\mathbf{IS}_6$.

Proof: If B is $\mathbf{IS}_6$-valid then $\neg B$ is not satisfied in any $\mathbf{IS}_6$-model, by the valuation rule for '$\neg$' and definitions. Hence, by the previous theorem, $\{\neg B\}$ is not a con set of $\mathbf{IS}_6$. Thus for some E, $\neg B \vdash E$ and $\neg B \vdash \neg E$; whence B is a theorem of $\mathbf{IS}_6$.

Theorem (3) *continued:*

(2) B is a theorem of $\mathbf{MS}_6$ iff B is $\mathbf{MS}_6$-valid.

(4) If Γ is a con set of $\mathbf{MS}_6$ then there is an $\mathbf{MS}_6$ model which satisfies Γ.

Proof of one half of (2) is by induction over the length of the proof of wff B. The details are like those given in *Theorem* 5. For the converse, it suffices to prove (4). A canonical model is defined, as follows:

G is an $\mathbf{MS}_6$ maximal extension of Γ; K_2 is the set of mac sets of $\mathbf{MS}_6$; K is the power set of the set of wff of $\mathbf{MS}_6$; $I(A, H) = t$ iff $A \in H$ and $I(A, H) = n$ iff $\sim SA \in H$, for $A \in At$ or $H \in (K - K_2)$; for $\alpha_i \in \mathscr{P}K$, $\langle \alpha_1, ..., \alpha_m \rangle \mathbf{R}_F{}^m H$ iff $(PA_1 ... A_m)$ $[F^m(\tau A_1, ..., \tau A_m) \in H$ & $\langle \{H \in K_\tau: A_1 \in H\},. .., \{H \in K_\tau: A_m \in H\} \rangle = \langle \alpha_1, ..., \alpha_m \rangle]$, where K_τ is K or K_2 according as τA_i is $qu(A_i)$ or $\S A_i$.

Then, as in *Theorem* 6, (β) and (λ) follow, once it is shown

($\in$) $\langle \{H \in K_\tau: A_1 \in H\}, ..., \{H \in K_\tau: A_m \in H\} \rangle \mathbf{R}_F{}^m H$ iff $F^m(\tau A_1, ..., \tau A_m) \in H$.

One half follows by quantification logic. For the converse, suppose for some $A'_1, ..., A'_m$, $F^m(\tau A'_1, ..., \tau A'_m)$ and for every i and $H \in K_\tau$, $A'_i \in H$ iff $A_i \in H$. Then where K_τ is K, $A'_i = A_i$, and where K_τ is K_2, $\vdash A'_i \rightharpoonup A_i$ and $\vdash A_i \rightharpoonup A'_i$ (otherwise mac sets would separate A_i and A'_i). Hence by substitutivity rules, $\vdash F^m(\tau A'_1, ..., \tau A'_m) \rightharpoonup F^m(\tau A_1, ..., \tau A_m)$; so $F^m(\tau A_1, ..., \tau A_m) \in H$.

Where a mixed method is adopted ($\in$) can be used in defining the canonical model.

§6 *Decidability*

The decidability of many multiply modal systems can be readily established through use of an equivalence class technique which facilitates proofs that the systems have the finite model property. Where Δ is a class of wff of $\mathbf{IS}_6$ closed under the operation of taking subformulae define the following relation on K: $H_1 \equiv_\Delta H_2 =_{Df} (UB \in \Delta)(v(B, H_1) = v(B, H_2))$.

Then $\equiv_\Delta$ is an equivalence relation, and equivalence classes can be defined:

$$(\hat{H})_\Delta =_{Df} \{H' : H' \equiv_\Delta H\}.$$

Relative to given class Δ of wff and a given $\mathbf{IS}_6$-model $\mathcal{M}$, a new $\mathbf{IS}_6$-model $\hat{\mathcal{M}}$ is defined, using these equivalence classes. Define a *filtration* $\hat{\mathcal{M}} = \langle \hat{G}, \hat{K}, \hat{N}_1, ..., \hat{N}_m, \hat{R}_1, ..., \hat{R}_m, \hat{v} \rangle$ of $\mathcal{M}$ through Δ, written $\hat{\mathcal{M}} = \mathcal{M}/\Delta$, (following Segerberg [1]) as follows:

$\hat{K} = \{\hat{H} : H \in K\}$; $\hat{N}_i = \{\hat{H} : H \in N_i\}$;
$\hat{H}_1 \hat{R}_i \hat{H}_2 \equiv .(UB \in \Delta)(v(\Phi_i B, H_1) = t \supset v(B_1, H_2) = t)$;
$\hat{v}(p, \hat{H}) = t \equiv .v(p, H) = t \;\&\; p \in \Delta$, and $\hat{v}(p, \hat{H}) = n \equiv .v(p, H) = n \;\&\; p \in \Delta$.

Lemma:
(i) $H_1 R_i H_2 \supset . \hat{H}_1 \hat{R}_i \hat{H}_2$. (ii) $\hat{\mathcal{M}}$ is an $\mathbf{IS}_6$-model.

Proof: (i) follows by quantification logic using the definitions of '$\hat{R}_i$' and '$v(\Phi_i B, H) = t$'. As to (ii), it is immediate from definitions that $\hat{G} \in \hat{K}$, that $\hat{N}_i \subseteq \hat{K}$ and that for $j < i \leqslant m$, $\hat{G} \in \hat{N}_i$.

What is more problematic, and proved in a case by case way, is that $\hat{\mathcal{M}}$ satisfies the modelling conditions corresponding to the further axioms that $\mathbf{IS}_6$ may satisfy. However the proof that $\hat{\mathcal{M}}$ satisfies condition $\hat{\mathcal{M}}\alpha$ in case $\mathcal{M}$ satisfies $\mathcal{M}\alpha$, frequently parallels the proof that where a given connective satisfies $A\alpha$ the corresponding modelling condition $\mathcal{M}\alpha$ is met. Again we simply illustrate elementary soluble cases:

(a) If R_i is reflexive then $\hat{R}_i$ is reflexive, follows at once from (i),
(b) If $N_i = K$ then $\hat{N}_i = \hat{K}_i$.

Lemma: For every wff $A \in \Delta$, $\hat{v}(A, \hat{H}) = v(A, H)$.
Proof establishes by induction over the connectives in A, using as a basis the specification of $\hat{v}(p, H)$ for initial wff,

(i) $\hat{v}(A, \hat{H}) = n$ iff $v(A, H) = n$. (ii) $\hat{v}(A, \hat{H}) = t$ iff $v(A, H) = t$.

The induction steps for '$\neg$', '$\rightharpoondown$' and 'D' are straightforward, since no evaluation leads outside H or $\hat{H}$; *e.g.*

$$\hat{v}(DA, \hat{H}) = t \equiv \hat{v}(A, \hat{H}) = n \equiv v(A, H) = n \equiv v(DA, H) = t.$$

Similarly for (i) in the case of $\Phi_i A$, where this is not initial:

$$\hat{v}(\Phi_i A, \hat{H}) = n \equiv \hat{v}(A, \hat{H}) = n \equiv v(A, H) = n \equiv (\Phi_i A, H) = n.$$

That $\hat{v}(\Phi_i A, \hat{H}) = t \supset v(\Phi_i A, H) = t$ follows from the induction hypothesis using $H_1 R_i H_2 \supset \hat{H}_1 \hat{R}_i \hat{H}_2$ and $\hat{H} \in \hat{N}_i \equiv H \in N_i$.

For the converse suppose $\hat{v}(\Phi_i A, \hat{H}) \neq t$. If $\hat{H} \notin \hat{N}_i$ then $H \notin N_i$ whence $v(\Phi_i A, H) \neq t$ as required. Otherwise for some $\hat{H}_1$, $\hat{H}\hat{R}_i\hat{H}_1$ & $\hat{v}(A, \hat{H}) \neq t$. Then by induction hypothesis $v(A, H_1) \neq t$, and by $\hat{H}\hat{R}_i\hat{H}_1$, $v(\Phi_i A, H) \neq t$.

Theorem (8):

(i) If a wff A is not true in $\mathbf{IS}_6$-model $\mathcal{M}$ then, where Δ is the subformulae closure of $\{A\}$, A is not true in finite $\mathbf{IS}_6$-model $\mathcal{M}/\Delta$

(ii) $\mathbf{IS}_6$ has the finite model property and accordingly is decidable.

Proof: By previous lemmata A is not true in $\mathcal{M}/\Delta$, and $\mathcal{M}/\Delta$ is finite since Δ is finite. Since, then, every unprovable wff can be shown to be not true, and therefore unprovable, in a finite countermodel, $\mathbf{IS}_6$ has, by definition, the finite model property. From this the decidability of $\mathbf{IS}_6$ readily follows (see Harrop [1]).

§7 *Generalizations, and the intensionality of connectives*

Analogous modelling and proof techniques work in the case of multiply modal systems based on weaker sentential bases, and in particular for systems $\mathbf{IS}_2$, $\mathbf{IS}_3$, $\mathbf{IS}_4$ and $\mathbf{IS}_5$, systems obtained by replacing the sentential basis of $\mathbf{IS}_6$. The salient differences are discussed in the quantificational case (§7.8), and relevant details are easily transferred to the modal case. Analogous methods also work for certain **C**-systems, *e.g.* for $\mathbf{ID}_6$, and the methods may be generalized to certain many-valued multiply modal logics. The methods break down however where the systems are not strong enough to define the connectives '*T*' '*F*' and '*S*', or in the many-valued case the Rosser-Turquette J-functors. In certain cases of breakdown a maximal set construction may still be applied, by using more than one set to determine values in the canonical model. (For an example of this technique, applied in significance tense logic, see Rennie [2]).

In view of our case for claiming that many significance assertions are a logical matter, and hence the need for theses such as: $Sp \rightharpoonup \Box Sp$ and $\sim Sp \rightharpoonup \Box \sim Sp$ (suggested in §4.8), it may seem surprising that the significance connective '*S*' which we have used in $S5S_6$ and $\mathbf{IS}_6$ does not conform to these theses. Does this not nullify our claims? Not at all; it is simply the case that in the presence of intensional connectives, the definition of '*S*' in terms of '$\neg D$' is no longer adequate. For when significance, like modality, is an intensional matter (*cf.* the issue of significance essentialism), the assessment of whether A is significant, or significantly holds in G, depends on assignments not just in G but also in other set-ups. If we consider the semantics of $S5S_6$, with a single relation R and with K = N, the contrast between contingent significance 'S' and non-contingent, or logical, significance '$\mathscr{S}$', is best brought out by comparing their holding functions. For 'S' we have:

$$v(SA,\mathrm{H}) = t \text{ iff } v(A,\mathrm{H}) \neq n$$
$$v(SA,\mathrm{H}) = f \text{ iff } v(A,\mathrm{H}) = n;$$

whereas for '$\mathscr{S}$' we should have:

$$v(\mathscr{S}A,\mathrm{H}) = t \text{ iff } (U\mathrm{H}')(\mathrm{HRH}' \supset v(A,\mathrm{H}') \neq n)$$
$$v(\mathscr{S}A,\mathrm{H}) = f \text{ iff } (P\mathrm{H}')(\mathrm{HRH}' \,\&\, v(A,\mathrm{H}') = n).$$

Given the holding function for '$\mathscr{S}$' it is not difficult to see that there is no need to introduce it as a new primitive and that it can be defined by $\mathscr{S}A =_{\mathrm{Df}} \Box SA$. That the defined connective has the right properties can be shown through theorems like the following:

$\vdash \mathscr{S}p \rightharpoondown \Box \mathscr{S}p$; $\vdash \sim \mathscr{S}p \rightharpoondown \Box \sim \mathscr{S}p$;
$\vdash \triangle \mathscr{S}p$ (where '$\triangle$' reads 'it is non-contingent (that)');
$\vdash \mathscr{S}\mathscr{S}p$; $\vdash \mathscr{S} \sim p \equiv \mathscr{S}p$; $\vdash \mathscr{S}(p \,\&\, q) \equiv \mathscr{S}p \,\&\, \mathscr{S}q$;
$\vdash \mathscr{S}\Box p \equiv \mathscr{S}p$.

It is assumed in the proofs that logical necessity satisfies S5 principles.

The order of intensionality of '$\mathscr{S}$' is that of a modal connective, as its definition makes plain. That it is not of higher intensionality is important; otherwise we should be in difficulties over such claims as that Pythagoras may have mistakenly believed that virtue is triangular though the assertion that virtue is triangular is not significant. For highly intensional connectives Φ_i, the principle $\mathscr{S}\Phi_i p \rightharpoondown \mathscr{S}p$, and so also J1 from which it follows, is not valid. In fact for many connectives the principles J1 and J2 are not valid; they have to be supplanted by other principles and the holding condition $v(\Phi_i A,\mathrm{H}) = n$ iff $v(A,\mathrm{H}) = n$ accordingly modified. Consider first J2. Though J2 holds for a large and important class of intensional connectives, significance-preserving connectives, it by no means holds for all such English connectives. For instance, it is true, and accordingly significant, that the tallest Eucalypt in the world is in the Styx valley, but the assertions "Rover smells that the tallest Eucalypt in the world is in the Styx valley" and "It is obligatory that the tallest Eucalypt in the world is in the Styx valley" are not. The second, and more debatable, of these prompts the observation that connectives may *transform* the sentence to which they apply internally as well as attaching further symbols, for example, the deontic connective 'It is obligatory (that)' which simply attached to indicative sentences differs from the more familiar English connective 'It is obligatory (that + subjunctive)' which both attaches a functor to a sentence and transforms to the subjunctive the mood of the sentence to which it attaches. These two obligation connectives appear to have different significance ranges; to suppose that the second satisfies J2 is not as implausible as supposing that the first does. In fact finding the appropriate evaluation for v(Rover smells that A,H) $= n$ is a matter of determining the significance range of 'Rover smells (that)' (and accordingly the matter is postponed until volume 2). Connectives like $D\Phi_i$ which are not significance-preserving are of course definable in $\mathbf{IS}_6$ though no such connectives are primitive. Similarly connectives which infringe J1 are definable, *e.g.* universal connectives such as $\Psi_i =_{\mathrm{Df}} T\Phi_i$ which are significant of every sentence; for whatever $A, v(\Psi_i A, \mathrm{H}) \neq n$. Such connectives have the right sort of significance features (though not the right distribution features) to represent formally universal connectives like 'Someone may think (that)', 'It is an interesting assertion (that)' and 'It is told (that)'.

No general method is available for proving that the F- and L-connectives of $\mathbf{IS}_6$ are, as intended, intensional. For it is consistent with most of these connectives—though not all, *e.g.* those satisfying the axiom $\neg\Phi\Phi p$—that $\Phi p \simeq p$ should hold, in which event it is perfectly possible that Φ is extensional or value-functional. On the other hand one knows from the semantics of $\mathbf{IS}_6$ that, with the exception of connectives which satisfy special reducing conditions, the value-functionality of F- and L-connectives is not derivable, since it is easy to design countermodels to their value-functionality. Though $\mathbf{IS}_6$ is complete relative to an $\mathbf{IS}_6$-modelling, it is not complete under its intended interpretation as an intensional system. For there are no pure variable axioms which can be added to $\mathbf{IS}_6$ to ensure just that a given F- or L-connective is not value-functional, *i.e.* without adding at the same time a stronger and more debatable thesis. What needs to be guaranteed, for a given Φ_i, is that each $\mathbf{IS}_6$-model satisfies the condition $(P\mathrm{H})(\mathrm{GR}_i\mathrm{H}\ \&\ \mathrm{G} \neq \mathrm{H})$. Such a condition does indeed follow from the modelling condition for $\neg\Phi_i\Phi_i p$, *viz.* $(P\mathrm{H})(\mathrm{GR}_i\mathrm{H}\ \&\ \mathrm{H} \notin \mathrm{N}_i)$, but undesirable extras also result, since this condition, unlike the weaker, is incompatible with the S5-modelling. We take $\mathrm{S5S}_6$ as an example, and show that, using just the wff of $\mathrm{S5S}_6$, there is no way of consistently adding to $\mathrm{S5S}_6$ so as to ensure that '$\Box$' is non-value-functional (Scroggs' result [1] shows that there is no consistent way of extending $\mathrm{S5S}_6$ which does not make the connectives matrix-accessible). The modelling condition corresponding to the non-value-functionality of '$\Box$' is $(P\mathrm{H})(\mathrm{GRH}\ \&\ \mathrm{G} \neq \mathrm{H})$, and in the use of $\mathrm{S5S}_6$ this condition can be replaced, using connected models (see Kripke [2]), by the condition that $(P\mathrm{H})(\mathrm{H} \neq \mathrm{G})$, *i.e.* there are at least two distinct worlds, G and H_1 say. Now $\mathrm{S5S}_6$ can be translated into a single variable monadic quantification logic $\mathbf{DS}_6$ (a subsystem of $\mathbf{QS}_6$ of §7.7), and the modelling condition into a condition on $\mathbf{DS}_6$-models. For convenience let *the* variable of $\mathbf{HS}_6$ be 'H', the atomic one-place wff be 'p(H)', 'q(H)' *etc.*, and the only quantifier 'U'. The $\mathbf{DS}_6$-translation A^* of wff A of $\mathrm{S5S}_6$ is obtained by replacing variables, *e.g.* 'p', of A by corresponding atomic wff, *e.g.* 'p(H)', and replacing connective '$\Box$' by quantifier 'UH'. Thus, for example, the translation of $\Box 2$ is $(U\mathrm{H})(p(\mathrm{H}) \rightarrow q(\mathrm{H})) \rightarrow .(U\mathrm{H})p(\mathrm{H}) \rightarrow$

$(UH)q(H)$. Then it may be shown by induction over proofs, or model-theoretically, that

(†) A is a theorem of $S5S_6$ iff A^* is a theorem of $\mathbf{DS}_6$. The modelling condition becomes the condition on $\mathbf{DS}_6$-models that the domain of interpretation should have at least two distinct elements, G and H_1 say. But this modelling condition cannot be expressed using just wff of $\mathbf{DS}_6$:
for

(††) Every simultaneously satisfiable set of wff of $\mathbf{DS}_6$ is simultaneously satisfiable in a one-element domain, G say. (Proof of (††) can be obtained as a special case of the Skolem-Lowenheim theorem in §7.7. In the main metatheorem of §7.7 it suffices, since only one variable occurs in $\mathbf{DS}_6$ wff, to take K_0 as $K \cup \{b_1\}$, and then a one-element model results).

Now if the intensional modelling conditions could be expressed in $S5S_6$, the $\mathbf{DS}_6$-translations of the wff which express the condition would express the two-element condition, which is impossible by (††). Incidentally (†) provides an alternative decision procedure for $S5S_6$ in terms of decision procedures for monadic quantification logic.

There are at least three connected ways of extending $\mathbf{IS}_6$ to guarantee the intensionality of its connectives: by adding sentence constants and axioms on them to the logic; by adding rejection axioms; or by introducing assertional quantifiers and simply asserting non-value-functionality. The last strategy is a more general technique which effectively includes the other moves. But it means that in order to complete our investigation of modal significance logics we shall have to look at their enrichment by quantifiers.

§8 *The context interpretation of set-ups and the addition of context*

The model $\mathscr{M}$ for multiply modal logics can be illuminatingly interpreted from the perspective of context logic. For the set K may be interpreted as a class of world contexts, R_i as a relation between contexts, and v as specifying whether atomic wff hold or hold significantly in given contexts. All the contexts considered are S_6 contexts in as much as the theses of S_6 hold in every context H in K: thus the modelling is obtained by considering a restricted class of contexts, but it would be possible alternatively to build the restrictions into the valuations for the connectives Φ_i. Instead of beginning with contexts K and a valuation function v indicating whether atomic wff hold or hold significantly in contexts, we could, to explain just one approach, begin with atomic contexts, build contexts from these and define v as a characteristic function indicating context membership. An *atomic context* H_A is an ordered pair $\langle H_A^1, H_A^2 \rangle$, where H_A^1 and H_A^2 are disjoint sets of atomic wff; the elements of H_A^1 are the atomic wff that hold in H_A and those of H_A^2 the atomic wff that do not significantly hold in H_A. In terms of atomic contexts, contexts may be specified by a recursion which replaces the recursive specification of function v. Thus for example we say:

$$\neg A \text{ holds in H iff } A \text{ fails to hold in H}$$
$$\neg A \text{ does not hold in H iff } A \text{ holds in H.}$$

A context H is then defined as the ordered pair $\langle H^1, H^2 \rangle$ where H^1 is the class of wff that hold in H and H^2 is the class of wff that do not significantly hold in H. In the case of all modal significance logics except those built on the most inadequate of sentential bases—those weaker than 1-systems—H may be reduced to a single set H^1, with H^2 defined as the class of wff B such that $\sim SB$ holds in H. (Hence the canonical model used in the completeness proof). In modal sentential logics with stronger significance bases the characteristic functions TA, FA, $\sim SA$ are replaced by the context relativized functions $TA(\text{H})$, $FA(\text{H})$ and $\sim SA(\text{H})$. Thus using the notion of context, slightly warped, the modelling can be interpreted without the introduction of states-of-affairs or worlds.

The context interpretation of modal logic semantics in terms of contexts satisfying the worlds assumption (of §2.3), suggests a method of obtaining a more general theory which takes account of both intensionality and context. For each sentence A may be assessed with respect both to a context $\boldsymbol{\theta}$ and to a situation **H**, in short with respect to a pair $\text{H} = \langle \boldsymbol{\theta}, \mathbf{H} \rangle$. Under this elaboration of H, $v(A, \langle \boldsymbol{\theta}, \mathbf{H} \rangle) = t$ reads: A, as assessed with respect to context $\boldsymbol{\theta}$, holds in situation **H**. The multiply modal logics developed may of course be interpreted in this way: but the context so far plays no independent part; and the semantical notation simply writes in explicitly an interpretational assumption we have been making throughout, that sentences are assessed with respect to a background parametric context. The connexion between the semantics so interpreted and the earlier theory (of chapter 2) where context is treated explicitly, is simply the contextual relativisation of the familiar relation between the valuation function and truth and significance evaluations, namely, $v(A, \langle \boldsymbol{\theta}, \mathbf{G} \rangle) = t$ iff $TA(\theta)$, *i.e.* A is true in context θ, and $v(A, \langle \boldsymbol{\theta}, \mathbf{G} \rangle) = n$ iff $\sim SA(\theta)$.

But context should be assigned a larger independent role. For very often we want to assess whether A holds or holds significantly not, or not merely, with respect to the actual situation **G** picked out by the predicate 'is true', but whether A holds, or holds significantly, with respect to other situations and cases: in the case (or cases) determined by, or in the context of, someone's dream, Newton's beliefs, Conan Doyle's detective novels, a make-believe case, naive set theory, and so forth. Such cases or contexts can be taken, as we have occasionally done, as part of what goes into the context θ, but it is advantageous, especially where intensional functors have to be assessed, to separate out *this* component of context and write it explicitly: thus to write θ

more fully as $\theta = \langle \boldsymbol{\theta}, \mathbf{H} \rangle$. We read $A\langle \boldsymbol{\theta}, \mathbf{H} \rangle$ as: A with respect to context $\boldsymbol{\theta}$ and case (situation) **H**, or: A with respect to context $\boldsymbol{\theta}$ and in the context of **H**, or: A at, or in, $\langle \boldsymbol{\theta}, \mathbf{H} \rangle$. Context-situations represented by the pair, context and situation—with respect to which sentences are assessed, for example as to whether they hold or not—provide *a* formal analogue of Wittgenstein's language-games. Holding, and holding significantly, are by no means the only properties of sentences in language-games: consider, for example, the various semantical properties of the order 'Get me a cab to Waterloo Station' given by Holmes in the context of a Doyle story, or of the question 'Is the Russell class self-membered?' set in the context of a certain set theory. Moreover the language-game $\langle \boldsymbol{\theta}, \mathbf{H} \rangle$ figures not only in the semantical evaluation of sentences, but also in the preliminary determination of logical form and in the resolution of various ambiguities.

Context-situations enable the implicit relativization of certain assertions to a given background to be represented semantically. Consider, for example, obligation claims which, though absolute in form, are understood, in their context, as relative to or in the context of a given moral or legal code c. Then what we shall very often want to evaluate is not $v(OA, G)$, whether 'it is obligatory that A' is true (or holds good) or is significant, but $v(OA, \langle c, \mathbf{G} \rangle)$, *i.e.* whether it is true (or holds good) within, or as regards, code c. So too where c and d are different codes, *e.g.* a legal code as distinct from a moral code, c-validity will not contradict d-validity; and under given modelling conditions OA may be c-valid but not d-valid. Of course where confusion is likely to result some of the background may be transferred to the sentence, *e.g.* where OA is assessed relative to a legal code c by one speaker but not by others, his 'O' may be expanded to read 'It is legally obligatory that'.

Context-situations can be profitably applied to increase the range of discourse the semantics can treat. Firstly, coding can be admitted. Consider, for example, a context c where 'Gladway' is code for 'It is raining'; then $v(\text{Gladway}, \langle c, \mathbf{G} \rangle) = t$ iff rain is now falling. Where non-standard contexts like coding contexts are not involved, it is however unnecessary to introduce new components of context or situation to cope with sentences at times, because the multiple method already given suffices for the combined treatment of tense, considered as a modality, with other modal functors, given that the actual situation G is construed as the situation obtaining now. There are, however, alternatives to this method of combination; for instance, in the case of a combined treatment of tense and necessity, H may be replaced by the pair $\langle \mathbf{H}, t \rangle$ where t is a time and **H** is a supposedly timeless world or else is the case for any time. Such a more complicated alternative is pursued by Montague [1] who introduces an elaborate, but otiose, restriction-of-possible-interpretations method; but in general it is much simpler to allow each function to determine, through its corresponding semantical relation, the class of situations over which it is evaluated.

Secondly, context-situations enable the semantical assessment, but without the usual distortions, of confident truth-claims such as "Sherlock Holmes lived at 221b Baker St." (p_0) said—to supply part of the context c for the assertion p_0—in answer to the question "Where did Sherlock Holmes live?" This claim is true, in its context, even though it clashes with other factual information we have about Baker St.; yet it is not equivalent to the claim "It is written in Conan Doyle's novels that Sherlock Holmes lives at 221b Baker St.". Such claims as p_0 can however be construed as involving a shift of the base situation with respect to which truth assessment is made, from the usual **G** to a new situation **H**, determined by c, *i.e.*, $\mathbf{H}_1 = f(c, \mathbf{G})$: roughly $\mathbf{H}_1$ is the situational closure of the set-up selected by the "It is asserted in Conan Doyle, or Sherlock Holmes', detective novels that". Then $v(p_0, \langle c, \mathbf{G}\rangle) = t$ because $v(p_0, \langle c, \mathbf{H}_1\rangle) = t$, *i.e.* because according to the novels Holmes did live at that address; and this claim is not undermined by objections such as "221 Baker St. was not a residential site" (q_0) because the context of 'q_0' does not shift the base situation **G**. How does one tell when the base is shifted? A non-trivial answer is: the context supplies a base-shifting function, and this figures as part of the modelling. For example, the modelling can, under this extension, associate with each $\boldsymbol{\theta}$ a base-shifting function θ_b from context-situations to situations such that $\theta_b^2 = \theta_b$, normally a projection function which makes no shift. Then where part of the context sets the sentence in another context-situation,

$$v(A, \langle \boldsymbol{\theta}, \mathbf{H}\rangle) = t \text{ iff } v(A, \langle \boldsymbol{\theta}, \theta_b\langle \boldsymbol{\theta}, H\rangle\rangle) = t.$$

It is important for the operation of this theory that incomplete situations be admitted, that not all situations be worlds.

Several important applications of context-situations require, however, analyses of sentences into at least subject-predicate parts, *e.g.* the semantical treatment of egocentric particulars of indexical subjects such as 'This', 'I' and 'Charlie' in sentences 'This is unfortunately very bad', 'I am sad' and 'I believe I am not BR', said by BR while suffering from amnesia. Such examples, where context-situations are essential to the evaluation, will be discussed briefly after subject-predicate analysis has been combined with intensionality (in §7.11).

7.3 Quantified Significance Logics: The Interpretation of Quantifiers.

Quantification logics are distinguished from sentential logics, inessentially by their analyses of sentences into subject and predicate parts, and essentially through the introduction of certain quantifiers, notably the quantifiers 'for every ...' and 'for some ...'. If we consider quantificational logics obtained from sentential S-logics simply by admitting decomposition of sentences into subject-predicate form and by adding quantifiers, then, since subject-predicate decomposition does not alter value assignments (though it may explain them), it is clear that the main new logical problem concerns the behaviour of quantifiers, their valuational analysis, and their axiomatisation in significance logics. As with connectives, so with quantifiers, it is easier and more illuminating to tackle such problems semantically rather than postulationally; for of course the semantics given should tie intimately with the meaning of quantifiers.

Because quantified sentences and their components may be neither true nor false but nonsignificant for some values of their variables, we cannot adopt an analysis of them in terms of truth and falsity alone.

If, however, we begin by adopting the standard account of quantifiers over finite domains of values, *e.g.* of the universal quantifier 'U' as a conjunction over a finite domain, then we can derive the properties of quantifiers from the expansions and the special features of the sentential connectives used in the expansions. Let us suppose, here, that the operator variable ranges over its full constants range, *i.e.* is unrestricted. Then, given that the constants range of a variable 'x' is just the finite set {'x_1', ..., 'x_m'}, then, where 'f_0' is a predicate constant, we may put,

$$(Ux)f_0x \simeq f_0x_1 \,\&\, f_0x_2 \,\&\, \ldots \,\&\, f_0x_m.$$

That is, '$(Ux)f_0x$' is significant if, and only if, '$f_0x_1 \,\&\, f_0x_2 \,\&\, \ldots \,\&\, f_0x_m$' is significant. If the conjunction is classical, it follows that '$(Ux)f_0x$' is significant if, and only if, 'f_0x_1' is significant and 'f_0x_2' is significant ... and 'f_0x_m' is significant. Extending this account now to an infinite domain, we may say that '$(Ux)f_0x$' is significant if, and only if, it is significant for all values in the primary range of the operator variable, *i.e.*

$$S(Ux)fx \simeq (Ux)Sfx.$$

On the assumption, then, that the quantifier 'U' as well as being unlimited or unrestricted, in that the total range of the operator variable is admitted, is appropriately *classical*, the sentence '(Ux)(x is happy)' expresses neither a truth nor a falsehood. It is nonsignificant since, for some values of the operator variable, such as 'Wittgenstein's stove' and 'the number seven', we have nonsignificant instantiation cases. This feature is expressed formally by the principle: $\sim Sfy \supset \sim S(Ux)fx$. Clearly, by taking non-classical conjunctions as the basis other universal quantifiers could be defined which do not satisfy these principles. But the question of providing an account of such quantifiers need not detain us, since given stronger S-logics they can be defined in terms of the quantifiers we shall consider.

In general, much as non-classical connectives and restricted variables can be introduced, so non-classical quantifiers and restricted quantifiers can be explained. The quantifiers of *PM*, for example, though classical, are restricted. For the quantifier in '$(x)f_0x$' is such that the range of the variable 'x' is restricted to values for which $qu(f_0x)$ is significant. Hence the sentence

'$(x)(x$ is happy)' is significant since $qu(x$ is happy) is significant for all values taken from its s-range.

Though we can make use of expansions over finite domains in devising quantificational evaluations, we cannot *define* quantifiers in this way in a finitary language since not all domains are finite. Although a theory of limits could presumably be set up, and quantifiers accounted for in terms of limits of progressively larger finite domains (*cf.* frequency evaluations), the account would only succeed in explaining the less obscure in terms of the more obscure, as multiple quantification is used in the theory of limits.

Given, however, that we confine attention to sentences or assertions, which are either true, false or nonsignificant and omit incomplete sentences as in the sentential systems, then we may explain the interpretation of quantifiers by providing valuation rules which determine the conditions under which quantified expressions are true, false or nonsignificant. The rules which we adopt for 'U' are such as to determine it as the unrestricted classical operator illustrated above, namely:

$v((Ux)f_0x) = t$ iff $v(f_0x) = t$ for all (primary) values of argument 'x';
$v((Ux)f_0x) = f$ iff $\begin{cases} v(f_0x) = f \text{ for some (primary) value of '}x\text{' and} \\ v(f_0x) \neq n \text{ for any (primary) value of '}x\text{';} \end{cases}$
$v((Ux)f_0x) = n$ iff $v(f_0x) = n$ for some (primary) value of 'x'.

The classical particular quantifier 'P' is then defined by:

$$(PX)A =_{\mathrm{Df}} \sim(UX) \sim A;$$

from which it follows, given the definition of the classical '$\sim$',

$v((Px)f_0x) = t$ iff $\begin{cases} v(f_0x) = t \text{ for some (primary) value of '}x\text{' and} \\ v(f_0x) \neq n \text{ for any (primary) value of '}x\text{';} \end{cases}$
$v((Px)f_0x) = f$ iff $v(f_0x) = f$ for all (primary) values of 'x';
$v((Px)f_0x) = n$ iff $v(f_0x) = n$ for some (primary) value of 'x'.

That is, '$(Px)f_0x$' reduces to the classical disjunction '$f_0x_1 \vee f_0x_2 \vee \ldots \vee f_0x_m$' over the finite constants range {'x_1', 'x_2', ..., 'x_m'}.

The point of introducing a second particular quantifier 'q' should now be clear. For since '$(Px)f_0x$' is nonsignificant if $qu(f_0x)$ is nonsignificant for some value of 'x', it is impossible to say significantly that some item is happy since, *e.g.*, 'The number seven is happy' is nonsignificant. This problem could be overcome by means of a restricted quantifier, but we prefer to introduce a non-classical quantifier which satisfies the conditions:

$v((qx)f_0x) = n$ iff $v(f_0x) = n$ for *all* (primary) values of 'x';
$v((qx)f_0x) = t$ iff $v(f_0x) = t$ for some (primary) value of 'x';
$v((qx)f_0x) = f$ iff $\begin{cases} v(f_0x) = f \text{ for some (primary) values of '}x\text{';} \\ v(f_0x) \neq t \text{ for any (primary) value of '}x\text{'.} \end{cases}$

The two particular quantifiers '*P*' and '*q*' coincide for significant values. Over the total range, their different roles are sufficiently marked by reading '*q*' as 'for some' and '*P*' as 'for some out of all'.

Significance logics with quantifiers '*U*' and '*q*' as primitive may be compared with intuitionist logic with its distinctive quantifiers '*V*' and '*Ǝ*'. In quantified logics with stronger sentential bases it is, however, unnecessary to take both '*U*' and '*q*' as primitive. For example, in quantified $\mathbf{S}_6$, '*q*' may be defined contextually thus: $(q\mathrm{X})A =_{\mathrm{Df}} D(U\mathrm{X})DA$. The adequacy of this definition is easily verified and it can be checked over finite domains of values.

The interpretation of quantifiers that we have stated, for a given predicate constant 'f_0', is a substitution interpretation of quantifiers inasmuch as it assesses values of quantified wff in terms of substitution values of operator variables. But it is not a substitution interpretation in the sense recently canvassed (for details see Dunn and Belnap [1]) that the values of variables are confined to closed terms or constants of the logic in which the quantifiers occur. For an assessment of a quantified wff to be made the constants-range of variables should be given, or stated, in advance, just as in classical logic a domain of interpretation has to be given. To distinguish the substitution interpretation of quantifiers we have given, we call it a *value-interpretation* of quantifiers (the label is suggested by the related account in Carnap [2]).

Now the values of the variables can be contracted (in the way explained in §3.6); and where a quantification logic contains only a restricted class of predicates, contractions may be made without affecting the truth value of quantified assertions. For example, if all the predicates of a two-valued logic are referentially transparent, then we shall not go wrong in value assessments by contracting using contingent identity of subject items. The classical semantics for two-valued quantification logic in fact assumes that the contraction relation is a reference relation (as previously explained in §3.6), that contingently identical items are identified and that each such item of the contracted domain exists. But for many purposes—almost any treatment of items which do not exist, or of opaque contexts—this assumption is far too restrictive; and it is quite unnecessary, for the logical apparatus of quantification theory applies much more generally and holds over non-classical domains. Whereas there is a loss of generality in using the classical semantics, since a quantification logic constrained by the requirements of the classical semantics can no longer be applied generally outside certain reference domains, there is no such loss in using the value interpretation.

Using a contraction function I, a more general style of semantics which includes, in effect, both the value interpretation and the classical interpretation of quantifiers as special cases, can be provided. Where the unrestricted constants-range of the variables is the class {'x_1','x_2', ...}, let $\mathrm{d} = \{\mathrm{I}(x_1), \mathrm{I}(x_2), ...\}$; *i.e.* the domain d of interpretation is the contraction, under I, of the constants-range. The contraction function may be extended to apply to predicates using the following connection:

$$\mathrm{I}(f^n)(\mathrm{I}(x),\mathrm{I}(x'),\ldots,\mathrm{I}(x'^{\cdots\prime})) \simeq f^n(x,x',\ldots,x'^{\cdots\prime}).$$

It is here that the assumption that contracting makes no difference to value assignments is required; for otherwise 'I(f^n)' is not well-defined. Given that the assumption is satisfied however—we shall fix the formal semantics so that it is—the quantificational conditions for 'U' can be replaced by these:

$$v((Ux)f_0(x)) = t \text{ iff } v(f_0(x)) = t \text{ for every contraction I}(x)\text{(on values of '}x\text{')};$$
$$v((Ux)f_0(x)) = n \text{ iff } v(f_0(x)) = n \text{ for some contraction I}(x).$$

The value interpretation of quantifiers is tantamount to that case where function I makes no reduction on the constants-range, and this amounts to the condition that $\mathrm{I}(x_i) = \mathrm{I}(x_j)$ iff $qu(x_i) = qu(x_j)$, for any two constants x_i, x_j. For if x_i, x_j are constants, $qu(x_i) =$ 'x_i' and $qu(x_j) =$ 'x_j'. But in general, 'x_i' $\neq$ 'x_j'; hence $qu(x_i) \neq qu(x_j)$, therefore $\mathrm{I}(x_i) \neq \mathrm{I}(x_j)$. Equivalently, $\mathrm{I}(x_i) = \mathrm{I}(x_j)$ iff $i = j$; in which case function I makes no reduction on the constants-range. Likewise, using contractions which do make a reduction on the constants-range corresponds to replacing primary values in the value interpretation by secondary values. Where I makes no reduction, the quantification rule for 'U' can become:

$$v((Ux)f_0(x)) = t \text{ iff } v(f_0(x)) = t \text{ for every } x;$$
$$v((Ux)f_0(x)) = n \text{ iff } v(f_0(x)) = n \text{ for some } x.$$

These are the controversial rules for domainless semantics; but they have the advantage of simplicity without causing loss of generality, and they are valuable when quantification and modal logics are combined. For when these logics are combined a separate domain is usually required for each set-up. Since, however, no allusion to domains, whether of values or contractions, is made in domainless quantification rules, and none occurs of course in valuation rules for connectives, domains can be eliminated altogether. Hence the name 'domainless semantics'.

It follows from the valuations which have been given that the following principle holds:

$$(\alpha)\ \ S(\mathfrak{q}x)fx \supset (\mathfrak{q}x)Sfx.$$

It is apparent from this that the particular quantifier '$\mathfrak{q}$' must not be identified with its existential mate 'E'$((E\mathrm{X}) =_{\mathrm{Df}} (\mathfrak{q}\mathrm{X})(E(\mathrm{X})\&A))$; otherwise ($\alpha$) would amount to,

$$(\beta)\ \ S(Ex)fx \supset (Ex)Sfx,$$

which is analogous to the Barcan formula of modal logic,

$$(\gamma)\ \ \Diamond(Ex)fx \supset (Ex)\Diamond fx,$$

and raises similar difficulties of interpretation. (α) not only follows immediately from the valuation rules, it is also needed to guarantee substitutivity and instantiation schemata essential for an adequate theory (*cf.* its role in the completeness proofs for $\mathbf{QS}_6$ below, §7.7). But interpreted as (β), the thesis is highly objectionable, for the following reasons (pointed out by V. Routley). (β) leads directly to Platonically swollen universes. For any predicate, it is significant to say that there exists something which has the property it determines (this is in fact a theorem of the logic $\mathbf{QS}_6\mathbf{E}$ below, §7.8). Therefore by the Barcan analogue (β), there actually exists something which significantly has the property. But if this were so, the mere significance of certain discourse would imply the existence of items of every sort. Similarly, the contraposed form of (β), *viz.*,

$$(\delta) \quad \sim (Ex)Sfx \supset \sim S(Ex)fx$$

is closely related to the Platonic doctrine that of what does not exist nothing meaningful can be said. For it says that, for any predicate, if it is not the case that the predicate is significant of anything existent, then it is not significant to say that something exists which has the property which it determines. Thus, for example, if no material object (number) existed, it would not be significant to say that something was heavy (prime, divisible by a prime), or that it was a material object: indeed no purely material-object predicates would be significant. The thesis consequently has the undesirable result that all categories must have some existent members: every category must be actually exemplified. Short of high Platonism, then, categories could be changed by catastrophes and other contingencies.

None of these difficulties arise in the case of (α). For taking '$\mathcal{P}$' as a non-ontological quantifier which has no existential implications, existence does not follow from mere significance. Moreover, the adoption of (α) does not lead to versions of (β). Since $S(x$ exists) does not imply that x exists, (β) is independent of (α). This will become clear in the formal theory.

A quantified significance logic which only contained the existentially-loaded quantifiers 'E' and '$\forall$' (read 'for every entity') would be singularly inadequate: it would apply only to a fraction of the claims where significance and absurdity can normally be assessed. For it is perfectly possible to assess the significance and absurdity of claims about non-existent items as well as those about entities. The sentences 'Unicorns are prime', 'The number seven dislikes dancing' and 'The president's sciatica addressed the meeting' are nonsignificant irrespective of whether unicorns exist, of whether the number seven even significantly exists and of whether the president has sciatica. Likewise 'A red unicorn is an animal' and 'The president's sciatica was treated by a doctor' are significant even though neither unicorns nor the president's sciatica exist. Similarly 'What I am thinking about is sad' is absurd if what I am thinking about is a certain square circle, and so is 'It has four feet' where 'it' is s-about some sequence, say the least rapidly converging sequence. The assessment of such claims is made possible by the fact that *qu*(z) may have a clear sense even though z does not exist, and that some item may be identical with z even though it does not exist.

An upshot of this is that the historically important project of using significance and category theories as a major tool in ontology, as a means of determining what exists, is mistaken. (For a recent strong statement of this thesis see Sommers [1]).

Significance theories have nothing to do with what sorts of things exist, since they assess significance independently of whether items exist or not; they have to do with the sorts of things which can be talked about. And that an item can be talked about does not imply that it exists: to think so is to adopt the quite erroneous Ontological Assumption: that x has features implies that x exists.

These points also serve to show the falsity of the view that significance and nonsignificance reveal "the way the world is" or "the real nature of things". For the nonsignificance of 'Unicorns are prime' and the significance of 'Unicorns are animals' can have little to do with the way the world is if unicorns do not exist. These points indicate, too, that significance and category theories have tended to confuse two different theses, the first of which is true, the second false. These are: that there is a transparent sense of 'significant'; and that significance is therefore connected with existence. That is, the theories confuse identifying talk about items with actual reference, and so with the existence of the items talked about. Nor would a significance theory with only existentially-loaded quantifiers provide an ontology—rather it presupposes one; for it must be known what items exist in order to quantify over them. In any case such a theory faces, as we have seen, formidable interpretational and formal problems. For example, how can it allow for significance and non-absurdity claims concerning expressions which lack actual referents, or are s-about non-existent items? Such claims cannot be satisfactorily dismissed as meaningless. Yet devices like theories of descriptions and fictions have no obvious extensions which satisfactorily allow for such claims.

7.4 Kinds of Quantified Significance Logics

The first quantified significance logic was the predicate theory of *PM* *9–*11. However, the significance features of *PM* predicate theory are not especially striking; indeed so little so that subsequent versions of it simply omit the significance riders which are imposed on theorems of *10 and drop the special significance-oriented interpretation. The reason for this, as we have seen, is that the significance features are regarded as metatheoretic by later writers, and since the only formal connectives and operators are classical, the presupposed significance principles can be reflected back into the formation rules as principles of elimination and the only admitted formulae are then two-valued. In the next section we examine various ways in which *10–*11 of *PM* can be represented as an explicit significance logic. Because it is so weak, however, it has little intrinsic interest as a *significance* logic. Indeed, if we take the sentential basis to be $\mathbf{S}_0$, then as Russell himself noticed, and as we stressed in the discussion of $\mathbf{S}_0$, no genuine *principles* of significance can formulated. The result is a logic which is structurally isomorphic to the classical two-valued quantification theory (**RQL**) and differs only in its interpretation. This explains why it is so easy, and indeed "natural", to take the predicate theory of *PM* as if it were a two-valued theory, with explicitly stated formation rules which eliminate nonsignificant formulae as non-wff. But this is not in fact the way it is presented in *PM*. Formulae which violate type principles are admissible, and their occurrence demands special restrictions on the application of certain theorems. By taking $\mathbf{S}_1$ as a basis, instead of $\mathbf{S}_0$, some of these restrictions can be made explicit as part of the formal theory and the resulting system then differs both in structure and interpretation from the classical **RQL**. It is still very weak, however, considered as a significance logic.

To build stronger and more adequate quantified significance logics, it is necessary to add quantifiers and other operators to stronger sentential bases. Various kinds of logics result according to the strength of the base. If a quantifier is added to a weak sentential system such as $\mathbf{S}_1$ or $\mathbf{S}_2$, analogues of many-sorted logics are the obvious outcome. Thus, suppose that we have the unrestricted classical quantifier 'U' and some one-place predicate constant 'f_0' such that the significance range of 'f_0' is a proper subclass of its primary (grammatical) range. Then '$(Ux)f_0x$' is nonsignificant since there are primary values for which $qu(f_0x)$ is nonsignificant. Accordingly, vital predicate logic principles such as universal instantiation formulated with '$\supset$', e.g. $(Ux)f_0x \supset f_0x$, are also nonsignificant. To salvage the logic, restricted variables, restricted to the significance ranges of predicates of the system, are wanted. Suppose that 'x/Sf_0' is a variable restricted to the s-range of 'f_0'. Then such principles as $((Ux/Sf_0)f_0(x/Sf_0) \supset f_0(x/Sf_0))$ are once again significant and true, and a logic of the expected sort can get off the ground. What is really done is to extend to predicate logics the idea already embodied in two-sorted sentential systems. Now, however, to arrive at significant principles, account must be taken of s-ranges.

The logics which emerge in this way are like many-sorted predicate systems. But they also have certain distinctive features of their own. Most important, there is no simple reduction of them to one-sorted logics as there is in the case of the usual many-sorted systems. We cannot for example affirm, as an equivalent of, say, the formula '$((Ux/Sf_0)f_0(x/Sf_0) \supset f_0(x/Sf_0))$', a formal version of: 'If $qu(x)$ is restricted to the s-range of 'f_0' then $(Ux)f_0x$

$\supset f_0x$; where $qu(x)$ is now *un*restricted. For since the only implication connective in S_1 is the classical '$\supset$', the formal representation of this would have to take the form: $Sf_0x \supset [(Ux)f_0x \supset f_0x]$. But the expression in the square brackets is now nonsignificant, and by the properties of '$\supset$' so is the whole formula. Exactly this problem, as we saw, arises at the sentential level if we try to reduce the two sorts of sentential variable to one by means of antecedent conditions on implicational formulae expressed in terms of '$\supset$'. To achieve a reduction of the sorts in significance logics, it is necessary to move beyond S_1 in order to take advantage of the properties of '$\rightharpoondown$' or some other non-classical implication.

Given stronger sentential bases in which *S*-restricted sentential variables can be eliminated (e.g. S_3 or stronger systems), so stronger quantificational logics without sorts can be developed. The resulting logics are then like many-valued predicate theories, and many of the techniques which apply generally to such systems can be taken over. These stronger logics are much more flexible than the many-sorted variety, and in any case, features of the many-sorted systems can be reflected in them by the introduction of restricted variables. Not only this, but many principles which cannot be expressed in the many-sorted systems can be affirmed in the stronger logics. Given a many-sorted logic, it is not possible to affirm relations between the sorts: consider for example assertions such as 'Provided the s-ranges of f_0 and g_0 overlap then . . .'. Such principles can however be expressed in predicate systems based on S_3 or higher logics. It is for reasons such as this that after a preliminary discussion of the *PM* predicate theory based on S_0, or at most S_1, we move immediately to the basis S_6 and derive weaker systems as special cases.

7.5 On the Quantification Logic of *Principia Mathematica*

Two predicate theories (theories of apparent variables so-called) are presented by Whitehead and Russell. They differ in the method by means of which quantifiers are introduced and the way in which the theory of (propositional) deduction is extended. The second theory, presented in *10–*11, is used throughout the remainder of *PM* and is explicitly limited by significance conditions. We therefore restrict attention to it. Essentially the same theory appears in the earlier Russell [2], and in a clearer form. In particular, many of the significance principles which are assumed are independent of the particular details of type-theory, and would hold even if different criteria of significance were adopted. This is made clear in Russell [2]. Thus,

> "In the above propositions our *x* and *y* may be of any type whatever. The only way in which the theory of types is relevant is that (11) only allows us to identify real variables occurring in different contexts when they are

shown to be of the same type by both occurring as arguments to the same function, and that in (7) and (9), y and a must respectively be of the appropriate type for arguments to $\phi\hat{z}$."

Now although significance riders are added to a number of *PM*-theorems in *10–*11, essential riders on other theorems are often not stated. This is not unusual. A great many rules and conditions, not only significance conditions, which are used and assumed in *10–*11 are not stated explicitly. This is so, for example, with the rule for changing bound variables used in *10.27 and *9.21, and the rule of substitution for propositional variables used, *e.g.* in *10.22.

As an example of a "missing" significance condition, consider

$$*10.14 \quad \vdash :. (x).\phi x : (x).\psi x : \supset . \phi y . \psi y$$

In the course of proving this, *10.13 is applied; and this thesis requires the assumption that the functions '$\phi\hat{x}$' and '$\psi\hat{x}$' take arguments of the same type. Accordingly *10.14 should be stated:

*10.14 If '$\phi\hat{x}$' and '$\psi\hat{x}$' take arguments of the same type, then $\vdash :. (x).\phi x : (x).\psi x : \supset . \phi y . \psi y$.

It might perhaps be said that in view of the interpretational conventions of *PM*, *10.14 can be stated without restriction. But this is far too liberal a view. For *10.14 is derived using *10.13, and *10.13 only holds subject to a condition on sameness of type.

Whitehead and Russell do comment on *10.14 as follows:

"This proposition is true whenever it is significant, but it is not always significant when its hypothesis is significant. For the thesis demands that ϕ and ψ should take arguments of the same type, while the hypothesis does not demand this. Hence if it is to be applied when ϕ and ψ are given, or when ψ is given as a function of ϕ or vice versa, we must not argue from the hypothesis to the thesis unless, in the supposed case, ϕ and ψ take arguments of the same type."

This comment, however, though correct, and important for clarifying the interpretation, is not enough. For detachment can be, and is, made on the strength of *10.14: for example in the proof of $\vdash :. (x).\phi x : (x).\psi x : \supset : (y).\phi y . \psi y$ in *10.22. But if we detach in cases where ϕ and ψ take arguments of different types, we shall end up with nonsense (*cf.* unrestricted detachment in **C**-logics). For example, if f is a function of ϕ, then applying *10.14 we have $\vdash . fy \vee \sim fy . \phi y \vee \sim \phi y$, which crosses types; an example would be: y has members or not and is red or not.

One way of rectifying this omission is to reformulate *10.14 in the way suggested. But very many theorems would have to be qualified in similar ways and some in more complicated ways. Examples in *10 alone are *10.22, *10.27, *10.271, *10.281, *10.29, *10.3–*10.32, *10.39, *10.41, *10.42. In some of these the required restrictions are not due to *10.13 but rather to what accounts for the qualification on *10.13, namely the introduction of two functions which may or may not carry different types. For example, in *10.41, *2.2 is used to obtain $\vdash:(x).\psi x.\supset.\phi y \vee \psi y$. Quite apart from the complication of introducing all such restrictions explicitly, however, they do not fit uniformly with the adoption of conventions that is made elsewhere: *e.g.* in asserting *10.1 $\vdash:(x).\phi x.\supset.\phi y$, where it is simply *taken* that y is of the type appropriate to ϕ. But if *10.1 can be guaranteed by such a metatheoretic convention, why should not *10.13, *10.14, *10.22, and so on, be likewise controlled by suitable conventions? In fact they can.

First, we clarify the interpretation of various expressions in *PM*.

The quantifiers are restricted, in fact significance restricted. Thus, "We are not asserting ϕx is true for all values of x, because, in accordance with the theory of types, there are values of x for which "ϕx" is meaningless; for example the function $\phi\hat{x}$ itself cannot be such a value" (*PM*, p. 127; *cf.* also p. 161). In terms of this, '$(x)\phi x$' reads 'for all significant (possible) values of $\phi\hat{x}$, ϕx (is true); and '$(\exists x)\phi x$' reads 'there exists a value from the s-range of $\phi\hat{x}$ such that ϕx (is true)'. Consequently, '$(x)\phi x$' and '$(\exists x)\phi x$', for given ϕ, are always true or false and never nonsignificant. Just this is in effect asserted by such theorems as $\vdash:\sim(x).\phi x \vee (x).\phi x$, a version of *9.23. Nonsignificance is thus strained out of the closed wff of the logic.

Free variables are not, however, so restricted in range that 'y' in $qu(\phi y)$ takes only values from the s-range of $qu(\phi)$. But it is assumed in *assertions* such as *10.1 (universal instantiation) that values of $qu(y)$ in $qu(\phi y)$ are limited to values from the s-range of $qu(\phi)$. And this convention can be adopted generally for assertions of predicate wff. Moreover, exactly this convention is effected if we follow out Russell's later suggestion and eliminate free variables from theorems by reading '$\vdash A$' as 'the universal closure of A is a theorem' (see Church [4], p. 46, footnote 107). Given either the convention that theorems are asserted for free variables only where these are restricted to the s-ranges of the predicates in which they occur, or the alternative convention that theorems are asserted only for the universal closures of wff, the predicate logic assumes modern standard formulations. That is, it has the same syntax, though by no means the same interpretation, as classical predicate logic (**RQL**).

Consider, then, the following formulations, where we change the vocabulary to fit with other systems subsequently investigated.

§1 *The Logic* $\mathbf{PM_1}$.

Vocabulary:

Improper primitives: $\sim$, $\vee$, (,).
Sentential variables: $r, r', r'', \ldots$
Individual variables: $x, y, z, x', y', \ldots$
Predicate variables: $f, g, h, f', g', \ldots$

Atomic wff:

A is an atomic wff iff A is a sentential variable or A is $F(u_1, ..., u_m)$ where F is a m-place predicate and $u_1, ..., u_m$ are m individual variables.

Well-formed formulae:

(1) An atomic wff is a wff;
(2) If A and B are wff, so are $\sim A$ and $(A \vee B)$;
(3) If A is a wff and u is an individual variable, then $(u)A$ is a wff.
(4) Defined equivalents of wff are wff. Thus Fu is a wff where F(u) is.

Free and bound variables:

An occurrence of a variable u in a wff A is a bound occurrence of u in A iff it is an occurrence in a well-formed part of A of the form $(u)B$; otherwise the occurrence of u is free in A.

Substitution notation:

$\check{S}^u_v A \mid$ is the wff resulting from the substitution of variable v for all free occurrences of variable u in A, provided that no free occurrence of u is in a well-formed part $(v)C$ of A; and otherwise is A.

Postulates:

1. $\vdash r \vee r \supset r$
2. $\vdash r' \supset . r \vee r'$
3. $\vdash r \vee r' \supset . r' \vee r$
4. $\vdash r' \supset r'' \supset . r \vee r' \supset r \vee r''$

R0. $\vdash A \twoheadrightarrow \vdash S^r_B A \mid$, where r is a sentential variable and A and B are wff.

If A is a consequence of 1–4 and R0, then A is a *substitution-instance* of an S_0-tautology. Thus, 1–4 and R0 can be replaced by the equivalent:

P1. If A is a substitution-instance of an S_0-tautology, then $\vdash A$.

To make the quantification axioms adequate for the full theory, without the complications of substitution rules, we use axiom schemata:

P2. $\vdash (u)(A \vee B) \supset . A \vee (u)B$, where u is a variable which does not occur free in A

P3. $\vdash (u)A \supset \check{S}^u_v A \mid$
R1. $\vdash A, \vdash A \supset B \twoheadrightarrow \vdash B$
R2. $\vdash A \twoheadrightarrow \vdash (u)A$

The notation '$\vdash A$' is read 'the significance restriction of A is a theorem', where the significance restriction of A limits those variables which have free occurrences in A to their appropriate types (more generally, if the details of type-theory are rejected, to their appropriate s-ranges). Hence A in '$\vdash A$' is always significant and we may instead read '$\vdash A$' as 'A is a theorem when A is significant.' An applied calculus based on **PM**$_1$ will have *vacuous* theorems which would be true if significant but which are never significant: *e.g.* y has six individual members or not and is red or not.

In $\mathbf{PM}_1$ the analogue of *10.14, i.e. $\vdash(x)fx \mathbin{\&} (x)gx \supset fy \mathbin{\&} gy$, is derivable without qualification. Whitehead and Russell's remarks on this are now apposite, except for the remark on detaching from the hypothesis. For the same interpretational conventions now apply to consequences like

$$\vdash(fy \vee \sim fy) \mathbin{\&} (gy \vee \sim gy),$$

as apply to *10.14. Similarly, *10.22 and the others can stand without significance riders. Adjunction, the principle which really brings out the unsatisfactoriness of the interpretational conventions of *PM* must, however, be amended. In particular, the following analogue of *10.13 follows at once in $\mathbf{PM}_1$:

$$Adj: \ \vdash fx, \vdash gx \longrightarrow \vdash fx \mathbin{\&} gx.$$

But the requirement that the conclusion is significant, and hence that 'f' and 'g' take arguments of the same type (if a type theory of significance is adopted) is now guaranteed by the intended interpretation of '$\vdash$', just as the requirements that 'x' belongs to the type of 'f' and 'x' (or 'y') belongs to the type of 'g' are guaranteed by the respective premiss assertions.

It is inevitable that our reformulation differs from the logic of *PM* *10. For the formulation of *PM* *10 is defective. Either *10.13 is used in the derivation of *10.14, in which case *10.14 is incompletely stated, or *10.13 is not used, in which case a more general adjunction rule (which incorporates the more liberal of the interpretational conventions adopted in *10) is necessary.

§2 *The Logic* $\mathbf{PM}_2$

$\mathbf{PM}_1$ is by no means the only possible reformulation of *PM* *10–11. Another would be obtained by qualifying the modus ponens rule, but Russell effectively rejects this line (see Russell [2], p. 87, (12)). A reformulation closely related to $\mathbf{PM}_1$ is obtained by eliminating free variables altogether. Then appropriate significance restrictions follow immediately from the adoption of significance-restricted quantifiers. We call this logic $\mathbf{PM}_2$. It corresponds to Russell's own suggested reformulation in the introduction to the second edition of *PM* (p. xviii).

$\mathbf{PM}_2$ has the same morphology (vocabulary, formation rules etc.) as $\mathbf{PM}_1$.

In place of the metalogical symbol '$\vdash$' we use '$\in$'; '$\in A$' reads 'the universal closure of A is a theorem', where the universal closure of a wff A is the wff $(u_1)(u_2) \ldots (u_m)A$, $u_1, u_2, \ldots, u_m$ being the m free individual variables in A in lexicographical order (*i.e.* the order given in the morphology).

The following economical formulation of $\mathbf{PM}_2$ is due essentially to Wang [1].

Postulates:

P1′. If A is a substitution instance of an $\mathbf{S}_0$-tautology, then $\in A$.

P2′. $\in (u)(A \vee B) \supset A \vee (u)B$, where u is a variable which does not occur free in A.

P3′. $\in (u)A \supset \check{\mathrm{S}}^u_v A \mid$

R1′. $\in A$, $\in A \supset B \twoheadrightarrow \in B$.

$\mathbf{PM}_2$ has advantages over $\mathbf{PM}_1$ in the interpretational details. For significance features are all taken up in $\mathbf{PM}_2$ by (i) the use of *S*-restricted sentential variables, and (ii) the use of *S*-restricted quantifiers. $\mathbf{PM}_2$ makes clear that the use of free variables in *PM*, *as used in theorems*, are really significance restricted.

§3 *The Criteria of Significance in Principia Mathematica*

There is an important distinction to be made between the criteria by means of which significance is assessed, which in *PM* are given by the theory of types, and the principles of significance which constitute the predicate theory. This distinction is not of course made in *PM*. Significance conditions which restrict the interpretation or application of the predicate theorems (principles) contain frequent references to types. In many cases, however, this is unnecessary and the explicit riders or interpretational conventions could quite well be expressed by conditions such as 'If 'ϕx' is significant ...' instead of 'If x is of a type appropriate to ϕ ...'. From this point of view, both $\mathbf{PM}_1$ and $\mathbf{PM}_2$ have an advantage over *PM* *10–*11 since they are not geared to particular criteria of significance such as the theory of types. Different criteria which would be consistent with the predicate principles could be adopted. However, the special features of some of the principles do imply that whatever criteria are adopted must satisfy some of the basic assumptions of type theory. In particular, unless the criteria are such that the types or categories are exclusive, and do not include or overlap one another, certain principles of the predicate theory are false. So the apparent independence of principles and criteria, in the case of $\mathbf{PM}_1$ and $\mathbf{PM}_2$, is to some extent deceptive; and it is an unsatisfactory feature of both, considered simply as significance logics, that they cannot be adopted consistently with more general and liberal criteria of significance. They are thus both too weak and too strong. Too weak, because a great many principles cannot be formulated in them, since their sentential basis is so weak, and too strong, because they are criteria bound.

The principles which fail over non-exclusive s-ranges, called by Halldén ([1], p. 11), *Behmann formulae*, are as follows:

(1) $[(x)(fx \supset gx) \;\&\; (x)(gx \supset hx)] \supset (x)(fx \supset hx)$

(2) $[(x)fx \;\&\; (x)(fx \supset gx)] \supset (x)gx$

(3) $(x)(fx \;\&\; gx) \supset [(x)fx \;\&\; (x)gx]$.

These are all false if overlapping s-ranges are admitted. For example, (3) may be falsified under the following conditions: *qu*(*f*) and *qu*(*g*) have overlapping significance ranges such that everything in the overlap is indeed both *f* and *g*, but an individual x_0, such that 'x_0' is not in the s-range of *qu*(*f*), is not *f*. In such a case, the antecedent of (3) is true but the consequent is false. Similarly, (1) may be falsified by a suitable predicate *qu*(*g*) such that the s-range of *qu*(*g*) is properly included in the s-ranges of both *qu*(*f*) and *qu*(*h*) (see Halldén [13] pp. 11–12 and 78–79).

In general, we may say, a wff of $\mathbf{PM}_2$ (or a closed wff of $\mathbf{PM}_1$) is a *Behmann wff* if, and only if, it is a theorem of $\mathbf{PM}_2$ (or $\mathbf{PM}_1$) but is false under a 3-valued significance interpretation which freely admits inclusive and overlapping s-ranges. And since there are such formulae, the intended 3-valued significance interpretation of $\mathbf{PM}_1$ and $\mathbf{PM}_2$ has to be underpinned by some theory of exclusive s-ranges, though not necessarily exactly that which is elaborated in the theory of types. However, since s-ranges are not always exclusive, and in any case must certainly be such that it is logically possible that they are not exclusive, $\mathbf{PM}_1$ and $\mathbf{PM}_2$ cannot stand as general significance logics.

In sum, $\mathbf{PM}_1$ and $\mathbf{PM}_2$, considered simply as quantified significance logics are radically unsatisfactory for several reasons. First, because they are largely independent of significance considerations; for since every closed wff is significant, significance features have been virtually eliminated, and it is not surprising that they should be interpretable as 2-valued systems and so lead directly to the standard elimination of nonsignificance from classical formal logic. Secondly, because they do not allow us to state nearly enough; this is partly due to the fact that they have such a weak sentential basis and partly due to the use of *S*-restricted quantifiers and variables, but it means that fundamental laws which express the inter-relations between s-ranges cannot be stated formally. What they really provide is the logic of *a type*, and, by way of the systematic ambiguity of all types, of each type separately. Thirdly, because they are criteria bound.

§4 *A Significance Semantics for* PM$_2$

Such a semantics may be constructed as follows:
m non-null domains of individuals, or types, $\{d_m\}$ are introduced for $m>0$: *m* need not be finite. The sum domain of individuals is $d = \bigcup_m d_m$. Individual variables range (designation-wise) over d. For each place *i* of a *m*-place predicate '*g*', there is exactly one domain d_i such that, for significant replacements of the remaining places of '*g*', $Sg(\ldots x \ldots) \equiv x \in d_i$; more explicitly, using Russellian significance restricted quantifiers,

$$(\exists y_1) \ldots (\exists y_{i-1})(\exists y_{i+1}) \ldots (\exists y_m)Sg(y_1, \ldots, y_{i-1}, x, y_{i+1}, \ldots, y_m) \equiv x \in d_i.$$

An interpretational assignment, as well as assigning to each m-place predicate an ordered set of types, also assigns to each such predicate a truth set. A truth set for an m-place predicate 'g' is a set of ordered m-tuples such that, for each m-tuple, the first element belongs to the first type, that is to the type of the first place of 'g', and generally, the ith element belongs to the ith type, and the m-tuple satisfies 'g' (*cf.* p. 205). Finally, the interpretational assignment assigns a secondary range $\{t, f\}$ to each sentential variable. Then a valuation function v is defined recursively for each interpretational assignment with respect to each set of domains as follows:

(i) where 'p' is a sentential variable, $v(p)=t$ or f according as 'p' is assigned t or f;

(ii) where 'g' is a m-place predicate and 'x_1', 'x_2', ..., 'x_m' are m individual variables,
$v(g(x_1, x_2, ..., x_m)) = n$
if some x_i does not belong to the ith type of 'g';
$v(g(x_1, x_2, ..., x_m)) = t$
if the ordered set $\langle x_1, ..., x_m\rangle$ belongs to the truth set of 'g';
$v(g(x_1, x_2, ..., x_m)) = f$
otherwise: *i.e.* if $v(g(x_1, ..., x_m)) \neq n$ and $\langle x_1, ..., x_m\rangle$ does not belong to the truth set of 'g'.

The inductive clauses for connectives '$\sim$' and '$\vee$' are those provided for $\mathbf{S_0}$: for example,

(iii) $v(\sim A) = t, f$ or n according as $v(A) = f$, t, or n; *etc.*

Finally, for an assignment to the free variables of $(x)A$,

(v) $v((x)A) = t$ or f according as $v(A) \neq f$ for all assignments from d to x, or not.

A more interesting rule runs: for given assignments to free variables, $v((x)A) = t$ iff $v(A) = t$ for every assignment to x for which $v(A) \neq n$ and $v(A) \neq n$ for some assignment to x; and $v((x)A) = n$ iff $v(A) = n$ for every assignment to x. But the rule is only sound for $\mathbf{PM_2}$ under restrictive assumptions, *e.g.*, under the unrealistic assumption, which reduces the rule to (v), that every predicate is significant of some one item, *i.e.* $(Px)(Uqu(f))Sfx$.

A wff A of $\mathbf{PM_2}$ is $\mathbf{PM_2}$-valid if, and only if, $v(A) = t$ for every interpretational assignment. A closed wff is one containing no free individual variables.

Lemma: No theorem or closed wff of $\mathbf{PM_2}$ has value n.

Theorem (9): Every theorem of $\mathbf{PM_2}$ is $\mathbf{PM_2}$-valid.

Proof is as usual by induction over the length of the proofs of $\mathbf{PM_2}$-theorems.

Theorem (10): Every $\mathbf{PM_2}$-valid wff is a theorem of $\mathbf{PM_2}$.

Proof: (I) Every classically valid closed wff is a theorem of $\mathbf{PM_2}$ (for details see Wang [1]).
Thus it suffices to show:
(II) Every $\mathbf{PM_2}$-valid wff is a classically valid closed wff.

If a wff A is not closed there are interpretational assignments such that $v(A)=n$ for assignments to its free variables, so it remains to consider closed ($\mathbf{PM_2}$-valid) wff, and hence wff which do not have value n. Suppose then that the closed wff A is not classically valid. Then there is a classical interpretation such that $v(A)=f$. But then too there is an interpretation, with $m=1$, so that $d=d_1$, such that $v(A)=f$. Hence A is not $\mathbf{PM_2}$-valid.

§5 *The Logic* $\mathbf{PS_1}$

Despite the above formal completeness result, $\mathbf{PM_2}$ is of course seriously incomplete from a more general point of view since it is not possible to express within it features of the intended interpretation which it should be possible to express. We cannot, for example, express formally the fact that its theorems are significant.

A first step in the direction of a more adequate logic consists in adding the quantificational superstructure of $\mathbf{PM_2}$ to $\mathbf{S_1}$ instead of to $\mathbf{S_0}$. In such a system, which we call $\mathbf{PS_1}$, we shall be able to affirm explicitly for each theorem that it is significant, and we shall be able to affirm also $\vdash S(x)fx$ and $\vdash S(\exists x)fx$. In general, some of the interpretational assumptions of the predicate theory of *PM* follow in $\mathbf{PS_1}$, but many of the special assumptions of type theory of course do not. A number of important principles which cannot be formulated in $\mathbf{PM_2}$ can be expressed in $\mathbf{PS_1}$: *e.g.* the principle of homogeneity (which is in fact false—see Halldén [1]):—

$$(\exists z)S(fz \,\&\, gz) \supset (y)(Sfy \equiv Sgy).$$

Again, significance riders on *PM* theorems can be formalized: *e.g.*, the required condition on *10.14 is:—$(\exists z)S(\phi z \& \psi z)$.

7.6 Many-sorted Significance Logics

A first step in improving the logic at least to the extent where many (non-exclusive) types or sorts can be treated at once and interlaced is provided by a many-sorted or many-typed development of the predicate logic of *PM*. The term 'sort' is, however, not particularly apposite here. For one thing, it suggests things in categories whereas our sorts are s-ranges, *i.e.* classes of expressions not designation-ranges. For another, the logics differ in important respects from many-sorted logics. A many-sorted significance logic cannot be set up in the free and easy way that 2-valued many-sorted logics have often been developed—so that in each argument place of primitive predicates any sort of variable may occur: (for such many-sorted systems, see *e.g.* Gilmore [1].) Suppose, for example, that some one-place predicate 'f' has its argument place occupied by a variable that ranges (designation-wise) over sorts different from that to which it is restricted for significance. Then its variable will have values such that $qu(f(x))$ is not significant and hence such that $qu(f(x) \vee \sim f(x))$ is not significant. For a *significance* many-sorted logic based on a weak sentential logic such as $\mathbf{S}_0$ or $\mathbf{S}_1$ we shall want, then, sortal or typed variables as well as unrestricted variables, if we are to have any predicate theorems. A many-sorted predicate system extends to predicate logic the kind of solution that was given at the sentential level to the no-theorems problem by introducing two sorts of sentential variable.

In order to extend $\mathbf{S}_1$ it is essential to have expressions of the form $f(x_1, ..., x_n)$ which are significant without qualification. Such expressions can be used to replace the S-restricted sentential variables of $\mathbf{S}_1$.

Many-sorted significance logics have the same large proper vocabulary as ordinary many-sorted logics (for details see Church [4], exercise 55.24, pp. 339–340). An n-sorted predicate logic without constants has the following sorts of variables:

(i) n sorts of individual variables: For each i, $1 \leqslant i \leqslant n, {}^ix, {}^iy, {}^iz, {}^ix'$, ... are individual variables of the ith sort;

(ii) n^m sorts of m-place predicate variables, for each $m \geqslant 1$.

For each i_j, $1 \leqslant i_j \leqslant n$ and $1 \leqslant j \leqslant m$, $({}^i{}_1, {}^i{}_2, ..., {}^i{}_m)f$; $({}^i{}_1, {}^i{}_2, ..., {}^i{}_m)g$; ... are m-place predicate variables such that the jth argument place of each is of the i_jth sort.

In ordinary many-sorted logics $({}^i{}_1, ..., {}^i{}_m)f(u_1, ..., u_m)$ is a wff iff u_1 is of the i_1th sort, etc., and generally u_k is of the i_kth sort. But in many-sorted significance logics based on (systems including) $\mathbf{S}_1$ this stipulation is not made: instead it is *postulated* that $S({}^i{}_1, ..., {}^i{}_m)f(u_1, ..., u_m)$ when u_k is of the i_kth sort for each k, $1 \leqslant k \leqslant m$. It does not follow conversely however from $S^{(i)}f(u)$, for example, that u is of the ith sort, for there is nothing to stop sorts including or overlapping one another. The requirement that if $S({}^i{}_1, ..., {}^i{}_m)f(u_1, ..., u_m)$ then, for each k, u_k is of the i_kth sort, is characteristic of an *exclusive* many-sorted theory. Henceforth '$({}^i)f$' is written '${}^if$'.

It is often convenient to have a universal sort, *i.e.* a sort that includes all other sorts. When the first sort is a universal sort, $S\,{}^1f(u)$, where '${}^1f$' is any predicate of the first sort and u is an individual variable of *any* sort. Predicates like '${}^1f$' are called *unlimited* predicates, since they are significant of everything. Sample unlimited predicates of English include 'is an item', 'is thought of' and 'exists' (the predicate 'E' of §7.9). Admitting unlimited predicates does not impose a restriction on the class of subjects admitted. Only the admission of unlimited predicates together with the common assumption, that a subject is significant iff it combines with some predicate to make a significant sentence, restricts the class of subjects. But it is difficult, and unreasonable, to stop the introduction of nonsignificant subjects such as 'the licentious number seven', so only the common assumption need be abandoned, and this causes no hardship (see §7.9, §2).

The many-sorted significance logic $\mathbf{QS}_1$ has as vocabulary the vocabulary of $\mathbf{S}_1$, the variables introduced above under headings (i) and (ii) and the quantifier symbol: U.

Formation rules of $\mathbf{QS}_1$

1. An atomic wff is a wff, where the following are atomic wff:
   (1a) Sentential variables standing alone;
   (1b) Formulae of the form $({}^{i}_{1},\ldots,{}^{i}_{m})f(u_1,\ldots,u_m)$, where $({}^{i}_{1},\ldots,{}^{i}_{m})f$ is an m-place predicate and $u_1,\ldots,u_m$ are m individual variables (of *any* sort).
2. If A and B are wff then $\sim A$, SA, and $(A \supset B)$ are wff.
3. If A is a wff and u is an individual variable (of any sort) then $(Uu)A$ is wff.

An atomic wff is *well-sorted* iff it is either an S-restricted sentential variable or is of the form $({}^{i}_{1},\ldots,{}^{i}_{m})f({}^{i}_{1}u_1,\ldots,{}^{i}_{m}u_m)$ for some m.

An atomic wff which is not well-sorted is *ill-sorted*: thus 'p' and '${}^{2}f({}^{3}x)$' are ill-sorted.

Definitions of further connectives and bracketing conventions as for $\mathbf{S}_1$ are adopted. Futhermore Church's account in [4] of substitution for predicate letters is adopted. The axiomatisation can be compared with a many-sorted extension of Church's second predicate logic F^2.

Axioms of $\mathbf{QS}_1$

S1. $\vdash r \supset . r' \supset r$

S2. $\vdash r'' \supset (r \supset r') \supset . r'' \supset r \supset . r'' \supset r'$

S3. $\vdash \sim r \supset \sim r' \supset . r' \supset r$

S4. $\vdash Sr$

S5. $\vdash S{\sim}p \equiv Sp$

S6. $\vdash S(p \supset q) \equiv . Sp \;\&\; Sq$

S7. $\vdash (U^{i}x)(r \supset {}^{i}f({}^{i}x)) \supset . r \supset (U^{i}x){}^{i}f({}^{i}x)$

S8. $\vdash (U^{i}x){}^{i}f({}^{i}x) \supset {}^{i}f({}^{i}y)$

S9. $\vdash S{}^{i}f({}^{j}x) \equiv S(U^{j}x){}^{i}f({}^{j}x)$.

S10. $\vdash SSp$

Transformation rules

R1. $\vdash A, \vdash A \supset B \longrightarrow \vdash B$

R2. $\vdash A, \vdash SB \longrightarrow \vdash \check{\mathrm{S}}^{R}_{B}A|$,

where R is a well-sorted atomic wff. This rule combines a rule of S-restricted sentential substitution and a rule of S-restricted predicate substitution.

R3. $\vdash A \longrightarrow \vdash \check{\mathrm{S}}^{P}_{B}A|$, where *P* is an unrestricted sentential variable; or an *S*-restricted sentential variable and *B* is a well-sorted atomic wff.

R4. $\vdash A \rightarrow \vdash B$,

where *B* results from *A* by substituting $\check{S}^u_v C|$ for a particular occurrence of *C* in *A*, and where *u* is an individual variable which is not free in *C* and *v* is an individual variable of the same sort which does not occur in *C*. (Sortal rule of change of bound variables).

R5. $\vdash A \rightarrow \vdash \check{S}^u_v A|$,

where *u* and *v* are individual variables of the same sort. (Sortal rule of substitution for individual variables).

R6. $\vdash A \rightarrow \vdash (Uu)A$
(Rule of generalisation).

This postulation set provides *a* fairly simple solution to the problem of axiomatising weak many-sorted significance logics. In particular by using axioms S7 and S8, instead of attempting to qualify more usual quantificational axiom-schemata, many difficulties are avoided. Nevertheless it would seem feasible to reaxiomatise **QS**$_1$ using axiom schemata. Then S7 for example would be replaced by

S7A $(Uu)(A \supset B(u)) \supset . A \supset (Uu)B(u)$, provided
(i) *A* does not contain *u* free, and
(ii) *A* and *B*(*u*) are significant wff.

But the second proviso makes the schema non-effective; and rendering it effective converts S7A into a further rule. There are, however, other ways out of this difficulty. For example, noting $\vdash Sp \equiv . Tp \simeq p$, we may replace S7A by

S7B $(Uu)(TA \supset TB(u)) \supset . TA \supset (Uu)TB(u)$, provided *u* is not free in A.

But this procedure leads beyond **QS**$_1$ to **QS**$_2$.

Ordinary many-sorted logics can be reduced to one-sorted classical logic—at worst after the introduction of sortal predicates for each sort, in case these cannot be defined in the logic. But for weaker many-sorted significance logics, like **QS**$_1$, an analogous reduction to one sort fails. To make the reduction problem definite, we consider necessary conditions for an adequate translation. To carry out a reduction, *n* sortal predicates, s_i, for each *i*, $1 \leqslant i \leqslant n$, are required. It is also required that $(Uu)Ss_i(u)$, *i.e.* all assertions as to the sort of a given item are significant. Now a reduction eliminates every closed wff of the many-sorted logic by a strongly equivalent closed wff of an unsorted (or one-sorted) logic. For by taking the universal closure of wff the problem can be reduced to one of eliminating closed wff. Hence for a reduction there must be some way of eliminating closed wff of the form $(U^i u)B(^i u)$. How could this be done? The obvious and standard way is to replace $(U^iu)B(^iu)$ by $(Uu)(s_i(u)IB(u))$, where 'I' is some sort of implication connective of $\mathbf{QS}_1$. This moreover appears to be the only way variable '^iu' can be eliminated, preserving truth and significance, in favour of 'u'—without transgressing the means of $\mathbf{QS}_1$. For otherwise '^iu' cannot be eliminated in favour of 'u' without making use of the *condition* that u is of sort s_i. Indeed in order to state that 's_i' is the sortal predicate of sort i such a condition needs to be satisfied.

Accordingly we say $\mathbf{QS}_1$ is *sort-reducible* only if

(i) there is available in $\mathbf{QS}_1$ for each sort i $(1 \leqslant i \leqslant n)$ a sortal predicate 's_i' (such that $(Uu)Ss_i(u)$), and

(ii) there is a connective 'I' of $\mathbf{QS}_1$ such that, for every closed wff $(U^iu)B(^iu)$, the following holds:

$$(\alpha)\quad (U^iu)B(^iu) \simeq (Uu)(s_i(u)IB(u)).$$

Actually the bracketed condition in requirement (i) can be dropped without loss.

Theorem (11): $\mathbf{QS}_1$ is not sort-reducible.

Proof: Every theorem of $\mathbf{QS}_1$ is $\mathbf{QS}_1$-valid in an interpretation in which each sort has just one element. Accordingly if $\mathbf{QS}_1$ is sort-reducible then the reduction is possible also in the case where each sort has just one element. Let the element of sort s_i be ^iu. Then for a reduction to be possible there must be a connective 'I' such that

$$(\alpha')\quad B(^ia) \simeq s_i(^1a)IB(^1a) \;\&\; \ldots \;\&\; s_i(^na)IB(^na),$$

upon expanding the quantifiers over their respective domains. This connection must hold also in the simpler case where there are just two sorts, *i.e.* I satisfies

$$(\alpha'')\quad \begin{aligned} &B(^1a) \simeq s_1(^1a)IB(^1a) \;\&\; s_1(^2a)IB(^2a)\\ &B(^2a) \simeq s_2(^1a)IB(^1a) \;\&\; s_2(^2a)IB(^2a),\\ &\text{where } s_1(^1a) \;\&\; s_2(^2a). \end{aligned}$$

The connection must also hold in the limiting case where there is only one sort. So I must also satisfy

$$(\alpha''')\quad \begin{aligned} &B(^1a) \simeq .\; s_1(^1a)IB(^1a),\\ &\text{where } s_1(^1a). \end{aligned}$$

Since $s_1(^1a)$ has value t, and $B(^1a)$ may have any of the values $\{t,f,n\}$, it follows since 'I' is matrix-accessible that part of its matrix is thus:

I	t	f	n
t	t	f	n .

Now apply (α''), whence it follows that '*I*' has this matrix:

I	*t*	*f*	*n*
t	*t*	*f*	*n*
f			*t*
n			*t*

But it is easily shown by induction over the connectives of $\mathbf{QS_1}$ that no such connective is definable in $\mathbf{QS_1}$.

Hence no sortal-reduction of $\mathbf{QS_1}$ is possible. Such a proof would break down at once in quantification extensions of stronger systems such as $\mathbf{S_3}$ where a suitable connective, '$\rightharpoondown$' is available. However the proof does extend to a many-sorted quantification extension, $\mathbf{QS_{2'}}$, of $\mathbf{S_2}$ obtained from $\mathbf{S_2}$ by adding just the quantifier '*U*', and postulates on it such that it has an interpretation as a conjunction over the finite domains of items. Then

Theorem (12): $\mathbf{QS_{2'}}$ is not sort-reducible.

However a reformulation of $\mathbf{QS_{2'}}$ without sorts which we call $\mathbf{QS_2}$ is possible, and is presented in §7.8, §3. Such a reformulation of $\mathbf{QS_1}$ appears impossible.

To stop with $\mathbf{QS_1}$ is not good enough. In particular $\mathbf{QS_1}$ does not contain the vital significance connective '*T*', and is therefore inferior to $\mathbf{QS_{2'}}$ in just the way that $\mathbf{S_1}$ is inferior to $\mathbf{S_2}$. Accordingly we consider briefly the system $\mathbf{QS_{2'}}$, obtained by an axiomatisation like that of $\mathbf{QS_1}$ except that sentential base ${}_3\mathbf{S_2}$ displaces ${}_1\mathbf{S_1}$. $\mathbf{QS_2}'$ has as axioms the axioms of $\mathbf{QS_1}$ except that S 10 is replaced by S 10′: $\vdash STp$. Also $\mathbf{QS_{2'}}$ has the additional axioms

S11. $\vdash r \supset Tr$

S12. $\vdash \sim T(p \,\&\, \sim p)$.

$\mathbf{QS_2}$ has all the transformational rules of $\mathbf{QS_1}$ together with the *T*-rule of ${}_3\mathbf{S_2}$.

Semantics for $\mathbf{QS_1}$ and $\mathbf{QS_{2'}}$ are now outlined. (This sketch should be compared with the detailed semantics of $\mathbf{QS_6}$ presented in §7·7). A *semantical basis* S for $\mathbf{QS_1}$ (or $\mathbf{QS_{2'}}$) is an ordered set $S = \langle d_1, d_2, \ldots, d_n, e\rangle$ where the individual domains $d_1, \ldots, d_n$ are each non-null classes of items (or expressions), and e is a class of relations over $d = \bigcup_i d_i$. A basis is *usual* if e is the class of all relations over d. An *m*-place relation of e is $(i_1, \ldots, i_m)$-*ic* iff for each j, $1 \leqslant j \leqslant m$, its jth place belongs to (is restricted to) domain d_{i_j}.

An interpretation I of $\mathbf{QS_1}$ ($\mathbf{QS_{2'}}$) over S is an assignment I which assigns to each *S*-unrestricted sentential constant one of the values *t*, *f*, or *n*, to each *S*-restricted sentential constant one of the values *t* or *f*; to each individual constant ${}^iu_0$ an element of d_i; to each *m*-place predicate constant $({}^i{}_1, \ldots, {}^i{}_m)f_0$ an $(i_1, \ldots, i_m)$-ic relation of e.

Given I and S, values for all wff are defined inductively just as for $\mathbf{QS_6}$

below, except that connectives and quantifiers of $\mathbf{QS_1}$ ($\mathbf{QS_{2'}}$) replace those of $\mathbf{QS_6}$. To be specific in the case of $\mathbf{QS_1}$ the definition of truth-value is as follows:

(i) $v(P) = \mathrm{I}(P)$ where P is a sentential variable
(ii) $v(({}^{i}{}_{1},\ldots,{}^{i}{}_{m})f(u_1, \ldots, u_m)) = t$ iff $T\mathrm{I}(({}^{i}{}_{1},\ldots,{}^{i}{}_{m})f)(\mathrm{I}(u_1), \ldots, \mathrm{I}(u_m))$

This does not require that $\mathrm{I}(u_j) \in \mathrm{d}_{i_j}$.

$$v(({}^{i}{}_{1},\ldots,{}^{i}{}_{m})f(u_1, \ldots, u_m)) = f \text{ iff } F\mathrm{I}(({}^{i}{}_{1},\ldots,{}^{i}{}_{m})f)(\mathrm{I}(u_1), \ldots, \mathrm{I}(u_m))$$
$$v(({}^{i}{}_{1},\ldots,{}^{i}{}_{m})f(u_1, \ldots, u_m)) = n \text{ iff} \sim S\mathrm{I}(({}^{i}{}_{1},\ldots,{}^{i}{}_{m})f)(\mathrm{I}(u_1), \ldots, \mathrm{I}(u_m))$$

This case does *not* occur if for each j, $1 \leqslant j \leqslant m$, $\mathrm{I}(u_j) \in \mathrm{d}_{i_j}$.

(iii)–(v) These clauses give the value of $\sim B$, $B \supset C$ and SB in terms of values for B and C; they are the same as those given in $\mathbf{S_1}$.
(vi) For a wff $(Uu)B$ for a given assignment to the free variables of $(Uu)B$, $v((Uu)B) = t$ iff $v(B) = t$ for all assignments to u; $v((Uu)B) = f$ iff $v(B) = f$ for some assignment to u and $v(B) \neq n$ for any assignment to u; and $v((Uu)B) = n$ iff $v(B) = n$ for some assignment to u.

A wff A is $\mathbf{QS_1}$-*valid for* I over S iff $v(A) = t$ for every assignment to its free variables with respect to I over S. A wff A is $\mathbf{QS_1}$-valid iff A is $\mathbf{QS_1}$-valid for every interpretation over usual bases.

It may be objected that clause (ii) of the inductive specification is insufficiently clear; and in particular it may be claimed that the interpretation should specify matters so that $({}^{i}{}_{1},\ldots,{}^{i}{}_{m})f({}^{i}{}_{1}u_1, \ldots, {}^{i}{}_{m}u_m)$ is automatically significant. This is not essential; the specification in (ii) is adequate, but the interpretation can be employed to clarify matters. We define a new interpretation I′ over S as an interpretation which makes the same assignments as I over S and these additional assignments: to each well-sorted atomic predicate wff $(({}^{i}{}_{1}\ldots,{}^{i}{}_{m})f({}^{i}{}_{1}u_1, \ldots, {}^{i}{}_{m}u_m))$ I′ assigns value t or f according as $\mathrm{I}'(({}^{i}{}_{1},\ldots,{}^{i}{}_{m})f)(\mathrm{I}'({}^{i}{}_{1}u_1), \ldots, \mathrm{I}'({}^{i}{}_{m}u_m))$ holds or not; and to each ill-sorted atomic predicate wff $({}^{i}{}_{1}\ldots,{}^{i}{}_{m})f(u_1, \ldots, u_m)$ I′ assigns value t, f, or n, according as $\mathrm{I}'(({}^{i}{}_{1},\ldots,{}^{i}{}_{m})f)(\mathrm{I}'(u_1), \ldots, \mathrm{I}'(u_m))$ is true, false or nonsignificant, i.e. according as the m-tuple $(\mathrm{I}'(u_1), \ldots, \mathrm{I}'(u_m))$ satisfies, does not satisfy, or does not significantly satisfy $\mathrm{I}'(({}^{i}{}_{1},\ldots,{}^{i}{}_{m})f)$. Then clause (ii) of the inductive definition of truth-value is replaced by

(ii′) $v(A) = \mathrm{I}(A)$, where A is an atomic predicate wff.

The way is now cleared for consistency and completeness proofs for $\mathbf{QS_1}$ and $\mathbf{QS_{2'}}$. But we do not develop these here because we think, for reasons we shall now explain, that many-sorted significance logics are far from being the quantified significance logics of most interest, scope and flexibility.

Stronger many-sorted significance logics can be reduced (using (α)) to significance logics without sorts, much as usual many-sorted logics are reduced to one-sorted logics. For in these latter logics what amounts to sorts can be introduced by use of restricted variables. Thus, using (α) above, (U^iu) $B(^iu)$ can be represented by (Uu) $(s_i(u) \rightharpoonup B(u))$, where the variable '$u$' is restricted by the antecedent to sort s_i.

Restricted-variable logics have big advantages over many-sorted significance logics. They are much more flexible; they have very much simpler axiomatisations and simpler semantics. The elaborate notational apparatus required in setting up the morphology of many-sorted logics is avoided, and thereby the logical impurity got by importing numerical apparatus into the very bottom of the logic is avoided. As with type theories, so in many-sorted logics, types or sorts are not explicitly dealt with in the logic, but only, as it were, at the periphery. For many purposes, furthermore, quantifiers which are *not* significance restricted are wanted; for example to state the relations between sorts and of items of different sorts. Now such quantifiers may be available in many-sorted significance logics, if the logic has a universal sort. But in this case if the logic is as strong as it ought to be, and contains '$\rightharpoonup$', then the sorts can be eliminated from the logic, usually *within* the logic itself. So such a logic does not differ appreciably from a significance logic with restricted variables.

Because of the vast preferability of logics without sortal variables and with restricted variables instead, we do not develop many-sorted significance logics based on stronger sentential bases. Such logics are, however, easily designed by analogy with those we have set up. They all consist in adding a suitable quantificational superstructure to the chosen sentential base. It is worth observing, however, that when stronger sentential bases are used, simple postulate sets for many-sorted significance logics can be obtained. We take as an example conversion of the system $\mathbf{QS_6}$ to a many-sorted significance logic; a similar conversion can be applied to other significance systems developed below. Let $\mathbf{QS_6M}$ have the individual, predicate and sentential letters of $\mathbf{QS_1}$, and otherwise let its morphology be that of $\mathbf{QS_6}$ (§7.7). Define a *well-sorted atomic wff*, as before, as an atomic wff which is either an S-restricted sentential variable or as an atomic predicate wff each of whose individual letters is of the sort specified by the predicate letter. Then $\mathbf{QS_6M}$ has the same postulates as $\mathbf{QS_6}$ except that the provisos on A4 and A5 are amended to read 'where u is an individual variable and t is an individual letter of the same sort' and the following postulate is added

A10. $\vdash SA$,

where A is a well-sorted atomic wff.

There is one reduction of notation, sometimes regarded as a sortal reduction that we have not considered; namely an elimination of distinctive subject and predicate variables in favour of a single neutral variable. Where 'ω' and

'ω'' are such neutral variables and '*Sub*' and '$Pred^1$' are (quotational) sortal predicates, then such a law as '$f(x) \rightharpoondown f(x)$' is reduced to the law:

$Sub(\omega) \rightharpoondown . Pred^1(\omega') \rightharpoondown . \omega'(\omega) \rightharpoondown \omega'(\omega)$.

Unlike levels of language theorists we have no objection to this procedure: it amounts to including some of what are normally formation rules within the logic. The procedure has some advantages, *e.g.* one can state in the logic that no expression is at once a subject and a predicate, and the main significance logics we go on to develop can be amended to include it. It has some drawbacks, though, from the significance point of view; for it tends to erode the subject-predicate distinction, a distinction which is all-important for the assessment of significance.

7.7 The Quantificational System $\mathbf{QS_6}$

§1 *The Syntax and Contractional Semantics*

Since $\mathbf{QS_6}$ is the main quantificational logic we shall develop, we give a full discussion of the system.

The Morphology of $\mathbf{QS_6}$

Vocabulary:

Improper primitive symbols: $\neg$, $\rightharpoondown$, D, (,), U, q

Sentential variables: p, q

Symbol rules:

(i) If P is a sentential parameter, then so is P'

(ii) If P is a sentential parameter and i a numerical subscript then P_i is a sentential constant.

Subject variables: x, y, z.

Symbol rules (i) and (ii): If X is a subject variable then X$'$ is a subject variable and X_i is a subject constant.

Function parameters: d, e

Symbol rules:

(i) If E is a function parameter and n a numeral then E^n is an n-place function parameter.

(ii) If E is a function parameter (of n places) then E$'$ is a function parameter (of n places).

(iii) If E is a function parameter (of n places) then E_i is a function constant (of n places).

Predicate parameters: f, g, h

Symbol rules: (i), (ii) and (iii) as for functions.

Formation rules:

(a) Well-formed terms (Wft): exhaustive specification:

(1) Subject variables and constants are wft.

(2) If E is a n-place function parameter or constant and $t_1, ..., t_n$ are terms, then $E(t_1, ..., t_n)$ is a wft.

(b) Atomic wff:

A is an atomic wff iff (i) A is $F(t_1, ..., t_n)$ and F is a n-place predicate parameter or constant and $t_1, ..., t_n$ are terms, or (ii) A is a sentential parameter or constant.

(c) Well-formed formulae: exhaustive specification:

(1) An atomic wff is a wff.

(2) If A and B are wff then so are $\neg A$, DA and $(A \rightharpoondown B)$.

(3) If A is a wff and u a subject variable then $(Uu)A$ and $(qu)A$ are wff.

Free and bound variables:

An occurrence of a variable u in a wff A is a *bound occurrence* of u in A iff it is an occurrence in a wf part of A of the form $(Uu)B$ or $(qu)B$; otherwise the occurrence is a *free occurrence* of u in A. The *bound variables* of A are the variables which have bound occurrences in A; the free variables of A are those which have free occurrences in A. In $(Uu)B$ [or $(qu)B$], B is the *scope* of quantifier (Uu) [or (qu)].

Substitution notation:

The substitution notation, over and above that already introduced, is explained as follows:

1. $\overset{\vee}{S}{}^u_N\, A|$ is the wff resulting from substituting variable, wft or wff N for all free occurrences of variable u in A, provided that no free occurrence of u in A is in a wf part $(Uw)C$ or $(qw)C$ or ξwC of A, where w is a free variable of N (or is N if N is a variable); and otherwise is A.
2. $S1^u_v A|$ is the wff resulting from substituting variable v for all (free) occurrences of variable u in wff A, provided v is free in the resulting wff where and only where u is free in A; and otherwise is A.

These explanations of the substitution notation could be broken down into recursive specifications, in much the way that the explanation of $S^P_B\, A|$ is replaced (at the beginning of chapter 6) by a recursive specification.

Definitions: To the definitions adjoined to S_6 the following is added:

$$(Pu)A =_{Df} \sim (Uu) \sim A.$$

In $\mathbf{QS}_6$ the quantifier 'q' is not independent: it may, as remarked, be eliminated definitionally thus: $(qu)A =_{Df} D(Uu)DA$. Accordingly our specification of $\mathbf{QS}_6$ and proofs concerning it could be simplified by omitting rules, postulates and inductions concerning 'q'. But in weaker quantified significance logics this definitional elimination is no longer possible, and to avoid long repetitions in dealing with these systems we shall carry 'q' in $\mathbf{QS}_6$. We shall, however, mark eliminable postulates and steps concerning 'q' with a dagger (†)

to indicate the simplifications made by defining '*q*'. Proof of the † postulates will be sketched in §7.11.

The Semantics of $\mathbf{QS_6}$

We develop a contractional semantics (in the sense explained in §7.3) for $\mathbf{QS_6}$. A *semantical basis* for $\mathbf{QS_6}$ is a triple S = ⟨d, e, f⟩, where the *item domain* d is a (non-null) set of contractions, e is a class of relations over d, and f is a class of operations in d (i.e. functions from the Cartesian product d × d × ... × d into d). A basis S is *usual* iff e is the class of all relations over d, and f is the class of all operations in d. The classes e and f can be restricted in various ways, *e.g.* they can be restricted to the classes of functions and relations in extension, in which case an extensional interpretation will result.

An *interpretation* I *of* $\mathbf{QS_6}$ *over basis* S is a contraction function I which assigns to each sentential constant one of the values *t*, *f* or *n*, to each subject constant an element of d, to each *n*-place predicate constant an *n*-place relation of e, and to each *n*-place function constant an *n*-place operation of f. In other words, I is a function which contracts sentences to values *t*, *f* or *n*, subjects to items of d, and predicates to relations between elements of d. Then, given I, secondary ranges of variables and parameters are defined as follows: sentential parameters have values *t*, *f* and *n* as their range, individual variables have items of d as their range, *n*-place predicate parameters have *n*-place relations of e as their range, and *n*-place functions have *n*-place operations of f as their range.

For each constant and variable *u*, I(*u*) is the *assignment-value* of *u* under I over S. As before '*A* has value *a*' is written '*v*(*A*) = *a*'. Value assignments are now relative to a given interpretation.

Assignments for all *wft* are defined inductively, as follows:

If E is a *n*-place function parameter or constant and $t_1, t_2, ..., t_n$ are terms, then $I(E(t_1, t_2, ..., t_n)) = I(E)(I(t_1), ..., I(t_n))$.

For example $qu(d_0(x_1, x_2))$ is about the result, of the operation 'd_0' is about, on the contractions of $qu(x_1)$ and $qu(x_2)$, in case the contraction is an aboutness function.

Given I and S, *values* for all *wff* are defined inductively as follows:

(i) For a wff consisting of a sentential parameter *P* alone, *v*(*P*) = I(*P*). Hence *v*(*P*) = *t* iff I(*P*) = *t*.

(ii) For a wff $F(t_1, ..., t_n)$ consisting of *n* terms $t_1, ..., t_n$ and *n*-place predicate F,

$v(F(t_1, ..., t_n)) = t$ iff $T[(I(t_1), ..., I(t_n))\ i\ I(F)]$;
$v(F(t_1, ..., t_n)) = f$ iff $F[(I(t_1), ..., I(t_n))\ i\ I(F)]$;
$v(F(t_1, ..., t_n)) = n$ iff $\sim S[(I(t_1), ..., I(t_n))\ i\ I(F)]$,

i.e. $F(t_1, ..., t_n)$ has value *n* iff $(I(t_1), ..., I(t_n))$ does not significantly instantiate relation I(F).

(iii), (iv), (v). These clauses, which give the values of $\neg B$, DB and $B \rightharpoonup C$ respectively, in terms of values of B and C, coincide with those already given for S_6.

(vi) For a wff $(Uu)A$, where u is a subject variable, for a given assignment to the free variables of $(Uu)A$,

$v((Uu)A) = t$ iff $v(A) = t$ for every assignment to u;

$v((Uu)A) = f$ iff $v(A) = f$ for some assignment to u, and $v(A) \neq n$ for any assignment to u;

$v((Uu)A) = n$ iff $v(A) = n$ for some assignment to u.

Thus $(Ux)f_0(x)$ has the same values as a (perhaps infinite) classical conjunction of values of $f_0(x)$.

(vii)† For a wff $(\text{ꟼ}u)A$, where u is an individual variable, for a given assignment to the free variables of $(\text{ꟼ}u)A$,

$v((\text{ꟼ}u)A) = t$ iff $v(A) = t$ for some assignment to u,

$v((\text{ꟼ}u)A) = f$ iff $v(A) = f$ for some assignment to u, and $v(A) \neq t$ for any assignment to u;

$v(\text{ꟼ}u)A) = n$ otherwise, *i.e.* iff $v(A) = n$ for every assignment to u.

Thus $(\text{ꟼ}x)f_0(x)$ has the same values as a (possibly infinite) Brady disjunction ('V' of §5.12) of values of $f_0(x)$.

A wff A is *valid for* I *over* S iff $v(A) = t$ for every assignment to its free variables, with respect to I over S; *satisfiable for* I *over* S iff $v(A) = t$ for some assignment to the free variables of A, with respect to I over S. A wff A of $\mathbf{QS_6}$ is *valid* iff it is valid for every interpretation over usual bases; *satisfiable* iff it is satisfiable for some interpretation over some usual basis.

Theorem (13): If A is valid for I over usual S then $\neg A$ is not satisfiable for I over usual S; if A is valid then $\neg A$ is not satisfiable.

Proof: From value rule (iii), if $v(A) = t$ always then $v(\neg A) \neq t$ always.

Theorem (14): If A is valid, A is satisfiable.

§2 *The Postulational Theory*

The postulate set of $\mathbf{QS_6}$ is as follows:

A1. If A is a substitution instance of an S_6-tautology, then $\vdash A$.

A2. $\vdash (Uu)(A \rightharpoonup B) \rightharpoonup . A \rightharpoonup (Uu)B$, where u is a (subject) variable which does not occur free in A.

A3.† $\vdash (Uu)(A \rightharpoonup B) \rightharpoonup . (\text{ꟼ}u)A \rightharpoonup B$, where u is a variable which does not occur free in B.

A4. $\vdash (Uu)A \rightharpoonup \check{S}^u_t A|$, where u is a variable and t is a wft.

A5.† $\vdash \check{S}^u_t A| \rightharpoonup (\text{ꟼ}u)A$, where u is a variable and t is a wft.

A6. $\vdash (Uu)SA \rightharpoonup S(Uu)A$

A7. $\vdash S(Uu)A \rightharpoonup (Uu)SA$

A8.† $\vdash (\mathcal{Q}u)SA \rightharpoondown S(\mathcal{Q}u)A$
A9.† $\vdash S(\mathcal{Q}u)A \rightharpoondown (\mathcal{Q}u)SA$
R1. $\vdash A, \vdash A \rightharpoondown B \twoheadrightarrow \vdash B$ (modus ponens)
R2. $\vdash A \twoheadrightarrow \vdash (Uu)A$ (generalisation).

A finite sequence of wff of $\mathbf{QS}_6$ is a *proof* iff each member of the sequence is either an axiom of $\mathbf{QS}_6$ or else is derived from preceding wff of the sequence by means of R1 or R2. Such a sequence is a proof of its last member. A is a theorem of $\mathbf{QS}_6$ iff *some* sequence is a proof of A. The quantifier 'some' used in this definition of 'theorem' is, of course, not existentially loaded. It is not required that the sequence actually exists, otherwise we should be committed to an obnoxious Platonism or else theorems would grow as this book is written.

The first connection between the syntactic notion of theoremhood and the semantical notion of validity is given in the following result:

Theorem (15): Every theorem of $\mathbf{QS}_6$ is valid.

Proof is by induction over proofs of theorems of $\mathbf{QS}_6$, and for this it suffices to show that each axiom scheme is valid and that the rules preserve validity.

ad R1. If for all I on all S, $v(A) = t$ and $v(A \rightharpoondown B) = t$ then for all I on all S, $v(B) = t$ by valuation rules for '$\rightharpoondown$'.

ad R2. If for all assignments to free variables of A for all I on all S, $v(A) = t$, then for all I on all S, $v((Uu)A) = t$ by the valuation rules for quantifier 'U'.

ad A1. If A is a substitution instance of a tautology of $\mathbf{S}_6$, then $v(A) = t$ by the valuation rules for sentential connectives, no matter what values the components of A assume as a result of assignment I over S; hence A is valid.

ad A2. If (a), for arbitrary I on S, $v((Uu)(A \rightharpoondown B)) \neq t$, then $v(\text{A2}) = t$ by the valuation rules for '$\rightharpoondown$'. Suppose then (b), for arbitrarily selected I on S and for arbitrary assignments to the free variables of A2, $v((Uu)(A \rightharpoondown B)) = t$. Then $v(A \rightharpoondown B) = t$ for every assignment to u. If further (b1), $v(A) \neq t$, then $v(A \rightharpoondown (Uu)B) = t$, so that $v(\text{A2}) = t$. Suppose then (b2), $v(A) = t$. Now if (c1), $v((Uu)B) = t$, then $v(A \rightharpoondown (Uu)B) = t$, so that $v(\text{A2}) = t$. If (c2), $v((Uu)B) = n$, then for some assignment to u, $v(B) = n$. But as u is not free in A, $v(A) = t$ for this assignment to u. Hence $v(A \rightharpoondown B) = n$ and hence $v((Uu)(A \rightharpoondown B)) = n$ contradicting (b). Similarly if (c3), $v((Uu)B) = f$ then (b) is contradicted. Since, finally, $v(\text{A2}) = t$ for arbitrary assignments to its free variables for arbitrarily selected I over S, A2 is valid. The extrasystematic notion of an arbitrarily selected I, upon which generalisation to all I can be made, can be taken up formally by a selection operator (as explained in §7.11).

ad A3–A9 The proofs of validity are similar in principle, though not in detail, to that of A2.

Theorem (16):

1. If A is a tautology then A is valid.
2. A is valid iff $(Uu)A$ is valid.
3. A is valid iff its universal closure is valid, where as before *the universal closure* of wff C is the wff $(Uu_1)(Uu_2) \ldots (Uu_n)C$, where $u_1, u_2, \ldots, u_n$ are all the free variables in C in lexicographical order.

Proof: 1 and 2 are immediate corollaries of *Theorem* (15).
3 follows from 2.

Theorem (17) (first method): $\mathbf{QS_6}$ is simply and absolutely consistent.

Proof: If A is a theorem of $\mathbf{QS_6}$ then A is valid. Therefore $\neg A$ is not satisfiable, and so $\neg A$ is not valid. Hence $\neg A$ is not a theorem. Therefore $\mathbf{QS_6}$ is absolutely consistent. If $\mathbf{QS_6}$ were simply inconsistent then, for some B, $\vdash B$ and $\vdash \sim B$; hence $\vdash \neg B$. Then $\vdash \neg A$, by A1. Contraposing, $\mathbf{QS_6}$ is simply consistent.

A more syntactical consistency proof, which presupposes less, is obtained as follows:

A wff is *quantifier-free* if it contains no quantifiers ('U' or '$\mathcal{I}$' in $\mathbf{QS_6}$). An *associated wff* is obtained from wff A by first replacing each term in A which is not a variable uniformly by a new free variable not occurring in A, second taking the universal closure A' of the resulting wff, and third replacing all occurrences of $(Ux)B(x)$ in A' by the conjunction $B(x_1) \& B(x_2)$, and all occurrences of $(\mathcal{I}x)B(x)$ in A' by the Brady disjunction $B(x_1) \mathrm{V} B(x_2)$. The *alf* f(A) of a wff A is obtained by taking its associated wff, and then in this replacing every wf part $\mathrm{F}(u_1, \ldots, u_n)$ by a sentential variable not previously occurring, according to the prescription: two wf parts $\mathrm{F}(u_1, \ldots, u_n)$ and $\mathrm{G}(v_1, \ldots, v_n)$ are replaced by the same sentential variable iff F and G are the same predicate parameter or constant and u_i is the same subject constant as v_i for each $i = 1, \ldots, n$. Semantically the alf f(A) of A is its interpretation over a two-element domain $\{x_1, x_2\}$.

Theorem (18): If A is a theorem of $\mathbf{QS_6}$ then its alf f(A) is an $\mathbf{S_6}$-tautology.

Proof is by induction over the proof of A in $\mathbf{QS_6}$.

ad A1. If A is an $\mathbf{S_6}$-tautology then f(A) is a conjunction of $\mathbf{S_6}$-tautologies, hence f(A) is an $\mathbf{S_6}$-tautology.

ad A2. f(A2) is a conjunction of expressions of the form

$$(A \rightharpoondown B_1) \& (A \rightharpoondown B_2) \rightharpoondown .\ A \rightharpoondown B_1 \& B_2.$$

But each such expression is a $\mathbf{S_6}$-tautology; hence the conjunction is also.

ad A3. Conjunctions of expressions of the form $(A_1 \rightharpoondown B) \& (A_2 \rightharpoondown B) \rightharpoondown .\ A_1 \mathrm{V} A_2 \rightharpoondown B$ are $\mathbf{S_6}$-tautologies.

ad A4. f(A4) is a conjunction of expressions of the form, $A_1 \& \ldots A_r \rightharpoondown A_1 \& \ldots A_r$. Thus f(A4) is an $\mathbf{S_6}$-tautology.

A5–A9 are similar in principle.

ad R1. The alfs of the premises are of the form

$$\vdash \underset{i}{\&}\, A_i \text{ and } \vdash \underset{ij}{\&}\, (A_j \rightharpoonup B_i).$$

Since these are tautologies it follows that $\vdash \underset{j}{\&}\, B_j$ is an $\mathbf{S}_6$-tautology.

R2 is immediate.

Theorem (17) (second method): $\mathbf{QS}_6$ is simply and absolutely consistent.

Proof: Let A be a universally closed theorem of $\mathbf{QS}_6$ which contains only variables as terms. $(Uy).\ (Ux)f(x) \rightharpoonup f(y)$ is such a theorem. By the previous result f(A) is an $\mathbf{S}_6$-tautology. However f($\neg A$)[$= \neg$ f(A)] is not an $\mathbf{S}_6$-tautology, by valuation rules for '$\neg$', hence by the previous results $\neg A$ is not a theorem of $\mathbf{QS}_6$. Hence $\mathbf{QS}_6$ is absolutely consistent. Simple consistency follows as in the first method.

Theorem (19): Classical restricted predicate logic, with '$\rightharpoonup$' symbolising material implication, '$\neg$' negation, and 'U' universal quantification, is a sublogic of $\mathbf{QS}_6$.

Proof: A1, A2, A4, R1 and R2 provide a standard postulate set for classical predicate logic so symbolised.

We now establish some formal theorems. These are numbered consecutively T1, T2, *etc.*

T1. $\vdash \neg D(Uu) \neg DA$; $\vdash S(Uu)SA$

Proof: $\neg D \neg DA$	A1
$(Uu) \neg D \neg DA$	by generalisation
$(Uu) \neg D \neg DA \rightharpoonup \neg D(Uu) \neg DA$	from A6
$\neg D(Uu) \neg DA$	by modus ponens

T2. $\vdash (Uu)SA \simeq S(Uu)A$

Proof: $p \rightharpoonup q \rightharpoonup.\ \ q \rightharpoonup p \rightharpoonup.\ \ Sp \rightharpoonup.\ \ Sq \rightharpoonup.\ \ p \simeq q$ — by A1

Result then using A6, A7, T1, *SSp*.

T3. $\vdash (Uu)SA(u) \supset SA(u)$; $\vdash SA(u) \supset (\mathit{q}u)SA(u)$

Proof: $p \rightharpoonup q \rightharpoonup.\ \ Sp \rightharpoonup.\ \ Sq \rightharpoonup.\ \ p \supset q.$

First result then using A1, T1, *SSp*. Second similarly.

T4. $\vdash S(Pu)A \simeq S(Uu)A$; $\vdash S(Pu)A \simeq S(Uu)A$.

Proof: $S(Pu)A \simeq S \sim (Uu) \sim A$	Definition
$\simeq S(Uu) \sim A$	A1.
$\simeq (Uu)S \sim A$	by T2.
$\simeq (Uu)SA$	A1, Theorem

This result depends crucially on the definition of 'P' as '$\sim U \sim$'. The particular quantifier 'P', in contrast to 'q', comprehends all subject expressions; hence the suggested reading of '*P*' as 'some out of all'. Other particular quantifiers may be defined in the theory, *e.g.* as $\neg(Uu)\neg A$, $\neg(Uu)\sim A$. Because of the properties of '$\neg$' these quantifiers are always significant.

T5. $\vdash (Uu)A \rightharpoonup (qu)A$ — by A4, A5

T6. $\vdash S(Uu)A \rightharpoonup S(qu)A$; $\vdash S(Uu)A \supset S(qu)A$.

Proof: $S(Uu)A \rightharpoonup (Uu)SA$ — A7
$\rightharpoonup (qu)SA$ — T5
$\rightharpoonup S(qu)A$ — A8

T7. $\vdash S(Pu)A \rightharpoonup S(qu)A$; $\vdash S(Pu)A \supset S(qu)A$ — by T4, T6

Theorem (20): The converse relations to T6 and T7,
(a) $S(qu)A \rightharpoonup S(Uu)A$
(b) $S(qu)A \rightharpoonup S(Pu)A$
are not theorems.

Proof: Consider a finite model with domain $d = \{x_1, x_2\}$ such that $Sf(x_1)$ and $\sim Sf(x_2)$, *e.g.* I(f) is a property significant just of x_1 (*cf.* §7.14).
Then $S(qx)f(x)$ but $\sim S(Px)f(x)$ and $\sim S(Ux)f(x)$.
Hence (a) and (b) are not valid, so they are not theorems.

T8. $\vdash (Uu)SA \rightharpoonup. \;A \supset (Pu)A$; $\vdash (Uu)SA \rightharpoonup. \;A \rightharpoonup (Pu)A$

Proof: $Sp \rightharpoonup. \;Sq \rightharpoonup. \;p \rightharpoonup q \rightharpoonup. \;\sim q \supset \sim p$ — A1
Take $p \rightharpoonup q$ as $(Uu)A \rightharpoonup A$. Then
$S(Uu)A \rightharpoonup. \;SA \rightharpoonup. \;\sim A \supset \sim (Uu)A$
$SA \rightharpoonup. \;(Uu)SA \rightharpoonup. \;\sim A \supset \sim (Uu)A$ — by A1, A6
Since $(Uu)SA \rightharpoonup SA$,
$(Uu)SA \rightharpoonup. \;\sim A \supset \sim (Uu)A$
$(Uu)SA \rightharpoonup. \;\sim\sim A \supset \sim (Uu) \sim A$ — A1
$(Uu)SA \rightharpoonup. \;A \supset (Pu)A$ — A1, Definition
Finally $p \supset q \rightharpoonup. \;p \rightharpoonup q$. — A1

T9. $\vdash (Uu)SA \rightharpoonup. \;(qu)A \rightharpoonup (Pu)A$

Proof: $(Uu)((Uu)SA \rightharpoonup. \;A \rightharpoonup (Pu)A)$ — generalising T8
$(Uu)SA \rightharpoonup (Uu)(A \rightharpoonup (Pu)A)$ — modus ponens, A2
$(Uu)SA \rightharpoonup. \;(qu)A \rightharpoonup (Pu)A$ — modus ponens, A1, A3

T10. $\vdash (Uu)A \rightharpoonup. \;\neg \sim (Uu) \sim \neg A$

Proof: $A \rightharpoonup \sim \neg A$ — A1
$(Uu). \;A \rightharpoonup \sim \neg A$ — generalising
$(Uu)A \rightharpoonup (Uu) \sim \neg A$ — distributing
$(Uu)A \rightharpoonup \neg\neg (Uu) \sim \neg A$ — using A1
$S \sim \neg A$ — A1
$S(Uu) \sim \neg A$ — generalising, A6
$Sp \rightharpoonup. \;\sim p \simeq \neg p$ — A1
Hence result.

T11. $\vdash \neg (qu) \neg\neg A \rightharpoonup \neg (qu)A$

Proof:	$\neg\neg A \rightharpoonup (\exists u)\neg\neg A$	A5
	$A \rightharpoonup (\exists u)\neg\neg A$	by A1
	$(Uu)(A \rightharpoonup (\exists u)\neg\neg A)$	generalising
	$(\exists u)A \rightharpoonup (\exists u)\neg\neg A$	by A3
	Result by A1.	
T12.	$\vdash T(\exists u)A \simeq (\exists u)TA$	
Proof:	$A \rightharpoonup TA$	A1
	$A \rightharpoonup (\exists u)TA$	by A5
	So, $(\exists u)A \rightharpoonup (\exists u)TA$	R2, A3
	$T(\exists u)A \rightharpoonup (\exists u)TA$	A1
	Similarly $(\exists u)TA \rightharpoonup T(\exists u)A$.	
	Also $ST(\exists u)A$ and STA	A1
	$(\exists u)\ STA$	A5
	$S(\exists u)TA$	A8
	The result now follows using	
	$\vdash Sp \rightharpoonup .\ \ Sq \rightharpoonup .\ \ p \rightharpoonup q \rightharpoonup .\ \ q \rightharpoonup p \rightharpoonup .\ \ p \simeq q.$	
T13.	$\vdash (Uu)TA \simeq T(Uu)A$	
Proof:	Similar to T12.	
T14.	$\vdash (Ux)A \rightharpoonup (Uy)\ \mathrm{S}^{1x}_{y}A\vert$; $\vdash (Uy)\ \mathrm{S}^{1x}_{y}A\vert \rightharpoonup (Ux)A$,	

where y is not free in A and x is not free in $(Uy)\ \mathrm{S}^{1x}_{y}A|$

Proof:	$(Ux)A \rightharpoonup \mathrm{S}^{1x}_{y}A\vert$	by A4
	Then $(Ux)A \rightharpoonup (Uy)\ \mathrm{S}^{1x}_{y}A\vert$	by R2, A2
	$(Uy)\ \mathrm{S}^{1x}_{y}A\vert \rightharpoonup A$	by A4
	$(Uy)\ \mathrm{S}^{1x}_{y}A\vert \rightharpoonup (Ux)A$	by R2, A2
T15.	$\vdash (\exists x)A \rightharpoonup (\exists y)\ \mathrm{S}^{1x}_{y}A\vert$; $\vdash (\exists y)\ \mathrm{S}^{1x}_{y}A\vert \rightharpoonup (\exists x)A$,	

where y is not free in A and x is not free in $(\exists y)\ \mathrm{S}^{1x}_{y}A|$. *Proof* as for previous theorem but using A5, A3, R2.
These theorems licence change of initial bound variables.

Theorem (21) (*Rule of substitution for subject variables*):

$\vdash A \longrightarrow \vdash \overset{\vee}{\mathrm{S}}{}^{u}_{t}A|$

Proof from R2 and A4.

A finite sequence $B_1, \ldots, B_m$ of $\mathbf{QS}_6$ is a *proof from hypotheses* $A_1, \ldots, A_n$ *of* B_m iff for each i either (1) B_i is one of the hypothesis $A_1, \ldots, A_n$, or (2) B_i is an axiom, or (3) B_i is derived from B_j and B_k where $j < i$, $k < i$ by R1, or (4) B_i is derived by R2 from B_j where $j < i$ and where the variable that is generalised upon does not occur as a free variable in A_n, or (5) B_i is derived by change of initial bound variable from B_j according to T14 or T15 where $j < i$, or (6) B_i is derived from B_j where $j < i$ by substitution for a subject variable, where the variable substituted for does not occur as a free variable in $A_1, \ldots, A_n$.

$A_1, ..., A_n \vdash B_m$ condenses: some sequence is a proof of B_m from hypotheses $A_1, ..., A_n$.

Theorem (22) (*The deduction theorem*):

If $A_1, ..., A_n \vdash B$ then $A_1, ..., A_{n-1} \vdash A_n \rightharpoondown B$.

Proof: The proof parallels that in Church [4] pp. 197–198 except for the following case: Case 5: B_i is derived from B_j by change of initial bound variable according to T14 or T15. But then by A1, $\vdash A_n \rightharpoondown B_j \rightharpoondown . A_n \rightharpoondown B_i$, and the result follows by R1.

A wff is *closed* iff it has no free subject variables. In the case that A_n is a closed wff the requirement upon the application of R2, generalisation, in a proof from hypothesis is automatically satisfied.

T16. $\vdash (Ux)A \simeq (Uy)\ \mathsf{S}^{1x}_{y}A|$; $\vdash (\mathcal{q}x)A \simeq (\mathcal{q}y)\ \mathsf{S}^{1x}_{y}A|$;
where y is not free in A and x is not free in $(Uy)\ \mathsf{S}^{1x}_{y}A|$.

Proof: As in T14 and T15 it follows, under the same conditions, applying A6–A9: $\vdash S(Ux)A \rightharpoondown S(Uy)\ \mathsf{S}^{1x}_{y}A|$, $\vdash S(Uy)\ \mathsf{S}^{1x}_{y}A| \rightharpoondown S(Ux)A$,
$\vdash S(\mathcal{q}x)A \rightharpoondown S(\mathcal{q}y)\ \mathsf{S}^{1x}_{y}A|$, $\vdash S(\mathcal{q}y)\ \mathsf{S}^{1y}_{x}A| \rightharpoondown S(\mathcal{q}x)A$.
The results then follow using these theorems: T14, T15, A1 and $\vdash p \rightharpoondown q \rightharpoondown .\ q \rightharpoondown p \rightharpoondown .\ Sp \rightharpoondown Sq \rightharpoondown .\ Sq \rightharpoondown Sp \rightharpoondown .\ p \simeq q$.

Theorem (23): If B results from A by substitution of B_1 for A_1 at zero or more places in A and if the variables $u_1, u_2, ..., u_n$ include at least those free variables of A_1 and B_1 which also occur as bound variables of A, then

$$\vdash (Uu_1) ... (Uu_n)(A_1 \simeq B_1) \rightharpoondown .\ A \simeq B.$$

Proof is by induction.

Theorem (24) (*Rule of change of bound subject variable*): If u is a subject variable which is not free in A_1 and v is a variable which does not occur in A_1, and if B results from A by substituting $\mathsf{S}^{u}_{v}A_1|$ for a particular occurrence of A_1 in A, and if $\vdash A$, then $\vdash B$.
Proof from T16, previous theorem, $A \simeq B \rightharpoondown . A \rightharpoondown B$ and R1.

Postulate A7 may be replaced by the equivalent A7′. $S(Uu)A \rightharpoondown SA$.
A7′ follows from A7 using A4 and A1; conversely A7 follows from A7′ using R2 and A2. Similarly A8 may be replaced by A8′. $SA \rightharpoondown S(\mathcal{q}u)A$.

§3 *Skolem-Löwenheim and Completeness Theorems*

Theorem (25): The set of wff of a formal system S with countable (*i.e.* at most denumerable) vocabulary, whose wff are finite, is denumerable.
This follows from the set-theoretic theorem that the number of selections of finitely many items from a countable set is denumerable. For any many-

valued predicate logic with finitely many connectives and quantifiers a Gödel numbering which provides the enumeration can be given.

For system S, class Γ of wff of S and wff B of S,

$\Gamma \vdash_S B$ iff, for some finite class

$\{A_1, \ldots, A_n\} \subseteq \Gamma$, there is a proof of B in S from $A_1, \ldots, A_n$.

Where 'N' is a negative symbol of S and Γ is a class of wff of S, Γ is *N-inconsistent* iff for some wff B of S, $\Gamma \vdash_S B$ and $\Gamma \vdash_S NB$. Γ is *N-consistent* iff it is not N-inconsistent.

C is *N-consistent with* Γ iff $\Gamma \cup \{C\}$ is N-consistent.

C is *N-inconsistent with* Γ iff $\Gamma \cup \{C\}$ is N-inconsistent.

Class Γ of S is a *maximal N-consistent* class iff Γ is N-consistent and every wff C of S consistent with Γ belongs to Γ.

Theorem (26): Every consistent class Γ of wff of a system S with denumerably many wff and negative 'N' can be extended to a maximal N-consistent class $\overline{\Gamma}$, *i.e.* for some class $\overline{\Gamma}$ of S, $\overline{\Gamma}$ is maximal consistent and $\Gamma \subseteq \overline{\Gamma}$.

Proof generalises that in §7.2.

A class Γ of wff of $\mathbf{QS}_6$ is *simultaneously satisfiable for* I *over* S iff for some uniform assignment to the free variables of all wff of Γ, $v(A) = t$ wrt I over S, for each A in Γ. Γ is *simultaneously satisfiable in a denumerable domain* iff for some I over S such that the domain of S is denumerable Γ is simultaneously satisfiable for I over S. Γ is *simultaneously satisfiable* iff for some I over S, Γ is simultaneously satisfiable for I over S.

Theorem (27): Every ¬-consistent class of wff of $\mathbf{QS}_6$ is simultaneously satisfiable in a denumerable domain.

Proof: Let Γ be a ¬-consistent class of wff of $\mathbf{QS}_6$. Let K be the $\mathbf{QS}_6$ quantificational logic obtained by adding to $\mathbf{QS}_6$ denumerably many new subject constants $b_1, b_2, \ldots$ not occurring in $\mathbf{QS}_6$ (or Γ). Let K_0 be the class of wff provable from Γ under proof from hypotheses in system K, *i.e.* $K_0 = \{B : \Gamma \vdash_K B\}$.

(1) K is ¬-consistent, by *Theorem* (17).

(2) K_0 is ¬-consistent.

For if K_0 is ¬-inconsistent then for some wff A, $K_0 \vdash A$ and $K_0 \vdash \neg A$, where '⊢' abbreviates '$\vdash_K$'. Replace each b_i appearing in the proof by a subject variable which does not appear in the proof. In this way a proof of ¬-inconsistency of Γ in $\mathbf{QS}_6$ results, contradicting assumptions.

The set of all wff of K_0 having at most one subject variable free is denumerable. For the set of all wff of K_0 is denumerable by a previous theorem. Delete from this set all wff which contain more than one subject variable free. Let the resulting denumerable set of wff with at most one subject variable free be represented:

$$A_1(x_{i_1}), A_2(x_{i_2}), \ldots, A_n(x_{i_n}), \ldots$$

Choose sequences $b_{j_1}, \ldots, b_{j_k}, \ldots; c_{j_1}, \ldots, c_{j_k}, \ldots;$

of new subject constants such that b_{j_k} and c_{j_k} are not contained in

$$A_1(x_{i_1}), ..., A_k(x_{i_k})$$

and such that b_{j_k} and c_{j_k} are distinct and different from each of

$$b_{j_1}, ..., b_{j_{k-1}}, c_{j_1}, ..., c_{j_{k-1}}.$$

Let K_n be the system obtained by adding to the axioms of K_0,

$$S_1, ..., S_n, T_1, ..., T_{n-1},$$

where S_k and T_k are the following formulae:

S_k: $A_k(b_{j_k}) \rightharpoonup (Ux_{i_k})A_k(x_{i_k})$
T_k: $(\text{ꟼ}x_{i_k})A_k(x_{i_k}) \rightharpoonup A_k(c_{j_k})$.

Let L_n be the system obtained by adding T_n to K_n.
Let L_∞ be the system obtained by adding S_i and T_i, for every i, to K. Where 'ꟼ' is defined the construction can be much simplified: there is no need to consider any T_i.

(3) If L_{n-1} is $\urcorner$-consistent, K_n is $\urcorner$-consistent. Suppose K_n is $\urcorner$-inconsistent. Then, for some wff B, both $L_{n-1}, S_n \vdash B$ and $L_{n-1}, S_n \vdash \urcorner B$.
Since S_n is a closed wff, by the deduction theorem for $\mathbf{QS}_6$, $L_{n-1} \vdash S_n \rightharpoonup B$ and $L_{n-1} \vdash S_n \rightharpoonup \urcorner B$.
It follows, using $p \rightharpoonup q \rightharpoonup . \, p \rightharpoonup \urcorner q \rightharpoonup . \urcorner p$, $L_{n-1} \vdash \urcorner S_n$,
i.e. $L_{n-1} \vdash \urcorner(A_n(b_{j_n}) \rightharpoonup (Ux_{i_n})A_n(x_{k_n}))$.
Now, by $\urcorner(p \rightharpoonup q) \rightharpoonup . \urcorner q$,
(i) $L_{n-1} \vdash \urcorner(Ux_{i_n})A_n(x_{i_n})$
and by $\urcorner(p \rightharpoonup q) \rightharpoonup p$
(ii) $L_{n-1} \vdash A_n(b_{j_n})$.
Let x_q be a variable which does not occur in proof (ii); and replace b_{j_n} throughout the proof (ii) by x_q.
So results: $L_{n-1} \vdash A_n(x_q)$.
Generalising on x_q, as is permissible since b_{j_n} does not occur in L_{n-1}, it follows:
(ii)′ $L_{n-1} \vdash (Ux_q)A_n(x_q)$.
It follows, after change of bound variable, from (i) and (ii)′ that L_{n-1} is inconsistent. Contraposing, (3) follows.

(4) If K_n is $\urcorner$-consistent, L_n is $\urcorner$-consistent.
Suppose L_n is $\urcorner$-inconsistent. By steps like those in (3) it follows

(i) $K_n \vdash (\text{ꟼ}x_{i_n})A_n(x_{i_n})$
(ii) $K_n \vdash \urcorner A_n(c_{j_n})$
(ii)′ $K_n \vdash (Ux_r)\urcorner A_n(x_r)$.

Since $\vdash S\urcorner p$, $\vdash (Ux_r)S\urcorner A_n(x_r)$. Hence by T9,
$\vdash (\text{ꟼ}x_r)\urcorner A_n(x_r) \rightharpoonup (Px_r)\urcorner A_n(x_r)$. Hence, contraposing,
$\vdash \urcorner \sim (Ux_r) \sim \urcorner\urcorner A_n(x_r) \rightharpoonup \urcorner(\text{ꟼ}x_r)\urcorner\urcorner A_n(x_r)$. So, by T10,
$\vdash (Ux_r)\urcorner A_n(x_r) \rightharpoonup \urcorner(\text{ꟼ}x_r)\urcorner\urcorner A_n(x_r)$; so $(Ux_r)\urcorner A_n(x_r) \rightharpoonup \urcorner(\text{ꟼ}x_r)A_n(x_r)$ by T11.

Thus it follows from (ii)′

(ii)″ $K_n \vdash \neg (\iota x_r) A_n(x_r)$.

The result now follows from (i) and (ii)″, as in (3).

(5) L_n is $\neg$-consistent, for each n.

This is proved by induction over n from (2), (3) and (4).

(6) L_∞ is $\neg$-consistent.

For any proof of the $\neg$-inconsistency of L contains only a finite set of wff—by the definition of $\neg$-inconsistency—and hence only a finite set of the S_i and T_i. Hence any proof of $\neg$-inconsistency of L_∞ is a proof of $\neg$-inconsistency of L_n for some finite n. So the result follows from (5).

Let J be the maximal $\neg$-consistent extension of L_∞. There is such a $\neg$-consistent set by a previous theorem. Every theorem of $\mathbf{QS_6}$ is in J.

A denumerable interpretation I_1 is now specified. Let S_1 be the usual basis over the domain d of all subject constants of K_0 (i.e. of J). Since the subject constants of K_0 are denumerable,

(7) d is denumerable.

A term is *closed* iff it contains no free subject variables. I_1 *over* S_1 is defined as follows:

(i) If X_0 is a subject constant of K_0, then $I_1(X_0) = X_0$.

(ii) If E_0^n is a functional constant of K_0 then, for closed terms $t_1, ..., t_n$, $I_1(E_0^n)(t_1, ..., t_n) = E_0^n(t_1, ..., t_n)$.

It follows by induction

(8) $I_1(t') = t'$, where term t' is closed.

(iii) If P_0 is a sentential constant, then
$I_1(P_0) = t, f$ or n according as
TP_0, FP_0 or $\neg SP_0 \in J$

(iv) If F_0^n is a predicate constant, and $t_1, ..., t_n$ are closed terms, then, abbreviating $(I(t_1), ..., I(t_n)) i I(F_0^n)$, using (8) and converting, to $I(F_0^n)(t_1, ..., t_n)$,

$$TI_1(F_0^n)(t_1, ..., t_n) \equiv TF_0^n(t_1, ..., t_n) \in J$$
$$FI_1(F_0^n)(t_1, ..., t_n) \equiv FF_0^n(t_1, ..., t_n) \in J$$
$$\neg SI_1(F_0^n)(t_1, ..., t_n) \equiv \neg SF_0^n(t_1, ..., t_n) \in J$$

(9) For every closed wff A of K_0, for valuation v_1, with respect to I_1,

(a) $v_1(A) = t \equiv TA \in J$

(b) $v_1(A) = f \equiv FA \in J$

(c) $v_1(A) = n \equiv \neg SA \in J$.

Proof of (9) is by induction over the number of connectives and quantifiers in wff A.

1. Induction basis for atomic wff.
   If P is a sentential variable or constant
   $v_1(P) = t \equiv I_1(P) = t \equiv TP \in J$.
   Similarly for (b) and (c).

If F^n is a predicate variable or constant, and $t_1, t_2, ..., t_n$ are closed terms, then

$$v_1(F^n(t_1, ..., t_n)) = t \equiv T[(I_1(t_1), ..., I_1(t_n))\iota I_1(F^n)]$$
$$\equiv T I_1(F^n)(t_1, ..., t_n), \quad \text{by (8)}$$
$$\equiv T F^n(t_1, ..., t_n) \in J.$$

Similarly for (b) and (c).

2. Induction step. The following preliminaries are used:

(10) $B \in J$ or $\neg B \in J$.

Suppose $B \notin J$ and $\neg B \notin J$. Then for some C and some E; $J, B \vdash C$; $J, B \vdash \neg C$; $J, \neg B \vdash E$; $J, \neg B \vdash \neg E$, since J is a mac set. By the deduction theorem, $J \vdash B \rightharpoonup C$; $J \vdash B \rightharpoonup \neg C$; $J \vdash \neg B \rightharpoonup E$, $J \vdash \neg B \rightharpoonup \neg E$. Since $\vdash p \rightharpoonup q \rightharpoonup . p \rightharpoonup \neg q \rightharpoonup . \neg p$, $J \vdash \neg B$ and $J \vdash B$, *i.e.* J is $\neg$-inconsistent. Contraposing, (10) follows.

(11) $B \in J$ iff $J \vdash B$.

If $B \in J$, $J \vdash B$ using $\vdash p \rightharpoonup p$. If $B \notin J$ then by (10) $\neg B \in J$, so $J \vdash \neg B$. Hence, since J is $\neg$-consistent, not-($J \vdash B$).

(12) If $A \in J$ and $A \rightharpoonup B \in J$ then $B \in J$.

For given the premises, by (11), $J \vdash A$ and $J \vdash A \rightharpoonup B$, whence $J \vdash B$ by $\mathbf{QS_6}$. Thus, by (11), $B \in J$.

(13) $A \mathbf{V} B \in J$ iff $A \in J$ or $B \in J$

If $A \in J$ or $B \in J$ then $A \mathbf{V} B \in J$ by (12). Conversely if $A \notin J$ and $B \notin J$, $\neg A \in J$ and $\neg B \in J$, by (10). Hence since $\vdash \neg p \rightharpoonup . \neg q \rightharpoonup \neg (p \mathbf{V} q)$ by A1, $\neg (A \mathbf{V} B) \in J$ by (12). Contraposing if $\neg (A \mathbf{V} B) \notin J$ then $A \in J$ or $B \in J$, and the result follows by (10). Hence $A \vee B \in J$ iff $A \in J$ or $B \in J$, for two-valued A and B.

(14) $A \& B \in J$ iff $A \in J$ and $B \in J$.

If $A \& B \in J$ then $A \in J$ and $B \in J$ by (12). If $A \in J$ and $B \in J$ then $J \vdash A$ and $J \vdash B$ by (11). Hence by $\vdash p \rightharpoonup . q \rightharpoonup . p \& q$, $J \vdash A \& B$, whence $A \& B \in J$ by (11).

Case 1. A is $\neg B$.

(a) $v_1(A) = t \quad \equiv v_1(\neg B) = t$
$\equiv v_1(B) = f \vee v_1(B) = n$
$\equiv FB \in J \vee \neg SB \in J$, by induction hypothesis
$\equiv FB \vee \neg SB \in J$, by (13)
$\equiv T\neg B \in J$, by (12) and $\vdash Fp \vee \neg Sp \rightharpoonup . T\neg p$ and $\vdash T\neg p \rightharpoonup . Fp \vee \neg Sp$
$\equiv TA \in J$

(b) $v_1(A) = f \quad \equiv v_1(\neg B) = f$
$\equiv v_1(B) = t$
$\equiv TB \in J$, by induction hypothesis
$\equiv F\neg B \in J$, by (12) and $\vdash Tp \simeq F\neg p$.

(c) $v_1(\neg B) \neq n$, by valuation rules for '$\neg$'.

Since $\vdash S\neg B$ by A1, $S\neg B \in J$; so by (10), $\neg S \neg B \notin J$.
Thus $v_1(\neg B) = n \quad \equiv \neg S \neg B \in J$.

Case 2. A is $B \rightharpoonup C$.

(a) $v_1(B \rightharpoonup C) = t \equiv v(C) = t \vee v(B) \neq t$
$\equiv TC \in J \vee TB \notin J$, by hypothesis
$\equiv TC \in J \vee \neg TB \in J$, by (10)
$\equiv TC \vee \neg TB \in J$, by (13)
$\equiv T(B \rightharpoonup C) \in J$, by (12) and
$\vdash Tq \vee \neg Tp \rightharpoonup . T(p \rightharpoonup q)$ and $\vdash T(p \rightharpoonup q) \rightharpoonup . Tq \vee \neg Tp$

(b) $v_1(B \rightharpoonup C) = f \equiv v_1(B) = t \ \& \ v_1(C) = f$
$\equiv TB \in J \ \& \ FC \in J$, by hypothesis
$\equiv TB \ \& \ FC \in J$ by (14)
$\equiv F(B \rightharpoonup C) \in J$ by (12) and
$\vdash Tp \ \& \ Fq \rightharpoonup F(p \rightharpoonup q)$ and $\vdash F(p \rightharpoonup q) \rightharpoonup . Tp \ \& \ Fq$

(c) $v_1(B \rightharpoonup C) = n \equiv v_1(B) = t \ \& \ v_1(C) = n$
$\equiv TB \in J \ \& \ \neg SC \in J$, by hypothesis
$\equiv TB \ \& \ \neg SC \in J$, by (14)
$\equiv \neg S(B \rightharpoonup C) \in J$, by (12) and
$\vdash Tp \ \& \neg Sq \rightharpoonup . \neg \ S(p \rightharpoonup q)$ and $\vdash \neg S(p \rightharpoonup q) \rightharpoonup . Tp \ \& \neg Sq$

Case 3. A is DB

(a) $v_1(DB) = t \quad \equiv v_1(B) = n$
$\equiv \neg SB \in J$, by induction hypothesis
$\equiv \neg \neg DB \in J$, by (12)
$\equiv TDB \in J$ by (12)

(b) $v_1(DB) = f \quad \equiv v_1(B) = f$
$\equiv FB \in J$, by hypothesis
$\equiv FDB \in J$, by (12)

(c) $v_1(DB) = n \quad \equiv v_1(B) = t$
$\equiv TB \in J$, by hypothesis.
$\equiv \neg SDB \in J$, by (12) and $\vdash Tp \simeq \neg SDp$

Case 4. A is $(Ux_n)B$. Then B is $A_k(x_n)$ for some k; for if B were to contain a variable distinct from x_n, A would not be closed, contrary to hypothesis.

Thus A is $(Ux_n)A_k(x_n)$.

(a) $v_1((Ux_n)A_k(x_n)) = t$
$\equiv v_1(A_k(b_j)) = t$, for every $b_j \in d$
$\equiv TA_k(b_j) \in J$, for every $b_j \in d$, by induction hypothesis.

(15) $\vdash p \rightharpoonup q \rightharpoonup . Tp \rightharpoonup Tq$ by A1

Now $TA_k(b_j) \in J$, for every $b_j \in d$, implies $TA_k(b_k) \in J$, and
$\vdash TA_k(b_j) \rightharpoonup T(Ux_{i_k})A_k(x_{i_k})$, by S_k and (15).

Hence $T(Ux_{i_k})A_k(x_{i_k}) \in J$, by (12)

and so by change of bound variable $T(Ux_n)A_k(x_n) \in J$. Conversely, if $T(Ux_n)A_k(x_n) \in J$, $TA_k(b_j) \in J$ for every $b_j \in d$, by A4, (15) and (12). Consequently, $v_1((Ux_n)A_k(x_n)) = t \equiv T(Ux_n)A_k(x_n) \in J$.

(b) $v_1((Ux_n)A_k(x_n)) = f$
$\equiv v_1(A_k(b_j)) = f$, for some $b_j \in$ d,
and $v_1(A_k(b_m)) \neq n$ for any $b_m \in$ d
$\equiv$ J$\vdash FA_k(b_j)$, for some $b_j \in$ d, and
J$\vdash SA_k(b_m)$, for every $b_m \in$ d, by induction hypothesis.

(16) $\vdash Sp \rightharpoonup . Fp \simeq \sim p$ by A1

Since $SA_k(x)$ is $A_e(x)$, for some e, J$\vdash SA_k(b_m)$ is J$\vdash A_e(b_m)$. Since this holds for every $b_m \in$ d, J$\vdash A_e(b_{j_e})$, whence, by S_e and change of bound variable, J$\vdash (Ux_n)A_e(x_n)$, *i.e.* J$\vdash (Ux_n)SA_k(x_n)$. From J$\vdash FA_k(b_j)$ for some $b_j \in$ d, and (16) it then follows, J$\vdash \sim A_k(b_j)$ for some $b_j \in$ d. Thus by A4 and A1 since J$\vdash (Ux_n)SA_k(x_n)$, J$\vdash \sim (Ux_n)A_k(x_n)$. Hence by (16) and (11), $F(Ux_n)A_k(x_n) \in$ J. Conversely if $F(Ux_n)A_k(x_n) \in$ J, then J$\vdash S(Ux_n)A_k(x_n)$ since $\vdash Fp \rightharpoonup Sp$. Hence J$\vdash (Ux_n)SA_k(x_n)$ by A7, so J$\vdash SA_k(b_m)$ for every $b_m \in$ d, by A4. Next, since J$\vdash SA_k(b_{j_k})$, J$\vdash F(Ux_{i_k})A_k(x_{i_k}) \rightharpoonup FA_k(b_{j_k})$. So, by (12), J$\vdash FA_k(b_{j_k})$. Thus $FA_k(b_j) \in$ J for some b_j (namely b_{j_k})$\in$ d.

(c) $v_1((Ux_n)A_k(x_n)) = n \equiv v_1(A_k(b_j)) = n$, for some $b_j \in$ d
$\equiv$ J$\vdash \neg SA_k(b_j)$, for some $b_j \in$ d.

Since $S(Ux_n)A_k(x_n) \rightharpoonup (Ux_n)SA_k(x_n)$ by A7
$\rightharpoonup SA_k(b_j)$ by A4
$\vdash \neg SA_k(b_j) \rightharpoonup \neg S(Ux_n)A_k(x_n)$ by A1
Hence $\neg SA_k(b_j) \in$ J $\supset . \neg S(Ux_n)A_k(x_n) \in$ J, by (12)
Conversely, $\neg S(Ux_n)A_k(x_n) \rightharpoonup \neg (Ux_n)SA_k(x_n)$ by A6, A1
But $\neg (Ux_{i_e})SA_k(x_{i_e}) \rightharpoonup \neg SA_k(b_{j_e})$, for some element of d by S_e
(using the same assignment as in (b)). Thus
$\neg S(Ux_n)A_k(x_n) \in$ J $\supset \neg SA_k(b_j) \in$ J for some $b_j \in$ d.

†*Case* 5. A is $(\text{ꝗ}x_n)B$. Then as in case 4, A is $(\text{ꝗ}x_n)A_k(x_n)$.

(a) $v_1((\text{ꝗ}x_n)A_k(x_n)) = t \equiv v_1(A_k(b_j)) = t$, for some $b_j \in$ d
$\equiv TA_k(b_j) \in$ J, for some $b_j \in$ d, by hypothesis.

Now $\vdash TA_k(b_j) \rightharpoonup T(\text{ꝗ}x_n)A_k(x_n)$, by A5 and (15); whence from $TA_k(b_j) \in$ J, by (12), $T(\text{ꝗ}x_n)A_k(x_n) \in$ J. Conversely, from J$\vdash T(\text{ꝗ}x_n)A_k(x_n)$, J$\vdash TA_k(c_{j_k})$, by T_k and (16), so $TA_k(b_j) \in$ J for some $b_j \in$ d.

(b) $v_1((\text{ꝗ}x_n)A_k(x_n)) = f \equiv v_1(A_k(b_j)) = f$ for some $b_j \in$ d
and $v_1(A_k(b_m)) \neq t$ for any $b_m \in$ d.
$\equiv FA_k(b_j) \in$ J for some $b_j \in$ d
and $\neg TA_k(b_m) \in$ J for every $b_m \in$ d.

Now $F(\text{ꝗ}x_n)A_k(x_n) \rightharpoonup S(\text{ꝗ}x_n)A_k(x_n)$ by A1
$\rightharpoonup (\text{ꝗ}x_n)SA_k(x_n)$ by
Hence J$\vdash F(\text{ꝗ}x_n)A_k(x_n) \supset$ J$\vdash (\text{ꝗ}x_n)SA_k(x_n)$ by (12)
$\supset$ J$\vdash SA_k(c_{i_e})$, for some e, by T_e
$\supset$ J$\vdash SA_k(b_j)$, for some $b_j \in$ d.

Now $\vdash \neg (\text{ꝗ}x_n)A_k(x_n) \rightharpoonup \neg A_k(b_j)$, by A5, A1
So by (16), J$\vdash \sim (\text{ꝗ}x_n)A_k(x_n) \rightharpoonup \sim A_k(b_j)$, for some $b_j \in$ d.
Thus J$\vdash T \sim (\text{ꝗ}x_n)A_k(x_n) \rightharpoonup T \sim A_k(b_j)$, for some $b_j \in$ d, by (15).

Hence $J \vdash F(\mathfrak{q}x_n)A_k(x_n) \supset . J \vdash FA_k(b_j)$, for some $b_j \in d$.
Also $\vdash \neg (\mathfrak{q}x_n)A_k(x_n) \rightharpoondown \neg A_k(b_m)$, for every $b_m \in d$, by A5, A1.
So as above, $J \vdash F(\mathfrak{q}x_n)A_k(x_n) \rightharpoondown T \neg A_k(b_m)$, for every b_m.
Since $T \neg A_k(b_m) \rightharpoondown \neg TA_k(b_m)$, by A1
$J \vdash F(\mathfrak{q}x_n)A_k(x_n) \supset . J \vdash \neg TA_k(b_m)$ for every $b_m \in d$.
Conversely if $J \vdash FA_k(b_j)$ for some $b_j \in d$ and $J \vdash \neg TA_k(b_m)$ for every $b_m \in d$, then first $J \vdash \neg A_k(c_{jk})$.
Hence by T_k and change of variable $J \vdash \neg (\mathfrak{q}x_n)A_k(x_n)$.
Thus since $\vdash \neg p \rightharpoondown . p \rightharpoondown q$ by A1, $J \vdash (\mathfrak{q}x_n)A_k(x_n) \rightharpoondown A_k(b_j)$.
Further, since $\vdash Fp \rightharpoondown Sp$, for some $b_j \in d$, $J \vdash SA_k(b_j)$ and by A5, A8,
$J \vdash S(\mathfrak{q}x_n)A_k(x_n)$. Thus $J \vdash \neg A_k(b_j) \rightharpoondown \neg (\mathfrak{q}x_n)A_k(x_n)$
leads to $J \vdash FA_k(b_j) \rightharpoondown F(\mathfrak{q}x_n)A_k(x_n)$, as $\vdash Sp \rightharpoondown . Fp \simeq \neg p$.
So by (12), $J \vdash F(\mathfrak{q}x_n)A_k(x_n)$.

(c) $v_1(\mathfrak{q}x_n)A_k(x_n)) = n \equiv v_1(A_k(b_j)) = n$ for every $b_j \in d$
$\equiv J \vdash \neg SA_k(b_j)$, for every $b_j \in d$.

Now $J \vdash (\mathfrak{q}x_n)SA_k(x_n) \rightharpoondown SA_k(c_{j_e})$, by T_e for some e.
Thus $J \vdash \neg SA_k(c_{j_e}) \rightharpoondown \neg S(\mathfrak{q}x_n)A_k(x_n)$, by A1, A9.
If then, $J \vdash \neg SA_k(b_j)$ for every $b_j \in d$, $J \vdash \neg S(\mathfrak{q}x_n)A_k(x_n)$.
Conversely, since $\vdash \neg S(\mathfrak{q}x_n)A_k(x_n) \rightharpoondown \neg SA_k(b_j)$ for every $b_j \in d$ by A1, A5, A9, if $J \vdash \neg S(\mathfrak{q}x_n)A_k(x_n)$ then by (12), $J \vdash \neg SA_k(b_j)$ for every $b_j \in d$.

(17) Γ is simultaneously satisfiable for I_1 over S_1. For proof of (17), first substitute uniformly for the free variables of members of Γ subject constants of K_0. This substitution provides a uniform assignment to free variables of members of Γ. Once the assignment is made each resulting wff B of Γ is a closed wff of K_0. Since $B \in K_0$, $B \in J$ and so, by (12) and $\vdash p \rightharpoondown Tp$, $TB \in J$. As B is closed, by (9), $v_1(B) = t$. Since then $v(B) = t$ wrt I_1 over S_1 for each B in Γ, Γ is simultaneously satisfied for I_1 over S_1.

(18) Γ is simultaneously satisfiable in a denumerable domain, by (7) and (17).

Theorem (28): Every simultaneously satisfiable class of wff of $\mathbf{QS}_6$ is $\neg$-consistent.

Proof: Let Γ be a $\neg$-inconsistent class of wff of $\mathbf{QS}_6$. Then for some finite subset $\{A_1, ..., A_m\}$ of Γ and for some wff B, $A_1, ..., A_m \vdash B$ and $A_1, ..., A_m \vdash \neg B$. Since $\mathbf{QS}_6$ is $\neg$-consistent the subset is not null. By the deduction theorem,

$\vdash A_1 \rightharpoondown . A_2 ... \rightharpoondown . A_m \rightharpoondown . B$ and $\vdash A_1 \rightharpoondown . A_2 ... \rightharpoondown . A_m \rightharpoondown \neg B$.

Hence where $A \,\&_1\, B =_{Df} \neg (A \rightharpoondown \neg B)$, ('$\&_1$' is treated in §6.7), by A1, $\vdash ((A_1 \&_1 A_2) ... A_m) \rightharpoondown B$ and $\vdash ((A_1 \&_1 A_2) ... A_m) \rightharpoondown \neg B$. By A1 again, $\vdash \neg ((A_1 \&_1 A_2) ... A_m)$. Consequently $\neg ((A_1 \&_1 A_2) ... A_m)$ is valid. Hence $v((A_1 \&_1 A_2) ... A_m) \neq t$ for any assignment. So by the derived valuation rules for '$\&_1$', $A_1, ..., A_m$ are not simultaneously satisfiable, and neither therefore is Γ.

Theorem (29) (*Skolem-Löwenheim*): Every simultaneously satisfiable class of wff of $\mathbf{QS}_6$ is satisfiable in a denumerable domain.

Proof by the preceding two theorems.

Theorem (30) (*Completeness*): Every valid wff of $\mathbf{QS}_6$ is a theorem.

Proof: Let A be a valid wff. Then $\neg A$ is not satisfiable, so $\{\neg A\}$ is not simultaneously satisfiable. Therefore $\{\neg A\}$ is $\neg$-inconsistent. Hence for some wff B, $\neg A \vdash B$ and $\neg A \vdash \neg B$. By the deduction theorem $\vdash \neg A \rightharpoonup B$ and $\vdash \neg A \rightharpoonup \neg B$; so by A1, $\vdash A$, *i.e.* A is a theorem.

T17. $\vdash (Pu)A \rightharpoonup (9u)A$

Proof: Under arbitrarily selected assignments to the free variables of A, $v((Pu)A) = t$ iff $v(A) = t$ for some assignment to u and $v(A) \neq n$ for some assignment to u. But in this case, $v((9u)A) = t$, since $v(A) = t$ for some assignment to u. It follows that $v((Pu)A \rightharpoonup (9u)A) = t$ always, so that $(Pu)A \rightharpoonup (9u)A$ is valid, and therefore a theorem.

T18. $\vdash (Uu)SA \rightharpoonup . (Pu)A \simeq (9u)A$

Proof: $(Uu)SA \rightharpoonup S(Pu)A$ T4, A6

$(Uu)SA \rightharpoonup S(9u)A$ T5, A8

Result then by T9, T17, A1.

Thus the particular quantifiers of the logic coincide under a significance proviso.

T19. $\vdash A \longrightarrow \vdash (Pu)A \simeq (9u)A$

Proof: $\vdash A \longrightarrow \vdash SA$ A1

$\longrightarrow \vdash (Uu)SA$ R2

§4 *Compatibility and Interpolation Theorems*

The Craig-Lyndon interpolation theorem for $\mathbf{QS}_6$ requires for its statement appropriate definitions of positive and negative occurrences of predicate parameters. In order to define modes of occurrence of predicate parameters in formulae—our next task—it is advantageous to re-express wff of $\mathbf{QS}_6$ using symbol set $G = \{f_0, {}^-, D, \vee, U\}$ where 'f_0' is a sentential constant which always takes value f, and '$^-$' is an irregular negation which has the matrix:

A	t	f	n
$\bar{A}$	n	t	t

Thus $\bar{p} \simeq Q\neg p$. A formula of $\mathbf{QS}_6$ is in *new* (*primitive*) *form* when it is written in terms of set G with individual constants construed as zero-place function parameters and sentence parameters as zero-place predicate parameters. Since, given G, '$\rightharpoonup$' and '$\neg$' may be defined $A \rightharpoonup B =_{\mathrm{Df}} \bar{A} \vee B$ and $\neg A =_{\mathrm{Df}} A \rightharpoonup f_0$, every wff of $\mathbf{QS}_6$ can be expressed in new form, and conversely every formula in new primitive form can be re-expressed in terms of the primitives of $\mathbf{QS}_6$. In the definition of modes of occurrence we consider throughout formulae expressed in new primitive form. A formula A is in set S, of *semi-atomic* formulae, iff it is one of B, $\overline{B}$, $\overline{\overline{\overline{B}}}$, $\overline{\overline{\overline{DB}}}$, $\overline{DB}$, DB where B is f_0 or an atomic wff.

The set $P_i(A)$—of predicate parameters which *occur in mode i in* formula A, for $0 \leqslant i \leqslant 5$—is defined by induction. The basis, for semi-atomic formula $A \in S$, is in two parts:

(i) if f_0 occurs in A, then for each $i<6$, $P_i(A) = \Lambda$;

(ii) otherwise, if A is constructed from atomic wff $B = F(t_1, ..., t_n)$, where F is an n-place predicate and $t_1, ..., t_n$ are terms, $P_i(A)$ is defined, for each $i<6$, according to the following table:—

A	$P_0(A)$	$P_1(A)$	$P_2(A)$	$P_3(A)$	$P_4(A)$	$P_5(A)$
B	{F}	Λ	Λ	Λ	Λ	Λ
$\overline{B}$	Λ	{F}	Λ	Λ	Λ	Λ
$\overline{\overline{\overline{B}}}$	Λ	Λ	{F}	Λ	Λ	Λ
$\overline{\overline{\overline{DB}}}$	Λ	Λ	Λ	{F}	Λ	Λ
$\overline{DB}$	Λ	Λ	Λ	Λ	{F}	Λ
DB	Λ	Λ	Λ	Λ	Λ	{F}

A classification of remaining types of new formulae, needed to extend the definition of mode of occurrence to all formulae, is provided in the following table:

A	C	D	U	J	$C_1(A)$	$C_2(A)$	$C_t(A)$
$\overline{\overline{\overline{B}}}$	x	x			$\overline{B}$	$\overline{B}$	
$\overline{\overline{\overline{D(\overline{B})}}}$	x	x			$\overline{\overline{\overline{B}}}$	$\overline{\overline{\overline{B}}}$	
$\overline{\overline{\overline{DDB}}}$	x	x			$\overline{\overline{\overline{B}}}$	$\overline{\overline{\overline{B}}}$	
$\overline{\overline{\overline{D(B \vee C)}}}$	x				$\overline{\overline{\overline{DB}}}$	$\overline{\overline{\overline{DC}}}$	
$\overline{\overline{\overline{D(Uu)B(u)}}}$				x			$\overline{\overline{\overline{DB^+(t)}}}$
$\overline{\overline{\overline{B \vee C}}}$		x			$\overline{\overline{B}}$	$\overline{\overline{C}}$	
$\overline{\overline{\overline{(Uu)B(u)}}}$			x				$\overline{\overline{\overline{B^+(t)}}}$
$\overline{D(\overline{B})}$	x	x			$\overline{B}$	$\overline{B}$	
$\overline{DDB}$	x	x			$\overline{B}$	$\overline{B}$	
$\overline{D(B \vee C)}$		x			$\overline{DB}$	$\overline{DC}$	
$\overline{D(Uu)B(u)}$			x				$\overline{DB^+(t)}$
$\overline{B \vee C}$	x				$\overline{B}$	$\overline{C}$	

$\overline{(Uu)B(u)}$				x			$\overline{B^{+}(t)}$
$D(\bar{B})$	x	x			$\bar{\bar{B}}$	$\bar{\bar{B}}$	
DDB	x	x			B	B	
$D(B \vee C)$	x				DB	DC	
$D(Uu)B(u)$				x			$DB^{+}(t)$
$B \vee C$		x			B	C	
$(Uu)B(u)$			x				$B^{+}(t)$

For each formula A, of the form indicated in the leftmost column, the table specifies, by a cross in the appropriate column, that the formula is in the set C (of *conjunctive* formulae), D (*disjunctive* formulae), U (*universal* formulae) or J (*particular* formulae). Furthermore each formula A in $C \cup D$ is assigned, as indicated, a *first component*, $C_1(A)$, and a *second component*, $C_2(A)$; and each formula A in $U \cup J$ is assigned, for each term t, a *t-component*, $C_t(A)$, given in the final column. $B^{+}(t)$ is $\check{S}^{u}_{t}B^{+}(u)|$, *i.e.* the result of substituting t for all free occurrences of u in $B^{+}(u)$, where $B^{+}(u)$ is the first formula, under some given well-ordering of new formulae, congruent to $B(u)$ such that t is free for u in $B^{+}(u)$, *i.e.* no free occurrences of u in $B^{+}(u)$ are within the scope of a quantifier (Uv) where v is a variable in t.

In view of the,

Lemma: A is a formula iff $A \in S \cup (C \cup D) \cup (U \cup J)$,

the following two clauses suffice to complete the definition of $P_i(A)$ for formulae:

(iii) if $A \in C \cup D$, then, for each $i<6$, $P_i(A) = P_i(C_1(A)) \cup P_i(C_2(A))$;

(iv) if $A \in U \cup J$, then, for each $i<6$, $P_i(A) = P_i(C_u(A))$, where u is the left-most subject variable occurring in A.

Where Γ is a set of formulae, $P_i(\Gamma) =_{Df} \bigcup_{A \in \Gamma} P_i(A)$, for each $i<6$; and

$P_{+}(\Gamma) =_{Df} P_0(\Gamma) \cup P_2(\Gamma) \cup P_4(\Gamma)$ and

$P_{-}(\Gamma) =_{Df} P_1(\Gamma) \cup P_3(\Gamma) \cup P_5(\Gamma)$.

$P_{+}(\Gamma)$ [$P_{-}(\Gamma)$] is the set of all predicate parameters which occur *positively* [*negatively*] in some formulae of Γ.

These definitions, which generalise well-known definitions for 2-valued quantification logic, and also the main results we prove using them are due to Rockingham Gill [1]. We follow his lucid presentation.

Lemmata:

(1) Where F is a predicate parameter, F occurs in some formula of set Γ iff $F \in \bigcup_{i<6} P_i(\Gamma)$.

(2) For each $i<6$, for terms t_1 and t_2, $P_i(A(t_1)) = P_i(A(t_2))$.

(3) $P_{+}(A) = P_{-}(\bar{A}) = P_{-}(DA) = P_{-}(\neg A)$ and
$P_{-}(A) = P_{+}(\bar{A}) = P_{+}(DA) = P_{+}(\neg A)$.

Result (2) shows that the definition is appropriately independent of terms and function symbols, and (3) the point of the labels 'positive' and 'negative'.

Where 'N' is a negative symbol (such as '$-$', '$\neg$', 'D', '$\sim$'), Γ is *N-incompatible* with Δ iff for some formula B, $\Gamma \vdash B$, $\Delta \vdash NB$, $P_+(B) \subseteq P_+(\Gamma) \cap P_-(\Delta)$ and $P_-(B) \subseteq P_-(\Gamma) \cap P_+(\Delta)$.

Γ is *N-compatible* with Δ iff Γ is not N-incompatible with Δ. An *open interpretation* I over basis S differs from an interpretation over S only in that I makes assignments to every term, whether closed or not; thus for new formulae of $\mathbf{QS}_6$ an open interpretation assigns elements of d to each subject variable as well as to each function and predicate parameter. A set of formulae is simultaneously satisfiable under open interpretation I iff it is simultaneously satisfiable under an (associated) interpretation. This result, which is easily proved directly, is also a corollary of earlier results together with the following theorem where we treat satisfaction in terms of an open interpretation.

Theorem (31) (*Compatibility*): For negative symbols '$-$', '$\neg$' and 'D',

(i) Γ is N-compatible with Δ.

iff (ii) $\Gamma \cup \Delta$ is N-consistent, and

iff (iii) $\Gamma \cup \Delta$ is simultaneously satisfiable in a denumerable domain.

Proof: By a previous theorem, (iii) guarantees that $\Gamma \cup \Delta$ is $\neg$-consistent; the converse also has been proved but we shall establish it afresh. Next $\neg$-consistency implies D-consistency and $\neg$-compatibility D-compatibility, since $DA \rightharpoondown \neg A$; and D-consistency implies $^-$-consistency and D-compatibility $^-$-compatibility, since $A \rightharpoondown \overline{\overline{A}}$, $A \rightharpoondown D\overline{\overline{A}}$, $P_+(\overline{\overline{A}}) = P_+(A)$ and $P_-(\overline{\overline{A}}) = P_-(A)$. Since, further, $^-$-consistency implies $\neg$-compatibility, because $\neg A \rightharpoondown \overline{A}$, it remains to show, given the $^-$-compatibility of Γ with Δ, that (iii) holds.

The first stage of the proof is the following maximal set construction. Add denumerably many new subject variables $u_0, u_1, \ldots$; let E be the set of all formulae A of this new system such that $P_+(A) \subseteq P_+(\Gamma)$ and $P_-(A) \subseteq P_-(\Gamma)$, and Z the set such that $P_+(A) \subseteq P_+(\Delta)$ and $P_-(A) \subseteq P_-(\Delta)$. Let A_n be the nth formula in an enumeration of $E \cup Z$. Finally,

(1) let $\Gamma_0 = \Gamma$ and $\Delta_0 = \Delta$:

(21) if $A_n \in E$ and $\Gamma_n \cup \{A_n\}$ comp Δ_n, *i.e.* is $^-$-compatible with Δ_n, then

(211) if A_n is $(\mathfrak{q}u)B(u)$, let $\Gamma_{n+1} = \Gamma_n \cup \{A_n\} \cup \{B(u_m)\}$ and $\Delta_{n+1} = \Delta_n$, with u_m the first new variable which occurs free in no formula of $\Gamma_n \cup \Delta_n \cup \{A_n\}$;

(212) otherwise, let $\Gamma_{n+1} = \Gamma_n \cup \{A_n\}$ and $\Delta_{n+1} = \Delta_n$;

(22) otherwise,

(221) if $A_n \in Z$ and Γ_n comp $\Delta_n \cup \{A_n\}$, then

(2211) if A_n is $(qu)B(u)$, let $\Delta_{n+1} = \Delta_n \cup \{A_n\} \cup \{B(u_m)\}$ and $\Gamma_{n+1} = \Gamma_n$, with u_m as before; (2212) otherwise, let $\Gamma_{n+1} = \Gamma_n$ and $\Delta_{n+1} = \Delta_n \cup \{A_n\}$;

(222) otherwise, let $\Gamma_{n+1} = \Gamma_n$ and $\Delta_{n+1} = \Delta_n$.

(3) let $\Gamma_\omega = \bigcup_n \Gamma_n$ and $\Delta_\omega = \bigcup_n \Delta_n$.

Then, for integers $m \leqslant n$, $\Gamma \subseteq \Gamma_m \subseteq \Gamma_n \subseteq \Gamma_\omega \subseteq E$ and $\Delta \subseteq \Delta_m \subseteq \Delta_n \subseteq \Delta_\omega \subseteq Z$.

We next establish, in the form of lemmata, requisite properties of Γ_ω and Δ_ω.

Lemma A: Γ_ω comp Δ_ω.
Proof is by induction over the set construction. The only cases that are not immediate are the induction steps corresponding to (211) and (2211). For (211), suppose $\Gamma_n \cup \{(qu)B(u)\}$ comp Δ_n but Γ_{n+1} incomp Δ_{n+1}. By the last, for some formula $C(u_m)$, $\Gamma_{n+1} \vdash C(u_m)$, $\Delta_{n+1} \vdash \overline{C(u_m)}$, $P_+(C(u_m)) \subseteq P_+(\Gamma_{n+1}) \cap P_-(\Delta_{n+1})$ and $P_-(C(u_m)) \subseteq P_-(\Gamma_{n+1}) \cap P_+(\Delta_{n+1})$. Since $\Gamma_{n+1} \vdash C(u_m)$, $\Gamma_n, A_n, B(u_m) \vdash C(u_m)$ with u_m free in no formula of $\Gamma_n \cup A_n$; whence, generalising and distributing, $\Gamma_n, A_n, (qu_m)B(u_m) \vdash (qu_m)C(u_m)$, and changing the bound variable, (α): $\Gamma_n \cup \{A_n\} \vdash (qu)C(u)$. Since $\Delta_{n+1} = \Delta_n$, $\Delta_n \vdash \overline{C(u_m)}$; whence generalising, since u_m is free in no formula of Δ_n, and changing the bound variable, $\Delta_n \vdash (Uu)\overline{C(u)}$. But $(Uu)\bar{A} \rightharpoonup \overline{(qu)A}$, so ($\beta$): $\Delta_n \vdash \overline{(qu)C(u)}$. Since, for each $i < 6$, $P_i(B(u_m)) = P_i(A_n)$ are $P_i((qu)C(u)) = P_i(C(u_m))$, ($\gamma$): $P_+((qu)C(u)) \subseteq P_+(\Gamma_n \cup \{A_n\}) \cap P_-(\Delta_n)$ and $P_-((qu)C(u)) \subseteq P_-(\Gamma_n \cup \{A_n\}) \cap P_+(\Delta_n)$. By ($\alpha$), ($\beta$) and ($\gamma$), $\Gamma_n \cup \{A_n\}$ incomp Δ_n, contradicting our supposition. The case for (2211) is similar.

Lemma B:

(i) $A \in \Gamma_\omega$ iff $A \in E$ and $\Gamma_\omega \cup \{A\}$ comp Δ_ω

(ii) $A \in \Delta_\omega$ iff $A \in Z$ and Γ_ω comp $\Delta_\omega \cup \{A\}$.

Proof of (ii), in this and succeeding lemmata, is similar to proof of (i). For (i) suppose, first, $A \in E$, so $A = A_n$ for some n, but $A \notin \Gamma_\omega$. Then by (21) $\Gamma_n \cup \{A_n\}$ incomp Δ_n, and so $\Gamma_\omega \cup \{A\}$ incomp Δ_ω. The converse is immediate using lemma A.

Lemma C:

(i) $A \in \Gamma_\omega$ iff $A \in E$ and $\Gamma_\omega \vdash A$

(ii) $A \in \Delta_\omega$ iff $A \in Z$ and $\Delta_\omega \vdash A$.

Proof: For the non-immediate half of (i) suppose $A \notin \Gamma_\omega$ though $A \in E$ and $\Gamma_\omega \vdash A$. Then by lemma B (i), for some formula B, $\Gamma_\omega, A \vdash B$, $\Delta_\omega \vdash \bar{B}$, $P_+(B) \subseteq P_+(\Gamma_\omega \cup \{A\}) \cap P_-(\Delta_\omega)$ and $P_-(B) \subseteq P_-(\Gamma_\omega \cup \{A\}) \cap P_+(\Delta_\omega)$. Since $A \in E$, by construction, $P_+(A) \subseteq P_+(\Gamma) \subseteq P_+(\Gamma_\omega)$ and $P_-(A) \subseteq P_-(\Gamma_\omega)$, hence $P_+(B) \subseteq P_+(\Gamma_\omega) \cap P_-(\Delta_\omega)$ and $P_-(B) \subseteq$

$P_-(\Gamma_\omega) \cap P_+(\Delta_\omega)$. But since $\Gamma_\omega \vdash A$ and $\Gamma_\omega, A \vdash B$, $\Gamma_\omega \vdash B$; whereas $\Delta \vdash_\omega \bar{B}$. Therefore Γ_ω incomp Δ_ω, contradicting lemma A.

Lemma D:

(i) If $A \vee B \in \Gamma_\omega$ then $A \in \Gamma_\omega$ or $B \in \Gamma_\omega$

(ii) If $A \vee B \in \Delta_\omega$ then $A \in \Delta_\omega$ or $B \in \Delta_\omega$.

Proof of (i). Suppose $A \vee B \in \Gamma_\omega$, $A \notin \Gamma_\omega$, $B \notin \Gamma_\omega$. Since $\Gamma_\omega \subseteq \mathrm{E}$, $A \vee B \in \mathrm{E}$; it follows since $A \vee B$ is a disjunctive formula that $A \in \mathrm{E}$ and $B \in \mathrm{E}$. Applying lemma B (i), $\Gamma_\omega \cup \{A\}$ incomp Δ_ω and $\Gamma_\omega \cup \{B\}$ incomp Δ_ω. Accordingly for some formulae C and D, $\Gamma_\omega, A \vdash C$, $\Delta_\omega \vdash \bar{C}$, $P_+(C) \subseteq P_+((\Gamma_\omega) \cup \{A\}) \cap P_-(\Delta_\omega)$ and $P_-(C) \subseteq P_-(\Gamma_\omega \cup \{A\}) \cap P_+(\Delta_\omega)$; and similarly with B and D replacing A and C respectively. Since $A, B \in \mathrm{E}$, $P_+(A), P_+(B) \subseteq P_+(\Gamma) \subseteq P_+(\Gamma_\omega)$ and $P_-(A)$, $P_-(B) \subseteq P_-(\Gamma_\omega)$. Therefore $P_+(C) \subseteq P_+(\Gamma_\omega) \cap P_-(\Delta_\omega)$ and $P_-(C) \subseteq P_-(\Gamma_\omega) \cap P_+(\Delta_\omega)$, and similarly for D; so $P_+(C \vee D) \subseteq P_+(\Gamma_\omega) \cap P_-(\Delta_\omega)$ and $P_-(C \vee D) \subseteq P_-(\Gamma_\omega) \cap P_+(\Delta_\omega)$. Furthermore, since $A \vee B$, $A \rightharpoonup C$, $B \rightharpoonup D \vdash C \vee D$ and $\bar{C}, \bar{D} \vdash \overline{C \vee D}$, it follows that $\Gamma_\omega \vdash C \vee D$ and $\Delta_\omega \vdash \overline{C \vee D}$; and so Γ_ω incomp Δ_ω.

To show that $\Gamma \cup \Delta$ is simultaneously satisfiable in a denumerable domain we show that its superset $\Gamma_\omega \cup \Delta_\omega$ is a Hintikka set on the set of terms and that all such sets are satisfiable. A set Γ of formulae is a *Hintikka set on a non-null set of terms* T iff Γ and T are such that

A. (i) if $t \in \mathrm{T}$ and term s occurs in t, then $s \in \mathrm{T}$; and
   (ii) if $A \in \mathrm{S} \cap \Gamma$ and term t occurs in A, then $t \in \mathrm{T}$.

B. (i) none of f_0, $\bar{\bar{f}}_0$, $\overline{\overline{Df_0}}$ and Df_0 belong to Γ
   (ii) if $A \in \Gamma$ or $\bar{\bar{A}} \in \Gamma$ with A atomic, then $\bar{A}$, $\overline{\overline{DA}}$, $DA \notin \Gamma$; and
   (iii) if $DA \in \Gamma$ or $\overline{\overline{DA}} \in \Gamma$ with A atomic, then $\overline{DA} \notin \Gamma$.

C. (i) if $A \in \mathrm{C} \cap \Gamma$ then $C_1(A) \in \Gamma$ and $C_2(A) \in \Gamma$;
   (ii) if $A \in \mathrm{D} \cap \Gamma$ then $C_1(A) \in \Gamma$ or $C_2(A) \in \Gamma$;
   (iii) if $A \in \mathrm{U} \cap \Gamma$ then for all $t \in \mathrm{T}$, $C_t(A) \in \Gamma$; and
   (iv) if $A \in \mathrm{J} \cap \Gamma$ then, for some $t \in \mathrm{T}$, $C_t(A) \in \Gamma$.

Lemma E: $\Gamma_\omega \cup \Delta_\omega$ is a Hintikka set on the set of all terms.

Proof: A is immediate, since all terms are taken. We illustrate the proof of B by proving some fairly representative cases. For the first case of B(i) suppose $f_0 \in \Gamma_\omega \cup \Delta_\omega$; then $f_0 \in \Gamma_\omega$ or $f_0 \in \Delta_\omega$. Suppose $f_0 \in \Gamma_\omega$; then $\Gamma_\omega \vdash f_0$. Also since $\vdash \bar{f}_0$, $\Delta_\omega \vdash \bar{f}_0$. Since $P_i(f_0) = \Lambda$, $P_+(f_0) = \Lambda$ and $P_+(f_0) \subseteq P_+(\Gamma_\omega) \cap P_-(\Delta_\omega)$; as $P_-(f_0) = \Lambda$, $P_-(f_0) \subseteq P_-(\Gamma_\omega) \cap P_+(\Delta_\omega)$. Hence Γ_ω incomp Δ_ω, contradicting lemma A. For the first case of B(ii), suppose A, $\bar{A} \in \Gamma_\omega \cup \Delta_\omega$. Then, to take one case, $A \in \Gamma_\omega$ and $\bar{A} \in \Delta_\omega$. (If A, $\bar{A} \in \Gamma_\omega$ then $f_0 \in \Gamma_\omega$ contradicting B(i)). Hence $\Gamma_\omega \vdash A$ and $\Delta_\omega \vdash \bar{A}$. Since for the predicate F of A,

$P_+(A) = P_+(\bar{A}) = \{\mathrm{F}\} = P_-(A) = P_-(\bar{A})$,

and since $A \in \mathrm{E}$ and $\bar{A} \in \mathrm{Z}$, remaining conditions for Γ_ω incomp Δ_ω

are met. For the last case of B(iii) suppose $\overline{\overline{A}}$, $DA \in \Gamma_\omega \cup \Delta_\omega$. Suppose next $DA \in \Gamma_\omega$ and $\overline{\overline{A}} \in \Delta_\omega$. (Since $\vdash \overline{\overline{A}} \rightarrow . DA \rightarrow f_0$, B(i) excludes the case DA, $\overline{\overline{A}} \in \Gamma_\omega$). Then since $\vdash DA \rightarrow \bar{A}$, $\Gamma_\omega \vdash \bar{A}$ and $\Delta_\omega \vdash \overline{\overline{A}}$, and as before lemma A is contradicted. Under C(i) there are 10 cases. We consider, to illustrate, the case where A is $\overline{\overline{\overline{DDB}}}$, and suppose $\overline{\overline{\overline{DDB}}} \in \Gamma_\omega$. Then by lemma C(i), $\Gamma_\omega \vdash \overline{\overline{\overline{DDB}}}$, whence $\Gamma_\omega \vdash \overline{\overline{B}}$, and $\overline{\overline{\overline{DDB}}} \in \mathrm{E}$, whence $\mathrm{P}_+(\overline{\overline{\overline{DDB}}}) \subseteq \mathrm{P}_+(\Gamma)$ and $\mathrm{P}_-(\overline{\overline{\overline{DDB}}}) \subseteq \mathrm{P}_-(\Gamma)$. But $\mathrm{P}_i(\overline{\overline{B}}) \subseteq \mathrm{P}_i(\overline{\overline{\overline{DDB}}})$; thus $\overline{\overline{B}} \in \mathrm{E}$. Hence, by lemma C(i), $\overline{\overline{B}} \in \Gamma_\omega$; so $\mathrm{C}_1(A)$, $\mathrm{C}_2(A) \in \Gamma_\omega \cup \Delta_\omega$. Under C(ii) there are also 10 initial cases. Suppose, to illustrate, A is $\overline{\overline{B \vee C}}$ and $\overline{\overline{B \vee C}} \in \Gamma_\omega$; then by lemma D(i), $\overline{\overline{B}} \in \Gamma_\omega$ or $\overline{\overline{C}} \in \Gamma_\omega$, so $\mathrm{C}_1(A) \in \Gamma_\omega \cup \Delta_\omega$ or $\mathrm{C}_2(A) \in \Gamma_\omega \cup \Delta_\omega$. Cases under C (iii) and C (iv) introduce no new features.

Lemma F (*Hintikka's Lemma*): If Γ is a Hintikka set on T, then Γ is simultaneously satisfiable in an open interpretation with domain T.

Proof: Suppose Γ is a Hintikka set on T; let t_0 be an arbitrarily selected term in T. Define a usual open interpretation I over domain d = T thus:

(i) for subject variable u, $\mathrm{I}(u) = u$;

(ii) for n-place function parameter E and $t_1, \ldots, t_n \in \mathrm{T}$,

$$\mathrm{I(E)}(t_1, \ldots, t_n) = \begin{cases} \mathrm{E}(t_1, \ldots, t_n), \text{ if } \mathrm{E}(t_1, \ldots, t_n) \in \mathrm{T} \\ t_0 \text{ otherwise;} \end{cases}$$

(iii) for n-place predicate parameter F and $t_1, \ldots, t_n \in \mathrm{T}$,

$$\mathrm{I(F)}(t_1, \ldots, t_n) = \begin{cases} t, \text{ if } \mathrm{F}(t_1, \ldots, t_n) \in \Gamma, \text{ or } \overline{\overline{\mathrm{F}(t_1, \ldots, t_n)}} \in \Gamma, \\ \quad \text{or } \overline{\mathrm{F}(t_1, \ldots, t_n)} \notin \Gamma \text{ and} \\ \quad \overline{\mathrm{DF}(t_1, \ldots, t_n)} \in \Gamma; \\ f, \text{ if } \overline{\mathrm{F}(t_1, \ldots, t_n)} \in \Gamma \text{ and} \\ \quad \overline{\mathrm{DF}(t_1, \ldots, t_n)} \in \Gamma; \\ n, \text{ otherwise.} \end{cases}$$

(ii) is well-defined since for every $t_i \in \mathrm{T}$, $\mathrm{I}(t_i) = t_i$; B(ii) and B(iii) ensure that (iii) is consistently specified.

Let m(A) be the number of occurrences of improper symbols of set $\{f_0, ^-, D, \vee, U\}$ in formula A, counting each occurrence of 'D' twice. Suppose, for an inductive proof, that for $A \in \Gamma$ and m(A)$<n$, $v(A) = t$: to show that for $A \in \Gamma$ and m(A) $= n$, $v(A) = t$ there are, by a previous lemma, just these cases:—

$A \in \mathrm{S}$; then each of the 12 cases is taken care of either by B or by specification (iii) for I; *e.g.* if A is f_0 or Df_0 then $A \notin \Gamma$; if A is $\overline{\overline{\mathrm{F}(t_1, \ldots, t_n)}}$ then using (iii), $v(A) = t$.

$A \in C$; then by C(i), $C_1(A)$, $C_2(A) \in \Gamma$, whence by induction hypothesis since $m(C_1(A)), m(C_2(A)) < n$, $v(C_1(A)) = t = v(C_2(A))$ and $v(C_1(A) \& C_2(A)) = t$. But, applying S_6, for $A \in C$, $C_1(A) \& C_2(A) \rightharpoondown A$; so $v(A) = t$.

$A \in D$; similarly, using the fact that for $A \in D$, $\vdash C_1(A) \vee C_2(A) \rightharpoondown A$.

$A \in U$; then, by C(iii), for every $u \in T$, $C_u(A) \in \Gamma$; whence, as $m(C_u(A)) < n$, $v(C_u(A)) = t$. Accordingly for every assignment under I to u, $v(C_u(A)) = t$, and so $v((Uu)C_u(A)) = t$. But for $A \in U$ and subject variable $u \vdash (Uu)C_u(A) \rightharpoondown A$; whence $v(A) = t$.

$A \in E$; similarly.

Lemma F, and the compatibility theorem, now follow.

Corollaries:

(1) Skolem-Löwenheim theorem

(2) Gödel completeness theorem

(3) Craig-Lyndon interpolation theorem:

$\Gamma \vdash A$ iff for some formula B such that $P_+(B) \subseteq P_+(\Gamma) \cap P_+(A)$ and $P_-(B) \subseteq P_-(\Gamma) \cap P_-(A)$, $\Gamma \vdash B$ and $B \vdash A$.

Proof: If $\Gamma \vdash A$ then $\Gamma \cup \{\bar{A}\}$ is $^-$-inconsistent; thus Γ incomp $\{\bar{A}\}$. Therefore for an appropriate B, $\Gamma \vdash B$ and $\bar{A} \vdash \bar{B}$; thus $\Gamma \vdash B$ and $B \vdash A$.

Other corollaries also follow, *e.g.* a significance version of Beth's definability theorem.

Another result which can be given a sharper formulation using the definitions of modes of occurrence of predicates is the theorem asserting that for every wff there is strongly equivalent wff in prenex normal form.

Theorem (32) (*Normal Form*): A is a wff of $\mathbf{QS_6}$ iff there is a wff B of $\mathbf{QS_6}$ such that (i) B is in prenex normal form (with respect to quantifiers 'U' and '$\mathcal{q}$'); (ii) the matrix of B is a conjunction (by '&') of disjunctions (by '$\vee$') of semi-atomic formulae; (iii) for each $i < 6$, $P_i(A) = P_i(B)$; and (iv) $\vdash A \simeq B$.

§5 *Generalised Quantified Significance Theories*

A generalised **QST** differs from an applied $\mathbf{QS_6}$-system only in that the theory may have a non-countable number of predicate symbols, of function symbols and of individual constants, and may have non-countably many axioms. Many of the results established for generalised first-order theories (see, e.g. Mendelson [1], p. 92 *ff*) extend to generalised **QST** (and indeed to generalised many-valued theories). To illustrate,

Theorem (33): If the class of symbols of a consistent generalised **QST** can be well-ordered and has cardinality $\aleph_\alpha$ then the theory has a model of cardinality $\aleph_\alpha$ (i.e. the theorems of the theory are simultaneously satisfiable in a domain of cardinality $\aleph_\alpha$).

7.8 Quantification Logics with Alternative Sentential Bases

We examine certain complete variations on $\mathbf{QS_6}$. The variations are obtained by weakening or replacing the connective set or quantifiers $\mathbf{QS_6}$ uses. In this way we obtain quantified significance logics analogous to $\mathbf{QS_6}$ but based on sentential logics $\mathbf{S_2}$, $\mathbf{S_3}$, $\mathbf{S_4}$, $\mathbf{S_5}$ and **L**.

§1 *The quantification logic* $\mathbf{QS_4}$

$\mathbf{QS_4}$ is a quantification theory with quantifiers '*U*' and '*q*' based on sentential logic $\mathbf{S_4}$ with connective set $\{\&, \sim, \rightharpoonup\}$, or with redundant sets such as $\{\supset, \sim, T, \rightharpoonup\}$ or $\{\&, \sim, V, \rightharpoonup\}$. Apart from connectives, $\mathbf{QS_4}$ has precisely the same vocabulary as $\mathbf{QS_6}$ and the formation rules parallel those of $\mathbf{QS_6}$.

The semantics of $\mathbf{QS_4}$ is also exactly the same as that of $\mathbf{QS_6}$, except for the inductive clauses defining values for sentential connectives in terms of those for components: here the clauses for $\mathbf{S_4}$ are used. To avoid confusion the central notion of validity is explicitly relativised, in much the way we have already system-relativised the notion of a tautology. Semantical bases and interpretations for $\mathbf{QS_4}$ are defined exactly as for $\mathbf{QS_6}$. Thus an interpretation I of $\mathbf{QS_4}$ over basis S is identical with an interpretation of $\mathbf{QS_6}$ over basis S. Given I and S, $\mathbf{QS_4}$-*values for all wff* of $\mathbf{QS_4}$ are defined inductively. Clauses (i) (ii) (vi) and (vii) of the inductive specification for $\mathbf{QS_6}$ values are taken over intact. The remaining clauses, which specify *values of* wff compounded using sentential connectives coincide with those used in defining $\mathbf{S_4}$-tautologies. A wff *A* of $\mathbf{QS_4}$ is $\mathbf{QS_4}$-*valid for* I *over* S iff under a $\mathbf{QS_4}$ valuation *v*, $v(A) = t$, for every assignment to its free variables wrt I over S. $\mathbf{QS_4}$-validity, $\mathbf{QS_4}$-satisfiability, $\mathbf{QS_4}$-simultaneous satisfiability, and so on, are defined analogously.

The postulates of $\mathbf{QS_4}$ are the same as those of $\mathbf{QS_6}$ except that in A1 '$\mathbf{S_6}$-tautology' is replaced by '$\mathbf{S_4}$-tautology'.

Theorem (34): Every theorem of $\mathbf{QS_4}$ is $\mathbf{QS_4}$-valid.
Proof is similar to that for $\mathbf{QS_6}$ (*Theorem* 15).

Theorem (35): $\mathbf{QS_4}$ is simply and absolutely consistent.
Proof: For $\mathbf{QS_4}$ is a subsystem of $\mathbf{QS_6}$. Hence the result follows from *Theorem* 17.

All the formal theorems actually proved in $\mathbf{QS_6}$ are also theorems, reached by the same proofs, of $\mathbf{QS_4}$, with the exception of the first formulation, using '*D*', of T1. The theorem of substitutivity of strong ($\simeq$) equivalents is likewise a *Theorem* of $\mathbf{QS_4}$; but the induction is over a different connective set. The deduction theorem and rules of substitution for free and bound variables follow just as for $\mathbf{QS_6}$.

Theorem (36): Every $\neg$-consistent class of wff of $\mathbf{QS_4}$ is $\mathbf{QS_4}$-simultaneously satisfiable in a denumerable domain.

Proof varies from the parallel theorem for $\mathbf{QS}_6$ (*Theorem* 27) only in the main induction step [(9)2. of the proof for $\mathbf{QS}_6$]. To complete the proof it suffices to consider these cases:—

Case a. A is $\sim B$

(a) $v_1(A) = t \equiv v_1(\sim B) = t$
$\equiv v_1(B) = f$
$\equiv FB \in \mathrm{J}$ by induction hypothesis
$\equiv T\sim B \in \mathrm{J}$ by definition of 'F'

(b) $v_1(A) = f \equiv v_1(B) = t$
$\equiv TB \in \mathrm{J}$
$\equiv T\sim\sim B \in \mathrm{J}$
$\equiv F\sim B \in \mathrm{J}$

(c) $v_1(A) = n \equiv v_1(B) = n$
$\equiv \sim SB \in \mathrm{J}$
$\equiv \sim SA \in \mathrm{J}$

Case b. A is $B \& C$

(a) $v_1(B \& C) = t \equiv v_1(B) = t \& v_1(C) = t$
$\equiv \mathrm{J} \vdash TB \& \mathrm{J} \vdash TC$
$\equiv \mathrm{J} \vdash TB \& TC$
$\equiv \mathrm{J} \vdash T(B \& C)$

(b) $v_1(B \& C) = f \equiv v_1(B) = f \& v_1(C) \neq n \vee . v_1(B) \neq n \& v_1(C) = f$
$\equiv \mathrm{J} \vdash FB \& \mathrm{J} \vdash SC \vee . \mathrm{J} \vdash SB \& \mathrm{J} \vdash FC$
$\equiv \mathrm{J} \vdash FB \& SC \vee . SB \& FC$
$\equiv \mathrm{J} \vdash F(B \& C)$

(c) $v_1(B \& C) = n \equiv v_1(B) = n \vee v_1(C) = n$
$\equiv \mathrm{J} \vdash \sim SB \vee \sim SC$
$\equiv \mathrm{J} \vdash \sim S(B \& C)$

Case c. A is $B \vee C$

(a) $v_1(B \vee C) = t \equiv v_1(B) = t \vee v_1(C) = t$
$\equiv \mathrm{J} \vdash TB \vee TC$
$\equiv \mathrm{J} \vdash T(B \vee C)$

(b) $v_1(B \vee C) = f \equiv v_1(B) = f \& v_1(C) \neq t \vee . v_1(B) \neq t \& v_1(C) = f$
$\equiv \mathrm{J} \vdash FB \& \sim TC \vee . \sim TB \& FC$
$\equiv \mathrm{J} \vdash F(B \vee C)$

(c) $v_1(B \vee C) = n \equiv v_1(B) = n \& v_1(C) = n$
$\equiv \mathrm{J} \vdash \sim SB \& \sim SC$
$\equiv \mathrm{J} \vdash \sim S(B \vee C)$.

Case d. A is TB

(a) $v_1(A) = t \equiv v_1(B) = t$
$\equiv TB \in \mathrm{J}$, *i.e.* $TTB \in \mathrm{J}$

(b) $v_1(TB) = f \equiv v_1(B) \neq t$
$\equiv \sim TB \in J$
$\equiv FTB \in J$

(c) $\sim (v_1(TB) = n)$ & $\sim (\sim STB \in J)$.

Theorem (37) (*Skolem-Löwenheim*): Every QS_4-simultaneously satisfiable class of wff is QS_4-satisfiable in a denumerable domain.

Theorem (38) (*Completeness*): Every QS_4-valid wff is a theorem of QS_4.

Proofs of these results are exactly the same as analogous results for QS_6 (*Theorems* 29 and 30).

For very many significance investigations QS_4-systems provide adequate logical bases. Complete quantificational extensions of S_3 (S_5) are obtained by combining S_3 and QS_4 (S_5 and QS_6); but the resulting systems are something of a mixture inasmuch as they contain two sorts of sentential expressions *S*-restricted and *S*-unrestricted, but only one sort of predicate expressions. To illustrate such mixed quantificational extensions of 2-sorted S-logics we investigate first a system QS_5.

§2 *The restricted quantificational logics* QS_5 *and* QS_3

The morphology of QS_5 is the same as that of QS_6 except for these features:

(1) QS_5 has these improper primitive symbols:

$$\sim \quad \supset \quad T \quad Q \quad (\quad) \quad U \quad q$$

(2) QS_5 has *S*-restricted sentential parameters:

$$r, \quad r', \quad r'', \quad \ldots$$

Further differences in formation rules just reflect these differences in primitive symbols; and the actual formation rules of QS_5 just appropriately combine those of S_5 with those of QS_6.

The semantics of QS_5 differ from those of QS_6 at the following points:

(1) An *interpretation* I *of* QS_5 *over* S is a contraction-function I which assigns to each *S*-unrestricted sentential constant one of the values *t*, *f* or *n*, and to each *S*-restricted sentential constant one of the values *t* or *f*, and otherwise makes the same assignments as an interpretation of QS_6. Then *S*-unrestricted sentential parameters have, as their range, values *t*, *f*, and *n*; and *S*-restricted parameters have as their (defined secondary) ranges, values *t* and *f*.

(2) Given I and S, QS_5-values for all wff are defined inductively thus: (i) For a wff consisting of an *S*-restricted or *S*-unrestricted sentential parameter *A* alone, $v(A) = I(A)$. Clauses (ii) (vi) and (iii) are the same as those for QS_6. Clauses for '$\sim$', '$\supset$', '*T*' and '*Q*' coincide with those given for the semantics of S_5.

The connectives '$\urcorner$' and '$\rightharpoonup$' are defined thus:

$$\urcorner A \quad =_{\mathrm{Df}} \sim TA$$
$$A \rightharpoonup B =_{\mathrm{Df}} \quad TA \supset Q(TA \supset QB).$$

As in $\mathbf{QS}_6$, 'q' can be eliminated definitionally.

$\mathbf{QS}_5$ has as postulates:

(1) The postulates (axioms and rules) of $\mathbf{S}_5$.

(2) Axiom schemes A1–A9 of $\mathbf{QS}_6$ and the rule R2 of generalisation.

The modus ponens rule for '$\rightharpoonup$' is derivable. Rule R2.1 for restricted substitution extends of course to all wff. Thus, since $S(Ux)SB$, $(Ux)SB \supset (Ux)SB$ follows by R2.1.

Theorem (39): Every theorem of $\mathbf{QS}_5$ is $\mathbf{QS}_5$-valid. *Proof* is analogous to the similar *Theorem* (15) for $\mathbf{QS}_6$.

Theorem (40): $\mathbf{QS}_5$ is simply and absolutely consistent. This follows either directly or from the fact that $\mathbf{QS}_5$ can be represented in $\mathbf{QS}_6$.

As was the case with $\mathbf{QS}_4$, all the theorems actually proved for $\mathbf{QS}_6$ (with the exception of one formulation of T1) are theorems of $\mathbf{QS}_5$. Since, further, every theorem of $\mathbf{S}_6$ is a theorem of $\mathbf{S}_5$, by a previous *Theorem*, the same proofs will suffice. The deduction theorem and the rules of change of free and bound variables, formulated as for $\mathbf{QS}_6$, are also *Theorems* of $\mathbf{QS}_5$.

Theorem (41): Every $\urcorner$-consistent class of wff of $\mathbf{QS}_5$ is $\mathbf{QS}_5$-simultaneously satisfiable in a denumerable domain.

Proof differs from the similar *Theorem* (27) for $\mathbf{QS}_6$ only in the specification of the denumerable interpretation I_1 and in the proof of its adequacy. Now S_1 is defined as for $\mathbf{QS}_6$ and I_1 over S_1 is also defined as for $\mathbf{QS}_6$, but clause (iii) is made more explicit, as follows:

(iii′) If P_0 is an S-unrestricted sentential constant, then $\mathrm{I}_1(P_0) = t, f$ or n according as TP_0, FP_0 or $\urcorner SP_0 \in \mathrm{J}$; and if P_0 is an S-restricted sentential constant then $\mathrm{I}_1(P_0) = t$ or f according as TP_0 or $FP_0 \in \mathrm{J}$. Note that if A is S-restricted then $\vdash SA$; and hence $\urcorner SA \notin \mathrm{J}$. Step (9) of the proof for $\mathbf{QS}_6$ is modified as follows:
Induction steps for '$\sim$' and 'T' are the same as those given for $\mathbf{QS}_4$; induction steps for 'U' and 'q' are the same as those for $\mathbf{QS}_6$. The remaining case is as follows:

Case A is QB

(a) $v_1(QB) = t \equiv v_1(B) = t$
$\equiv \mathrm{J} \vdash TB$
$\equiv \mathrm{J} \vdash TQB$, by T10 of Stage 5.

(b) $v_1(QB) = f \equiv v_1(B) = n$
$\equiv \mathrm{J} \vdash \sim SB$
$\equiv \mathrm{J} \vdash FQB$

(c) $v_1(QB) = n \equiv v_1(B) = f$
$\equiv \mathrm{J}\vdash FB$
$\equiv \mathrm{J}\vdash \sim SQB$

Theorem (42) (*Skolem-Löwenheim*): Every $\mathbf{QS}_5$-simultaneously satisfiable class of wff is $\mathbf{QS}_5$-satisfiable in a denumerable domain.

Theorem (43) (*Completeness*): Every $\mathbf{QS}_5$-valid wff is a theorem of $\mathbf{QS}_5$.

Let $\mathbf{QS}_3$ be a quantificational extension of S_3 extending the sentential system precisely as $\mathbf{QS}_5$ extends S_5. Hence $\mathbf{QS}_3 = \mathrm{S}_3 \cup \mathbf{QS}_4$, in *an* obvious sense of union of systems. Consistency, plausibility, Skolem-Löwenheim and completeness theorems then follow very much as for $\mathbf{QS}_5$: where the proofs differ, relevant details transfer from $\mathbf{QS}_4$ proofs.

These completeness results yield alternative proofs of the completeness of S_3, S_4 and S_5. Suppose, for instance, that A is an S_5-tautology. Then A is $\mathbf{QS}_5$-valid. Hence A is a theorem of $\mathbf{QS}_5$. But,

Theorem (44): For every quantifier-free theorem of $\mathbf{QS}_5$ there is a proof of A in which only quantifier-free wff occur. Hence A is a theorem of S_5.

The theorem needed to complete this argument may be proved along the lines of Church [4], pp. 182–184, with '$\rightharpoonup$' replacing '$\supset$' and '$\&_1$' representing conjunction. The theorem does, however, follow directly and easily from the completeness result already established for S_5.

§3 *The restricted quantification logic* $\mathbf{QS}_2$

The morphology of $\mathbf{QS}_2$ differs from $\mathbf{QS}_5$ only in respect of the primitive sentential connectives. $\mathbf{QS}_2$ has the primitive sentential connectives '$\sim$', '$\supset$' and 'T' of S_2.

The semantics of $\mathbf{QS}_2$ is the same as that of $\mathbf{QS}_5$ except that the inductive clauses in the definition of the valuation function are the connectives of $\mathbf{QS}_2$.

In the light of our axiomatisation of $\mathbf{QS}_6$, a simple strategy for obtaining a complete quantification system based on S_2 is to introduce a suitably classical "implication". Although '$\rightharpoonup$' cannot be defined in S_2 the connective '$\searrow$', already noticed (§5.11), can be,

$$A \searrow B =_{\mathrm{Df}} TA \supset TB.$$

$\mathbf{QS}_2$ has as postulates:

(1) The postulates of S_2, and
(2) Axiom-schemes A2–A5 of $\mathbf{QS}_6$ *but* with '$\rightharpoonup$' replaced throughout by '$\searrow$', and
(3) Axiom-schemes A6, A7, A8′, A9 of $\mathbf{QS}_6$ *but* with '$\rightharpoonup$' replaced by '$\supset$', and
(4) The rule R2 of generalisation.

To derive modus ponens for '$\mathrel{\smallsetminus\!\!-}$' we collect these results:

T1. $\vdash S(Ux)SA$
Proof: $(Ux)SSA$ by $\mathbf{S}_2$ and R2.
Result by A6 and $\supset$-detachment.
T2. $\vdash S(\text{ꟼ}x)SA$
Proof: $SSA \supset S(\text{ꟼ}x)SA$ from A8′.

Theorem (45): $\vdash A \rightarrow \vdash SA$ (S-rule)
Proof is by induction over proofs of $\mathbf{QS}_2$.
(1) If A is an axiom then $\vdash SA$, holds for the quantificational postulates, by *Theorem* (15) of §6.3, $\vdash STA$, $\vdash SSA$, $\vdash SA \;\&\; SB \supset .S(A \;\&\; B)$ and T1 and T2.
(2) R2 preserves significance. For if $\vdash A$ then, by induction hypothesis, $\vdash SA$. Hence, by R2, $\vdash (Ux)SA$, and, by A7, $\vdash S(Ux)A$.

Theorem (46): $\vdash A \rightarrow \vdash TA$
Proof as for *Theorem* (16) of §6.3.

Theorem (47): $\vdash A, \vdash A \mathrel{\smallsetminus\!\!-} B \rightarrow \vdash B$.
Proof: $\vdash A, \vdash A \mathrel{\smallsetminus\!\!-} B \rightarrow \vdash TA, \vdash TA \supset TB$
$\rightarrow \vdash TB$
$\rightarrow \vdash TB, \vdash SB$, since $Tp \supset Sp$
$\rightarrow \vdash B$, by axiom 2.4.

Theorem (48): Classical restricted predicate logic, with '$\mathrel{\smallsetminus\!\!-}$' representing material implication, '$\urcorner$' negation, and 'U' universal quantification, is a sublogic of $\mathbf{QS}_2$.

T3. $\vdash (Ux)SA \mathrel{\smallsetminus\!\!-} S(Ux)A$; $\vdash S(Ux)A \mathrel{\smallsetminus\!\!-} (Ux)SA$
T4. $\vdash S(\text{ꟼ}x)A \mathrel{\smallsetminus\!\!-} (\text{ꟼ}x)SA$; $\vdash (\text{ꟼ}x)SA \mathrel{\smallsetminus\!\!-} S(\text{ꟼ}x)A$
Proof: The first three follow using $A \supset B \mathrel{\smallsetminus\!\!-} . A \mathrel{\smallsetminus\!\!-} B$, a consequence of the completeness of $\mathbf{S}_2$. Similarly $\vdash SA \mathrel{\smallsetminus\!\!-} S(\text{ꟼ}x)A$ follows from A7′, and the last then follows by R2 and A3.

Theorem (49): $\mathbf{QS}_2$ is a sublogic of $\mathbf{QS}_5$, *i.e.* every theorem of $\mathbf{QS}_2$ is a theorem of $\mathbf{QS}_5$ (since the vocabulary of $\mathbf{QS}_2$ is definable in $\mathbf{QS}_5$).
Proof: Since it has been shown that $\mathbf{S}_2$ is a sublogic of $\mathbf{S}_5$ it suffices to show that the quantificational postulates of $\mathbf{QS}_2$ are derivable in $\mathbf{QS}_5$; and this follows using the $\mathbf{QS}_5$ theorems $A \mathrel{\smallsetminus\!\!-} B \rightharpoonup . A \rightharpoonup B$ and $A \rightharpoonup B \rightharpoonup . A \mathrel{\smallsetminus\!\!-} B$. It follows that $\mathbf{QS}_3$–$\mathbf{QS}_6$ could have been formulated using '$\mathrel{\smallsetminus\!\!-}$' instead of '$\rightharpoonup$'.
Corollary: $\mathbf{QS}_2$ is simply and absolutely consistent.

Theorem (50): Every theorem of $\mathbf{QS}_2$ is valid.
Proof: In the case of $\mathbf{S}_2$-postulates, the requisite results follow since

every S_2-postulate is an S_2-tautology or else preserves S_2-tautologousness. In the case of quantificational postulates *Theorem* (49) can be applied. Finally the result follows by induction over proofs.

Since theorems like the following hold for $\mathbf{QS}_2$,

$$A \leftharpoonup B \leftharpoonup . B \leftharpoonup A \leftharpoonup . SA \leftharpoonup . SB \leftharpoonup . A \simeq B,$$
$$A \leftharpoonup B \leftharpoonup . SA \leftharpoonup . SB \leftharpoonup . A \supset B,$$

particular theorems proved for $\mathbf{QS}_6$ carry over, with '$\rightharpoonup$' replaced by '$\leftharpoonup$', to $\mathbf{QS}_2$.

By the same techniques as used for $\mathbf{QS}_6$, a deduction theorem and rules of change of free and bound variables can be proved for $\mathbf{QS}_2$.

Theorem (51): Every $\urcorner$-consistent class of wff of $\mathbf{QS}_2$ is $\mathbf{QS}_2$-simultaneously satisfiable in a denumerable domain.

Proof: The proof of the parallel *Theorem* (27) for $\mathbf{QS}_6$ is changed in these respects:

(1) The connective '$\rightharpoonup$' is replaced throughout by '$\leftharpoonup$';
(2) The denumerable interpretation I_1 is specified as for $\mathbf{QS}_5$;
(3) The induction steps in the proof of stage (9) are those, already given, for '$\supset$', '$\sim$' and 'T', or if '$\leftharpoonup$' supplants 'T' this one:

Case A is $B \leftharpoonup C$

(a) $v_1(B \leftharpoonup C) = t \equiv v_1(B) \neq t \vee v_1(C) = t$
$\equiv J \vdash \sim TB \vee TC$
$\equiv J \vdash T(B \leftharpoonup C)$

(b) $v_1(B \leftharpoonup C) = f \equiv v_1(B) = t \;\&\; v_1(C) \neq t$
$\equiv J \vdash TB \;\&\; \sim TC$
$\equiv J \vdash F(B \leftharpoonup C)$

(c) $v_1(B \leftharpoonup C) \neq n$. Also since $\vdash S(B \leftharpoonup C)$, $J \vdash \sim \sim S(B \leftharpoonup C)$.

Theorem (52): Every $\mathbf{QS}_2$-simultaneously satisfiable class of wff is $\urcorner$-consistent.

Proof is as for the corresponding result for $\mathbf{QS}_6$ only '$\leftharpoonup$' replaces '$\rightharpoonup$'. Note that

$$\urcorner(A \leftharpoonup \urcorner B) \simeq . A \,\&_1 B$$

Theorem (53): Every $\mathbf{QS}_2$-simultaneously satisfiable class is satisfiable in a denumerable domain.

Theorem (54): Every $\mathbf{QS}_2$-valid wff is a theorem of $\mathbf{QS}_2$.

Theorem (55): The systems $\mathbf{QS}_2$, $\mathbf{QS}_3$, $\mathbf{QS}_4$, $\mathbf{QS}_5$ and $\mathbf{QS}_6$ are, necessarily, undecidable systems. For they each contain syntactically full classical predicate logic.

An inadequacy of $\mathbf{QS}_2$, as opposed to $\mathbf{QS}_3$, is that there is no general way of expanding wff containing the quantifier 'q' over finite domains in terms of the sentential connectives of $\mathbf{QS}_2$. For 'V' is not definable in S_2. In other words, even in the theory of $\mathbf{QS}_2$ restricted to finite domains, 'q' occurs uneliminably.

§ 4 *The Quantification Logic* **QL**, *with alternative quantifiers*

The 4-logic **L** may be quantificationally enriched in precisely the way we have treated $\mathbf{S}_4$. It is customary however to add to **L** primitive quantifiers which reduce over finite domains to strong normal connectives (see, *e.g.*, Kleene[2]). It suffices to take as primitive one of these quantifiers, say the universal quantifier 'A', for the other quantifier can then be defined in the usual way: $(\mathrm{S}u)A =_{\mathrm{Df}} \sim (\mathrm{A}u) \sim A$. The semantics for **QL** differ from those for $\mathbf{QS}_4$ only in the evaluation rules given for the primitive connectives, '$\sim$' and '$\supset_{\mathrm{L}}$', which are the same as those for **L**, and in the rule for the primitive quantifier 'A', which is as follows:

$v((\mathrm{A}u)B) = t$ iff $v(B) = t$ for every assignment to u;
$v((\mathrm{A}u)B) = f$ iff $v(B) = f$ for some assignment to u; and
$v((\mathrm{A}u)B) = n$ otherwise.
Expressed algebraically, the value of $(\mathrm{A}u)B$ is the minimum of the values of B for assignments of u, under the ordering $t > n > f$ of values. The morphology of **QL** differs from that of $\mathbf{QS}_4$ only so far as change of primitive connectives and quantifier affect it.

QL has these postulates:

L1. If B is a substitution instance of an **L**-tautology then $\vdash B$.
L2. $\vdash(\mathrm{A}u)(A \supset_{\mathrm{L}} B) \supset_{\mathrm{L}} . A \supset_{\mathrm{L}} (\mathrm{A}u)B$, provided u is not free in A.
L3. $\vdash(\mathrm{A}u)B \supset_{\mathrm{L}} \overset{\vee}{\mathrm{S}}{}^{u}_{t} B|$
L4. $\vdash(\mathrm{A}u)TB \supset_{\mathrm{L}} T(\mathrm{A}u)B$
L5. $\vdash T(\mathrm{S}u)B \supset_{\mathrm{L}} (\mathrm{S}u)TB$.
R1. $\vdash A, \vdash A \supset_{\mathrm{L}} B \longrightarrow \vdash B$
R2. $\vdash B \longrightarrow \vdash(\mathrm{A}u)B$.

We sketch derivations of some theorems needed for a completeness proof.

DR1. $\vdash A \supset_{\mathrm{L}} B \longrightarrow \vdash TA \supset_{\mathrm{L}} TB$
T1. $\vdash T(\mathrm{A}u)B \supset_{\mathrm{L}} (\mathrm{A}u)TB$

Proof by L3, DR1, R2, L2

T2. $\vdash F(\mathrm{A}u)B \supset_{\mathrm{L}} (\mathrm{S}u)FB$ from L5
T3. $\vdash(\mathrm{S}u)FB \supset_{\mathrm{L}} F(\mathrm{A}u)B$, as in T1

Theorem (56):

(1) B is a theorem of **QL** iff B is **QL**-valid.
(2) Every **QL**-simultaneously satisfiable class of wff is **QL**-satisfiable in a denumerable domain.
(3) B is a theorem of **QL** iff B is a theorem of $\mathbf{QS}_4$, using definitional substitutions. Hence, too, **QL** is a sublogic of $\mathbf{QS}_6$.

Proof: Instead of proving (1) and (2) directly, (3) and results on $\mathbf{QS}_4$ may be used to prove (1) and (2) (cf. §6.5). As to (3) it suffices (building on §6.5) to show that the quantifiers of **QL** and $\mathbf{QS_4}$ are interdefinable and that, using the definitions, the quantificational postulates of each system are derivable in the other. Suitable definitions are as follows:

$$(\mathrm{A}u)B =_{\mathrm{Df}} (Uu)B \cap \urcorner(\mathrm{E}u)FB \qquad \text{(in } \mathbf{QS_4})$$
$$(\mathrm{E}u)B =_{\mathrm{Df}} (\mathrm{A}u)B \equiv_{\mathrm{K}}.\sim (\mathrm{S}u)B \cup (\mathrm{A}u)B \qquad \text{(in } \mathbf{QL}).$$

$$(Uu)B =_{\mathrm{Df}}[(\mathrm{A}u)\sim FB \cup (\mathrm{A}u)\ SB \rightharpoonup .(\mathrm{A}u)B] \cap$$
$$[(\mathrm{S}u)FB \cap (\mathrm{S}u)\sim SB \rightharpoonup .(\mathrm{A}u)(B \cup \sim B)]$$

If the sentential basis of **QL** is made functionally complete, by adding an irregular connective such as 'D' or the Slupecki '$\dagger$' (which converts all values to n), then 'U' may be defined, somewhat more obviously, thus:

$$(Uu)B =_{\mathrm{Df}} (\mathrm{A}u)B \vee D \sim T(\mathrm{A}u)SB$$

Since $\vdash(\mathrm{A}u)(A \supset_{\mathrm{L}} B) \supset_{\mathrm{L}}.(\mathrm{A}u)A \supset_{\mathrm{L}}(\mathrm{A}u)B$, $\vdash(\mathrm{A}u)A \cap B \supset_{\mathrm{L}}(\mathrm{A}u)A$ and $\vdash(\mathrm{A}u)(A \cap B) \supset_{\mathrm{L}}.(\mathrm{A}u)A \cap (\mathrm{A}u)B$ are theorems of **QL**. Conversely $\vdash(\mathrm{A}u)A \supset_{\mathrm{L}}.(\mathrm{A}u)B \supset_{\mathrm{L}}(\mathrm{A}u)(A \cap B)$, but importation is rejected and only weakened imported forms hold, such as:
$\vdash S((\mathrm{A}u)A \cap (\mathrm{A}u)B) \supset_{\mathrm{L}}.(\mathrm{A}u)A \cap (\mathrm{A}u)B \supset_{\mathrm{L}}.(\mathrm{A}u)(A \cap B)$.
Also rejected are the following principles: $S(\mathrm{A}u)A \supset_{\mathrm{L}}(\mathrm{A}u)SA$, $S(\mathrm{A}x)f(x) \supset_{\mathrm{L}} Sf(y)$, $(\mathrm{S}u)\sim SA \supset_{\mathrm{L}} \sim S(\mathrm{A}u)A$, $(\mathrm{A}u)SA \supset_{\mathrm{L}} S(\mathrm{A}u)A$, $(\mathrm{A}u)(A \simeq B) \supset_{\mathrm{L}}.(\mathrm{A}u)A \simeq (\mathrm{A}u)B$.

The quantificational behaviour of 'A' and 'S' is very like the behaviour one would expect when the third value is interpreted as incompleteness, or paradoxicalness, instead of nonsignificance. For example, 'All Cretan assertions are false' is only paradoxical, when a Cretan assertion, if all other Cretan assertions are false; for presumably if one Cretan statement is true then the Cretan statement that all Cretan assertions are false is false, even if Cretan statements include such assertions as the paradoxical 'This very Cretan assertion is false'. **QL** is then a profitable system to examine in the study of incompleteness and paradoxicalness, though of course the relevant features can be studied by examining the strong regular, *i.e.* $\{\sim, \cup, \cap, \mathrm{A}, \mathrm{S}\}$, part of $\mathbf{QS_6}$. Since, moreover, Absorption, $p \supset_{\mathrm{L}}(p \supset_{\mathrm{L}} q) \supset_{\mathrm{L}}.p \supset_{\mathrm{L}} q$, and its equivalent Frege, $p \supset_{\mathrm{L}}(q \supset_{\mathrm{L}} q') \supset_{\mathrm{L}}.p \supset_{\mathrm{L}} q \supset_{\mathrm{L}}.p \supset_{\mathrm{L}} q'$, are not theorems of **QL**, **QL** has strong appeal as a system on which to base a set theory with a relatively unimpeded abstraction axiom; for the failure of Absorption stops Curry's paradox while 3-valued features eliminate Russell's paradox. The addition of an unqualified set-abstraction axiom of course leads to inconsistency by variations on Russell's paradox using non-classical connectives such as '$\urcorner$', and by extensions of Curry's paradox using higher-order absorption laws which are theorems of **QL** (see Shaw-Kwei [1]). The sorts of wff which can appear in the abstraction axiom have, then, to be restricted;

but the restrictions can be rather different from the usual ones. In particular, the class of connectives and quantifiers that can appear in the abstraction axiom may be restricted (as suggested, *e.g.* in Routley [3]). Brady set theory, which we consider briefly, makes just such a restriction. The theory is an applied logic obtained from **QL**, or $\mathbf{QS}_6$, by requiring that the two-place predicate constant '$\in$' satisfies these postulates:

(Abs.): $(\mathrm{S}y)(\mathrm{A}x).x \in y \simeq A(x, z_1, ..., z_n)$,

where wff $A(x, z_1, ..., z_n)$, which does not contain 'y' free, is one of the constants $\{t, f, n\}$ or is constructed from atomic wff of the form $u \in v$ using only connectives and quantifiers definable in terms of the strong normal set $\{\sim, \cap, \cup, \mathrm{A}, \mathrm{S}\}$;

(Ext): $(\mathrm{A}z)(z \in x \simeq z \in y) \rightharpoonup (\mathrm{A}z')(x \in z' \simeq y \in z')$.

The initial interest of this set theory is enhanced by the following important result (proved essentially in Brady [3]):

Theorem (57): Brady set theory is consistent (with respect to any set theory which provides adequately for transfinite induction).

Since among the connectives definable in terms of the strong normal set are the connectives of the classical set $\{\sim, \&, \vee, U, P\}$, the constructions and methods of classical "naive" set theory can be reflected, naturally in somewhat weakened form, in Brady set theory. (Compare the development in Fitch [1].) For instance union sets, power sets and so forth can be defined in much the usual way, with the connective '$\simeq$' taking up the usual 'iff'. Further such familiar theses as $\sim S(x \in x)$ and $\sim S \sim (x \in x)$ quickly follow. However Brady set theory is not adequate for the development of incompleteness theory or its analogue, significance range theory (or for the obvious extension of the virtual set theory developed in §7.9); in particular the theory does not permit full use of the expected definition of significance domains since the connective 'S' is not definable in terms of the strong normal set. If 'S' were definable then $t(P)$ which takes the value t whatever the value of its variable would be definable; but $t(P)$ is not definable since every connective definable in terms of strong normal connectives takes the value n when all its components take value n. It seems likely, however, that Brady's consistency theorem can be extended to show the consistency of an enlarged theory which admits the connective 'S' in (Abs). But the fact that such connectives as 'T', 'F' '$\supset$', '$\rightharpoonup$' and '$\simeq$' cannot be admitted at all in (Abs), on pain of inconsistency, means that for a sufficiently powerful set theory even the enlarged theory needs to be augmented by further principles. There are several ways in which Brady set theory can be extended: one (developed by Brady [3]) is to combine the theory with a standard set theory such as **ZF**; another is to supplement the theory by significance postulates, *i.e.* assumptions that certain membership assertions are significant; and a third is to use

(Abs) to define complete elements or membership-eligible sets and to add a further, unrestricted, abstraction postulate for complete elements (*i.e.* to exploit the von Neumann technique that Quine uses to inflate **NF** into **ML**; see Quine [1] and [4]. If non-sets are admitted, as they may be, as membership-eligible, then this theory includes the virtual set theory outlined in §7.9). We make no attempt to elaborate these or other methods here; the large issues of set theory touched upon, here and later in this chapter, belong to volume 2, and beyond, as a good set theory has yet to be devised.

A very different development based in **QL**, or in $\mathbf{QS}_6$, is the formulation of the theory of partial recursive functions of Kleene. (The theory is developed in a mixed semantical-syntactical way in Kleene [2] §§63–66.) But a formalisation within a restricted quantification logic such as **QL** requires first a formalisation of number theory, to which we now turn.

7.9 Developments of Quantified Significance Logics

§1 SA, *a significance arithmetic*; *and the theory of partial recursive functions*

By interpreting the subject arguments and constants of **QL**, or $\mathbf{QS}_6$, as natural number arguments and constants and by restricting the predicates and functions of the logic to natural number predicates and functions, the usual formal number theory (as presented for example in Kleene [2] or Mendelson [1]) can be taken over virtually intact. It suffices, given Mendelson's [1] presentation (p. 103), to replace '$\supset$' '$\sim$' and '(x)' systematically by '$\rightharpoonup$' '$\neg$' and '(Ux)' respectively. It is more illuminating, however, to insert significance features into the logic, rather than leaving the matter at the interpretational level; and such a strategy leads to a system more closely resembling the Peano postulates than the usual first-order arithmetic.

SA has as vocabulary the improper primitive symbols and individual variables of **QL**, or $\mathbf{QS}_6$, and these constant symbols: the individual constant '0', the 1-place function constant ''', the 2-place function constants '+' and '×', the 1-place predicate constant 'N' and the 2-place predicate constant '='.

The postulates of **SA** are those of $\mathbf{QS}_6$ together with the following:

P1. $\mathrm{N}0$

P2. $\mathrm{N}x \rightharpoonup \mathrm{N}x'$

P3. $\mathrm{N}x \rightharpoonup . \sim (x' = 0)$

P4. $\mathrm{N}x \rightharpoonup . x' = y' \rightharpoonup . x = y$

P5. For any wff $A(x)$ of **SA** with x free,
$A(0) \rightharpoonup . (Ux)(Nx \rightharpoonup . A(x) \rightharpoonup A(x')) \rightharpoonup . (Ux)(Nx \rightarrow A(x))$

P6. $\mathrm{N}x \rightharpoonup . x = y \rightharpoonup \mathrm{N}y$

P7. $\mathrm{N}x \rightharpoonup . x = y \rightharpoonup . x = z \rightharpoonup . y = z$

P8. $\mathrm{N}x \rightharpoonup . x = y \rightharpoonup x' = y'$

P9. $Nx \rightharpoonup .x+0=x$
P10. $Nx \rightharpoonup .Ny \rightharpoonup .x+y'=(x+y)'$
P11. $Nx \rightharpoonup .x\times 0=0$
P12. $Nx \rightharpoonup .Ny \rightharpoonup .x\times y'=(x\times y)+x$

Postulates P1–P5 are first-order formulations of Peano's postulates for arithmetic. P6–P8 furnish the requisite identity properties; P6 reads 'Provided x is a natural number, then provided x is identical with y, y is a natural number'. P9–P12 are significance-qualified recursion postulates for addition and multiplication.

The usual systemic results of first-order arithmetic follow, subject to qualifications of the form 'Provided Nx'.

In richer significance logics (*e.g.* **2QS**$_6$ below) the special primitives of **SA** may be defined, and arithmetic postulates deduced.

The theory of partial recursive functions can be given a more formal turn by using **QL** or **QS**$_6$ as the background logic; for Kleene's strong senses of connectives and quantifiers (Kleene [2], p. 334 ff) are given by the strong regular connectives and quantifiers. To make the tie-up we characterise *partial* functions using nonsignificance (or better, incompleteness) as the third value:—A function from any subset of n-tuples of natural numbers to the natural numbers is called a *partial* function. For an n-tuple $\underset{\sim}{x}$ for which partial function ϕ is assigned a natural number as function value, ϕ, or $\phi(\underset{\sim}{x})$, is *significant*; otherwise ϕ is nonsignificant (or indefinite). The theory of partial recursive functions may now be developed as in Kleene [2], p. 326 ff, but with greater formalisation of the intuitive theory and with a significance interpretation. Under this interpretation the Richard paradox (discussed in Kleene [2], p. 341) appears, as on Russell's account, as a significance fault; the paradox-generating function of the Richard paradox is not significant for all arguments. The partial recursive functions, including Richard functions, can be enumerated because partial functions may be non-significant for some arguments.

§2 *Significant subjects, special subjects and existence*

The quantified significance logic with existence, **QS**$_6$**E**, is obtained by adding to **QS**$_6$ the primitive one-place predicate constant 'E' read 'exists', together with an appropriate formation rule and these axioms:

A10. $\vdash(Px)E(x)$
A11. $\vdash(Px)\sim E(x)$

It follows:

T1. $\vdash(Ux)SE(x)$; $\vdash SE(x)$.

Proof:	$S(Px)E(x)$	by A1, A10
	$S(Ux)E(x)$	by T4
	$(Ux)SE(x)$	by A7

A semantical basis for QS_6E is a quadruple $S = \langle d_1, d_2, e, f \rangle$ where the entity and non-entity domains, d_1 and d_2 respectively, are both non-null. Then the domain d of QS_6-basis is defined: $d =_{Df} d_1 \cup d_2$; and e and f are defined as for the basis of QS_6. An *interpretation* I of QS_6E *over basis* S is defined exactly as for an interpretation of QS_6 except that the following requirements are added:

(1) I assigns to some subject an element of d_1 and to some subject an element of d_2; and

(2) $T(I(E)(I(x))) \equiv I(x) \in d_1$; $F(I(E)(I(x))) \equiv I(x) \in d_2$.

Since $I(x) \in d_1$ or $I(x) \in d_2$ it follows $v(E(x)) \neq n$; and it also follows $v(E(x)) = t$ or f according as $I(x) \in d_1$ or $I(x) \in d_2$.

Validity and satisfiability are defined as for QS_6; so also is satisfiability over a denumerable domain.

Theorem (58): Every theorem of QS_6E is valid.

To the proof of the analogous result on QS_6 (*Theorem* (15)) we add:

ad A10. For some assignment to variable x, $v(E(x)) = t$, since for some assignment $I(x)$ to x, $I(x) \in d_1$, and for every assignment to x, $v(E(x)) \neq n$. Hence $v((Px)E(x)) = t$. Since this is so for every interpretation over every basis A10 is valid.

ad A11. Similar to A10.

Theorem (59): QS_6E is simply and absolutely consistent.

Proof as for the corresponding *Theorem* (17) of QS_6.

The *alpe* g(*A*) of a wff *A* of QS_6E is obtained by taking its associated wff (defined as for QS_6) and then in this replacing every wf part $F(u_1, ..., u_n)$ *not* of the form $E(u)$ by a sentential variable not previously occurring according to the same prescription as for QS_6 and replacing every wf part $E(x_1)$ by t and every wf part $E(x_2)$ by f.

Theorem (60): If *A* is a theorem of QS_6E then its alpe g(*A*) is an S_6-tautology.

Proof extends that for QS_6 (*Theorem* (18)) thus:

ad A10. g(A10) is $t \vee \sim f$ *i.e.* t. Thus $v(g(A10)) = t$.

ad A11. Similar.

Corollary: QS_6E is simply and absolutely consistent.

Theorem (61): Every $\neg$-consistent class of wff of QS_6E is simultaneously satisfiable in a denumerable domain.

Proof: The proof adds to the proof of the corresponding theorem for QS_6 (*Theorem* (27)) the following details concerning predicate '*E*'. Define entity domain $d_1 =_{Df} \{t: E(t) \in J\}$ and non-entity domain $d_2 = \{t: E(t) \notin J\}$. Hence for closed term $t \in d$, $t \in d_2$ iff $\neg(E(t) \in J)$, and $d = d_1 \cup d_2$.

Since $\vdash (Px)E(x)$, $(\mathfrak{q}x)E(x) \in J$. Hence by some T_k, $E(b_1) \in J$ for some b_1. Hence d_1 is non-null. Similarly, using $\vdash (Px) \sim E(x)$,

$\sim E(b_2) \in J$ for some b_2; whence $\neg E(b_2) \in J$, $E(b_2) \notin J$, and d_2 is non-null.

Further $TI_1(E)(t) \equiv TE(t) \in J$
$\equiv t \in d_1$ and
$FI_1(E)(t) \equiv FE(t) \in J$
$\equiv t \in d_2$
$\equiv I(t) \in d_2$ as required.

Then, as before, I_1 over S_1 provides a denumerable interpretation for $\mathbf{QS_6E}$.

Theorem (62): Every simultaneously satisfiable class of wff of $\mathbf{QS_6E}$ is $\neg$-consistent.

Theorem (63) (*Skolem-Löwenheim*): Every simultaneously satisfiable class of wff of $\mathbf{QS_6E}$ is satisfiable in a denumerable domain.

Theorem (64) (*Completeness*): Every valid wff of $\mathbf{QS_6E}$ is a theorem of $\mathbf{QS_6E}$. Proofs as for corresponding results for $\mathbf{QS_6}$.

Existentially loaded quantifiers are defined:

$(\forall x)A =_{Df} (Ux)(E(x) \supset A)$; $(\forall_1 x)A =_{Df} (Ux)(E(x) \rightharpoonup A)$
$(\exists x)A =_{Df} (Px)(E(x) \& A)$; $(\exists_1 x)A =_{Df} (Px)(E(x) \&_1 A)$
$(Ex)A =_{Df} (\text{ᑫ}x)(E(x) \& A)$

It then follows:

$\vdash (\forall x)A \rightharpoonup (\forall_1 x)A$, but the converse $(\forall_1 x)A \rightharpoonup (\forall x)A$ is not valid.
$\vdash \sim(\exists x) \sim A \simeq (\forall x)A$; $\vdash \sim(\forall x) \sim A \simeq (\exists x)A$
$\vdash \sim(\forall_1 x) \neg A \simeq (\exists_1 x)A$; $\vdash S(\exists_1 x)A$.

Although both

$\vdash \sim(\exists_1 x) \neg A \rightharpoonup (\forall_1 x)A$ and $\vdash (\forall_1 x)A \rightharpoonup \sim(\exists_1 x) \neg A$,
$\sim(\exists_1 x) \neg A \simeq (\forall_1 x)A$ is not valid, because $S(\forall_1 x)A$ is not valid.

Finally the Barcan analogue, $S(Ex)A \supset (Ex)SA$, is not valid, as counter-models show.

The theorem $SE(x)$ which follows from A10, or A11, is not indisputable; after all, are sentences like 'The drunken numbet seven exists' significant? There appears indeed to be an English predicate 'exists', which we symbolise '$E!$', which is significant just of significant subjects. The predicate 'E' may be defined in terms of '$E!$' as '$TE!$'. Conversely, given that we can mark out the class of significant subjects, we can define $E!u =_{Df} E(u) \& QS(u)$, where '$S(u)$' reads '$qu(u)$ is a significant subject', and $SS(u)$.

In contrast to significant subjects like 'Pegasus' and 'the number seven', such subjects as 'a purple drunken number seven', 'this red round idea' and 'the person heavier than 10 o'clock' are nonsignificant and very often render sentences in which they occur nonsignificant. The initially appealing definition of a nonsignificant subject as one with no significant couplings breaks down, however, because not all sentences in which nonsignificant subjects occur are nonsignificant. For consider combinations of nonsignificant subjects with intensional predicates, *e.g.* 'We spent all afternoon thinking about the purple number seven' and 'John believes there is a purple number seven', with E or T-ed predicates, and with identifying and distinguishing predicates, *e.g.* 'the purple number seven is different from the orange number'. The initial definition is not easily repaired; for the predicates which can be significantly coupled with nonsignificant subjects are rather diverse, *e.g.* not all are intensional, many cannot be reconstrued as quotational, and some are not significant of every item. The attempt to deal with descriptions like 'the drunken hyperbolic functions' by noting that the predicate 'is a drunken hyperbolic function' is nonsignificant, likewise leads to an impasse, since in order to distinguish nonsignificant predicates we seem to need a way of separating off subjects like 'what we are writing about'. Finally most means for defining nonsignificant subjects which do suggest themselves go well beyond the resources of **QS₆E**. One method however which is scarcely more demanding, is to replace 'E' by '$E!$' and to define 'S' over subjects thus: $S(u) =_{\mathrm{Df}} SE!u$. What this definition assumes seems to be correct.

We sketch the theory **QS₆E!** of existence and significant subjects. We add to **QS₆** these axioms:

E1. $\vdash (\mathcal{q}x)E!x$

E2. $\vdash (\mathcal{q}x) \sim E!x$

E3. $\vdash (\mathcal{q}x) \sim SE!x$, *i.e.* some subject is not significant.

It follows at once that **QS₆E** is a subtheory of **QS₆E!** using the definition of 'E'. A semantical basis for **QS₆E!** is a quintuple $S = \langle d_1, d_2, d_3, e, f \rangle$ where d_1, d_2 and d_3 are all non-null. $d =_{\mathrm{Df}} d_1 \cup d_2 \cup d_3$. To the assignments (given for 'E', but taken over also for '$E!$') we add: $\sim SI(E!)(I(x)) \equiv . I(x) \in d_3$. Then $(E!x) = t, f$ or n according as $I(x) \in d_1$, d_2 or d_3 respectively; d_3 is the domain of nonsignificant contractions.

Theorem (65):

(1) Every con class of **QS₆E!** wff is simultaneously satisfiable in a denumerable domain.

(2) B is a theorem of **QS₆E!** iff B is **QS₆E!**-valid.

In the proof of (1) we define:

$d_1 = \{t \in d : TE!t \in J\}, d_2 = \{t \in d : FE!t \in J\}$ and
$d_3 = \{t \in d : \sim SE!t \in J\}$.

Significance logics such as **PM** exclude nonsignificant subjects and what these are about, absurd items, from logical treatment, just as they exclude non-entities; but as **QS₆E!** shows these restrictions are over-restrictive, since all such subjects and items can be treated logically. Logic does not stop with the possible, nor with the significant.

'Exists' is by no means the only special one-place predicate we shall eventually need. Predicates which classify the subjects we have admitted into their sorts are also required. For we want to be able to distinguish, as well as significant subjects, attributive subjects, assertoric subjects and so forth. Attributives, for example, satisfy further postulates which distinguish them logically from other subjects such as individual subjects. The predicate '*At*' read '*qu*( ) is an attribute (quality)', which is true of the subject 'redness' but not true of 'the author of Waverley', may be defined in second-stage logics (§7.12), by

$$At(u) =_{\mathrm{Df}} (\mathit{9}z)S(ziu).$$

In quantification logics like **QS$_6$E!** however, '*At*' has, it seems, to be introduced as a further primitive. About the basic semantics of *At* there are few problems; for, like '*E*!', '*At*' divides domain d into three divisions, items which are attributes, items which are not, and items like the author of Waverley which are not significantly attributes. The only new problem lies in formulating conditions which correspond semantically to the special postulates, if any, that '*At*' satisfies. In fact '*At*' will presumably satisfy further axioms exactly like those '*E*!' satisfies—which emphasizes the superficiality of the logical treatment of existence that we have so far provided. Thus we may go on to develop quantification logics with *n* special predicates, which provide an alternative to *n*-sorted logics, to which *n*-sorted logics are usually reduced.

Indeed, in our view, the main logical development of quantified significance logic (when intensionally enriched, in the style of **QMS$_6$** of §7.11 below) consists, not in the extension of the logic to higher-order logics, but in the introduction and analysis, both postulationally and semantically, of special subjects and classes of subjects, and of special predicates and classes of predicates. Special classes of subjects of importance are those formed by descriptors such as, '§', '*qu*', 'λ', 'ξ' and 'ι'. Our treatment of special subjects and predicates is very selective. For instance we examine only a tiny selection of predicates of logical importance, and omit altogether such predicates as 'is a part of' and 'precedes'.

§3 *Quantified significance logic with Leibniz identity*

The predicate '≈', read 'is (Leibniz) identical with' is to be so understood that $x \approx y$ is true iff *x* has all and only the properties that *y* has. To turn this criterion (traditionally ascribed to Leibniz) into a *definition* of identity would require logical resources not available in **QS$_6$**, *viz.* quantification using property terms. Indeed Leibniz identity cannot be defined in **QS$_6$**, as we shall show.

Thus to handle the logic of Leibniz identity we add to **QS$_6$**, **QS$_6$E** or **QS$_6$E**! the two-place predicate constant '≈'; we call the resulting logics **QS$_6$**≈, **QS$_6$E**≈ and **QS$_6$E**! ≈ respectively. For wft t_1 and t_2, $t_1 \approx t_2 =_{\mathrm{Df}} \approx(t_1, t_2)$.

It is a ready consequence of the interpretation of Leibniz identity that the relation $\approx$ should satisfy the following postulates. Note that unlike the postulates (P14) of §3.5, these are many-valued, not two-valued.

I1. $x \approx x$

I2. $u \approx v \rightarrow (A(u) \simeq A(v))$

where u and v are subject terms and $A(v)$ results from atomic wff $A(u)$ by replacing one occurrence of u by v, and this occurrence of u is not within the scope of quantifiers binding u or v.

These two postulates in fact suffice for completeness proofs for $\mathbf{QS_6}\approx$ and $\mathbf{QS_6E}!\approx$, and they suffice to show that identicals are indiscernible. It is hardly a deficiency of quantification logic that the converse principle, of the identity of indiscernibles, is not expressible within its resources since the logic makes no claim to comprehensiveness. Moreover a (trivial) indiscernibility principle is provable in rule form: *viz*:

$$\text{if } \vdash A(u) \simeq A(v) \quad \text{then } \vdash u \approx v$$

A systemic result which the full Leibnizian version of I2 would entail, $S(x \approx y)$, does not follow in $\mathbf{QS_6E}\approx$; should it be insisted that identity must be homogeneous the further postulate

I3. $S(x \approx y)$

can be added. (If it is added, minor changes are needed in the semantics given below.)

Theorem (66):

(i) For every term t, u and v,
$t \approx t$, $u \approx t \rightarrow . \; t \approx u$, and
$u \approx v \rightarrow . \; v \approx t \rightarrow . \; u \approx t$. Also $u \approx v \rightarrow . \; E(u) \equiv E(v)$.

(ii) Scheme I2 holds not just for atomic wff $A(u)$ but for every wff $A(u)$—call the general scheme I2$^+$.

Proof of (i) follows by quantification logic using I1 and I2, and proof of (ii) then follows using induction.

Proposition (ii) guarantees the full indiscernibility of identicals. I1 and I2 do not, however, guarantee the full principle if $\mathbf{QS_6}$ is enriched by intensional functors such as '$\Box$'. For the induction to go through the further principle

$$u \approx v \rightarrow . \; \Box(u \approx v)$$

correct for Leibniz identity though not for contingent identity, would be needed. I1 and I2$^+$ also characterise Leibniz identity in intensional logics. In contrast I1 and I2 do not provide an adequate characterisation of Leibniz identity in intensionally enriched logics; nor, however, do they provide a satisfactory characterisation of the usual, extensional, identity unless the primitive predicates of $\mathbf{QS_6}$ are limited—either explicitly or through the interpretation—to extensional predicates, and exclude such predicates as 'is necessarily red'. In case the predicates of $\mathbf{QS_6}$ are so limited the relation '$\approx$' can be replaced by the extensional identity '='.

The contractional semantics for $\mathbf{QS_6}\approx$ and $\mathbf{QS_6E}!\approx$ simply extend those for $\mathbf{QS_6}$ and $\mathbf{QS_6E}!$ by adjoining a valuation rule for $t_1 \approx t_2$, *viz*: $v(t_1 \approx t_2) = t, f$ or n according as it is true, false or nonsignificant that $I(t_1)$ is identical with $I(t_2)$. The rule is not circular, any more than quantificational evaluation rules are circular; and in the case of the identity rule the Leibniz identity used in the explanation can be explained, in turn, in terms of coincidence of all properties.

The main results for $\mathbf{QS_6}$ extend to $\mathbf{QS_6}\approx$, $\mathbf{QS_6E}\approx$ and $\mathbf{QS_6E}!\approx$.

Theorem (67): $\mathbf{QS_6E}\approx$ is simply and absolutely consistent.

Proof differs from the corresponding syntactical proof for $\mathbf{QS_6E}$ only in that wff of the form $x_2 \approx x_1$, $x_1 \approx x_1$ and $x_2 \approx x_2$ are, like $E(x_1)$ and $E(x_2)$, given special treatment, and are replaced respectively by f, t, and t. Then if A is a theorem of $\mathbf{QS_6E}\approx$ its alpe, g(A), as just defined, is an S_6-tautology.

Theorem (68a): Every $\neg$-consistent class of wff of $\mathbf{QS_6E}!\approx$, or of $\mathbf{QS_6}\approx$, is simultaneously satisfiable in a countable domain.

Proof adapts the Henkin-Gödel proof for quantification logic with identity. Given that Γ is a $\neg$-consistent class of wff of $\mathbf{QS_6E}\approx$, then $\Delta = \Gamma \cup I1' \cup I2'$ is a $\neg$-consistent class of wff of $\mathbf{QS_6E}$, where $I1'$ is the set of closures of I1 and $I2'$ of scheme I2. By a previous theorem Δ is simultaneously satisfiable, say by I_1 over S_1. Now the relation $E = I_1(\approx)$ assigned by I_1 to predicate '$\approx$' is an equivalence relation, since as $\Delta \subseteq J$, $(t \approx t) \in J$, and similarly for symmetry and transitivity of Leibniz identity. Using this equivalence relation equivalence classes of domain d may be formed, and used to provide a new interpretation I_2 on new basis S_2.

Explicitly, $[d] =_{Df} \{[b]: b \in d\}$ where $[b] =_{Df} \{b' \in d : b'Eb\}$. S_2 is the usual basis over domain [d]. Also $[d_1] =_{Df} \{[b]: b \in d_1\}$, $[d_2] =_{Df} \{[b]: b \in d_2\}$ and $[d_3] =_{Df} \{[b]: b \in d_3\}$, where d_1, d_2 and d_3 are as before.

I_2 over S_2 is defined as follows:

$$I_2(X_0) = [X_0] \text{ for subject constants of } K_0;$$
$$I_2(D^n)([t_1], \ldots, [t_n]) = [I_1(D^n)(t_1, \ldots, t_n)].$$

Hence, by induction, $I_2(t) = [t]$ for closed term t.

$$([t_1], \ldots, [t_n]) i I_2(F^n) \simeq I_1(F^n)(t_1, \ldots, t_n)$$
$$I_2(P) = I_1(P).$$

$I_2(D^m)$ and $I_2(F^m)$ are well defined since

$$\vdash X_1 \approx Y_1 \rightharpoonup . \ldots X_m \approx Y_m \rightharpoonup . F^m(X_1, \ldots, X_m) \simeq F_1{}^m(Y_1, \ldots, Y_m)$$

and

$$\vdash X_1 \approx Y_1 \rightharpoonup . \ldots X_m \approx Y_m \rightharpoonup . D^m(X_1, \ldots, X_m) \approx D^m(Y_1, \ldots, Y_m);$$

so it makes no difference which representatives of equivalence classes are selected.

For the result it suffices to show

(9_1). For every closed wff A, for valuation v_2 wrt I_2,

$$v_2(A) = t \equiv .TA \in J$$
$$v_2(A) = f \equiv .FA \in J$$
$$v_2(A) = n \equiv .\neg SA \in J$$

Once, however, the induction basis is established the induction steps follow as for $\mathbf{QS_6}$; and the basis is established, except for '$\approx$', using the basis for $\mathbf{QS_6E}$, thus:

$$v_2(P) = I_2(P) = I_1(P) = v_1(P)$$
$$v_2(F^n(t_1, ..., t_n)) = t \equiv .T(I_2(t_1), ..., I_2(t_n)) i I_2(F^n)$$
$$\equiv .TI_1(F^n)(t_1, ..., t_n)$$
$$\equiv .v_1(F^n(t_1, ..., t_n)) = t; \; etc.$$
$$v_2(E(t_1)) = t \equiv [t_1] \in [d_1] \equiv t_1 \in d_1 \equiv v_1(E(t_1)) = t$$
$$v_2(E(t_1)) = f \equiv .v_1(E(t_1)) = f$$
$$v_2(E(t_1)) = n \equiv .v_1(E(t_1)) = n$$

Finally, $v_2(t_1 \approx t_2) = t \equiv .T(I_2(t_1), I_2(t_2)) i I_2(\approx)$

$$\equiv .TI_1(\approx)(t_1, t_2)$$
$$\equiv .T(t_1 E t_2)$$
$$\equiv .T((t_1, t_2) i I_1(\approx)),$$
$$\equiv .T(t_1 \approx t_2) \in J, \text{ by definition of '}I_1\text{'}.$$

Similarly $v_2(t_1 \approx t_2) = f \equiv .F(t_1 \approx t_2) \in J$ and

$$v_2(t_1 \approx t_2) = n \equiv .\neg S(t_1 \approx t_2) \in J$$

Theorem (68b) (*Skolem Löwenheim*): Every simultaneously satisfiable class of wff of $\mathbf{QS_6E!\approx}$ ($\mathbf{QS_6\approx}$) is simultaneously satisfiable in a countable domain.

Theorem (69) (*Completeness, Consistency*): Every valid wff of $\mathbf{QS_6E!\approx}$ ($\mathbf{QS_6\approx}$) is a theorem of $\mathbf{QS_6E!\approx}$ ($\mathbf{QS_6\approx}$), and conversely.

Theorem (70): Leibniz identity, as characterised by postulates I1 and I2, is not definable in $\mathbf{QS_6E!}$ or in $\mathbf{QS_6E}$, and so not definable in $\mathbf{QS_6}$.

Proof: (1) Given any finite number $n \geqslant 3$ there is a simultaneously satisfiable class of wff Γ of $\mathbf{QS_6E!\approx}$ such that Γ is simultaneously satisfiable only in a domain of exactly n items. But, (2) any simultaneously satisfiable class of wff of $\mathbf{QS_6E!}$ is simultaneously satisfiable in a denumerably infinite domain, by the Skolem-Löwenheim theorem. It follows from (1) and (2) that Leibniz identity is not definable in $\mathbf{QS_6E!}$; for if it were we could form classes of wff which restrict domains to the finite. As to (1) define, by induction on positive n,

$$(\mathfrak{q}_1 x)B(x) =_{\mathrm{Df}} (\mathfrak{q} x)B(x) \; \& \; (Ux, y)(B(x) \; \& \; B(y) \rightharpoonup .x \approx y)$$
$$(\mathfrak{q}_{n+1} x)B(x) =_{\mathrm{Df}} (\mathfrak{q} y)[B(y) \; \& \; (\mathfrak{q}_n x)(\neg(x \approx y) \; \& \; B(x))].$$

Then $(\mathfrak{q}_n x)E(x)$ asserts that there are exactly n entities, in the sense that if $\Delta \cup \{(\mathcal{q}_n x)E(x)\}$ is simultaneously satisfiable it is only simultaneously satisfiable in domains with n entities. Likewise $(\mathcal{q}_m x) \sim E(x)$ asserts that there are exactly m non-entities. Now the set $\{(\mathcal{q}_k x)E(x), (\mathcal{q}_m x) \sim E(x)\}$ is simultaneously satisfiable in a domain of $n = k + m$ items, as may be laboriously verified, but it is not simultaneously satisfiable in domains of cardinality different from n (essentially because the interpretation of the identity relation in any model which satisfies the wff, in any *normal* model that is, is the Leibniz identity relation in the domain of the model).

§4 *Contextually defined definite descriptions in* $\mathbf{QS_6E}\approx$

The definite description operator 'the' symbolized' '$\imath$', may be defined contextually in $\mathbf{QS_6E}\approx$ in various ways. We shall examine a theory based on the definition:

$$(\imath u)(B(u), A(u)) =_{\mathrm{Df}} (\mathcal{q}u)\,[B(u)\ \&\ (Uv)(\urcorner(u \approx v) \rightharpoonup \urcorner B(v))\ \&\ A(u)].$$

$A(...)$ provides the *scope* of the description $\imath uB(u)$. Where $A(u)$ is the smallest sentence context in which a subject u may occur we write $(\imath u)(B(u), A(u))$ in the more familiar form $A(\imath uB(u))$. Thus

$$\vdash \mathrm{F}(\imath uB(u)) \simeq (\mathcal{q}u)\,[B(u)\ \&\ (Uv)(\urcorner(u \approx v) \rightharpoonup \urcorner B(v))\ \&\ \mathrm{F}(u)].$$

The advantages of this theory of descriptions over the Russellian theory on which it is based, are that it reduces the two basic definitions of the Russellian theory to a single unified treatment, that it allows for a more satisfactory treatment of items which do not exist, and that, by defining '$\imath$' as a binary quantifier, it minimizes some of the problems of scope that afflict (and aid) Russell's theory. To facilitate comparison with the usual theories we introduce the following definitions:

$$B!u =_{\mathrm{Df}} B(u)\ \&\ (Uv)(\urcorner(u \approx v) \rightharpoonup \urcorner B(v));$$
$$(\mathcal{q}!u)A(u) =_{\mathrm{Df}} (\mathcal{q}u)A!(u)$$
$$(E!u)A(u) =_{\mathrm{Df}} (Eu)A!(u)\ \ [\simeq (\mathcal{q}u)(A!u\ \&\ E(u))].$$

Then $\vdash E(\imath uB(u)) \simeq .(E!u)B(u)$, a suitable analogue of *PM**14.11. It is already evident from this result, however, that $\imath uB(u)$ is not a proper subject, or a value of subject variables. For if it were we should have $SE(\imath uB(u))$ (from $\vdash SE(x)$), whence $(\mathcal{q}u)SB(u)$ for every wff B; and hence Sp. In fact the treatment of the descriptive terms of this theory as complete subjects—the way that they should be treated, we have argued—leads to inconsistency (for details see the alternative, preferred, theory of descriptions in §7.10). The condition under which a description $\imath uB(u)$ may be treated as a genuine subject is given by the following instantiation theorem:

$$\vdash (\mathcal{q}!u)B(u) \rightharpoonup .(Uu)\mathrm{F}(u) \rightharpoonup \mathrm{F}(\imath uB(u));$$

it is the uniqueness condition, $(\mathcal{q}!u)B(u)$. On the theory this uniqueness condition is equivalent to the assumptibility condition $(\iota u)(B(u), B!u)$; for $\vdash(\iota u)(B(u), B!u) \simeq (\exists!u)B(u)$. In fact the slash may be omitted: $\vdash F(\iota u F(u)) \simeq (\exists!u)F(u)$. Given $(\exists!u)F(u)$ we may introduce $\iota u F(u)$ and let it satisfy all subject rules. For then all quantificational principles hold, and

$\iota u F(u) \approx \iota u F(u)$ holds, since

$\vdash \iota u F(u) \approx \iota u F(u) \simeq . (\exists!u)F(u)$.

A minor result, with an application to philosophical discussion of significance, is the theorem

$\vdash SF(\iota u B(u)) \equiv (\exists u)S(B(u) \,\&\, F(u))$.

Thus, for instance, the assertion that the seat of a bed is hard (Smart's [1] example) is significant iff something is significantly a seat of a bed and hard; similarly for such examples as 'the wheels of an elephant are robust'. The issue of the significance of sentences containing descriptions is thus reduced to the simpler issue of the significance of predications; and in our opinion such controversial examples as Smart's are resolved using this transformation, since it is significant, for instance, that some item is a seat of a bed and is hard.

§5 *The logic of classes and relations in* $\mathbf{QS_6E}\approx$

Most of the theory developed can, in fact, be carried out in $\mathbf{QS_6}$. The definitional schema on which the theory is based jointly defines membership ('$\in$') and class abstraction ('$\{\underset{\sim}{x}: \ldots \underset{\sim}{x} \ldots\}$') as follows:

(1) $\underset{\sim}{y} \in \{\underset{\sim}{x}: A(\underset{\sim}{x})\} =_{\mathrm{Df}} A(\underset{\sim}{y})$

This conversion principle, called the 'law of concretion' in Quine [4] (p. 16), has to be understood as accompanied by a relettering principle (as explained in Fitch [1] and Quine [4]) in order to count as a proper definition. Alternatively, we can stipulate (in the one-variable case)

$$y \in \{x: A(x)\} =_{\mathrm{Df}} \mathrm{S}^{1x}_{y} A(x)|,$$

provided $\mathrm{S}^{1x}_{y}A(x)\,|$ differs from $A(x)$ when y differs from x.

We abbreviate class abstracts by Greek letters 'α', 'β', 'γ' and 'δ'. A class identity relation '$=$' is defined:

(2) $\alpha = \beta =_{\mathrm{Df}} (Ux)(x \in \alpha \simeq x \in \beta)$.

The strong equivalences ($\simeq$) yielded by (1) and (2) provide the basis of a three-valued ideal class theory—except that quantification over class abstracts is not defined, *i.e.*, class abstracts are not values of subject variables of $\mathbf{QS_6E}\approx$. Because quantification over class abstracts and over relation abstracts is not defined, and cannot be defined, in the theory, much of the theory of numbers and of relations cannot be developed (for details, readily extended from the 2-valued to the 3-valued case, see Quine [4], and Hilbert and Ackermann [1], Ch. IV). Because quantification over class abstracts is not defined the theory is dubbed, by Quine, a *virtual* theory of classes; similarly the relation theory is a *virtual* theory of relations. This terminology is harmless enough so long as virtual classes are not contrasted with real or existent classes; that a class is virtual in a given logic has nothing to do with whether or not it exists.

The theory of virtual properties differs from the theory of virtual classes through (2): for identity of properties is not guaranteed by sameness of instances. The theory of virtual attributes is based on the conversion principle

(1′) $\underset{\sim}{y}\, i\, \lambda \underset{\sim}{x} A(\underset{\sim}{x}) =_{\mathrm{Df}} A(\underset{\sim}{y})$.

Anticipating later developments, we abbreviate attribute abstracts by Greek letters 'ψ', 'ϕ' and 'χ'.

Analogues of the main operations and relations of the Boolean algebra of classes may be defined in much the usual way:

$$\alpha \cup \beta =_{\mathrm{Df}} \{x : x \in \alpha \ \vee . x \in \beta\}$$
$$\alpha \cap \beta =_{\mathrm{Df}} \{x : x \in \alpha \ \&. \ x \in \beta\}$$
$$\bar{\alpha} =_{\mathrm{Df}} \{x : \sim x \in \alpha\}$$
$$\alpha \subseteq \beta =_{\mathrm{Df}} (Ux)(x \in \alpha \rightharpoonup . \ x \in \beta)$$

(Similarly for attributives; *e.g.*

$$\psi \cup \chi =_{\mathrm{Df}} \lambda x(x i \psi \vee x i \chi)).$$

But because the underlying logic is 3-valued not all the usual Boolean connexions hold, and operations and relations which cannot be defined at all in Boolean algebra can be defined, *e.g.* the range,

$$\hat{\alpha} =_{\mathrm{Df}} \alpha \cup \bar{\alpha}.$$

Further, where just one notion is definable given a 2-valued basis, several competing notions can be defined given a many-valued basis. In particular, along with union and intersection we have dot union and intersection, defined:

$$\alpha \dot{\cup} \beta =_{\mathrm{Df}} \{x : x \in^{+} \alpha \ \vee . \ x \in^{+} \beta\},$$
$$\alpha \dot{\cap} \beta =_{\mathrm{Df}} \{x : x \in^{+} \alpha \ \& \ x \in^{+} \beta\},$$
$$\text{where } x \in^{+} \alpha =_{\mathrm{Df}} T(x \in \alpha),$$
$$\alpha \doteq \beta =_{\mathrm{Df}} (Ux)(x \in^{+} \alpha \equiv . x \in^{+} \beta)$$
$$\dot{\hat{\alpha}} =_{\mathrm{Df}} \alpha \dot{\cup} \bar{\alpha}$$
$$\alpha^{+} =_{\mathrm{Df}} \{x : T(x \in \alpha)\}.$$

We state, without proof, sample theorems.

T. $x \in^{+} \{y : f(y)\} \simeq . Tf(x)$.

T. $x \in^{+} \overline{\{y : f(y)\}} \simeq . Ff(x)$.

T. $x \in^{+} \dot{\widehat{\{y : f(y)\}}} \simeq . Sf(x)$.

By defining truth-, falsity- and significance-domains,

$$TD\text{‘}f =_{\mathrm{Df}} \{x{:}Tf(x)\};\ FD\text{‘}f =_{\mathrm{Df}} \{x{:}Ff(x)\};\ SD\text{‘}f =_{\mathrm{Df}} \{x{:}Sf(x)\},$$

we obtain the now familiar forms:

T. $x \in TD\text{‘}f \simeq . Tf(x)$,
T. $SD\text{‘}f = TD\text{‘}f \cup FD\text{‘}f$ *etc.*

Much of the Boolean logic of classes follows:—

T. $\alpha \cap \alpha = \alpha$
T. $\alpha \cup \alpha = \alpha$
T. $\alpha \cap \beta = \beta \cap \alpha$
T. $\alpha \cup \beta = \beta \cup \alpha$
T. $\alpha \cap (\beta \cap \gamma) = (\alpha \cap \beta) \cap \gamma$
T. $\alpha \cup (\beta \cup \gamma) = (\alpha \cup \beta) \cup \gamma$
T. $\alpha \cap (\beta \cup \gamma) = (\alpha \cap \beta) \cup (\alpha \cap \gamma)$
T. $\alpha \cup (\beta \cap \gamma) = (\alpha \cup \beta) \cap (\alpha \cup \gamma)$
T. $\alpha = \beta \ \&\ \beta = \gamma \rightharpoonup . \gamma = \alpha$

Theorem (71):

$$\alpha = \beta \rightharpoonup . \Delta = \mathsf{S}^{\alpha}_{\beta}\, \Delta|,$$

where Δ is an expression built up from class-abstracts using '$\cup$', '$\cap$' and '$^-$'. *Proof* is by induction over the number of occurrences of class operations. Since however the absorption principles,

$$\alpha \cap (\alpha \cup \beta) = \alpha \quad \text{and} \quad \alpha \cup (\alpha \cap \beta) = \alpha,$$

do not hold generally, the theory of operations $\{\cap, \cup, ^-\}$ on classes does not form a lattice, let alone a Boolean algebra as in the two-valued case. Of course Boolean algebra does hold for a restricted class of abstracts, *viz.* for β such that $\beta = \beta^+$. For

T. $\alpha = \alpha^+ \equiv . S(Ux)(x \in \alpha)$.

In place of Boolean algebra new algebraic theories appear, *e.g.* one based on operations $\{^-, \cup, \cap\}$ and another based on $\{^-, \cup, \cap, V, \Lambda\}$ where $V =_{\mathrm{Df}} \{x{:}T(x \approx x)\}$ and $\Lambda = \{x{:}\neg(x \approx x)\}$.

Also theories can be built using dot operations. Dot operations and relations can however be eliminated using $^+$ abstracts.

T. $\alpha \doteq \beta \equiv . \alpha^+ = \beta^+$
T. $\alpha \mathbin{\dot\cup} \beta = \alpha^+ \cup \beta^+$
T. $\alpha \mathbin{\dot\cap} \beta = \alpha^+ \cap \beta^+ = (\alpha \cap \beta)^+$
T. $\hat{\dot{\alpha}} = \alpha^+ \cup \bar{\alpha}^+$
T. $\alpha = \alpha^+ \supset . \bar{\alpha}^+ = \bar{\alpha}^+$
T. $x \in^+ \hat{\dot{\alpha}} \equiv S(x \in \alpha) \equiv x \in \hat{\dot{\alpha}}$
T. $\hat{\dot{\alpha}} = \hat{\alpha}^+$
T. $Sf(x) = x \in^+ \widehat{\{y{:}f(y)\}}$
T. $\widehat{\alpha \mathbin{\dot\cap} \beta} = \hat{\dot{\alpha}} \cap \hat{\dot{\beta}} = \widehat{\alpha \mathbin{\dot\cup} \beta}$
T. $\hat{\bar{\alpha}} \doteq \hat{\alpha}$

The theory of virtual two-place relations (i.e. relations-in-extension) assumes a more familiar form if in place of (1) we use

$$z\{xy{:}A(x,y)\}w =_{\mathrm{Df}} A(z,w),$$

and abbreviate the defined abstracts with capitals 'R', 'S' and 'T'. Various analogues of the principal notions of the theory of relations may be defined: *e.g.*

$$\mathrm{R} \cup \mathrm{S} =_{\mathrm{Df}} \{xy : x\mathrm{R}y \vee x\mathrm{S}y\}$$
$$\overline{\mathrm{R}} =_{\mathrm{Df}} \{xy : \sim x\mathrm{R}y\}$$
$$\mathrm{R} = \mathrm{S} =_{\mathrm{Df}} (Ux,y) \,.\, x\mathrm{R}y \simeq x\mathrm{S}y$$
$$\alpha \times \beta =_{\mathrm{Df}} \{xy : x \in \alpha \;\&\; y \in \beta\}$$
$$\breve{\mathrm{R}} =_{\mathrm{Df}} \{xy : y\mathrm{R}x\}$$
$$\mathrm{R}|\mathrm{S} =_{\mathrm{Df}} \{xz : (\mathcal{q}y)(x\mathrm{R}y \;\&\; y\mathrm{S}z)\}$$
$$\mathrm{R}``\alpha =_{\mathrm{Df}} \{x : (\mathcal{q}y)(x\mathrm{R}y \;\&\; y \in \alpha)\}$$
$$\mathrm{D}`\mathrm{R} =_{\mathrm{Df}} \{x : (\mathcal{q}y)x\mathrm{R}y\};\ \mathrm{\Omega}`\mathrm{R} = \{y : (\mathcal{q}x)x\mathrm{R}y\}$$
$$\alpha 1 \mathrm{R} =_{\mathrm{Df}} \{xy : x \in \alpha \;\&\; x\mathrm{R}y\}$$

Such notions as these provide the basis for a virtual significance logic of relations (compare *PM*, section D). But though many of the equations of the classical algebra of relations carry over, many do not. The discrepancy leaves room for the development of the specifically significance features of relations (only touched on in *PM*, and excluded by the classical theory). Among new notions of importance are these:

$$S\mathrm{D}`\mathrm{R} =_{\mathrm{Df}} \{x : (\mathcal{q}y)S(x\mathrm{R}y)\};\ S\mathrm{\Omega}`\mathrm{R} =_{\mathrm{Df}} \{y : (\mathcal{q}x)S(x\mathrm{R}y)\}$$
$$\mathrm{Hom}\ (\mathrm{R}) =_{\mathrm{Df}} (Ux,y)(x \in S\mathrm{D}`\mathrm{R} \;\&\; y \in S\mathrm{\Omega}`\mathrm{R} \supset .S(x\mathrm{R}y))$$
$$\mathrm{Fam}\ (\mathrm{R}) =_{\mathrm{Df}} \mathrm{Hom}\ (\mathrm{R})\ \&.\ S\mathrm{D}`\mathrm{R} = S\mathrm{\Omega}`\mathrm{R}$$
$$\mathrm{Cat}\ (\mathrm{R}) =_{\mathrm{Df}} \mathrm{Hom}\ (\mathrm{R})\ \&.\ S\mathrm{D}`\mathrm{R} \cap S\mathrm{\Omega}`\mathrm{R} = \Lambda$$

(For a preliminary discussion see Goddard [2]).

§6 *Restricted variables and restricted quantifiers*

Earlier attempts to formulate systems of quantified significance logic based on classical or clean connectives all use restricted variables. For example, the individual variables and quantifiers Halldén and Russell use are both significance and existence restricted. As explained (in §3.2), there are various more and less satisfactory ways in which restricted variables and quantifiers can be introduced into the logic. Since a wff *A* is provable iff its universal closure is provable, the problem of introducing restricted variables and quantifiers can be limited to that of introducing restricted *bound* variables; thus the problems of restricting variables and quantifiers merge into one problem, that of restricting bound variables.

A restricted variable is a variable whose constants-range is restricted by a condition or proviso. Where '*u*' is restricted so that *u* satisfies *A*, we write '$u \ni A(u)$', a wft read '*u* such that *A*(*u*)'. The symbol '∋', read 'such that' or 'for which', is taken from Peano (but we give it a new setting); it is a restricting relative term. We have two cases to consider, variables bound by '*U*' and variables bound by '$\mathcal{q}$'. As before '$\mathcal{q}$' may be eliminated definitionally;

but the elimination is problematic where '$\mathcal{q}$' binds restricted variables. At the very least then we have to explain expressions of the form,

$$(Uu\ni B(u))C(u\ni B(u)),$$

read 'for every u for which u satisfies B, u such that $B(u)$ satisfies C'. We discuss briefly two methods for introducing such expressions into $\mathbf{QS_6E}\approx$. *Method* 1. The most satisfactory and comprehensive way of handling restricted variables and quantifiers is by introducing '$\ni$' as a further primitive, satisfying this formation rule: if $B(u)$ is a wff with u free and unrestricted, then $u\ni B(u)$ is a free, restricted variable.

This rule excludes iterations of '$\ni$'; it would be easy to amend the rule to admit iterations, but their force would be a bit obscure. Quantifiers may bind both restricted and unrestricted free variables. Were a descriptor 'τ' added as a primitive, a formation rule would be simply: where $u\ni B(u)$ is a restricted variable $\tau u\ni B(u)$ is a wft. Thus the more explicit '$\xi u\ni B(u)$', read 'for any u such that $B(u)$' would replace '$\xi uB(u)$'.

Apart from instantiation schemes, the postulates of $\mathbf{QS_6E}\approx$ can be simply taken over intact, permitting the general variables u, v, *etc.*, in terms of which the postulates are stated, to be either restricted or unrestricted. But instantiation (and particularisation) schemes demand special treatment; for substituted variables must carry every restriction that the variable upon which substitution is made carries. In place of the instantiation scheme, A4, we postulate, then:

RA4. $(Uu\ni B)A\rightarrow \check{S}^{u\ni B}_{v\ni B\,\&C}\, A\,|.$

(Where $u\ni B$ is u, $v\ni B\&C$ is $v\ni C$. For simplicity we exclude function parameters). From the theory so far presented we can obtain such theorems as:

$$(Ux)f(x)\rightarrow(Uy\ni g(y))f(y\ni g(y)); \quad (Ux)h(x)\rightarrow(Uy\ni E(y))h(y\ni f(y));$$
$$S(Ux\ni g(x))f(x\ni g(x))\simeq(Ux\ni g(x))Sf(x\ni g(x)).$$

Instances of the last theorem make it plain that the quantifiers are not transparent; it matters how the items talked about are described. More disputable are such theorems as:

$$Tf(x\ni(Tg(x)\ \&\ \neg g(x)))\vee\neg f(x\ni(Tg(x)\ \&\ \neg g(x))).$$

So far restricted variables only work like further, notationally complex, variables; and the semantics given for $\mathbf{QS_6E}\approx$ can be adapted. Given the intended interpretation of restricted variables it is clear, however, that the axiomatisation is incomplete. It should be a theorem, for example, that $Sf(x\ni Sf(x))$; but it is not, since $Sf(y)$ is not a theorem (and should not be a theorem; for consider such predicates as 'is a happy prime number'). $Sf(x\ni Sf(x))$ is but a special case of the unrestricted Assumption Postulate $B(x\ni B(x))$, which tempted Meinong. The abstraction axiom introduced below is also a special case of this principle. The unrestricted Assumption Postulate leads, however, to inconsistency; most simply it entails $(\mathcal{q}x)B(x)$ for *any* B.

We postpone the tasks of developing an adequate theory of *such that* and providing semantics for the rather incomplete R-(restricted variable) systems that we introduce. For the present volume it seems enough to assume the following unadventurous postulates:

R1. $E(x\ni B(x))\rightharpoondown B(x\ni B(x))$, i.e. entities are assumptible
R2. $SB(x\ni SB(x))$
R3. $(Ux\ni B(x))C(x\ni B(x))\rightharpoondown . B(y)\rightharpoondown C(y)$.

But the converse principle $(Uy)(B(y)\rightharpoondown C(y))\rightharpoondown .(Ux\ni B(x))C(x\ni B(x))$, is not assumed, and appears to be wrong (and wrong just where the Assumption Postulate fails). For the antecedent may hold vacuously, but the consequent entails $(\mathcal{q}x\ni B(x))C(x\ni B(x))$. All this suggests—what turns out to be the case in the two-valued subtheory—that the quantificational theory of restricted variables can reproduce and interpret in a straightforward way the full traditional theory of the syllogism. That traditional term logic emerges intact depends in part on the fact that, in contrast to the standard analysis, one and the same connective '∋' is used in restricting both universal and particular assertions.

Method 2. We consider, what is less satisfactory, the standard definitional elimination of restricted variables. Because only matrix-accessible connectives occur in $\mathbf{QS_6E}\approx$, only an extensional-style elimination is possible. In contrast, one of the advantages of introducing '∋' as a primitive is that '∋' is not so matrix-accessible; and consequently does not suffer the drawbacks of an extensional-style elimination—for example, that seemingly correct Aristotelian principles break down. (See §3.2).

There are various ways we can define '$(Uu\ni B(u))C(u\ni B(u))$' depending on which connective we use to eliminate '∋', and on whether we add weighting requirements, *e.g.* whether we require also $(Pu)TB(u)$, say. In view of what we have already said about expressing provisos in significance logic, one obvious way of eliminating restricted universal quantifiers is by using the connective '⇁'. Similarly restricted particular quantifiers can be eliminated extensionally using '↿' (see §5.13 ff). To avoid confusion we write the extensionally eliminable 'such that' '/', instead of '∋'. Whereas '/' is eliminable, '∋' is not. The basic definitions are as follows:—

$(Uu/B)C =_{\mathrm{Df}} (Uu)(B\rightharpoondown C)$; $(\mathcal{q}u/B)C =_{\mathrm{Df}} (\mathcal{q}u)(B\upharpoonleft C)$;
$(Pu/B)C =_{\mathrm{Df}} \sim(Uu/B)\sim C$.

Hence, since $A\rightharpoondown B\simeq D(A\upharpoonleft DB)$, $(\mathcal{q}u/B)C\simeq D(Uu/B)DC$. Thus, just as with unrestricted variables, where '$\mathcal{q}$' can be definitionally eliminated (in 6-logics) using the equivalence $(\mathcal{q}u)C\simeq D(Uu)DC$, so with restricted variables '$\mathcal{q}$' can be similarly eliminated. In fact the principles of quantified significance logic remain valid for variables restricted to any non-null domain d.

Where $u_1, \ldots, u_n$ are all the free variables of A and F is a one-place predicate, the F-*restriction* A^F of A is obtained by (i) replacing each wf part $(Uv)B$ of A by $(Uv)(F(v) \to B)$ and each wf part $(qv)B$ of A by $(qv)(F(v) \mid B)$ [a change of bound variables may also be required] and (ii) restricting the result A_F of (i) by $F(u_1), \ldots, F(u_n)$, *i.e.* A^F is $F(u_1) \to . F(u_2) \ldots \to . F(u_n) \to A_F$.

Theorem (72): (1) Where F is a one-place predicate, if $\vdash A$ then $\vdash A^F$; provided that if any term t occurs in A, $\vdash F(t)$.

(2) Where F is a one-place predicate such that $(qv)F(v)$ is valid, then if A is valid so is A^F.

In other words if the domain d is restricted to the non-null class of t such that $F(t)$ then the principles of quantified significance logic remain valid.

Proof of (2) follows from (1) using the completeness of QS_6.

Proof of (1) is by induction over the proof of A. Since 'q' can be definitionally eliminated it suffices to consider postulates not involving 'q'.

ad A1. If A is a substitution instance of an S_6-tautology, so is A_F, whence $\vdash A^F$ follows using A1 (by applying $p \to . q \to p$ repeatedly to restrict free variables of A).

ad A2. Apply A1 to the theorem:

$$(Uu)(F(u) \to . A_F \to B_F) \to . A_F \to (Uu)(F(u) \to B_F), \text{ with } u \text{ not free in } A_F.$$

ad A4. Case 1, where a term t instantiates u.

Since $(Uu)(F(u) \to A_F) \to . \check{S}^u_t(F(u) \to A_F) \mid$

$$\to . F(t) \to \check{S}^u_t A_F \mid,$$

$\vdash F(t) \to . (Uu)(F(u) \to A_F) \to \check{S}^u_t A_F \mid$. Detach $F(t)$ and apply A1.

Case 2, where a variable v instantiates u. Apply A1 to

$$F(v) \to . (Uu)(F(u) \to A_F) \to \check{S}^u_v A_F \mid .$$

ad A6. Since $\vdash p \to Sq \to . S(p \to q)$,

$$(Uu)(F(u) \to SA_F) \to (Uu)S(F(u) \to A_F)$$
$$\to S(Uu)(F(u) \to A_F).$$

Then apply A1.

ad A7. Similarly, using $\vdash S(p \to q) \to . p \to Sq$.

ad R1. Use A1 and R1.

ad R2. Use A1 and R2.

Where the restricting predicate is 'E' or '$E!$' the theorem provides conditions under which the logic of existence-restricted quantifiers is valid, and this points the way to an axiomatisation of free quantified significance logic. But the theorem provides more generally for the application of quantified significance logic to any special subject matter: not just to entities, but to numbers, properties, farms, and so forth.

An important application, for significance theory, is in the matter of significance restricted quantifiers and variables. Once again use of connectives like '$\rightharpoonup$' and '$\urcorner$' is quite essential for an extensional analysis; use of classical connectives would be disastrous (see §5.11). The basic extensional definitions are there:

$(USu)C =_{\mathrm{Df}} (Uu/SC)C$; $(\mathcal{9}Su)C =_{\mathrm{Df}} (\mathcal{9}u/SC)C$;
$(PSu)C =_{\mathrm{Df}} (Pu/SC)C$.

'$(USu)C$' can be read 'for every u for which C is significant, C(is true)'. It follows, $\vdash S(USu)C$, $\vdash S(PSu)C$, $\vdash (USu)(B \supset C) \supset .(USu)(\sim C \supset \sim B)$, $\vdash (USu)C \supset (PSu)C$, $\vdash (USu)C(u) \supset C(u)$, provided $\vdash SC(u)$.
Although $(USu)(B \supset C) \supset .(USu)B \supset (USu)C$ is not a theorem, we have $\vdash (USu)(B \supset C) \supset .\ (Uu/S(B \,\&\, C))B \supset (Uu/S(B \,\&\, C))C$.

Using S-restricted variables, conditions under which Behmann formulae (defined in §7.5) hold can be worked out. Though
$(USu)(f(u) \supset g(u)) \,\&\, (USu)(g(u) \supset h(u)) \supset .\ (USu)(f(u) \supset h(u))$ —(α)
is invalid, the wff holds for example if the s-range of 'f' is included in that of 'g' or the s-range of 'h' is included in that of 'g'. (So too the theory vindicates Halldén's counterexample to Behmann formula (1) discussed in §7.5). More generally, $\vdash (Uu)(Sf(u) \supset Sg(u) \vee .Sh(u) \supset Sg(u)) \supset .(\alpha)$. Similarly, though Behmann formula $(USu)f(u) \,\&\, (USu)(f(u) \supset g(u)) \supset .(USu)g(u)$ is not valid, it is a theorem under the condition that $(Uu)(Sg(u) \supset Sf(u))$, that is provided the S-range of 'g' is included in or coincides with that of 'f'. Again though $(USu)(f(u) \,\&\, g(u)) \supset .(USu)f(u) \,\&\, (USu)g(u)$ is not valid, it is valid under the condition for example that $(Uu)(Sf(u) \equiv Sg(u))$, and thus holds for an exclusive type theory.

In so assessing Behmann wff we have taken it for granted that the quantifiers of these wff are S-restricted. That quantifiers are taken as so S-restricted by previous writers on significance theory, such as Russell, Behmann and Halldén, is clear from the interpretation they regularly provide for their quantifiers. Thus S-restricted quantifiers give us a *formal* way of bringing out interpretational features of earlier significance theories.

7.10 Quantified Multiply Modal Significance Logics

These logics, which combine quantification and modal operators on a significance basis, simply amalgamate multiply modal system $\mathbf{IS_6}$ with the various quantified significance logics we have studied—apart from one controversial detail, the Barcan formulae. A combination of the syntax of $\mathbf{IS_6}$ with that of $\mathbf{QS_6}$ results in a system without such Barcan wff as

B. $(Uu)\Phi_i B \rightharpoonup \Phi_i (Uu)B$;

but the expected combination of the semantics of $\mathbf{IS_6}$ and $\mathbf{QS_6}$ validates B, for each L- and F-connective. The semantics seems a better guide to correctness than the syntax. We shall take the simpler course of adding Barcan wff B, for each *i*, as postulates of the combined system **QIS**$_6$, since we think that the modal connectives that satisfy the postulates of **IS**$_6$ mostly satisfy B. Certainly such central intensional connectives as logical necessity, probability, rational belief and verifiability, do. The main objections to Barcan wff, like the analogous objections to the scheme $(\mathcal{q}u)SB \rightharpoondown S(\mathcal{q}u)B$, have in fact depended on an existential construal of quantifiers. Those objections break down where non-ontological quantifiers like '*U*' and '$\mathcal{q}$' are used. The existential construal also threatens the converse Barcan wff

CB. $\Phi_i(Uu)B \rightharpoondown (Uu)\Phi_i B$,

principles which are provable for F- and L-connectives just from theorems of **IS**$_6$ and **QS**$_6$. If Φ_i is an L-connective then $\Phi_i(Uu)B \rightharpoondown \Phi_i B$ follows by RB from the instantiation scheme; then CB follows by generalisation and distribution of '*U*'. Thus the existential construal of quantification forces a modification in quantification logic—in the direction of free logic (see Kripke [2]). We take the position that both B and CB fail for existentially-loaded quantifiers, but that the principles hold, at least for central necessity-type connectives, with our non-ontological quantifiers. Existentially-loaded quantifiers may however be defined in **QIS**$_6$**E**; with these quantifiers both B and CB break down, and the free logic theory emerges as a sublogic of **QIS**$_6$**E**. It is possible, however, to provide a formal semantical analysis of a system which simply amalgamates **QS**$_6$ and **IS**$_6$ systematically, and accordingly has CB but not B as theorems. But in the case of classical quantified modal logic the semantics is got in effect by reinterpreting classical quantification logic as a quantified **C**-logic with basis **C**$_0$; thus the semantics is open to the objections we have already made to **C**-logics.

We form system **QIS**$_6$ by combining the morphology of **QS**$_6$ (with '$\mathcal{q}$' defined) and **IS**$_6$; and we add to the postulates of **QS**$_6$, *viz.* A1, A2, A4, A6, A7 and R2, the modal postulates of **IS**$_6$ (reformulated as schemata in the case of D, J1, J2 and supplementary postulates) except that RT is replaced by the scheme:

RTQ. $\vdash A \twoheadrightarrow \vdash \Phi_i A$, where A is a theorem of **QS**$_6$ or an instance of scheme D or B, for each i such that $j \leqslant i \leqslant m$. Finally we add Barcan wff B for each $1 \leqslant i \leqslant g$, *i.e.* for each L-connective.

By combining the contractional semantics for **QS**$_6$ with the set semantics for **IS**$_6$, we can provide a contractional semantics for **QIS**$_6$. A **QIS**$_6$-model $\mathcal{M}$ is a structure $\mathcal{M} = \langle G, K, N_1, ..., N_g, N_j, ..., N_m, R_1, ..., R_g, R_j, ..., R_m, d, e, f, I\rangle$, where $\langle G, ..., R_m, I\rangle$ is an **IS**$_6$-model (with assignment I replacing valuation function *v*), and where d is a function which associates with each H in K a non-null set d(H) of contractions; e(H) is a class of relations over d(H), and f(H) is a class of operations over d(H). The class of all contractions d $=_{\text{Df}} \bigcup_{H \epsilon K}$ d(H); e is a class of relations over d (which includes $\bigcup$e(H)) and f is a class of operations on d. As before a model is *usual* iff e is the class of all operations on d. I makes the same assignments to atomic wff of $\mathbf{S_6}$ and to wff of the form $\Phi_i A$ with $g<i<j$ (*initial* wff of $\mathbf{IS_6}$) as valuation function v of $\mathbf{IS_6}$-models, *i.e.* for such a wff B, I(B, H) is one of values t, f, n. For subject constant X_i, I(X_i, H) is an element of d; I(F_i^n, H) is an n-place relation of e; and I(E_i^n, H) is an n-place operation of f. The *invariance* requirement that I(X_i, H) is independent of H, *i.e.* that for H_1, H_2 in K, $I(X_i, H_1) = I(X_i, H_2) = I(X_i)$, makes d(H) otiose: this, together with the quantification rule, chiefly accounts for the validity of B and CB. Assignments for wft are defined inductively thus:

$$I(E(t_1, ..., t_n), H) = I(E, H)\,[I(t_1, H), ..., I(t_n, H)].$$

Given a model, a two-place valuation function v is defined inductively thus:
For initial wff of $\mathbf{IS_6}$, $v(P, H) = I(P, H)$.
Where $t_1, ..., t_n$ are terms,
$v(F^n(t_1, ..., t_n), H) = t$ iff $T[(I(t_1, H), ..., I(t_n, H))\, i I(F^n, H)]$ *etc.*

The recursion rules for sentential connectives are the same as those given in $\mathbf{IS_6}$ semantics. Finally, given assignments from d to the free variables of $(Uu)A$,
$v((Uu)A, H) = t$ iff $v(A, H) = t$
for every assignment of an element of d to u, and
$v((Uu)A, H) = n$ iff $v(A, H) = n$
for some assignment of an element of d to u.

An alternative quantification rule—$v((Uu)A, H) = t$ iff $v(A, H) = t$ for every assignment of an element of d(H) to u—falsifies, in the style of free logic, both B and A3. (See Kripke [1]). Typically, invariance is also dropped. Truth, validity, satisfaction, *etc.*, are defined as for corresponding $\mathbf{IS_6}$ notions, *e.g.* B is true in $\mathbf{QIS_6}$-model $\mathcal{M}$ iff $v(B, G) = t$.

Theorem (73a): Every theorem of $\mathbf{QIS_6}$ is $\mathbf{QIS_6}$-valid.

A short completeness theorem can once again be obtained using a maximal set construction; more constructive but longer proofs may be obtained, for example using a triple semantic tableaux method where each tableau has three columns, roughly a truth column, a falsity column and a nonsignificance column. The maximal set construction we give employs one new strategy beyond the proof given for $\mathbf{IS_6}$, *viz.* each maximal set used in the canonical model for $\mathbf{IS_6}$ is completed in the way that the con set of $\mathbf{QS_6}$ is completed by adding instances of $A_k(b_{jk}) \rightarrow (Ux_{ik})A_k(x_{ik})$. To ensure that the completion can be effected we need the following details, which are like those in the completeness lemma proved for $\mathbf{QS_6}$.

Where ∇ is a con set of $\mathbf{QIS_6}$, let K be the system got by adding a denumerable set $\{b_1, ..., b_j, ...\}$ of new subject constants to $\mathbf{QIS_6}$, and let ∇B_0 be the wff provable in K from ∇, *i.e.* $\nabla B_0 = \{C : \nabla \vdash_K C\}$: B is the set of all subject constants of K. Let the denumerable set of wff of ∇B_0 with at most one subject variable free be represented: $A_1(x_{i1})$, ..., $A_n(x_{i_n})$, ... Choose a sequence b_{j_1}, ..., b_{j_k}, ... of new subject constants of B such that b_{j_k} is distinct from each of b_{j_1}, ..., $b_{j_{k-1}}$. Let ∇B_n be the system obtained by adding to ∇B_{n-1} the wff

$$S_n : A_n(b_{j_n}) \rightarrow (Ux_{i_n})A_n(x_{i_n});$$

and let ∇B be the union of these systems, $\bigcup_{j<\omega} \nabla B_j$. ∇B is the *ω-completion* or the *ω-complete extension* of ∇ wrt. B.

Lemma. ∇B is a con set.

Proof is as in the completeness proof for $\mathbf{QS_6}$.

Lemma. Every con set ∇ of $\mathbf{QIS_6}$ has wrt a denumerable set B a maximal ¬-consistent ω-complete extension (a *Macω* set) $\overline{\overline{\nabla}} = (\nabla B)^+$.

Theorem (73b): If Γ is a con set of $\mathbf{QIS_6}$ then there is a $\mathbf{QIS_6}$-model $\mathcal{M}$ with denumerable domain d which satisfies Γ.

Proof: Define a canonical $\mathbf{QIS_6}$-model $\mathcal{M}$ as follows: G is the macω extension of Γ wrt some set of constants d(G). K is the set of macω sets of filter system $\mathbf{QIS_6}^*$ wrt d(G); for H in K, d(H) is the denumerable set of constants of H. Each N_i and R_i is defined as for a canonical $\mathbf{IS_6}$ model. e(H) and f(H) are determined by taking the usual model. Lastly assignment I is defined by relativising the assignment defined for the canonical model for $\mathbf{QS_6}$ to H. Thus, where X_i is a subject constant $I(X_i, H) = X_i$; and where E_i^n is a functional constant then, for closed terms t_1, ..., t_n, $I(E_i^n, H)(t_1, ..., t_n) = E^n(t_1, ..., t_n)$. It follows that $I(E^n(t_1, ..., t_n), H) = E^n(t_1, ..., t_n)$; and hence that for every closed term $t_1, I(t_1, H) = t_1$. Thus we may rewrite $(I(t_1), ..., I(t_n)) i I(F^n)$ as $I(F^n)(t_1, ..., t_n)$ for closed terms. Where, further, F^n is a predicate constant, I is defined,

$T[I(F^n, H)(t_1, ..., t_n)] \equiv T(F^n(t_1, ..., t_n)) \in H$

$TD[I(F^n, H)(t_1, ..., t_n)] \equiv TD(F^n(t_1, ..., t_n) \in H$

For initial wff P of $\mathbf{IS_6}$, $I(P, H) = t, f$ or n according as TP, FP, or $TDP \in H$.

Then the canonical model $\mathcal{M}$ is a $\mathbf{QIS_6}$-model.

It remains to show, by induction, that for every closed wff A and every H, $v(A, H) = t$ iff $TA \in H$, $v(A, H) = f$ iff $FA \in H$ and $v(A, H) = n$ iff $TDA \in H$. The induction basis and the induction step for quantifier '*U*' are like the analogous steps for $\mathbf{QS_6}$, and the induction steps for sentential connectives are the same as the steps for $\mathbf{IS_6}$. Since d(G) is denumerable, so is d.

Theorem (73c):

(1) Every simultaneously satisfiable set of wff of $\mathbf{QIS_6}$ is satisfiable in a denumerable domain.

(2) Every $\mathbf{QIS_6}$-valid wff is a theorem of $\mathbf{QIS_6}$.

Further principles formalisable within $\mathbf{QIS_6}$ such as von Wright's principle of predication and Pollock's principle $\neg\Box((\mathit{q}x)f(x)\ \&\ (\mathit{q}x)\sim f(x))$ (essentially that proposed in Pollock [2]), have been advanced as valid. In our view, however, these particular principles are certainly false. But once existence and identity are added to $\mathbf{QIS_6}$, questions as to which logical principles are correct become more vexed. We shall, in developing $\mathbf{QIS_6E}$, $\mathbf{QIS_6E!}$ and $\mathbf{QIS_6E!=}$ in the way we do, make plain, without much argument, the accounts of existence and identity we believe correct.

The logic $\mathbf{QIS_6E}$ adds the primitive 'E' to $\mathbf{QIS_6}$. To axiomatise existence satisfactorily in a modal context we need to single out a logical necessity connective, say Φ_m; we write $\Box A$ in place of $\Phi_m A$. Then we assert

E1. $\vdash (Px)E(x)$

E2. $\vdash (Px)\Box\sim E(x)$

E3. $\vdash\sim\Box(Px)E(x)$.

E2 is true because such items as Meinong's round square necessarily do not exist. E1 and E3 taken together assert that it is contingently true that something exists, but E3 goes on to assert, what is more controversial, that nothing exists of logical necessity. It follows:

$\vdash SE(x)$. Indeed $\vdash(Ux)\mathscr{S}E(x)$, where $\mathscr{S}A =_{\mathrm{Df}} \Box SA$.

For $\Box(Px)\sim E(x) \rightharpoondown \Box S(Px)\sim E(x)$ $\qquad$ $\mathrm{S_6}$, RD

$\rightharpoondown \Box S(Ux)\sim E(x)$

$\rightharpoondown \Box(Ux)SE(x)$

$\rightharpoondown (Ux)\Box SE(x)$

$\vdash\Box(Px)\sim E(x)$; $\qquad \vdash(Ux)\sim\Box E(x)$

$\vdash(Px)(E(x)\ \&\sim\Box E(x))$. Thus since, $\vdash\Box(Ux)(E(x)\supset E(x))$,

$\vdash\sim(\Box(\forall x)E(x)\rightharpoondown(\forall x)\Box E(x))$; the converse Barcan wff is demonstrably false where existentially loaded quantifiers are used. Rule RTQ is *not* extended to apply to E1–E3; if it were, inconsistency would result from $\Box$E1 and E3.

A $\mathbf{QIS_6E}$-model differs from a $\mathbf{QIS_6}$-model only in replacing function d by a pair of functions $\mathrm{d_1}$ and $\mathrm{d_2}$. Given $\mathrm{d_1}$ and $\mathrm{d_2}$, $\mathrm{d(H)} =_{\mathrm{Df}} \mathrm{d_1(H)}\cup\mathrm{d_2(H)}$. For $\mathrm{d_1}$ and $\mathrm{d_2}$ we require: $\mathrm{d_1(G)}$ is non-null, but for some H such that $\mathrm{GR}_m\mathrm{H}$, $\mathrm{d_1(H)} = \Lambda$; and for every H such that $\mathrm{GR}_m\mathrm{H}$, $\mathrm{d_2(H)}$ is non-null. Since $\mathrm{GR}_m\mathrm{G}$, $\mathrm{d_2(G)}$ like $\mathrm{d_1(G)}$ is non-null. Assignment function I assigns thus:

$T(\mathrm{I}(t,\mathrm{H})\ i\mathrm{I}(E,\mathrm{H})) \equiv . \mathrm{I}(t,\mathrm{H})\in\mathrm{d_1(H)}$;

$F(\mathrm{I}(t,\mathrm{H})\ i\mathrm{I}(E,\mathrm{H})) \equiv . \mathrm{I}(t,\mathrm{H})\in\mathrm{d_2(H)}$.

Theorem (74a):

(1) If Γ is a con set of $\mathbf{QIS_6E}$ then there is a $\mathbf{QIS_6E}$-model $\mathscr{M}$ with a denumerable domain d which satisfies Γ.

(2) Every simultaneously satisfiable set of wff of $\mathbf{QIS_6E}$ is satisfiable in a denumerable domain.

(3) B is a theorem of $\mathbf{QIS_6E}$ iff B is $\mathbf{QIS_6E}$-valid.

Proof: The new details for (1) are as follows:—In terms of the canonical model $\mathscr{M}$,

$d_1(H) =_{Df} \{t \in d(H): E(t) \in H\}$ and

$d_2(H) =_{Df} \{t \in d(H): \neg E(t) \in H\}$. Hence

$d(H) = d_1(H) \cup d_2(H)$ as required. Also,

$T(I(E,H)(t_1)) \equiv TE(t_1) \in H$

$\equiv TE(t_1) \in H \;\&\; t_1 \in d(H)$, since $TE(t_1) \in H \supset . t_1 \in d(H)$

$\equiv t_1 \in d_1(H)$

$F(I(E,H)(t_1)) \equiv FE(t_1) \in H \equiv FE(t_1) \in H \;\&\; t_1 \in d(H)$

$\equiv t_1 \in d_2(H)$.

Since $\vdash (Px)E(x)$, as for $\mathbf{QS_6E}$, $E(b_1) \in G$ for some $b_1 \in d(G)$; hence $d_1(G)$ is non-null. Because of E2, $\Box \sim E(b) \in G$ for some b. If $GR_m H$ then $\Box \sim E(b) \in G \supset . \sim E(b) \in H$, from which it follows that $d_2(H)$ is non-null. Because of E3, $\Box (Px)E(x) \notin G$; hence for some H, $GR_m H$ and $(Px)E(x) \notin H$. But $(Px)E(x) \notin H$ implies $d_1(H) = \Delta$.

$\mathbf{QIS_6E}$ lumps together, what it is important to distinguish, inconsistent subjects such as 'The husband of Socrates', 'Meinong's round square' and 'the successfully married bachelor' and nonsignificant subjects such as 'the colour of Time', 'a colourless green idea' and 'the wheels of happiness'. To allow for these differences we introduce system $\mathbf{QIS_6E!}$, with these axioms on $E!$:

E!1. $\vdash (\text{\textquestiondown}x)E!x$

E!2. $\vdash (\text{\textquestiondown}x)\Box \sim E!x$

E!3. $\vdash \sim \Box (\text{\textquestiondown}x)E!x$

E!4. $\vdash (\text{\textquestiondown}x) \sim SE!x$

$S(u) =_{Df} SE!u$; $\mathscr{S}(u) =_{Df} \mathscr{S}E!u$, *i.e.* $\Box SE!u$; $E(u) =_{Df} TE!u$.

Using the last definition it follows that every theorem of $\mathbf{QIS_6E}$ is a theorem of $\mathbf{QIS_6E!}$ It also follows that the nonsignificance of some subjects is not just a contingent matter: for

$\vdash (\text{\textquestiondown}x) \sim \mathscr{S}E!x$, $\vdash (\text{\textquestiondown}x)\Box \sim \mathscr{S}E!x$ and $\vdash \Box (\text{\textquestiondown}x) \sim \mathscr{S}E!x$.

A $\mathbf{QIS_6E!}$-model replaces the function d of $\mathbf{QIS_6}$-models by three functions d_1, d_2 and d_3. $d(H) = d_1(H) \cup d_2(H) \cup d_3(H)$. d_1 and d_2 satisfy the same conditions as for a $\mathbf{QIS_6E}$-model; for d_3 we require that $d_3(G)$ is non-null. I assigns so that $\sim S(I(t_1,H)\, i I(E!,H)) \equiv . I(t_1,H) \in d_3(H)$.

Theorem (74b): (1), (2) and (3) as stated for **QIS$_6$E** holds for **QIS$_6$E**!

In canonical model,

$d_3(H) =_{Df} \{t \in d(H) : \sim SE(t) \in H\}$. That $d_3(G)$ is non-null follows from E!4.

Although we can, in **QIS$_6$E**!, exclude nonsignificant subjects from various logical roles by using *S*-restricted subject variables, we should not exclude them outright from the theory. For we should, if we did, fail to reflect in the theory the evident distinction we find between sentences such as 'Saturday is in bed' and 'the number seven is pregnant' on the one hand, and 'the man taller than 10 o'clock lives next door' and 'the purple number seven is prime'. A distinguishing feature of proper names is that they are always significant.

The identity relation '$=$' of **QIS$_6$E**! $=$ satisfies the postulates I1. $x = x$ and I2. $u = v \rightarrow . A(u) \simeq A(v)$, where u and v are subject terms and $A(v)$ is an atomic wff which results from atomic wff $A(u)$ by replacing one occurrence of u by v, where this occurrence of u is not within the scope of quantifiers binding u or v.

Rule RTQ is not extended to apply to I1 and to instances of I2. Though identity is, much as before, an equivalence relation in terms, it no longer guarantees full intersubstitutivity of identicals. For example $x = y \rightarrow . \Phi_i(x = x) \rightarrow \Phi_i(x = y)$ is not provable. Stronger identity relations may be defined in terms of '$=$'; for instance strict identity, defined:

$$u \equiv v =_{Df} \Box(u = v).$$

Were '$\Box$' the sole modal connective of **QIS$_6$E**! $=$ then full intersubstitutivity would follow, by induction, for strict identities. There is however no guarantee that strict identities are intersubstitutable in other intensional sentence contexts. It is not possible to define Leibniz identity $u \approx v$ as a conjunction of the form $\Phi_1(u = v)$ & ... $\Phi_m(u = v)$, for there is a counter model that refutes

$\Phi_i(x = y) \rightarrow . \Phi_i f(x) \rightarrow \Phi_i f(y)$, for $g < i < j$.

Leibniz identity can of course be introduced as a further primitive using postulate $x \approx x$ and scheme I2$^+$; but the semantics previously given breaks down for connectives like Φ_{g+1}. To treat Leibniz identity modally one simple course is to set $j = g + 1$ and to define $u \approx v$ as a conjunction: to prove the full scheme I2$^+$ further postulates $\Phi_i(p \simeq q) \rightarrow . S\Phi_i p \rightarrow S\Phi_i q$ are needed for certain i, namely those for which $GR_i G$ does not hold. An alternative course which avoids these additional axioms is simply to set $j = g + 1$ and to adjoin '$\approx$' as a further primitive satisfying I1 and I2$^+$. Then the semantics previously given, relative to H, can be taken over: $(I(t_1, H), I(t_2, H)) i I(\approx, H) \simeq . I(t_1, H) \approx I(t_2, H)$, *i.e.* there items are Leibniz identical.

In the system **QIS$_6$E**! $= \approx$, $t_1 \approx t_2 \rightarrow t_1 = t_2$ of course, but not conversely.

A $\mathbf{QIS_6E}$!=-model adds to a $\mathbf{QIS_6E}$!-model the following assignment for identity:

$(I(t_1, H), I(t_2, H)) i I(=, H) \simeq . I(t_1, H) =_H I(t_2, H)$,

where H-identity ($=_H$) is coincidence with respect to all properties which can be evaluated purely in H, *i.e.* without proceeding off H, via a relation R_i say, to another set-up H_j for the evaluation of properties. H-identity is a generalisation of extensional identity, which, as often understood, is G-identity. The valuation rule for identity follows: *viz.*
$v(t_1 = t_2, H) = t, f$ or n according as $T\Theta$, $F\Theta$ or $TD\Theta$ where 'Θ' abbreviates '$I(t_1, H) =_H I(t_2, H)$'. A countermodel to the wff $\Gamma: t_1 = t_2 \rightharpoonup . \Box f(t_1) \rightharpoonup \Box f(t_2)$, is worth examining. If $v(\Gamma, G) \neq t$ then $v(t_1 = t_2, G) = t$, $v(\Box f(t_1), G) = t$ and $v(\Box f(t_2), G) \neq t$.
Thus for some H, GRH and $\neg T(I(t_2, H)\, i I(f, H))$ and $T(I(t_1, H)\, i I(f, H))$, and also $I(t_1, G) =_G I(t_2, G)$. But a G-identity does not ensure an H-identity; and even though in the canonical model $I(t_1, G) = t_1$, $t_1 =_G t_2$ is compatible with $t_1 \neq_H t_2$ for $H \neq G$. Though Scott $=_G$ the author of Waverley, for certain set-ups H selected by the necessity connective, Scott $\neq_H$ the author of Waverley.

Theorem (75):

(1) If Γ is a con set of $\mathbf{QIS_6E}$!= then there is a $\mathbf{QIS_6E}$!=-model $\mathcal{M}$ with a countable domain which satisfies Γ

(2) Every simultaneously satisfiable set of wff of $\mathbf{QIS_6E}$!= is satisfiable in a countable domain.

(3) Every $\mathbf{QIS_6E}$!=-valid wff is a theorem of $\mathbf{QIS_6E}$!=, and conversely.

Proof: The only new points are in the proof of (1), which relativizes to H the steps in the proof of the corresponding theorem for $\mathbf{QS_6E}$!=. As before, $\Delta = \Gamma \cup I1' \cup I2'$ is simultaneously satisfiable, say by a canonical model $\mathcal{M}$ for Δ with interpretation I_1, domain d_1 *etc.* The relation $E(H) = I_1(=, H)$ is an equivalence relation on H, since $t = t, t_1 = t_2 \rightharpoonup . t_2 = t_1$ and $t_1 = t_2 \rightharpoonup . t_2 = t_3 \rightharpoonup . t_1 = t_3$ belong to H, for each H. Using E(H) we may form equivalence classes of pairs of the form (t_1, H):

$[t_1, H] =_{Df} \{(t_2, H) : (t_1, H)\, E(H)\, (t_2, H)\}$, where
$(t_1, H)\, E(H)\, (t_2, H)$ iff $(I_1(t_1, H), I_1(t_2, H))\, i I_1(=, H)$.

Then $[d(H)] =_{Df} \{[t, H] : t \in d(H)\}$; $[d] =_{Df} \bigcup_{H \in K} [d(H)]$;

$[d_1(H)] =_{Df} \{[t, H] : t \in d_1(H)\}$; $[d_2(H)] =_{Df} \{[t, H] : t \in d_2(H)\}$;
$[d_3(H)] =_{Df} \{[t, H] : t \in d_3(H)\}$.

Domain [d] is countable since d is denumerable.A new interpretation I_2 over model structure $\mathcal{M}$ is specified as follows:

$I_2(X, H) = [X, H]$ for subject constant X;
$I_2(E^n, H)([t_1, H], \ldots, [t_n, H]) = [I_1(E^n, H)(t_1, \ldots, t_n), H]$.

Hence for closed term t, $I_2(t, H) = [t, H]$. For closed terms $t_1, ..., t_n$, $([t_1, H], ..., [t_n, H])\, i I_2(F^n, H) \simeq I_1(F^n, H)(t_1, ..., t_n)$; and $I_2(P, H) = I_1(P, H)$ for initial wff of $\mathbf{IS}_6$. Then

$$v_2(F^n(t_1, ..., t_n), H) = t \simeq (I_2(t_1, H), ..., I_2(t_n, H))\, i I_2(F^n, H)$$
$$\simeq I_1(F^n, H)(t_1, ..., t_n), \text{ etc.; whence}$$
$$v_2(F^n(t_1, ..., t_n), H) = v_1(F^n(t_1, ..., t_n), H)$$
$$v_2(P, H) = v_1(P, H), \text{ for initial } P.$$
$$v_2(E!t_1, H) = t \equiv T(I_2(t_1, H)\, i I_2(E!, H))$$
$$\equiv [t_1, H] \in [d_1(H)]$$
$$\equiv t_1 \in d_1(H)$$
$$\equiv v_1(E!t_1, H) = t.$$

Similarly,

$$v_2(E!t_1, H) = f \equiv . v_1(E!t_1, H) = f;\ v_2(E!t_1, H) = n \equiv . v_1(E!t_1, H) = n.$$
$$v_2(t_1 = t_2, H) = t \quad \equiv T(I_2(t_1, H), I_2(t_2, H))\, i I_2(=, H)$$
$$\equiv T([t_1, H], [t_2, H])\, i I_2(=, H)$$
$$\equiv T I_1(=, H)(t_1, t_2)$$
$$\equiv T(t_1 = t_2) \in H$$

Similarly,

$$v_2(t_1 = t_2, H) = n \quad \equiv TD(t_1 = t_2) \in H.$$

Quantified modal logics are often criticised as involving, in one way or another, essentialism. But the main problems of modal essentialism result not from the quantification but, like those of significance essentialism (§3.8), from referential insistence that Leibnitz identity is referential identity and on taking designation-values of constants and designation-ranges of variables as primary. These problems vanish when Leibnitz identity and extensional identity are properly distinguished and the value-interpretation of variables is taken, where variables are kept track of through their constants-range and where constants do not vary as between situations, though they may be assigned different "designations" with respect to different situations. Thus, too, no insurmountable problems concerning transworld identifications arise; in fact, with respect to a given model an inter-situation identity relation can be determined thus: if for every extensional predicate $qu(f)$,

$$v(f(x_1), H_1) = v(f(x_2), H_2)$$

then $(x_1, H_1) = (x_2, H_2)$ is an extensional inter-situation identity.

7.11 Further Enlarged Quantified Significance Logics

§1 *Logics based on an any-operator*

A more comprehensive logic results if, instead of introducing quantifiers as primitive, the logic is based on the *any*-operator 'ξ' and quantifiers are defined. The symbol 'ξ', a many-valued version of Hilbert's operator, is subject-forming on wff. Like our basic quantifiers, 'ξ' must be construed non-existentially, or interpretational difficulties like the Barcan analogues reappear. Thus '$\xi xf(x)$' reads, 'an arbitrarily selected x which is f' or in most contexts, 'any x which is f': it does not require that there exists an item which is f. The operator 'ξ' is not definable in $\mathbf{QS}_6$ (it provides an alternative to such adjuncts to quantification logic as Skolem functions). Thus we add 'ξ' to the improper symbols of $\mathbf{S}_6$ as a new primitive. We consider systems $\xi\mathbf{S}_6$, $\xi\mathbf{S}_6\mathbf{E}$, $\xi\mathbf{S}_6\approx$ and $\xi\mathbf{S}_6\mathbf{E}\approx$, each system being obtained from the corresponding $\mathbf{QS}_6$ by replacing quantifiers 'U' and 'q' by 'ξ'. In place of the formation rules for quantifiers we adopt the rule:

(c) (3). If A is a wff and u a subject variable then ξuA is a wf term.

In order to state the postulates of $\xi\mathbf{S}_6$ we introduce the notation $B_x(\xi yC)$ where

$$B_x(\xi yC) =_{\mathrm{Df}} \check{\mathrm{S}}^{x}_{\xi y\, \mathrm{S}^{1x}_{y}C} B \,|$$

$\check{\mathrm{S}}^{x}_{\xi y \mathrm{S}^{1z}_{y}C} A\,|$ is called a *ξ-variant* of $\check{\mathrm{S}}^{x}_{\xi zC} A\,|$.

We extend the notion of *substitution instance* to admit ξ-variants: thus, *e.g.*

$$f_x(\xi yg(y)) \rightarrow f_x(\xi zg(z))$$

is a substitution instance, in the extended sense, of '$p \rightarrow p$'. $\xi\mathbf{S}_6$ has just these postulates:

ξ1. If A is a substitution instance of an $\mathbf{S}_6$-tautology, or of a theorem, then $\vdash A$.

ξ2. $\vdash \check{\mathrm{S}}^{u}_{t} B \,| \rightarrow B_u(\xi yB)$, where t is a wf term.

ξ3. $\vdash \check{\mathrm{S}}^{u}_{t} SB \,| \rightarrow SB_u(\xi yB)$.

R1. $\vdash A \vdash A \rightarrow B \twoheadrightarrow \vdash B$.

According to ξ2, if B is true of some arbitrarily chosen term t then B is true of any term which is B; and according to ξ3, if B is significant of some arbitrary term t then any selected item which is B is significantly B.

The quantifiers of $\mathbf{QS}_6$ are defined as follows in $\xi\mathbf{S}_6$:

$$(qu)B =_{\mathrm{Df}} B_u(\xi yB)$$
$$(Uu)B =_{\mathrm{Df}} B_u(\xi yDB).$$

ξ1 may be replaced by A1 (of $\mathbf{QS}_6$) and the rule

$\vdash B \twoheadrightarrow \vdash B_u(\xi uDB)$ (generalisation).

Systems $\xi\mathbf{S}_6\mathbf{E}$, $\xi\mathbf{S}_6\mathbf{E}!$, $\xi\mathbf{S}_6\approx$ and $\xi\mathbf{S}_6\mathbf{E}!\approx$ result from $\xi\mathbf{S}_6$ in precisely the way the corresponding $\mathbf{Q}$-systems are obtained from $\mathbf{QS}_6$.

Theorem (76a): Every theorem of **QS**$_6$ (**QS**$_6$**E**, ...) is a theorem of ξ**S**$_6$ (ξ**S**$_6$**E**, ...).

Proof: We first derive the postulates of **QS**$_6$ containing '$\mathfrak{q}$' essentially; then we derive the equivalence $D(\mathfrak{q}u)DB \simeq (Uu)B$; and finally using these results we sketch the derivation of remaining postulates of **QS**$_6$, *i.e.* those that contain 'U' essentially. Thus the proof shows at the same time the eliminability of one of the two quantifiers of **QS**$_6$.

ad A3.

$$\check{S}^u_{\xi y A} S[D(A \rightharpoondown B)] \,|\!\rightharpoondown . SD(A_u(\xi y D(A \rightharpoondown B)) \rightharpoondown B),$$

by $\xi 3$ using the assumption that u is not free in B. Hence using **S**$_6$ theorems, in particular $\vdash SA \simeq \neg DA$ and $\vdash DDA \simeq A$,

$A_u(\xi y D(A \rightharpoondown B)) \rightharpoondown B \rightharpoondown . A_u(\xi y A) \rightharpoondown B,$

i.e. $(Uu)(A \rightharpoondown B) \rightharpoondown . (\mathfrak{q}u)A \rightharpoondown B$, where u is not free in B.

ad A5. Immediate from $\xi 2$ and the definition of '$\mathfrak{q}$'.

ad A8′. Immediate from $\xi 3$ and the definition of '$\mathfrak{q}$'.

Then A8 follows by $\xi 1$ and A3.

ad A9. $\check{S}_{\xi y B} SB \,|\!\rightharpoondown . SB_u(\xi y\, SB)$, by $\xi 2$.

Thus $SB_u(\xi y\, B) \rightharpoondown (\mathfrak{q}u)SB$

ad $D(Uu)DB \simeq (\mathfrak{q}u)B$.

$D(Uu)DB \simeq DDB_u(\xi y DDB)$ by definition of 'U'

$\simeq (\mathfrak{q}u)DDB$ by definition of '$\mathfrak{q}$'

But (a) $(\mathfrak{q}u)B \rightharpoondown (\mathfrak{q}u)DDB$, either by ξ schemes or thus:

$B \rightharpoondown DDB \rightharpoondown (\mathfrak{q}u)DDB$, whence (a) by $\xi 1$ and A3.

(b) $(\mathfrak{q}u)DDB \rightharpoondown (\mathfrak{q}u)B$, similarly.

(c) $S(\mathfrak{q}u)B \rightharpoondown S(\mathfrak{q}u)DDB$

For

$\check{S}^u_{\xi y B} S(DDB) \,|\!\rightharpoondown SDDB_u\, (\xi y DDB)$ by $\xi 3$

(d) $S(\xi u)DDB \rightharpoondown S(\mathfrak{q}u)B$

For $\check{S}^u_{\xi y DDB}\, SB \,|\!\rightharpoondown S(\mathfrak{q}u)B$ by $\xi 3$

By (a)–(d) $(\mathfrak{q}u)DDB \simeq (\mathfrak{q}u)B$

ad $D(\mathfrak{q}u)DB \simeq (Uu)B$

$(\mathfrak{q}u)DB \simeq DDB_u(\xi y DB)$, using the definition of '$\mathfrak{q}$'

$\simeq B_u(\xi y DB)$.

*ad*A4. $\check{S}^u_t \neg A \,|\!\rightharpoondown (\mathfrak{q}u) \neg DDA$ A5, $\xi 1$

$\neg(\mathfrak{q}u)SDA \rightharpoondown \check{S}^u_t A \,|$ $\xi 1$

$\neg S(\mathfrak{q}u)DA \rightharpoondown \check{S}^u_t A \,|$, using A8

$D(\mathfrak{q}u)DA \rightharpoondown \check{S}^u_t A \,|$, whence A4 follows.

$$\begin{array}{lll} \textit{ad}\text{A3.} & \neg DD(\neg B \rightharpoonup \neg A) \rightharpoonup \neg DD(A \rightharpoonup B) & \xi 1 \\ & \quad \rightharpoonup (\mathit{9}u)\neg DD(A \rightharpoonup B) & \text{A5} \\ & (\mathit{9}u)\neg DD(\neg B \rightharpoonup \neg A) \rightharpoonup (\mathit{9}u)\neg DD(A \rightharpoonup B) & \xi 1, \text{A3} \\ & \neg D(\mathit{9}u)D(\neg B \rightharpoonup \neg A) \rightharpoonup \neg D(\mathit{9}u)D(A \rightharpoonup B) & \text{A8, A9} \\ & (Uu)(A \rightharpoonup B) \rightharpoonup (Uu)(\neg B \rightharpoonup \neg A) & \xi 1 \\ & (Uu)(A \rightharpoonup B) \rightharpoonup (Uu)(\neg DDB \rightharpoonup \neg A) & \text{similarly} \\ & \quad \rightharpoonup (\mathit{9}u)(\neg DDB \rightharpoonup \neg A), & u \text{ not free in } A, \text{ by A3} \\ & \quad \rightharpoonup . A \rightharpoonup D(\mathit{9}u)DB & \xi 1, \text{A9.} \end{array}$$

A3 may be proved directly from the following case of $\xi 3$:

$$\check{\mathrm{S}}^{u}_{\xi y BD} SD(A \rightharpoonup B) \mid\rightharpoonup SD(A \rightharpoonup B_u(\xi y D(A \rightharpoonup B))), \qquad u \text{ not free in } A$$

$$\begin{array}{lll} \textit{ad}\text{ A6.} & B \rightharpoonup \neg DD\neg B \\ & \quad \rightharpoonup (\mathit{9}u)\neg DD\neg B \\ \text{Hence} & (\mathit{9}u)B \rightharpoonup \neg D(\mathit{9}u)D\neg B \\ \text{Thus} & D(\mathit{9}u)D\neg DA \rightharpoonup \neg(\mathit{9}u)DA \\ & \quad \rightharpoonup \neg DD(\mathit{9}u)DA. \\ \textit{ad}\text{ A7.} & (\mathit{9}u)\neg DD\neg B \rightharpoonup (\mathit{9}u)B \\ & \neg D(\mathit{9}u)D\neg DA \rightharpoonup (\mathit{9}u)DA \\ & \neg DD(\mathit{9}u)DA \rightharpoonup D(\mathit{9}u)DSA. \end{array}$$

We extend the contractional semantics given for $\mathbf{QS}_6$ ($\mathbf{QS}_6\mathbf{E}$, *etc.*) to a contractional semantics for $\xi\mathbf{S}_6$ ($\xi\mathbf{S}_6\mathbf{E}$, *etc.*) by adding to the semantical basis S for $\mathbf{QS}_6$ a choice function c. Thus a semantical basis for $\xi\mathbf{S}_6$ is a structure $\mathrm{S} = \langle \mathrm{d}, \mathrm{e}, \mathrm{f}, \mathrm{c} \rangle$ where d, e, and f are as before and c is a choice or selection function defined on d such that for each non-null subset d′ of d, c(d′) is an element of d′, and such that $\mathrm{c}(\Lambda) = \mathrm{c}(\mathrm{d})$. An *interpretation* I of $\xi\mathbf{S}_6$ *over* basis S is defined in exactly the same way as an interpretation I of $\mathbf{QS}_6$.

Given I and S, we use a joint inductive definition to define a valuation function v and to extend interpretation I to ξ-terms. The inductive definition of v is precisely the same as that given in the semantics of $\mathbf{QS}_6$, except that clauses (vi) and (vii) for quantifiers are deleted, and the following added:

(viii) If for some $(\mathit{9})$ u, $v(A) = t$,
$\mathrm{I}(\xi u A) = \mathrm{c}(\{\mathrm{I}(u) : v(A) = t\})$;

otherwise, $\mathrm{I}(\xi u A) = \mathrm{c}(\{\mathrm{I}(u) : v(A) \neq n\})$.

In other words $\mathrm{I}(\xi u A)$ is a selection from the truth-domain of A (under the interpretation) if this is non-null, otherwise a selection from the significance-domain of A if this is non-null; and otherwise a selected item from d.

Validity, satisfiability, and so forth, are defined just as for $\mathbf{QS}_6$.

Theorem (76b):

(1) Every $\neg$-consistent class of wff of $\xi\mathbf{S}_6$, ($\xi\mathbf{S}_6\mathbf{E}$, $\xi\mathbf{S}_6\mathbf{E}!$ $\xi\mathbf{S}_6\!\approx$, $\xi\mathbf{S}_6\mathbf{E}!\!\approx$) is simultaneously satisfiable in a denumerable (countable) domain.

(2) Wff A is a theorem of $\xi\mathbf{S}_6$ ($\xi\mathbf{S}_6\mathbf{E}$, *etc.*) iff A is $\xi\mathbf{S}_6$ ($\xi\mathbf{S}_6\mathbf{E}$, *etc.*)-valid.

Proof of (1) is a simplification of corresponding theorems for $\mathbf{QS}_6$ (*etc.*). We prove the result for $\xi\mathbf{S}_6$; extensions to further systems are then obtained as before. Let Γ be a $\urcorner$-consistent class of wff of $\xi\mathbf{S}_6$; and let J be the maximal $\urcorner$-consistent extension of Γ. A semantical basis S_1 for interpretation I_1 is specified as follows: d_1 is the (denumerable) domain of subject constants and closed ξ-terms of $\xi\mathbf{S}_6$. $\langle \mathrm{d}_1, \mathrm{e}_1, \mathrm{f}_1\rangle$ is the usual basis over d_1. The function c_1 is so specified in terms of I_1 over S_1 that

$$\mathrm{c}_1(\{\mathrm{I}_1(u): v_1(A) = t\}) = \mathrm{c}_1(\{\mathrm{I}_1(u): v_1(A) \neq n\}) = \xi u A$$

where A contains at most u free.

I_1 over S_1 is defined exactly as for the corresponding theorem for $\mathbf{QS}_6$. It follows

(i) $\mathrm{I}_1(\xi u A) = \xi u A$, and so
(ii) $\mathrm{I}_1(t_i) = t_i$, for every closed wf term t_i.

As before the central part of the proof consists in showing that for every closed wff A,

$$v_1(A) = t \equiv. \quad TA \in \mathrm{J}$$
$$v_1(A) = n \equiv. \urcorner SA \in \mathrm{J}.$$

The induction basis and the induction steps for sentential connectives are as for the corresponding theorem for $\mathbf{QS}_6$. It remains to consider the case where A is of the form $\mathrm{F}^n(t_1, ..., t_{i-1}, \xi u B, ..., t_n)$ where B contains at most u free. Then

$$\begin{aligned} v_1(\mathrm{F}^n(t_1, ..., t_{i-1}, \xi u B, ..., t_n)) = t &\equiv T[(\mathrm{I}_1(t_1), ..., \mathrm{I}_1(t_{i-1}), \mathrm{I}_1(\xi u B), ..., \mathrm{I}_1(t_n))\, i\, \mathrm{I}_1(\mathrm{F}^n)] \\ &\equiv T[\mathrm{I}_1(\mathrm{F}^n)(t_1, ..., t_{i-1}, \xi u B, ..., t_n)] \\ &\equiv T\mathrm{F}^n(t_1, ..., t_{i-1}, \xi u B, ..., t_n) \in \mathrm{J}. \end{aligned}$$

Similarly,

$$v_1(\mathrm{F}^n(t_1, ..., t_{i-1}, \xi u B, ..., t_n)) = n \equiv \urcorner S\mathrm{F}^n(t_1, ..., t_{i-1}\, \xi u B, ..., t_n) \in \mathrm{J}.$$

The remainder of the proof of (1) is as before; and the only new steps in (2) are the verifications of ξ-postulates $\xi 1$ and $\xi 2$.

Corollaries

(1) $\xi\mathbf{S}_6$ ($\xi\mathbf{S}_6\mathbf{E}$, *etc.*) are consistent. But, much as before, syntactical consistency proofs can also be given.
(2) Where A is a wff of $\mathbf{QS}_6$ (and this is a wff of $\xi\mathbf{S}_6$, with quantifiers defined in terms of 'ξ') (of $\mathbf{QS}_6\mathbf{E}$, *etc.*), $\vdash_{\xi S_6 (etc.)} A$ iff $\vdash_{QS_6 (etc.)} A$.

Proof: (*) If A is a wff of $\mathbf{QS}_6$ then, if A is $\xi\mathbf{S}_6$-valid then A is $\mathbf{QS}_6$-valid. It is enough to show that quantifiers have their usual interpretations when defined in $\xi\mathbf{S}_6$, and it suffices to show this in the case of '$\mathcal{q}$'. But

$v((\mathcal{9}u)A) = t$ iff $v(A_u(\xi y A)) = t$
iff $v(A) = t$ for some u; and
$v((\mathcal{9}u)A) \neq n$ iff $v(A_u(\xi y A)) \neq n$
iff $v(A) \neq n$ for some u, by the interpretation of 'ξ'.

Now, applying (*) to a wff A of $\mathbf{QS}_6$, if $\vdash_{\xi S_6} A$ then A is $\xi\mathbf{S}_6$-valid; so A is $\mathbf{QS}_6$-valid, whence $\vdash_{\mathbf{QS}_6} A$.

Corollary (2) provides the main case (and other cases may be derived from this) of Hilbert's second ξ-theorem for quantified significance logic. It shows that $\xi\mathbf{S}_6$ is a conservative extension of $\mathbf{QS}_6$, $\xi\mathbf{S}_6\approx$ of $\mathbf{QS}_6\approx$, *etc.* (when definitions of quantifiers are taken as an integral part of ξ-systems).

§2 *Definite descriptions in* $\xi\mathbf{S}_6\mathbf{E}\approx$

Using the ξ-operator a theory of definite descriptions, superior in several respects to that developed in $\mathbf{QS}_6\mathbf{E}\approx$ can be designed. A definite description operator 'ι' can be defined in $\xi\mathbf{S}_6\mathbf{E}\approx$ without the need for scoping devices thus:

$$\iota u A(u) =_{\mathrm{Df}} \xi u(A(u) \,\&\, (Uv)(\neg(u \approx v) \rightarrow . \neg A(v))).$$

In other words, the item which is A is an *unique* item which is A. In contrast to Russell's theory, '$\iota x f(x)$' is here a complete symbol; and in particular $\iota u A(u)$ fulfils all subject roles, *e.g.* instantiation holds unrestrictedly:

$$\vdash (Uv)A(v) \rightarrow . A_v(\iota u B(u));$$

and reflexivity of identity holds unrestrictedly:

$$\vdash \iota u B(u) \approx \iota u B(u).$$

Nevertheless the theory has important features in common with a significance version of Russell's theory. Where uniqueness is defined, as before (§7.8):

$$(\mathcal{9}!u)A(u) =_{\mathrm{Df}} (\mathcal{9}u)(A(u) \,\&\, (Uv)(\neg(u \approx v) \rightarrow \neg A(v))),$$
$$B!(u) =_{\mathrm{Df}} B(u) \,\&\, (Uv)(\neg(u \approx v) \rightarrow \neg B(v)),$$

then $\vdash \iota u A(u) \simeq \xi u A!(u)$; $\vdash (\mathcal{9}!u)A(u) \simeq (\mathcal{9}u)A!(u)$, and
$\vdash A!(\iota u A(u)) \simeq (\mathcal{9}!u)A(u)$.

For $(\mathcal{9}!u)A(u) \simeq A(\xi u A!(u)) \,\&\, (Uv)(\neg(\xi u A!(u) \approx v) \rightarrow \neg A(v))$ (a)
$\simeq A!(\iota u A(u))$.

$\vdash SA(\iota u A(u)) \equiv . (\mathcal{9}v)SA(v)$, from (a) and the definition of '$\mathcal{9}!$' since $\vdash S(Uu)(\neg(u \approx v) \rightarrow \neg A(v))$.

Thus it is significant that the seat of a bed is a bed-seat (an adaption of an example used by Smart [1]) iff it is significant that some item is a bed-seat; and it presumably is significant that some item is a bed-seat.

$\vdash E(\iota u A(u)) \,\&\, TA!(\iota u A(u)) \simeq (E!u)TA(u)$,
where of course $(E!u)B(u) =_{\mathrm{Df}} (\mathcal{9}!u)(B(u) \,\&\, E(u))$.

Thus where $\iota u A(u)$ is assumptible, *i.e.* $A!(\iota u A(u))$ holds,

$\vdash E(\iota u A(u)) \simeq (E!u)TA(u)$ (A)

Proof: Since $SE(u)$ and STp,

$$S[E(\iota u A(u)) \,\&\, TA(\iota u A(u))] \simeq S[(E!u)TA(u)].$$

Further, $(E!u)TA(u) \rightharpoondown .(\text{ᑫ}!u)A(u)$

$$\rightharpoondown .A!(\iota u A(u)) \; [\rightharpoondown A(\iota u A(u))]$$

Also $(E!u)TA(u) \rightharpoondown (\text{ᑫ}u)\,[TA(u) \,\&\, (Uv)(TA(v) \rightharpoondown T(u \approx v)) \,\&\, E(u)]$

i.e. $(\text{ᑫ}u)\beta(u)$

$\rightharpoondown (Uv)\,[TA(v) \rightharpoondown T(v \approx \xi u \beta(u))] \,\&\, E(\xi u \beta(u))$

$\rightharpoondown \iota u A(u) \approx \xi u \beta(u) \,\&\, E(\xi u \beta(u))$

$\rightharpoondown E(\iota u A(u))$

Conversely, $E(\iota u A(u)) \,\&\, TA!(\iota u A(u))$

$\rightharpoondown . E(\iota u A(u)) \,\&\, TA(\iota u A(u)) \,\&\, (Uv)(\neg(v \approx \iota u A(u)) \rightharpoondown \neg TA(v))$

$\rightharpoondown (\text{ᑫ}!w)(E(w) \,\&\, TA(w)).$

$\vdash B(\iota u A(u)) \,\&\, A!(\iota u A(u))$

$\rightharpoondown (\text{ᑫ}v)\,[A(v) \,\&\, (Uw)(TA(w) \rightharpoondown T(w \approx v)) \,\&\, B(v)]$ (B)

Proof is like the last part of the previous theorem.

$\vdash (\text{ᑫ}v)\,[A(v) \,\&\, (Uw)(TA(w) \rightharpoondown T(w \approx v)) \,\&\, B(v)] \rightharpoondown . B(\iota u A(u)).$ (C)

Proof is as in the proof of (A), but with $B(u)$ replacing $E(u)$.

Hence where $\iota u TA(u)$ is assumptible,

$\vdash TB(\iota u A(u)) \,\&\, TA!(\iota u TA(u))$

$\simeq (\text{ᑫ}v)\,[TA(v) \,\&\, (Uw)(TA(w) \rightharpoondown T(w \approx v)) \,\&\, TB(v)]$ (D)

But the T-qualifications cannot be removed without changing the logic by adding the postulate,

$\xi 4.$ $\vdash SB(\xi u A(u)) \simeq (\text{ᑫ}u)S(A(u) \,\&\, B(u)).$

For $\xi 4$ is not a theorem schema of $\xi \mathbf{S_6 E} \approx$, as may be shown by countermodels of the following sort: Consider cases where it is false that $(\text{ᑫ}u)SA(u)$. Then the choice for $\xi u A(u)$ may be made arbitrarily; so select a b such that $SB(b)$. Since models can readily be designed suitable for wff B, $\xi 4$ is not $\xi \mathbf{S_6 E} \approx$-valid. Countermodels to the other half of $\xi 4$ may be constructed by ensuring that the significance ranges of $A(u)$ and $B(u)$ just overlap and by letting the b chosen from the s-range of $A(u)$ for $\xi u A(u)$ be such that $\sim S(b)$. By appropriately restricting the semantics these countermodels could, of course, be eliminated; but the restrictions appear both peculiar—roughly, a nonsignificant item is wanted—and unwarranted. Moreover properly qualified forms of $\xi 4$ can be derived:

$\vdash SB(\xi y A(y)) \supset . (\text{ᑫ}u)S(A(u) \,\&\, B(u))$, *provided* $\vdash SA(\xi y A(y))$.

$\vdash (\text{ᑫ}u)SA(u) \,\&\, (Uv)(SA(v) \supset SB(v)) \supset . SB(\xi u A(u)),$

i.e. $(\text{ᑫ}u)SA(u) \supset SB(\xi u A(u))$ provided B's significance-range includes A's.

In principles (A)–(D) we collect qualified analogues of Russell's two contextual definitions of definite descriptions. The assumptibility restrictions on these principles are quite crucial. Without them inconsistency would quickly result, *e.g.* as follows:

Let $B(u)$ be a wff containing u free such that $\vdash B(u)$. Then instantiating $(Uu)B(u)$, $\vdash TB(\iota v(TC(v) \,\&\, \neg C(v)))$; whence, by an unqualified form of (B) or (D), $(\mathcal{9}v)(TC(v) \,\&\, \neg C(v))$.

The main features of the Hilbert-Bernays' theory of descriptions follow, in a more satisfactory form, in this theory. For it follows that whenever (and only when) $(\mathcal{9}!u)A(u)$ is true, one is entitled to take $A!(\iota u A(u))$ as true. Thus given the uniqueness condition $(\mathcal{9}!u)A(u)$ one can introduce a description $\iota u A(u)$ such that it satisfies the assumptibility postulate $\vdash A!(\iota u A(u))$, and such that

$$\vdash B(\iota u A(u)) \rightharpoonup (\mathcal{9}v)\,[A(v) \,\&\, (Uw)(TA(w) \rightharpoonup T(w \approx v)) \,\&\, B(v)]$$

and conversely. Moreover the serious disadvantage of the Hilbert-Bernays' theory that $B(\iota u A(u))$ is not defined when the uniqueness condition is not satisfied is removed.

Descriptive functions may be defined, following *PM*, for two-place predicates, as follows:

$$\mathrm{F}`v =_{\mathrm{Df}} \iota u \mathrm{F}(u, v).$$

But in contrast to *PM*, $\vdash \mathrm{F}`v \approx \mathrm{F}`v$ and $\vdash (Uu)A(u) \rightharpoonup A(\mathrm{F}`v)$. Generally, theorems for descriptions carry over to theorems for descriptive functions.

§3 *Modalisation: ξ*$\mathbf{IS}_6$ *etc.*

System $\xi\mathbf{IS}_6$ adds to $\mathbf{IS}_6$ the ξ-schemes $\xi 2$ and $\xi 3$ and, in place of RT, the rule RTξ. $\vdash A \twoheadrightarrow \vdash \Phi_i A$, where A a theorem of $\xi\mathbf{S}_6$ or an instance of (D) (for $j \leqslant i \leqslant m$). Systems $\xi\mathbf{IS}_6\mathbf{E}$ $\xi\mathbf{IS}_6\mathbf{E}!$ and $\xi\mathbf{IS}_6\mathbf{E}!=$ result from $\xi\mathbf{IS}_6$ just as corresponding **Q**-systems result from $\mathbf{QIS}_6$.

One striking feature of $\xi\mathbf{IS}_6$ is the derivability of Barcan wff B (§7.10), as follows:—

Define, $\Theta_i A =_{\mathrm{Df}} D\Phi_i DA$. Then by $\xi 2$,
$\vdash \check{S}^{u}_{\xi y B} S\Theta_i B(u) \mid\rightharpoonup S\Theta_i B_u(\xi y \Theta_i B)$; whence,
$\vdash D(\mathcal{9}u)\Theta_i B \rightharpoonup D\Theta_i(\mathcal{9}u)B$. Hence,
$\vdash DD(Uu)DD\Phi_i DB \rightharpoonup DD\Phi_i DD(Uu)DB$; so taking
DB for B, $\vdash (Uu)\Phi_i B \rightharpoonup \Phi_i(Uu)B$, which is the Barcan wff B for Φ_i. Also, for $j \leqslant i \leqslant m$, Φ_iB follows using RTξ. Thus, $\xi\mathbf{IS}_6$ is superior to $\mathbf{QIS}_6$ since the Barcan wff do not have to be assumed as further postulates. To sum up:

Theorem (77): Every theorem of $\mathbf{QIS}_6$ ($\mathbf{QIS}_6\mathbf{E}$, *etc.*) is a theorem of $\xi\mathbf{IS}_6$ ($\xi\mathbf{IS}_6\mathbf{E}$, *etc.*).

The converse fails, however, since the ξ-schemes facilitate the derivation of further schemes which are much more questionable than the Barcan wff: for example, the provably equivalent,

(a) $(\mathcal{q}u)\Box[(\mathcal{q}v)A(v) \rightharpoonup A(u)]$, and

(b) $\Box(B \rightharpoonup (\mathcal{q}u)A) \rightharpoonup (\mathcal{q}u)\Box(B \rightharpoonup A)$, for u not free in B.

(a) follows from (b) by taking B to be $(\mathcal{q}u)A$. Conversely, it follows from (a), by distributing '$\Box$' and a change of bound variable: $(\mathcal{q}u)\Box(\mathcal{q}u)A \rightharpoonup \Box A$, whence by quantifier distribution, as u is not free in $(\mathcal{q}u)A$,

(c) $\Box(\mathcal{q}u)A \rightharpoonup (\mathcal{q}u)\Box A$.

Since (a) in turn follows from (c), (a) and (c) are provably equivalent. Now, $\vdash \Box(B \rightharpoonup (\mathcal{q}u)A) \rightharpoonup \Box(\mathcal{q}u)(B \rightharpoonup A)$, when u is not free in A. Hence (b) follows upon applying (c).

(a) and (b) are readily proved in $\xi\mathbf{IS}_6$. For instance,

$\vdash \Box((\mathcal{q}v)A(v) \rightharpoonup A(\xi u A(u)))$,

using the definition of '$\mathcal{q}$'. But (a) now follows by particularising.

To show, on the other hand, that (a) is not a theorem of $\mathbf{QIS}_6$, and hence that $\xi\mathbf{IS}_6$ is not a conservative extension of it, consider the expansion of (a) over a two-element domain $\{x_0, x_1\}$, namely,

$$\Box(A(x_0)\mathrm{V}A(x_1) \rightharpoonup .\, A(x_0))\mathrm{V}.\, \Box(A(x_0)\mathrm{V}A(x_1) \rightharpoonup .\, A(x_1)).$$

This formula is not a theorem of $\mathbf{IS}_6$, hence not of $\mathbf{QIS}_6$, since a counter-model with $\mathrm{K} = \{\mathrm{G}, \mathrm{H}_1, \mathrm{H}_2\}$ is readily designed. But were (a) a theorem of $\mathbf{QIS}_6$ it would hold in such a finite domain, in which case the above formula would hold; hence (a) is not a theorem.

It follows, too, that quantified wff of $\xi\mathbf{IS}_6$ cannot be expanded in the usual way over finite domains. That is, the quantifiers 'U' and '$\mathcal{q}$' of $\xi\mathbf{IS}_6$ are not value-functional, since even in finite cases they do not reduce respectively to conjunctions and Brady disjunctions. The descriptor 'ξ' in intensional contexts thus gives rise to intensional quantifiers.

It will be objected that some highly undesirable results hold for these intensional quantifiers, notably (c) and its equivalents. Indeed there are well-known counter-examples to (c) designed to unmask it as a kind of modal fallacy. A typical example is the following:

> It is logically necessary that some number n is the number that numbers the planets, but it is not the case that there is some number n which is necessarily the number that numbers the planets.

This is a doubtful counterexample, however. It seems to us that there is such a number, namely the number that numbers the planets. For it is logically necessary that the number that numbers the planets, numbers the planets; and hence, by particularisation, for some number n it is logically necessary that n numbers the planets.

In general, alleged counterexamples to (c), like counterexamples to the Barcan wff, and like the case for the value-functional expansion of quantified expressions over finite domains, are convincing enough (and indeed correct) for existential (referential) quantifiers. But they are neither correct nor convincing when non-referential quantifiers, which take up indeterminate or intensionally specified descriptions and names, are admitted. Since such quantifiers and descriptions are an integral part of discourse, they ought to be encompassed in, or at least not excluded from, the general logical theory of quantification.

There is, besides, a powerful direct argument in favour of (a). For (a) to be true, it is sufficient that,

(d) $\Box((\mathcal{q}u)A(v) \rightharpoondown A(\tau uB))$

be true for *some* description or name τuB, not necessarily a ξ-description. For then (a) follows by particularisation. Now,

(e) $(\mathcal{q}v)A(v) \rightharpoondown A(\tau uB)$

is true—a point which can be admitted without extending quantification logic—but it is not simply a contingent truth. For if $A(v)$ for some v, then we can always and easily find, by purely logical means, a name or description which we can correctly use to instantiate $(\mathcal{q}v)A(v)$; that is, (e) is necessarily true. It does not matter how hazy or indeterminate the item described is, or whether it is intensionally specified. Such a "new-name" move is regularly made in both mathematical and ordinary argumentation: *e.g.* $A(v)$ holds for some v, say (call it) Charlie; then, necessarily A(Charlie). Thus, to drop (d) and (e) would eliminate an important and seemingly correct method of reasoning which employs names and descriptions as instantiations of particular quantifiers in cases where the quantifiers are not existential and the names are not referential.

We therefore propose the system **TQIS**$_6$, formed by adding (a) to **QIS**$_6$, as worth serious consideration. **TQIS**$_6$ differs markedly from **QIS**$_6$. In contrast to **QIS**$_6$, every wff of **TQIS**$_6$ has a prenex-normal equivalent; quantified wff of **TQIS**$_6$ cannot be expanded truth-functionally over finite domains; and ξ**IS**$_6$ is a conservative extension of **TQIS**$_6$. The interpretational problem for the quantifiers of **TQIS**$_6$ is solved by way of the connection with the ξ-system. Thus, the quantifiers are interpreted in terms of 'ξ': *e.g.* $(\mathcal{q}u)A$ holds in situation H iff A holds in H for an arbitrarily selected item which is A. It follows that the semantics for **TQIS**$_6$ is a case of the semantics for ξ**IS**$_6$.

A simple amalgam of modal and ξ modellings provides a semantics for ξ**IS**$_6$. Accordingly, a ξ**IS**$_6$ model $\mathcal{M}$ is a structure,

$\mathcal{M} = \langle G, K, ..., d, e, f, c, I \rangle$ where $\langle G, K, ..., d, e, f, I \rangle$

is a **QIS**$_6$-model and c is a choice function defined on properties on d. The quantificational rules for **QIS**$_6$ are deleted and replaced by:

If, for some u, $v(A, H) = t$,
$I(\xi uA, H) = c(\lambda I(u)(v(A, H) = t))$; and otherwise
$I(\xi uA, H) = c(\lambda I(u)(v(A, H) \neq n))$.

ξ**IS**$_6$**E** and ξ**IS**$_6$**E**! = similarly modify corresponding E-models.

Theorem (78):

(1) If Γ is a con class of ξ**IS**$_6$ (*etc.*), there is a ξ**IS**$_6$ (*etc.*) model with a countable domain which satisfies Γ.

(2) *B* is ξ**IS**$_6$ (*etc.*)-valid iff *B* is a theorem of ξ**IS**$_6$ (*etc.*).

In the proof of (1) for ξ**IS**$_6$ we define K as the class of mac sets of ξ**IS**$_6$* and G as the ξ**IS**$_6$ mac set of Γ. For closed term $\xi u A$, we define function c thus:

$$\mathrm{c}(\lambda u \in \mathrm{d}(TA \in \mathrm{H})) = \mathrm{c}(\lambda u \in \mathrm{d}(\sim SA \in \mathrm{H})) = \xi u A$$

It then follows canonically that $\mathrm{I}(\xi u A, \mathrm{H}) = \xi u A$; and since, by inductive proof $v(A, \mathrm{H}) = t$ iff $TA \in \mathrm{H}$, and $v(A, \mathrm{H}) = n$ iff $\sim SA \in \mathrm{H}$, that

$$\mathrm{c}(\lambda \mathrm{I}(u)(v(A, \mathrm{H}) = t)) = \mathrm{c}(\lambda \mathrm{I}(u)(v(A, \mathrm{H}) \neq n)) = \xi u A.$$

Remaining proof details are like those for ξ**S**$_6$ and **QIS**$_6$.

Property abstraction replaces set abstraction to ensure that c is a function; alternatively properties may be eliminated in favour of pairs of sets.

On the basis of the definition:

$$\iota u A(u) =_{\mathrm{Df}} \xi u(A(u) \ \& \ (Uv)(\neg(u = v) \rightharpoonup \neg A(v))),$$

a theory of descriptions may be developed. The item which is *A* is an extensionally unique item which is *A*. Using this theory the modal paradoxes are simply resolved; for it does not follow from the truth that

$$\iota x \text{ morning star}(x) = \iota x \text{ evening star}(x)$$

that the statement is necessary, known to the rational man, *etc.*

'Any' and 'the' are by no means the only operators that can profitably be investigated in operator-enriched non-extensional logics and whose behaviour cannot be fully analysed in terms of the quantifiers 'every' and 'some', others of importance are 'a certain' and 'each'.

§4 *General intensional logics, context and pragmatics*

We consider briefly a set of general intensional logics obtained by adding quantifiers, descriptors, and existence and identity predicates to systems **MS**$_6$ (of §7.2) instead of to **IS**$_6$. For axiomatic formulations **QMS**$_6$, *etc.*, we make, in each case, exactly the same additions to **MS**$_6$ as are made to **S**$_6$ for corresponding systems **QS**$_6$, *etc.* Thus **QMS**$_6$, unlike **QIS**$_6$, does not have the Barcan wff,

B. $(Uu)\mathrm{F}\S B \rightharpoonup \mathrm{F}\S(Uu)B$

as a thesis. Of course when modal constraints are imposed, *e.g.* on R_{f_0}, the corresponding Barcan wff, for '□' will be valid. In each case the semantics too are an amalgam of those for system **MS**$_6$ with the semantics for the extension that is made; *e.g.* for **QMS**$_6$ the semantical rules for quantifiers, and associated semantical bases, are added to the semantics for **MS**$_6$. Then a cluster of adequacy results follows:—

Theorem (79): *B* is a theorem of **QMS**$_6$, *etc.*, iff *B* is **QMS**$_6$-valid, *etc.*

Systems like $\mathbf{QMS}_6$ are amenable to several philosophically important modifications and extensions. For example the syntax and semantics may be enlarged to provide for a more adequate treatment of the range of intensional functions, of modifying phrases and restricted variables, and of special subjects and descriptors. A first way of extending the semantics is by generalizing the model sets or worlds to arbitrary set-ups, by defining situations and worlds in terms of the rules satisfied by extensional connectives, and by letting each intensional connective state as part of its evaluative rule, or associated relation, which set-ups it is assessed over. For instance, the rule for '$\Box$' will begin 'for every world', whereas that for entailment will begin 'for every situation', and that for Pythagoras's belief functor will select at least some set-ups where nonsignificant sentences hold.

The semantics can accommodate special subjects such as assertoric subjects, quotational subjects and attributive subjects. Thus §*A*, like other subjects, has an interpretation which is invariant with respect to situations, namely $I(\S A) = \{H \in K_1 : v(A, H) = t\} = I(\S A, H')$ for arbitrary H'. Similarly $I(qu(A)) = \{H \in K : v(A, H) = t\}$; accordingly sentence predicates like 'contains five words' may be evaluated in terms of appropriate sets of set-ups. The semantical apparatus may also be enlarged to cater for quotation of the usual constant-forming sort where quotation is an operation which forms a subject from any part of speech; but some analogue of quotation, *e.g.* quotation itself or concatenation, seems needed among the building materials of set-ups. If, for example, concatenation and blanks are available, then I ('Socrates') = S⁀o⁀c⁀r⁀a⁀t⁀e⁀s; I('I am') = I⁀⁀a⁀m, *etc.*

The semantics of intensional systems may be illuminatingly interpreted in terms of context-situations or language-games (*cf.* §7.2). Such an interpretation has the advantage of allowing several more general matters of context to be treated, including matters that are commonly reckoned to belong to the discipline of pragmatics—that study of linguistic expressions which takes account of the users of the expressions and of the contexts in which they are used. For instance, now that subject-predicate analysis is available, sentences containing indexical expressions can be evaluated in the enlarged semantics. (These points are explained in Montague [1], but the analyses provided appear inadequate.)

There are several components of context (see §1.7, §2 and §2.3), *e.g.* the topic of conversation, the identifications made (including names of those discussed and of the speakers and audience), special assumptions made in terms of which non-standard senses can be determined and ambiguities removed, the time and place of the utterance, the description of relevant physical circumstances in which the utterance takes place including prominent items, *etc.* Thus, for the context for 'Charlie is standing between you and me' we should have at least to say who Charlie is, where he is, and who the speaker and audience are. Not all of these components will be needed in general, and in certain cases further components will be necessary. For example, in 'This is larger than this, but this is smaller than this' it is not good enough to specify what *this* is; the context has to tell us what the first *this* is and what each later *this* is, and these may all differ since four items may be pointed to in turn. We have already taken for granted that the context supplies as many contextual details as are needed for the assessment at hand, and this flexibility and openness affects the final details. It counts against the sequential method of treating context, which supposes that we make an advance listing of all components which matter, *e.g.* the time, the place, the speaker, *etc.*, to a predetermined level of detail and then specify context by a sequence which has a place held for each of these components, *e.g.* the first place in the context sequence is occupied by a time component, the second by a place component, *etc.*; in other words, each context is a point in context space. The alternative which we have adopted is to let the context be specified by a set (of sentences) each element of which indicates which component is being selected and includes the levels of detail, *e.g.* the time is 10 o'clock July 15, 1917, in a case where minutes and seconds do not matter. We indicate how our account can be applied to the problem of indexical expressions.

Indexical subjects like 'I' and 'you', 'this' and 'those' are subject parameters. But while 'I' is always evaluated in the same way throughout a sentence (though not throughout a stretch of discourse with several speakers), a new occurrence of 'this' in a sentence provides a new parameter *when* a new indication of aboutness is made. To distinguish different indications syntactically, 'this' about item 1, 'this' about item 2, *etc.*, we shall subscript 'this', 'this_1' in the first case, 'this_2' in the second, *etc.*; the notation does not of course exclude, say, '$\text{this}_2 = \text{this}_6$', where a return is made to an earlier subject. Resulting parameters 'I' 'this_1', 'this_2' *etc.* are parameters which are constant in each particular context, but which may vary, by making different identifications, from one context to another. Thus, with respect to a given context Θ, 'I', 'me', 'you', *etc.*, behave in many logical respects like a constant, 'x_0' say; and just as in standard contexts $\text{I}(x_0, \langle\Theta, \text{H}\rangle) = x_0$, so in standard contexts $\text{I}(\text{me}, \langle\Theta, \text{H}\rangle) = \text{me}$. But in the case of indexical expressions there remain further logical details. For one, we can identify I and me with the speaker of the sentence in which 'I' or 'me' occurs; for another, the context Θ supplies further negotiable information in the form of extensional identifications. For example if the speaker of the sentence 'I am tall' is *BR* then the context c of the sentence is the context $c \ni (I = BR)$; and the sentence may be evaluated as follows:

$$v(\text{I am tall}, \langle c \ni (I = BR), \text{G}\rangle) = v(BR \text{ is tall}, \langle c, \text{G}\rangle),$$

which holds true, given that c is standard, iff *BR* is tall. The contextual identification is discharged in making the evaluation by a substitution, but the original sentence is only contextually equivalent to the resulting sentence. In the case of certain sentences, even where the speaker is say *BR*, a fuller relevant description of the speaker may be applied by the context; consider, *e.g.*, the sentence 'I believe I am not *BR*' said by *BR* suffering from amnesia. Abbreviate the sentence 'p_0', and '*BR* suffering from amnesia', 'BR_1'. Then the context c_1 of 'p_0' is not adequately specified by $c_1 \ni (I = BR)$; what is required is $c_1 \ni (I = BR_1)$. On the assumption that c_1 is otherwise logically standard we can shorten the evaluation to the following: $v(p_0, \langle I = BR_1, \text{G}\rangle)$, that is $v(BR_1$ believes BR_1 is not $BR, \text{G})$, which holds iff, for every $\text{H} \in \text{K}_1$ such that $\text{GR}_{\text{BR}_1}\text{H}, v(BR_1 \neq BR, \text{H}) = t$; that is if $BR_1 \neq BR$ is part of BR_1's belief set. Consider as an example of simultaneous use of indexical expressions, the sentence 'I am the author of this but I am not the author of this' said by *BR* pointing first to *PM* and then to the *Tractatus*. Then, given an otherwise standard context c_2, the context is discharged in,

$$v(\text{I am the author of } \mathit{this}_1 \text{ but I am not the author of } \mathit{this}_2, \langle c_2 \ni (I = BR) \;\&\; (\mathit{this}_1 = PM) \;\&\; (\mathit{this}_2 = \text{the } \mathit{Tractatus}), \text{G}\rangle)$$

to give $v(BR$ is the author of PM & BR is not the author of the *Tractatus*, G), which is significant but presumably false since Whitehead was an author of *PM*. The contextual dependence of certain definite descriptions can be dealt with similarly; consider, *e.g.*, the sentence 'The man had red hair' said in a context where the man = the man who stole car COP 001.

This account of the semantical evaluation of sentences containing expressions whose aboutness varies on different occasions of use, *i.e.* in different contexts, naturally does not preclude a fuller account of the behaviour of indexical expressions.

7.12 Second-stage Quantified Significance Logics

§1 *Interpretational Problems*

Developments of the logic of attributives, *i.e.* of property and relation abstracts, beyond that definable in $\mathbf{QS_6E!}\approx$ (§7.9) fall into one of two groups:

(1) the logic adds to quantified logic at least quantification over predicates, and typically a theory of logical types;

(2) a new primitive predicate '*i*' (or '∈'), and perhaps also an abstraction operator, is added to quantified logic, and the logic of instantiation and abstraction is axiomatised.

We have argued that the first course is thoroughly confused (Ch. 3). But although we have ruled out developments of kind (1), and their stages, such as second-order logic, these developments have, of course, analogues in developments of kind (2). One analogue, worth separate discussion, is that of second-order quantification theory which we call *second-stage* logic.

The wff of a second-*order* significance logic would differ from those of

$\mathbf{QS_6}$ only by admission of a more comprehensive rule for quantified formulae, *viz.*

(c)(3′). If A is a wff and u is a variable (subject, predicate or sentential), then $(Uu)A$ and $(\mathit{9}u)A$ are wff.

In terms of such an approach, both predicates and sentences are treated as subjects since both stand in quantifiable positions, but neither are allowed as values of the subject arguments 'x', 'y', *etc*. Hence the quantifiable variables of second-order logic are accordingly all restricted subject variables.

We have already given in detail our objections to treating predicates and sentences as subject factors, and in terms of these objections we find standard second-order theory unintelligible. In formulating a second-*stage* logic, $\mathbf{2QS_6}$, then, we shall have to proceed rather differently. But we shall, so far as is possible, parallel the development of second-order logic in order to mark the essential differences and similarities. In particular, we shall begin by adopting the convention that all the quantifiable variables are *restricted* subject variables. Later, however (§7.13, §3), we drop this requirement.

So far as an alternative to predicate quantification is concerned, we have indicated our solution in Chapter 3. There we allowed attributives to be formed from predicate wff by means of full λ-conversion. The new formulae so obtained are then subject expressions which can stand in quantifiable positions. In Chapter 3, however, we went further and allowed attributive variables and constants formed by conversion transformations of predicate wff to be taken as values of the subject arguments. Here, to keep the different kinds of subjects distinct, as in second-order theory, we shall not admit attributive subject factors as instantiations of subject arguments. They do, however, stand in quantifiable positions since they are subjects, though restricted subjects.

Thus, where $F^n(X_1, ..., X_n)$ is an atomic wff (of $\mathbf{QS_6}$) with free subject arguments $X_1, ..., X_n$, then $\lambda X_1 ... X_n F^n(X_1, ..., X_n)$, shortened to λF^n, is an attributive variable with bound components $X_1, ..., X_n$. Then, under the limitation that all variables are properly restricted subject variables, the λ-conversion scheme:

$$F^n(t_1, ..., t_n) \simeq (t_1, ..., t_n)\ i\ \lambda F^n$$

is valid. This scheme allows quantified significance logic to be reformulated so that predicates are uniformly replaced by attributives which occupy quantifiable places. This suggests a compact and unequivocal formulation of second-stage logic in which 'i' is introduced as a further primitive and attributive variables are abbreviated by new single letters. Thus, where 'ϕ' is such a letter, we may replace formulae of the sort 'fx' uniformly by formulae of the sort '$xi\phi$'. The clear difference is that while in the former 'f' is not quantifiable, in the latter 'ϕ' is.

[illegible] is, therefore, a straightforward alternative to predicate quantification [illegible]ing from the simple transformation of predicative formulae to attributive formulae. Indeed we could retain the standard syntactical formulation of second-order theory and regard second-stage logic (so far as its predicate-quantificational features are concerned) as arising from an alternative interpretation of formulae of the sort $(Uf)A$. The matter of sentential quantification, however, is much more difficult. Here, as we shall see presently, there is no such simple transformation. We shall, nevertheless, digress to develop systems which contain an analogue of sentential quantification; and this for three reasons. First, because the difficulties encountered are themselves instructive and suggest ways of getting to more adequate logics; secondly, because consistency proofs for systems which contain analogues of sentential quantification are simpler than corresponding proofs for systems which do not; thirdly, because the main system we develop, $\mathbf{2QS_6}$, though given a non-standard interpretation, in fact a clause interpretation, is syntactically equivalent to standard second-order theory (with attributive variables) and so provides a "normal" second-order significance logic for those who, unlike us, find no difficulty in accepting sentential quantification. We therefore carry an analogue of sentential quantification until §7.13, §5, where we finally discard it and revert to a more satisfactory two-sorted system $\mathbf{A2QS_6}$ which has a standard sentential interpretation and contains attributive quantification but no sentential quantification. $\mathbf{A2QS_6}$ can in fact be picked up at any stage of the development of $\mathbf{2QS_6}$ simply by dropping the clause quantifiers and clause interpretation, and making appropriate adjustments in the proofs of various metatheorems.

As an alternative to sentential quantification, there are two courses open to us, both of which were discussed in Chapter 3. We may derive subject factors from sentences by means of either §-conversion or *qu*-conversion. For although a sentential parameter P is not a subject, both $\S P$ and $qu(P)$ are.

Let us begin, then, by considering first whether we can provide an analogue of sentential quantification by introducing the descriptor '§', as the introduction of 'λ' enables us to provide an analogue of predicate quantification. Suppose we add the rule:

If A is a wff, $\S A$ is a clause.

This would give us formulae such as $\S P$ which, being subjects, are quantifiable; but again, being subjects, they stand in need of predicates. Now in Chapter 3, as with attributives, we allowed that-clauses as values of the subject arguments in predicate formulae of the sort FX. Predicates were thus automatically supplied. Here, however, if we are to regard clauses as a restricted class of subjects, as the quantifiable sentential variables of second-order logic are, they must be kept distinct from subject arguments and cannot be allowed as instantiations of them. Special predicates have therefore to be added to the logic so that wff can be constructed with clauses as subjects, just as the special two-place predicate '*i*' has to be added to enable wff to be constructed with attributives as (one of the) subjects. What is necessary is a set of predicates analogous to '**T**', '**F**', '**imp**' *etc.* of **SL**. These particular predicates, however, are at present inappropriate since they are only defined for use in conjunction with wfc (*i.e.* clauses constructed from statement-yielding sentences), but here, since the logic is three-valued, the clauses which are formed from the three-valued wff are not simply the earlier wfc; they include also clauses formed from nonsignificant sentences. Hence we should need versions of '**T**', '**F**', '**imp**', *etc.* (say '**T***', '**F***', '**imp***' *etc.*), applicable to nonsignificant clauses as well as to wfc. In general, therefore, as part of this way of developing $\mathbf{2QS_6}$, it would be necessary to construct a full three-valued analogue of **SL**, in effect the logic of three-valued assertions, extended to include quantification over clauses and linked to the sentential formulae of $\mathbf{QS_6}$ by the rule: if *A* is a (sentential) wff, §*A* is a clause.

To illustrate, we might introduce,

assertoric arguments: **p**, **q**, *etc.* (read 'that-*p*', 'that-*q*')
and assertoric predicates: **T***, **F***, **imp***, *etc.*

in addition to the usual,

sentential parameters: *p*, *q*, *etc.*
and sentential connectives: $\neg$, *D*, $\rightharpoonup$;

where the assertoric arguments (*i.e.* assertion variables or clause variables) are three-valued analogues of the statement variables of **SL** and the assertoric predicates are analogues of the **SL** predicates. Of course, it may be possible to develop the logic using the two-valued **SL** predicates themselves, instead of the special assertoric predicates, since formulae such as '**T*p**', say, may be two-valued even though '**p**' is three-valued, but we cannot prejudge the answer to this question at this stage.

Given this basis, wffs and clauses could then be defined as follows:

p, *q*, *etc.*, are wff;
p, **q**, *etc.* are clauses;
If α and β are clauses, $\mathbf{T^*}\alpha$, $\mathbf{F^*}\alpha$, $\alpha\,\mathbf{imp^*}\,\beta$, *etc.* are wff;
If *A* is a wff, §*A* is a clause;
If *A* and *B* are wff, so are $\neg A$, *DA* and $A \rightharpoonup B$.

Within such a system we could develop the theory of s-p transforms (§7.2, §1).

Quantification over the clause variables is then permissible, but not quantification over the sentential parameters. This, together with an appropriate extension to the usual predicate theory and the introduction of attributives would give us a three-valued second-stage logic containing an analogue of sentential quantification, but would take us a long way from the formal structure of second-order logic. For we now have two sets of variables in place of one and two sets of corresponding connectives. It is in fact possible to do without the assertoric variables, given '§' as a primitive, since '§*p*', '§*q*', *etc.*, do the same job as '**p**', '**q**', *etc.*, though their use makes for some notational simplification, especially in the quantifiers since we may write '(*U***p**)' in place of '(*U*§*p*)'. Given the distinction between clauses and sentences, however, it is not possible to manage with only one set of connectives. For since the ordinary connectives '$\neg$', '*D*' and '$\rightharpoondown$' are sentence-forming on sentences, they cannot be used to combine clauses; and since the assertoric predicates are sentence-forming on clauses, they cannot be used to combine sentences. Again, '§' is an essential operator in a logic in which a distinction is made between sentences and clauses since it is clause-forming on sentences, and therefore does a different job from each of the other two sets of connectives. Unless we have all these logical constants, therefore, it would not be possible to construct either a full set of sentential wff or a full set of clauses.

Much more important than this formal divergence from second-order theory, however, is the fact that by proceeding in this way we do not achieve what we set out to achieve. The logic which we have sketched will contain an analogue of sentential quantification but it will still contain sentential parameters which are unquantified, and indeed unquantifiable. We therefore achieve nothing in this way, so far as the present problem is concerned, and we might instead take the much simpler system $\mathbf{A2QS_6}$ which arises by adding attributives (and attributive quantification) to $\mathbf{QS_6}$ but which contains no sentential quantification.

The difficulty is that any system which contains a sentential theory, whether or not it also contains clauses, cannot be a fully closed system if sentential quantification is regarded as improper. Hence the addition of clause variables, either explicitly, or by way of the rule: if *A* is a wff, §*A* is a clause, is irrelevant to the problem of providing a system which is fully closed and in which sentential quantification is displaced by clause quantification. The solution is to develop a system in which clauses wholly displace sentences; that is, a system in which all wff stand for clauses.

This suggests an alternative approach which does not make use of '§' and which requires only one set of variables and connectives. A logic so developed will have the formal appearance of a standard extended sentential logic, and in general, when the predicate theory is added, it will have the formal appearance of a standard second-order theory (with attributives in place of predicates), but its interpretation will be very different. In fact, the simplest such system arises from a re-interpretation of second-order theory. Thus:—

First regard all atomic wff of $\mathbf{QS_6}$ as standing for clauses. For example, the sentential parameters '*p*', '*q*', *etc.*, are now to be regarded as assertoric arguments and, if desired, written '**p**', '**q**', *etc.* Since these variables cannot be combined using the ordinary sentential connectives, these will have to be re-interpreted also. It is not adequate, however, to re-interpret them as the assertoric predicates '**T***', '**imp***' *etc.* For although this would allow us to form sentential wff from clauses, we should then have no way of combining the wff so constructed: *e.g.*, we could not combine sentential wff such as '**T*p**' and '**pimp*q**' to yield further sentential wff (*e.g.* '**T*p⇀(pimp*q)**') unless we *also* introduce the sentential connectives. Such compounds cannot arise using the assertoric connectives, for formulae such as

(T*p) imp* (p imp* q)

are ill-formed since '**imp***' combines clauses but '**T*p**' and '**pimp*q**' stand for sentences, not clauses. In general, iteration is not possible with assertoric predicates, *e.g.* '**T*T*p**' stands neither for a sentence nor a clause (by contrast, '**T*§T*p**' would be well-formed). The solution is to re-interpret the connectives so that they are clause-forming on clauses. To mark this, we might label them with a subscript '*c*' and read them as follows:

$\neg_c$: that it is not true
D_c: that *D*
$\rightharpoonup_c$: that (... implies ...).

Since these combine with clauses to yield further clauses, we have formulae such as,

$\neg_c \mathbf{p}$; $\mathbf{p} \rightharpoonup_c \mathbf{q}$; *etc.*,

which, since '**p**', '**q**', *etc.*, are to be interpreted as 'that-*p*', 'that-*q*', *etc.*, read,

that it is not true that-*p*; that (that-*p* implies that-*q*);

and so on. Moreover, since these are clauses, further clauses can be constructed from them using the connectives. In particular, iteration is possible. Thus, for example, '$\neg_c \neg_c \mathbf{p}$' reads 'that it is not true that it is not true that-*p*'.

In general, therefore, if the formation rules of $\mathbf{QS}_6$ are retained, and this special interpretation is put on the sentential parameters and connectives, every wff stands for a clause. To achieve this interpretation fully, it is only necessary to specify appropriate interpretations for the quantifiers and the instantiation connective:

$(U_c u)$**A** where **A** is a clause (wff), reads: that for every *u*, that-*A* is true;
$(\mathcal{q}_c u)$**A** reads: that for some *u*, that-*A* is true;
$\mathrm{X} i_c \Psi$ reads: that X instantiates Ψ.

Hence the logic becomes a clause logic, and the "sentential" variables are now quantifiable since they are no longer sentential but assertoric.

The advantage of this approach is that the logic has the formal structure of second-order logic since it has only one set of "sentential" connectives, it does not contain '§' explicitly, and it can be fully closed since "sentential" quantification is in fact clause quantification. But it might be thought that the interpretation necessary to achieve this is too strained. For since the theses are clauses, they are not assertions. We recall, however, that if we take a Boolean interpretation of sentential logic (or more generally a cylindrical algebraic interpretation of quantification logic) by stipulating that the variables '*p*', '*q*', ..., range over sets α, β, ..., and interpreting '$\sim$' as the complement connective '$-$' and '$\vee$' as the union connective '$\cup$', then every wff stands for a *set* and is not a statement or assertion about sets: *e.g.* a formula of the form '$p \supset q$' stands for a set of the form $\bar{\alpha} \cup \beta$. In particular, every thesis stands for the universal set. The theses of the logic so interpreted are only made into assertions by adding to each the complex connective '$= \mathbf{U}$' ('is equal to the universal set'). Similarly the theses of the clause logic we propose could be made into assertions by prefixing each by '**T**' ('is true'), or simply by regarding the gate sign '$\vdash$' as the connective 'is true', and therefore as part of the thesis rather than simply a thesis-marker.

In general, standard quantification logic is characterised by the fact that all its wff are sentence-wff. This means that only one set of connectives is necessary—those which are sentence-forming on sentences—but it means, too, that the extension to a second-order theory, which includes an extended sentential logic, involves quantification over sentences and the acceptance of formulae such as '$(Up)p$' and '$(Up)(p \,\&\, q)$' which are without sense since the sentences over which the quantification is made, and which are now therefore to be conceived as subjects, have no predicates. If, on the other hand, we introduce clauses as quantifiable subjects, and retain sentential wff, then the logic becomes complicated by two sets of connectives, and in any case still contains unquantified sentential parameters or unacceptable sentential quantification. A logic which contains clauses and only one set of connectives cannot also contain sentences; it must therefore be a clause logic. These are the only possibilities. Other variants, which may differ in detail from the logics we have sketched, must fall into one or other of these broad types. Suppose, for example, we retain the sentential variables but re-interpret the connectives by suffixing them with 'that'. This kind of interpretation is suggested by W. and M. Kneale [1] (p. 539 ff). Thus, '$\neg$' reads 'it is not the case that'; '$\rightharpoondown$' reads 'provided that ... then that ...'; and so on. The formation rules remain the same and each wff (apart from sentential variables standing alone) appears to be an expression which can be interpreted as a remark involving clauses, not sentences: for example, '$\neg p$' and '$p \rightharpoondown q$' read, respectively, 'it is not the case that *p*' and 'provided that *p* then that *q*'. We therefore have a similar interpretation to that which we have put on '**F*p**' and '**p imp* q**' with the advantage that iteration is now possible (since all wff stand for sentences and the connectives are sentence-forming on sentences), and the apparent advantage that the variables are quantifiable since they are effectively clause-variables. This last, however, is an illusion. For under this interpretation the variables must remain sentential variables (otherwise we have a stuttering 'that' in formulae like '$\neg p$') and formulae such as '$(Up)p$' and '$(Up)(p \,\&\, q)$' contain no predicate to go with the quantified subjects. Such an approach therefore represents no advantage over ordinary extended logic. What we achieve in this way is simply a variant of the first broad type mentioned above, namely a logic in which all wff are sentential and in which quantification over sentences is illegitimate.

It should now be clear that although it is relatively easy to provide an analogue of predicate quantification, it is by no means easy to provide an analogue of sentential quantification by introducing '§', explicitly or implicitly, and constructing clauses as properly quantifiable subjects to take the place of the improperly quantified sentences.

It may seem, however, that there is a much simpler solution if we construct subjects by means of '*qu*' instead of '§'. For '*qu*' is already implicit in a standard interpretation of $\mathbf{QS}_6$, and so does not have to be introduced as a further primitive since, given this interpretation, formulae such as '*Tp*' and '*Sq*' are abbreviations for '*tru qu*(*p*)' and '*sig qu*(*q*)'. Here, therefore, we already have appropriate quantifiable subjects, '*qu*(*p*)', '*qu*(*q*)', *etc.*, and predicates '*tru*', '*sig*', *etc.* which go with them. Hence we may introduce the restricted subject quantifiers '(*Uqu*(*p*))', (*9qu*(*q*))', *etc.* and form formulae of the sort '(*Uqu*(*p*))*Tp*' in which the quantification is over acceptable subjects which are associated with appropriate predicates. Moreover, we may allow, as well formed, quantified sentential formulae in which no predicate appears, for example '(*Uqu*(*p*))*p*', '(*Uqu*(*p*))(*p* & *q*)'; for here, it seems, the quantification is merely vacuous not, as in extended formulae such as '(*Up*)*p*', improper (*cf.* §3.5 R1 (xiii)). If we then write '(*Up*)' as an (admittedly misleading) abbreviation for '(*Uqu*(*p*))' and similarly, '(*9p*)' as an abbreviation for '(*9qu*(*p*))', we obtain a logic which has exactly the same syntax as extended sentential theory but which differs in interpretation. In particular, quantification over any sentential variable which is not associated with an appropriate predicate appears to be vacuous. Thus, the quantification over '*p*', but not '*q*', is vacuous in '(*Up*)(*p*⇀(*9q*)*Tq*)' and its consequence '(*Up*)(*p*)⇀(*9q*)*Tq*'; both, however, are wffs, and the latter remains a consequence of the former as in standard quantification theory. By contrast, the quantification over '*p*' is not vacuous in those formulae which express explicitly what these try unsuccessfully to express, *viz.*: '(*Up*)(*Tp*⇀(*9q*)*Tq*)' and '(*Up*)*Tp*⇀(*9q*)*Tq*'. Most important, the quantification is always vacuous over the variables in sentential wff which contain no predicates: *e.g.* '(*Up*)*p*'.

There is, however, a serious difficulty present in such a system if we allow the usual rule that vacuous quantifiers can be deleted. For we should then be able to derive non-theses from theses (*cf.* §7.2,§1). As follows:

⊢*p*⇀*Tp*, $\mathbf{S}_6$;
⊢(*Up*)(*p*⇀*Tp*), generalisation;
⊢(*Up*)*p*⇀(*Up*)*Tp*, distribution;
⊢*p*⇀(*Up*)*Tp*, deletion of vacuous quantifier.

Clearly this last should not be a thesis since, given any true sentence, it would allow us to conclude, by detachment, that every sentence is true. In fact, if it were a thesis, it would not only raise interpretational problems but also formal ones since it implies the absolute (and simple) inconsistency of the system. For we have:

$\vdash (Up)Tp \rightharpoonup Tq$, instantiation;
$\vdash Tq \rightharpoonup q$, S_6;
So, from $\vdash p \rightharpoonup (Up)Tp$ and the transitivity of '$\rightharpoonup$',
$\vdash p \rightharpoonup q$.
Hence, $\vdash (p' \rightharpoonup p') \rightharpoonup q$;
therefore, $\vdash q$.

It is not obvious why this difficulty arises. If we expand the symbolism and write the formulae in primitive notation, then the second line of the "proof", for example, becomes $\vdash (Uqu(p))(p \rightharpoonup tru\ qu(p))$ and there is a case for saying that since '*p*' and '*qu*(*p*)' are different variables, so the quantifier is vacuous over the antecedent. This, apparently, justifies the later deletion. But of course, since *qu*(*p*) is a function of *p*, the two variables, though different, are not independent. There is, for example, a clear difference between
'$(Uqu(p))(p \rightharpoonup tru\ qu(p))$' and '$(Uqu(p))(q \rightharpoonup tru\ qu(p))$'.
If we remove the quantifiers from these, then in the latter we may substitute for '*q*' independently of substitutions for '*p*' in '*qu*(*p*)'; but in the former, any substitution for '*p*' will control the value of '*qu*(*p*)'. This means, of course, that if we had ordinary sentential quantifiers, then in an expression like '$(Up)(p \rightharpoonup tru\ qu(p))$', the variable '*p*' would be bound at both occurrences in the scope, just as each occurrence of '*x*' is bound in the scope of '$(Ux)(x < sin\ x)$', since the fact that the second occurrence appears as part of a function does not affect the binding. But here the question is rather different, namely whether '$(Uqu(p))$' binds '*p*' in '$(Uqu(p))(p \rightharpoonup tru\ qu(p))$', given that *qu*(*p*) is a function of '*p*' (*cf.* $(U\ sin\ x)(x < sin\ x)$). Perhaps one may say that though the quantifier '$(Uqu(p))$' does not bind '*p*' in the way that it binds '*qu*(*p*)', neither is it vacuous over '*p*'. Related to this are other problems to which we have not yet given clear answers: whether, for example, connectives such as '$\rightharpoonup$' absorb '*qu*'.

Such considerations make it clear that the proposal to provide an analogue of sentential quantification by falling back on *qu*-transformation is by no means straightforward. Far from providing an exact analogue of extended sentential logic, the proposal would in fact lead to a new logic with very special features.

We do not regard any of the courses we have indicated as being satisfactory, though we consider that each is viable in a way that ordinary extended logic is not. The least unsatisfactory is that which arises by reinterpreting QS_6 as a clause logic, and this is the course we propose to adopt.

It has limitations, however, and causes special difficulties when reduction definitions of sets and numbers are introduced. At that point we revert to a second-stage theory which contains no analogue of sentential quantification. Thus the clause logic which we develop, and which contains assertoric quantification, is a digression, but an important one because it seems to us to represent the only possible fully closed system which is interpretationally sound and which corresponds to standard second-order theory. In effect, our conclusion is that standard second-order theory fails at the level of interpretation unless it is interpreted as a clause logic.

We therefore set up the second stage logic **2QS$_6$** in a normal second-order way but regard the sentential parameters as assertoric variables and interpret the connectives and quantifiers in the way already described, *i.e.* as if with subscript '*c*'. To make these semantical conventions explicit, the assertoric variables should be expressed in Clarendon type and the clause connectives and quantifiers should be explicitly *c*-subscripted. We choose not to do this in order to emphasise that the major difference between second-stage logic so conceived and second-order logic with sentential quantification, is interpretational, not formal. This allows us to present **2QS$_6$** as a simple formal extension of **QS$_6$** in the way that second-order theory (with attributive variables) is presented as a formal extension of standard quantification theory.

System **2QS$_6$** has the same vocabulary as **QS$_6$** except that the symbol '*i*' is added to the improper primitive symbols; predicate parameters are replaced by:

attributive variables: ψ, ϕ, χ, θ;

and sentential parameters are replaced by (re-interpreted as):

assertoric variables: *p*, *q*, *r*.

Note: '*r*' is not here a restricted (two-valued) variable as in the previous chapters. We introduce it (briefly) as a three-valued variable to avoid complicating the symbolism by an excessive use of primes in some of the proofs.

There are three symbols rules, analogous to those for functions.

2QS$_6$ has three sorts of quantifiable variables: assertoric; attributive; and restricted subject, or *individual* variables; where clauses and attributives are not ($\neg$) individuals.

The formation rules (for well formed *clauses*) are like those for **QS$_6$** except at the following points:

(b) *A* is an atomic wff (read: well formed clause) iff (i) *A* is $(t_1, ..., t_n)\ i\ \Theta$, where Θ is an *n*-place attributive variable or constant and $t_1, ..., t_n$ are terms, or (ii) *A* is an assertoric variable or constant.

(c) (3) If *A* is a wff (read: well formed clause) and *u* is an individual, attributive or assertoric variable then $(Uu)A$ and $(\mathcal{G}u)A$ are wff.

System **A2QS$_6$** results from **2QS$_6$** by deleting the assertoric variables and quantifiers. The system may or may not have sentential parameters (*e.g.* it need not have if the sentential theory is set up in terms of schemata). All the results which we derive for **2QS$_6$** extend to **A2QS$_6$** provided assertoric quantifiers do not occur in the proofs. **A2QS$_6$** need not be interpreted as a clause logic, and in fact we shall take a normal sentential interpretation for **A2QS$_6$**.

§2 *A Semantics for* 2QS$_6$

A semantics for **2QS$_6$** may be specified in the same way as the semantics of **QS$_6$** except that assignments are made for new variables in terms of corresponding assignments for constants, and that clause (ii) is replaced by the clause:

$$v((t_1, ..., t_n)\ i\ \Theta) = t \text{ iff } T[(\mathrm{I}(t_1), ..., \mathrm{I}(t_n))\ i\ \mathrm{I}(\Theta))]\ etc.$$

But the semantics, like the syntax, is re-interpreted. The valuation function v is re-interpreted as a function from clauses to values from $\mathrm{V} = \{t,f,n\}$. The values of V may be construed either as clause values or, at the cost of still further notational ambiguity, as sentential values. In either case '$v(A) = t$' may be read 'that-A is true'.

As before a wff A is *valid for* I *over* S iff $v(A) = t$, for every assignment to the free variable of A, with respect to I over S; and A is *satisfiable for* I *over* S iff $v(A) = t$ for some assignment to the free variables of A with respect to I over S. A is *valid* iff it is valid for every interpretation over usual bases; *satisfiable* iff it is satisfiable for some interpretation over some usual basis.

A basis S is *countable* iff its domain d is countable. When d is countable so is e.

Interpretation I over S is *normal* iff all the axioms of **2QS$_6$** are valid for I over S and the transformation rules of **2QS$_6$** preserve validity wrt I over S, *i.e.* when the premisses of the rules are valid for I over S so are the conclusions.

Normality introduces a serious impurity into the semantics insofar as it makes subsequent semantical definitions depend on the axiomatisation of the logic; however this dependence can sometimes be removed.

A wff A of **2QS$_6$** is *secondarily valid* (*2-valid*) iff A is valid for every normal interpretation I over S; *secondarily satisfiable* (*2-satisfiable*) iff it is satisfiable for some normal interpretation.

A class Γ of wff of **2QS$_6$** is *simultaneously satisfiable for* I over S iff $v(A) = t$ for some uniform assignment to the free variables of each wff A of Γ.

Definitions:

$t =_{\mathrm{Df}} (Up)(p \rightharpoondown p)$; $\quad f =_{\mathrm{Df}} \neg t$; $\quad n =_{\mathrm{Df}} Dt$.

Theorem (80): A is satisfiable for I over S iff its q-closure A' is satisfiable for I over S.

Proof: If A is satisfiable then $v(A) = t$ for some assignment to the free variables of A. But then $v((9u)A) = t$; so $(9u)A$ is satisfiable. Iterate then until A' is obtained. Conversely, if A' is satisfiable (for I over S) then $v(A') = t$. So since A' is, in the main case, $(9u)A''$ (in the other case A' is A), $v((9u)A'') = t$. Hence $v(A'') = t$ for some value of u, so that A'' is satisfiable. Then, after iteration, A is satisfiable.

Theorem (81): A is valid for I over S iff its universal closure (U-closure) is valid for I over S. *Proof* is along the same lines as the preceding proof.

Theorem (82): If A is 2-valid then $\neg A$ is not 2-satisfiable.

Theorem (83): A finite set Γ of wff of $\mathbf{2QS_6}$ is simultaneously satisfiable for I over S iff $\Gamma_{\&}$, *i.e.* the conjunction by '&' of all wff in Γ, is satisfiable for I over S.

§3 *The Postulational Theory of* $\mathbf{2QS_6}$

$\mathbf{2QS_6}$ has as postulates the postulates A1, A2, A3, A6, A7, A8, A9, R1 and R2 of $\mathbf{QS_6}$, with just this difference: the restriction that variable u be a subject variable is amended. Thus A2, formulated for $\mathbf{2QS_6}$, reads:

A2. $\vdash (Uu)(A \rightharpoonup B) \rightharpoonup . A \rightharpoonup (Uu)B$, where u is a variable (individual, attributive or assertoric) which is not a free variable of A. Postulates A4 and A5 of $\mathbf{QS_6}$ are replaced, in $\mathbf{2QS_6}$, by the following postulates:

2A4. $\vdash (Uu)A \rightharpoonup \check{S}^{u}_{N} A\,|$, where either

(i) u is an individual variable and N is an individual term; or

(ii) u is an assertoric variable and N is a wff; or

(iii) u is an n-place attributive variable and N is an attributive term (*i.e.* given the vocabulary of $\mathbf{2QS_6}$, an attributive variable or constant). More explicitly, 2A4(iii) is the scheme:

$$\vdash (U\Psi)A \rightharpoonup \check{S}^{(u_1, \ldots, u_n)i\Psi}_{zi\Phi} A\,|,$$

where $u_1, \ldots, u_n$ are distinct individual variables and Φ is an attributive term. (The substitution notation in this case is effectively that of Church [4], p. 193).

2A5. $\vdash \check{S}^{u}_{N} A\,| \rightharpoonup (9u)A$, with the same conditions as 2A4.

In view of 2A4, A1 can be weakened to: if A is a tautology of $\mathbf{S_6}$ then $\vdash A$. As in the case of $\mathbf{QS_6}$, so here '9' and postulates concerning it can be definitionally eliminated.

T1. $\vdash (Uu)(A \rightharpoonup B) \rightharpoonup . (Uu)A \rightharpoonup B$

Proof: $(Uu)(A \rightharpoonup B) \rightharpoonup . A \rightharpoonup B$; $(Uu)A \rightharpoonup A$: both by 2A4

Result by A1.

T2. $\vdash(Uu)(A\rightarrow B)\rightarrow.(Uu)A\rightarrow(Uu)B$

Proof from T1, R2, A2.

T3. $\vdash(Uu)(A\rightarrow B)\rightarrow.A\rightarrow(\exists u)B$

T4. $\vdash(Uu)(A\rightarrow B)\rightarrow.(\exists u)A\rightarrow(\exists u)B$

T5. $\vdash(Uu)(A\simeq B)\rightarrow.(Uu)A\simeq(Uu)B$.

Proof: $A\simeq B\rightarrow.A\rightarrow B$ by A1

(a) $(Uu)(A\simeq B)\rightarrow.(Uu)A\rightarrow(Uu)B$ using R2, A2, T2

$(Uu)(A\simeq B)\rightarrow.(Uu)B\rightarrow(Uu)A$ similarly

$A\simeq B\rightarrow.SA\simeq SB$ by A1

$(Uu)(A\simeq B)\rightarrow.(Uu)SA\rightarrow(Uu)SB$ as in proof of (a)

$(Uu)(A\simeq B)\rightarrow.S(Uu)A\rightarrow S(Uu)B$ using A1, A6, A7

$(Uu)(A\simeq B)\rightarrow.S(Uu)B\rightarrow S(Uu)A$ similarly.

Result then using A1.

T6. $\vdash(Uu)(A\simeq B)\rightarrow.(\exists u)A\simeq(\exists u)B$

Proof parallels proof of T5.

T7. $\vdash T(Uu)A\simeq(Uu)TA$

T8. $\vdash T(\exists u)A\simeq(\exists u)TA$.

Proofs as for $\mathbf{QS_6}$.

T9. $\vdash(Uu)A\simeq(Uv)\,\mathrm{S}^{1u}_{v}A\,|$, where v is not free in A and u is not free in $(Uv)\,\mathrm{S}^{1u}_{v}A\,|$.

Proof: As in T14 for $\mathbf{QS_6}$ it follows:
$\vdash(Uu)A\rightarrow(Uv)\,\mathrm{S}^{1u}_{v}A\,|$ and $\vdash(Uv)\,\mathrm{S}^{1u}_{v}A\,|\rightarrow(Uu)A$.

By similar arguments it follows $\vdash(Uu)SA\rightarrow(Uv)\,\mathrm{S}^{1u}_{v}SA\,|$ and $\vdash(Uu)\,\mathrm{S}^{1u}_{v}SA\,|\rightarrow(Uu)SA$; and thence $\vdash S(Uu)A\rightarrow S(Uv)\,\mathrm{S}^{1u}_{v}A\,|$ and $\vdash S(Uv)\,\mathrm{S}^{1u}_{v}A\,|\rightarrow S(Uu)A$. The result then follows using A1 and $p\rightarrow q\rightarrow.q\rightarrow p\rightarrow.Sp\rightarrow Sq\rightarrow.Sq\rightarrow Sp\rightarrow.p\simeq q$.

T10. $\vdash(\exists u)A\simeq(\exists v)\,\mathrm{S}^{1u}_{v}A\,|$, where v is not free in A and u is not free in $(\exists v)\,\mathrm{S}^{1u}_{v}A\,|$.

Theorem (84): If B results from A by substitution of B_1 for A_1 at zero or more places in A and if the variables $u_1, u_2, ..., u_n$ include at least those free variables of A_1 and B_1 which also occur as bound variables of A, then

$$\vdash(Uu_1), ..., (Uu_n)(A_1\simeq B_1)\rightarrow.A\simeq B.$$

Proof is by induction over the number of occurrences of connectives and quantifiers in A.

Theorem (85): If u is a variable which is not free in A_1 and v is a variable of the same kind as u which does not occur in A_1, if B results from A by substituting $\mathrm{S}^{1u}_{v}A_1\,|$ for a particular occurrence of A_1 in A, and if $\vdash A$, then $\vdash B$. (*Rule of change of bound variable*).

Proof: By T9 and T10, $\vdash(Uu)A' \simeq (Uv)\,\mathrm{S}'^{u}_{v}A' \mid$ and $\vdash(\mathfrak{q}u)A' \simeq (\mathfrak{q}v)\,\mathrm{S}'^{u}_{v}A' \mid$.

The result then follows by applying the previous theorem.

Theorem (86): $\vdash A \twoheadrightarrow \vdash \check{\mathrm{S}}^{u}_{N}A \mid$, where u is a variable and N is an expression of the same kind as u (*Rules of substitution for variables*).

Proof from R2 and 2A4.

Proof from hypotheses is defined as for $\mathbf{QS_6}$ except that under clause 5 any change of bound variable conforming with the rule of change of bound variable is admitted. Then, as for $\mathbf{QS_6}$ this deduction theorem follows:

Theorem (87): If $A_1, A_2, ..., A_n \vdash B$ then $A_1, A_2, ..., A_{n-1} \vdash A_n \rightharpoondown B$.

T11. $\vdash \mathrm{S}^{p}_{t}TA \mid \& \; \mathrm{S}^{p}_{f}TA \mid \& \; \mathrm{S}^{p}_{n}TA \mid \rightharpoondown .(Up)TA$, where assertoric variable p is not bound in A.

Proof: $\vdash p \simeq t \rightharpoondown .A \simeq \mathrm{S}^{p}_{t}A \mid$ by the substitution theorem

$\rightharpoondown .TA \equiv \mathrm{S}^{p}_{t}TA \mid$

Similarly $\vdash p \simeq f \rightharpoondown .TA \equiv \mathrm{S}^{p}_{f}TA \mid$

and $\vdash p \simeq n \rightharpoondown .TA \equiv \mathrm{S}^{p}_{n}TA \mid$

$\vdash p \simeq t \vee p \simeq f \vee p \simeq n$ by A1

Hence by A1 again,

$$\vdash TA \equiv \mathrm{S}^{p}_{t}TA \mid \quad \vee .TA \equiv \mathrm{S}^{p}_{f}TA \mid \quad \vee .TA \equiv \mathrm{S}^{p}_{n}TA \mid$$

Since $\vdash STA$ and $\vdash TA_1 \equiv TB \vee TA_2 \equiv TB \vee TA_3 \equiv TB \rightharpoondown .\, TA_1 \,\&\, TA_2 \,\&\, TA_3 \rightharpoondown TB$,

$$\vdash \mathrm{S}^{p}_{t}TA \mid \& \; \mathrm{S}^{p}_{f}TA \mid \& \; \mathrm{S}^{p}_{n}TA \mid \rightharpoondown TA.$$

The result then follows by generalisation and distribution.

T12. $\vdash \mathrm{S}^{p}_{t}SA \mid \& \; \mathrm{S}^{p}_{f}SA \mid \& \; \mathrm{S}^{p}_{n}SA \mid \rightharpoondown (Up)SA$, where assertoric variable p is not bound in A.

Proof: Similar to previous theorem, with 'SA' replacing 'TA'.

T13. $\vdash(\mathfrak{q}p)TA \rightharpoondown .\mathrm{S}^{p}_{t}TA \mid \vee \mathrm{S}^{p}_{f}TA \mid \vee \mathrm{S}^{p}_{n}TA \mid$, where assertoric variable p is not bound in A.

Proof: As in T11,

$$\vdash TA \equiv \mathrm{S}^{p}_{t}TA \mid \vee .TA \equiv \mathrm{S}^{p}_{f}TA \mid \vee .TA \equiv \mathrm{S}^{p}_{n}TA \mid .$$

Hence since $\vdash STA$ and

$\vdash TA_1 \equiv TB \vee TA_2 \equiv TB \vee TA_3 \equiv TB \rightharpoondown .TB \rightarrow TA_1 \vee TA_2 \vee TA_3$;

$\vdash TA \rightharpoondown .\mathrm{S}^{p}_{t}TA \mid \vee \mathrm{S}^{p}_{f}TA \mid \vee \mathrm{S}^{p}_{n}TA \mid .$

The result then follows by generalisation and distribution of '$\mathfrak{q}$'.

T14. $\vdash(\mathfrak{q}p)SA \rightharpoondown .\mathrm{S}^{p}_{t}SA \mid \vee \mathrm{S}^{p}_{f}SA \mid \vee \mathrm{S}^{p}_{n}SA \mid$, where p is not bound in A.

§4 *Soundness, Completeness and Skolem-Löwenheim Theorems for* $\mathbf{2QS_6}$

Theorem (88): Every theorem of $\mathbf{2QS_6}$ is 2-valid, and therefore valid.

But unlike $\mathbf{QS_6}$ the converse, that every valid wff is a theorem, does not hold (in view of Gödel's incompleteness theorem, since arithmetic can be developed in $\mathbf{2QS_6}$); only the weakened converse, that every 2-valid wff is a theorem, holds.

Theorem (89): Every $\neg$-consistent class of wff of **2QS$_6$** is simultaneously satisfiable wrt some normal interpretation over a countable basis.

Proof parallels in many respects the corresponding *Theorem* (27) for **QS$_6$**. Let Γ be a $\neg$-consistent class of wff of **2QS$_6$**. Let K be the **2QS$_6$** logic obtained by adding to **2QS$_6$** the symbol set $\{b_1^0, b_2^0, ...; b_1^1, b_2^1, ...; ...; b_1^k, b_2^k, ...; ...\}$, where $b_1^0, b_2^0, ...$ are denumerably many new individual constants and for each positive k, $b_1^k, b_2^k, ...$ are countably many attributive constants. (Exactly how many are required in each class will depend on Γ).

Let K_0 be the class of wff provable from Γ under proof from hypotheses in system K. Then

(1) K is $\neg$-consistent.

(2) K_0^- is $\neg$-consistent. As for *Theorem* (27).

Let the denumerable set of wff of K_0^- with at most one individual or attributive variable free be represented:

$A_1(u_{i_1}), A_2(u_{i_2}), ..., A_n(u_{i_n}), ...$

Much as for *Theorem* (27) choose sequences $b_{j_1}, ..., b_{j_k}; ... c_{j_1}, ..., c_{j_k}, ...$ of new constants, such that b_{j_k} and c_{j_k} are n-place iff i_{i_k} is n-place. Systems K_n^-, L_n^- and L_ω are formed, where S_k and T_k are these formulae:

$S_k: A_k(b_{j_k}) \rightarrow (Uu_{i_k})A_k(u_{i_k})$

$T_k: (\exists u_{i_k})A_k(u_{i_k}) \rightarrow A_k(c_{j_k})$.

(6) L_ω is $\neg$-consistent. As for *Theorem* (27).

Let J be the maximal $\neg$-consistent extension of L_ω. A *countable basis* S_1 is now specified.

d is the domain of all individual constants of K_1 (*i.e.* of J),

e is the class of attributive constants of K_1 (*i.e.* of J),

f is the class of all operations on d.

(7) S_1 is countable.

An individual term is *closed* iff it contains no free variables. *Interpretation* I_1 over S_1 is defined as for *Theorem* (27). Note that where Ψ^n is an attributive constant of K_0,

(8′) $I_1(\Psi^n) = \Psi^n$

(8″) $I_1(t) = t$; $I_1(f) = f$; $I_1(n) = n$.

For $\vdash Tt$, $\vdash Ff$, $\vdash \neg Sn$; and hence $Tt, Ff, \neg Sn \in J$.

(9) For every closed wff A of K_0, for valuation v_1 w.r.t. I_1 on S_1,

(a) $v_1(A) = t \equiv . TA \in J$

(*b*) $v_1(A) = f \equiv . FA \in J$

(c) $v_1(A) = n \equiv . \neg SA \in J$

Proof of (9):

1. Induction basis, where P is a sentential constant or Ψ^n is an attributive constant, is much as for *Theorem* (27).
2. Induction step.

Cases 1, 2, 3 as for *Theorem* (27).

Case 4. A is $(Uu_n)B$, where u_n is an individual or attributive variable. Then B is $A_k(u_n)$ for some k. If u_n is an individual variable then the case is as for *Theorem* (27); if u_n is an attributive variable, then the case is a slight modification of that of *Theorem* (27). The modification consists in replacing 'for every (some) $b_j \in$d', *e.g.* by 'for every (some) $b_j \in$e of the same number of places as the variable in question'. For example,

4(a) begins:

$v_1((Uu_n)A_k(u_n)) = t \equiv . v_1(A_k(b_j)) = t$, for every $b_j \in$e of the same number of places as u_n, *i.e.* b_j is n-place iff u_n is n-place.

Case 5. Similarly modifies case 5 in *Theorem* (27).

Case 4.1 A is $(Uu)B$, where u is an assertoric variable. Then B is of the form $C(u)$ with at most u free. Thus A is $(Uu)C(u)$. Now change the initial bound variable of A to a new variable, r say, such that r is not bound in $C(u)$. Thus A is the same as $(Ur)C(r)$.

(a) $v_1((Ur)C(r)) = t \equiv . v_1(C(r)) = t$ for assignments t, f, and n to r.

$\equiv v_1(C(t)) = t \;\&\; v_1(C(f)) = t \;\&\; v_1(C(n)) = t$

$\equiv TC(t) \in \mathrm{J} \;\&\; TC(f) \in \mathrm{J} \;\&\; TC(n) \in \mathrm{J}$

$\equiv TC(t) \;\&\; TC(f) \;\&\; TC(n) \in \mathrm{J}$

$\supset (Ur)TC(r) \in \mathrm{J}$, by T11.

Also $\vdash T(Ur)C(r) \rightharpoonup (Ur)TC(r)$

$\rightharpoonup TC(t)$.

Similarly $\vdash T(Ur)C(r) \rightharpoonup TC(f)$; $\vdash T(Ur)C(r) \rightharpoonup TC(n)$.

Since $\vdash STA$, by A1, $\vdash T(Ur)C(r) \rightharpoonup . TC(t) \;\&\; TC(f) \;\&\; TC(n)$.

Hence $T(Ur)C(r) \in \mathrm{J} \supset . TC(t) \;\&\; TC(f) \;\&\; TC(n) \in \mathrm{J}$.

(b) $v_1((Ur)C(r)) = f \equiv . v_1(C(r)) = f$ for t, f or n

and $v_1(C(r)) = n$ for none of t, f and n;

$\equiv \mathrm{J} \vdash FC(t) \vee FC(f) \vee FC(n)$

$\&. \; SC(t) \;\&\; SC(f) \;\&\; SC(n)$.

Now $\vdash S(Ur)C(r) \rightharpoonup . SC(t) \;\&\; SC(f) \;\&\; SC(n)$ by A1, A4.

Since $\vdash F(Ur)C(r) \rightharpoonup S(Ur)C(r)$, $\vdash F(Ur)C(r) \rightharpoonup SC(t) \;\&\; SC(f) \;\&\; SC(n)$.

Also $\vdash F(Ur)C(r) \rightharpoonup \sim T(Ur)C(r)$

$\rightharpoonup \sim (Ur)TC(r)$

$\rightharpoonup \sim (TC(t) \;\&\; TC(f) \;\&\; TC(n))$ by T11.

Combining results,

$\vdash F(Ur)C(r) \rightharpoonup . \sim (TC(t) \;\&\; TC(f) \;\&\; TC(n)) \;\&\; SC(t) \;\&\; SC(f) \;\&\; SC(n)$

$\rightharpoonup (FC(t) \vee FC(f) \vee FC(n)) \;\&\; SC(t) \;\&\; SC(f) \;\&\; SC(n)$ by A1.

Conversely, the preceding line, by A1,

$\rightharpoonup \sim (TC(t) \;\&\; TC(f) \;\&\; TC(n)) \;\&\; SC(t) \;\&\; SC(f) \;\&\; SC(n)$

$\rightharpoonup \sim T(Ur)C(r) \;\&\; S(Ur)C(r)$ by A4, T12, A1

$\rightharpoonup F(Ur)C(r)$.

(c) $v_1((Ur)C(r)) = n \equiv v_1(C(r)) = n$ for t, f or n

$\equiv \mathrm{J} \vdash \neg SC(t) \vee \neg SC(f) \vee \neg SC(n)$.

But $S(Ur)C(r) \rightarrow (Ur)SC(r)$

$\rightarrow SC(t) \,\&\, SC(f) \,\&\, SC(n).$

Also $SC(t) \,\&\, SC(f) \,\&\, SC(n) \rightarrow S(Ur)C(r)$ by T12.

Contraposing both these results, and using $\vdash SSP$, it follows

$$J\vdash \neg SC(t) \vee \neg SC(f) \vee \neg SC(n) \equiv . J\vdash \neg S(Ur)C(r).$$

Strictly, it suffices to prove just two of cases (a) (b) and (c) since the third case then follows: thus, for example, the complexities of (b) can be avoided. *Case* 5.1 A is $(\mathfrak{q}u)B$ where u is an assertoric variable. As in case 4.1, after change of initial bound variable, A is deductively the same as $(\mathfrak{q}r)C(r)$.

(a) $v_1((\mathfrak{q}r)C(r)) = t \equiv v_1(C(r)) = t$, for t, f or n

$\equiv J\vdash TC(r)$, for t, f or n

$\equiv J\vdash TC(t) \vee TC(f) \vee TC(n).$

Since $TC(t) \rightarrow (\mathfrak{q}r)TC(r)$

$\rightarrow T(\mathfrak{q}r)C(r)$ *etc.*

$TC(t) \vee TC(f) \vee TC(n) \rightarrow T(\mathfrak{q}r)C(r),$ (ϵ).

Conversely, $T(\mathfrak{q}r)C(r) \rightarrow (\mathfrak{q}r)TC(r)$

$\rightarrow TC(t) \vee TC(f) \vee TC(n)$ by T13.

(b) $v_1((\mathfrak{q}r)C(r)) = f \equiv v_1(C(r)) = f$, for some t, f, or n

$\equiv J\vdash FC(t) \vee FC(f) \vee FC(n) \,\&. \sim TC(t) \,\&\, \sim TC(f) \,\&\, \sim TC(n).$

Now $F(\mathfrak{q}r)C(r) \rightarrow . \sim T(\mathfrak{q}r)C(r)$

$\rightarrow . \sim TC(t) \,\&\, \sim TC(f) \,\&\, \sim TC(n)$ by (ϵ).

Also $F(\mathfrak{q}r)C(r) \rightarrow . S(\mathfrak{q}r)C(r)$

$\rightarrow . SC(t) \vee SC(f) \vee SC(n)$ by T14

$\rightarrow . FC(t) \vee FC(f) \vee FC(n),$ by A1 using the other consequent of $F(\mathfrak{q}r)C(r)$ as well.

Conversely, first,

$\sim TC(t) \,\&\, \sim TC(f) \,\&\, \sim TC(n) \rightarrow , \sim (\mathfrak{q}r)TC(r)$ by T13

$\rightarrow . \sim T(\mathfrak{q}r)C(r).$

Second, $FC(t) \vee FC(f) \vee FC(n) \rightarrow . SC(t) \vee SC(f) \vee SC(n)$

$\rightarrow (\mathfrak{q}r)SC(r)$ by T14

$\rightarrow S(\mathfrak{q}r)C(r).$

Combining these and using $\vdash \sim Tp \,\&\, Sp \rightarrow Fp$, the converse follows.

(c) $v_1((\mathfrak{q}r)C(r)) = n \equiv v_1(C(r)) = n$, for t, f, and n

$\equiv J\vdash \sim SC(t) \,\&\, \sim SC(f) \,\&\, \sim SC(n)$

$\equiv J\vdash \sim S(\mathfrak{q}r)C(r)$ by A5 and T14.

(17) Γ is simultaneously satisfiable for I_1 over S_1. For proof, a simultaneous substitution for the values of the free variables of Γ is made. By (8), (8′) and (8″) this will provide a uniform assignment to free variables of members of Γ. The substitution is as follows: if X is an individual variable substitute

a new individual constant of K_0^- in such a way that a different individual constant is substituted for the free occurrences in Γ of each different individual variable; if Ψ is an n-place attributive variable in a similar way substitute a new distinct n-place attributive constant of K_0^-; and if P is an assertoric variable substitute t, f, or n according as TP, FP or $\neg SP \in J$. Let the set resulting from Γ by substitution be Γ'; and let the wff resulting from member B of Γ by the substitution be B'. For every $B' \in \Gamma'$, $B' \in J$, by rules for substitution of constants upon variables.

Since $B' \in J$, $TB' \in J$. Since B' is closed, by (9), $v_1(B') = t$. Since for every B' in Γ', $v_1(B') = t$ wrt I_1 over S_1, Γ' is simultaneously satisfied for I_1 over S_1. So too then is Γ.

(17′) In case Γ contains only finitely many wff, a simpler procedure is this: Γ is satisfiable iff $\Gamma_{\&}$, the conjunction of wff of Γ is satisfiable. $\Gamma_{\&}$ is satisfiable iff the 𝑞-closure of $\Gamma_{\&}$ is satisfiable. Also the 𝑞-closure of $\Gamma_{\&} \in J$; hence this 𝑞-closure has value t for I_1 over S_1, and so is satisfiable for I_1 over S_1. Hence Γ is satisfiable for I_1 over S_1.

(18) Γ is simultaneously satisfiable over a countable basis.

(19) I_1 over S_1 is normal. First, an axiom B of $\mathbf{2QS_6}$ is valid for I_1 over S_1 iff its universal closure c(B) is valid for I_1 over S_1, by a previous theorem. But c(B)$\in$J, hence Tc(B)$\in$J. Thus since Tc(B) is closed, $v_1(\text{c}(B)) = t$ always. Since then c(B) is valid for I_1 over S_1, so is B. Second, if the premisses of a rule are valid, then the universal closures of these premisses are valid. Since these closures are closed and have value t always they belong to J.

ad R1. If c(A)$\in$J and c($A \rightharpoonup B$)$\in$J then c(A)$\rightharpoonup$c(B)$\in$J by distribution. Hence c(B)$\in$J, so Tc(B)$\in$J. Then $v_1(\text{c}(B)) = t$ always, so B is valid.

ad R2. c(A) coincides, perhaps after change of initial bound variables, with c((Uu)A). Hence the conclusion is valid if the premiss is.

Theorem (90): Every simultaneously satisfiable class of wff of $\mathbf{2QS_6}$ is $\neg$-consistent.

Proof as for $\mathbf{QS_6}$, *Theorem* (28).

Theorem (91) (*Skolem-Löwenheim*): Every simultaneously satisfiable class of wff of $\mathbf{2QS_6}$ is simultaneously satisfiable wrt some normal interpretation over a countable basis.

Proof by the two preceding theorems.

Theorem (92) (*Completeness*): Every secondarily valid wff of $\mathbf{2QS_6}$ is a theorem.

Proof: Let A be a 2-valid wff. Then $\neg A$ is not 2-satisfiable, so $\{\neg A\}$ is not simultaneously satisfiable for any normal interpretation. Therefore $\{\neg A\}$ is $\neg$-inconsistent. Hence, as in completeness theorem for $\mathbf{QS_6}$, *Theorem* (30), A is a theorem.

§5 *The Extended Assertoric Logic* **ES$_{6'}$** *and the Consistency of* **2QS$_6$**

The extended assertoric logic **ES$_6$** is that subsystem of **2QS$_6$** where the only variables and constants are assertoric variables and constants. The extended logic **ES$_{6'}$** is obtained from **ES$_6$** by adding, as primitives, the assertoric constants '*t*', '*f*' and '*n*' and the axioms on these constants $\vdash Tt$, $\vdash Ff$, $\vdash \sim Sn$. To be explicit **ES$_{6'}$** has as vocabulary:

The improper clause symbols: $\rightharpoonup$, $\urcorner$, D, (,), U, q

Assertoric variables: $p, q, p', q', p'', \ldots$

Assertoric constants: $t, f, n, p_0, q_0, p_1, q_1, p_2, \ldots$

The postulates of **ES$_{6'}$** are these:

A1. If A is an **S$_6$**- tautology, then $\vdash A$ (*i.e.* $\vdash_{ES_{6'}} A$)
A2. $\vdash (Up)(A \rightharpoonup B) \rightharpoonup . A \rightharpoonup (Up)B$, where p is not free in A.
A3. $\vdash (Up)(A \rightharpoonup B) \rightharpoonup . (qp)A \rightharpoonup B$, where p is not free in B.
2A4. $\vdash (Up)A \rightharpoonup \check{S}^{p}_{B}A|$
2A5. $\vdash \check{S}^{p}_{B}A| \rightharpoonup (qp)A$
A6. $\vdash (Up)SA \rightharpoonup S(Up)A$
A7. $\vdash S(Up)A \rightharpoonup (Up)SA$
A8. $\vdash (qp)SA \rightharpoonup S(qp)A$
A9. $\vdash S(qp)A \rightharpoonup (qp)SA$
A10. $\vdash Tt \;\&\; Ff \;\&\; \sim Sn$
R1. $\vdash A, \vdash A \rightharpoonup B \longrightarrow \vdash B$
R2. $\vdash A \qquad \longrightarrow \vdash (Up)A$.

By defining 'q' in the usual way postulates A3, 2A5, A8 and A9 may be eliminated.

The semantics of **ES$_{6'}$** developed is a specialisation of that of **2QS$_6$**: the semantical basis S for **2QS$_6$** is simply omitted. Thus, an *interpretation* I of **ES$_{6'}$** is an assignment I which assigns to each assertoric constant one of the values t, f or n, and to constant t value t, constant f value f and constant n value n. The *range* of an assertoric variable is then: $\{t,f,n\}$. Given I, **ES$_{6'}$**-values for all wff are defined inductively by clauses (i), (iii)–(vii) of those given for **2QS$_6$** (but only assertoric variables are considered).

A wff A is **ES$_{6'}$**-*valid for* I iff $v(A) = t$, for every assignment to the free variables of A, with respect to I. A wff is **ES$_{6'}$**-valid iff it is valid for every interpretation. The definitions of **ES$_{6'}$**-satisfiability and **ES$_{6'}$**-simultaneous-satisfiability are similarly related to definitions for **2QS$_6$**.

Theorem (93): Every theorem of **ES$_{6'}$** is **ES$_{6'}$**-valid. *Proof* is by induction over the length of proofs.

Corollary: **ES$_{6'}$** is absolutely and simply consistent. For n is not **ES$_{6'}$**-valid.

Theorems T1–T14 of **2QS$_6$** are also theorems, for assertoric variables, of **ES$_{6'}$**. Also proved as for **2QS$_6$** are theorems permitting inter-substitution of strong equivalents, change of bound variables and substitution for variables, and a deduction theorem.

$\mathbf{ES_6}$ and $\mathbf{ES_{6'}}$ are equivalent systems.

Theorem (94):

(1) Every wff of $\mathbf{ES_6}$ is a wff of $\mathbf{ES_{6'}}$; and every wff of $\mathbf{ES_{6'}}$ is, using definitions of 't', 'f', and 'n' for $\mathbf{ES_6}$, a wff of $\mathbf{ES_6}$.

(2) Every theorem of $\mathbf{ES_6}$ is a theorem of $\mathbf{ES_{6'}}$, and every theorem of $\mathbf{ES_{6'}}$ is, using definitional replacement, a theorem of $\mathbf{ES_6}$.

Theorem (95): Every $\neg$-consistent class of wff of $\mathbf{ES_{6'}}$ is $\mathbf{ES_{6'}}$-simultaneously satisfiable.

Proof is a special case of the corresponding theorem for $\mathbf{2QS_6}$. Let Γ be a $\neg$-consistent class of wff of $\mathbf{ES_{6'}}$; and let J be the maximal-consistent extension of Γ. Then an *interpretation* I_1 is defined by the requirement:

(iii) If P is an assertoric constant then $I_1(P) = t, f,$ or n according as TP, FP, or $\sim SP \in J$. Then

(9) For every closed wff A of Γ, for valuation v_1 wrt I_1

(a) $v_1(A) = t \equiv . TA \in J$

(b) $v_1(A) = f \equiv . FA \in J$

(c) $v_1(A) = n \equiv . \sim SA \in J$

is proved by induction, precisely as for $\mathbf{2QS_6}$. It also follows as a special case of the result for $\mathbf{2QS_6}$, that Γ is $\mathbf{ES_{6'}}$-simultaneously satisfiable w.r.t. to I_1.

Theorem (96): Every $\mathbf{ES_{6'}}$-valid wff is a theorem of $\mathbf{ES_{6'}}$.

Corollary: $\mathbf{ES_{6'}}$ is complete.

T15. $(Up)A \simeq . S^p_t A | \,\&\, S^p_f A | \,\&\, S^p_n A |$,

where assertoric variable p is not bound in A.

Proof: From T11, $Tp \,\&\, Tq \simeq T(p \,\&\, q)$ and $(Up)TA \rightharpoonup T\, S^p_t A |$, *etc.* it follows,

(1) $T(Up)A \rightharpoonup T(S^p_t A | \,\&\, S^p_f A | \,\&\, S^p_n A |)$, and

(2) the converse of (1).

Similarly from T12 and $Sp \,\&\, Sq \simeq S(p \,\&\, q)$ and 2A4 it follows

(3) $S(Up)A \rightharpoonup S(S^p_t A | \,\&\, S^p_f A | \,\&\, S^p_n A |)$, and

(4) the converse of (3).

From (1), (2), (3), (4) and

$\vdash (Tp \rightharpoonup Tq) \,\&\, (Tq \rightharpoonup Tp) \,\&\, (Sp \rightharpoonup Sq) \,\&\, (Sq \rightharpoonup Sp) \rightharpoonup . p \simeq q$,

the result follows. The restriction is a consequence of T11 and T12.

T16. $(\exists p)A \simeq . S^p_t A | \vee S^p_f A | \vee S^p_n A |$,

where assertoric variable p is not bound in A.

Proof: The method is similar to T15, but T13 and T14 are used in place of T11 and T12, and the following connections are then taken advantage of:

$$\vdash T(p \vee q) \simeq Tp \vee Tq; \vdash S(p \vee q) \simeq Sp \vee Sq$$

Theorem (97): $\mathbf{ES}_{6'}$ is a decidable system.

Proof: Given any wff A of $\mathbf{ES}_{6'}$ an equivalent associated wff A^* of $\mathbf{S}_{6'}$, a system obtained from $\mathbf{S}_6$ by adding constants t, f, and n and the postulate $Tt \& Ff \& \sim Sn$, may be obtained as follows:

Replace each well-formed part of A of the forms $(Up)B$ and $(\exists q)C$ respectively according to T15 to T16. The resulting quantifier-free wff A^* is also a wff of $\mathbf{S}_{6'}$. (Note here that in passing from $\mathbf{ES}_{6'}$ to $\mathbf{S}_{6'}$, the interpretation changes from well-formed clauses to wff). Moreover, by the theorem on substitutivity of strong equivalents $\vdash A \simeq A^*$. Thus A is a theorem iff A^* is a tautology of $\mathbf{S}_{6'}$, a matter decidable by a matrix decision procedure.

The method provides a decision procedure for provability, and hence by the completeness theorem, for validity for $\mathbf{ES}_{6'}$.

As far as quantification goes, $\mathbf{ES}_{6'}$ is really a theory of values t, f, and n, not of assertions. For in view of T15 and T16 every assertion about all assertions and some assertions is tantamount to assertions about t, f, and n. To obtain a quantified theory of assertions that does not so reduce we shall have to enrich $\mathbf{ES}_{6'}$ with intensional functors (as is done below with $\mathbf{EIS}_{6'}$).

For any wff A of $\mathbf{2QS}_6$ an associated wff h(A) of $\mathbf{ES}_{6'}$ is obtained as follows:

(1) Delete all expressions of the form (Uu) and $(\exists u)$, where u is an individual variable.

(2) Where $\Psi_1, \ldots, \Psi_m$ are the distinct attributive variables and constants in A, and $P_1, \ldots, P_m$ are distinct assertoric variables not occurring in A, replace every well-formed part $(u_1, \ldots, u_n)i\Psi_j$ of A by P_j and each quantifier $(U\Psi_j)$ and $(\exists\Psi_j)$, respectively, by (UP_j) and $(\exists P_j)$.

Theorem (98): If A is a theorem of $\mathbf{2QS}_6$ then h(A) is a theorem of $\mathbf{ES}_{6'}$.

Proof: For each axiom scheme Ai of $\mathbf{2QS}_6$, h(Ai) is an axiom scheme of $\mathbf{ES}_{6'}$. Further the rules R1 and R2 preserve the mapping.

Theorem (99): $\mathbf{2QS}_6$ is simply consistent.

Proof: If $\mathbf{2QS}_6$ is simply inconsistent then for some wff B, both B and $\sim B$ are theorems. Thence, by the previous theorem,

h(B) and $\sim$ h(B) [= h($\sim B$)]

are theorems of $\mathbf{ES}_{6'}$, and

(h(B))* and ($\sim$ h(B))* [= $\sim$ (h(B))*]

are theorems of $\mathbf{S}_{6'}$. That is $\mathbf{S}_{6'}$ is inconsistent. The result follows from the consistency of $\mathbf{S}_{6'}$ by contraposition.

Corollary: $\mathbf{2QS}_6$ is absolutely and Post consistent.

7.13 Developments of Second-stage Quantified Significance Logics

§1 *Further second-stage logics*

First, second-stage logics may be built on weaker bases. Logic $\mathbf{2QS_4}$, for example, extends $\mathbf{QS_4}$ to second stage in precisely the way that $\mathbf{2QS_6}$ extends $\mathbf{QS_6}$ to second stage. The main metatheorems proved for $\mathbf{2QS_6}$ all go over to $\mathbf{2QS_4}$. Further, as we shall see presently, $\mathbf{2QS_6}$ can be enlarged by the addition of various axioms and in exactly the same way so, too, can $\mathbf{2QS_4}$. Second-stage versions of $\mathbf{QS_2}$, $\mathbf{QS_3}$ and $\mathbf{QS_5}$ are also readily designed, and extended assertoric logics resembling $\mathbf{ES_6}$ can be constructed on top of any S-logic which includes $\mathbf{S_2}$.

Secondly, second-stage logics may be built using a choice operator which is clause-forming on clauses, in place of the quantifiers. Thus, in place of assertoric quantificational schemata, we should require:

$\xi 2.\quad \vdash \check{\mathsf{S}}^{p}_{B} A \,|\!\rightharpoonup A_p(\xi q A)$ and

$\xi 3.\quad \vdash \check{\mathsf{S}}^{p}_{B} SA \,|\!\rightharpoonup SA_p(\xi q A)$

and we could then define: $(\mathsf{q} p)A =_{\mathrm{Df}} A_p(\xi p A)$. The clause formula '$(\mathsf{q} p)p$' is thus defined by the clause 'ξpp' or else is not well defined, according as degenerate cases of the definition are sanctioned or not.

Logics with assertoric terms such as $\xi q A$ bring up in a dramatic way the interpretational problems generated by second-stage logics. In fact, ξ has three distinct roles in these systems: (i) It combines with individual variables to form individuals from clauses (wff). Thus $\xi x A$ is, as in §7.11, a well-formed term, and hence is an admissible replacement for the individual variables 'x', 'y', *etc.*; it is not an admissible replacement for the assertoric variables 'p', 'q', *etc.* (ii) It combines with attributives to form attributives from clauses (wff). For example, $\xi \Psi A$ is an attributive and hence an admissible replacement for attributive variables but, again, not for assertoric variables. (iii) It combines with assertoric variables to form clauses from clauses. Thus, $\xi p A$ is itself a clause and therefore an admissible value for the assertoric variables. In this last role, but not the others, its interpretation is prefixed by 'that', *i.e.* '$\xi p A$' reads 'that for any p ...'. It is this use of 'ξ' which is expressed in the schemata $\xi 2$ and $\xi 3$ above. Hence in subsequent theses such as $\vdash \xi q A \vee \neg \xi q A$, the connectives have to be regarded as clause-forming on clauses, as in $\mathbf{2QS_6}$, and understood as being c-subscripted. Such theses could in fact be written more revealingly as $\vdash \xi \mathbf{q} \mathbf{A} \vee_c \neg_c \xi \mathbf{q} \mathbf{A}$. In general, then, '$\mathbf{q}$' and '$\xi \mathbf{q} \mathbf{A}$' are expressions of the same kind, namely clauses (wff), and assertoric terms such as '$\xi \mathbf{q} \mathbf{A}$' can be taken as values of the variables by means of which they are formed.

Thirdly, second-stage logics may be enriched using special predicates such as '*E*', and they may be modalised. Modalisation of $\mathbf{2QS_6}$, to obtain $\mathbf{2QIS_6}$, say, is obtained simply by extending $\mathbf{QIS_6}$ to second stage: in particular, the Barcan wff are now assumed for all variables, individual, assertoric and attributive.

Although the syntax of $\mathbf{2QIS_6}$ is got by a straightforward extension of $\mathbf{QIS_6}$, its semantics requires new details; for assertions can no longer be contracted to values in the presence of quantification. Since two assertions with the same value may have different modal properties, theorems T11–T14 break down, and a contraction to values would not be appropriately invariant under intersubstitution of assertions with the same value in wff. The interpretation of assertions over domain $V = \{t, f, n\}$ has to be replaced by an interpretation of assertions over a (possibly infinite) domain δ—of modal values, assertions, closed wff, or some such, so long as sufficiently many appropriate discriminations are made, and each element of δ has a value from V associated with it. Then for assertoric constant P_0, $I(P_0, H) \in \delta$, δ being non-null; and $v(P_0, H)$ is the value of $I(P_0, H)$, *e.g.* $v(P_0, H) = n$ iff the value from V of $I(P_0, H)$ is *n*. The quantifier rules are adjusted to the new domain; thus for a given assignment to the free variables of *B*,

$$v((UP)B, H) = t \text{ iff } v(B, H) = t \text{ for every assignment of an element of } \delta \text{ to } P.$$

Validity, and other semantical notions, are defined as for $\mathbf{QIS_6}$. Then theoremhood and 2-validity coincide for $\mathbf{2QIS_6}$. The soundness half of the proof requires an extension of lemma (A) of §6.3 in order to verify instantiation of assertoric variables by wff. In the completeness half of the proof we ω-complete each con set H by adding S_k and T_k (of the corresponding proof for $\mathbf{2QS_6}$) not only for individual and attributive variables but for assertoric variables as well.

As part of second-stage multiply modal logics we have system $\mathbf{IS_6}$ extended by assertoric quantifiers or choice operators. We consider briefly a system $\mathbf{EIS_6}$ obtained by adding 'ξ' to $\mathbf{IS_6}$. Where *A* is a (clause) wff ξpA is an assertoric term. In the formation rules we treat assertoric terms like assertoric variables. $\mathbf{EIS_6}$ adds to $\mathbf{IS_6}$ the schemes $\xi 2$ and $\xi 3$ for assertoric terms and the rules:

RTξ. $\vdash A \twoheadrightarrow \vdash \Phi_i A$, where *A* is a theorem of $\mathbf{S_6}$ or an instance of D or $\xi 2$ or $\xi 3$;

RG. $\vdash A \twoheadrightarrow \vdash (U\phi)A$.

An $\mathbf{EIS_6}$-model $\mathscr{M}$ is a structure $\mathscr{M} = \langle G, K, ..., R_m, \delta, c, I \rangle$ where

$\langle G, ..., R_m \rangle$ is an $\mathbf{IS_6}$ model structure (*i.e.* model less its valuation function), δ is a non-null domain of assertions each with exactly one of values *t*, *f* and *n*, c is a choice function, and I is an assignment function which assigns to each initial wff *P* of $\mathbf{EIS_6}$ with respect to each set-up $H \in K$ an element of δ. Then

$v(P,\mathrm{H})$ is the value of $\mathrm{I}(P,\mathrm{H})$. The valuation rules for connectives of $\mathbf{IS}_6$ are as for $\mathbf{IS}_6$-semantics. If for some variable P, $v(A,\mathrm{H}) = t$,

$$\mathrm{I}(\xi PA,\mathrm{H}) = \mathrm{c}(\lambda \mathrm{I}(P,\mathrm{H})(v(A,\mathrm{H}) = t)); \text{ and otherwise}$$
$$\mathrm{I}(\xi PA,\mathrm{H}) = \mathrm{c}(\lambda \mathrm{I}(P,\mathrm{H})(v(A,\mathrm{H}) \neq n)).$$

Theorem (100):

(1) If Γ is a con class of $\mathbf{EIS}_6$ there is an $\mathbf{EIS}_6$-model with a denumerable domain δ which satisfies Γ.
(2) B is $\mathbf{EIS}_6$-valid iff B is a theorem of $\mathbf{EIS}_6$.

In $\mathbf{EIS}_6$ we can formulate explicitly many features intensional connectives may have, but which are inexpressible in $\mathbf{IS}_6$; in particular we can add to $\mathbf{EIS}_6$ theses asserting that modal connectives are not value-functional. For example we can add to $\mathbf{EIS}_6$ the postulate:

$$(\mathit{9}p,q)\,.\,Tp \simeq Tq \;\&\; \neg(\Box Tp \simeq \Box Tq).$$

For the semantics we then add the expected modelling condition

$$(P\mathrm{H})(\mathrm{G}\,\mathrm{R}_{\Box}\mathrm{H} \;\&\; \mathrm{H} \neq \mathrm{G}),$$

which entails that the actual world is not the only modal set-up, that—what is true—there are worlds beyond G.

§2 *The weakness of completeness: extensions of* $\mathbf{2QS}_6$ *by abstraction.*

The logic $\mathbf{2QS}_6$ may well be regarded as unnecessarily weak, because of the restricted forms of the axiom scheme 2A4(iii). For example, instead of 2A4(iii), a standard second stage scheme would be

$$2\mathrm{A}4(t)\ \vdash (U\psi)A \rightarrow \check{\mathrm{S}}^{(u_1,\ldots,u_n)i\psi}_{B} A\,|,$$

where ψ is an n-place attributive variable and $u_1, \ldots, u_n$ are distinct individual variables. (Compare Church [1], p. 297. The substitution notation in this scheme is that of Church.) In fact it is not possible to prove in $\mathbf{2QS}_6$ such (apparent) truths as $(\mathit{9}\psi)(Ux)T(xi\psi)$ and $(Ux)(\mathit{9}\psi)(xi\psi)$; for countermodels to these wff can be constructed. With the adoption of 2A4(t), and its mate 2A5(t), such results are readily proved, using such theorems as
$\sim T(xi\psi) \vee T(xi\psi)$ or $SS(xi\psi)$.
However a postulate much weaker than 2A4(t) also suffices for these particular results.

With respect to two-valued systems an alternative to the introduction of 2A4(t) with advantages—it eliminates a complex substitution notation and it provides an intuitively more accessible axiom—is the adoption of an abstraction axiom, written for one place attributives:

$(P\psi)(Ux)(xi\psi \equiv B(x))$, where $B(x)$ is a wff which does not contain ψ free. Various many-valued analogues of such an axiom can be designed. For our investigations the most attractive abstraction scheme is that which replaces '$\equiv$' by '$\simeq$'. In the general sense, where no further restrictions are made on wff B, the scheme will then be:

(A) $(\mathcal{I}\psi)(U\underset{\sim}{x})(\underset{\sim}{x}i\psi \simeq B)$,

where ψ is not free in wff B. The restriction, that ψ is not free in B, appears at once when (A) is deduced from 2A5(t). And were the restriction not imposed inconsistency would result immediately, e.g. by taking B as $\urcorner(\underset{\sim}{x}i\psi)$. But the need for the restriction perhaps emerges best from a consideration of what such an abstraction axiom is supposed to do, and why it has some plausibility. It is supposed to provide a single new attributive equivalent to a compound expression (which is in turn usually taken to be the same as a compound predicate). In terms of our analysis it combines the claims (1) that every sentential (clause) form provides a predicate, and (2) that every predicate yields by λ-conversion an attributive. Thus the postulate is supposed to capture the notion that given any sentential (clause) form a single equivalent attributive can be abstracted or defined. For the introduced attributive to fulfil these roles it must be a new distinct attributive, otherwise the definition will suffer from circularity, and the attributive will not be eliminable.

Various grades of attributive abstraction can be admitted, these varying according as different conditions are imposed on the sort of wff B can be. Simple restrictions can be imposed on B by limiting the kind of vocabulary that can be used in it; more complex restrictions would also limit the ways in which B is built up out of its given vocabulary. Contrary to popular logical opinion, an assumption of unlimited attributive abstraction does not strike us as intuitively obvious. The abstraction assumption is not "obvious" when B contains such non-classical connectives as '$\urcorner$' and 'D'. In terms of our diagnosis the trouble is in the λ-conversion principle, with (2) not (1). For

$$\underset{\sim}{x}i(\lambda \underset{\sim}{y} f(\underset{\sim}{y})) \simeq f(\underset{\sim}{x})$$

does not hold universally, as applications of impredicativity paradoxes readily show.

We consider stronger systems $\mathbf{2Q_2}$–$\mathbf{2Q_5}$, obtained by adding to $\mathbf{2QS_6}$ stronger abstraction schemes. Thus $\mathbf{2Q_2}$ is obtained by adding (A) subject to the further condition: and B contains only classical connectives and the universal quantifier 'U'; $\mathbf{2Q_3}$ by the slightly more liberal condition: and B contains only clean connectives and 'U'; $\mathbf{2Q_4}$ by the requirement that B contains only strong normal connectives and the corresponding quantifiers. Finally $\mathbf{2Q_5}$ adds (A) to $\mathbf{2QS_6}$. Given the restriction on subject variables to individual variables which do not take attributives as values the syntactic consistency proof extends to $\mathbf{2Q_5}$.

Theorem (101): Each of the systems $\mathbf{2Q_2}$–$\mathbf{2Q_5}$ is simply and absolutely consistent.

Proof: It suffices to show that

h($(\mathcal{q}\psi)(Ux_1) \ldots (Ux_n).(x_1, \ldots, x_n)i\psi \simeq B$) with ψ not free in B, is a theorem of $\mathbf{ES}_{6'}$, that is that $(\mathcal{q}p)(p \simeq$ h$(B))$, with p not free in h(B) (an alphabetic change of bound variable may be needed in the transformation if ψ occurs bound in B), is a theorem of $\mathbf{ES}_{6'}$. It is; for when p is not free in h(B),

$\vdash$h$(B) \simeq$ h$(B) \rightarrow .(\mathcal{q}p)(p \simeq$ h$(B))$.

This proof breaks down, however, as soon as attributive variables are equated with subject variables, *i.e.* if second-stage restrictions are lifted. The completeness proof is not so sensitive. Completeness theorems can be established for each of $\mathbf{2Q}_2$–$\mathbf{2Q}_5$ (and indeed even if second-stage restrictions are removed). The relevant semantical notions are relativised as follows: Interpretation I over S is $\mathbf{2Q}_i$-normal (where $i = 1$–5) iff all the axioms of $\mathbf{2Q}_i$ are valid for I over S and the transformation rules of $\mathbf{2Q}_i$ preserve validity w.r.t. I over S. A wff A of $\mathbf{2Q}_i$ is $\mathbf{2Q}_i$-2-*valid* iff A is valid for every $\mathbf{2Q}_i$-normal interpretation; $\mathbf{2Q}_i$-2-*satisfiable* iff it is satisfiable for some $\mathbf{2Q}_i$-normal interpretation. (Validity and satisfiability *for* are those already defined for $\mathbf{2QS}_6$).

Theorem (102): Every theorem of $\mathbf{2Q}_i$ is $\mathbf{2Q}_i$-2-valid (where $i = 1$–5).

Proof is by induction over the length of proofs of theorems.

But an axiom A of $\mathbf{2Q}_i$ is $\mathbf{2Q}_i$-2-valid iff it is valid for every normal interpretation I, that is iff it is valid for every interpretation for which it is valid. Similarly the rules trivially preserve 2-validity.

Less trivial is the next result,

Theorem (103): Every theorem of $\mathbf{2Q}_i$ is $\mathbf{2Q}_i$-valid.

Proof is a direct extension of the analogous *Theorem* (88) for $\mathbf{2QS}_6$.

We consider briefly the unrestricted abstraction axiom:

ad (A). It suffices to show that, for arbitrarily selected I on S,

$v((\mathcal{q}\psi)(Ux_1) \ldots (Ux_n)(x_1, \ldots, x_n)i\psi \simeq B) = t.$

Suppose, for a reduction that it does not; then, since

$v(A \simeq B) \neq n, v((Ux_1) \ldots (Ux_n)((x_1, \ldots, x_n)i\psi \simeq B)) \neq t$

for *any* assignment to attributive ψ. But ψ may be so assigned that

$v((x_1, \ldots, x_n)i\psi) = v(B).$

Theorem (104): Every $\neg$-consistent class of wff of $\mathbf{2Q}_i$ is $\mathbf{2Q}_i$-simultaneously satisfiable wrt some normal interpretation over a countable basis.

Proof is exactly as for $\mathbf{2QS}_6$, *Theorem* (89).

Theorem (105): Every $\mathbf{2Q}_i$-simultaneously satisfiable class is $\mathbf{2Q}_i$-simultaneously satisfiable wrt some normal interpretation over a countable basis.

Theorem (106) (*Weak completeness*): Every $\mathbf{2Q}_i$-2-valid wff is a theorem of $\mathbf{2Q}_i$.

Corollary: Every $\mathbf{2Q}_i$-2-valid wff is $\mathbf{2Q}_i$-valid.

Proof: Every $\mathbf{2Q}_i$-2-valid wff is a $\mathbf{2Q}_i$ theorem, and every $\mathbf{2Q}_i$ theorem is $\mathbf{2Q}_i$-valid.

That these different second-stage systems are all complete underscores the weakness of the completeness results. In fact completeness using secondary validity is something of a formal swindle; for secondary validity is deliberately aligned using the axiomatisation of the logic.

The addition of abstraction axioms is by no means the only way of enlarging $\mathbf{2QS}_6$. Significance analogues of axioms of choice and infinity, for instance, may also be added.

§3 *Extensions of second-stage logics*

Many grammatically correct sentences, some significant and some not, cannot be directly or satisfactorily expressed, and many valid arguments cannot be formulated, in the framework of second-stage logic or its reductions. Consider, first of all, such examples as 'The property of roundness, as distinct from Meinong's round square, is not an individual', 'John thought about his uncle and duplicity'. 'Triangularity is abstract but neither red nor triangular', 'That happiness is felt entails that some property is felt, which in turn entails that some property has some property.' More generally, although certain properties of properties and attributes of classes can be expressed in second-stage logics, it is not possible within them to generalise about properties of properties, and so on. In consequence such theories as those of the identity of attributes and of real numbers (see Hilbert & Ackermann [1], p. 158 ff) cannot be fully treated. (Of course identity of attributes, *e.g.*, can be treated using a second-stage postulate theory, much as identity can be added postulationally to quantification logic: but an explicit definition yielding the requisite properties—a logicistic treatment—cannot be provided).

Within a significance framework there is no very good reason—Russell's theory of types included—for adopting the usual way of extending second-stage logic to cope with (some of) these relational sentences, namely adopting a still higher order (type) theory. For an adequate assessment of significance and truth a much more flexible logic than a rigid type hierarchy is wanted. (Russell's theory of types, and fixed type theories of significance generally, will be discussed in volume 2).

Many of the inadequacies of second-stage logic result from the grammatical exclusion of attributives from various subject positions. Yet since they are subjects, there is no reason why they should not occupy these positions. We shall extend $\mathbf{2QS}_6$ to a system $\mathbf{HQS}_6$ (shortened to **HQ**) which admits attributives to subject positions. It is claimed, by proponents of type theory, that such a move cannot be made consistently, because it allows the formulation of such logical paradoxes as the impredicativity paradox in the resulting system. But in this case the danger of encountering paradoxes is exaggerated: in fact the system **HQ** is consistent, and versions of Russell's impredicativity paradox appear simply as formal fallacies.

HQ, like $\mathbf{2QS_6}$, is given a clause interpretation; and in general it has the same vocabulary and the same formation rules. There are, however, two important differences.

The first difference is purely notational and technical, carrying no semantic overtones: we replace assertoric terms (variables and constants) by zero-place attributive symbols. Thus, in place of P, Q, P_0, Q_0, *etc.*, we write λF^0, λG^0, λF^0_0, λG^0_0, *etc.*, which are then abbreviated Φ^0, Ψ^0, Φ^0_0, Ψ^0_0, *etc.* We wish to emphasise, however, that terms such as F^0 are not predicate terms, and similarly, those such as Ψ^0 are not attributive terms. They are assertoric terms distinguishable from predicate and attributive terms by the zero superscript. Predicate and attributive terms continue to have (implicitly or explicitly) non-zero superscripts which indicate the number of subjectival places associated with them. In particular, therefore, for reasons given in Chapter 3, we are not identifying clauses with predicates, not even with degenerate predicates, at the semantic level.

The second syntactic difference is a consequence of our intention to admit attributives to positions at present restricted to individual terms, and this does, of course, have semantic implications. From the point of view of the background language, since attributives (*e.g.* abstract nouns) are subject terms, *all* the variables in basic $\mathbf{2QS_6}$-wff of the form $(X_1, ..., X_n)i\Psi$ are subject variables. But since individual and attributive terms, being restricted terms, are formally distinguishable, we may (misleadingly) distinguish *subjectival positions*, namely those reserved for individual terms, from *predicative positions* reserved for attributive terms. (The tags are of course suggested by the equivalent subject-predicate form $\Psi^*(X_1, ..., X_n)$ to which $(X_1, ..., X_n)i\Psi$ converts (see §7.13, §4)). Using this nomenclature, the characteristic feature of **HQ** is that it opens subjectival positions to attributive terms. This brings us nearer to the logic of Chapter 3. Further, since clauses are subject terms, we may allow these too in subjectival positions. This explains why it is convenient to represent assertoric terms as zero-place attributives, for we may simply say that all attributive symbols of zero or more places are admissible in subjectival positions. Thus, the formation rules of **HQ** differ from those of $\mathbf{2QS_6}$ only in the following particulars:

(1) Where Ψ is a zero-place attributive term (*i.e.* a variable or a constant), and B is a wff, Ψ, $D\Psi$, $\neg\Psi$, $(\Psi \rightharpoonup B)$ and $(B \rightharpoonup \Psi)$ are wff.

(2) Where Ψ is an attributive term of zero or more places and Φ is an attributive term which is not zero-place, $\Psi i\Phi$ is a wff.

So, for instance, expressions like $\sim(\Psi i\Psi)$ are well-formed where Ψ is an attributive of more than zero places. As before, the connectives have a clause interpretation, *i.e.* as D_c, $\neg_c$, $\rightharpoonup_c$ and i_c. Individual variables and constants are perhaps better renamed (purely) *subjectival* variables and constants in **HQ**.

The postulates of **HQ** are the same as those of **2QS**$_6$ except that the instantiation schemes 2A4 and 2A5 have to be enlarged, and the substitution notation used in the new schemes has to be redefined in order to avoid variable clashes generated by the new symbolism. The instantiation schemes of **HQ** are as follows:—

HA4. $\vdash (Uu)B \rightharpoondown \check{S}^u_v B \mid,$

where (i) u is a subjectival variable and v is a subjectival term or an attributive variable which does not occur in B, or (ii) u is an attributive variable and v is an attributive term of the same number of places.

HA4(n) $\vdash (U\Psi)B \rightharpoondown \check{S}^{(X_1, \ldots, X_n)i\Psi}_A B \mid,$

where Ψ is an n-place attributive variable and $X_1, \ldots, X_n$ are distinct subjectival variables.

HA5 and HA5(n) are formulated analogously. The substitution notation used in HA4(n) is extended to cope with attributives occurring in subjectival positions by tightening the second proviso in Church's account ([4], pp. 192–3) of predicate substitution notation: Church's "(2). for each ordered n-tuple $a_1, \ldots, a_n$ of individual variables or individual constants ..." is replaced by "(2′). for each ordered n-tuple $u_1, \ldots, u_n$ of terms, subjectival or attributive ...".

Theorem (107): **HQ** is consistent.

The same syntactical proof that shows that **2QS**$_6$ is consistent, by mapping it into **ES**$_{6'}$, also shows that **HQ** is consistent. This observation and the above formulation of **HQ** are due, in essentials, to N. Cocchiarella [1]; (an early version of this section was completely rewritten in the light of this article). In fact Cocchiarella introduces a two-valued version **T*** of **HQ**, and argues that **T*** provides, effectively, the original logical context within which the reasoning of Russell's impredicativity paradox was codified, and that accordingly the paradox argument involves but a simple formal fallacy. But the claim that Russell's reasoning can be satisfactorily codified in **T*** seems to us decidedly suspect, since, for example, **T*** does not admit versions of the abstraction axiom which Russell certainly regarded as admissible.

Theorem (108): **2Q**$_5$ is a proper sublogic of **HQ**.

For the following abstraction principle is derivable in **HQ**:

(B) $\vdash (\mathfrak{q}\Psi)(UX_1, \ldots, X_n)((X_1, \ldots, X_n)i\Psi \simeq A)$

where A is a wff, containing all (or some) of the subjectival variables $X_1, \ldots, X_n$, but not containing Ψ free. In view of the consistency theorem a paradox cannot be got using this principle; but it is worth seeing why versions of the impredicativity paradox are not derivable. In significance logic the usual impredicativity paradox, generated by the formula $\sim (u i u)$, is, of course, not a problem. Given the naive abstraction principle for attributives, an attempt to derive the paradox simply shows, what Russell concluded, that it is not significant to assert that impredicativity is impredicative and not significant to assert that impredicativity is not impredicative. However the naive abstraction principle does lead to inconsistency using variants of the impredicativity paradox formulated using non-classical connectives; a simple example is furnished by the formula $\neg(u i u)$. Then by the abstraction principle: $(q\Psi)(U\mathrm{X})(\mathrm{X}i\Psi \simeq \neg\, \mathrm{X}i\mathrm{X})$, whence $(q\Psi)(\Psi i\Psi \simeq \neg\,(\Psi i\Psi))$, contradicting the theorem $(U\Psi^{0}) \sim (\Psi^{0} \simeq \neg\,\Psi^{0})$. In **HQ** such a derivation is prevented because $\neg(\mathrm{X}i\mathrm{X})$ is not a wff; it has an individual term in a predicative position.

It is at this point that the categorial distinction between individual subjects and attributives, reflected in **HQ** in the *grammatical* requirement that individual expressions may not occupy predicative positions, assumes fundamental importance. This version of the impredicativity paradox is blocked then in the characteristic way that modern type-logics block logical paradoxes —by syntactical division of the class of subjects. In the case of **HQ** this is simply the same bifurcation of subjects into individual and attributive subjects that is adopted in second-stage logics. This simple bifurcation, unlike the elaborate classifications of type theory, has considerable initial appeal. For we know, more or less, how to divide subjects up into individual and attributive subjects; indeed our tests for subjects and predicates (in §3.2) can, combined with conversion, be extended to make such a distinction. And after all, are not sentences such as '..... instantiates Socrates', where an individual expression occupies a predicative position, ungrammatical? (If instead we say that such sentences are grammatical but nonsignificant, then it is a comparatively straightforward matter to reformulate **HQ** to mirror this feature. This reformulation would accord better with the significance way of doing things.) Not too much hangs on the term 'individual'; the only metaphysical force that 'individual subject' need carry, in distinction from 'attributive', is that individual subjects cannot occupy predicative positions. But this limitation is enough to diminish the importance of individuals in the formulation of impredicativity paradoxes.

For in fact Russell's impredicativity paradox is typically derived by considering properties of *properties*: it is properties which are not (truly) properties of themselves which are impredicative, and accordingly impredicativity is a property of properties. Thus the expected paradox-generating formulae in significance logic is not $\sim T(u i u)$, but is instead of the form $\sim T(\Phi i \Phi)$. What then prevents the derivation of the paradox-producing principle

(C) $(q\Psi)(U\Phi)(\Phi i\Psi \simeq \neg\,(\Phi i\Phi))$,

and of the corresponding general abstraction scheme

$(q\Psi)(U\Phi_1 \ldots \Phi_n)((\Phi_1, \ldots, \Phi_n)i\Psi \simeq A(\Phi_1, \ldots, \Phi_n))$?

We explain briefly why (C) and the one-place case of the general scheme, *viz.*

(D) $(q\Psi)(U\Phi)(\Phi i\Psi \simeq . A(\Phi))$,

with Ψ not free in $A(\Phi)$, are not forthcoming. First, neither HA4(i) nor its consequence, $\vdash(Uu)B \rightarrow (U\Phi)\check{S}^u_\Phi B \,|$, with Φ not occurring in B, applied to scheme (B) yield (D) in cases where $A(\Phi)$ has type-violating components; for example to obtain $A(\Phi)$ in the form ... $(\Phi i\Phi)$ ... it would be necessary to begin with the ill-formed (or nonsignificant) form ... (XiX) ... which (B) of course excludes. Secondly, HA5(n) applied to the thesis $(U\Phi)(A(\Phi) \simeq A(\Phi))$ does not lead to (D) because in whichever way the substitution step is carried out conditions on legitimate substitution are violated: hence the formal fallacy in the paradox argument. Consider the use of HA5(n) needed for (C), *viz.*

$(U\Phi)(\neg\Phi i\Phi \simeq . \neg\Phi i\Phi) \rightarrow (q\Psi)(U\Phi)(\Phi i\Psi \simeq . \neg\Phi i\Phi)$,

where the antecedent $(U\Phi)(\neg\Phi i\Phi \simeq . \neg\Phi i\Phi)$ *should* be

$\check{S}^{Xi\Psi}_{\neg Xi\Phi}\,(U\Phi)(\Phi i\Psi \simeq . \neg\Phi i\Phi)\,|$.

It is not however; for Φ occurs free in $\neg(Xi\Phi)$ but would become bound upon substitution, thereby violating Church's condition (1) for non-vacuous substitution. Moreover the substitution cannot be alternatively construed as a case of

$\check{S}^{\Phi i\Psi}_{\neg\Phi i\Phi}\,(U\Phi)(\Phi i\Psi \simeq \neg\Phi i\Phi)\,|$

since HA5(n) is not asserted for substitutions on wff with attributives occupying subjectival places. It is not asserted in such cases for the reason that, if it were, clashes of free and bound variables upon substitution, which should be ruled out as illegitimate, would slip through in a disguised form. In the above substitution, for example, the differential treatment of subjectival and predicative occurrences of Φ in $\neg(\Phi i\Phi)$ with respect to bondage, clear enough intuitively, would be covered up completely.

The technique by which **HQ** successfully avoids variable clashes suggests that a satisfactory unrestricted quantification logic may be got by using the separation features of type theory to avoid variable clashes and vicious vicious circles. Such an unrestricted theory **H**, reminiscent of Quine's system **NF**, admits wff of the form $(u_1, \ldots, u_n)iv$ with unrestricted subject variables (of $\mathbf{QS}_6$) on both the left and right of 'i'. Like **HQ** the postulates of **H** differ from those of $\mathbf{QS}_6$ (or $\mathbf{2QS}_6$, with attributive variables construed either as further general variables or as restricted variables) only as regards instantiation schemes. The schemes replacing A4, for example are as follows:

H4. $\vdash(Uu)B \rightarrow \check{S}^u_v B\,|$,

where u is a variable and v a term (of the same number of places—if grammar rather than significance is to determine this matter).

$$\text{H4}(n) \quad \vdash(Uv)A \rightarrow \check{\mathsf{S}}_{B}^{(u_1, \ldots, u_n)\, iv} A \,|,$$

with $u_1, \ldots, u_n, v$ distinct variables and A and B well-typed, where a wff C is *well-typed* iff there is a way of supplying type classifications to components of C so that C is well-formed according to the theory of types (using, say, Carnap's [3] classification of types, p. 84 ff). Many of the arguments for a type-theoretic logic (especially those in *PM*), can be recast as arguments for amending substitution principles in the style of H4–H4(n) or two-valued versions of these schemes. The following abstraction principle follows at once in **H**:

$$\vdash(qv)(Uu_1 \ldots u_n)((u_1, \ldots, u_n)iv \simeq A),$$

where A is a well-typed wff not continuing v free. This principle includes a 3-valued version of the stratified scheme of **NF** as a special case. In view of the known disadvantages of **NF**, it would seem advisable to advance beyond **H** to a system which includes 3-valued unextensionalised **ML** rather than **NF** as a special case. But we postpone this matter, and the main issues concerning choice of an unrestricted quantificational logic until volume 2. One central point we should however anticipate: the big issue concerning choice of an unrestricted quantification logic comes down, for significance logics as for 2-valued systems, to the question of how the logical paradoxes are to be *convincingly* resolved. Now although a significance resolution of the paradoxes is certainly worth considering carefully, we do not believe that a significance resolution is really satisfactory. More promising are resolutions using incompleteness either as a further value or as introduced through new connectives (see §3.9). In the first case the development of specifically 3-valued systems is of course very much to the point; in the second case, however, where paradoxical items such as impredicativity and the Russell class turn out to be impossible items, intensionally enriched two-valued systems provide the main object of study. But whichever case obtains, when a logic which provides a satisfactory resolution of the paradoxes is achieved, it should be "significancized".

Though **H** exploits the technique of **HQ** for avoiding variable clashes, it is a very different theory from **HQ**; for **HQ** is but a conservative extension of $\mathbf{2Q_5}$ with some curious theorems in the enriched notation, whereas **H** is an extension of **HQ** only on pain of inconsistency.

Theorem (109):

(i) **HQ** is a conservative extension of $\mathbf{2Q_5}$, *i.e.* every theorem of $\mathbf{2Q_5}$ is a theorem of **HQ**, and every theorem of **HQ** which is a wff of $\mathbf{2Q_5}$ is a theorem of $\mathbf{2Q_5}$.

(ii) If **H** is consistent, **H** is not an extension of **HQ**.

Proof of (i) is like that of Cocchiarella [1], §6; it differs only in that 's-transform' is alternatively defined representationwise on the connectives of $\mathbf{2Q_5}$; thus $s(\neg A) = \neg s(A)$; $s(DA) = Ds(A)$; *etc.*

Proof of (ii) adapts Cocchiarella [1], §5. Thus where,

$$\Psi \overset{\text{H}}{\approx} \Phi =_{\text{Df}} (U\Theta)(\Psi i\Theta \simeq \Phi i\Theta),$$

define $\text{Id} =_{\text{Df}} (U\Psi,\Phi)(\Psi \overset{\text{H}}{\approx} \Phi) \rightharpoonup . (U\Theta)(\Theta i\Psi \simeq \Theta i\Phi)$.

Since $\vdash_{\mathbf{HQ}} \neg \Psi i\Psi \rightharpoonup . (q\Phi)(\Psi \overset{\text{H}}{\approx} \Phi \;\&\; \neg \Psi i\Phi)$, and since, assuming Id, $(q\Phi)(\Psi \overset{\text{H}}{\approx} \Phi \;\&\; \neg \Psi i\Phi) \rightharpoonup . \neg \Psi i\Psi$, then, $\neg \Psi i\Psi \simeq . (q\Phi)(\Psi \overset{\text{H}}{\approx} \Phi \;\&\; \neg \Psi i\Phi)$.

But by (B), $(q\Psi)(UX) . Xi\Psi \simeq (q\Phi)(\Phi \overset{\text{H}}{\approx} X \;\&\; \neg Xi\Phi)$, whence by HA4(ii), $(q\Psi) . \Psi i\Psi \simeq (q\Phi)(\Phi \overset{\text{H}}{\approx} \Psi \;\&\; \neg \Psi i\Phi)$. Thus, assuming Id, for some Ψ, $\Psi i\Psi \simeq \neg \Psi i\Psi$; hence $\neg$Id. Next,

since $\vdash_{\mathbf{HQ}} (U\Psi)(UX)[(\Psi,X)i\Theta \simeq Xi\Psi] \rightharpoonup . (\Psi,X)i\Theta \simeq Xi\Psi \;\&\; (\Phi,X)i\Theta \simeq Xi\Phi$,
$\vdash_{\mathbf{HQ}} (U\Psi)(UX)[(\Psi,X)i\Theta \simeq Xi\Psi] \rightharpoonup . \Psi \overset{\text{H}}{\approx} \Phi \rightharpoonup (UX)[Xi\Psi \simeq Xi\Phi]$. Thus
$\vdash_{\mathbf{HQ}} (q\Theta)(U\Psi)(UX)[(\Psi,X)i\Theta \simeq Xi\Psi] \rightharpoonup \text{Id}$; so,
$\vdash_{\mathbf{HQ}} \neg (q\Theta)(U\Psi)(UX)[(\Psi,X)i\Theta \simeq Xi\Psi]$. But since $Xi\Psi$ is well-typed
$\vdash_{\mathbf{H}} (q\Theta)(U\Psi)(UX)[(\Psi,X)i\Theta \simeq Xi\Psi]$.

Therefore if **H** extends **HQ**, **H** is inconsistent.

In fact $(q\Phi)(\Phi \overset{\text{H}}{\approx} X \;\&\; \neg Xi\Phi)$ is *not* well-typed, so the abstraction postulate cannot be applied in **H** in the way it is in **HQ** to contradict Id.

§4 *Reductions of second-stage logics to many-sorted significance logics*

Clearly, quantification logic with ordinary predicate terms and only individual quantifiers, can be formulated as part of second-stage logic. We use the converse of the concretion principle to define predicates in terms of attributives:

$$\Psi^{*}\underset{\sim}{X} =_{\text{Df}} \underset{\sim}{X}\, i\Psi$$

Now consider the sublogic **Q** of $\mathbf{2QS_6}$ whose wff contain at most individual quantifiers and whose attributives are replaced uniformly by starred predicates. Then the wff and theorems of **Q** are the same, after predicate re-lettering, as those of $\mathbf{QS_6}$. Similarly, $\mathbf{QIS_6}$ is a sublogic of a definitionally enriched $\mathbf{2QIS_6}$. There remains, however, a difference in interpretation since the quantificational sublogics inherit the clause interpretation of their parental systems.

Much more interesting is the fact that $\mathbf{2QS_6}$ can be regarded as a sublogic of an enriched $\mathbf{QS_6}$. Enrichment is essential for, as an outcome of Gödel's incompleteness theorem, $\mathbf{2QS_6}$ cannot be re-expressed just in $\mathbf{QS_6}$. In particular, by an extension of Ackermann's [1] proof that the elimination problem for second-order theory is not solvable, not every wff of $\mathbf{2QS_6}$ can be eliminated in favour of an equivalent ($\simeq$) wff of $\mathbf{QS_6}$.

To see what enrichment is necessary, we note first that second-stage and higher-order logics can be regarded as dividing the subjects of first-order quantification theory (and in the case of most higher-order logics, the predicates of first-order quantification theory) into a number of notationally distinguished classes. For example, second-stage logics divide subjects into three classes: individual, attributive and assertoric (or sentential). The division is not compulsory. An escape from the logical paradoxes does not show, as Hilbert & Ackermann ([1], pp. 151–2), for example, suppose, that the "undifferentiated" notions of subject and predicate we introduced in Chapter 3 cannot be used consistently. For there are other suspect moves in the paradox arguments (see *e.g.* Rosser's [1] discussion). Rather, the division of subjects is made in second-stage theory because each of the different kinds of subjects has a slightly different characteristic logic. But it is also apparent from the setting up of second-stage logic, that despite some differences, there is a common shared logic of subjects, namely that which is featured in (first-order) quantification theory, and the logical differences between different subjects can be isolated as special assumptions for each different subject-sort. These special assumptions can then be re-cast as postulates on specific predicates such as '*i*'. Subjects are not, therefore, completely undifferentiated. Though there are some logical distinctions to be made between different sorts of subjects, the sorts are, in a clear sense, logically united.

In the case of second-stage logics, the differences in the sorts appears primarily in the instantiation axioms, though they could, equivalently, have been introduced in other ways, *e.g.* in the case of attributives by an abstraction scheme. Now special postulates for certain sorts of subjects can be stated, without introducing new variables, simply by using restricted variables, given that the logic has the apparatus for making the restriction. To illustrate, consider first a second-stage system without assertoric quantification and with a normal sentential interpretation, *e.g.* **A2QS$_6$**. Then, instead of assuming

$$\vdash (U\Psi')A \rightarrow \check{S}^{\underset{\sim}{u}i\Psi'}_{\underset{\sim}{z}i\Phi} A \,|,$$

we can require, more explicitly,

$$\vdash (Uv \ni At(v))A \rightarrow \check{S}^{\underset{\sim}{u} \ni In(\underset{\sim}{u})\; i\; v \ni At(v)}_{\underset{\sim}{z} \ni In(\underset{\sim}{z})\; i\; w \ni At(w)} A \,| \qquad \text{provided we have}$$

available the predicates '*In*' and '*At*' (read respectively as 'each component of *qu*( ) is a subject' and '*qu*( ) is an attributive'). The upshot is that we can reformulate such second-stage logics as applied **QS$_6$ R** systems given that we add to them, or define using their existing resources, such special predicates as '*In*' and '*At*'. Since **QS$_6$R** systems can in turn be approximated by **QS$_6$** systems (using method 2 of §7.9, §6), second-stage and higher-order logics can be reformulated without loss of expressive power as applied first-order theories, such that each system, the original and the reformulated, is consistent with respect to each other. Thus, whatever can be expressed and asserted in **A2QS$_6$** can be expressed and asserted in its reformulation **RS$_6$**; and **RS$_6$** is consistent iff **A2QS$_6$** is.

In the case of second-stage systems which contain assertoric quantification, however, *e.g.* $\mathbf{2QS_6}$, special difficulties arise. Provided the first-order quantificational theory contains a predicate '*As*' (read '*qu*(that ...) is an assertion (clause)') there is no problem in reformulating the instantiation scheme. Thus, we could affirm,

$$\vdash (Uu \ni As(u))A \rightarrow \check{S}_{B}^{u \ni As(u)} A \,|$$

in place of,

$$\vdash (UP)A \rightarrow \check{S}_{B}^{P} A \,|.$$

But since assertoric quantification imposes a clause interpretation on the second-stage theory, this will be inherited by the first-order system which results. Secondly, suppose the stock of quantifiable variables in the first-order theory is '*x*', '*y*', '*z*', *etc.*, then these will have to stand both as subject variables *and* as well-formed formulae (clauses). The reformulating system will therefore transgress the limits of ordinary quantified significance logic, and indeed, of admissible discourse.

There are steps which can be taken to meet these difficulties, but none is wholly satisfactory. If, for example, we have the predicate '*Wff*' (read '*qu*( ) is a well-formed formula'), then we could make the required distinction within the formal system by, in effect, writing the formation rules as theses. For example, we could affirm:

$\vdash As(u) \rightarrow Wff(u)$,
$\vdash In(u) \vee At(u) \rightarrow \sim Wff(u)$,
$\vdash In(u) \;\&\; At(v) \rightarrow Wff(u\,i\,v)$, *etc.*;

but these of course are, in an important sense, ineffective since they presuppose formation rules of the ordinary kind. Alternatively, an intensional extension of first-order theory with the descriptor '§' included could be taken as the base logic of the reformulating system, but once again we meet the difficulty of §7.12, §1, that such a logic must also contain unquantified sentential parameters.

The simplest way of avoiding, rather than solving, these problems is to drop assertoric arguments and quantifiers from any second-stage system before reducing it. The clause interpretation can then either be retained or not, but in fact there is then no longer any point in keeping it and it is simpler to regard it as a secondary interpretation and to revert to a normal sentential interpretation as primary. The reduction may then be accomplished in two steps: first, a reformulation and reinterpretation of the second-stage logic as a many-sorted quantified significance logic; and second, a reduction of this many-sorted logic, after the introduction or definition of special predicates such as '*At*', to an applied $\mathbf{QS_6}$ by one of the methods of §7.9, §6 for treating restricted variables. For the special many-sorted logics we reach, the second of these stages, reduction to a one-sorted applied $\mathbf{QS}_6$ system using '$\rightharpoondown$' to eliminate bound restricted variables, is standard (see §7.6).

To illustrate the first stage of the reduction, we reformulate second-stage logic $\mathbf{2Q}_5$ *without assertoric arguments*, as an applied many-sorted logic $\mathbf{MQS}_6$, with '*U*' as the only primitive quantifier, with '*i*' as the only predicate symbol, and with two sorts of arguments: subjectival variables, '*x*', '*y*', '*z*', *etc.*; and attributive variables, 'ϕ', 'ψ', *etc.*, of finitely many places. The formation rule for predicate '*i*' is as follows:

$(u_1, \ldots, u_n)iv$ is a wff iff $u_1, \ldots, u_n$ are *n* subject variables and *v* is an attributive variable of *n*-places.

Otherwise wff are built as for $\mathbf{QS}_6$, but with quantification over both sorts of variables introduced. The postulates of $\mathbf{MQS}_6$ are the same as those of $\mathbf{QS}_6$ except that A4 is replaced by:

M4. $\vdash (Uu)A \rightharpoondown \check{S}^u_t A\,|$,

where *t* is a term (variable) of the same sort, subjectival, or attributive of the same number of places, as *u*. The sole special postulate for '*i*' is the scheme:

2A4(iii), *viz.*: $\vdash (Uv)A \rightharpoondown \check{S}^{(u_1, \ldots, u_n)\, iv}_{B} A\,|$,

where *v* is an *n*-place attributive variable and $u_1, \ldots, u_n$ are *n* distinct subjectival variables. Now, by inspection, $\mathbf{2Q}_5$ (without assertoric variables) is the same system as $\mathbf{MQS}_6$. Similarly, **HQ** and **H**, postulated without assertoric variables, may be reduced.

The reduction thus indicates how developments of second- and higher-stage logics which do not use assertoric quantification essentially, may be more satisfactorily represented as applications of enriched quantified significance logics. In fact our second-stage logics, in contrast to second-order theories, are really nothing but applied many-sorted systems; for the reduction simply consists in explicating bound restricted variables.

Since individual and attributive quantification alone suffice for the later applications which we wish to make, we re-set the direction of the development at this point to restore normality of interpretation. In particular, we take $\mathbf{A2QS}_6$ as the basic second-stage logic and work either with it or its reduction $\mathbf{RS}_6$. Both $\mathbf{A2QS}_6$ and $\mathbf{RS}_6$ are given a standard sentential interpretation (though they can still, of course, be given a secondary clause interpretation). Under the primary sentential interpretation, the logics are effectively straightforward extensions of $\mathbf{QS}_6$, *i.e.* first-order sentential theories with special predicates and restricted-role attributives added.

§5 *Applications of normally interpreted second-stage logics and their extensions: class theory and number theory*

Standard applications and developments of second-order theory—for example postulate theory in its various forms (see Church [4]), identity theory, the theory of numbers, and much of the theory of classes and relations (see *e.g.* Hilbert and Ackermann [1])—may be paralleled in $\mathbf{A2QS_6}$ and its extensions, *i.e.* in sententially-interpreted second-stage logics without assertoric quantification. Some of these applications, in particular the theory of natural numbers, are in fact better carried out in second-stage significance logics; and there are other developments which are not possible in second-order theories. For example, second-stage significance logics, unlike classical second-order systems, provide a suitable framework for the development of category theory.

We first sketch the theory of classes and relations, then numbers, before going on to a more detailed treatment of identity and categories. For these applications, however, a sentential interpretation of second-stage logic is virtually essential. That is why we take $\mathbf{A2QS_6}$ rather than $\mathbf{2QS_6}$. Under a clause interpretation some of the definitions have unusual features.

An important feature of the second-stage elaboration of the theories of classes and relations is that, unlike the development in $\mathbf{QS_6E}\approx$ (§7.9, §5), they are no longer virtual theories. For quantification over abstracts can be defined in terms of quantification over attributives. Just as there are various more or less satisfactory methods by which descriptions can be introduced into a quantification logic (see Carnap [2], pp. 32–39), so there are various methods by which class abstracts can be introduced into second-stage logics and their extensions. In particular, class abstracts may be defined explicitly in terms of descriptors such as '℩' or 'ξ', as in,

$$\{z{:}A(z)\} =_{\mathrm{Df}} \xi\phi(Ux)(xi\phi \simeq A(x)),$$

or may be introduced contextually, as in **PM** (the system of the first edition of *PM*) or, differently, as in Hilbert & Ackermann [1].

In order to exploit the detailed and philosophically sophisticated treatment in **PM**, and to pave the way for a discussion of the significance features of type theory (as developed most fully in *PM* *63ff), we introduce class and relation abstracts contextually by taking significance analogues of the fundamental **PM** definitions. The technique we adopt may be carried out in several of the systems we have introduced, *e.g.* in $\mathbf{A2QS_6}$, in extensions of it such as **HQ** and **H** (all without assertoric quantifiers, and interpreted sententially), and in Brady's set theory (Brady [3]) and a property theory derivable from it (by taking '*i*' in place of '∈' and dropping the axiom of extensionality). What differences there are between the theories can be largely isolated in the different abstraction principles which hold for the different systems. To avoid repetition of qualifications we simply use expressions of the form $\alpha(u)$ and $\beta(u)$ to represent wff for which abstraction is permissible in the system in question—what we call *assumptible* schemes of the system. Thus, in Brady property theory the class abstraction scheme

$$v \in \hat{u}\alpha(u) \simeq \alpha(v)$$

holds for $\alpha(u)$ with improper symbols definable in a strong normal set; but in **HQ** the scheme holds for wff $\alpha(u)$ of **HQ** with u an individual variable; and in **H** for well-typed wff $\alpha(u)$ of **H**. In order to follow **PM** we present the theory using, as with second-stage logics, distinctive symbolism for attributive subjects; from the viewpoint of Brady property theory, attributive variables such as 'ϕ' and 'ψ' may simply be regarded as further subject variables.

The fundamental definition of the derived theory of classes (*cf. PM* *20.01), with scope explicitly featured, is:

$$[\hat{z}A(z)]\,[B(\hat{z}A(z))] =_{\mathrm{Df}} (\mathcal{g}\phi).(Uv)(vi\phi \simeq A(v))\ \&\ B(\phi).$$

It turns out that scope matters only in the presence of intensional predicates, so that for usual applications of the theory of classes scope indications can be omitted (and the definiendum written $B(\hat{z}A(z))$, and restored according to the conventions given in **PM**—essentially that each abstract has the smallest scope (previously defined symbols uneliminated) compatible with its position and that if an abstract $(\hat{z}A(z))$ is to the right of another abstract in a wff it takes a smaller scope. Instead of *showing* scope by square bracketed abstracts, class abstract symbols may be introduced not as descriptors, but as binary quantifiers (as in §7.9, §3 for descriptions), thus:

$$\hat{z}(A(z), B(z)) =_{\mathrm{Df}} (\mathcal{g}\phi).(Uv)(vi\phi \simeq A(v))\ \&\ B(\phi);$$

then $\hat{z}(A(z), B(\hat{z}A(z))$ may be introduced in terms of the definiendum; and finally the abbreviation $B(\hat{z}A(z))$, with the class symbol now as a descriptor, may be introduced when the conventions for omitting scope are satisfied. Reasons for adopting the fundamental definition are well explained by Whitehead and Russell (see *PM*, pp. 71–84 and pp. 187–190).

The determining features of classes follow thus:

$\vdash v\, i\, \hat{z}\alpha(z) \simeq \alpha(v)$ [*cf. PM* *20.3];

one part of the proof uses the abstraction theorem.

I. $\vdash A(u) \simeq_u B(u) \rightarrow . C(\hat{z}A(z)) \simeq C(\hat{z}B(z))$,
where $(A \simeq_u B) =_{\mathrm{Df}} (Uu)(A \simeq B)$; [*cf. PM* *20.11].

This result reveals that classes are determined solely through their argument-values (or extensions), since the classes abstracted from coincident sentence frames have the same truth and significance values in all sentence contexts (where scope is immaterial). Such a result is often said to show that classes are extensional; but we prefer to say that it shows that classes are argument-functional, so as to avoid confusion of the sense of 'extensional' we have used elsewhere with the connected but nevertheless *different* sense of 'extensional' used in connection with classes. In the light of this argument-functionality result we adopt the definition

$u\in\hat{z}A(z) =_{Df} u\, i\, \hat{z}A(z)$; thence

II. $\vdash u\in\hat{z}\alpha(z) \simeq \alpha(u)$ (*class abstraction scheme*).

To derive appropriate conditions for class identity we first define coincidence of attributes: $\Phi \simeq \Psi =_{Df} ui\Phi \simeq_u ui\Psi$.

Then $\vdash \hat{z}A(z) \simeq \hat{z}B(z) \rightharpoonup . A(u) \simeq_u B(u)$ [*cf. PM* *20.14],

and conversely $\vdash \alpha(u) \simeq_u \beta(u) \rightharpoonup . \hat{z}\alpha(z) \simeq \hat{z}\beta(z)$ [*cf. PM* *20.13; note that features of coincidence suffice for the proof];

whence

III. $\vdash \alpha(u) \simeq_u \beta(u) \simeq . \hat{z}\alpha(z) \simeq \hat{z}\beta(z)$.

Since $\hat{z}\alpha(z) \simeq \hat{z}\alpha(z)$ and, by I and III,

$\hat{z}\alpha(z) \simeq \hat{z}\beta(z) \rightharpoonup . (U\psi). \hat{z}\alpha(z) i\psi \simeq \hat{z}\beta(z) i\psi$,

'$\simeq$' is an identity relation on assumptible class abstracts.

Accordingly we define $\hat{z}A(z) = \hat{z}B(z) =_{Df} \hat{z}A(z) \simeq \hat{z}B(z)$;

whence abbreviating assumptible class abstracts using small Greek letters 'α', 'β', 'γ', 'δ' (much as in §7.9, §5, following *PM*, p. 188),

IV. $\vdash \alpha = \beta \simeq (Uu)(u\in\alpha \simeq u\in\beta)$.

Also

V. $\vdash \alpha = \beta \rightharpoonup . \alpha\in\gamma \simeq \beta\in\gamma$.

Thus coincident properties determine identical classes; a class on this theory is an abstraction from coincident properties: it is the invariant (argument) under coincidence. (A different class theory can be derived using dot coincidence, which requires only preservation of truth.)

That such results as I–V hold is what justifies us in calling the theory we have derived a *class* theory, and what vindicates the claim that class theory can be represented in significance logics. In the theory we have sketched, but by no means in all theories of classes, classes are effectively argument-functional properties. The fact is that *class* like many of the other important logical notions we have examined, is a logical determinable. The determining features of a class logic are that class abstraction and class argument-functionality principles should hold for sufficiently many sentence frames $\alpha(\ldots)$; but the particular form these principles take (which frames are assumptible, which equivalence connectives are used), and the further principles elucidating the nature of the classes, vary with the determinate class theory. Thus we adopt weaker conditions of adequacy for a theory of classes than Russell and Whitehead (*PM* pp. 76–77), and drop altogether their over-restrictive requirement (5) that it is nonsignificant for a class to be identical with one of its own members. Because we only impose weak conditions of adequacy many very weak theories and non-standard theories, *e.g.* non-Cantorian theories, will rate as class theories, alongside the theory we have sketched and alongside such others as Zermelo-Fraenkel class theory—a better known but philosophically less well motivated theory than Russell's—which has recently become entrenched as *the* theory of sets according to mathematicians.

Schemes II and III provide the basic schemes of the virtual theory of classes (of §7.9, §5); hence we can take over at once the developments of the virtual theory of classes based on these schemes. In fact for theories based on **A2QS**$_6$, **HQ** and **H**, we can subsume at once the whole of virtual class (and relation) theory, since to every class-determining wff of quantification logic corresponds an assumptible scheme of the larger systems. That the present theory is not virtual, and its enhanced richness, is already apparent through definitions like these:

$$\mathrm{Cls} =_{\mathrm{Df}} \hat{\alpha}((9\phi).\,\alpha = \hat{z}(zi\phi));$$
$$(U\alpha)C(\alpha) =_{\mathrm{Df}} (U\phi)C(\hat{z}(zi\phi));$$
$$[\hat{\alpha}\beta(\alpha)]\,[C(\hat{\alpha}\beta(\alpha))] =_{\mathrm{Df}} (9\phi).\,\alpha i\phi \simeq \hat{\alpha}\beta(\alpha)\ \&\ C(\phi).$$

The class theory thus far developed, along with an analogous redevelopment of the basic theory of relations (*i.e.* argument-functional attributes) places us in a position to rework in a significance way, a substantial fragment of **PM**.

Especially important properties of classes are numbers; hence our next application, that of outlining a significance theory of cardinal and ordinal numbers (the controversial thesis that numbers are non-distributive properties of manifolds is defended in Routley [1], where a more detailed treatment of cardinal numbers is given). To define numbers as properties we introduce (weak) attribute descriptions, in a way paralleling our introduction of descriptions in quantification logic, *viz.*:

$$(\text{℩}\phi)(A(\phi), B(\phi)) =_{\mathrm{Df}} (9\psi).\,(A(\theta) \simeq_{\theta}.\,\psi \simeq \theta)\ \&\ B(\psi).$$

Coincidence of attributes is the appropriate condition to define uniqueness of these descriptions, since classes which are identical have the same cardinality. Just as with the analogous contextual definition of class abstracts, so in this case, the definition may be written more revealingly

$$(\text{℩}\phi)(A(\phi), B(\text{℩}\phi A(\phi)),$$

and then the scope omitted under specified conditions where scope ambiguities are avoided. Alternatively, and more satisfactorily, '℩' could be added as a primitive descriptor or defined in terms of some other descriptor such as 'ξ' or 'λ'.

We adapt, to ontologically neutral significance logics, the classical Frege-Russell definitions of cardinal and ordinal numbers. The cardinal n is the property of all and only those classes which correspond one-one to a given paradigmatic class, *pd*(*n*), with *n* members: in symbols

$$n =_{\mathrm{Df}} (\imath\phi).\alpha\ sm\ pd(n) \simeq_{\alpha} \alpha i\phi.$$

The ordinal $\bar{n}$ is the property of all and only those classes which are ordinally similar to a given paradigmatic ordered class *opd*(*n*) with *n* members: in symbols

$$\bar{n} =_{\mathrm{Df}} (\imath\phi).\alpha\ smor\ opd(n) \simeq_{\alpha} \alpha i\phi.$$

'*sm*' and '*smor*' can both be defined in second-stage logic in much the usual way; *e.g.* $\alpha\ sm\ \beta$ iff some (q) one-one relation R has α as (truth-) domain and β as counter-domain, where R is one-one iff both R and $\breve{R}$ are functions (under a suitable identity). How the paradigmatic classes are chosen depends rather more critically on the resources of the logic. In particular in $\mathbf{2Q_5}$ and **HQ** there is no way of selecting a paradigmatic infinite class (the negation of the axiom of infinity—where the axiom is taken in the form, *some* class is infinite—is consistent with **HQ**; see Cocchiarella [1]). For finite numbers, one simple way of selecting paradigm classes, for both cardinals and ordinals, is given by the following recursion scheme:

(P). $pd(0) = \Lambda$

$pd(n') = \{pd(0), \ldots, pd(n)\}.$

Then $pd(1) = \{\Lambda\}, pd(2) = \{\Lambda, \{\Lambda\}\}$, *etc.*

Alternatively the recursion clause could be replaced by

$$pd(n') = \{0, 1, \ldots, n\}.$$

Or, as in **PM**, unit classes may be used to provide paradigms; *e.g.* *pd*(2) is the union of the unit classes of distinct elements *x* and *y*.

Given the finite numbers, addition and multiplication may be defined, in pretty much the usual logicist way, and finite cardinal and ordinal arithmetic adduced in significance logics. (At this stage the theory can encompass **SA** of §7.9, §1). Once a suitable recursive progression of paradigms such as (P), on which to fix (quasi-ostensively) the equipollence relations, is selected, the *successor* relation '*Su*' can be defined in such a way that property ϕ succeeds property ψ iff ϕ and ψ are manifold properties and if ψ is the property of all and only those manifolds similar with some member of the progression then ϕ is the property of all and only those possible manifolds similar to its successor in the progression. Finally the *positive integers* can be defined as the set of property ancestors of 1 under the successor relation, and the positive integer property 'N' can be defined. Thus positive integers are finite cardinals ordered under the successor relation.

§6 *On Identity*

The argument (of §7.9, §3) that Leibniz identity cannot be defined in quantified significance logic $\mathbf{QS_6}$ breaks down for second- and higher-stage logics:

these do provide the resources to define various identity relations. Thus second-stage logics provide sufficient logical apparatus to define identity relations between individuals, though not between attributes; and higher logics enable definitions of identity relations of both individuals and attributes. Given that identity relations on individuals can be defined in second-stage logics, it follows (using §7.13, §2) that these relations can be defined in applied quantified significance logics enriched by further predicates; in other words that what is needed for definitions of identity is the enrichment of quantified significance logic by appropriate axiomatically-constrained new predicates. Once this is realised some of the resistance to defining ordinary, non-Leibnizian, identity relations by introducing appropriate new predicates (such as 'extensional' below) is removed.

Identity is always a matter of the communion of features. Both what is identified, whether items or entities or symbols, and how they are identified, whether contingently, strictly or strongly, whether typewise or tokenwise, depends on the shared features required for the particular identity. Identity is not a unique relation but (as explained in §3.5) a determinable under which various determinate relations fall. It is the reference theory of meaning and truth, the resources of which permit only the characterisation of the weakest of identity relations, identity of reference (defined in terms of coincidence of extensional properties of entities) that has led to the idea that there is a *unique* identity relation and to standard puzzles about identity, *e.g.* how can *two* things ever be identical?, how is identity compatible with change? Once out of the grips of the reference theory the problem as to how there can be several identity relations dissolves—there are several determinates of the one determinable—and traditional puzzles about identity generated by the reference theory can be straightforwardly resolved (see, *e.g.* Routley [9]).

Strong or Leibnizian identity relations place no restrictions on the class of predicates required to transfer over equivalences which characterise the identities; *i.e.* they require coincidence of all attributes of particulars identified. (In calling strong identities *Leibnizian* we are not committing ourselves to an historical thesis as to Leibniz's position on identity). More familiar identity relations such as contingent identity and identity of reference do however curtail the class of transferring predicates; *i.e.* coincidence of all attributes is not necessary for an identity. We treat strong identity of items first; then we give a parallel classification of extensional identity relations. We assume, as always, that the predicates of the logic are non-quotational and that they provide attributives on conversion: otherwise we should arrive, not with identity of particulars, but with some sort of identity of subject-expressions, *e.g.* type identity.

Strong homogeneous identity is defined thus:

$$X \overset{\text{H}}{\approx} Y =_{\text{Df}} (U\Psi).Xi\Psi \simeq Yi\Psi.$$

Then $\vdash S(X \overset{H}{\approx} Y)$; $\vdash X \overset{H}{\approx} X$; $\vdash X \overset{H}{\approx} Y \equiv . Y \overset{H}{\approx} X$; and $\vdash X \overset{H}{\approx} Y \,\&\, Y \overset{H}{\approx} Z \supset . X \overset{H}{\approx} Z$; *i.e.* $\overset{H}{\approx}$ is an equivalence relation on individuals which is always significant. Further, $\overset{H}{\approx}$ guarantees full intersubstitutivity in second-stage logics **A2QS$_6$–2Q$_5$** (without assertoric quantification).

Theorem (110): $X \overset{H}{\approx} Y \rightharpoonup . A \simeq B$, where B results from A by substitution of Y for zero or more free occurrences of X, no one of which is in a wf part $(UY)C$ or $(\mathcal{q}Y)C$ in A. So, in particular, $\vdash X \overset{H}{\approx} Y \rightharpoonup . A \simeq \check{S}^{X}_{Y} A\,|$.

Proof: For second-stage logics without the full abstraction postulate the proof is by induction over the number of primitive connectives and quantifiers occurring in A. We present the proof in the case of **A2QS$_6$**.

Induction basis: $X \overset{H}{\approx} Y \rightharpoonup . (U\Psi). Xi\Psi \simeq Yi\Psi$
$\rightharpoonup . (X_1, ..., X_{i-1}, X, ..., X_n)i\Psi^n \simeq . (X_1, ..., X_{i-1}, Y, ..., X_n)i\Psi^n$,
by applying 2A4.

Induction steps. Those for connectives '$\neg$', 'D' and '$\rightharpoonup$' follow immediately from previously established sentential theorems. For the quantificational cases to be considered suppose first A is of the form $(Uu)A_1$; then B will be of the form $(Uu)B_1$. By induction hypothesis $X \overset{H}{\approx} Y \rightharpoonup . A_1 \simeq B_1$. Then $X \overset{H}{\approx} Y \rightharpoonup . A_1 \rightharpoonup B_1$; $X \overset{H}{\approx} Y \rightharpoonup . B_1 \rightharpoonup A_1$; $X \overset{H}{\approx} Y \rightharpoonup . SA_1 \rightharpoonup SB_1$ and $X \overset{H}{\approx} Y \rightharpoonup . SB_1 \rightharpoonup SA_1$. Then, since '$u$' is distinct from 'X' and 'Y', $X \overset{H}{\approx} Y \rightharpoonup . (Uu)(A_1 \rightharpoonup B_1)$, *etc.*; hence $X \overset{H}{\approx} Y \rightharpoonup . (Uu)A_1 \rightharpoonup (Uu)B_1$, *etc.*, also $X \overset{H}{\approx} Y \rightharpoonup . S(Uu)A_1 \rightharpoonup S(Uu)B_1$, *etc.* Finally the result follows by combining all these cases using **S$_6$**. The proof in the case that A is of the form $(\mathcal{q}u)A_1$ is similar except that distribution and interchange principles for '$\mathcal{q}$' are used.

For **2Q$_5$**, however, the induction steps are unnecessary. The result follows at once upon applying the abstraction axiom to the induction basis.

Since homogeneous identities of items preserve significance features, homogeneous identities only hold between items of the same minimal significance categories, where sameness of minimal significance category is defined thus: $X \overset{M}{=} Y =_{Df} (U\Psi)(SXi\Psi \equiv SYi\Psi)$. (The importance of minimal categories is explained in R. and V. Routley [1].) It is a fact however (see §7.14) that identity relations hold between items of different minimal categories: *i.e.* not all true identities are fully homogeneous. But there is not a single strong heterogeneous identity; rather there are various related notions some of which are important for the development of category theories.

Once significance homogeneity is abandoned the principle of full intersubstitutability of identity, often mistakenly regarded as essential for an identity relation, breaks down. For example it breaks down for dot identity which only requires coincidence as to truth: $X \dot{\approx} Y =_{Df} (U\Psi). TXi\Psi \equiv TYi\Psi$. Though dot identities are not intersubstitutable preserving significance, dot identify is nonetheless an equivalence relation which is always significant.

Every homogeneous identity is of course a dot identity; and a sufficient condition that $\overset{H}{\approx}$ and $\dot{\approx}$ coincide is that the items they relate are in the same minimal category. Dot identity belongs among the dot notions discussed in §7.9, §5.

A more drastic way of deleting significance features uses wide identity defined thus:

$$X \overset{W}{\approx} Y =_{Df} (U\Psi) SXi\Psi \; \& \; SYi\Psi \supset . Xi\Psi \simeq Yi\Psi.$$

Since $\vdash X \overset{W}{\approx} Y \equiv . (U\Psi). SXi\Psi \; \& \; SYi\Psi \supset . TXi\Psi \equiv TYi\Psi$,

dot identity implies wide identity; but not conversely. Wide identity, too, is a two-valued equivalence relation. Moreover it is intersubstitutable in all significant sentence contexts.

Theorem (111): $X \overset{W}{\approx} Y \; \& \; SA \; \& \; SB \rightharpoonup . A \simeq B$, where A and B are as in the previous theorem.

Proof is, as before, by induction.

A very different heterogeneous identity is inclusive identity, defined:

$$X \overset{I}{\approx} Y =_{Df} (U\Psi) . (SXi\Psi \supset SYi\Psi) \vee (U\Psi)(SYi\Psi \supset SXi\Psi) \; \& . \; X \overset{W}{\approx} Y.$$

$\overset{I}{\approx}$ is not an equivalence relation; for though it is reflexive and symmetric it is not transitive. It can however be broken down into two directed identities which keep transitivity features. Thus

$$X \precsim Y =_{Df} (U\Psi)(SXi\Psi \supset SYi\Psi) \; \& \; X \overset{W}{\approx} Y.$$

$\precsim$ is reflexive and (*one*-way) transitive, *i.e.* $\vdash X \precsim Y \; \& \; Y \precsim Z \supset . X \precsim Z$; but it is not symmetric. Inclusive identities such as $\precsim$ are basic relations in the theory of absurdity and inclusive categories (see §7.14).

The identity notions so far defined are always significant. Yet not all identity assertions are significant (see §3.6). For example, assertions such as 'Julius Caesar = the number one' and 'This chair = President Nixon' are often cited as nonsignificant, not merely false, claims. (That they are nonsignificant of course does not imply that they cannot occur as components of significant sentences, *e.g.* as antecedents of subjunctive conditionals, unless we restrict ourselves to classical significance connectives). It is, however, easy to provide among identity determinants relations which are sometimes nonsignificant. Thus one restrictive identity which, like the identity relation of **PM**, allows only items of the same minimal category, or type, to be significantly identified is defined as follows:

$$X \overset{M}{\approx} Y =_{Df} X \overset{H}{\approx} Y \; \& \; Q(X \overset{M}{=} Y).$$

Since S(Julius Caesar likes soup) and $\sim S$(The number one likes soup), $\sim S$(Julius Caesar $\overset{M}{\approx}$ the number one). Though $\overset{M}{\approx}$ is not always significant, it is nonetheless an equivalence relation. (Proof of transitivity uses the theorem: $(p \mathbin{\&} q \rightharpoonup p') \rightharpoonup (Qp \mathbin{\&} Qq \rightharpoonup Qp')$). It may be objected that $\overset{M}{\approx}$ is too restrictive, because some important cross-type identifications are significant, and, as science shows, sometimes true. (For example theoretical identifications concerning temperatures, colours, smells, sensations, *etc.*) But all this reveals is that where these identities are not wide identities there remain further identity determinates to be taken account of. Satisfactory accounts of such identities do however appear to depend on the development of significance-range theory.

The bedrock identity relation is that of extensional or contingent identity, not Leibniz identity. (The word 'contingent' is somewhat misleading, since extensional identities may be non-contingent. Contingency enters as follows: where a Leibniz identity is true it is necessarily true, but where an extensional identity is true it may be contingently true.) As remarked, Leibniz identities are those limiting identities where every non-quotational predicate transfers; but for an identity relation it is sufficient that only a certain subclass of predicates transfer across the defining equivalence, the so-called extensional predicates. Extensional identity is the bedrock relation because weaker relations, which do not transfer every extensional predicate, do not provide identity relations but only similarity and correspondence relations at best.

Extensionality of predicates and attributes is a primitive notion not so far introduced notationally in the many-valued logics studied. It can be clarified, to begin with, through examples of predicates which are extensional and of predicates which are not. All empirical predicates, such as '... is round', '... is hard', 'It is true that ... is hard', '... is between ... and ...' and '... dissolves in aqua regia' are, in their standard senses, extensional. But predicates such as 'Bill believes that ... is red', 'It is widely known that ... is hard', 'It is logically necessary that ... is greater than ...' and 'I am expecting ...' are not extensional, but intensional. Since very many intensional predicates are compounded from sentence connectives, it is tempting, but not satisfactory, to define an intensional predicate simply as one which can be analysed into a non-value-functional connective combined with some other predicate, *e.g.* thus:

$$\textit{Intensional}\ (\mathrm{F}) =_{\mathrm{Df}} (\mathcal{q}\Phi_i)(\mathcal{q}\Psi)(U\mathrm{X})\,[\mathrm{F(X)} \simeq \Phi_i(\mathrm{X}i\Psi)] \mathbin{\&} {\sim}\,\textit{Val}(\Phi_i).$$

It is, however, easy to concoct artificial predicates Φ_i and Ψ which cancel out one another's intensional features. A simple example is provided by re-entrant non-value-functional predicates of period 2, *i.e.* with Φ_i such that ${\sim}\,\textit{Val}(\Phi_i)$ and $\Phi_i\Phi_i P \simeq P$. Though such predicates are "unnatural", at least in the sense that they do not seem to occur in natural languages, formally nothing excludes them, and some of them can be provided with a standard semantical treatment. The definition can of course be patched up—by requiring further that Ψ be extensional—but only at the cost of serious circularity. The definition can alternatively be progressively complicated to exclude new cases of re-entrant predicates; but so far we have not been able to devise a satisfactory and workable complication of the original definition. In any case such a definition is not essential, since a semantical account can be given of extensionality.

A predicate F is intensional when for some situation H and some subject vector $\underset{\sim}{X}$, the evaluation of whether F($\underset{\sim}{X}$) holds in situation H is a function of evaluations in situations other than H (and H*—in case inconsistent situations are considered). Thus, to illustrate the point, empirical-style predicates such as '... is hard' are extensional, because whether Tom, say, is hard can be evaluated just in the actual case G without recourse to possible situations. But 'It is logically necessary that ... is hard' is not extensional for its evaluation even in G leads beyond G to all other possible situations.

Thus, instead of trying to reduce intensionality to non-value-functionality by syntactical analysis we may generalise the semantical equivalent of non-value-functionality, *viz.* leading to a new situation, to give a semantical account of intensionality.

Given such a semantical explanation there is no reason why we should not introduce the predicate 'is extensional' as a primitive predicate. The position of 'is extensional' is in fact somewhat like that of the logical necessity connective '□'; *viz.* it cannot be adequately defined within the resources of extensional logic or *its* semantics, but it can be explained initially through examples in a case by case way, and it can be fully explained semantically.

Were we to introduce 'extensional' as a predicate of predicates we should have to allow for quotation of predicates in the logic or else that 'extensional' absorbs a quotation function. For developments of **A2QS**$_6$ *etc.* it is however simpler to introduce 'extensional' as a primitive predicate (of attributives) conforming to the following rule:

If Ψ is an attributive then *Extensional* (Ψ) is a wff.

We shall immediately eliminate '*Extensional*' and work with a two-valued predicate '*ext*' defined:

$$ext\ (\Psi) =_{Df} T\ Extensional\ (\Psi).$$

We can, in the usual ways, form restricted variables using extensionality.

To each strong identity relation corresponds an extensional identity relation, obtained by replacing '$(U\Psi)$' suitably by '$(U\ ext\ \Psi)$'. Thus homogeneous extensional identity, defined

$$X \overset{H}{=} Y =_{Df} (U\ ext\Psi)(Xi\Psi \simeq Yi\Psi),$$

parallels strong homogeneous identity. $\overset{H}{=}$ is, once again, an equivalence relation which is always significant; but, unlike strong homogeneous identity, homogeneous identicals are not fully intersubstitutable. They are only intersubstitutable in extensional sentence contexts; they are not intersubstitutable in places controlled either by a non-extensional attributive or by a non-value-functional connective.

Theorem (112): For **A2QS$_6$** and **HQS$_6$** systems (without intensional sentence functors),

$X \overset{H}{=} Y \rightharpoondown . A \simeq B,$

where B results from A by substitution of Y for zero or more occurrences of X, no one of which is in a wf part $(\mathcal{U}Y)C$ or $(\mathcal{q}Y)C$ of A and no one of which is in a wf part $(..., ,...)i\Psi$ of A where Ψ is not extensional.

Proof is, as before, by induction, and the induction steps are as before. In the induction basis the restriction to extensional attributives is essential.

The theorem breaks down when the logic is enriched by non-value-functional connectives such as '□', 'It is probable (that ...)' and 'Bill believes (that ...)'. However the result holds in the following modified form:

Theorem (113): $X \overset{H}{=} Y \rightharpoondown . A \simeq B$, where B results from A as in the previous theorem and, moreover, no occurrence of X upon which substitution is made is within the scope of a non-value-functional operator.

The usual identity substitution principle is of course a special case of this result; it is that special case where (a) all sentence connectives are value-functional, and (b) all attributives are extensional.

But the special case does not hold generally since, as examples readily show, both (a) and (b) are false. A favourite hard-line way around the problems is first to restrict attributes to say traits, traits being so determined that (b) does hold for traits, and, second, to discount, as unintelligible or in need of analysis, connectives which are not value-functional. As to the first move, if traits can be suitably distinguished then so can extensional attributes, and in the same way. Needless to say, we find the second move completely unacceptable: if the analysis or explanation demanded is to conform to classical (extensional) requirements, then an adequate analysis is impossible; while if the analysis does not have to conform then, at least in the case of some modal connectives, a satisfactory semantical explanation can already be given (along the lines of §7.2).

A more reasonable move is to reject (a), but to retain (b) on the assumption that intensional features can always be isolated, after analysis, in a non-value-functional connective. So, for example, 'Bill believes that ... is red' can be separated into the connective 'Bill believes (that ...)' and the extensional predicate '... is red'. Despite the many clear cases where separation can be effected, it is by no means an uncontroversial matter whether the separation assumption holds generally. Consider the problems in recasting such sentences as 'Bob expects ...', 'Harry is looking for ...' and 'Tom sees ...', and recall familiar disputes in epistemology over whether such sentences can be analysed satisfactorily in terms of a perhaps new verb followed by a that-clause, *e.g.* whether 'Tom sees ...' can be replaced, preserving values, by a sentence of the form 'Tom sees that ... is so-and-so'. Retaining (a) as well as (b) enables us to avoid these difficulties in the separation assumption. (The same problem arises for the related moves made in §3.6 where it is assumed that all intensional predicates will, after logical analysis, contain an occurrence of 'that'. This is not to say that extensionality, or—what is distinct—transparency, of a predicate amounts to the absence of '§'. It simply does not matter that extensional predicates such as 'It is true that ... is red' also contain an occurrence of '§', for an inductive argument shows that such predicates do transfer over extensional identities. The approach of §3.6 cannot be used in connexion with second-stage logics which do not contain '§').

Just as '$\overset{H}{=}$' corresponds to '$\overset{H}{\approx}$', so we use connectives '$\overset{W}{=}$' and '$\mathrel{\overset{-}{\leftharpoondown}}$' corresponding, respectively to '$\overset{W}{\approx}$' and '$\mathrel{\overset{\sim}{\leftharpoondown}}$'.

$$X \overset{W}{=} Y =_{\mathrm{Df}} (U\, ext\, \Psi).SXi\Psi \;\&\; SYi\Psi \supset . Xi\Psi \simeq Yi\Psi;$$
$$X \mathrel{\overset{-}{\leftharpoondown}} Y =_{\mathrm{Df}} (U\Psi)(SXi\Psi \supset SYi\Psi) \;\&\; X \overset{W}{=} Y$$

$\mathrel{\overset{-}{\leftharpoondown}}$ is the basic relation in the theory of absurdity, *i.e.* in the theory of context-dependent nonsignificance where context is taken up into the sentence through extensional identity. Thus the absurdity connective '*Ab*' is defined:

$$Ab\,B(X_1, \ldots, X_n) =_{\mathrm{Df}}$$
$$(qY_1, \ldots, Y_n)\,[Y_1 \mathrel{\overset{-}{\leftharpoondown}} X_1 \;\&\; \ldots\; Y_n \mathrel{\overset{-}{\leftharpoondown}} X_n \;\&\; \sim S\check{S}^{X_1 \ldots X_n}_{Y_1 \ldots Y_n}\, B(X_1, \ldots, X_n)|\,].$$

For example, $Ab\,f(x) \equiv (qy)(y \mathrel{\overset{-}{\leftharpoondown}} x \;\&\; \sim Sfy)$; so, too, 'What I am thinking about is happy' is absurd where what I am thinking about is Wittgenstein's stove. Wide identities include not only inclusive identities but also exclusive identities, defined

$$X \overset{e}{=} Y =_{\mathrm{Df}} (q\Psi)(SXi\Psi \;\&\sim SYi\Psi) \;\&\; (q\Psi)(SYi\Psi \;\&\sim SXi\Psi) \;\&\; X \overset{W}{=} Y.$$

Exclusive identities occur when items of exclusive minimal categories are correctly identified. Such scientific identities as: The odour of violets = the molecule irone (an example from the Feynman lectures pointed out to us by V. Routley) appear to be true exclusive identities. For the identity statement is true (though strictly the non-synthetic odour of violets consists of α-, β- and γ-irone), yet exclusive. For predicates like '... has shape', 'weighs ...' are certainly significant of the right-hand subject of the identity—since it is precisely the shape and structure of the molecule which helps explain the distinctive smell of violets—but are not significant of 'the odour of violets'. Yet predicates like 'is sweet', 'is sharp' and 'is pungent' are significant of the left-hand subject but not of the right.

Such extensions of **A2QS**$_6$ as **HQ** provide the logical resources to define identity relations for attributes. In fact in **HQ**, and enrichments of it by predicates such as '*ext*', earlier definitions can simply be generalised by using general variables in place of individual variables. Then identity of attributes with particulars can also be considered; but, it follows, if some predicate is a purely individual predicate, that an individual is never homogeneously identical with an attribute. Homogeneous identity of attributes, defined

$$\Psi \overset{\text{H}}{\approx} \Phi =_{\text{Df}} (U\Theta)(\Psi i\Theta \simeq \Phi i\Theta),$$

provides as before an equivalence relation, from which intersubstitutivity conditions can be derived: it provides what is sometimes taken to be *the* identity relation for attributes. Coincidence of attributes does not provide an identity relation on attributes, though it does for argument-functional attributes (§7.13, §5).

7.14 The Structure of Categories

Second-stage significance logics provide much of the apparatus required to formulate a theory of categories. For a full account we need also a detailed analysis of significance ranges, but a number of conclusions can be drawn at this point. In the discussion which follows we assume as the background logic a sententially interpreted second-stage system with full conversion of predicates to attributives, *i.e.* **A2QS**$_6$ or one of its extensions, enriched by restricted quantifiers over quoted sentential variables.

The problem of defining categories and the relations between them is often taken to be the basic problem of significance, the solution to which will lead to an adequate significance theory. In fact, however, significance theory is the richer and more fundamental discipline. One can no more get quantified significance logic from a category theory than one can get quantification logic from Boolean algebra, or predicate theory from an elementary class theory. Significance theory is thus not reducible to category theory, and the traditional approach to it through categories has to be ruled out. It might, however, be thought that category theory is reducible to significance theory. For given an adequate significance logic, we can define significance ranges and then identify these as the categories. But this, too, is inadequate. Every category is ultimately determined by a significance range, but in general not every significance range can be taken to determine a category. For in a sufficiently rich logic, as we shall show, predicates can be constructed which determine as distinct significance ranges the unit classes which consist of each separate item in the range of the subject arguments; but it cannot be accepted that every item constitutes a unique category. There are, therefore, certain special problems which vex category theory but which do not affect significance logic; in particular, the problem of characterising naturally-occurring or intuitively well-behaved predicates.

Both significance ranges and categories are classes of certain sorts. (We use 'category' in the well-established philosophical sense, not in the sense recently introduced in mathematics). A significance range is the class of descriptions, or derivatively the class of s-items, for which, or over which, a given predicate is significant. If incomplete sentences are neglected, this reduces to the class of descriptions for which a given predicate is true-or-false. Every predicate determines a significance range; and conversely, every significance range is relative to a given predicate. The concept of category, however, is less comprehensive.

It is interesting to compare the difference between a significance range and a category with the difference between the wide concept of a class and the narrower concept of a natural kind. Indeed the concept of a category may be looked at, from one point of view, as an extension of the concept of a natural kind to the concept of a logical kind.

If two items belong to the same natural kind, they each have certain characteristic properties. Two items are *birds* iff both have wings, beaks, feathers, *etc*; more particularly, they are *finches* iff they are birds, seed-eating, *etc*; more particularly still, they are *yellowhammers* iff they are finches, have yellow underparts, *etc*. In general, two items belong to the same natural kind if, and only if, certain characteristic predicates are true of them. The word 'characteristic' is important. Not all predicates true of one item are true of another of the same kind (otherwise the items would be identical): many things true of one yellowhammer will not be true of another, though both are yellowhammers. Natural kinds, then, are classes determined by sets of characteristic attributes which occur together naturally, *i.e.* in nature. The notion may be extended to kinds which are not natural, to classes of artefacts, for example, but even so, not every class is a kind. The class of items which constitute the rubbish in someone's dustbin at a particular time is not a kind. In general, any given predicate determines a class, but not every predicate, nor yet every set of predicates, determines a kind.

Now whereas two items belong to the same natural kind or artefact-kind iff certain characteristic predicates are true of them, two items belong to the same logical kind (category) iff certain characteristic predicates are significant of them. Thus, the difference between a natural kind or artefact-kind, and a logical kind, is that membership of the latter is determined not by truth conditions but by significance conditions (though this is not to deny that a logical kind and a natural kind, though generated differently, may be co-extensive: for example, the predicate 'is jovial' determines the category of human beings since it is significant of these and these alone, but the category of human beings is also a natural kind specifiable in terms of truth conditions: *e.g.* as the class of rational animals). In spite of this difference, however, the structural relationship between truth-determined classes and truth-determined kinds is paralleled by the structural relationship between significance-determined classes (significance ranges) and significance-determined kinds (logical kinds, or categories). Any given predicate determines a significance range, but not every predicate, nor yet every set of predicates, determines a category. A category is determined by a characteristic predicate or set of predicates, as a natural kind is; and the main problem of a category theory is to determine what these predicates are. There are, too, other parallels. For example, not all Boolean operations apply to natural kinds: the complement of a natural kind is not normally a natural kind (nor is the complement of an artefact-kind itself an artefact-kind), and the union and intersection of two natural kinds are not usually natural kinds; but the complement of a class, and the union and intersection of two classes, are always classes. Similarly, it seems clear (though we discuss this in more detail below), the complement of a category, and the union and intersection of two categories, are not in general categories; but it is at least arguable, though it may seem doubtful, that Boolean operations do apply to significance ranges. Again, natural kinds can be ordered in genus-species trees, though classes in general cannot; and it seems possible that categories are orderable while significance ranges are not.

In making this comparison we have of course neglected the special problem of whether significance ranges and categories are best regarded as classes of items or classes of descriptions. We take up this question below and it is discussed fully in R. and V. Routley [1] (much of the present section is based on work done with Valerie Routley in this article and other, unpublished, papers). Because of this, and for other reasons, we do not wish to place too much emphasis on the similarity between the logical structure imposed on items by applying truth conditions and the structure imposed by applying significance conditions, but the comparison does serve to highlight the important differences between significance ranges and categories. A theory which identified categories as significance ranges would be as inadequate as a theory which identified natural kinds and simple classes. Kinds are, of course, classes; and categories are significance ranges (which are, in turn, classes). But just as kinds are special sorts of classes, so categories are special sorts of significance ranges.

We take for granted the thesis that categories, whether classes of descriptions or classes of items, are not exclusive; they have an inclusive structure. For example, natural numbers are included in numbers and these in abstractions; people are material things; persons and numbers are however exclusive categories. What is meant by an inclusive category theory, therefore, is that at least some categories include others. The theory must concede that some categories exclude others, but it can leave open, to begin with, the issues of whether some categories are properly overlapping and of whether some category is universal.

Given that categories are inclusive then an absolute sameness-of-category relation cannot be defined. However an unqualified sameness-of-minimal category relation can be defined since minimal categories partition their domain (the class of expressions, or of items) into exclusive and exhaustive subclasses. Thus the problem of defining categories under an inclusive category theory can conveniently be divided into two parts, that of defining minimal categories, *i.e.* categories which properly contain or overlap no further categories; and that of defining categories proper, as certain combinations of minimal categories. Both parts raise problems.

Minimal categories can be defined as the classes into which s-items (or descriptions) are partitioned by the sameness-of-minimal category, $\overset{\mathrm{M}}{=}$, relation. An initially appealing definition of $\overset{\mathrm{M}}{=}$ is that given in §7.13, §6:

$$x \overset{\mathrm{M}}{=} y =_{\mathrm{Df}} (U\psi)\,[S(xi\psi) \equiv S(yi\psi)],$$

a prescription adopted in fact by exclusive category theories to define sameness-of-category in general, and thereby exclusive categories. Though $\overset{\mathrm{M}}{=}$ does provide a partition of the desired sort, there are, as we shall soon see, troublesome attributives and predicates which collapse the notion of minimal category so defined.

Given minimal categories there is still the problem of defining categories. If, of course, Boolean operations on minimal categories provided categories then the remainder of the theory would coincide with Boolean algebra (or with the calculus of individuals if a null category is excluded) with minimal categories as the atoms (or "individuals"). On such a Boolean category theory the only real problem would be that of characterising minimal categories. But it is commonly contended that not all combinations of (minimal) categories provide categories, *e.g.* there is no intuitively discerned category of prumbers, *i.e.* of persons and natural numbers only.

In terms of significance theory the classical solution to the problem of defining categories is very simple—as we have noted, far too simple—*positive categories* are the significance ranges of predicates, thus:

$$p\text{-}cat\ (w) \equiv . w \approx \{x : (\mathcal{9}\psi) S(xi\psi)\}.$$

Then, either all categories are positive categories, or else there are more categories than p-categories, such as those obtained by complementation and union. In the second case *category* can be defined by recursion; *e.g.* a p-category is a category and the non-null complement of a category is a category. If union, say, were also permitted as a category-determining operation then we should be back, as before with prumbers and with the Boolean theory built on minimal categories.

A category theory is not unique; what theory emerges depends in particular on three factors:

(I) what significance notion is taken as fundamental in defining categories;

(II) which predicates are finally admitted as category determining;
(III) which combinations of initial categories or p-categories are counted as categories.

We discuss these in turn.

What a category theory is a theory *of* depends on the basic significance notion used to define categories and minimal categories. If the notion is that of significance, connoted by '*S*', then the categories are initially classes of descriptions and only derivatively classes of s-items; but if a more transparent significance notion, say the absurdity notion '*Ab*' is used (§7.13, §6), then the categories resemble rather more classes of things, The categories of these theories are not isomorphic; hence the theories are different (see R. and V. Routley [1]). Though we shall concentrate on categories defined using '*S*' as the basic significance connective, there is no great loss of generality; for the theory based on '*S*', which we call when we need to be specific *theory 1*, can be formally paralleled by theories with more thingy categories simply by using other, identity-transparent, significance connectives in place of '*S*'. For example, by using '*Ab*' in place of '$\sim S$' in definitions of 'category' *etc.*, a different theory, *theory 2*, which takes account of contingent identifications, emerges. Since the different significance notions reflect different identity notions, in effect different category theories result as different identity relations are used at the base. Put differently, categories are determined by significance-weighted contractions; with different identities used in the contractions, different categories result.

Part of the point of a category theory based on an identity-transparent significance notion is, then, to escape from descriptions to the things described. In theory 2, for example, many descriptions are discounted by use of directed identity; only the more specific descriptions of an item count in its categorisation. But is even theory 2 a theory in which description is inessential; is it a purely thing theory? Apparently not. For as exclusive identities occur, the categories of theory 2 cannot really be categories of things, in the sense that the same thing appears in just the one minimal category. For minimal categories are exclusive, and yet distinct categories contain the same *thing*. Therefore, although theory 2 is a thingy theory, it is not a purely thing theory.

The clear candidate for a "pure thing" theory would, we suggest, be *theory 3* based on wide identity, $\overset{W}{=}$; for it does not rule out the important exclusive identities. But such a theory would collapse categories—at least for every such exclusive identity which is true, formerly distinct categories which include identified individuals would collapse into one another. For, if exclusive identities were added to theory 2, things in distinct minimal categories would be identified. So the categories would not be exclusive, insofar as distinct categories include the same *thing*. In other words, if we tried to define an absurdity relation, and thereby minimal categories, using '$\overset{W}{=}$', things of different minimal categories of theory 2 would be identified:

thus the minimal categories of theory 3 would collapse locally and would tend to collapse outwards (*i.e.* all categories would reduce to one). A satisfactory category theory based on wide identity seems therefore to be ruled out. Exclusive identities are indeed cross-category identities, and they tend to disregard usual ways of conceptualising items; for conceptualising or describing items places them in one category or another. For these reasons, exclusive identification provides an escape from conceptualisation or categorisation—not by the use of purely referential expressions such as logically proper names (for there can be no purely referential referring expressions, we should argue), but by allowing us to ignore or bypass those features which result from a particular categorisation. This effect is achieved by the cumulative force of categorially different descriptions: given one description of a thing we cancel the effect of the description by another categorially distinct description of the thing. Wide identities are not designed to change categories, or to eliminate our ordinary ways of talking, but rather to escape the restrictions usually imposed by them. Wide identity is so characterised that significance considerations, like intensional features, are simply discounted. We appear to have, then, through exclusive and wide identifications, a way of doing that apparently impossible thing, sometimes called "peeling language off the world" or "removing the veneer of language". Of course exclusive identities do not presuppose that there are pure things and other, semantical things, as separate items.

A category theory like theory 1, which can be interpreted as categorizing concepts, is directly tied with ways of conceptualising. It now appears that even a theory like theory 2 also ties up with, and perhaps sometimes imposes, ways of conceptualising. For theory 2 is not a purely thing theory: reference in theory 2 is circumscribed by significance. Theory 2, unlike theory 3, is partly semantical. That it is semantical is a consequence of the identity on which it is based: sense and significance still feature. Thus categorising ties with conceptualising. Suppressing the role of significance, as in theory 3, means virtually abandoning categories.

Because of exclusive identities there can be no pure thing theory of categories which does not tend to collapse outwards, *i.e.* to reduce to a no-category theory. Hence any genuine category theory, *i.e.* one such that some categories remain distinct from others, since it cannot be wholly a thing theory, must be partly semantic in the sense that it will depend to some extent on the way things are described and not simply on the things themselves. Categorisation depends on limiting what can count as an alternative description or identification of a thing. To this extent categories limit our ways of "seeing" the world.

In explaining factor (II), and the need to restrict category-determining predicates, we concentrate on theory 1. Theory 1 has both universal and null categories. Because of predicates like 'is an s-item' and 'may be thought about' the universal class consisting of all s-items should be a p-category. Since predicates such as 'is heavier than 10 o'clock' have, in standard contexts, no significant couplings the null class comprising no s-items should also be a p-category. One ragbag category worth noticing consists of the class of nonsignificant subjects, *i.e.* of s-items such as a purple drunken number and the happy toffee axe; the category is a ragbag because, though seemingly minimal, it includes such apparently diverse s-items as kinds of numbers, ideas, axes, and so forth. Both categories and p-categories are inclusive under the intended interpretation since predicates such as 'is prime', 'is a natural number', and 'is a number', which specify inclusive categories, are supplied. Nor are properly overlapping categories excluded, as construction of a suitable finite model will show. But though it appears that, under the intended interpretation of the theory, sufficiently many categories will result, it is by no means clear that too many categories have not been provided.

A surfeit of categories does result from category determination by ill-behaved predicates. But there is no viable alternative to the determination of categories by predicates. For any theory which classifies individuals must group them in terms of certain properties they exhibit, and these properties are specified by predicates. Category theories classify individuals in terms of their significant or absurd couplings, and hence a crucial problem is raised by the sorts of predicates which are admitted. The problem is enhanced because formal theories are conventionalistic in that any predicate which is not proscribed and which can be defined in the theory, may be introduced; but it is not just a problem for formalised category theories. The same difficulty arises for theories which are not formalised, where ill-behaved predicates may also be introduced, since natural languages present few barriers to the introduction of new predicates.

Many of the major objections designed to show that some specific category theory fails, or that no category theory is possible, have in fact depended on the introduction of ill-behaved predicates which create category differences where none should be, or obliterate those there are.

In view of the possibility of defining ill-behaved predicates, conditions of adequacy are required to show that the theory has an intuitively adequate interpretation.

Three conditions of adequacy for any formal category theory are that the following theses are consistent with it:

(1) The categories should not amount to or reduce to just one category.
(2) The categories should be such that it is not the case that each item belongs to its own category.
(3) Categories should not just coincide with classes.

If (1) can be shown to fail the categories of the theory are said to *collapse outwards*; if (2) does not hold the categories of the theory *collapse inwards*.

These conditions are rather weak, for a Boolean category theory with a minimal category containing several elements would satisfy them. But given that every predicate determines an attributive, a conversion principle with considerable appeal at least in the case of predicates of individuals, the conditions are not so easily satisfied. For there are certain sorts of predicates which raise difficulties for any category theory, particularly with respect to (2) above. One such is 'QIx_0', where 'Ix_0' is an identity predicate; another is 'DIx_0'. It can be shown that any category theory in which such predicates are definable, and freely admitted to all roles, collapses inwards. Consider, for example, the effect of 'QIx_0' on the theory, where, *e.g.*, $Ix_0(x) =_{\mathrm{Df}} (x \overset{\mathrm{H}}{\approx} x_0)$. Then $QIx_0(x)$ is true if x is identical with x_0 and is nonsignificant otherwise: *i.e.*, 'QIx_0' is significant of just one item, x_0, so that x_0 by itself forms a category. Similarly 'QIy_0' is significant just of y_0, and so on for each item. Hence for each item x there is a predicate $qu(QIx)$ which is significant of just that item, and nonsignificant of anything else. This can be shown formally as follows:

$$\begin{aligned} y \overset{\mathrm{M}}{=} x_0 \supset . & SQIx_0(y) \equiv SQIx_0(x_0) \\ \supset . & SQIx_0(y), \text{ since } SQIx_0(x_0) \\ \supset . & T(y = x_0)(\text{or } y \overset{\mathrm{H}}{\approx} x_0). \end{aligned}$$

Thus y is of the same minimal category as x_0 only if y is identical with x_0. Consequently the minimal category to which x_0 belongs is just the unit class consisting of x_0, for each x_0, under the given identity relation. Hence the theory collapses inwards if 'QIx_0' is admitted for each x_0. The argument is readily generalised to other category theories, since the only property of categories presupposed is that if items are in the same category, they have the same significance properties, a feature of Leśniewski's and Russell's theories, for example.

By considering such expressions as $qu(Q(x = x_1 \vee x = x_2 \vee \ldots x = x_n))$ and $qu(QTf)$ it can be shown that (3) also fails to hold; *i.e.* every class is a category if ill-behaved predicates like 'Q' are admitted as category-determining. Further, any category u of theory 2 can be split into two sub-categories u_1 and u_2, which consist respectively of the true and false ranges of u under a given category predicate $qu(f^u)$, by predicates $qu(QTf^u)$ and $qu(QFf^u)$. For example, the category of material objects, which can be specified by the predicate 'is heavy' can be split into the categories of heavy material objects and objects which are not heavy. Thus ill-behaved predicates carve up categories, including minimal categories, which are supposed to be irreducible. They provide means of getting inside any category. It follows that not all predicates are category determining—on pain of violating conditions of adequacy.

Given an adequate significance logic on which to base a category theory, collapse outwards is not a worry. Theory 1 collapses outwards iff all individual expressions have the same significance couplings; that is, iff all individual expressions are in the same minimal category. Thus the categories collapse outwards iff the following condition is satisfied:

$$(Ux, y, \psi).(S(xi\psi) \equiv S(yi\psi)).$$

Supposing it is, it follows, choosing a *y* appropriately free,

$$(Ux, \psi).S(xi\psi) \supset (Uy)S(yi\psi), \text{ and hence}$$
$$(Ux).(U\text{non-}S\text{-null } \psi)S(xi\psi) \supset .(Uy, \text{non-}S\text{-null } \psi)S(yi\psi),$$

where an attributive is non-*S*-null iff it is significant of some s-item. Then, since *x* is not free in the consequent,

$$(Px)(U \text{ non-}S\text{-null } \psi)S(xi\psi) \supset .(Uy, \text{non-}S\text{-null } \psi)S(yi\psi).$$

But the antecedent is easily verified; *e.g.* for non-*S*-null ψ, $S(\xi xIn(x)i\psi)$, since $SIn(x)$, *i.e.* since it is significant that any *x* is a subject item. Detaching, $(Uy)(U \text{ non-}S\text{-null } \psi)S(yi\psi)$ follows; that is, effectively, *Sp* for any *p*. Any significance theory for which such a result holds is *S*-inconsistent. Consequently theory 1 collapses outwards only if the underpinning significance logic is inadequate. Alternatively it can be argued: since all individual expressions have the same significance couplings iff all predicates are unlimited, the categories collapse outwards iff all predicates are unlimited; that is, only if *Sp*.

These points can be argued, as well, from the interpretation of the underlying significance logic. If this logic is adequate, predicates which are not significant over all items can be consistently added to it. But, using the predicates supplied in this extension of the significance logic, categories distinct from the category of items can be distinguished. Therefore the categories do not collapse outwards.

But to prevent collapse inwards category-reducing predicates must be somehow excluded. This can generally be done by some combination of these methods: either the interpretation provided for predicates is restricted, or formal restrictions are made ensuring that delinquent predicates play a limited role. The first course alone is not open to us in this theory, since, unless formal restrictions are imposed, predicates can be so constructed within the theory that the collapse of categories is demonstrable.

The predicates which cause trouble, '*QI*' and '*DI*', and the connectives '*Q*', '*D*' and '*QT*', exhibit unusual features with respect to negation. They display a certain negation asymmetry, not found in the case of more ordinary predicates and connectives. To illustrate: when *P* is true, *QP* is true; but then $\sim P$ is false, so $Q{\sim}P$ is nonsignificant. Similarly *QP* is nonsignificant when *P* is false; but then $\sim P$ is true, so that $Q{\sim}P$ is true. A similar asymmetry affects '*D*'. Of normal and natural-language connectives, by contrast, it is true that if a connective Φ is significant of P, it is also significant of its classical negation $\sim P$, *i.e.*

$$S\Phi p \equiv S\Phi \sim p$$

holds. To say that normal connectives have this feature is just to say that a sentence and its classical negation belong to the same category. To allow connectives without this regularity feature a role in determining categories would be to assume in certain cases that a sentence and its negation belong to different categories. For to assign the value n to DP when P is true, and the value f to DP when P is false, is to assume that there is some connective which is significant of a false sentence P', but not significant of its negation $\sim P'$; and this is to assume a category difference between P' and $\sim P'$, Such an assumption is surely false. For in view of the matrix for classical negation, the significance properties of P and $\sim P$ should coincide; and all normal connectives yield this result.

We define *irregular* connectives and predicates in terms of this feature (*cf.* §5.13). Thus, where Φ is a one-place sentence connective (*e.g.* '$\sim$') which is related by an absorption principle to a predicate coupled with a descriptor $\Phi'\tau$ (see §7.12, §1), which for clarity we may write as Gτ, where G is a predicate,

$$reg(\mathrm{G}) =_{\mathrm{Df}} (U\tau p)(S\mathrm{G}\tau p \equiv S\mathrm{G}\tau \sim '\tau p);$$

where '$\sim '\tau p$' is the s-p transform of '$\sim p$' (in terms of $\mathbf{2Q}_i$,
$reg(\Phi) \simeq (Up)S\Phi p \equiv S\Phi \sim p)$);

$$reg(\mathrm{G}\tau) =_{\mathrm{Df}} reg(\mathrm{G}); \qquad reg(\Phi) =_{\mathrm{Df}} reg(\mathrm{G}\tau);$$
$$irreg(\mathrm{G}) =_{\mathrm{Df}} \sim reg(\mathrm{G}); \qquad irreg(\Phi) =_{\mathrm{Df}} irreg(\mathrm{G}).$$

Here '*reg*' and '*irreg*' absorb '*qu*'. The restricted quantifiers are acceptable since τ converts 'p' to a subject. Since every connective can be resolved into a predicate coupled with a descriptor, it follows immediately from these definitions:

$irreg(Q)$; $reg(QQ)$; $reg(T)$; $reg(\sim)$; $(U\,\tau p)S\mathrm{G}\tau p \supset reg(\mathrm{G})$.

The following metatheorem also results:

G is irregular iff G converts *one* of the values t and f of P to n, *but not both*, for some P.

This theorem explains how it is that the use of irregular predicates provides a means of getting inside categories and subdividing them.

Irregular connectives are not the only ones which collapse categories. There are some individual predicates which have exactly the same effect. An example is the individual identity predicate '$\mathscr{I}x_0$', interpreted as 'is identical with x_0', defined in such a way that it is true, hence significant, of the individual x_0, but not significant of anything else. This predicate behaves just like 'QIx_0', but since it does not contain a distinct connective explicitly, the above definition of 'regular', which applies only to sentence predicates and connectives, does not apply to it. Clearly, if the predicates '$\mathscr{I}x_0$', '$\mathscr{I}y_0$', *etc.* for each individual, were permitted a role in forming categories, each individual would be in its own category.

Individual predicates like '$\mathscr{I}x_0$' which do not contain connectives explicitly, create special problems. '$\mathscr{I}x_0$' is like 'grue'—Goodman's disjunctive predicate—in that it cannot, as it stands, be split up to obtain the appropriate negation for the definition of 'irregular'. However, since it behaves like an irregular connective together with an individual predicate, it might be thought appropriate to seek an extension of the definition of 'irregular' along the following lines:

Where 'f_0' and 'h_0' are individual predicates and Φ is some irregular connective:

$$irreg(f_0) \text{ iff } (Ux)(f_0x \simeq \Phi h_0 x) \;\&\; irreg(\Phi).$$

To obtain a more general characterisation, it is necessary to introduce attributives since predicate parameters are not quantifiable. Thus,

$$irreg(f) \text{ iff } irreg(\lambda f), \text{ and}$$
$$irreg(\lambda f) \text{ iff } (P\lambda \mathrm{G})\,[(P\lambda h)(Ux)(xi\lambda f \simeq \tau(xi\lambda h)i\lambda\mathrm{G} \;\&\; irreg(\lambda\mathrm{G})].$$

The last condition may be stated more succinctly as:

$$irreg(\psi) \text{ iff } (P\chi)\,[(P\phi)(Ux)(xi\psi \simeq \tau(xi\phi)\, i\chi) \;\&\; irreg(\chi)].$$

Such a definition expresses the fact that an individual predicate which behaves just like a compound predicate consisting of some individual predicate and an irregular connective, is itself irregular. Unfortunately, however, the definition is unsatisfactory. For the regular identity predicate 'Ix_0' is equivalent to '$QQIx_0$'; hence, if we take 'h_0' to be 'QIx_0', we can show that the regular predicate 'Ix_0' is irregular, by choosing 'Q'' for G; and in general, it follows that *any* individual predicate is irregular.

What seems to be required is a restriction to regular predicates in the definition, thus ensuring that 'h_0' is itself regular and so preventing the choice of 'QIx_0' for it (or more generally, a requirement that $qu(h)$ (or $qu(\lambda h)$) is regular to avoid instantiation cases such as 'QIx_0'). But such a requirement would make the definition circular. In fact, an explicit characterisation of regularity for individual predicates cannot be obtained within the scope of the logic so far developed. For a little reflection will show that connectives and individual predicates combine as follows:

Where Φ is a sentence connective and F is an individual predicate:

(i) If *irreg*(Φ) and *irreg*(F) then ΦF may be either regular or irregular: *i.e.* the irregularity of ΦF is not determined by the irregularity of Φ and F. For if we choose Φ as '*D*' and F as '*Q*H', where H is some intuitively regular individual predicate which takes all values, the outcome is an irregular predicate: *DQ*H, since we have *irreg*(*DQ*). On the other hand, if we take Φ to be '*Q*', and again F to be '*Q*H', the outcome is regular since we have *reg*(*QQ*).

(ii) If *irreg*(Φ) and *reg*(F), then ΦF may be either regular or irregular. This is easily seen as a consequence of the property that irregular connectives convert *one* of the values *t* and *f* to *n*, but not both. If, now, our regular F takes just one value for any X, whether it be *t*, *f* or *n*, ΦF may not be irregular even though Φ is. For F may not take for any X the value which Φ, if irregular, converts to *n*. For example, *Q*FX is equivalent, for all X, to FX in the case that FX is true for any X. On the other hand, even though F is regular and for some X takes each value, ΦF may still be irregular when *irreg*(Φ).

(iii) If *reg*(Φ) and *irreg*(F), then ΦF may be regular or irregular. Let Φ be an unlimited connective, or, for example, a connective which converts every value to *n*. Then Φ is not irregular. But then, even if F is irregular, ΦF does not behave irregularly for any X, for either none of the values of ΦFX are *n* or all are. But, if, say, Φ is the echo connective, ΦF behaves just like F and is therefore irregular, given that F is.

(iv) Finally, if *reg*(Φ) and *reg*(F), then *reg*(ΦF). This, then, is the only general case in which the regularity of ΦF is determined just by the regularity features of Φ and F.

Now an explicit definition of the kind suggested depends on the regularity features of ΦF being a function of the regularity features of Φ. But in view of the properties (i)–(iv) above, this is not in general the case and the definition must therefore fail. Not only this, but it is clear from (i)–(iv), and especially (ii), that the regularity features of ΦF are not even a function of the regularity features of Φ for some determined class of individual predicates F. That is, even if a class of regular F can be guaranteed (for example, unlimited predicates), in view of (ii) the irregularity of ΦF for these particular F would still not be a function of the regularity features of Φ. Hence no explicit definition for individual predicates can be obtained from the definition of irregularity for sentence connectives alone.

The property of regularity set out under (iv) does, however, allow us to use an inductive definition or axiomatic characterisation of a regular predicate. Thus, an individual predicate is *regular* iff either it is a primitive regular predicate or it is obtainable by abstraction, from primitive regular predicates using only regular sentence connectives and quantifiers. Now the assumption that all primitive individual predicates of a category theory are drawn from an initial class Ω of intuitively regular predicates does not appear over-restrictive; for, so it seems (see §5.13), irregular predicates lack adequate interpretation and are never needed to specify intuitively given categories or minimal categories. But if there is no adequate interpretation of irregular predicates in natural language, then all natural-language predicates are intuitively regular, and irregular individual predicates are concocted from them using irregular sentence functors. Thus, primitive regular predicates can be *interpreted* over some subclass of natural language predicates. This procedure is not, however, completely satisfactory; for it inevitably puts a heavy load on the interpretation of the system, yet without giving much guide as to which interpretations are admissible.

One way of lightening the burden on the interpretation is by enlarging the logic. An obvious way of doing this is by introducing a new negation, a predicate negation, so that a definition of *irregularity* for individual predicates which parallels that for sentence connective, can be obtained. But without a good interpretation of the new negation this development is little better than introducing a new primitive, '*reg*' say. At this stage we shall simply take '*reg*' as a primitive predicate, of attributives, which is characterised inductively so that predicates obtained from primitive predicates using only regular connectives and quantifiers are regular. It follows that all predicates of 4-logics are regular. We restrict the attributives admitted in determining minimal categories and p-categories to regular attributives (where of course a predicate is regular iff its full conversion is). So, for example, the definition of sameness-of-minimal-category now takes the form:

$$x \overset{M}{=} y =_{\text{Df}} (U\, reg\, \psi)(S(xi\psi) \equiv S(yi\psi)).$$

Both minimal categories and p-categories are categories.

We now argue that a category theory in which the role of irregular attributives is so limited does not collapse formally, either inward or outward. Theory 1 collapses inwards iff for each item x, there is an attributive $qu(\psi)$ such that $qu(\psi)$ is significant over x, and not significant of anything not identical with x; and furthermore, if $qu(\psi)$ is an attributive not excluded by the definitions of the theory from specifying categories, *i.e.* if $qu(\psi)$ is regular. The first part of these conditions for category collapse inwards can be represented:

$$(Ux)(P\psi)\,[S(xi\psi)\ \&\ (Uy)(\neg(x = y) \supset \sim S(yi\psi))],$$

for an appropriate identity relation =. But then $qu(\psi)$ would surely be irregular. For in terms of it, given any x_0, $qu(\psi)$ would be circumscribed by the condition,

$$S(x_0 i\psi)\ \&\ (Uy)(\neg(y = x_0) \supset \sim S(yi\psi)).$$

There are two cases to consider: (i) $T(x_0 i\psi)$; (ii) $F(x_0 i\psi)$. For each z, if $z = x_0$ then in case (i) $T(zi\psi)$ and $S(zi\psi)$ hold, and in case (ii) $F(zi\psi)$ and $S(zi\psi)$ hold; on the other hand if $\neg(z = x_0)$ then $\sim S(zi\psi)$ holds in both cases. Thus for each z, either $zi\psi \simeq Q(z = x_0)$ or $zi\psi \simeq \sim Q(z = x_0)$. Since so constructed, using the irregular connective 'Q', we should dismiss $qu(\psi)$ as intuitively irregular. We do not however pretend that this argument is conclusive; it is at best persuasive, until intuitively regular predicates are better characterised.

Given that initial categories are determined by well-behaved predicates, what form should the recursion rules which facilitate the construction of further categories take? This brings us to factor (III). Are categories so structured that the following conditions of adequacy on inclusive category theories are satisfied?

(4) The difference of any two categories is a category;

(5) The union of any two (or more) categories is a category?

The issue is a conflict one. There are category theories, see *theory 1A* below, in which both (4) and (5) are demonstrable; there are theories, like *theory 1C* (below), in which neither (4) nor (5) hold for any general class of cases; and there are intermediate theories, of which *theory 1B* (below) is an example, for which (4) holds in a qualified form. Deciding which of these theories to adopt provides a conflict issue, since there are reasons for and against adopting each.

Conditions (4) and (5) are equivalent. First, *(5) implies (4)*. For consider the difference u–v of two categories u and v, where u–v is non-null. Since u–v is non-null, it can be partitioned into minimal categories. If there is just one such minimal category, *viz.* u–v itself, then, since a minimal category is a category, (4) follows. If u–v consists of two or more minimal categories, then, by (5), since these minimal categories are non-inclusive their union, *i.e.* u–v, is a category. So (4) follows. Secondly, *(4) materially implies (5)*. Consider two categories u and v. We need not consider further the case (i) where one includes the other. It remains to consider the case (ii) where u and v are non-inclusive. Consider the difference of these categories with respect to the item category I, which may be specified by the predicate $qu(Tf)$, where $qu(f)$ is some predicate of the system. Both u and v are properly contained in I; otherwise the result would follow by case (i). Since I-u is non-null, by (4), I-u is a category. Since u and v are non-inclusive (I-u)-v is not null; so, by (4), (I-u)-v is a category. Similarly I-((I-u)-v) is a category, that is $u \cup v$ is a category, and (5) follows. If, in place of categories, positive categories had been considered, the first argument, that (5) implies (4) would fail; for the minimal categories obtained may not be p-categories. Consideration of p-categories leads to further theories, like *theory 1D*, where (5) holds generally but (4) does not.

When either (4) or (5) holds, categories are Boolean; they are closed under all Boolean operations with minimal categories as atoms. Categories are then treated exactly like classes with respect to Boolean intersection, union and complementation; and they are so treated in theory 1A.

Theory 1A has these features. It is built on a significance logic such as **A2QS**$_6$ with a functionally complete sentential basis. No restrictions are imposed on category-determining predicates beyond those already imposed, namely restrictions to regular predicates. Since the sentential basis of 1A is functionally complete, the following regular sentence connectives, here defined by cases, are explicitly definable:

$$\begin{aligned} v(K_+p) &= n \text{ when } v(p) \neq n \\ &\quad t \text{ when } v(p) = n \\ v(K_-p) &= n \text{ when } v(p) \neq n \\ &\quad f \text{ when } v(p) = n \end{aligned}$$

These sentence connectives 'K_+' and 'K_-' are not significance-preserving, and accordingly are not definable explicitly in 4-logics; they are inversion connectives, where an *inversion connective* is defined thus:

$$Inv(\mathrm{G}) =_{\mathrm{Df}} (U\tau p)(S'\tau p \equiv \sim S\mathrm{G}\tau p), \text{ where } Sp \simeq S'\tau p \text{ (§7.2, §1).}$$

Hence where Φ is $\mathrm{G}\tau$,

$$Inv(\Phi) \text{ iff } Inv(\mathrm{G}).$$

An important feature of 1A is that every category has a minimal spanning predicate (or attributive), *i.e.*

$$cat(w) \supset .(P\psi)(S(xi\psi) \equiv .x \in^{+} w).$$

Thus every category of 1A is a positive category. This condition can be demonstrated if the theory contains only finitely many predicate constants or if sentential connectives can be iterated infinitely many times. Then a minimal spanning predicate $qu(f^v)$ for a minimal category v, which otherwise lacks such a spanning predicate, can be defined by the relation,

$$f^v(x) \simeq .f_1(x)\mathrm{W}.f_2(x)\mathrm{W} \ldots \mathrm{W}.K_+g_1(x)\mathrm{W}.K_+g_2(x) \ldots,$$

where 'f_1', 'f_2', ... are all predicates significant over v and 'g_1', 'g_2' ... are all predicates nonsignificant over v, and where the (*non-U*) connective 'W' is defined:

$$A \mathrm{W} B =_{\mathrm{Df}} K_+(K_+A \;\&\; K_+B).$$

'W' may also be defined in 3-logics using Brady's disjunction 'V' as follows:

$$A \mathrm{W} B =_{\mathrm{Df}} SA \mathrm{V} SB.$$

A minimal spanning predicate $qu(f^u)$ for a non-minimal category u, which would otherwise lack such a predicate, can be defined by the relation,

$$f^u(x) \simeq . f_1(x) \mathrm{W} . f_2(x) \mathrm{W} \ldots$$

where 'f_1', 'f_2' ... are all the predicates which span u. Generally, finitely many predicates would suffice.

Conditions (4) and (5) are theorems of theory 1A. It suffices to establish (4), since (5) follows. To show that 1A yields condition (4), suppose that categories u and v are such that $u-v$ is non-null. Now each category w has a minimal spanning predicate, $qu(f^w)$ say, which distinguishes the category. It is minimal because all other predicates which span the category apply to classes which include (properly or improperly) the class or category to which it applies. Such a minimal spanning predicate can always be obtained in 1A. Using 'K_+' or 'K_-' and minimal spanning predicates of categories u and v we can define a minimal spanning predicate of class $u-v$ as follows:

$$f^{u-v}(x) =_{\mathrm{Df}} K_+ f^v(x) \,\&\, f^u(x).$$

It is readily verified that $qu(f^{u-v})$ has the desired feature that it is nonsignificant of $qu(x)$ when $x \in^+ v$ and when $x \in^+ (-u)$, where $-u$ is the complement of category u, but it is significant of $qu(x)$ when $x \in^+ (u-v)$. Finally the predicate $qu(f^{u-v})$ and all predicates which are significant over all elements of $u-v$ can be used, in accordance with the definition of *category*, to define the category $u-v$. For $qu(f^{u-v})$ is significant, since $u-v$ is non-null, and is regular, since $qu(f^u)$, $qu(f^v)$, '&' and 'K_+' are regular. Hence the difference $u-v$ of u and v is a category.

The intersection of u and v also forms a category. The predicate $qu(f^{u \cap v})$, defined

$$f^{u \cap v}(x) =_{\mathrm{Df}} f^u(x) \,\&\, f^v(x),$$

provides a minimal spanning predicate. For $qu(f^{u \cap v})$ is significant when coupled with $qu(x)$, under theory 1, when $x \in^+ (u \cap v)$, but is nonsignificant when coupled with $qu(x)$ if $x \in^+ (u-v)$ or $x \in^+ (v-u)$. The category relations of inclusion, intersection, difference, and so on, that we are working with are, of course, the usual Boolean relations, *e.g.* u includes v if every element of category v is an element of category u.

Theory 1A, indeed any theory which satisfies (5), is inconsistent with Sommers' [1] category theory. For as the union of any two categories is a category, overlapping categories are generated. Consequently Sommers' theory cannot be consistently erected on the basis of a functionally complete logic, or a 3-logic, without severe curtailment of the class of category-determining predicates. We show later how leading features of Sommers' theory can be deduced given the assumption that no categories properly overlap.

Why not adopt theory 1A? The usual objection is that it admits too many categories. But the objection has different—inconsistent—developments. It is contended first, that 1A is unsatisfactory because it does not limit (starting) categories to p-categories as it should. In particular the complement of a p-category should not always be a category; for such classes as the class of items that are not natural numbers are not spanned by natural-language predicates. But the use of this natural-predicate criterion, according to which compounding rules should only supply categories spanned by natural-language predicates, also counts against retention of (5), as we shall see. Indeed application of the natural-predicate criterion would appear to lead, owing to the unsystematic design of natural languages, to the general breakdown of compounding rules for categories (and thus in the direction of theory 1C).

The *theory 1D*, which we shall consider briefly, replaces the Boolean theory 1A by a lattice category theory. Theory 1D concedes that the union and intersection of any two categories is a category and that there is a universal category, and a null category, but denies that the complement of every category is a category. It follows that not every minimal category is a category: otherwise the complement of each category would be a category upon union of minimal categories. Hence, too, the theory is compatible with Sommers' theory. A Sommers' 1D theory is a 1D theory where no categories properly overlap; in such a case the union operation is really the method of category combination, and a tree structure, determined by the lattice union operation, results.

Given a class Ω of well-behaved primitive predicates, categories of 1D may be defined by abstraction using the connectives and quantifiers of system $A2QS_{21}$, *i.e.* using the set $\{\sim, T, \&, V, U, q\}$ of a quantified S_{2I} (§5.12) extended to second stage (the connectives and quantifiers admitted in category formation may be alternatively characterised using the notion of directness: see §5.12). Since classical negation does not permit the definition of new categories, because $Sp \equiv S \sim p$, and since the connective '*T*' does not provide categories not already obtained by union, or by the use of arbitrary significant wff, all category construction in theory 1D is in terms of (finite and infinite) intersection, through '&' and '*U*', and union, through V and '*q*' (this theorem is proved in detail in Brady [3], p. 355 ff.). From the point of view of lattice category theory 1D then, the fault with Boolean theory 1A is that it permits the use of inversion connectives in the formation of categories; for given an inversion connective all complement categories can be generated.

A second, quite different objection to theory 1A is that it permits the unification in one category of items which are too heterogeneous. Categories should not be combinable to form a further category unless they are neighbouring and sufficiently homogeneous. Thus the category of persons and the category of numbers should not be combinable, as they are in 1A, to form the further category of prumbers. When two categories are sufficiently homogeneous to form a third, there is a predicate which is significant of just the members of both. Since in 1A this condition is automatically guaranteed for any two categories at all, either homogeneity is violated, or the union of any two categories is counted as sufficiently homogeneous, so that homogeneity loses its point. Finally, categories, which should be very special sorts of classes, are treated under 1A just as ordinary classes are.

However, the notion of homogeneity used in this argument is suspect. The homogeneity criterion, if used consistently, would lead to an exclusive category theory. If an inclusive theory is accepted, some categories containing quite diverse items—the universal category, for example, which contains both numbers and persons—must be accepted also. Since homogeneity is always considered with respect to some property, a given group of items may be homogeneous with respect to one property (for example, that of being heavy or not heavy) and heterogeneous with respect to another (thinking or not thinking). Hence homogeneity is a relative notion, just as 'in the same category' is, and the only categories which should be homogeneous are minimal categories, which are not relative and which are exclusive. To ask that all inclusive categories be homogeneous is to try to have both an inclusive and an exclusive theory at the same time. Under theory 1A categories are still special sorts of classes, for not every class forms a category. But it does not follow from the fact that categories are special sorts of classes that they must have a special sort of union.

Even if this argument succeeds in showing the inadequacy of the homogeneity criterion as a principle for combining categories, there may still be a case for combining only certain sorts of categories. Although all categories can be combined in the universal category, unintuitive results are obtained by combining some exclusive categories to form a further category consisting just of these exclusive categories, as in the case of prumbers. On the other hand, we might still want to obtain as categories certain difference classes, such as the class of non-persons, and the class of everything which is a material object but not a person. A theory intended to meet these conditions would satisfy only a qualified form of (4), as well as,

(6) Not any two non-inclusive categories are integrable by means of a connective combining their characterising predicates.

Condition (6) cannot be guaranteed if irregular predicates are admitted as category determining predicates, *i.e.* as what we shall distinguish as *U-predicates*. A vital question here is whether the restriction of *U*-predicates to regular predicates is sufficient. It is not, because of *K*-predicates. Alternatively, there are certain 2-place connectives which, for any two given predicates, permit the construction of a third predicate which is significant of just the items of which either of the given predicates is significant. Such connectives, whether two or more place, we shall call *non-U connectives*. With a 2-place classical connectives Φ we have, quite generally,

$$\sim Sf(x) \;\&\; \sim Sg(x) \supset . \sim S\Phi(f(x), g(x)).$$

Non-U connectives, however, must also satisfy the converse condition:

$$\sim S\Phi(f(x), g(x)) \supset . \sim Sf(x) \;\&\; \sim Sg(x).$$

which is not satisfied by classical connectives.

The use of *non-U* connectives is both necessary and sufficient for the integration of any two categories by means of a *compound predicate*, *i.e.* a predicate which in primitive form contains a 2-place connective.

Consider two categories u and v, with minimal spanning predicates $qu(f^u)$ and $qu(f^v)$.

We show that the condition,

$$(\alpha): \; \sim S\Phi(f(x), g(x)) \equiv . \sim Sf(x) \;\&\; \sim Sg(x)$$

is necessary and sufficient for it to be true that $\Phi(f^u(x), f^v(x))$ is significant of just the members of u and v, *i.e.* for

$$x \in^+ u \;\&\; x \in^+ v \equiv . S\Phi(f^u(x), f^v(x)).$$

Sufficiency:

$$x \in^+ u \equiv Sf^u(x); \; x \in^+ v \equiv Sf^v(x).$$

Hence,

$$\begin{aligned} x \in^+ u \vee x \in^+ v &\equiv . Sf^u(x) \vee Sf^v(x) \\ &\equiv . \sim (\sim Sf^u(x) \;\&\sim Sf^v(x)) \\ &\equiv . S\Phi(f^u(x), f^v(x)), \qquad \text{by } (\alpha). \end{aligned}$$

Necessity:

$$\begin{aligned} \sim Sf^u(x) \;\&\sim Sf^v(x) &\equiv . \sim (x \in^+ u \vee x \in^+ v) \\ &\equiv . \sim S\Phi(f^u(x), f^v(x)). \end{aligned}$$

In the light of this result we define *non-U connective* Φ^2, using (α), where Φ^2 is related by an absorption principle to $G^2(\tau \ldots, \tau \ldots)$, as follows:

$$non\text{-}U(\Phi^2) =_{\mathrm{Df}} (U\tau p, \tau q)(SG^2(\tau p, \tau q) \equiv S'\tau p \vee S'\tau q).$$

Φ^2 can be iterated to connect n-places; or, generalising,

$$non\text{-}U\,(\Phi^n) =_{\mathrm{Df}} (U\tau p, \ldots, \tau p_n)(SG^n(\tau p_1, \ldots \tau p_n) \equiv S'\tau p_1 \vee S'\tau p_2, \vee \ldots \vee S'p_n).$$

If either $u \subseteq v$ or $v \subseteq u$ there is always a predicate $qu(f)$, *e.g.* $qu(f^u)$ or $qu(f^v)$, such that, $(Ux)(x \in^+ u \vee x \in^+ v \equiv Sf(x))$. Hence inclusive categories can always be combined. We have that the use of *non-U* connectives is necessary for the integration of any two categories by means of compound predicates.

Therefore if the use of *non-U* compound predicates, (*i.e.* of predicates compounded by using *non-U* connectives), is suitably limited, for example is prohibited in category determination, it is not the case that any two categories can be combined by a compound predicate. Since any two inclusive categories can be combined, it follows, if use of *non-U* compound predicates for category determination is prohibited, that there are some non-inclusive categories that cannot be combined by means of compound predicates. Therefore condition (6) is established, provided the use of *non-U* compound predicates in the determination of category extensions, and thereby in the determination of categories, is prevented.

The only definitions we need to revise are the definitions of minimal and positive category. The effect of theories 1B and 1C is to revise these definitions so as to exclude at least determining use of *non-U* predicates and thereby of *non-U* compounded predicates. By requiring, in theory 1B, that determining predicates are *U*-predicates, condition (6) is guaranteed. Accordingly the usual characterisation of p-categories is revised so that they are only determined by regular *U*-predicates.

It remains to underpin theory 1B by a characterisation of *U* individual predicates. An explicit characterisation of *non-U* individual predicates in terms of individual predicates and *non-U* sentence predicates appears to be excluded, for reasons similar to those given in arguing that irregular individual predicates are not explicitly definable in analogous fashion. But once again a characterisation can be given relative to some initial class Ω of primitive individual predicates, where elements of Ω are interpreted over intuitively well-behaved predicates.

We outline an argument to show that it is not demonstrable that

(6′) Any two categories of theory 1B are combinable to form a further category.

The argument simply consists in the construction of a finite set-theoretic model for theory 1B such that (6′) is false under the model. The model has exclusive minimal categories u_1 and u_2 specified by the primitive predicates 'h^{u_1}' and 'h^{u_2}' which belong to Ω. Both u_1 and u_2 are properly included in the item category which is specified by the predicate 'Th^{u_1}'. 'h^{u_1}' and 'h^{u_2}' are the only primitive predicates. It can be verified that the model has just 3 minimal categories and just $6(=2^3-2)$ categories. Since the theory is consistent over this finite model, (6′) is not demonstrable since it is falsified in the model.

Interpretations of 1B which violate (6′) have not been formally excluded: for the class Ω can always be enriched. The theory could, however, be extended so as to ensure (6′) by addition of an axiom of limitation. But such an axiom by no means guarantees intuitive requirements.

Theory 1B allows use of inversion predicates in determining categories provided that these do not, through iteration, amount to *non-U* predicates.

For example, the predicate '$K_+h^{u_1}$' is admitted as category-determining in the model, but the attributive '$\lambda x K_+(K_+h^{u_1}(x) \& K_+h^{u_2}(x))$' is not because it is equivalent to the combination '$\lambda x(h^{u_1}(x) \mathrm{W} h^{u_2}(x))$', and 'W' is a *non-U* connective. The outcome is this: within limits the complements and differences of categories form categories in 1B, but the union of non-inclusive categories does not form a category.

Although 1B appears to achieve what it is designed to do, it is formally a very cumbersome theory. This might be anticipated: for 1B is a compromise position, based on the conviction that while 1A sanctions undesirable categories, 1C sacrifices too many categories and does not even yield certain harmless difference categories.

In theory 1C, in contrast to 1A, there is no formal method for the unification of categories; and, in contrast to both 1A and 1B, there is no formal method for obtaining the difference of any categories. For 1C excludes all *non-U* and inversion predicates from category-determining roles. There are, once again, alternative ways of formulating theory 1C. Either we can base it on a weak functionally incomplete significance logic, which does not contain inversion and *non-U* connectives, or on a more extensive logic but restrict the role of undesirable connectives in category determination. The first method can be effected by basing the theory on sentential logic S_2. Then only S_2-connectives, *i.e.* connectives definable only in terms of classical connectives and 'T', are available: inversion connectives, irregular connectives, and *non-U* connectives are not available. Under the second method, category-determining predicates are restricted to those obtainable from predicates of initial class Ω using only S_2-connectives and quantifiers. The second method is preferable. Although irregular and *non-U* connectives create difficulties and must play a restricted role, it is really not a satisfactory solution to take as a base logic a weaker functionally incomplete system which suppresses difficult connectives by prohibiting their definition. For the chief problems are simply covered up in this way. Moreover there are heavy technical and expressive sacrifices, with no real gain in solving the problems. If a connective is definable, its role can be restricted in just those places where it needs to be restricted: but not, as in a weaker system, restricted to the point of disappearance. In any case predicates such as 'prumber' and '$\mathscr{I}x_0$' may be *introduced* into natural languages, and hence should be definable in systems designed to reflect these languages and to treat the philosophical problems they raise.

1C does not have formal means of unifying such categories as persons and numbers; nor does it have formal means of constructing from the category of persons such non-minimal categories as that of non-persons, for there is no formal way to construct their spanning predicates from primitives, as there is in 1B. Some categories, such as the minimal category of material objects which are not persons, which are positive in 1B where there is a predicate which spans them, are negative in 1A. In contrast to 1A, there is no formal means in 1C for obtaining inclusive categories from minimal categories—except in the case of the universal category which is spanned by defined predicates such as '*Tf*'.

The question of unification of categories is entirely a matter for the interpretation under 1C. Hence it is possible to interpret 1C so that all categories are minimal and exclusive with the exception of the universal category. Under 1C categories, and under 1B exclusive categories, can be combined only by the introduction of suitably interpreted primitives. Further, 1B and 1C are more interpretation-dependent than 1A, since they can be interpreted so as to satisfy (4) and (5), whereas 1A cannot be interpreted so as to satisfy (6).

The fact remains that in all these theories the primitive predicates can be interpreted perversely so as to bring out undesirable results, and it is impossible to avoid some result of this sort for primitive predicates. Constructed predicates are constructed from primitives using just admitted resources of the system, and hence it is both possible and necessary to control their behaviour in the system. But we cannot so control the choice of primitive predicates syntactically except by devices such as the filter suggested for irregular predicates, or by introducing a class Ω of intuitively well-behaved primitive predicates. Since, however, the choice of primitive predicates remains a matter of interpretation there is at least as much room for a perverse choice of predicates as there is for perverse decisions about the significance ranges of predicates. We can spoil the interpretation of 1B and 1C by introducing the primitive 'prinks', which behaves just like the *non-U* predicate 'is prime or thinks', to obtain from primitives the category of prumbers.

Although the theories have unintuitive interpretations, they also have more intuitive ones, which seem to be yielded by predicates actually occurring in natural language. This suggests an initial way of selecting the class Ω of primitive predicates, and of cashing in "intuitive well-behaviour", by talking of the p-categories of a given (natural) language and by taking Ω as interpretationally determined by the dictionary of the language. The naturalness of a set of categories is then the naturalness of a natural language. "Natural" categories are fixed by semantic, and perhaps also pragmatic features of a natural language, much as the natural species of taxonomy are fixed by the world. A category theory, like a taxonomical theory, requires underpinning by factual data; in the case of a category theory, by semantical data in particular. In view of the complexity of the data it would be rash indeed to expect a syntactical characterisation of the class Ω of naturally-occurring predicates. The problem is not one that can be solved by syntax alone, nor is it peculiar to category theory.

Bibliography

This is not a complete bibliography; only those works to which reference is made in the text are noted. Works are listed serially under authors' names usually in the order in which they are referred to in the text rather than chronologically. Where articles have been published more than once, only one source, not necessarily the original, is given. Unless otherwise stated, all book references are to the first edition.

The following abbreviations are used:

In the text:

OED	*Oxford English Dictionary*
PM	Whitehead and Russell [1], *Principia Mathematica*

In the bibliography:

AJP	*Australasian Journal of Philosophy*
APF	*Acta Philosophica Fennica*
BJPS	*British Journal for the Philosophy of Science*
CM	*Compositio Mathematica*
J.Phil	*Journal of Philosophy*
JPL	*Journal of Philosophical Logic*
JSL	*Journal of Symbolic Logic*
LA	*Logique et Analyse*
MA	*Mathematische Annalen*
NDJFL	*Notre Dame Journal of Formal Logic*
PAAAS	*Proceedings of the American Academy of Arts and Science*
PCPS	*Proceedings of the Cambridge Philosophical Society*
PR	*Philosophical Review*
PS	*Philosophy of Science*
RM	*The Review of Metaphysics*
Rev. IP	*Revue Internationale de Philosophie*
ZML	*Zeitschrift für Mathematische Logik und Grundlagen der Mathematik*

Ackermann, W. [1]:'Untersuchungen über das Eliminationsproblem der mathematischen Logik', *MA*, vol. 110, 1935.

Anderson, A. R. [1]: 'Some open questions concerning the system **E** of entailment', *Proceedings of a Colloquium on Modal and Many-Valued Logics*, *APF*, vol. 16, 1963.

[2]: 'Completeness theorems for the systems **E** of entailment and **EQ** of entailment with quantification', *ZML*, vol. 6, 1960.

Angell, R. B. [1]: 'A propositional logic with subjunctive conditionals', *JSL*, vol. 27, 1962.

Åqvist, L. [1]: 'Reflections on the logic of nonsense', *Theoria*, vol. 28, 1962.

Aristotle, [1]: *Categories*

[2]: *Metaphysics*

[3]: *Physics*

Austin, J. L. [1]: *How to do things with Words* (ed. Urmson, J. O.), Clarendon Press, Oxford, 1962.

[2]: *Philosophical Papers* (*ed.* Urmson, J. O. and Warnock, G. J.), Clarendon Press, Oxford, 1961.

Ayer, A. J. [1]: *Language, Truth and Logic*, Gollancz, second edn. 1946.

Behmann, M. [1]: 'Der Prädikatenkalkül mit limitierten Variablen: Grundlegung einer natürlichen exacten Logik', *JSL*, vol. 24, 1959.

Belnap, N. D. [1]: 'Conditionally restricted assertions', *Nous*, vol. 4, 1970.

Black, M. [1]: 'Bertrand Russell's philosophy of language', in Schilpp, P. A. [1].

Brady, R. [1]: 'The consistency of the axioms of abstraction and extensionality in a three-valued logic', *NDJFL*, vol. 12, 1971.

[2]: 'The relative consistency of the class axioms of abstraction and extensionality and the axioms of NBG in a three-valued logic', *NDJFL*, vol. 13, 1972.

[3]: *A 4-valued Theory of Classes and Individuals*, doctoral dissertation, 1971, deposited in the University Library, St Andrews.

Carnap, R. [1]: 'Testability and meaning', *PS*, vol. 3, 1936 and vol. 4, 1937.

[2]: *Meaning and Necessity*, University of Chicago Press, Chicago, 1948.

[3]: *Logical Syntax of Language* (trans. Smeaton, A.), Routledge & Kegan Paul, London, 1949.

[4]: *Introduction to Symbolic Logic* (trans. Meyer, W. H. and Wilkinson, J.), Dover, New York, 1958.

[5]: *Introduction to Semantics*, Studies in Semantics, vol. 1, Cambridge, Mass., 1942.

[6]: *Formalisation of Logic*, Studies in Semantics, vol. 2, Cambridge, Mass., 1943.

[7]: *Foundations of Logic and Mathematics*, International Encyclopaedia of Unified Science, vol. 1, no. 3, Chicago, 1939.

[8]: *Logical Foundations of Probability*, University of Chicago Press, Chicago, second edn. 1962.

Chomsky, N. [1]: *Language and Mind*, Harcourt Brane, New York, 1968.

Church, A. [1]: *The Calculi of Lambda-conversion*, Princeton University Press, Princeton, 1941.

[2]: 'On Carnap's analysis of statements of assertion and belief', *Analysis*, vol. 10, 1950.

[3]: 'The need for abstract entities in semantic analysis', *PAAAS*, vol. 80, 1951.

[4]: *Introduction to Mathematical Logic*, Princeton University Press, Princeton, 1956.

Cocchiarella, N. B. [1]: 'Whither Russell's paradox of predication' in *Identity and Individuation* (ed. Munitz, M. K.), New York University Press, 1971.

Copi, I. M. [1]: 'Language analysis and metaphysical inquiry', *PS*, vol. 16, 1949.

Curry, H. B. [1]: *Foundations of Mathematical Logic*, McGraw Hill, New York, 1963.

[2]: 'Generalizations of the deduction theorem', *Proceedings of International Congress of Mathematician*, 1954.

Davidson, D. [1]: 'Truth and Meaning', in *Philosophical Logic* (ed. Davis, J. W., *et al.*), Dordrecht, Holland, 1969.

Drange, T. [1]: *Type Crossings*, Moulton, The Hague, 1966.

Dugundji, J. [1]: 'Note on a property of matrices for Lewis and Langford's calculi of propositions', *JSL*, vol. 5, 1940.

Dunn, J. M. and Belnap, N. D. [1]: 'The substitution interpretation of the quantifiers', *Nous*, vol. 2, 1968.

Feys, R. [1]: *Modal Logics*, (ed. Dopp, J.) Louvain; Paris, 1965.

Fitch, F. [1]: *Symbolic Logic*, Ronald Press Co., New York, 1962.

Fraassen, B. van [1]: 'Singular terms, truth-value gaps and free logic', *J. Phil*, vol. 63, 1966.

Frege, G. [1]: 'On concept and object' in *Translations from the Philosophical Writings of Gottlob Frege* (ed. Geach, P. and Black, M.), Blackwell, 1960.

Geach, P. [1]: *A History of the Corruptions of Logic*, Inaugural Lecture, University of Leeds, 1968.

[2]: *Reference and Generality*, Cornell University Press, New York, 1962.
Gilmore, P. C. [1]: 'An addition to "logic of many-sorted theories"', *CM*, vol. 13, 1956–58.
[2]: 'Attributes, sets, partial sets and identity', *CM*, vol. 20, 1968.
Goddard, L. [1]: 'Sense and nonsense', *Mind*, vol. 73, 1964.
[2]: 'Predicates, relations and categories', *AJP*, vol. 44, 1966.
[3]: 'Towards a logic of significance', *NDJFL*, vol. 9, 1968.
[4]: 'Nonsignificance', *AJP*, vol. 48, 1970.
[5]: 'The existence of universals' in *Contemporary Philosophy in Australia* (ed. Brown, R. and Rollins, C. D.), Allen and Unwin, 1969.
Goddard, L. and Routley, R. [1]: 'Use, mention and quotation', *AJP*, vol. 44, 1966.
Halldén, S. [1]: *The Logic of Nonsense*, Uppsala Universitets Årsskrift, 1949.
Harrop, R. [1]: 'On the existence of finite models and decision procedures for propositional calculi', *PCPS*, vol. 54, 1958.
Hempel, C. G. [1]: 'Problems and changes in the empiricist criterion of meaning', *Rev. IP*, vol. 4, 1950.
Hilbert, D. and Ackermann, W. [1]: *Principles of Mathematical Logic* (trans. Hammond, L. M. *et al.*), Chelsea, New York, 1950.
Hintikka, J. J. K. [1]: *Knowledge and Belief*, Cornell University Press, 1962.
[2]: 'The modes of modality', *Proceedings of a Colloquium on Modal and Many-valued Logics*, *APF*, vol. 16, 1963.
Hughes, G. E. and Cresswell, M. J. [1]: *An Introduction to Modal Logic*, Methuen, 1968.
Hume, D. [1]: *Enquiries* (ed. Selby-Bigge, L. A.), second edn., Clarendon Press, Oxford, 1902.
Husserl, E. [1]: *Logical Investigations* (trans. Findlay, J. N.), Routledge & Kegan Paul, 1970.
Kant, I. [1]: *Critique of Pure Reason* (trans. Smith, N. K.), Macmillan, 2nd impression 1933.
Kleene, S. C. [1]: 'On notation for ordinal numbers', *JSL*, vol. 3, 1938.
[2]: *Introduction to Metamathematics*, North-Holland, Amsterdam, 1952.
Kneale, W. and Kneale, M. [1]: *The Development of Logic*, Clarendon Press, Oxford, 1962.
Kripke, S. [1]: 'Semantical analysis of modal logic I', *ZML*, vol. 9, 1963.
[2]: 'Semantical considerations on modal logic', *Proceedings of a Colloquium on Modal and Many-Valued Logics*, *APF*, vol. 16, 1963.
Lambert, K. [1]: 'On the no type theory of significance', *AJP*, vol. 46, 1968.
Lewis, C. I. [1]: *A Survey of Symbolic Logic*, Dover edn., Dover, New York, 1960.
Linsky, L. [1]: *Referring*, Routledge & Kegan Paul, 1967.
Lyons, J. [1]: *Introduction to Theoretical Linguistics*, Cambridge University Press, 1968.
Mendelson, E. [1]: *Introduction to Mathematical Logic*, Van Nostrand, Princeton, New Jersey, 1964.
Mill, J. S. [1]: *A System of Logic*, eighth edn., Longmans Green, 1900.
Montague, R. [1]: 'Pragmatics' in *Contemporary Philosophy: A Survey* (ed. Klibansky, R.), Firenze, La Nuova Italia Editrice, 1968.
Pap, A. [1]: 'Types and Meaninglessness', *Mind*, vol. 69, 1960.
Passmore, J. [1]: 'Russell and Bradley' in *Contemporary Philosophy in Australia* (ed. Brown, R. and Rollins, C. D.), Allen & Unwin, 1969.
Plato [1]: *Sophist*
Pollock, J. L. [1]: 'Model theory and modal logic', *LA*, vol. 9, 1966.
[2]: 'Logical validity in modal logic', *Monist*, vol. 51, 1967.
Popper, K. R. [1]: 'The demarcation between science and metaphysics', in Schilpp, P. A. [2].
[2]: *The Logic of Scientific Discovery*, Hutchinson, 1959.
Presley, C. F. [1]: 'Arguments about meaninglessness', *BJPS*, vol. 12, 1961.
Prior, A. N. [1]: 'Entities', *AJP*, vol. 32, 1954.

[2]: *Formal Logic*, second edn., Clarendon Press, Oxford, 1962.
[3]: *Time and Modality*, Clarendon Press, Oxford, 1957.
Quine, W. V. O. [1]: *Mathematical Logic*, revised edn., Harvard University Press, Cambridge, Mass., 1955.
[2]: *Word and Object*, Wiley and Technology Press of M.I.T., Cambridge, Mass., 1960.
[3]: *From a Logical Point of View*, second edn., Harper & Row, Cambridge, Mass., 1961.
[4]: *Set Theory and its Logic*, Belnap Press, Cambridge, Mass., 1965.
[5]: *Methods of Logic*, revised edn., Holt, Rinehart and Winston, New York, 1966.
Reichenbach, H. [1]: *Elements of Symbolic Logic*, Macmillan, New York, 1952.
Rennie, M. [1]: 'Models for multiply modal systems', *ZML*, vol. 16, 1970.
[2]: 'Semantics for RK_t', *JSL*, vol. 36, 1971.
Rockingham Gill, R. R. [1]: 'The Craig-Lyndon interpolation theorem in 3-valued logic', *JSL*, vol. 35, 1970.
Rosser, J. B. [1]: *Logic for Mathematicians*, McGraw-Hill, New York, 1953.
Rosser, J. B. and Turquette, A. [1]: *Many-valued Logics*, North-Holland, Amsterdam, 1952.
Routley, R. [1]: 'What numbers are', *LA*, vol. 8, 1965.
[2]: 'Some things do not exist', *NDJFL*, vol. 7, 1966.
[3]: 'On a significance theory', *AJP*, vol. 44, 1966.
[4]: 'The need for nonsense', *AJP*, vol. 47, 1969.
[5]: 'Identity and existence in quantified modal logic', *NDJFL*, vol. 10, 1969.
[6]: 'Extensions of Makinson's completeness theorems in modal logic', *ZML*, vol. 16, 1970.
[7]: 'Exploring Meinong's jungle', *NDJFL* (forthcoming).
[8]: 'Domainless semantics for free quantification and significance logics', *LA*, vol. 14, 1971.
[9]: 'Non-existence does not exist', *NDJFL*, vol. 11, 1970.
Routley, R. and Routley, V. [1]: 'Categories–expressions or things?', *Theoria*, vol. 35, 1969.
[2]: 'The semantics of the first degree of entailment', *Nous*, vol. 6, 1972.
[3]: 'A fallacy of modality', *Nous*, vol. 3, 1969.
[4]: 'Rehabilitating Meinong's theory of objects', *Rev. IP*, vol. 26, 1972.
Routley, R. and Meyer, R. K. [1]: 'Semantics of entailment II', *JPL*, vol. 1, 1972.
Russell, B. W. [1]: *The Principles of Mathematics*, second edn., Allen & Unwin, 1937.
[2]: 'Mathematical Logic as based on the theory of types' in Russell, B. W. [6].
[3]: 'The philosophy of logical atomism', in Russell, B. W. [6].
[4]: 'Critical notice of R. MacColl's *Symbolic Logic and its Applications*', *Mind*, vol. 15, 1906.
[5]: *A Critical Exposition of the Philosophy of Leibniz*, second edn., Allen & Unwin, 1949.
[6]: *Logic and Knowledge* (ed. Marsh, R. C.), Allen & Unwin, 1956.
Ryle, G. [1]: 'Systematically misleading expressions' in *Logic and Language*, first series (ed. Flew, A. G. N.), Blackwell, Oxford, 1951.
[2]: 'Categories' in *Logic and Language*, second series (ed. Flew, A. G. N.), Blackwell, Oxford, 1953.
[3]: *Philosophical Arguments*, Inaugural Lecture, *OUP*, London, 1945.
[4]: *The Concept of Mind*, Hutchinson, 1949.
Schilpp, P. A. (ed.) [1]: *The Philosophy of Bertrand Russell*, Tudor, New York, 1946.
[2]: *The Philosophy of Rudolf Carnap*, Open Court, New York, 1963.
Scroggs, S. J. [1]: 'Extensions of the Lewis System S5', *JSL*, vol. 16, 1951.
Segerberg, K. [1]: 'Propositional logics related to Heyting's and Johansson's', *Theoria*, vol. 34, 1968.
Sellars, W. [1]: 'Abstract entities', *RM*, vol. 16, 1962–63.

Shaw Kwei, Moh. [1]: 'Logical paradoxes for many-valued systems', *JSL*, vol. 19, 1954.
Shorter, M. [1]: 'Meaning and grammar', *AJP*, vol. 34, 1956.
Słupecki, J. [1]: 'The full three-valued propositional calculus' in *Polish Logic 1920–1939* (ed. Storrs McCall), Clarendon Press, Oxford, 1967.
Smart, J. J. C. [1]: 'A note on categories', *BJPS*, vol. 4, 1953–54.
Sommers, F. [1]: 'Types and ontology', *PR*, vol. 72, 1963.
[2]: 'The ordinary language tree', *Mind*, vol. 68, 1959.
Strawson, P. F. [1]: 'On referring' in *The Theory of Meaning* (ed. Parkinson, G. H. R.), Oxford University Press, 1968.
[2]: *Introduction to Logical Theory*, Methuen, 1952.
[3]: *Individuals: an essay in descriptive metaphysics*, Methuen, 1959.
Tarski, A. [1]: 'The concept of truth in formalized languages' in Tarski, A. [3].
[2]: 'Investigations into the sentential calculus' (with Łukasiewicz, J.) in Tarski, A. [3].
[3]: *Logic, Semantics and Metamathematics* (ed. Woodger, J. H.), Clarendon Press, Oxford, 1956.
Toms, E. [1]: *Being, Negation and Logic*, Blackwell, Oxford, 1962.
Wang, H. [1]: 'A note on Quine's principles of quantification', *JSL*, vol. 12, 1947.
[2]: *A Survey of Mathematical Logic*, Science Press, Peking, and North-Holland, Amsterdam, 1964.
Whitehead, A. N. and Russell, B. W. [1]: *Principia Mathematica*, first and second edns., Cambridge University Press, 1910 and 1925.
Wittgenstein, L. [1]: *Tractatus Logico-Philosophicus*, Routledge & Kegan Paul, 1922.
[2]: *Philosophical Investigations* (ed. Anscombe, G. E. M. and Rhees, R.; trans. Anscombe, G. E. M.), second edn., Blackwell, Oxford, 1953.
[3]: *Remarks on the Foundations of Mathematics*, (ed. von Wright, G. H., Rhees, R. and Anscombe, G. E. M.; trans. Anscombe, G. E. M.), Blackwell, Oxford, 1956.

